NUMERICAL METHODS FOR HYPERBOLIC EQUATIONS: THEORY AND APPLICATIONS

PROCEEDINGS OF THE INTERNATIONAL CONFERENCE ON NUMERICAL METHODS FOR HYPERBOLIC EQUATIONS: THEORY AND APPLICATIONS, SANTIAGO DE COMPOSTELA, SPAIN, 4–9 JULY 2011

Numerical Methods for Hyperbolic Equations: Theory and Applications

An international conference to honour Professor E.F. Toro

Editors

Elena Vázquez-Cendón
Department of Applied Mathematics, University of Santiago de Compostela, Spain

Arturo Hidalgo
Department of Applied Mathematics and Computational Methods, Technical University of Madrid, Spain

Pilar García-Navarro
Fluid Mechanics, EINA, University of Zaragoza, Spain

Luis Cea
Environmental and Water Engineering Group, University of A Coruña, Spain

CRC Press
Taylor & Francis Group
Boca Raton London New York Leiden

CRC Press is an imprint of the
Taylor & Francis Group, an **informa** business
A BALKEMA BOOK

Cover photo: "Simulations of the magnetized Kelvin-Helmholtz instability in neutron-star mergers". By M. Obergaulinger, M.A. Aloy & E. Müller

CRC Press/Balkema is an imprint of the Taylor & Francis Group, an informa business

Typeset by V Publishing Solutions Pvt Ltd., Chennai, India
Printed and bound in Great Britain by CPI Group (UK) Ltd, Croydon, CR0 4YY

Published by: CRC Press/Balkema
P.O. Box 447, 2300 AK Leiden, The Netherlands
e-mail: Pub.NL@taylorandfrancis.com
www.crcpress.com – www.taylorandfrancis.com

ISBN: 978-0-415-62150-2 (Hbk)
ISBN: 978-0-203-56233-8 (eBook)

Numerical Methods for Hyperbolic Equations – Vázquez-Cendón et al. (eds)
© 2013 Taylor & Francis Group, London, ISBN 978-0-415-62150-2

Table of contents

Numerical Methods for Hyperbolic Equations – Vázquez-Cendón et al. (eds)
© 2013 Taylor & Francis Group, London, ISBN 978-0-415-62150-2

Preface

These Proceedings of the International Conference on Numerical Methods for Hyperbolic Equations, held in July 2011 in Santiago de Compostela, Spain, present some of the most recent advances in the theory and the computation of solution of the equations of compressible flows, including reactive flows. A wide variety of applications are presented: climatology, heat transfer, magnetohydrodynamics, tsunami modeling and biomedical problems such as blood flow in various regimes.

This volume is a good source for learning the mathematical theory of compressible flows, an array of numerical methods for solving the partial differential equations that express the laws governing these flows, and a variety of real world phenomena described by such equations.

Peter D. Lax
New York, 2012

Numerical Methods for Hyperbolic Equations – Vázquez-Cendón et al. (eds)
© 2013 Taylor & Francis Group, London, ISBN 978-0-415-62150-2

Foreword

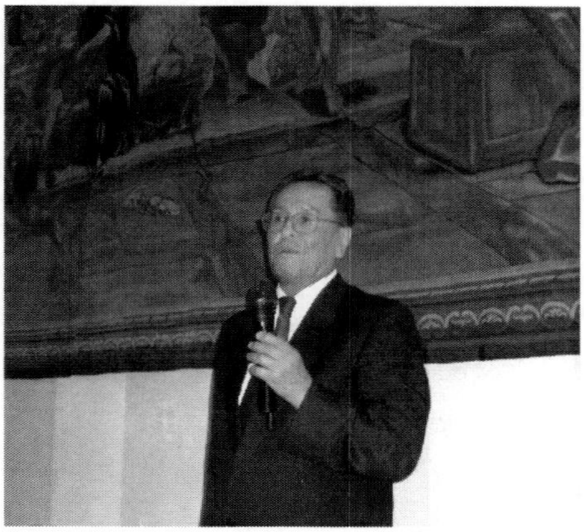

It was a dream birthday party, in a dream place and attended by 130 wonderful human beings and scientists from twenty countries. I am so much indebted to so many people who made this extraordinary event possible, from the local organizing committee, the international scientific committee to each individual who contributed to the full, one-week top-class scientific programme. We worked hard but we also played hard. The full range of social and cultural activities held in this historical city *Santiago de Compostela* were a most wonderful experience, not to mention that *Cena de Gala* in the *Hostal de los Reyes Católicos* and our made-to-measure tour along the *Camino de Santiago*. My very special thanks go to *Universidad de Santiago de Compostela* and the main organizer of the event, Professor M. Elena Vázquez Cendón.

Eleuterio F. Toro
Santiago de Compostela, Spain, July 2011

All the contributors to this book hope that Toro enjoys reading it as much as we loved his birthday celebrations. Our work is our gift to you. Thank you for your knowledge and friendship through the years.

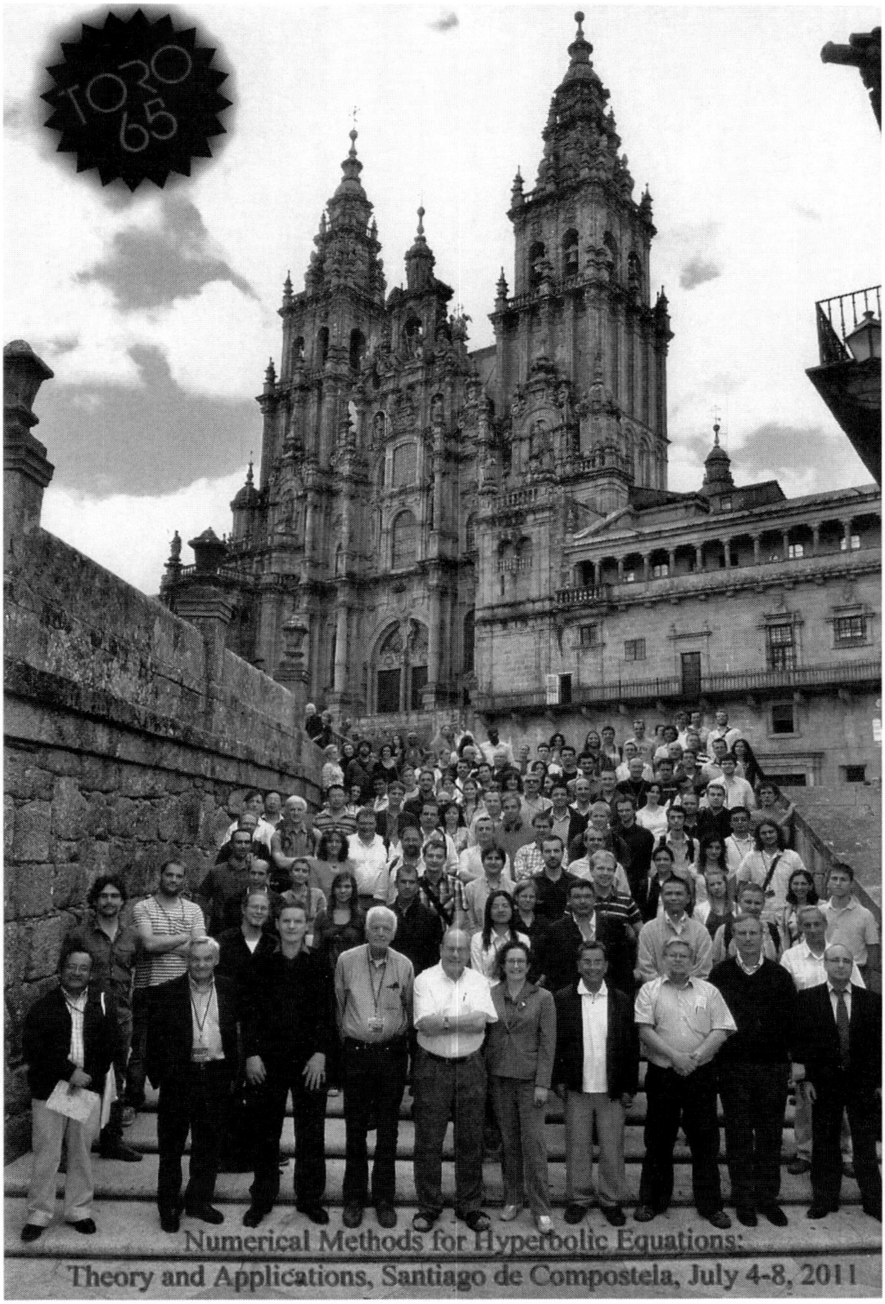

Numerical Methods for Hyperbolic Equations – Vázquez-Cendón et al. (eds)
© 2013 Taylor & Francis Group, London, ISBN 978-0-415-62150-2

Scientific committee

Honour president: Peter Lax, *Courant Institute of Mathematical Sciences, USA.*

Rémi Abgrall, *INRIA and Institut Polytechnique de Bordeaux, France.*
Ángel Aloy, *University of Valencia, Spain.*
Blanca Ayuso, *Centre de Recerca Matemática CRM, Spain.*
Alfredo Bermúdez, *University of Santiago de Compostela, Spain.*
Luis Cea, *University of A Coruña, Spain.*
Tomás Chacón Rebollo, *University of Sevilla, Spain.*
Juan Cheng, *Institute of Applied Physics and Computational Mathematics, China.*
Frédéric Coquel, *Centre de Mathématiques Appliquées, UPMC Paris 06, CNRS, France.*
Rosa Donat, *University of Valencia, Spain.*
Michael Dumbser, *University of Trento, Italy.*
Enrique D. Fernández-Nieto, *University of Sevilla, Spain.*
Pilar García-Navarro, *University of Zaragoza, Spain.*
James Glimm, *University at Stony Brook, USA.*
Arturo Hidalgo, *Technical University of Madrid, Spain.*
José María Ibañez, *University of Valencia, Spain.*
Smadar Karni, *University of Michigan, USA.*
Philippe LeFloch, *University of Paris 6, France.*
Amable Liñan, *Technical University of Madrid, Spain.*
Nikos Nikiforakis, *University of Cambridge, UK.*
Carlos Parés, *University of Málaga, Spain.*
Jerónimo Puertas, *University of A Coruña, Spain.*
Alfio Quarteroni, *Swiss Federal Institute of Technology, Switzerland.*
Philip L. Roe, *University of Michigan, USA.*
Chi-Wang Shu, *Brown University, USA.*
Eleuterio F. Toro, *University of Trento, Italy.*
Elena Vázquez-Cendón, *University of Santiago de Compostela, Spain. (Coordinator)*

Numerical Methods for Hyperbolic Equations – Vázquez-Cendón et al. (eds)
© 2013 Taylor & Francis Group, London, ISBN 978-0-415-62150-2

Sponsors

Numerical Methods for Hyperbolic Equations – Vázquez-Cendón et al. (eds)
© 2013 Taylor & Francis Group, London, ISBN 978-0-415-62150-2

Thanks!

This book contains the contributions of all the authors present in the Author Index as well as the work of the Scientific Committee via proposals, comments and revisions of the papers and sessions. Also the financial support of the Sponsers and the collaboration of efficient people from different academic institutions and international scientific societies. The International Conference Numerical Methods for Hyperbolic Equations: Theory and Applications, held in Santiago de Compostela in July 2011, gathered a great number of personalities and students of the field to pay tribute to a vital scientist, Tito Toro. It also counted with a renowned figure of this scientific family, Peter Lax, as honorary president.

The editors gratefully acknowledge all the people that in one way or another left a personal mark both in the Conference and the Book. Thank you for the great work and energy provided.

Elena Vázquez-Cendón, Arturo Hidalgo, Pilar García-Navarro and Luis Cea
Santiago de Compostela, Spain, September, 2012

I Invited lectures

Numerical Methods for Hyperbolic Equations – Vázquez-Cendón et al. (eds)
© 2013 Taylor & Francis Group, London, ISBN 978-0-415-62150-2

Santiago and the Jacobean pilgrimage at the height of the Middle Ages: Mentality and culture

Francisco Singul
S.A. de Xestión do Plan Xacobeo, Santiago de Compostela, Spain

ABSTRACT: The consolidation of the Way of St. James pilgrimage route mass as for the central centuries of the Middle Ages, is due to the particular worldview of medieval society in the Christian West. All culture generated in this European pathway responds to the intellectual and spiritual demands of a civilization in search regeneration.

1 INTRODUCTION

The 11th century in Western Europe was a time marked by a burst of activity along the pilgrimage ways leading to the holy sepulchre of St. James the Apostle. The popularisation of the pilgrimage was largely due to the fact that the year 1000 had come and gone, and with it the fear that it would bring about the end of the world—a superstitious belief widely held by the rural, and generally uneducated society of the time. These routes whose destination was the Hispanic *finisterre* or land's end, based on the foundations of Roman Christianity, were consolidated in the late 11th century and continued during the following century with the spontaneous pilgrimages undertaken by thousands of believers who were backed on the peninsula by the monarchies of Aragón and Navarra as well as Castile and León. The central period of the Middle Ages was, in fact, the golden age of the Jacobean pilgrimage. Under the episcopate of Diego Gelmírez (1100–1140), SantiagodeCompostelawasthecentreofa thriving world, at once devout, cosmopolitan, eclectic and Babelic, devoted to a collective experience, blending—in this Jacobean cocktail—Christian piety, faith in the intervention of the Apostle and the generous practice of hospitality—channelled through a way/umbilical cord that connected the universal city to the West—with the advantages of exchanging news, ideas, goods, and knowledge, that were circulating along these devotional routes for the purpose of spreading out, opening up new markets and participating in the sheer creative possibilities offered by the Way of St. James.

2 THE CAMINO DE SANTIAGO

This pilgrimage route crossed the Christian Kingdoms of the Northern Spain from East to West, connecting the capitals and major cities in the area (Jaca, Pamplona, Nájera, Burgos, León, Astorga and Santiago de Compostela) with the roads, lands, towns and cities of the rest of Europe, until it became a ritual, indeed a sacred place, closely linked to the final goal of Santiago de Compostela. All kinds of stimuli—cultural, demographic, urbanistic, political and diplomatic—disseminated through this communication channel, took on greater force thanks to the pilgrims themselves with their capacity to convey information. So much so that this dynamism spurred the creation of a special place for devotion and culture in which the great works of Romanesque art make up an enormous artistic compendium that portray, unlike any other cultural expression, the conquests of the Medieval Christian community. One of the decisive factors was the support of the ecclesiastic elite and the contribution made by the Benedictine order. This new monastic ambience backed by the Cluny Reform and the liturgical innovations put forth by the papacy through the Gregorian reformist program was introduced on the peninsula, reaching Aragón in 1072 and shortly after appearing in Castile and León, particularly after the Council of Burgos of 1080. These processes proved to be an important stimulus in the creation of great monastic buildings and cathedrals integrated into a rural world that was undergoing major changes—thanks to the cultivation of new plots of land and innovative farming techniques introduced by the monks—and in the cities located along the Way of St. James. The entire monastic and hospitality dispensing structure organized by the Cluny order along the Way of St. James favoured the introduction of the Roman liturgy, the evolution of the worship of the Apostle and the rendering of care to the pilgrims. In addition to these achievements, stemming from the Reform of the Church, expressed through the Gregorian chant and the common liturgy, the need for larger and safer churches triggered the dissemination of the Romanesque style all along the route, thus forming what we might call the first European style.

These events, which took place during the reigns of Alfonso VI of Castile and León (1072–1109) and Sancho Ramírez of Aragón and Navarra (1063–1094), led to a dynamism of a social, economic, spiritual and cultural nature, that was accentuated by the pilgrimage phenomenon. At the time the route to Santiago de Compostela was one of the most powerful means of communication of the masses ever known, and it witnessed the spread of all types of commerce initiated in the Christian kingdoms. Pilgrims of all nationalities, masterfully reflected in Book V of the "Codex Calixtinus" circulated in two opposing directions—towards the goal represented by the Apostle and the return trip heading towards the countries located overseas. These heteroclite and devout masses constituted a group of people who readily consumed both cultural and commercial products, while at the same time they served as a channel through which privileged information flowed at the service of creativity in multi-faceted form.

3 IN THE BEGINNING

The first pilgrimages to Santiago started in the 9th century, becoming increasingly popular in the 10th century and reached their heyday in halfway through the Middle Ages. From the very beginning, the phenomenon of the Jacobean Way was associated with a Christian sentiment of the worship of relics. In Hispania, after the end of the Visigoth State and the spread of Islam on the peninsula, 8th century Christian society—all but limited to Asturias and Galicia—held St. James the Great as their patron saint and protector of the kingdom. From as far back as the 7th century, thanks to the dissemination of the *Breviarum Apostolorum*—a text indicating the places that had been evangelised by the apostles after the Pentecost, Europe was already familiar with the tradition of the preaching of the Apostle Zebedee, on the peninsula. The oral transmission of these teachings was followed by a small set of texts disseminated in the West during the 7th–9th centuries. These reports insisted that St. James had preached the gospel in the western confines of the known world in an effort to spread the word of the Redemption of mankind through the sacrifice of Jesus on the cross. Another different tradition and other written sources also dating from before the discovery of the tomb of St. James, informed that this tomb was located in Galicia in a place "very near the British Sea", quoting from a text written by the English monk, the Venerable Bede, in the late 8th century. The *Revelatio* of the Jacobean relics confirmed the veracity of these suppositions, traditions and written information. The Bishop from Iria, Theodomir (+847) had the honour of finding the tomb. As a result of the discovery of the tomb, between the years 820–830, in an almost deserted location in the diocese of Iria, the Asturian King Alfonso II (791–842) and Bishop Theodomir sponsored the creation of the basic infrastructure to build the sanctuary. This decision would extend far and wide, owing particularly to the donation of lands to the holy place, granted by the King, who also ordered the construction of the first basilica and set up a monastic community in charge of the cult of the Apostle. As time passed, this foundation became something much more complex. The centre of worship, consisting of a church housing the Apostle's tomb, a place of residence for the Bishop and his servants and a small monastery, gave rise to an urban structure with the capacity to offer lodging and services, as well a sizeable population and social substance suitable to the development of a walled city of some importance in the 10th–11th centuries. During the early Middle Ages, Compostela was the spontaneous product of the pilgrimage, the fruit of the devotion of pilgrims who came from the most diverse corners of the kingdom, from other parts of the peninsula and even from abroad. People, who, out of their love for St. James, and faithful to his power of intervention, were drawn to the sanctuary with the desire to live and die in shadow of the holy sepulchre.

The urban dynamism of this bustling Jacobean city was in full force by the year 1000, with the pastoral aspects being the concern of the Bishops of Iria-Compostela, Theodomir, Sisnando I, Sisnando II and St. Peter of Mezonzo, along with the support of the Asturian Monarchs Alfonso II and his successor, Alfonso III. The Monarchs donated lands and granted privileges, in addition to promoting the construction of the first two Jacobean basilicas housing the tomb. The first church, erected by the Alfonso II, the "Chaste" King must be understood as a kind of emergency action, built with poor quality material. However, the second pre-Romanesque church, donated by Alfonso III and consecrated in 899, was the largest building of the Christian religion on the Iberian Peninsula. This endeavour by Alfonso III the Great, to present St. James with a magnificent church—larger than any other in Oviedo—would appear to indicate, in addition to the symbolic and representative value of the gift to the patron saint, the projection of the pilgrimage and the hopes for the future held by the promoters of the sanctuary. After the year 1000 was over, with the self-affirmation of the Christian North and the end of al-Andalus as the most important power of the peninsula, the pilgrimages to Santiago continued to grow during the 11th century. This period gave rise to the idea of building a new cathedral with a greater capacity and more impressive presence in the city. The construction of the Romanesque basilica got underway in 1075, under the auspices of Bishop Diego Peláez and sponsored by Alfonso VI of Castile and León. This building served to bind and stimulate the medieval urban structure of the city that was the goal of the Way of St. James. From a typological and stylistic standpoint, it must be said, in short, that the new European artistic order achieved a perfect example in 12th century Compostela: the most exquisitely finished model of the European Romanesque style, a building that was planned after a century of trials and errors, a building in which the architecture, monumental sculpture, and mural painting—today either lost or hidden—reached artistic heights of almost unsurpassable perfection and beauty.

4 THE GOLDEN AGE

During the episcopate of Gelmírez there was a flurry of activity of a constructive and cultural nature to serve the cult of St. James and the pilgrimage. A municipal program under which artists carried out the work on the cathedral, clergymen and intellectuals worked on the *scriptorium* drafting the books of the History of Compostela and the "Codex Calixtinus" as well as copying the most valuable historical and legal documents of the Church of Santiago in order to compile them into a register artistically decorated in vermilion. The period of Gelmírez was rich in achievements for the Archdiocese of Santiago de Compostela, especially in the Hispanic context, with the Church of Santiago and the pilgrimage way closely linked to it, acquiring great prestige inside and outside the borders of the kingdom. In this representative and symbolic program, the new metropolitan see was embodied through a city-sanctuary open to all the peoples of the West—a spiritual and commercial centre of international renown, a model reference for the entire Christian world. A framework and a situation that were at once, the consequence and the source of the vibrant development of the Jacobean pilgrimages. Twelfth-century Compostela would become a fruitful cultural synthesis, a product of the exchange of technical ideas and artistic processes at the service of a political, religious and social project. The demographic, economic and cultural boom taking place in the north of the peninsula prompted King Alfonso VI and King Sancho to promote in their respective kingdoms, the construction of a series of religious buildings which synthesised the theoretical and empirical accomplishments of various generations of European artists, concerned, in turn, with the different solutions and systems in terms of construction—the layout and elevation of stone domes able to cover large spaces was, for centuries, the bone of contention of architects—that would be fitting to convey the symbolic universe embodied in the great abbatial churches and in the new urban cathedrals that would be built in the 12th and 13th centuries. The prolonged period of the early Middle Ages in the West witnessed the study and practical application of late medieval artistic traditions, forming a common architectural and symbolic order with local peculiarities that would serve to enrich its splendour, inspiring what has come to be known on the peninsula as the art of the Way of St. James. In this physical and symbolic space the monarchs demonstrated, on the artistic front, the same decision they had made when they created the basic infrastructure for the pilgrimage, repairing old roads, planning new stretches, building bridges, doing away with taxes and tolls and fomenting the creation of a network to serve the pilgrim. In this collective project sponsored by the royal power, a decisive factor was the economic contribution provided by the taxes levied on the weakened Islamic Taifa kingdoms.

The French Way was built up during the short period from 1075 to 1125. A half century that witnessed the construction of its most important churches and cathedrals, which are related in style, giving rise to a certain homogeneity—a kind of family resemblance. An appearance of unity stemming from the fact that they belong to the same culture and to an integrated artistic and spiritual space. This group of buildings comprising the artistic backbone of the Way of St. James, is made up of the Cathedral of Santiago, the Collegiate Church of San Isidoro de León, the Church of San Martín de Frómista (Palencia), the Cathedral of Jaca and the French churches of Santa Fe de Conques and Saint-Sernin de Toulouse. This ensemble also would have included the Romanesque cathedrals of Pamplona, Burgos, León and Astorga, no longer in existence. A group of buildings which indeed shine in splendour along the French Way, like a trick of mirrors, captivating us with the stylistic similarity of their architecture and the sculptures of their porches. The flow and reflection of these relationships are directed towards the goal of the Way of St. James whose Jacobean Basilica evokes solutions and forms taken from many of the buildings mentioned earlier, projecting once again—but this time in other directions—to the East and other areas, the magnificence of a unique building which may be explained as the synthesis and recapitulation of the art along the pilgrimage ways.

At the beginning of the 12th century, the promotion of the Cathedral of Santiago was backed by the Mitre of Compostela, with Bishop Gelmírez being the principal driving force. For the purpose of achieving his political and religious objectives, which boiled down to enhancing the prestige of Santiago as the Apostolic See, a metropolitan centre and goal of a universal pilgrimage travelled by throngs of pilgrims, the bishop cultivated the friendship of the powerful Cluny abbots as well as the influential members of the Roman Curia. Thanks to his tenacity, in 1105 he succeeded in obtaining the privilege of the pallium from the Pope in the Basilica of Letrán; he organised the Church of Santiago and its temporary domain; he rebuilt churches, monasteries and fortresses in his territory; he boosted the cultural and pastoral prestige of the clergy of Santiago de Compostela and strengthened the importance of the diocese in the context of the Kingdom of Castile and León, and lastly, in 1120, thanks to his efforts, the Church of Santiago was granted the ranks of metropolitan see, and Gelmírez became its first archbishop. A complex promotional project that had a positive effect on the success of the pilgrimage. The spiritual seed that grew under the protection of the pilgrimages to Santiago fomented the productive and collective creation of a Christian culture with deep-rooted values and a universal vocation—a product of the religious experience and the creativity of a pluralistic society, which was not as hermetic as it was commonly thought to be, open to a variety of cultural proposals of Judaic and Islamic origin. The sheer scope of this unique synthesis was beneficial in the North of Spain, especially during the 12th and 13th centuries, in that it provided the necessary vitality to generate the creative experience of the art of the Way of St. James. A new world of forms, symbols, spaces and sensations expressed in artistic projects—integrated into a synthesis of architecture, sculpture, painting and gold and silver work—renewing the cultural roots of Europe with the visualisation of the Scriptures through art, in a proposal enriched through a new way of understanding and visualising the legacy of the past. What is achieved, then, is a new visual, symbolic and spatial order in the most important buildings where the greatest creative effort of the western world as a whole has been exerted. This happened in such a way that the new cathedrals, churches, and abbeys—in addition to civil and ecclesiastic palaces, hospitals, liturgical objects and symbols of power—correspond to an integrated model, highlighted by the Romanesque order and starting in the 13th century, by the Gothic experience, which was the result of the splendour of a culture that found one of the most vigorous channels for dissemination and creativity in the Way of St. James.

5 HOSPITALITY AND HOLY PLACES

Although the compelling presence of the cathedrals along the way—Jaca, Pamplona, Burgos, León, Astorga and Santiago de Compostela—was a determining factor in the development of the new channels for expression, it was the Cluny and Cistercian monasteries along the route—and the mendicant convents, which, starting in the 13th–14th centuries, were the most prominent institutions, where a large part of the creative backbone of this artistic and cultural experience can be traced to. This is also in keeping with the fact that these centres of religion and power were at the forefront of the Christian experience of charity and the rendering of hospitality to the pilgrims—a fundamental identity mark of the western pilgrimage. The hospitality offered along each of the Jacobean routes reflected a common practice in Medieval society—attending to the needs of the most destitute, in the spiritual realm as well as in terms of material and sanitary requirements, of pilgrims who, owing to the distances and the arduous journey, would often reach the end of the stretch battered, hungry and sick. This situation of gruelling hardship suffered by the pilgrims on a day to day basis motivated a zealous endeavour to provide charity and hospitality—a noble gesture practiced by monastic communities and convents, ecclesiastic chapters, towns and villages along the way, noblemen and kings. Civil and religious institutions and many devout private citizens with economic means founded hospitals, and helped to equip and maintain them. Depending on the origin of the foundation of these centres set up to provide care and assistance to the pilgrims, they may be classified into hospitals belonging to the episcopate, the cathedral, military orders, monastic, royal, noble, or parish hospitals and in the urban world, hospitals pertaining to guilds or associations. During the Middle Ages—particularly during the centuries in the middle of this period—the practice of hospitality offered by private citizens or monastic communities must be understood as a custom typical of the society and in keeping with the Christian faith. The will to practice the Christian virtue of charity and thus, serve God and St. James.

At the end of the way, in the city Santiago during medieval times charity and hospitality were dispensed at the doors of the cathedral in the middle of the Plaza del Paraíso, adjacent to the Puerta de Francia, the north façade of the Romanesque cathedral. The Hospital of Santiago operated throughout the Medieval period and was supported by the donations of the archbishops. The hospital was often overwhelmed during times of massive pilgrimages, and it was unable to cope with the great demand for beds. When this would happen, the aisles and possibly the galleries of the basilica of the Apostle were used to provide lodging and assistance. A fundamental aspect to understand the mentality that made this rich culture possible is the fact that the Way of St. James—a holy place par excellence, a way of asceticism and self-denial, protected at all times by Jesus, the Virgin Mary, the Pilgrim Apostle and other saints who mark this way of faith and devotion—was dotted with many secondary sanctuaries where important relics were worshiped. The reasons for the success attained by the pilgrimage to Santiago are difficult to rationalise. However, clearly of great importance are the aspects related to the religious mentality of Medieval Christianity and its zealous worship of relics: the experience of the miracle imparted in holy places, particularly in cases involving the holy remains of a saint, the symbolic image that was held of an ethereal world, and in short, the experience of spirituality and the psycho-social atmosphere inherent to the medieval world itself. In general, the motivations shared by the pilgrims were related to devotion, faith and/or atonement. Owing to the charismatic experience it represented, far beyond mere sensitivity, the phenomenon of the Jacobean pilgrimage transcended even the most prosaic meaning of the Way of St. James as a well-travelled communication route, to become a holy place where the pilgrim surrendered himself to self-discipline and penance and to the practice of charity and solidarity that would help him to achieve his encounter with uplifting spirituality. Throughout the entire Medieval period, the Way of St. James was thus, a kind of symbolic road of purification, where human beings put their lives in the hands of God and of his close friend and beloved apostle, St. James. All of their efforts were directed at obtaining the indulgences offered by the Church of Santiago and attaining the special grace for themselves or a loved one, living or dead. This was one of the other common motives. The Way of St. James was a road to purification for both the living and the dead, for an individual person or a group.

6 A CIVILIZATION IN SEARCH OF REGENERATION

The penitent could surrender himself to this experience of salvation in representation of a sick or deceased family member, making the pilgrimage for the suffrage of the soul of the invalid or deceased person, or even as a representative of a community, a parish or a city in need of supernatural help to combat a natural or man-made disaster, such as a plague or a war. Whatever his reasons for making the journey to Compostela, the pilgrim was given a farewell ceremony at his parish during which prayers were said for the cause and the symbolic elements of his ascetic journey, the staff and the leather pouch were blessed. On his return trip to his land of origin, he would buy a scallop shell in the Plaza del Paraíso in Santiago, at the entrance to the cathedral, and either hang it around his neck or pin it to his clothing. The theologians from Santiago de Compostela in the 12th century explained in the *Veneranda dies* sermon that this scallop shell symbolised good deeds, representing the acts of charity and love carried out, which would tip the scales in favour of the pilgrim on the Judgement day. The good works achieved in life and the spiritual purification attained through the gruelling pilgrimage were symbolised by the scallop shell, which served to identify and protect the pilgrim. Over the course of time with the spread of information on the pilgrimage ways, many pilgrims set out on the Way of St. James with scallop shell pinned to their clothing, pouch or brim of their hats. In this way the Jacobean pilgrims boasted a distinguishing mark, an element that symbolised and spoke for their pious, charitable and penitent intentions.

The impossibility of visiting the Holy Tomb of Jerusalem, cut off during the last thirty years of the 11th century by the religious intolerance of the Seljuk Turks, was an anxiety-provoking situation, difficult to overcome. As regards Rome, the Christian *caput mundi*, the pilgrimage there had become an unsafe prospect at that time as well, owing to the strong political and military tensions between the papacy and the empire. The dispute between the Investitures and the Church's resolute stance against the excesses associated with simony and Nicolaitan practices, left Italy in a state of turmoil, and its road network so unsafe as to discourage any type of pilgrimage or act of piety. In this historic context, with such adverse circumstances for journeys to Rome or Jerusalem, the peoples of the West looked to the Way of St. James to soothe their spiritual needs and to exercise an act of devotion and salvation by making the pilgrimage to a holy place. Beyond these horizons fraught with bellicosity and disagreement, the 11th century was a fruitful period for the Way of St. James. With the demise of the Caliphate of Córdoba and the excision of the Hispano-Islamic world into many different Taifa kingdoms, the Christian North strengthened its position, reaching the border marked by the Tagus River, and creating a safe realm that went form the Pyrenees to the Atlantic. Such a favourable situation led to the organisation and promotion of the physical infrastructure and the hospitality network of a pilgrimage route with a providential moral and cultural scope. A route that was laid out ex novo, protected by the Hispano-Christian sovereigns who endowed it with infrastructures to improve the roads and offer assistance, from the mountain passes of the Pyrenees to the Sanctuary of Santiago de Compostela, crossing Aragón, Navarra, La Rioja, Castile, León and Galicia. A route that would be a devotional and holy way, able to channel the devout faith to the man who preached the gospel in the West, as well as representing most of the spiritual and cultural values generated by Western Christian culture. Values and a heritage based on religiousness, hospitality and mutual generosity, as well as triggering a cultural and artistic explosion unprecedented in the north of Spain. Political and demographic interests, which originally gave rise to the design and the development of the Way of St. James and the establishment of its infrastructure for the dispensing of hospitality and buildings for support, gave official approval to this privileged space through which were channelled and spread different ideas and cultural forms arising from the different focal points of knowledge and experimentation in the West. This is, in short, the key to the cultural birth that was experienced in the shadow of the Jacobean pilgrimage. The western route became a line across which knowledge was conveyed, gaining more and more strength over time until it actually turned into a communication route that would open the doors of the peninsula to Europe, the embodiment of the universal ideals and values befitting a spiritual orientation whose fundamental distinguishing feature was the practice of hospitality with the pilgrims, giving charity and developing a positive awareness that led to the fruitful exchange of ideas and experiences whose virtues have been demonstrated for centuries. While in the last twenty-five years of the 11th century and over the course of the 12th century, the roads leading to Rome were insecure and the ways to the Holy Land by land or sea just as dangerous, the Way of St. James was characterised as a secure route, boasting the support of kings, noblemen, burgers, cathedral chapters and monastic communities during this is same period.

7 CONCLUSIONS

It would doubtless appear that Compostela and its Way, the cult of St. James and the success of the pilgrimage were the product of the medieval Christian mentality. The intensity of the religious experience and the other-worldly meaning found by the faithful in the sanctuaries where important relics were kept, changed reality, creating the necessary conditions—in terms of mentality and sensitivity—to start out on a way of atonement and self-denial where pilgrims could experience the miracle. A Way on which the devout aspired, thanks to the generous system of indulgences, to be forgiven for their sins. The cultural and symbolic knowledge of this route of universal values was brought to life with the creation and protection of the physical infrastructures offering assistance, with the construction of churches, monasteries and cathedrals, and above all, with the birth of a special awareness of charity and hospitality, which are considered as one of the hallmarks of the western pilgrimage.

REFERENCES

Constable, G., (1996). *The Reformation of the Twelfth Century*. Cambridge University Press.

Díaz y Díaz, M.C., (1997). La espiritualidad de la peregrinación en el siglo XII, De Santiago y de los Caminos de Santiago. *Colección de inéditos y dispersos reunida y preparada por Manuela Domínguez García*, Santiago, Xunta de Galicia, pp. 249–260.

Díaz y Díaz M.C., (1999). Las tres grandes peregrinaciones vistas desde Santiago, in P. Caucci Von Saucken (ed.), *Santiago, Roma, Jerusalén. Actas del III Congreso Internacional de Estudios Jacobeos*, Santiago, Xunta de Galicia, pp. 81–97.

Geary, P.J., (1994). *Living with the Dead in the Middle Ages*, Ithaca and London, Cornell University Press, pp. 163–176.

Haskins, Ch.H., (1927). *The Renaissance of the Twelfth Century*, Harvard University Press.

Hehl, E.-D., (1999). Cruzada y peregrinación bajo el signo de la *Imitatio Christi*, in P. Caucci Von Saucken (ed.), *Santiago, Roma, Jerusalén. Actas del III Congreso Internacional de Estudios Jacobeos*, Santiago, Xunta de Galicia, pp. 145–159.

Jacomet, H., (1995). Pèlerinage et culte de saint-Jacques en France: bilan et perspectives. *Pèlerinages et croisades. Actes du 118e congrès national annuel des sociétés historiques et scientifiques (Pau, octobre 1993)*, París, Éditions du CTHS.

Le Goff, J., (1981). *La Naissance du Purgatoire*, Paris, Gallimard.

López Alsina, F., (1999)., Años Santos Romanos y Años Santo Compostelanos. P. Caucci Von Saucken (ed.), *Santiago, Roma, Jerusalén. Actas del III Congreso Internacional de Estudios Jacobeos*, Santiago, Xunta de Galicia, pp. 216–219.

Márquez Villanueva, F., (2004). *Santiago: trayectoria de un mito*, Barcelona, Edicions Bellaterra.

Singul, F. (2008). La sacralidad del espacio en el Camino de Santiago. *La Corónica. A Journal of Medieval Spanish Language, Literature.*

Singul, F., (2010). Simbología y mentalidad. Dos referentes en el camino de Santiago durante la Edad Media. E. CORRAL DÍAZ (ed.), *In marsupiis peregrinorum. Circulación de textos e imágenes alrededor del camino de Santiago en la Edad Media, Actas del congreso internacional, Santiago, 24–28, marzo, 2008*, Firenze, Edizioni del Galluzzo, pp. 83–109.

Sumption, J., (2003). *The Age of Pilgrimage. The Medieval Journey to God*, London, Faber & Faber, 1975, New Jersey, Hidden Spring.

Swanson, R.N., (1995). *Religion and Devotion in Europe, c.1215–c.1515*, Cambridge University Press, 1995, pp. 191–199.

Swanson, R.N., (1999). *The Twelfth-Century Renaissance*, Manchester University Press.

Numerical Methods for Hyperbolic Equations – Vázquez-Cendón et al. (eds)
© 2013 Taylor & Francis Group, London, ISBN 978-0-415-62150-2

Recent developments on the Lagrangian and remapping methods for compressible fluid flows

Juan Cheng
National Key Laboratory of Science and Technology on Computational Physics, Institute of Applied Physics and Computational Mathematics, China

Chi-Wang Shu
Division of Applied Mathematics, Brown University, USA

ABSTRACT: The Lagrangian method and the Arbitrary Lagrangian-Eulerian method (ALE) are widely used in many fields, especially for multi-material flow simulations, due to their distinguished advantage in capturing material interfaces automatically. In this paper we give a short survey on our recent research on the Lagrangian scheme and the remapping method, both of which constitute essential and important components of the ALE method. Specifically, we first introduce a class of high order Lagrangian schemes which are based on Essentially Non-Oscillatory (ENO) and Weighted Essentially Non-Oscillatory (WENO) reconstructions. The schemes are constructed both on straight-line quadrilateral meshes and on curvilinear quadrilateral meshes. Meanwhile, two approaches for high order time discretization in the Lagrangian scheme are investigated, namely, the Total Variation Diminishing (TVD) Runge-Kutta time discretization and the Lax-Wendroff (LW) type time discretization. Next, we give a class of cell-centered Lagrangian schemes with the properties of both conservation and spherical-symmetry-preserving in two-dimensional cylindrical coordinates. After that, we summarize a high order conservative ENO remapping algorithm. At last, several numerical examples are shown to further demonstrate the performance of these schemes.

1 INTRODUCTION

In numerical simulations of multidimensional fluid flow, there are two typical formulations: a Lagrangian framework, in which the grid moves with the local fluid velocity, and an Eulerian framework, in which the fluid flows through a grid fixed in space. More generally, the motion of the grid can also be chosen arbitrarily, this method is called the Arbitrary Lagrangian-Eulerian method (ALE; cf. (Hirt, Amsden, and Cook 1974)). Most ALE algorithms consist of three phases, a Lagrangian phase in which the solution and the grid are updated, a rezoning phase in which the nodes of the computational grid are moved to a more optimal position and a remapping phase in which the Lagrangian solution is projected to the new grid.

Pure Lagrangian methods and certain ALE methods can capture material interfaces sharply and automatically, so they are widely used in many fields for multi-material flow simulations such as astrophysics and Inertial Confinement Fusion (ICF). In this paper, we give a brief overview of our recent research on the Lagrangian scheme and the remapping method, both of which constitute essential components of the ALE method.

In a Lagrangian or ALE simulation, critical issues for a scheme include conservation, accuracy, non-oscillation and other special properties such as the preservation of certain symmetry (for example, cylindrical or spherical symmetry) in a coordinate system distinct from that symmetry.

In recent decades, many Lagrangian schemes have been developed which are successful in simulating multi-material flows. Some of these algorithms are built on a staggered discretization in which velocity (momentum) is stored at vertices, while density and internal energy are stored at cell centers. The density/internal energy and velocity are solved on two different control volumes, see, e.g. (von Neumann and Richtmyer 1950; Caramana, Burton, Shashkov, and Whalen 1998). An alternative to the staggered discretization is to use a conservative cell-centered discretization in which density, momentum and total energy are all centered within cells and evolved on the same control volume, see, e.g. (Godunov, Zabrodine, Ivanov, Kraiko, and Prokopov 1979; Munz 1994; Maire, Abgrall, Breil, and Ovadia 2007). These schemes usually have first or at most second order accuracy. More recently, we have investigated a class of high order cell-centered Lagrangian type schemes. The schemes are based on the Essentially Non-Oscillatory (ENO) or simple Weighted Essentially Non-Oscillatory (WENO) reconstructions which were first introduced in the Eulerian formulation by Harten et al. (Harten and Osher 1987; Harten, Engquist, Osher, and Chakravarthy 1987) in 1987 and by Liu et al. (Liu, Osher, and Chan 1994) in 1994 respectively. Both ENO and WENO Eulerian schemes can achieve uniformly high-order accuracy with sharp, essentially non-oscillatory shock transitions. In (Cheng and Shu 2007a; Cheng and Shu 2008b), a class of ENO/WENO schemes in the Lagrangian framework for solving the Euler equations were designed both on quadrilateral meshes and curved quadrilateral meshes respectively. The schemes are conservative for the mass, momentum and total energy, and are essentially non-oscillatory. They can achieve at least uniformly second order accuracy on moving and distorted Lagrangian quadrilateral meshes and third order accuracy both in space and time on curved quadrilateral meshes. Both the TVD Runge-Kutta and Lax-Wendroff type time discretization methods were proposed for the ENO/WENO Lagrangian schemes as well, see (Cheng and Shu 2007a; Cheng and Shu 2008b; Liu, Cheng, and Shu 2009).

In a Lagrangian simulation, another crucial issue for a scheme is to preserve certain symmetry (for example, cylindrical or spherical symmetry) in a coordinate system distinct from that symmetry. In fact, the preservation of spherical symmetry is especially important for the simulation of the implosion problem with strong compressions, since the small deviation from spherical symmetry due to numerical errors may be amplified by Rayleigh-Taylor or other instabilities which may potentially produce unpredictably large errors. Many works have been done concerning the issue related to spherical symmetry preservation in two dimensional cylindrical coordinates, for example (Caramana, Burton, Shashkov, and Whalen 1998; Maire 2009; Browne 1986). Several different efforts were made to maintain such symmetry, which however also often brought some side effects such as the violation of conservation for momentum or total energy. In (Cheng and Shu 2010; Cheng and Shu 2011), we proposed a new class of cell-centered Lagrangian schemes for solving compressible Euler equations in cylindrical coordinates which possess the properties of both spherical symmetry and conservation.

The remapping algorithm is an essential and important component of ALE methods. In the past thirty years, many remapping algorithms were developed for the cell-centered and nodal variables, see for example (Dukowicz and Baumgardener 2000; Kucharik, Shashkov, and Wendroff 2003; Margolin and Shashkov 2003). However most of them are at most second order accurate, especially in two or higher spatial dimensions. In our recent papers (Cheng and Shu 2007b; Cheng and Shu 2008a), we developed a local remapping algorithm on two types of staggered grids in one and two dimensions, which is also applicable to non-staggered grids. The ENO reconstruction idea is used on the remapping strategy so that the algorithm has the framework of arbitrary order of accuracy and has the properties of producing essentially non-oscillatory output which is conservative for mass, total energy and momentum.

An outline of the rest of this paper is as follows. In Section 2, we review the ENO/WENO Lagrangian type schemes for Euler equations on both straight-line and curvilinear quadrilateral meshes. In Section 3, we give a brief overview of a class of the Lagrangian schemes with the spherical symmetry and conservation properties in two-dimensional cylindrical coordinates. In Section 4, we summarize a high order ENO remapping algorithm, while in Section 5 a few numerical examples are given to further demonstrate the performance of the methods under discussion. In Section 6 we give concluding remarks.

2 HIGH ORDER ENO/WENO CONSERVATIVE LAGRANGIAN TYPE SCHEMES

The Euler equations for unsteady compressible flow in the reference frame of a moving control volume can be expressed in integral form as

$$\frac{d}{dt}\int_{\Omega(t)} \mathbf{U}d\Omega + \int_{\Gamma(t)} \mathbf{F}d\Gamma = 0 \tag{1}$$

where $\Omega(t)$ is the moving control volume enclosed by its boundary $\Gamma(t)$. The vector of the conserved variables $\mathbf{U}$ and the flux vector $\mathbf{F}$ are given by

$$\mathbf{U} = \begin{pmatrix} \rho \\ \rho\mathbf{u} \\ E \end{pmatrix}, \qquad \mathbf{F} = \begin{pmatrix} (\mathbf{u}-\dot{\mathbf{x}})\cdot\mathbf{n}\rho \\ \rho\mathbf{u}(\mathbf{u}-\dot{\mathbf{x}})\cdot\mathbf{n} + p\cdot\mathbf{n} \\ (\mathbf{u}-\dot{\mathbf{x}})\cdot\mathbf{n}E + p\mathbf{u}\cdot\mathbf{n} \end{pmatrix} \tag{2}$$

where ρ is the density, $\mathbf{u}$ is the velocity, E is the total energy and p is the pressure, $\dot{\mathbf{x}}$ is the velocity of the control volume boundary $\Gamma(t)$, $\mathbf{n}$ denotes the unit outward normal to $\Gamma(t)$. The system (1) represents the conservation of mass, momentum and total energy.

The set of equations is completed by the addition of an Equation of State (EOS) with the following general form

$$p = p(\rho, e) \tag{3}$$

where $e = E/\rho - 1/2|\mathbf{u}|^2$ is the specific internal energy. Especially, if we consider the ideal gas, then the equation of state has a simpler form,

$$p = (\gamma - 1)\rho e$$

where γ is a constant representing the ratio of specific heat capacities of the fluid.

In a Lagrangian framework, we have $\dot{\mathbf{x}} = \mathbf{u}$, and then the vectors $\mathbf{U}$ and $\mathbf{F}$ take the simpler form

$$\mathbf{U} = \begin{pmatrix} \rho \\ \rho\mathbf{u} \\ E \end{pmatrix}, \qquad \mathbf{F} = \begin{pmatrix} 0 \\ p \cdot \mathbf{n} \\ p\mathbf{u} \cdot \mathbf{n} \end{pmatrix}. \tag{4}$$

In this section we overview our recent development on a class of finite volume ENO/WENO Lagrangian type schemes for the Euler equations in the Lagrangian formulation (1) and (4).

Three steps are used to construct a high order ENO/WENO Lagrangian scheme, namely, spatial discretization, the determination of the vertex velocity and the time discretization. In the step of spatial discretization, to get a high order scheme in space, similarly to the Eulerian finite volume scheme, the high order ENO or WENO reconstruction procedure is used to obtain high order and non-oscillatory approximations to the solution at the Gaussian points along the cell boundary from the neighboring cell averages. The determination of velocity at the vertex is a special step for a Lagrangian scheme which decides how the grid moves at the next time. The reconstruction information of density and momentum from the four neighboring cells of the vertex is used in this step which guarantees the approximation of velocity also has the same order accuracy as the primitive conserved variables. In the step of time discretization, the TVD Runge-Kutta or Lax-Wendroff type time discretization methods are applied.

Next, as an example, we introduce the specific construction procedures of the scheme in one dimension.

The spatial domain Ω is discretized into N computational cells $I_{i+1/2} = [x_i, x_{i+1}]$ of sizes $\Delta x_{i+1/2} = x_{i+1} - x_i$ with $i = 1, \ldots, N$. The location of the cell center for Cell $I_{i+1/2}$ is denoted by $x_{i+1/2}$. The fluid velocity u_i is defined at the vertex of the grid. All the conserved variables are stored at the cell center $x_{i+1/2}$ in the form of cell averages and the cell is their common control volume. To be more specific, the values of the cell averages of density, momentum and total energy for Cell $I_{i+1/2}$, denoted by $\bar{\rho}_{i+1/2}$, $\bar{M}_{i+1/2}$ and $\bar{E}_{i+1/2}$, are defined as follows

$$\bar{\rho}_{i+1/2} = \frac{1}{\Delta x_{i+1/2}} \int_{I_{i+1/2}} \rho dx,$$
$$\bar{M}_{i+1/2} = \frac{1}{\Delta x_{i+1/2}} \int_{I_{i+1/2}} \rho u dx,$$
$$\bar{E}_{i+1/2} = \frac{1}{\Delta x_{i+1/2}} \int_{I_{i+1/2}} E dx.$$

Step 1: Spatial discretization

We first formulate the semi-discrete finite volume scheme of the governing equations (1) and (4) as

$$\frac{d}{dt}\begin{pmatrix} \bar{\rho}_{i+1/2}\Delta x_{i+1/2} \\ \bar{M}_{i+1/2}\Delta x_{i+1/2} \\ \bar{E}_{i+1/2}\Delta x_{i+1/2} \end{pmatrix} = -\begin{pmatrix} \hat{f}_D(\mathbf{U}_{i+1}^-, \mathbf{U}_{i+1}^+) - \hat{f}_D(\mathbf{U}_i^-, \mathbf{U}_i^+) \\ \hat{f}_M(\mathbf{U}_{i+1}^-, \mathbf{U}_{i+1}^+) - \hat{f}_M(\mathbf{U}_i^-, \mathbf{U}_i^+) \\ \hat{f}_E(\mathbf{U}_{i+1}^-, \mathbf{U}_{i+1}^+) - \hat{f}_E(\mathbf{U}_i^-, \mathbf{U}_i^+) \end{pmatrix} \tag{5}$$

where $\mathbf{U}_i^\pm$, $\mathbf{U}_{i+1}^\pm$ represent the left and right values of $\mathbf{U}$ at the cell's boundary x_i and x_{i+1} respectively. $\hat{f}_D$, $\hat{f}_M$, $\hat{f}_E$ are the numerical fluxes of mass, momentum and total energy across the boundary of its control volume $I_{i+1/2}$ respectively, which are consistent with the physical fluxes (4) in the sense that

$$\begin{pmatrix} \hat{f}_D(\mathbf{U},\mathbf{U}) \\ \hat{f}_M(\mathbf{U},\mathbf{U}) \\ \hat{f}_E(\mathbf{U},\mathbf{U}) \end{pmatrix} = \begin{pmatrix} 0 \\ p \\ pu \end{pmatrix}.$$

To determine the fluxes $(\hat{f}_D, \hat{f}_M, \hat{f}_E)$, we first identify the values of the primitive variables on each side of the boundary, that is $\mathbf{U}_i^\pm = (\rho_i^\pm, M_i^\pm, E_i^\pm)$, $i = 1, \ldots, N$. The information we have is the cell average values of the conserved variables $\bar{\mathbf{U}}_{i+1/2} = (\bar{\rho}_{i+1/2}, \bar{M}_{i+1/2}, \bar{E}_{i+1/2})$. To obtain uniformly second or higher order accurate schemes, the ENO/WENO idea (Harten, Engquist, Osher, and Chakravarthy 1987; Jiang and Shu 1996) is used to reconstruct polynomial functions on each $I_{i+1/2}$ by using the information of the cell $I_{i+1/2}$ and its neighbors, such that they are second or higher order accurate approximations to the functions $\rho(x)$, $M(x)$ and $E(x)$ etc. on $I_{i+1/2}$. Thus the approximate values of each conserved variable $(\rho_i^\pm, M_i^\pm, E_i^\pm)$ at both sides of the cell's boundary are obtained from its reconstructed polynomial.

Four typical numerical fluxes are provided in (Cheng and Shu 2007a) to compute the fluxes given the primitive states at each side of a control volume's boundary, namely, 1) The Godunov flux, 2) The Dukowicz flux, 3) The HLLC (Harten-Lax-van Leer contact wave) flux, 4) The L-F (Lax-Friedrichs) flux. We refer to (Cheng and Shu 2007a) for more details.

Step 2: The determination of vertex velocity
In the Lagrangian formulation, the grid moves with the fluid velocity which is defined at the vertex, thus we would need to know the velocity at the vertex to move the grid. Since the velocity is a derived quantity, it needs to be obtained it from the conserved variables. Specifically, for the Godunov flux and the Dukowicz flux, the vertex velocity is naturally obtained by the (exact or approximate) Riemann solver in the above procedure of flux determination. For the L-F flux and the HLLC flux, the velocity at the cell's vertex is defined as the Roe's average of velocities from both sides,

$$u_i = \frac{\sqrt{\rho_i^-}\,u_i^- + \sqrt{\rho_i^+}\,u_i^+}{\sqrt{\rho_i^-} + \sqrt{\rho_i^+}}. \tag{6}$$

Step 3: Time discretization
The TVD Runge-Kutta type method is used to discretize the time derivative term in the semi-discrete scheme (5). As the grid changes with the time marching in the Lagrangian simulation, the velocity, the position of each vertex and the size of each cell need to be updated at each Runge-Kutta stage. Thus the form of the Runge-Kutta method in our Lagrangian type schemes is as follows (the third-order case is taken here as an example).

Stage 1,

$$x_i^{(1)} = x_i^n + u_i^n \Delta t^n, \qquad \Delta x_{i+1/2}^{(1)} = x_{i+1}^{(1)} - x_i^{(1)},$$
$$\overline{\mathbf{U}}_{i+1/2}^{(1)} \Delta x_{i+1/2}^{(1)} = \overline{\mathbf{U}}_{i+1/2}^{n} \Delta x_{i+1/2}^{n} + \Delta t^n \mathbf{L}(\overline{\mathbf{U}}_{i+1/2}^{n});$$

Stage 2,

$$x_i^{(2)} = \frac{3}{4} x_i^n + \frac{1}{4}[x_i^{(1)} + u_i^{(1)} \Delta t^n],$$
$$\Delta x_{i+1/2}^{(2)} = x_{i+1}^{(2)} - x_i^{(2)},$$
$$\overline{\mathbf{U}}_{i+1/2}^{(2)} \Delta x_{i+1/2}^{(2)} = \frac{3}{4} \overline{\mathbf{U}}_{i+1/2}^{n} \Delta x_{i+1/2}^{n} + \frac{1}{4}\left[\overline{\mathbf{U}}_{i+1/2}^{(1)} \Delta x_{i+1/2}^{(1)} + \Delta t^n \mathbf{L}(\overline{\mathbf{U}}_{i+1/2}^{(1)})\right];$$

Stage 3,

$$x_i^{n+1} = \frac{1}{3} x_i^n + \frac{2}{3}[x_i^{(2)} + u_i^{(2)} \Delta t^n],$$
$$\Delta x_{i+1/2}^{n+1} = x_{i+1}^{n+1} - x_i^{n+1},$$
$$\overline{\mathbf{U}}_{i+1/2}^{n+1} \Delta x_{i+1/2}^{n+1} = \frac{1}{3} \overline{\mathbf{U}}_{i+1/2}^{n} \Delta x_{i+1/2}^{n} + \frac{2}{3}\left[\overline{\mathbf{U}}_{i+1/2}^{(2)} \Delta x_{i+1/2}^{(2)} + \Delta t^n \mathbf{L}(\overline{\mathbf{U}}_{i+1/2}^{(2)})\right],$$

where $\mathbf{L}$ is the numerical spatial operator representing the right hand of the scheme (5). Here the variables with the superscripts n and $n+1$ represent the values of the corresponding variables at the n-th and $(n+1)$-th time steps respectively.

For the construction of the scheme in two-dimensional cartesian and cylindrical coordinates, we refer to (Cheng and Shu 2007a) for more details.

In the paper (Liu, Cheng, and Shu 2009), we also explore the Lax-Wendroff (LW) type time discretization as an alternative procedure to the high order Runge-Kutta time discretization for the above described high order ENO/WENO Lagrangian schemes. The LW time discretization is based on a Taylor expansion in time, coupled with a local Cauchy-Kowalewski procedure to utilize the Partial Differential Equation (PDE) repeatedly to convert all time derivatives to spatial derivatives, and then to discretize these spatial derivatives based on high order ENO reconstruction. Comparing with the Runge-Kutta time discretization procedure, an advantage of the LW time discretization is the saving in computational cost and memory requirement.

In the accuracy test in (Cheng and Shu 2007a), a phenomenon for the high order Lagrangian type scheme is observed, that is, the third order Lagrangian type scheme can only achieve second order accuracy on two dimensional distorted Lagrangian grids. This is analyzed to be due to the error from the mesh approximation. Since in a Lagrangian simulation, a cell with an initially quadrilateral shape may not keep its shape as a quadrilateral at a later time. It usually becomes a curved quadrilateral. Thus if during the Lagrangian simulation the mesh is always kept as quadrilateral with straight-line edges, this approximation of the mesh will bring second order error into the scheme. So for a Lagrangian type scheme in multi-dimensions, it can be at most second order accurate if curved meshes are not used. In (Cheng and Shu 2008b), We demonstrate the previous claim by developing a third order scheme on curved quadrilateral meshes in two space dimensions. The reconstruction is based on the high order WENO procedure. Each curvilinear cell consists of four quadratically-curved edges by the information of the coordinates of its four vertices and the four middle points of its four edges. The accuracy test and some non-oscillatory tests are presented to verify the properties of the scheme. The Lagrangian type scheme can also be extended to higher than third order accuracy if a higher order approximation is used on both the grids and the discretization of the governing equations.

3 THE CELL-CENTERED CONSERVATIVE LAGRANGIAN SCHEMES WITH THE PRESERVATION OF SPHERICAL SYMMETRY IN CYLINDRICAL COORDINATES

In the simulation of certain application problems such as implosion, there is a critical requirement for a Lagrangian scheme to keep spherical symmetry in a cylindrical coordinate system, since a small departure from spherical symmetry due to numerical errors may be magnified by some physical instabilities which may lead to possible large errors. In the past several decades, many research works have been performed concerning the spherical symmetry preservation in two-dimensional cylindrical coordinates. The most widely used method that keeps spherical symmetry exactly on an equal-angle-zoned grid in cylindrical coordinates is the area-weighted method, see e.g. (Caramana, Burton, Shashkov, and Whalen 1998; Maire 2009). In this approach one uses a Cartesian form of the momentum equation in the cylindrical coordinate system, hence integration is performed on area rather than on the true volume in cylindrical coordinates. However, these area-weighted schemes have a flaw in that they may violate momentum conservation. In our recent work (Cheng and Shu 2010), we have developed a new cell-centered control volume Lagrangian scheme for solving the compressible Euler equations in two-dimensional cylindrical coordinates. It is constructed on a genuine volume discretization formulation. By the compatible discretization of pressure in the source term and the flux term, the scheme is designed to be able to preserve one-dimensional spherical symmetry in a two-dimensional cylindrical geometry when computed on an equal-angle-zoned initial grid. A distinguished feature of our scheme is that it can keep both the symmetry and conservation properties on straight-line grids.

In this section, we will summarize the construction strategy of the scheme.

We consider the axisymmetric compressible Euler equations in the cylindrical coordinates which have the following integral form in the Lagrangian formulation

$$\begin{cases} \frac{d}{dt}\iint_{\Omega(t)} \rho r dr dz = 0 \\ \frac{d}{dt}\iint_{\Omega(t)} \rho u_z r dr dz = -\int_{\Gamma(t)} p n_z r dl \\ \frac{d}{dt}\iint_{\Omega(t)} \rho u_r r dr dz = -\int_{\Gamma(t)} p n_r r dl + \iint_{\Omega(t)} p dr dz \\ \frac{d}{dt}\iint_{\Omega(t)} E r dr dz = -\int_{\Gamma(t)} p u_n r dl \end{cases} \tag{7}$$

where z and r are the axial and radial directions respectively. $\mathbf{u} = (u_z, u_r)$ where u_z, u_r are the velocity components in the z and r directions respectively, $\mathbf{n} = (n_z, n_r)$ is the unit outward normal to the boundary $\Gamma(t)$ in the $z - r$ coordinates, and $u_n = (u_z, u_r)\cdot\mathbf{n}$ is the normal velocity at $\Gamma(t)$.

The 2D 1/4-circle-shaped spatial domain Ω is discretized into $K \times L$ computational cells. Figure 1 shows an equal-angle-zoned grid. $I_{k+1/2,l+1/2}$ is a quadrilateral cell constructed by the four vertices $\{(z_{k,l}, r_{k,l}), (z_{k+1,l}, r_{k+1,l}), (z_{k+1,l+1}, r_{k+1,l+1}), (z_{k,l+1}, r_{k,l+1})\}$. $S_{k+1/2,l+1/2}$ and $V_{k+1/2,l+1/2}$ denote the area and volume of the cell $I_{k+1/2,l+1/2}$ with $k = 1, \ldots, K$, $l = 1, \ldots, L$ respectively. The fluid velocity $((u_z)_{k,l}, (u_r)_{k,l})$ is defined at the vertex of the grid.

For the cell-centered scheme, all the variables except velocity are stored at the cell center of $I_{k+1/2,l+1/2}$ in the form of cell averages. To specific, the values of the cell averages of density, z-momentum, r-momentum and total energy for the cell $I_{k+1/2,l+1/2}$, denoted by $\bar{\rho}_{k+1/2,l+1/2}$, $\bar{M}^z_{k+1/2,l+1/2}$, $\bar{M}^r_{k+1/2,l+1/2}$ and $\bar{E}_{k+1/2,l+1/2}$, are defined as follows

$$\bar{\rho}_{k+\frac12,l+\frac12} = \frac{1}{V_{k+\frac12,l+\frac12}}\iint_{I_{k+\frac12,l+\frac12}} \rho r dr dz,$$

$$\bar{M}^z_{k+\frac12,l+\frac12} = \frac{1}{V_{k+\frac12,l+\frac12}}\iint_{I_{k+\frac12,l+\frac12}} \rho u_z r dr dz,$$

$$\bar{M}^r_{k+\frac12,l+\frac12} = \frac{1}{V_{k+\frac12,l+\frac12}}\iint_{I_{k+\frac12,l+\frac12}} \rho u_r r dr dz,$$

$$\bar{E}_{k+\frac12,l+\frac12} = \frac{1}{V_{k+\frac12,l+\frac12}}\iint_{I_{k+\frac12,l+\frac12}} E r dr dz,$$

where $V_{k+1/2,l+1/2} = \iint_{I_{k+1/2,l+1/2}} r dr dz$. In our scheme, the cell averages of the above conserved variables are evolved in time directly.

To construct a finite volume Lagrangian scheme for the governing equations (7), we first formulate the semi-discrete scheme as

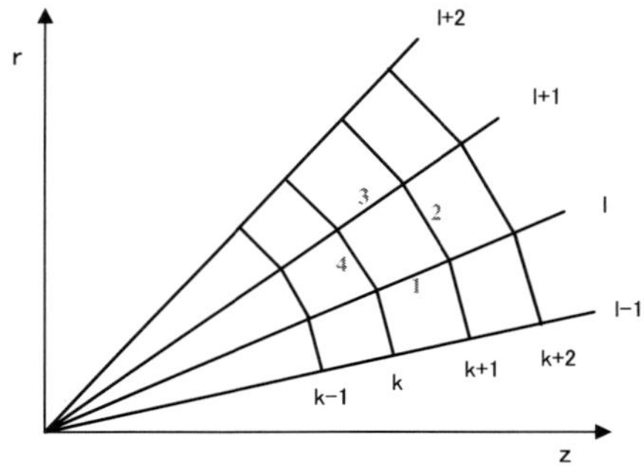

Figure 1. Equi-angular polar grid for cylindrical geometry.

$$\frac{d}{dt}\begin{pmatrix} \bar{\rho}_{k+\frac{1}{2},l+\frac{1}{2}} V_{k+\frac{1}{2},l+\frac{1}{2}} \\ \bar{M}^z{}_{k+\frac{1}{2},l+\frac{1}{2}} V_{k+\frac{1}{2},l+\frac{1}{2}} \\ \bar{M}^r{}_{k+\frac{1}{2},l+\frac{1}{2}} V_{k+\frac{1}{2},l+\frac{1}{2}} \\ \bar{E}_{k+\frac{1}{2},l+\frac{1}{2}} V_{k+\frac{1}{2},l+\frac{1}{2}} \end{pmatrix}$$

$$= -\int_{\partial I_{k+\frac{1}{2},l+\frac{1}{2}}} \hat{\mathbf{F}} dl + \begin{pmatrix} 0 \\ 0 \\ p_{k+\frac{1}{2},l+\frac{1}{2}} s_{k+\frac{1}{2},l+\frac{1}{2}} \\ 0 \end{pmatrix} \tag{8}$$

where

$$\int_{\partial I_{k+\frac{1}{2},l+\frac{1}{2}}} \hat{\mathbf{F}} dl = \int_{\partial I_{k+\frac{1}{2},l+\frac{1}{2}}} \begin{pmatrix} \hat{f}_D(\mathbf{U}_n^-, \mathbf{U}_n^+) \\ \hat{f}_{M^z}(\mathbf{U}_n^-, \mathbf{U}_n^+) \\ \hat{f}_{M^r}(\mathbf{U}_n^-, \mathbf{U}_n^+) \\ \hat{f}_E(\mathbf{U}_n^-, \mathbf{U}_n^+) \end{pmatrix} dl \tag{9}$$

and

$$\begin{cases} \hat{f}_D(\mathbf{U}_n, \mathbf{U}_n) = 0, \\ \hat{f}_{M^z}(\mathbf{U}_n, \mathbf{U}_n) = p n_z r \\ \hat{f}_{M^r}(\mathbf{U}_n, \mathbf{U}_n) = p n_r r \\ \hat{f}_E(\mathbf{U}_n, \mathbf{U}_n) = p u_n r \end{cases} \tag{10}$$

$\hat{f}_D$, $\hat{f}_{M^z}$, $\hat{f}_{M^r}$ and $\hat{f}_E$ are the numerical fluxes for mass, z-momentum, r-momentum and total energy across the cell boundary respectively. $\mathbf{U}_n^\pm = (\rho^\pm, (\rho u_n)^\pm, E^\pm)$ are the values of density, normal component of momentum and total energy at both sides of the cell boundary. $p_{k+1/2,l+1/2}$ is the pressure.

The first step for establishing the scheme is to determine the line integral term on the right side of Equation (8). As the cell boundary $\partial I_{k+1/2,l+1/2}$ consists of 4 edges, the line integral concerned with the flux in Equation (8) is discretized by the following formula,

$$\int_{\partial I_{k+\frac{1}{2},l+\frac{1}{2}}} \hat{\mathbf{F}} dl \approx \sum_{m=1}^{4} \hat{\mathbf{F}}(\mathbf{U}_n^{m+}, \mathbf{U}_n^{m-}) \Delta l^m \tag{11}$$

where Δl^m is the length of the cell edge m. $\hat{\mathbf{F}}(\mathbf{U}_n^{m+}, \mathbf{U}_n^{m-})$ is a numerical flux at the edge m.

As a first order accurate scheme, we just take $\mathbf{U}_n^{m\pm}$ as the left and right cell average values of the conserved variables, that is, for example, for Edge 1 in the cell $I_{k+1/2,l+1/2}$ in Figure 1, we take $\mathbf{U}_n^{1-} = (\bar{\rho}_{k+1/2,l+1/2}, (\bar{M}_n)_{k+1/2,l+1/2}, \bar{E}_{k+1/2,l+1/2})$ and $\mathbf{U}_n^{1+} = (\bar{\rho}_{k+1/2,l-1/2}, (\bar{M}_n)_{k+1/2,l-1/2}, \bar{E}_{k+1/2,l-1/2})$.

After getting $\mathbf{U}_n^{m\pm}$, $m = 1,4$, then we could use the numerical fluxes such as Godunov flux and Dukowicz flux mentioned in the previous section to compute the fluxes at the cell boundary and obtain the pressure p^m and normal velocity u_n^m at each boundary edge.

Thus finally we can get the fluxes $\hat{f}_D$, $\hat{f}_{M^z}$, $\hat{f}_{M^r}$ and $\hat{f}_E$ at Edge m as follows,

$$\begin{cases} \hat{f}_D(\mathbf{U}_n^{m-}, \mathbf{U}_n^{m+}) = 0, \\ \hat{f}_{M^z}(\mathbf{U}_n^{m-}, \mathbf{U}_n^{m+}) = p^m n_z^m r^m \\ \hat{f}_{M^r}(\mathbf{U}_n^{m-}, \mathbf{U}_n^{m+}) = p^m n_r^m r^m \\ \hat{f}_E(\mathbf{U}_n^{m-}, \mathbf{U}_n^{m+}) = p^m u_n^m r^m \end{cases} \tag{12}$$

where (n_z^m, n_r^m) and r^m are the unit outward normal direction and the r coordinate of the middle point of Edge m respectively.

As to the discretization of the source term in (8), commonly $p_{k+1/2,l+1/2}$ is taken as the cell-center value of pressure which is determined by the conserved variables at the cell by the EOS equation. Unfortunately, we can prove that the scheme can never preserve the spherical symmetry if $p_{k+1/2,l+1/2}$ is defined in this way. In order to keep the property of symmetry, we found $p_{k+1/2,l+1/2}$ should be discretized in a compatible way with that in the flux term. By the derivation, we finally choose it as

$$p_{k+\frac{1}{2},l+\frac{1}{2}} = \frac{1}{2}(p^1 + p^3), \tag{13}$$

where p^1 and p^3 are the values of pressure at Edges 1 and 3 of Cell $I_{k+1/2,l+1/2}$ (see Figure 1) which are obtained in the above procedure of flux determination.

For the determination of the vertex velocity, we use a similar method as that shown in the previous section and in (Cheng and Shu 2007a). The Euler forward method is used to discretize the time derivatives in the semi-discrete scheme (8).

Based on the above manipulation, the scheme can preserve the conservation for all the conserved variables, since it is discretized on the true volume and the numerical flux across each cell boundary is single-valued for the update of its two neighboring cells. Meanwhile, we can also prove that the scheme can keep the spherical symmetry property if equi-angular polar initial grids are used, see the paper (Cheng and Shu 2010) for more details.

The above introduced strategy can also be applied to other existing Lagrangian schemes to improve their symmetry property. For example, in the paper

(Cheng and Shu 2011), it is used to improve Maire's control volume scheme (Maire 2009) on its property of spherical symmetry without losing its original good properties such as the conservation of mass, momentum and total energy, the Geometric Conservation Law (GCL) and robustness.

4 THE HIGH ORDER ENO CONSERVATIVE REMAPPING METHOD

In the recent papers (Cheng and Shu 2007b; Cheng and Shu 2008a), we developed a local remapping algorithm on two types of staggered meshes in one and two dimensions where the ENO reconstruction idea is used on the remapping strategy so that the algorithm has the good properties such as uniformly high order accuracy, essential non-oscillation and the conservative for mass, momentum and total energy. The remapping algorithm does not require any relationship between the old and the new meshes. It is therefore suitable not only for continuous rezoning but also for occasional rezoning.

Specifically we consider a two-dimensional computational domain Ω which are covered by quadrilateral cells $I_{i+1/2,j+1/2}$, $i=1,\dots I$, $j=1,\dots J$ without gaps or overlaps. The cell $I_{i+1/2,j+1/2}$ is defined by a set of vertices $\{(i,j),(i+1,j),(i+1,j+1),(i,j+1)\}$, see Figure 2.

Two meshes are concerned with the same numbers of cells and vertices. The mesh that contains the cells $\{I_{i+1/2,j+1/2}\}$ is called the Lagrangian or old mesh. The second mesh, containing the cells $\{\tilde{I}_{k+1/2,l+1/2}\}$, is called the rezoned or new mesh, see Figure 2. In the following, all the quantities of the new mesh are distinguished with a "tilde". After rezoning, the solutions on the old mesh $\{I_{i+1/2,j+1/2}\}$ is mapped into the new mesh $\{\tilde{I}_{k+1/2,l+1/2}\}$. Each cell of the new mesh $\tilde{I}_{k+1/2,l+1/2}$ is formed from pieces of several cells of the old mesh $\{I_{i+1/2,j+1/2}\}$:

$$\tilde{I}_{k+\frac{1}{2},l+\frac{1}{2}} = \bigcup_{i=1,j=1}^{I,J} \left(\tilde{I}_{k+\frac{1}{2},l+\frac{1}{2}} \cap I_{i+\frac{1}{2},j+\frac{1}{2}}\right) \quad (14)$$

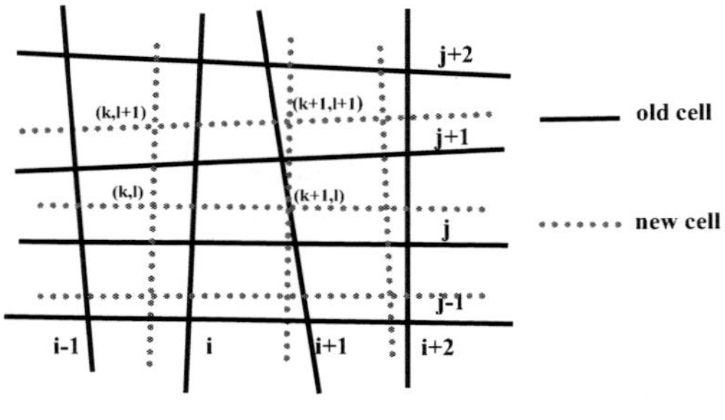

Figure 2. The sketch map of old and new grids.

In the papers (Cheng and Shu 2007b; Cheng and Shu 2008a), the remapping strategies both for the cell-centered quantities such as density and nodal quantity such as velocity are investigated. Here, to save space, we take the density as an example to illustrate the ENO remapping algorithm.

Before the remapping, the only information we know is the cell average of density in each of the cells of the old grid:

$$\bar{\rho}_{i+\frac{1}{2},j+\frac{1}{2}} = \frac{1}{S_{i+\frac{1}{2},j+\frac{1}{2}}} \iint_{I_{i+\frac{1}{2},j+\frac{1}{2}}} \rho(x,y)\,dxdy \quad (15)$$

where $S_{i+1/2,j+1/2}$ is the area of the cell $I_{i+1/2,j+1/2}$.

Next we will reconstruct a polynomial function on each $I_{i+1/2,j+1/2}$ by using the information of the cell $I_{i+1/2,j+1/2}$ and its neighbors, such that it is a second or higher order accurate approximation to the function $\rho(x,y)$ on $I_{i+1/2,j+1/2}$. The procedure allows us to obtain any high order accurate approximation by a proper reconstruction. For simplicity, in the following we will only give the details of the second order accurate remapping. we refer to the paper (Cheng and Shu 2007b) for details about the third order remapping.

Suppose that our reconstructed linear polynomial on the cell $I_{i+1/2,j+1/2}$ is of the form

$$\rho_{i+\frac{1}{2},j+\frac{1}{2}}(x,y) = a^1_{i+\frac{1}{2},j+\frac{1}{2}}(x - x_{i+\frac{1}{2},j+\frac{1}{2}}) + a^2_{i+\frac{1}{2},j+\frac{1}{2}}(y - y_{i+\frac{1}{2},j+\frac{1}{2}}) + a^3_{i+\frac{1}{2},j+\frac{1}{2}} \quad (16)$$

where $(x_{i+1/2,j+1/2}, y_{i+1/2,j+1/2})$ are the coordinates of the cell center and the coefficients $a^1_{i+1/2,j+1/2}$, $a^2_{i+1/2,j+1/2}$ and $a^3_{i+1/2,j+1/2}$ are determined by satisfying the given cell averages on a 3-cell stencil including $I_{i+1/2,j+1/2}$, for example by

$$\begin{aligned}
&\iint_{I_{i+\frac{1}{2},j-\frac{1}{2}}} \rho_{i+\frac{1}{2},j+\frac{1}{2}}(x,y)\,dxdy = \rho_{i+\frac{1}{2},j-\frac{1}{2}} S_{i+\frac{1}{2},j-\frac{1}{2}} \\
&\iint_{I_{i+\frac{1}{2},j+\frac{1}{2}}} \rho_{i+\frac{1}{2},j+\frac{1}{2}}(x,y)\,dxdy = \rho_{i+\frac{1}{2},j+\frac{1}{2}} S_{i+\frac{1}{2},j+\frac{1}{2}} \\
&\iint_{I_{i+\frac{3}{2},j+\frac{1}{2}}} \rho_{i+\frac{1}{2},j+\frac{1}{2}}(x,y)\,dxdy = \rho_{i+\frac{3}{2},j+\frac{1}{2}} S_{i+\frac{3}{2},j+\frac{1}{2}}
\end{aligned} \quad (17)$$

then it is a second-order accurate approximation to the function $\rho(x,y)$ on $I_{i+1/2,j+1/2}$.

In order to have an essentially non-oscillatory algorithm, we use the idea of an ENO reconstruction. For the density reconstruction on the cell

$I_{i+1/2,j+1/2}$, we consider the following four sets of nonsingular stencils:

$$\begin{aligned}&\left\{I_{i+\frac{1}{2},j-\frac{1}{2}}, I_{i+\frac{1}{2},j+\frac{1}{2}}, I_{i-\frac{1}{2},j+\frac{1}{2}}\right\},\\&\left\{I_{i-\frac{1}{2},j+\frac{1}{2}}, I_{i+\frac{1}{2},j+\frac{1}{2}}, I_{i+\frac{1}{2},j+\frac{3}{2}}\right\},\\&\left\{I_{i+\frac{3}{2},j+\frac{1}{2}}, I_{i+\frac{1}{2},j+\frac{1}{2}}, I_{i+\frac{1}{2},j-\frac{1}{2}}\right\},\\&\left\{I_{i+\frac{1}{2},j+\frac{3}{2}}, I_{i+\frac{1}{2},j+\frac{1}{2}}, I_{i+\frac{3}{2},j+\frac{1}{2}}\right\}.\end{aligned} \tag{18}$$

Finally we choose the function which has the least slope value ($|a^1_{i+1/2,j+1/2}|+|a^2_{i+1/2,j+1/2}|$) among four linear functions reconstructed by the above stencils as the reconstruction function $\rho_{i+1/2,j+1/2}(x,y)$ on the cell $I_{i+1/2,j+1/2}$.

The approximation to the average of the density $\tilde{\rho}_{k+1/2,l+1/2}$ on the new cell $\tilde{I}_{k+1/2,l+1/2}$ is then given by

$$\tilde{\rho}_{k+\frac{1}{2},l+\frac{1}{2}} = \frac{1}{\tilde{S}_{k+\frac{1}{2},l+\frac{1}{2}}}\iint_{\tilde{I}_{k+\frac{1}{2},l+\frac{1}{2}}} \rho(x,y)dxdy = \frac{\sum_{i=1}^{I}\sum_{j=1}^{J}\iint_{\tilde{I}_{k+\frac{1}{2},l+\frac{1}{2}}\cap I_{i+\frac{1}{2},j+\frac{1}{2}}} \rho_{i+\frac{1}{2},j+\frac{1}{2}}(x,y)dxdy}{\tilde{S}_{k+\frac{1}{2},l+\frac{1}{2}}}$$

In the paper (Cheng and Shu 2007b), the efficient algorithms for determining the intersection regions between the old cells and the new cells and calculating the integration of the reconstruction function on a polygon with any possible shape are also discussed, which are very helpful to the remapping algorithm being implemented efficiently, see (Cheng and Shu 2007b) for details.

Based on the above described procedure of remapping, the algorithm can keep the good properties of conservation, uniformly high order accuracy and essential non-oscillation which are also demonstrated by the numerical tests performed in (Cheng and Shu 2007b; Cheng and Shu 2008a).

5 NUMERICAL EXAMPLES

In our previous mentioned papers, many numerical tests have been shown to demonstrate the properties of the related schemes. Here in this section, we give a few more numerical experiments to further verify the performance of the schemes. Purely Lagrangian computation with the Dukowicz flux is used to do the following first two tests for the Lagrangian schemes.

Example 1 (Sod problem for the Lagrangian simulation) The initial data of this one dimensional shock tube problem are

$$(\rho,u,p)=\begin{cases}(1,0,1), & -10\le x\le 0\\(0.125,0,0.1), & 0<x\le 10.\end{cases}$$

$\gamma=1.4$. The reflective boundary condition is used at two sides. The ENO Lagrangian schemes introduced in Section 2 are applied to simulate this problem. Figure 3 plots the results with 200 initially uniform cells at $t=1.4$. From the figures, we can observe the front of the contact discontinuity is quite sharp which demonstrates the advantage

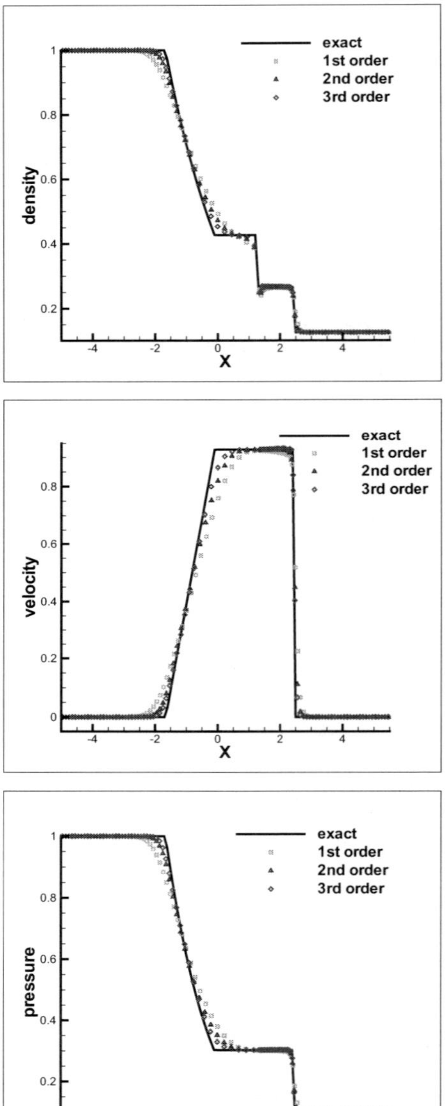

Figure 3. The results of the sod problem at $t=1.4$ with 200 cells. Top: density; Middle: velocity; Bottom: pressure. Solid line: exact solution; symbol: computational solution.

of the Lagrangian scheme. Also, it can be observed that more satisfactory resolution for all the waves is obtained by the higher order scheme.

Example 2 (One-dimensional spherical shock tube problem for the two-dimensional Lagrangian simulation)
This is a spherical shock tube problem. The initial condition is given by the following data

$$(\rho,u,e)=\begin{cases}(1,0,0.1), & 0\le radius<3,\\(0.01,0,0.01), & 3<radius\le 9.\end{cases}$$

$\gamma=5/3$. We test it by the cell-centered Lagrangian scheme described in Section 3 in two-dimensional cylindrical coordinates.

The equi-angular polar grid is applied in the $\frac{1}{4}$-circle computational domain defined in the polar coordinates by $[0,9]\times[0,\pi/2]$. The reflective boundary condition is applied at the outer boundary. Figure 4 shows the initial grid and the final grid at $t=6$ with 100×50 cells and internal energy as a function of radial radius with 400×100 and 1000×400 cells at $t=6$ respectively. In the plot of the grid, we observe the symmetry is perfectly preserved. In the plot of internal energy, we see that the numerical solution especially of the contact is in good agreement with the reference solution which is computed using a one-dimensional spherical third order Eulerian WENO scheme with 10000 cells. The location and magnitude of the shock are also closer to the reference solution with the refinement of the grid which reflects the convergence trend of the numerical solution.

Example 3 (The smooth problem for remapping)
For the following two remapping tests, we choose a sequence of random meshes which are obtained by an independent random perturbation from a uniform mesh as the moving meshes.

$$x^n_{i,j}=\xi_i+cr^n_{i,j}\Delta x,\qquad y^n_{i,j}=\eta_j+cs^n_{i,j}\Delta y$$

where $-0.5\le r^n_{i,j},s^n_{i,j}\le 0.5$ are random numbers. c is the parameter taken as 0.5 in the following two tests. We will show the results on the random mesh after 200 remapping steps. The computational domain is $[0,1]\times[0,1]$ with 40×40 cells.

The smooth test function is chosen as follows:

$$\rho(x,y)=1+0.6(\sin(\pi x)^2+\sin(\pi y)^2).$$

In Figure 5 we show the surface plots of density for the exact solution and the results of remapping obtained by the method described in Section 4. The figures demonstrate that the higher order remapping method produces results with better resolution.

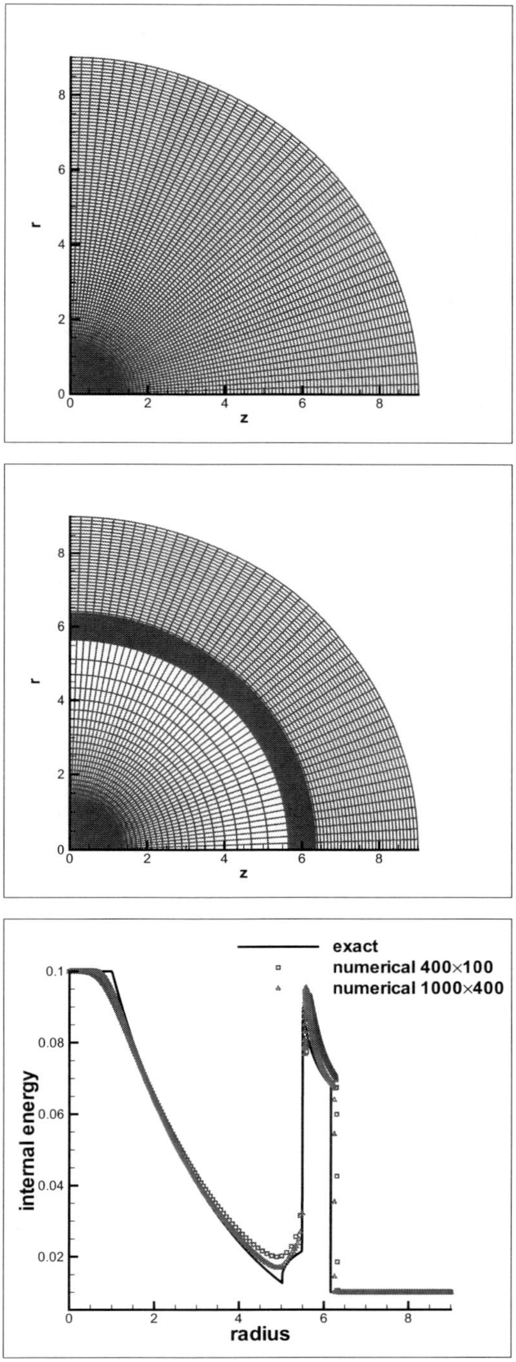

Figure 4. The results of the spherical shock tube problem. Top: initial grid; Middle: final grid at $t=6$; Bottom: internal energy vs radial radius with 400×100 and 1000×400 cells at $t=6$. Solid line: reference solution; symbol: computational solution.

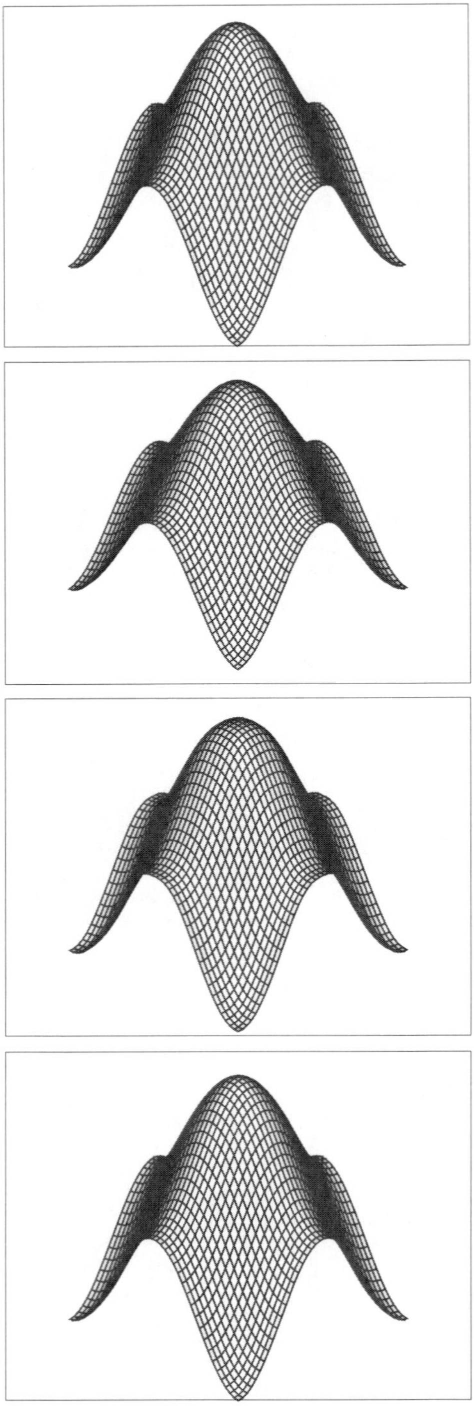

Figure 5. The results of the smooth problem with 200 remapping steps. First: exact; Second: first order; Third: second order; Last: third order.

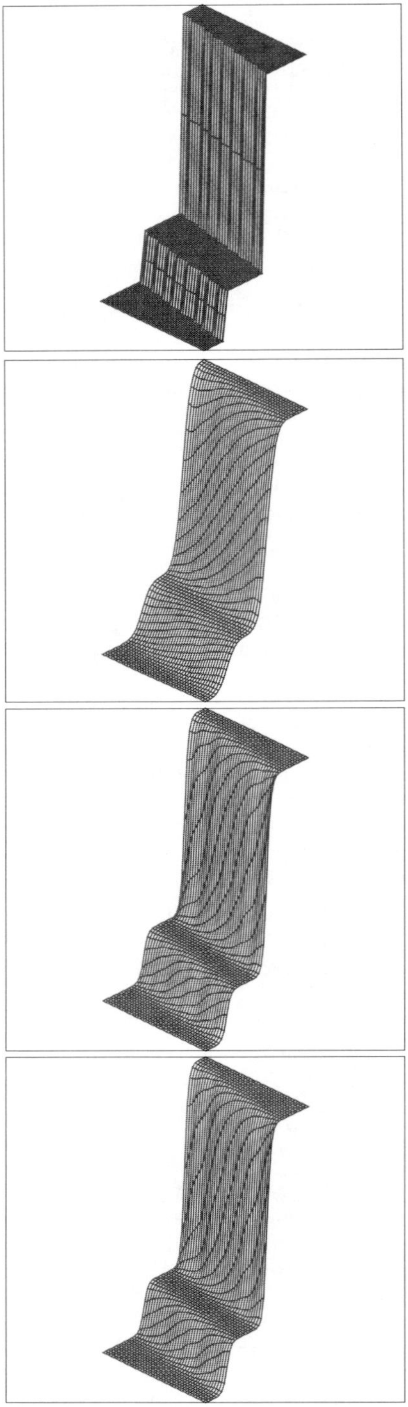

Figure 6. The results of the two-shock interaction problem with 200 remapping steps. First: exact; Second: first order; Third: second order; Last: third order.

Example 4 (The two-shock problem for remapping) We investigate profiles associated with two shocks which are defined as:

$$\rho(x,y) = \begin{cases} 1, & y \le 5(x-0.6), \\ 200, & 5(x-0.6) < y \le 5(x-0.2), \\ 1000, & else. \end{cases} \quad (19)$$

The remapping results for this two-shock case are shown in Figure 6 where we find that the shock profile for the first order remapping method smears a quite lot while the ones for the second and third order remapping methods are sharp and non-oscillatory. The higher order ENO scheme obtains more precise solution than the lower order one.

6 CONCLUSIONS

In this paper, we have reviewed our recent research works on the Lagrangian method and remapping method which include cell-centered high order ENO/WENO Lagrangian type schemes both on straight-line quadrilateral meshes and curvilinear quadrilateral meshes, the cell-centered conservative Lagrangian scheme with the preservation of spherical symmetry in cylindrical coordinates and a high order conservative ENO remapping algorithm. Several new numerical tests are given to further verify the performance of the methods. Our future work will involve the improvement of the schemes in accuracy and robustness and more application in the simulation of multi-fluid problems.

REFERENCES

Browne, P. (1986). Integrated gradients: a derivation of some difference forms for the equation of motion for compressible flow in two-dimensional lagrangian hydrodynamics, using integration of pressures over surfaces. *Los Alamos National Laboratory Report LA-2105872-MS*.

Caramana, E., D. Burton, M. Shashkov, and P. Whalen (1998). The construction of compatible hydrodynamics algorithms utilizing conservation of total energy. *Journal of Computational Physics 146*, 227–262.

Cheng, J. and C.-W. Shu (2007a). A high order eno conservative lagrangian type scheme for the compressible euler equations. *Journal of Computational Physics 227*, 1567–1596.

Cheng, J. and C.-W. Shu (2007b). High order eno conservative remapping method on staggered grids for ale methods: a review and an alternative momentum remapping. pp. 40–56.

Cheng, J. and C.-W. Shu (2008a). A high order accurate conservative remapping method on staggered meshes. *Applied Numerical Mathematics 58*, 1042–1060.

Cheng, J. and C.-W. Shu (2008b). A third order conservative lagrangian type scheme on curvilinear meshes for the compressible euler equation. *Communications in Computational Physics 4*, 1008–1024.

Cheng, J. and C.-W. Shu (2010). A cell-centered lagrangian scheme with the preservation of symmetry and conservation properties for compressible fluid flows in two-dimensional cylindrical geometry. *Journal of Computational Physics 229*, 7191–7206.

Cheng, J. and C.-W. Shu (2011). Improvement on spherical symmetry in two-dimensional cylindrical coordinates for a class of control volume lagrangian schemes. *Communications in Computational Physics*.

Dukowicz, J. and J. Baumgardener (2000). Incremental remapping as a transport/advection algorithm. *Journal of Computational Physics 160*, 318–335.

Godunov, S., A. Zabrodine, M. Ivanov, A. Kraiko, and G. Prokopov (1979). Résolution numérique des problémes multidimensionnels de la dynamique des gaz. *Edition Mir, Moscou*.

Harten, A., B. Engquist, S. Osher, and S. Chakravarthy (1987). Uniformly high order accurate essentially non-oscillatory schemes, iii. *Journal of Computational Physics 71*, 231–303.

Harten, A. and S. Osher (1987). Uniformly high-order accurate nonoscillatory schemes i. *SIAM Journal on Numerical Analysis 24*, 279–309.

Hirt, C., A. Amsden, and J. Cook (1974). An arbitrary lagrangian-eulerian computing method for all flow speeds. *Journal of Computational Physics 14*, 227–253.

Jiang, G. and C.-W. Shu (1996). Efficient implementation of weighted eno schemes. *Journal of Computational Physics 126*, 202–228.

Kucharik, M., M. Shashkov, and B. Wendroff (2003). An efficient linearity-and-bound-preserving remapping method. *Journal of Computational Physics 188*, 462–471.

Liu, W., J. Cheng, and C.-W. Shu (2009). High order conservative lagrangian schemes with lax-wendroff type time discretization for the compressible euler equations. *Journal of Computational Physics 228*, 8872–8891.

Liu, X.-D., S. Osher, and T. Chan (1994). Weighted essentially non-oscillatory schemes. *Journal of Computational Physics 115*, 200–212.

Maire, P.-H. (2009). A high-order cell-centered lagrangian scheme for compressible fluid flows in two-dimensional cylindrical geometry. *Journal of Computational Physics 228*, 6882–6915.

Maire, P.-H., R. Abgrall, J. Breil, and J. Ovadia (2007). A cell-centered lagrangian scheme for compressible flow problems. *SIAM Journal of Scientific Computing 29*, 1781–1824.

Margolin, L. and M. Shashkov (2003). Second-order sign-preserving conservative interpolation (remapping) on general grids. *Journal of Computational Physics 184*, 266–298.

Munz, C. (1994). On godunov-type schemes for lagrangian gas dynamics. *SIAM Journal on Numerical Analysis 31*, 17–42.

von Neumann, J. and R. Richtmyer (1950). A method for the calculation of hydrodynamics shocks. *Journal of Applied Physics 21*, 232–237.

Numerical Methods for Hyperbolic Equations – Vázquez-Cendón et al. (eds)
© 2013 Taylor & Francis Group, London, ISBN 978-0-415-62150-2

Coupling of nonlinear hyperbolic systems: A journey from mathematical to numerical issues

Frédéric Coquel
CNRS, UMR 7641 and Centre de Mathémathiques Appliquées, École Polytechnique, France

ABSTRACT: The present paper proposes an overview of some recent results devoted to the coupling of nonlinear hyperbolic PDEs. It reviews several distinct but complementary mathematical formalisms for which existence of a solution for the coupledproblem is proved under fairly general assumptions. In some instances, resonance phenomena may occur at the expense of multiple discontinuous solutions and suitable selection criteria for restoring uniqueness are investigated. Several illustrative examples of importance in the applications are discussed. Generic well-balanced numerical techniques are described and the known convergence results are stated.

1 INTRODUCTION

The coupling of non-linear hyperbolic equations is motivated by the modeling of complex industrial setups under the form of a partition of components. The partition under consideration results from the fact that the phenomena taking place in a given component can be characterized by a typical range of physical time and space scales which are in addition well separated. To the resulting hierarchy of characteristic scales is associated a corresponding hierarchy of PDEs models that are naturally arranged from the coarsest to the finest ones according to the characteristic scales they are supposed to resolve.

The large size of the complete device generally prevents from modeling each component using the finest model, the CPU effort would be indeed too large. The simulation of the whole operating system therefore requires to solve a Cauchy problem for a collection of hyperbolic PDEs formulated on a given partition of the physical domain, separated by interfaces at which transient exchange conditions have to be prescribed. For simplicity, we only address the case of given fixed interfaces, whose locations are decided *ab initio* by the user.

We describe in the present survey two complementary mathematical formalisms for handling the coupling of hyperbolic systems. Each supports the property that it is possible to prescribe infinitely many distinct coupling conditions. The first framework relies on the use of infinitely thin interfaces that allows to model the coupling on the ground of coupled boundary conditions. Such conditions are formulated so as to promote some continuity properties for the unknown across the coupling interface, the choice of which is left to the user. Existence of a solution for the coupled problem is achieved under fairly general assumptions. But in some instances, resonance phenomena may occur at the expense of loosing the expected continuity properties and generally yields multiple discontinuous solutions. Despite that resonance is not expected to occur in several industrial settings, this nevertheless rises the natural question of designing suitable selection criteria to restore uniqueness.

The second framework intends to tackle this issue. Roughly speaking, it consists in modeling the existence of the coupling interfaces under the form of standing waves for an augmented PDEs model which is set over the entire physical domain. Each standing wave is designed so as to exhibit a complete set of Riemann invariants (away from resonance) in agreement with the expected continuity properties in the coupled solutions. The coupling problem thus takes the form of a standard Cauchy problem which can in turn support various regularization mechanisms. In a single space dimension, we adopt the viscous regularization *á la Dafermos*. Existence of self-similar weak solutions for the coupling of two hyperbolic systems can be obtained under fairly general conditions. However, in the limit of a vanishing viscosity, the lack of uniqueness of resonant solutions can be still observed. We will explain in which sense multiplicity in the resonant solutions sounds natural. This will naturally lead us to promote another regularization mechanism based on thickened coupling interfaces resulting from the use of smooth transition profiles in between two domains. The proposed framework naturally allows for the definition of multidimensional and multicomponent couplings with possible recovering. Within this frame, existence and uniqueness of the coupled solution is proved in the setting of scalar conservations laws thanks to a convenient well-balanced finite volume strategy in several space variables. The format of the paper closely follows the argumentation developed along the introduction.

2 THIN INTERFACE COUPLING: A DOUBLE IBVP FORMALISM

For simplicity, we restrict ourselves to PDEs in a single space dimension when addressing first the coupling of nonlinear hyperbolic systems with unknown of the same size. We investigate the setting of an infinitely thin coupling interface, located say at $x = 0$ for definiteness. Let us consider the coupling of two strictly hyperbolic systems of conservation laws posed in half-spaces:

$$\begin{aligned} &\partial_t w + \partial_x f_-(w) = 0, \qquad x<0, \quad t>0, \\ &\partial_t w + \partial_x f_+(w) = 0, \qquad x>0, \quad t>0, \end{aligned} \tag{1}$$

where the flux-functions $f^{\pm}$ are given smooth maps defined on open subsets $\Omega_{\pm} \subset \mathbb{R}^N$. The unknown of the problem is $w = w(t,x) \in \Omega_- \cup \Omega_+$. Prescribing the initial data $w(0,x) = w_0(x)$ at time $t = 0$ does not suffice, an additional condition for modeling the transient exchange of informations at the (fixed) coupling interface $x = 0$ must be supplemented. Such an additional closure is called hereafter the *coupling condition*, which is also referred to as a *transmission condition*. Focusing at this stage on piecewise Lipschitz-continuous solutions with bounded left and right traces, a coupling condition expresses most of the time a continuity property at $x = 0$ for the unknown w or some nonlinear transformation of it. The key feature stems from the possibility to impose infinitely many distinct coupling conditions. Indeed, the coupling condition may model the conservation of the unknown when requiring the continuity of the fluxes

$$f^-(w(t,0^-)) = f^+(w(t,0^+)), \quad t>0. \tag{2}$$

Such a continuity condition seems natural in several settings and has received a considerable attention over the past decade. Let us quote for instance (Adimurthi et al. 2005a; Audusse and Perthame 2005; Bachmann and Vovelle 2006; Bürger and Karlsen 2008; Seguin and Vovelle 2003). However, there exists many other situations from the Physics of compressible materials, according to which distinct continuity properties have to be preferred. We refer the reader to the recent works (Ambroso et al. 2007a; Ambroso et al. 2008a; Ambroso et al. 2008b; Ambroso et al. 2008c; Ambroso et al. 2005b; Ambroso et al. 2007b; Chalons et al. 2008) devoted to the mathematical modeling of several complex media. In these works, the transmission conditions rely on two monotonic (say increasing) functions θ_- and θ_+, $\theta_\pm : winIR^N \mapsto \theta_\pm(w) \in \mathbb{R}^N$, for expressing the following continuity property:

$$\theta_-(w(t,0^-)) = \theta_+(w(t,0^+)), \quad t>0. \tag{3}$$

As clearly seen hereafter, the reported monotonicity condition actually preserves the time arrow in the coupled problem. Besides we will show hereafter that the framework for 3 also allows to handle the conservative coupling setting 2 in the case of pairs of general non-invertible fluxes $(f^-;f^+)$ thanks to convenient relaxation approximation procedures (see also (Ambroso et al. 2007a)).

Promoting a given transmission condition 3 allows to regard the coupling problem as a boundary and initial value problem within each half-domain, and the fundamental question is how to formulate physically relevant boundary conditions so that the coupled problem is well-posed. Boundary conditions must be formulated so that to understand 3 in a weaker form. Indeed, it is well-known that prescribing boundary conditions for hyperbolic PDEs, generally requires a weak formalism. In the pioneering contributions (Godlewski and Raviart 2004) and (Godlewski and Raviart 2004; Godlewski et al. 2005), the formalism by Dubois and LeFloch (Dubois and LeFloch 1988) is used.

Following these authors, Ambroso *et al* (Ambroso et al. 2006) impose that the interface condition is satisfied *in a weak sense* and takes the form

$$\begin{aligned} &w(0_+,t) \in \mathcal{O}_+(\theta_+ \circ \theta_-^{-1}(w(0-,t))), \\ &w(0_-,t) \in \mathcal{O}_-(\theta_- \circ \theta_+^{-1}(w(0+,t))). \end{aligned} \tag{4}$$

where $\mathcal{O}_+(b_+)$ (resp. $\mathcal{O}_-(b_-)$) is the Dubois-LeFloch's set of admissible traces of the associated Riemann solutions

$$\begin{aligned} &\mathcal{O}_+(b_+) := \{\mathcal{W}_+(0+;b_+,a), \quad a \in \Omega_+\}, \\ &\mathcal{O}_-(b_-) := \{\mathcal{W}_-(0-;a,b_-), \quad a \in \Omega_-\}. \end{aligned}$$

Here, $R = \mathcal{W}_+(x/t;b_+,a)$ denotes the solution of the Riemann problem

$$\begin{aligned} &\partial_t R + \partial_x f_+(R) = 0, \qquad x \in \mathbb{R}, \quad t>0, \\ &R(x,0) = \begin{cases} b_+, & x<0, \\ a, & x>0, \end{cases} \end{aligned}$$

and similarly $R = \mathcal{W}_-(x/t;a,b_-)$ is the self-similar solution of

$$\partial_t R + \partial_x f_-(R) = 0, \qquad x \in \mathbb{R}, \quad t > 0,$$
$$R(x,0) = \begin{cases} a, & x < 0, \\ b_-, & x > 0. \end{cases}$$

The question of the existence and uniqueness of weak solutions satisfying 4 is a challenging issue. We refer the reader to (Boutin et al. 2010) for general existence results in the scalar setting (see also (Godlewski and Raviart 2004)). General existence results can be also found in (Boutin et al. 2009a), it is shown in particular that the weak form 4 for the coupling condition resumes to the strong version 3 as long as some resonance phenomenadoes not take place at the interface. By resonance, it is meant that waves from either the left and/or the right problems do interfere with the interface, the reported phenomena will be seen hereafter to coincide with the usual notion of resonance between nonlinear waves and a standing one. We refer to the pioneering paper (Isaacson and Temple 1992). In such cases, the expected continuity property 3 is lost in general and in addition, multiple discontinuous solutions are in order (see also (Boutin et al. 2010). The natural question of designing suitable selection criteria to restore uniqueness will be addressed later on.

2.1 *The case of hyperbolic systems with distinct phase space dimension*

We briefly address the coupling of hyperbolic systems with distinct size. Similar transmission condition can be expressed as soon as there exists a lift operator L from the smaller set of variables w to the larger one W, and conversely a projection operator P from the larger to the smaller set of variables. In practice, the definition of these two operators readily follows from underlying relaxation procedures. We refer to (Ambroso et al. 2006), (Ambroso et al. 2008a), and hereafter forexamples. Assuming without restriction that the system with the larger set of variables is posed on the half-domain $\mathbb{R}^-$, coupling conditions then generically write:

$$\begin{aligned} &w(0^-,t) \in \mathcal{O}_-(\theta_-(\mathcal{L}W(0^+,t))), \\ &W(0^+,t) \in \mathcal{O}_+(\theta_+(\mathcal{P}w(0^-,t))). \end{aligned} \tag{5}$$

2.2 *About the choice of the transmission conditions*

The particular choice of the two mappings $\theta_\pm$ entering the definition of the coupling condition 3 and their weak form 4 should be performed according to the Physics underlying the PDE models to be coupled. Partial guidelines for promoting a particular pair can be anyhow formulated on the ground of the following two observations.

First and importantly, the choice of a given pair (θ_-, θ_+) actually dictates the definition of the constant in space and time solutions of the coupled problem, namely piecewise constant solutions made of two constant states w_- and w_+ with a jump $x = 0$, such that

$$u \equiv \theta_-(w_-) = \theta_+(w_+). \tag{6}$$

From the standpoint of the Physics, such constant solutions actually model all the states at thermo-mechanical rest. For instance, focusing on the Euler equations with two distinct pressure laws, namely p_- and p_+, one could promote the conservation of mass, momentum and total energy $(\rho, \rho u, \rho E)^T$ choosing

$$\theta_-(w) = \theta_+(w) = w,$$

but one could instead favor the constance of the density, velocity and pressure $(\rho, u, p)^T$ with

$$\theta_-(w) = (\rho, u, p_-(w))^T, \quad \theta_+(w) = (\rho, u, p_+(w))^T.$$

Obviously these two choices are distinct. The last one 6 preserves the conservation of the density and of momentum but not of the total energy.

Next and if a particular pair (θ_-, θ_+) prescribes the definition of stationary constant solutions, it does also obviously affect the transient behavior of the coupling interface. A particular attention has been paid to the study of the long time behavior of the solutions of the general Cauchy coupled problem. The analysis of the solutions of the coupled Riemann problem, that is the Cauchy problem 1–4 with initial data

$$w_0(x) = \begin{cases} w_L, & x < 0, \\ w_R, & x > 0, \end{cases} \tag{7}$$

plays a central role in that direction, since it is known after Dafermos (Dafermos 1973), that the self-similar solution coming with 7 actually represents the long time behavior of the solution of the Cauchy problem with initial data $w_0(x)$ satisfying $\lim_{x\to-\infty} w_0(x) = w_L$ and $\lim_{x\to+\infty} w_0(x) = w_R$.

Besides, self-similar solutions are also a well-known major ingredient in the assessment of the relevance of numerical methods.

3 NUMERICAL COUPLING: A FIRST WELL-BALANCED APPROACH

We briefly describe a generic finite volume strategy for approximating the solutions of the coupled Cauchy problem 1–4. Introduced in the case 2.2 in (Godlewski and Raviart 2004), this strategy has been extended in (Ambroso et al. 2006) to general pairs 3 and used in (Ambroso et al. 2008a) for instance. In order to ease the description of the method, we make use of dedicated notations. Being given a constant space step Δx, a given mesh cell $\mathcal{C}_{j+1/2}$, with $j \in Z$, is defined by its center $x_{j+1/2} = (j+1/2)\Delta x$ and its two interfaces $x_j = j\Delta x$ and x_{j+1}. The discrete solution $w_{\Delta x}(t,x)$ is sought under the form of a piecewise constant function:

$$w_{\Delta x}(t,x) = w^n_{j+1/2},\quad (t,x) \in \mathcal{C}_{j+1/2} \times [t^n \equiv n\Delta t, (n+1)\Delta t],$$

with $n \in N$, $j \in Z$. Here and for simplicity, Δt denotes a constant time step. The initial data w_0 is classically discretized according to:

$$w^0_{j+1/2} = \frac{1}{\Delta x}\int_{C_{j+1/2}} w_0(x)dx, \quad j \in Z.$$

Being given a numerical flux function g_- (respectively g_+): $\Omega \times \Omega \to \mathbb{R}^N$, consistent with the exact flux function f_- (respectively with f_+), the discrete solution is evolved in time thanks to the following finite volume formulae:

$$\begin{aligned} w^{n+1}_{j-1/2} &= w^n_{j-1/2} - \frac{\Delta t}{\Delta x}(\{g_-\}^n_j - \{g_-\}^n_{j-1}), \\ &\qquad j \le 0, n \ge 0, \\ w^{n+1}_{j+1/2} &= w^n_{j+1/2} - \frac{\Delta t}{\Delta x}(\{g_+\}^n_{j+1} - \{g_+\}^n_j), \\ &\qquad j \ge 0, n \ge 0, \end{aligned} \tag{8}$$

where for $j \neq 0$ (namely except at the coupling interface $j = 0$), the numerical fluxes are naturally defined by $\{g\}_{\pm j}^{\,n} = g_\pm(w^n_{j-1/2}, w^n_{j+1/2})$. The transmission condition 4 being chosen, the numerical fluxes at the coupling interface $j = 0$ are defined as follows:

$$\begin{aligned} \{g_-\}^n_0 &= g_-(w^n_{-1/2}, \theta_- \circ \theta_+^{-1}(w^n_{1/2})), \\ \{g_+\}^n_0 &= g_+(\theta_+ \circ \theta_-^{-1}(w^n_{-1/2}), w^n_{1/2}). \end{aligned} \tag{9}$$

This two numerical fluxes approach can be understood as a direct extension to the coupling framework detailed in the first section of a ghost fluid approach due to Fedkiw *et al* (Fedkiw et al. 1999) (see also Abgrall-Karni (Abgrall and Karni, 2001)). By construction,the proposed finite volume method is well-balanced in the sense that it exactly preserves stationary solutions of the coupled Riemann problem, namely solutions under the form:

$$R(x/t; w_L, w_R) = \begin{cases} w_L, & x < 0, \\ w_R, & x > 0, \end{cases} \tag{10}$$

with $\quad \theta_-(w_L) \equiv \theta_+(w_R)$.

4 EXAMPLES

4.1 *The scalar setting*

A first complete mathematical and numerical study has been performed in (Godlewski and Raviart 2004) for the continuity condition $w(0^-,t) = w(0^+,t)$, namely in the case $\theta^- = \theta^+ = Id$. The coupled Riemann problem in the setting of a general pair (θ^-, θ^+) of increasing functions is analyzed for general pair of fluxes (f^-, f^+) in (Boutin et al. 2010). A distinct approach based on the Dafermos self-similar regularization is proposed in (Boutin et al. 2008) (see also hereafter) under the same general assumptions. It avoids the intricate gluing analysis of distinct families of waves at the coupling interface in case of the resonance. It yields in addition useful representation formulae of the solutions. According to the quoted works, existence of self-similar solutions for the coupled Riemann problem is guaranteed under fairly general assumptions. Failure of uniqueness can be exhibited and corresponds to a change of sign in the wave velocities $f'^-(w)$ and $f'^+(w)$. This scenario coincides with the well-known resonance phenomena first analyzed by Isaacson-Temple (Isaacson and Temple 1992) in the scalar setting.

4.2 *Euler systems of gas dynamics in Lagrangian coordinates*

Let us consider the coupling of two p-systems

$$w = (\tau, u)^T, \quad f_\pm(w) = (-u, p_\pm(\tau))^T, \tag{11}$$

where the two isentropic pressure laws $p_\pm$ classically denote two strictly convex decreasing functions of the specific volume τ. In (Ambroso et al. 2008a) it is shown that the coupling condition 4 expressed for a pair (θ_-, θ_+) in the form $\theta_\pm(w) = (h_\pm(\tau), u)^T$, where $h_\pm$ denote two strictly (say) increasing functions, is always satisfied in the strong sense 3:

$$\begin{aligned} h_-(\tau(0^-,t)) &= h_+(\tau(0^+,t)), \\ u(0^-,t) &= u(0^+,t). \end{aligned} \tag{12}$$

In addition and excluding vacuum, the solution of the coupled Riemann problem 11–12 exists and is uniquely defined. It is worthy to briefly exemplify two distinct cases. The first one with $h_\pm = Id$ yields the continuity of the unknown $w(0^-,t) = w(0^+,t)$ while the second choice, $h_\pm(\tau) = p_\pm(\tau)$, preserves the conservation of the unknown w, since it ensures the continuity of the fluxes $f_-(w(0^-,t)) = f_+(w(0^+,t))$.

Let us now address the coupling of the 3 × 3 Euler equations in Lagrangian coordinates:

$$w = (\tau, u, e)^T, \quad f_\pm(w) = (-u, p_\pm(\tau,\varepsilon), p_\pm(\tau,\varepsilon)u)^T, \quad (13)$$

where each of the pressure laws $p_\pm(\tau,\varepsilon)$, with $\varepsilon = e - u^2/2$, gives rise to an hyperbolic PDE model whose two extreme fields are (genuinely) nonlinear. The very discrepancy with the previous isentropic setting comes from the fact that the coupling interface is now characteristic: 0 is always an eigenvalue of both left and right PDE models. Nevertheless, the coupling interface is never resonant in the sense that the other extreme eigenvalues never vanishes (excluding vacuum). These two properties are responsible for the well-posedness of the coupled problem with

$$\theta_\pm(w) = (\tau, u, p_\pm(w))^T. \quad (14)$$

Proposition (Ambroso et al. 2008a)
The coupling condition 14 yields the following continuity properties:

$$\begin{aligned} u(0^-,t) &= u(0^+,t), \\ p(0^-,t) &= p(0^+,t). \end{aligned} \quad (15)$$

Excluding vacuum, the coupled Riemann problem admits an unique solution.

Observe that the specific volume τ is not continuous, generally speaking, at $x = 0$ as a consequence of the characteristic nature of the coupling interface. The proposed coupling condition 14 nevertheless ensures the continuity of the fluxes $f_-(w(0^-,t)) = f_+(w(0^+,t))$ and hence the conservation of the unknown w. Distinct coupling conditions for 13 are studied in (Ambroso et al. 2008a). In this paper, it is proved in addition that the well-posedness of the purely conservative coupling extends to framework by Despres (Desprès 2001) for general fluid models in Lagrangian coordinates (existence and uniqueness of the del-similar solution to the coupled Cauchy problem for sufficiently flat data).

4.3 *Euler systems of gas dynamics in Eulerian coordinates*

Let us now examine the more delicate framework of two 3 × 3 Euler systems in Eulerian coordinates:

$$\begin{aligned} &w = (\rho, \rho u, \rho e)^T, \\ &\times f_\pm(w) = (\rho u, \rho u^2 + p_\pm, (\rho e + p_\pm)u)^T, \end{aligned} \quad (16)$$

where with little abuse in the notations, the pressure laws $p_\pm(\tau,\varepsilon)$ obey similar properties as the ones put forward in the latter Lagrangian setting. The difficulty stems from the property that the eigenvalues of the two extreme fields can now well vanish and interact with the coupling interface: this interface may thus become resonant. It is therefore mandatory to address the weak formulation 4 of the coupling condition and the study of the Riemann problem with general data turns rather involved. Chalons, Raviart and Seguin (Chalons et al. 2008) have proposed a direct derivation of some self-similar solutions illustrating that the resonancephenomena comes at the expense of uniqueness. In addition, they prove that uniqueness is guaranteed in the case of fully subsonic self-similar coupled solutions.

In the definition of natural transmission conditions for 16, preserving the conservation of the density ρ seems natural (but not mandatory). As a consequence, we will ask for the continuity of the mass flux $\rho u(0^-,t) = \rho u(0^+,t)$. Physicists usually promote the continuity of velocity and pressure at the interface: $u(0^-,t) = u(0^+,t)$ et $p(0^-,t) = p(0^+,t)$. A natural choice would thus read just like in 14:

$$\theta_\pm(w) = (\tau, u, p_\pm(w))^T \quad (17)$$

Obviously, such a choice promotes the conservation of mass and momentum since it asks for the continuity of ρu and $\rho u^2 + p$ at the coupling interface, but by contrast does not preserve the conservation of the total energy ρE. In (Ambroso et al. 2005b), it is proved that preserving the conservation of ρE precludes the constant in space profiles for velocity and pressure to be stationary solutions, as soon as $p_- \neq p_+$. In other words, a fully conservative coupling generates artificial acoustic waves at the interface when the latter is crossed by pure contact waves. Amplitude in the spurious acoustic waves increases with the strength of the contact wave. The work (Ambroso et al. 2005b) provides several numerical illustrations of the deeply different behavior of a fully conservative coupling with the coupling 16–14.

4.4 *Coupling of PDE models with distinct size*

In (Ambroso et al. 2008a), the coupling of 2 × 2 and 3 × 3 gas dynamics models in Lagrangian coordinates is investigated. Requiring the continuity of the velocity u and the pressure p seems natural as already underlined in the Lagrangian setting. Due to the characteristic nature of the coupling interface, due to the zero eigenvalue in the 3 × 3 PDE model, notice that one cannot ask for the continuity of the specific volume τ. This quantity jumps freely. Lift $\mathcal{L}$ and projection $\mathcal{P}$ operators can be simply defined when considering the inverse function $\tau(p)$ from the isentropic pressure law $p(\tau)$ (under the hyperbolicity condition $p'(\tau) < 0$).

Existence and uniqueness of the solution of the coupled Riemann problem is proved.

In (Ambroso et al. 2007b), the coupling of two simplified PDE models for multiphase flows in Eulerian coordinates is analyzed. Falling within the frame of the so-called homogeneous models, they differ from the modeling of the phase transition: the first model assumes instantaneous thermodynamic equilibrium (the so-called HEM 3×3 model) by opposition with the second finer model (the so-called HRM 4×4 model). These two models are naturally linked through a relaxation mechanism which in the limit of an infinite relaxation parameter restores the thermodynamic equilibrium.

Several coupling scenarii are explored in (Ambroso et al. 2007b).

4.5 *Coupling two phase flow and drift models*

The mathematical coupling of a two-fluid two-pressure model (a 7×7 PDE system referred as to the Baer-Nunziato model with a drift flux model (4 equations) is analyzed in (Ambroso et al. 2008c). These models come from the Physics of compressible twophase flows resulting from the mixture of a gas and a liquid. The so-called Baer Nunziato model makes use of two momentum equations, one per phase, while the Drift models only involve one momentum equation for the mixture, the difference in the phase velocities being modeled by an algebraic closure law (the so-called drift law). For simplicity, we address a simplified setting in which each of the phase's pressure laws is isentropic:

$$\begin{cases} \partial_t \alpha_1 + u_2 \partial_x \alpha_1 = \dfrac{v}{\varepsilon}(p1 - p2), \\ \partial_t \alpha_k \rho_k + \partial_x (\alpha_k \rho_l u_k) = 0, \\ \partial_t \alpha_k \rho_k u_k + \partial_x (\alpha_k \rho_k u_k^2 + \alpha_k p_k) - p_l \partial_x \alpha_k \\ = \alpha_k \rho_k f_k + \dfrac{\lambda}{\varepsilon}(u_{\hat{k}} - u_k), \end{cases} \tag{18}$$

where the index $k \in \{1,2\}$ allows to describe the time evolution of mass and momentum for the first and the second phases. Here α_k is the volume fraction of the phase k with $\alpha_1 + \alpha_2 = 1$. Drift models can be written under the following convenient form:

$$\begin{cases} \partial_t \rho Y + \partial_x (\rho Y u + \rho Y(1-Y)u_r) = 0, \\ \partial_t \rho + \partial_x \rho u = 0, \\ \partial_t \rho u + \partial_x (\rho u^2 + p + \rho Y(1-Y)u_r^2) \\ \quad = \rho(1-Y) f_1^0 + \rho Y f_2^0, \end{cases} \tag{19}$$

where the unknowns characterize the time evolution of the mixture of gas and liquid. Y denotes the mass fraction of phase 2 while $u_r = u_2 - u_1$, the relative velocity, is defined thank to the drift law. At last f_k and f_k^0 stand for external forces, the $p_k = p_k(\rho_k)$ in 18 are the pressure laws for each phase k while in 19 $p = p(\rho)$ denotes a pressure law for the mixture. The first mathematical question is to understand how to link the large PDE model 18 to the reduced one 19. This question has been investigated in (Ambroso et al. 2008c) on the basis of the relaxation mechanisms involved in 18: namely a pressure relaxation procedure which in the formal limit $\varepsilon \to 0$ yields equality in the phase pressures:

$$p_1\left(\frac{(\alpha_1 \rho_1)}{\alpha_1}\right) = p_1\left(\frac{(\alpha_2 \rho_2)}{1-\alpha_1}\right) \tag{20}$$

(which in turn reads as an algebraic closure equation for defining the volume fraction α_1 at pressure equilibrium), and then a velocity relaxation mechanism which gives in the limit of a vanishing relaxation parameter $\varepsilon \to 0$

$$u_1 = u_2. \tag{21}$$

In these limits, it can be seen that the mixture pressure $p(\rho)$ (respectively the mixture velocity u) coincides with 20 (respectively 21). Then the drift closure law for defining the relative velocity $u_r = u_2 - u_1$ can be regarded as a first order corrector in ε for modeling the local departure from equilibrium $u = u_1 = u_2$. This first order corrector acts like a Darcy law but for the relative velocity and using a Chapman-Enskog type of expansion allows to express the required closure law involving the external forces f_k in 18. We refer the reader to (Ambroso et al. 2008c) for the details.

5 FULLY CONSERVATIVE COUPLING AND RELAXATION APPROXIMATION PROCEDURES

Let us again investigate the coupling of two hyperbolic systems of conservation laws 1 of the same size under the condition of a fully conservative transmission:

$$f_-(w(0^-,t)) = f_+(w(0^+,t)), \quad t > 0. \tag{22}$$

It is useful to rephrase this question within the frame of systems of conservation laws with a non-homogeneous flux:

$$\partial_t w + \partial_x f(w, a_0(x)) = 0, \quad x \in \mathbb{R}, t > 0,$$

where the given scalar function $a_0(x)$ can be discontinuous. But, the PDEs 22, set over the whole real line $\mathbb{R}$, can be understood in the usual weak sense of the distributions. With this respect, the mathematical question of a fully conservative coupling can be reexpressed in the PDE framework 22 defining for instance:

$$f(w,a_0(x)) = (1-a_0(x))f_-(w) + a_0(x)f_+(w),\\ a_0(x) = H(x), \tag{23}$$

where $H(x)$ denotes the Heavyside's function. Systems of conservation laws with a discontinuous flux have received a considerable attention over the past years. Let us quote for instance (Bürger and Karlsen 2008) and the references therein (see also (Audusse and Perthame 2005), (Bachmann and Vovelle 2006), (Seguin and Vovelle 2003).

After (Isaacson and Temple 1992), equations 5–23 are commonly rewritten under the convenient form of an augmented PDE system:

$$\begin{cases} \partial_t w + \partial_x f(w,a) = 0, & x \in \mathbb{R}, \quad t>0, \\ \partial_t a = 0, \end{cases} \tag{24}$$

when choosing $a_0(x)$ in 23 as the initial data for the unknown a. The first order system 24 is hyperbolic provided that (generally speaking) the matrix $\nabla_w f(w,a)$, which we assume to be $\mathbb{R}$-diagonalizable, solely admits non-vanishing real eigenvalues, otherwise a resonance phenomena may take place. We refer the reader to (Isaacson and Temple 1992) for a mathematical analysis.

We show after (Ambroso et al. 2008b) that convenient relaxation approximation procedures for 5–23 may easily circumvent the resonance phenomena and turn to be very useful in the derivation of well-suited numerical methods in the setting of a fully conservative coupling. We first exemplify this claim on the ground of the generic Jin et Xin relaxation framework (Jin and Xin 1995). We then show how to handle a Suliciu-like relaxation procedure (Suliciu 1990) devoted to Eulerian gas dynamics equations in Eulerian coordinates. The main interest of this strategy stems from its lower numerical dissipation (contact waves at rest are exactly preserved).

5.1 *Jin and Xin relaxation procedure for a conservative coupling*

The Jin and Xin (Jin and Xin 1995) relaxation approximation of the system of conservation laws 23–24 formally writes:

$$\begin{cases} \partial_t w^\lambda + \partial_x v^\lambda = 0, & x \in \mathbb{R}, \quad t>0, \\ \partial_t a^\lambda = 0, \\ \partial_t v^\lambda + c^2 \partial_x w^\lambda = \lambda(f(w^\lambda,a^\lambda) - v^\lambda), \end{cases} \tag{25}$$

with a given real constant $c > 0$ large enough to meet the Whitham's sub-characteristic condition. Observe that the PDE model 25 is hyperbolic by contrast to the original equations 24. Then, invoking the classical splitting approach

$$\begin{cases} \partial_t w + \partial_x v = 0, & x \in \mathbb{R}, \\ \partial_t a = 0, \\ \partial_t v + c^2 \partial_x w = 0, \end{cases}$$
$$\text{followed by} \tag{26}$$
$$\begin{cases} \partial_t w^\lambda = 0, & x \in \mathbb{R}, \\ \partial_t a^\lambda = 0, \\ \partial_t v^\lambda = \lambda(f(w^\lambda,a^\lambda) - v^\lambda), & \lambda \to \infty, \end{cases}$$

obviously yields a consistent and conservative finite volume approach for 24, namely for the fully conservative coupled problem 1–22. In practice, such an approximation procedure may be localized at the coupling interface under a suitable CFL restriction (CFL $\le 1/2$), in order to produce a coupling numerical flux function at $x = 0$.

5.2 *Gas dynamics and Suliciu's relaxation procedure*

In (Ambroso et al. 2008b), it is shown how to perform the fully conservative coupling of to Euler gas dynamics models in Eulerian coordinates using Sulicu-like relaxation approximation. We refer the reader to (Coquel et al. 2001) and (Bouchut 2004) (see also (Chalons and Coquel 2008)) for a detailed presentation of this relaxation procedure. Such a strategy amounts to couple in a fully conservative manner the following two relaxation systems made only of linearly degenerate fields:

$$W = (\rho, \rho u, \rho s, \rho\mathcal{T})^T,\\ F_\pm(W) = (\rho u, \rho u^2 + \Pi_\pm, \rho s u, \rho\mathcal{T}u)^T, \tag{27}$$

where the two relaxation pressure laws read:

$$\Pi_\pm(\tau, s, \mathcal{T}) \equiv p_\pm(\mathcal{T}, s) + c^2(\mathcal{T} - \tau). \tag{28}$$

Here and with a little abuse in the notations, the two exact pressure laws $p_\pm(\tau,s)$ are function of the specific volume τ and of the specific entropy s. As they stand, the proposed relaxation in 27 do preserve the conservationof the mathematical entropy ρs, but such a property is natural in view of the property that all the fields are linearly degenerate (so that all additional non trivial conservation laws are true not only for smooth but also for discontinuous solutions in the usual sense of distribution). Obviously such a property fails at equilibrium and some (trivial !) correction procedure is in order (see again (Coquel et al. 2001) or (Bouchut 2004) for the details). In 28, the relaxation frozen (Lagrangian sound) speed $c > 0$ is chosen large enough according to the following Whitham-like sub-characteristic condition:

$$c^2 > -\partial_\tau p_\pm(\mathcal{T}, s), \tag{29}$$

for all the values of τ and s under consideration.

Let us underline that the relaxation frozen speed c is the same for both the left and right relaxation models, so that the two PDEs systems with flux F_- and flux F_+ do admit the same eigenvalues $u - c\tau$, u and $u + c\tau$. This is the very reason why the resonance phenomena is bypassed: the two family of waves cannot badly interact at the coupling interface $x = 0$.

Then, it is worth underlining that both relaxation models 27 admit an additional non trivial conservation law for governing a relaxation energy $\rho\Sigma_\pm(W)$ with energy flux $\rho H_\pm(W)u$:

$$\Sigma_\pm(\tau, u, s, \mathcal{T}) = u^2/2 + \varepsilon_\pm(\mathcal{T}, s) + (\Pi_\pm(\tau, s, \mathcal{T})^2 - p_\pm(\mathcal{T}, s)^2)/2c^2,$$
$$H_\pm = (\Sigma_\pm + \Pi_\pm \tau).$$

The relaxation energy obeys the equation:

$$\partial_t \rho\Sigma_\pm(W) + \partial_x \rho H_\pm(W)u = 0,$$

which we again stress to be valid in the sense of distributions for weak solutions. Under the Whitham's sub-characteristic condition 29 and the natural assumption of positive temperature $\partial_s(\rho E) < 0$, it is proved in (Chalons and Coquel 2008) that the mapping

$$W = (\rho, \rho u, \rho s, \rho\mathcal{T})^T \to V = (\rho, \rho u, \rho\Pi, \rho\Sigma)^T \tag{30}$$

actually defines a smooth admissible change of variable for both relaxation systems in 27. We can thus model the coupling of the relaxation systems 27, requiring a continuity condition in the V-variable: $V(0^-, t) = V(0^+, t), T > 0$. It can be proved that the proposed strong coupling condition (in the form 3 yields a well-posed coupled problem for 27:

Proposition
Being given two stares W_L and W_R evaluated at equilibrium, i.e. with $\mathcal{T} = \tau$, the coupled Riemann problem 27–30 admits a unique self-similar solution $R(\xi; W_L, W_R)$ which coincides with theusual Riemann solution in each half space:

$$R(\xi; W_L, W_R) = \begin{cases} R(\xi; W_L, \theta_-(V_R)), & \xi < 0, \\ R(\xi; \theta_+(V_L), W_R), & \xi > 0. \end{cases} \tag{31}$$

In addition, the fluxes ρu, $(\rho u^2 + \Pi)$ *and* $(\rho\Sigma + \Pi)u$ are continuous at $\xi = 0$ (namely at the coupling interface $x = 0$).

The reported flux continuity properties just express that the coupled Riemann problem 27–30 allows to define a (fairly simple) numerical Godunov type solver for handling the fully conservative coupling of two Euler gas dynamicsmodels in Eulerian coordinates (see again (Ambroso et al. 2008b) for the details and the consistency property with the entropy condition at equilibrium).

6 COUPLING VIA AN AUGMENTED PDE FORMULATION

The second framework for coupling problems we address intends to tackle the failure of uniqueness in the case of resonance by designing convenient formalisms. Introduced in (Boutin et al. 2008) in the scalar setting and extended to the case of systems in (Boutin et al. 2009a), it consists roughly speaking in modeling the existence of the coupling interface under the form of a standing wave for an augmented PDE model which is set over the whole real line. The standing wave is designed so as to exhibit a complete set of Riemann invariants (away from resonance) in agreement with the expected continuity property 3 in the coupled solutions. The coupling problem thus takes the form of a Cauchy problem which can in turn support various regularization mechanisms that are expected to restore uniqueness. The design principle is to consider a new unknown $u(t, x)$ defined by:

$$u_- := \theta_-(w), \qquad u_+ := \theta_+(w),$$
$$u(t, x) := \begin{cases} u_-(t, x), & x < 0, \quad t > 0, \\ u_+(t, x), & x > 0. \end{cases} \tag{32}$$

Observe that 32 is actually a well-defined change of variable since by assumption $d_w\theta_\pm(w) > 0$ for all $w \in \mathbb{R}$. On the ground of 32, the half-space problems in the (conservative) form are rewritten

$$\partial_t \gamma_\pm(u) + \partial_x f_\pm(\gamma_\pm(u)) = 0, \quad \pm x > 0, \tag{33}$$

or equivalently in the (nonconservative) form

$$(D_u\gamma_\pm(u))\partial_t u + (D_\gamma f_\pm)(\gamma_\pm(u))(D_u\gamma_\pm(u))\partial_x u = 0,$$

for $\pm x > 0$. The coupling condition becomes

$$u(0+, t) \in \mathcal{O}_+(u(0-, t)),$$
$$u(0-, t) \in \mathcal{O}_-(u(0+, t)). \tag{34}$$

In absence of a resonance phenomenon, this reformulation allows to simply impose the *continuity of u at interface*

$$u(0-,t) = u(0+,t). \tag{35}$$

In (Boutin et al. 2008) and (Boutin et al. 2009a), Boutin *et al* propose to replace the problem 33–34 by the augmented system posed over the whole real line:

$$\begin{cases} A_0(u,v)\partial_t u + A_1(u,v)\partial_x u &= 0, \\ \partial_t v &= 0, \end{cases} \tag{36}$$

which is a nonlinear first order system in nonconservative form where $v:[0,+\infty)\times\mathbb{R}\to[-1,1]$ will be referred to as the *color function.* By construction, time-space regions where $v=-1$ correspond to the left-hand half-problem while regions where $v=1$ correspond to the right-hand half-problem, by requiring the following consistency property on A_0, A_1:

$$\begin{aligned} A_0(u,\pm1) &= D_u\gamma_\pm(u), \\ A_1(u,\pm1) &= D_\gamma f_\pm(\gamma_\pm(u))\,D_u\gamma_\pm(u). \end{aligned} \tag{37}$$

The existence of a smooth function $C_0:\Omega_u\times[-1,1]\mapsto C_0(u,v)\in\mathbb{R}^N$ so that

$$\begin{aligned} A_0(u,v) &= D_u C_0(u,v), \\ Det(D_u C_0(u,v)) &\neq 0, \end{aligned} \tag{38}$$

with $C_0(u,\pm1)=\gamma_\pm(u)$, is assumed. The invertibility property stated in 38 obviously preserves the time arrow in the nonlinear first order augmented system 36. Similarly, the existence of a smooth function $C_1:\Omega_u\times[-1,1]\mapsto C_1(u,v)\in\mathbb{R}^N$ so that

$$A_1(u,v) = D_u C_1(u,v), \tag{39}$$

with $C_1(u,\pm1)=f_\pm(\gamma_\pm(u))$ is assumed.

For $j=0,1$, by definition, the matrices $A_j(u,v)$ smoothly connect $A_j(u,-1)$ to $A_j(u,1)$ as v describes the interval $[-1,1]$. Moreover A_0 must be invertible and $A_0^{-1}A_1$ have real and distinct eigenvalues, extending here the strict hyperbolicity of the original hyperbolic half-problems. In the scalar setting, the simplest example makes use of straight lines in the phase space

$$\begin{aligned} A_0(u,v) &= \frac{1-v}{2}\gamma_-(u) + \frac{1+v}{2}\gamma_+(u), \\ A_1(u,v) &= \frac{1-v}{2}f^-(\gamma_-(u)) + \frac{1+v}{2}f^+(\gamma_+(u)), \end{aligned} \tag{40}$$

but obviously many other definitions can be in order. We refer the reader to (Boutin et al. 2009a) for related examples in the framework of hyperbolic systems. Due to assumption 38–39, observe that the augmented PDEs 36 can be equivalently recast as

$$\begin{cases} \partial_t C_0(u,v) + \partial_x C_1(u,v) - \partial_v C_1(u,v)\partial_x v = 0, \\ \partial_t v = 0. \end{cases} \tag{41}$$

To make the coupling problem a Cauchy problem, the system 36 is then supplemented with the initial data

$$\begin{aligned} u_0(x,0) &= u_0(x) =: \theta_\pm(w_0(x)), \quad \pm x>0 \\ v_0(x,0) &= v_0(x) := \pm1, \quad \pm x>0, \end{aligned} \tag{42}$$

for some given data u_0.

It is worth underlining that as long as the matrix $A_0^{-1}(u,v)A_1(u,v)$ does not admit a vanishing eigenvalue, then the coupling problem 36–42 resumes to a Cauchy problem for an nonlinear hyperbolic system (in non-conservationform). The Riemann invariants associated with the standing wave modeling the coupling interface are then easily seen to obey

$$A_0^{-1}(u,v)A_1(u,v)Du = 0, \quad \text{namely } Du = 0,$$

so that as long as $A_1(u,v)$ stays invertible, the expected continuity condition

$$u(0-,t) = u(0+,t)$$

stated in 35 is restored. In other words, the strong form 3 of the coupling condition is satisfied.

From now on, we are especially interested in the case that the interface is resonant for some state value $(u^\star,v^\star)$, that is, when the matrix $A_1(u^\star,v^\star)$ admits the eigenvalue 0 and 35 need not be satisfied as an equality, in general. As already stressed, a weak formulation is necessary. As already emphasized, this situation is nothing but a classical resonance phenomena.

The objective of Boutin *et al* in (Boutin et al. 2008) is precisely to study this *resonant regime.* Note that this is a challenging issue since one first needs to give sense to weak solutions of the PDE model 36 which writesas a non-linear system but in non-conservation form. It is well known that entropy weak solutions for such PDEs may not be uniquely defined (see the pioneering work of LeFloch (LeFloch 1989)). In (Boutin et al. 2008) the definition of weak solutions for 41 in the resonant regime has been tackled on the ground of a vanishing viscosity analysis. In order to explicitly derive the limit solutions, the self-similar viscosity approach by Dafermos (Dafermos 1973) has been privileged. This analysis has been extended to the case of systems in (Boutin et al. 2009a) under fairly general assumptions.

Recall that Dafermos (Dafermos 1973) advocates the use of self-similar regularizations in order to capture the whole wave fan structure of weak solutions to the Riemann problem. This consists in searching for self-similar solutions depending only on the variable $\xi := x/t$ and, then, introducing a self-similar regularization of the given hyperbolic system. Specifically, the coupling problem 36 under consideration is regularized as follows:

$$\begin{cases} A_0(u^\varepsilon,v^\varepsilon)\partial_t u^\varepsilon + A_1(u^\varepsilon,v^\varepsilon)\partial_x u^\varepsilon = \varepsilon t\,\partial_x(B_0(u^\varepsilon,v^\varepsilon)\partial_x u^\varepsilon), \\ \partial_t v^\varepsilon = \varepsilon^2 t\,\partial_{xx} v^\varepsilon. \end{cases} \tag{43}$$

Here $\in > 0$ is a small parameter and $B_0 = B_0(u,v)$ is a given matrix referred to as the viscosity matrix. In the self-similar variable ξ, the equations satisfied by the viscous solutions $(u^\varepsilon,v^\varepsilon) = (u^\varepsilon(\xi),v^\varepsilon(\xi))$ read (with $\xi \in \mathbb{R}$)

$$\begin{cases} (-\xi A_0(u^\varepsilon,v^\varepsilon) + A_1(u^\varepsilon,v^\varepsilon))u^\varepsilon_\xi = \in (B_0(u^\varepsilon,v^\varepsilon)u^\varepsilon_\xi)_\xi, \\ -\xi v^\varepsilon_\xi = \varepsilon^2 v^\varepsilon_{\xi\xi}. \end{cases} \tag{44}$$

The self-similar initial data can be readily inferred from 42, choosing as expected constant states in the half and right domain.

Let us highlight that the ε^2 weight in the last equation reflects the linearly degenerate nature of the associated field. It intends to interplay in balance with the genuinely nonlinear phenomena taking place in the first equations.

The objective in (Boutin et al. 2008) and (Boutin et al. 2009a) is to study the existence and smoothness of solutions to 44 and to rigorously justify the passage to the limit $\in \to 0$. In that aim, uniform ($\in$-independent) bound on the total variation $TV(u^\epsilon)$ is derived. As a consequence, an existence theorem under fairly general structural assumptions on the hyperbolic system and its regularization, is inferred. More precisely, one has:

Theorem [(Boutin et al. 2009a)]

There exists a solution $u^\varepsilon \in \Omega$ of the problem 43–42 satisfying, for some constant K independent of $\in$,

$$TV(u^\epsilon) \le K\,|u_R - u_L|, \ \in |u^\epsilon_\xi| \le K. \tag{45}$$

This results asserts that up to an extracted subsequence, the family $\{u_\epsilon\}_{\epsilon>0}$ converges to a limit function u with bounded total variation. This limit function can be characterized as follows:

Theorem [(Boutin et al. 2009a)]

The sequence u^ϵ converges pointwise toward $u \in BV$, satisfying in the usual weak sense

$$-\xi\frac{d}{d\xi}\theta_\pm(u) + \frac{d}{d\xi}f_\pm(\theta_\pm(u)) = 0, \pm\xi > 0. \tag{46}$$

Let $(\eta_\pm, q_\pm)$ be an entropy pair for the PDES with flux $f_\pm$, where $\eta_\pm(w)$ is a convex entropy satisfying the compatibility condition

$$\nabla^2\eta_\pm(u)B_0(u) \ge 0, \qquad u \in U,$$

with the dissipative tensor $B_0(u)$. Then following entropy inequalities are satisfied,

$$-\xi\frac{d}{d\xi}\eta_\pm(\theta_\pm(u)) + \frac{d}{d\xi}q_\pm(\theta_\pm(u)) \le 0, \quad \pm\xi > 0. \tag{47}$$

In other words and as expected, the limit u is entropy weak solution of the left PDEs in $\mathbb{R}^-$ and right PDEs in $\mathbb{R}^+$. It remains to characterize the viscous inner profile taking place within the interface $\{\xi = 0\}$ so as to exhibit matching conditions in between the left and right half-problems. The required analysis has been performed in the scalar framework in (Boutin et al. 2009b) using a *blow-up* technic. Introducing the fast variable y with $\xi = \in y$, one considers the rescaled profiles:

$$U^\epsilon(y) = u^\epsilon(\in y), \quad V^\epsilon(y) = v^\epsilon(\in y). \tag{48}$$

It can be readily checked that as $\in$ goes to zero, the family of rescaled profiles $\{V^\epsilon\}_{\in>0}$ converges to the following limit profile:

$$V(y) := -1 + 2\int_{-\infty}^{y} e^{-s^2/2}\,ds \Big/ \int_{-\infty}^{+\infty} e^{-s^2/2}\,ds.$$

mapping monotonically $\mathbb{R}$ onto $(-1,1)$. For simplicity, we focus ourselves here on the straight-line closure 40. The following result is established in (Boutin et al. 2009b):

Theorem

1. *The sequence of rescaled profiles* $\{U^\epsilon\}_{\in>0}$ *48 strongly converges in the limit* $\in \to 0$ *to a smooth inner profile* $U \in C^2(\mathbb{R})$, *heteroclinic solution of the non-homogeneous second order ODE*

$$U_{yy} = \left(\frac{1-V(y)}{2}f'_-(U) + \frac{1+V(y)}{2}f'_+(U)\right)U_y \tag{49}$$

with the following asymptotic behaviour

$$\lim_{y\to-\infty} U(y) = U_{-\infty},$$
$$\lim_{y\to+\infty} U(y) = U_{+\infty}. \quad (50)$$

The viscous profile $U(y)$ may be trivial (i.e. uniformly constant in y) but is always monotonic with the following bounds valid for all $y \in \mathbb{R}$

$$\min(u_L, u_R) \le U(y) \le \max(u_L, u_R). \quad (51)$$

2. *Let (η, q_-) (respectively (η, q_+)) be an entropy pair for the PDEs with flux f_- (respectively f_+) where $\eta(w)$ is a convex entropy. Let $u \in L^\infty(\mathbb{R}) \cap BV(\mathbb{R})$ denotes the limit of the sequence of self-similar solutions $\{u_\epsilon\}_{\epsilon>0}$ of 48 as ϵ goes to zero. Then, the matching conditions between the limit u and the inner profile U read:*

$$\begin{aligned} & f_-(\theta_-(u(0^-))) = f_-(\theta_-(U_{-\infty})),\\ & q_-(\theta_(u(0^-))) \ge q_-(\theta_-(U_{-\infty})),\\ & \textit{and}\\ & f_+(\theta_+(U_{+\infty})) = f_+(\theta_+(u(0^+))),\\ & q_+(\theta_+(U_{+\infty})) \ge q_+(\theta_+(u(0^+))). \end{aligned} \quad (52)$$

3. *Assuming either $f'_-(U_{-\infty}) < 0$ or $f'_+(U_{+\infty}) > 0$, then the internal boundary layer U is necessarily trivial:*

$$U(y) = U_{-\infty} = U_{+\infty}, \quad \text{for all } y \in \mathbb{R}. \quad (53)$$

When the viscous profile U achieves a non trivial connection in between the two endpoints $U_{-\infty}$ and $U_{+\infty}$, observe that one cannot infer neither $u(0^-) = u(0^+)$ or $f_-(u(0^-)) = f_+(u(0^+)))$. Indeed the conditions 52 express that the inner boundary layer can be well connected to the outer domain, $\mathbb{R}^- \cup \mathbb{R}^+$ via a standing entropy shock either at $x = 0^-$ or $x = 0^+$, or even actually at both sides of the coupling interface $x = 0$.

In (Boutin et al. 2009b), the internal structure of the coupling interface $\{x = 0\}$ is studied in the scalar setting. The proposed analysis allows to build up to four distinct solutions in the case of a Riemann data leading to the resonance phenomena. Thus multiplicity of self-similar solutions does persist even though in the presence of underlying viscous regularizing mechanisms. This situation should not read too surprising. The origin for multiple solutions may be found after Dafermos in the deep property that Riemann solutions describe the time-asymptotic behavior of the Cauchy problem for parabolic perturbations of 41. It is useful to think of a Cauchy problem with regularized initial data v_η for the color function v (for some small regularization parameter $\eta > 0$). Here, taking for granted the reported asymptotic property, the precise definition of the regularization v_η can be understood to play a central role in the non-uniqueness of Riemann solutions. Indeed and focusing on the 2×2 system 41 for coupling scalar conservation laws, the regularized profile v_η does not generally weight equally the wave speeds $f^{-\prime}$ and $f^{+\prime}$ in the definition of the eigenvalue

$$\lambda(u, v_\eta) = \{\partial_u C_0(u, v_\eta)\}^{-1} \; ((1 - v_\eta)\gamma'_-(u) f^{-\prime}(\gamma_-(u)) + v_\eta \gamma'_+(u) f^{+\prime}(\gamma_+(u))). \quad (54)$$

Consequently, in the resonance phenomena more importance can be paid either to the left problem or the right one and this property is at the core of the failure of uniqueness for self-similar solutions. We refer the reader to (Boutin et al. 2009b) where up to four solutions can be built from the Dafermos analysis.

In (Boutin et al.), several numerical illustrations are provided. They strongly support that all the multiple solutions are stable in the sense that each of these multiple solutions can be captured numerically speaking.

7 COUPLING WITH THICK INTERFACES

In this paragraph, we investigate another regularization mechanism based on thick interfaces resulting from the use of smooth transition profiles

$$v : x \in \mathbb{R} \to v(x) \in [-1, 1] \quad (55)$$

in between the two half regions to be coupled. Such a strategy has been introduced and analyzed first in a single space dimension in (Boutin et al.) and then in several space variables in (Boutin et al.) within the setting of scalar conservation laws. This framework is strongly motivated by the root of the failure of uniqueness experienced in the viscous regularized coupling problem and it is shown to restore the expected uniqueness. To start with and without loss of generality, the invertibility condition 38 may be re-expressed as:

$$\partial_u C_0(u, v) > 0, \quad \text{for all } u \in \mathbb{R} \text{ and } v \in [0,1]. \quad (56)$$

In the setting of thick interfaces 55, it turns useful to extend the change of variable 32 to smooth color functions $v(x)$ when defining:

$$w(t, x) = C_0(u(t, x), v(x)), \quad t > 0, x \in \mathbb{R}. \quad (57)$$

In view of the monotonicity property 56 satisfied by $C_0(.,v)$, u can be conversely recovered from w, the color function v being fixed. For simplicity and with little abuse in the notations, one writes:

$$w = w(u,v), \qquad u = u(w,v). \tag{58}$$

The interest in the w change of variable stems from the smoothness assumption 67 verified by v. Indeed the first equation in 36 merely resumes to a scalar equation with source term we recast as

$$\begin{cases} \partial_t w + \partial_x f(w,v) = \ell(w,v)\partial_x v, \\ w(0,x) = w_0(x), \end{cases} \tag{59}$$

where in agreement with the augmented formulation 41, one has

$$\begin{aligned} f(w,v) &= \mathcal{C}_1(u(w,v),v), \\ \ell(w,v) &= \{\partial_v \mathcal{C}_1\}(u(w,v),v). \end{aligned} \tag{60}$$

The initial data in 59 is prescribed by $w_0 = w(u_0, v)$.

Importantly, the thick interface framework for coupling problems supports the classical notion of entropy pairs for scalar conservation laws with source terms. Any given convex function $U(w)$ is as usual a natural candidate for defining an entropy pair (U,F). To define the required flux, it turns useful to start from the augmented PDEs 41 so as to write for smooth solutions

$$\partial_t \mathcal{C}_0(u,v(x)) + \partial_u \mathcal{C}_1(u,v(x))\partial_x u = 0. \tag{61}$$

This leads to the identity

$$\partial_t U(\mathcal{C}_0(u,v)) + \partial_x \mathcal{Q}(u,v) - \partial_v \mathcal{Q}(u,v)\partial_x v = 0, \tag{62}$$

where by construction the entropy flux reads:

$$\mathcal{Q}(u,v) = \int^u \mathcal{U}'(\mathcal{C}_0(\theta,v))\partial_u \mathcal{C}_1(\theta,v)\,d\theta. \tag{63}$$

Let us re-express the identity 62 in terms of the unknown w:

$$\partial_t \mathcal{U}(w) + \partial_x \mathcal{F}(w,v) - \mathcal{L}(w,v)\partial_x v = 0, \tag{64}$$

with

$$\begin{aligned} \mathcal{F}(w,v) &= \mathcal{Q}(u(w,v),v), \\ \mathcal{L}(w,v) &= \{\partial_v \mathcal{Q}\}(u(w,v),v). \end{aligned} \tag{65}$$

Weak solutions of the scalar conservation law with smooth spatial inhomogeneities 59 are then naturally selected according to the standard entropy inequalities

$$\partial_t \mathcal{U}(w) + \partial_x \mathcal{F}(w,v) - \mathcal{L}(w,v)\partial_x v \le 0, \quad \mathcal{D}', \tag{66}$$

for any given entropy pair $(\mathcal{U},\mathcal{F})$ with $\mathcal{U}$ convex.

Here and again, $L(w,v)\partial_x v$ acts as a source term. The classical Kruzkov uniqueness theory (Kruzkov 1970) directly applies to the entropy weak solutions of 59–66 provided that the spatial inhomogeneities entering the flux and the source term are smooth enough, namely at least piecewise differentiable. The minimal smoothness property on the color function v to meet the Kruzkov assumptions is therefore

$$v \in W^{2,\infty}(\mathbb{R}_x, [-1,1]). \tag{67}$$

The main result of the present paper gives the existence and uniqueness of an entropy weak solution of 59–66 under this smoothness condition but for general initial data $w_0 \in L^\infty(\mathbb{R})$.

To motivate the numerical method developed in (Boutin et al.), it is worth considering the time independent solutions of the scalar conservation law 59, namely solutions of

$$\partial_x f(w,v) = \ell(w,v)\partial_x v, \tag{68}$$

or in the u variable

$$((1-v)f^{-\prime}(\gamma_-(u))\gamma'_-(u) + vf^{+\prime}(\gamma_+(u))\gamma'_+(u))\partial_x u = 0. \tag{69}$$

At the numerical level, the difficulty consists in capturing such steady solutions, especially when the coefficient $((1-v)f^{-\prime}(\gamma_-)\gamma'_-(u) + vf^{+\prime}(\gamma_+)\gamma'_+(u))$ vanishes, that is when the resonance phenomenon takes place. In agreement with the coupling condition imposing the continuity of u, Boutin *et al.* propose propose in (Boutin et al.) to enforce the solution w of 59–66 with constant u but variable v to be stable solutions, even if resonance occurs. The numerical scheme has been designed in a single space variable in (Boutin et al.) and extended to problems with arbitrary space dimensions in (Boutin et al.), according to this well-balanced criteria. To achieve the well-balanced property, the proposed finite volume method is non co-localized: approximate solutions are sought as usual to be piecewise constant functions but with the property that constant values in u and v are achieved on two staggered grids. For simplicity, the discrete method is described in one space dimension.

Let be given a constant time step $\Delta t > 0$ and a constant space step $\Delta x > 0$. Let $t^n = n\Delta t$, $n \in N$, be the time levels, $x_j = j\Delta x$ be the cell centers and $x_{j+1/2} = (j+1/2)\Delta x$ with $j \in Z$ be the cell interfaces.

The approximate solutions $u_{\Delta x}$ and $v_{\Delta x}$ are sought under the form:

$$\begin{cases} u_{\Delta x}(t,x) = u_j^n, & x \in (x_{j-1/2}, x_{j-1/2}), \\ v_{\Delta x}(t,x) = v_{j+1/2}, & x \in (x_j, x_{j+1}), \end{cases} \tag{70}$$

for time $t \in (t^n, t^{n+1})$. Notice that as already claimed, the numerical approach promote is not co-localized. Let us define from 57 the companion approximate solution $w_{\Delta x}(t,x)$:

$$w_{\Delta x}(t,x) = C_0(u_{\Delta x}(t,x), v_{\Delta_x}(t,x)). \tag{71}$$

Since the color function $v(x)$ in the Cauchy problem 59 is time independent, $v_{\Delta x}(t,x)$ is chosen to coincide at all time t with the discrete initial data $v_{\Delta x}^0(x)$ given by:

$$v_{\Delta x}^0(x) \equiv v_{j+1/2} = \frac{1}{\Delta x}\int_{x_j}^{x_{j+1}} v_0(x)dx, \tag{72}$$

for $x \in (x_j, x_{j+1})$. Then and as usual, the discrete initial data $u_{\Delta x}^0(x)$ reads:

$$u_{\Delta x}^0(x) = u_j^0 = \frac{1}{\Delta x}\int_{x_{j-1/2}}^{x_{j+1/2}} u_0(x)dx, \tag{73}$$

for $x \in (x_{j-1/2}, x_{j-1/2})$. The discrete solution $u_{\Delta x}$ is evolved in time thanks to a finite volume method which has to be consistent with:

$$\partial_t w(u,v) + \partial_x f(w(u,v),v) - (f_+(\gamma_+(u)) - f_-(\gamma_-(u)))\partial_x v = 0. \tag{74}$$

To that purpose, one observes that in each cell interface $x_{j+1/2}$ the discrete function $v_{\Delta x}$ stays locally constant in space, so that the above equation reduces at each interface $x_{j+1/2}$ to a standard scalar conservation law in the unknown $w = w(u, v_{j+1/2})$:

$$\partial_t w + \partial_x f(w, v_{j+1/2}) = 0, \tag{75}$$

for $(x,t) \in (x_j, x_{j+1}) \times (t^n, t^{n+1})$. This motivates the introduction at each interface of a two-point numerical flux function $g(\cdot,\cdot; v_{j+1/2}) : \mathbb{R} \times \mathbb{R} \to \mathbb{R}$ with the consistency property:

$$g(a,a; v_{j+1/2}) = f(a, v_{j+1/2}), \quad \text{for all } a \in \mathbb{R}. \tag{76}$$

For the sake of stability, this two-point numerical flux function is assumed to be monotone (see (Godlewski and Raviart 1996) for instance) for instance).

The discrete solution $u_{\Delta x}(t^n,.)$, being known at time t^n, is updated at the next time level t^{n+1} into two steps: namely, a subcell reconstruction step followed by an evolution step. The overall algorithm takes the form of a time explicit finite volume method consistent with 74, and for which a natural CFL restriction reads:

$$\frac{\Delta t}{\Delta x} \max_{j \in Z} \sup_{u \in [m,M]} \left| \frac{\partial f}{\partial w}(w(u, v_{j+1/2}), v_{j+1/2}) \right| \leq \frac{1}{2}, \tag{77}$$

with $m = \inf_{x \in \mathbb{R}} u_0(x)$ and $M = \sup_{x \in \mathbb{R}} u_0(x)$. The two steps read as follows.

- *Subcell reconstruction.* At time t^n, define in each cell $(x_{j-1/2}, x_{j+1/2})$ the two subcell values:

$$\begin{aligned} w_{j-1/2,+}^n &= w(u_j^n, v_{j-1/2}), \\ w_{j+1/2,-}^n &= w(u_j^n, v_{j+1/2}), \end{aligned} \tag{78}$$

  together with their averaged value

$$w_j^n = \frac{1}{2}(w_{j-1/2,+}^n + w_{j+1/2,-}^n). \tag{79}$$

- *Evolution in time.* At time t^{n+1}, define in each cell $(x_{j-1/2}, x_{j+1/2})$ the required update u_j^{n+1} as the unique solution of

$$\frac{1}{2}(w(u_j^{n+1}, v_{j-1/2}) + w(u_j^{n+1}, v_{j+1/2})) = w_j^{n+1}, \tag{80}$$

  where w_j^{n+1} is given by the following finite volume formula:

$$w_j^{n+1} = w_j^n - \frac{\Delta t}{\Delta x}(G_{j+1/2,-}^n - G_{j-1/2,+}^n). \tag{81}$$

  Here, the left and right numerical fluxes at each cell interface $x_{j+1/2}$ respectively write:

$$\begin{aligned} G_{j+1/2,-}^n &= g(w_{j+1/2,-}^n, w_{j+1/2,+}^n; v_{j+1/2}) \\ &\quad - f(w_{j+1/2,-}^n, v_{j+1/2}), \\ G_{j+1/2,+}^n &= g(w_{j+1/2,-}^n, w_{j+1/2,+}^n; v_{j+1/2}) \\ &\quad - f(w_{j+1/2,+}^n, v_{j+1/2}). \end{aligned} \tag{82}$$

This completes the description of the discrete method.

As expected, the scheme 78–82 can be proved (Boutin et al.) to be well-balanced:

Proposition

Let $u^\star$ be any given real number. Define the initial data u_0 of the Cauchy problem for 41 by

$$u_0(x) = u^\star, \quad x \in \mathbb{R}, \tag{83}$$

and v_0 by any given smooth function with value in $[-1,1]$. Then, the discrete solution $u_{\Delta x}$ of 78–82 stays constant in space with

$$u_{\Delta x}(t^n, x) = u^\star, \quad x \in \mathbb{R}, \quad \text{at all time level } t^n. \tag{84}$$

In addition, it is shown in (Boutin et al.) that the finite volume method under consideration does converge, more precisely:

Theorem. [Convergence of the well-balanced finite volume method]

Consider the Cauchy problem for 41 with initial data u_0 in $L^\infty(\mathbb{R})$ and v_0 in $W^{2,\infty}(\mathbb{R},[-1,1])$. Assume the monotonicity property 56 $\partial_u w(u,v) > 0$. Let $\{v_{\Delta x}\}_{\Delta x>0}$ be given by 70–72. Let $\{w_{\Delta x}\}_{\Delta x>0}$ be the sequence of approximate solutions where $w_{\Delta x}$ in 71 is defined from $u_{\Delta x}$ in 70 by the finite volume method 78–82 with monotone numerical functions.

Then under the CFL condition 77, the sequence $\{w_{\Delta x}\}_{\Delta x>0}$ stays uniformly bounded in $L^\infty(\mathbb{R}_+ \times \mathbb{R})$ and converges in the L^p_{loc} norm strongly, $1 \le p < \infty$, as $\Delta x \to 0$ to the unique Kruzkov solution w to the problem 59–66. Namely for all time $T > 0$ and for all compact $\mathcal{K}$ in $\mathbb{R}$

$$\|w - w_{\Delta x}\|_{L^p((0,T)\times K)} \le o(\Delta x), \tag{85}$$

for $1 \le p < \infty$, and where $o(\Delta x)$ is a function going to zero with Δx.

In the work (Boutin et al.), the framework is extended to coupling problems in several space dimensions. The formulation via augmented PDE's systems turns out to be fairly flexible in that it allows the coupling of several distinct conservation laws with possible covering. The existence and uniqueness of the solution of the coupled Cauchy problem (for initial data in L^∞) can be established thanks to the design and analysis of a well-balanced finite volume method which applies to general triangulations. Strong convergence is established using the DiPerna's uniqueness theorem in the class of entropy weak measure valued solutions (see (Coquel and LeFloch 1993) and the references therein). Numerical illustrations involving problems with covering highlight the interest of the proposed coupling strategy.

ACKNOWLEDGEMENTS

A significant part of the contributions discussed in the present survey has been carried out with co-workers of a joint research program devoted to the coupling of thermo-hydraulics models between CEA-Saclay and University Pierre et Marie Curie-Paris 6.

The author would like also to warmly thank the organizers of the International Conference to honor Professor E.F. Toro in the year of his 65th birthday. Special thanks to Prof. Elena Vázquez-Cendón, the Coordinator of the Organizing Committee, for thedelightful welcome and the deeply appreciated scientific event.

REFERENCES

Abgrall R. & S. Karni. 2001. Computations of compressible multifluids, *J. Comput. R, Phys.*, **169**, p. 594–623.

Adimurthi S., G.D. Mishra & V. Gowda. 2005. Optimal entropy solutions for conservation laws with discontinuous flux-functions, *J. Hyperbolic Differ. Equ.* **2** no. 4, p. 783–837.

Ambroso A., C. Chalons, F. Coquel, T. Galié, E. Godlewski, P.-A. Raviart & N. Seguin. 2006. Extension of interface coupling to general Lagrangian systems, *Numerical mathematics and advanced applications*, Springer, Berlin, pp. 852–860.

Ambroso A., C. Chalons, F. Coquel, E. Godlewski, F. Lagoutière, P.-A. Raviart & N. Seguin. 2007. A relaxation method for the coupling of systems of conservation laws, in *Hyp2006 Proceedings*, Springer, Berlin.

Ambroso A., C. Chalons, F. Coquel, E. Godlewski, F. Lagoutière, P.-A. Raviart & N. Seguin. 2008. Coupling of general Lagrangian systems, *Math. Comp.* **77**, no. 262, p. 909–941.

Ambroso A., C. Chalons, F. Coquel, E. Godlewski, F. Lagoutière, P.-A. Raviart & N. Seguin. 2008. Relaxation methods and coupling procedures, *Internat. J. Numer. Methods Fluids* **56**, no. 8, p. 1123–1129.

Ambroso A., C. Chalons, F. Coquel, T. Galié, E. Godlewski, P.-A. Raviart & N. Seguin. 2008. The drift-flux asymptotic limit of barotropic two-phase two-pressure models, *Commun. Math. Sci.* **6**, no. 2, p. 521–529.

Ambroso A., C. Chalons, F. Coquel, E. Godlewski, F. Lagoutière, P.-A. Raviart & N. Seguin. 2007. The coupling of homogeneous models for two-phase flows, *Int. J. Finite Vol.* **4** (2007), no. 1, p. 39.

Ambroso A., C. Chalons, F. Coquel, E. Godlewski & P.-A. Raviart. 2005. Couplage de deux systèmes de la dynamique des gaz, in *17e congrès francais de mécanique*, Université de technologie de Troyes, Troyes.

Ambroso A., J.M. Hérard & O. Hurisse. 2009. A method to couple HEM and HRM two-phase flow models, *Computers and Fluids* **38**, p. 738–756.

Audusse E. & B. Perthame. 2005. Uniqueness for scalar conservation laws with discontinuous flux via adapted entropies, *Proc. Roy. Soc. Edinburgh Sect. A* **135**, no. 2, p. 253–265.

Bachmann F. & J. Vovelle. 2006. Existence and uniqueness of entropy solution of scalar conservation laws with a flux function involving discontinuous coefficients, *Comm. Partial Differential Equations* **31**, no. 1–3, p. 371–395.

Baer M.R. & J.W. Nunziato. 1986. A two-phase mixture theory for the deflagration-to-detonation transition (DTR) in reactive granular materials, *International J. Multiphase Flows* **12**, p. 861–889.

Bouchut F. 2004. Nonlinear stability of finite volume methods for hyperbolic conservation laws and well-balanced schemes for sources, Birkhäuser.

Boutin B., C. Chalons & P.-A. Raviart. 1898. Existence result for the coupling problem of two scalar conservation laws with Riemann initial data, *Math. Models Methods Appl. Sci.* **20**, no. 10, p. 1859–1898.

Boutin B., F. Coquel & E. Godlewski. 2008. Dafermos regularization for interface coupling of conservation laws, in *Hyperbolic problems: Theory, Numerics, Applications*, Springer, Berlin, 2008, p. 567–575.

Boutin B., F. Coquel & P. G. LeFloch. 2011. Coupling techniques for nonlinear hyperbolic equations. I Self-similar diffusion for thin interfaces, *Proc. Roy. Soc. Edinburgh Sect. A* **141**, no. 5, p. 921–956.

Boutin B., F. Coquel & P. G. LeFloch. Coupling techniques for nonlinear hyperbolic equations. II Resonant interfaces with internal structure, submitted.

Boutin B., F. Coquel & P.G. LeFloch. Coupling techniques for nonlinear hyperbolic equations. III. A regularization method via thick interfaces, submitted.

Boutin B., F. Coquel & P.G. LeFloch. Coupling techniques for nonlinear hyperbolic equations. IV. A multidimensional finite volume framework, submitted.

Bürger R. & K.H. Karlsen. 2008. Conservation laws with discontinuous flux: a short introduction, *J. Engrg. Math.* **60**, no. 3–4, p. 241–247.

Chalons C., P.-A. Raviart & N. Seguin. 2008. The interface coupling of the gas dynamics equations *Quart. Appl. Math.* **66**, no. 4, p. 659–705.

Chalons C. & F. Coquel. 2005. Navier-Stokes equations with several independent pressure laws and explicit predictor-corrector schemes *Numer. Math.* **101(3)**, p. 451Ð478.

Coquel F., E. Godlewski, B. Perthame, A. In & P. Rascle. 2001. Some new Godunov and relaxation methods for two-phase flow problems. In Godunov methods (Oxford, 1999), Kluwer/Plenum, p. 179Ð188.

Coquel F. & P. LeFloch. 1993. Convergence of finite difference schemes for conservation laws in several space dimensions: a general theory, *SIAM J. Numer. Anal.* **30**, no. 3, p. 675–700.

Dafermos C. M. . 1973. Solution of the Riemann problem for a class of hyperbolic systems of conservation laws by the viscosity method, *Arch. Rational Mech. Anal.* **52** (1973), p. 1–9.

Desprès, B. 2001. Lagrangian systems of conservation laws. Invariance properties of Lagrangian systems of conservation laws, approximate Riemann solvers and the entropy condition, *Numer. Math.*, **89**, p. 99–134.

Diehl S. 1996. Scalar conservation laws with discontinuous flux function. I. The viscous profile condition, *Comm. Math. Phys.* **176(1)**, p. 23–44.

Dubois F. & LeFloch P.G. 1988. Boundary conditions for nonlinear hyperbolic systems of conservation laws, *J. Differential Equations* **71**, p. 93–122.

Eymard R., T. Gallouët & R. Herbin. 2000. Finite volume methods, in *Handbook of numerical analysis, Vol. VII*, Handb. Numer. Anal., VII, North-Holland, Amsterdam, p. 713–1020.

Fedkiw R., T. Aslam, B. Merriman & S. Osher. 1999. A non-oscillatory Eulerian approach to interfaces in multimaterial flows (the ghost fluid method), *Journal of Computational Physics*, **152**, p. 457–492.

Gallouet T., J.M. Hérard, 0. Hurisse & A.Y. LeRoux. 2006. Well balanced schemes versus fractional step method for hyperbolic systems with source terms, *Calcolo* **43 (4)**, p. 217–251.

Girault L. & J.M. Hérard J.M.. 2010. Multi-dimensional computations of a two-fluid hyperbolic model in a porous medium, *IJFV, (electronic: http://www.latp.univ-mrs.fr/IJFV/)* **7 (1)**, p. 1–33.

Girault L. & J.M. Hérard J.M.. 2010. A two-fluid model in a porous medium *Math. Mod. Numer. Anal.* **44 (6)**, p. 1319–1348.

Goatin P. & P.G. LeFloch. 2004.The Riemann problem for a class of resonant hyperbolic systems of balance laws, *Ann. Inst. H. Poincaré Anal. Non Linéaire* **21**, no. 6, p. 881–902.

Godlewski E. & P.-A. Raviart. 1996. *Numerical approximation of hyperbolic systems of conservation laws*, Applied Mathematical Sciences, vol. 118, Springer-Verlag, New York.

Godlewski E. & P.-A. Raviart. 2004. The numerical interface coupling of nonlinear hyperbolic systems of conservation laws. I. The scalar case, *Numer. Math.* **97**, no. 1, p. 81–130.

Godlewski E., K.-C. Le Thanh & P.A. Raviart. 2005. The numerical interface coupling of nonlinear hyperbolic systems of conservation laws. II. The case of systems, *M2AN Math. Model. Numer. Anal.* **39**, 649–692.

Hérard J.M. & O. Hurisse. 2010. Some attempts to couple distinct fluid models, *Networks and Heterogeneous Media* **5 (3)**, p. 649–660.

Hérard J.M. & O. Hurisse. 2007. Coupling of two and one dimensional unsteady Euler equations through a thin interface, *Computers and Fluids* **36(4)**, p. 651–666.

Hérard J.M. 2006. A rough scheme to couple free and porous media, *IJFV (electronic: http://www.latp.univ-mrs.fr/IJFV/)* **3(2)**, p. 1–28.

Isaacson E. & B. Temple. 1992. Nonlinear resonance in systems of conservation laws, *SIAM J. Appl. Math.* **52**, no. 5, p. 1260–1278.

Isaacson E. & E.B. Temple. 1995. Convergence of the 2×2 Godunov method for a general resonant nonlinear balance law, *SIAM J. Appl. Math.* **55**, p. 625–640.

Jin S. & Z. Xin. 1995. The Relaxation Scheme for Systems of Conservation Laws in Arbitrary Space Dimension, *Comm. Pure Appl. Math.* **45**, p. 235–276.

Kruzkov S.N. 1970. First order quasilinear equations with several independent variables, *Mat. Sb. (N.S.)* **81 (123)**, p. 228–255.

LeFloch P.G. Shock waves for nonlinear hyperbolic systems in nonconservative form, *Institute for Math. and its Appl., Minneapolis*, Preprint # 593.

Seguin N. & J. Vovelle. 2003. Analysis and approximation of a scalar conservation law with a flux function with discontinuous coefficients, *Math. Models Methods Appl. Sci.*, **13(2)**, p. 221–257.

Suliciu I. 1990. On modelling phase transitions by means of rate-type constitutive equations, shock wave structure, *Int. J. Engrg. Sci.*, **28**, p. 829–841.

Toro E.F. 1999. Riemann solvers and numerical methods for fluid dynamics. A practical introduction, Second edition. Springer-Verlag, Berlin.

Tzavaras A.E. 1996. Wave interactions and variation estimates for self-similar zero-viscosity limits in systems of conservation laws, *Arch. Rational Mech. Anal.*, **135(1)**, p 1–60.

Numerical Methods for Hyperbolic Equations – Vázquez-Cendón et al. (eds)
© 2013 Taylor & Francis Group, London, ISBN 978-0-415-62150-2

Stochastic convergence and the software tool W*

Ryan Kaufman, Tulin Kaman, Yan Yu & James Glimm
Department of Applied Mathematics and Statistics, Stony Brook University, USA

ABSTRACT: The concept of w* convergence of probability distribution functions to a probability distribution limit, also known as a Young measure, as a description of Large Eddy Simulation (LES) convergence has been proposed separately, and preliminary tests have been encouraging. Here we develop the practical side of implementation for this program.

The first step is a conceptual reinterpretation of simulation data, and in this sense the idea is basically one of a post processing. In a given simulation, the grid limited information is partitioned into (a) coarse grained grid information, i.e. geometrical or space-time localization relative to a coarse grid composed of supercells, and (b) stochastic information within a supercell, giving information on random or probabilistic solution state values. The second step in a convergence study is to compare the results of the first step, i.e. to compare the probabilistic state values (with their–limited resolution–geometrical localizations). This second step, comparison of grid defined Young measures, is the subject of the present paper.

1 INTRODUCTION

W* convergence and Young measures have been known in the theory of hyperbolic conservation laws as an intermediate step on the route to existence theories (Perna 1983; Perna 1985; Ding et al. 1985; Chen 1986) in 1D space. Here we base an interpretation of LES simulations of turbulent flows in the inertial range on this same concept. A recent existence theory for the 3D Euler equations was based on this set of ideas (Gangbo and Westerdickenberg 2009). In (Chen and Glimm 2010) we assumed Kolmogorov 1941 type bounds and derived as a consequence convergence (through a subsequence) to a strong solution for the incompressible Euler equation, but extension to passive scalars lead to w* convergence and Young measures.

The conceptual basis for use of w* convergence and Young measures has been explained in a series of papers (Kaman et al. 2011; Kaman et al. 2011; Lim et al. 2011; Kaman et al. 2011). Briefly, the method could be called: coarse-grain and sample. Given a mesh approximation, we combine all mesh values within a single supercell or single coarse grained cell. These values are neither averaged, nor are they given any geometrical meaning coming from their distinct grid locations. Rather, they are regarded as samples from a probability distribution (at the supercell location) defined by some measure (the Young measure) on the space of solution state variables. They thus define a finite ensemble which is a discrete approximation to a measure on the space of state variables, e.g. momentum, energy and density, species concentration, for the multifluid compressible Euler equation.

In Sec 2, we describe a software tool, W*, designed to aid in comparision of Young measures and to assess w* convergence rates. In the two following sections, Sec 3 and 4, we apply this tool to LES simulations of turbulent mixing and turbulent combustion. A key advantage of this formalism can be seen in Sec. 4: w* convergence of Young measures has completely different properties from those anticipated from conventual convergence methods. Nonlinear functions of the solution converge as well. The reason, basically, is that averages are never used. Fluctuations are preserved and shown to be convergent. Many nonlinear functionals of the solution (combustion being only one of these) respond to the fluctuations and not to the averages. Thus the power of w* convergence and Young measures is revealed in this example.

2 HE SOFTWARE TOOL W*

The goal is to compare discrete approximations to Young measures. The Young measures are defined in terms of Probability Distribution Functions (PDFs) and their indefinite integrals, the Cumulative Distribution Functions (CDFs). These are discretized relative to space-time and also discretized relative to the solution state variables over which the measure is defined. For the space time discretization, we use a grid, which will normally be a coarsening of some (simulation) grid on which the data is presented. The method is applicable as well to inertial range data of an experimental or observational nature. We refer to the cells of the coarsened grid as supercells. There is also a discretization of the solution state variables. These can be thought of as bins. The number of entries in a bin (the "counts") define an empirical frequency, and a discrete approximation to the Young measure at the supercell location. The degree of discretization of the Young measure, i.e. the bin size, is a convergence issue quite different from conventional grid resolution issues. There is a tradeoff between the loss of information caused by a coarse gridded bin size, as no knowledge of solution values internal to a bin is retained, and the loss of information due to the finite sample size. In view of the latter, the finite data approximation to the bin frequencies, these frequencies will be subject to random errors. Standard statistical tests can evaluate this tradeoff; here we choose a reasonable middle value without further analysis.

In comparing two discrete Young measures, we compare the L_1 norm of the difference of their CDFs. Each CDF is a monotone increasing step function. The location of the steps is determined by the bins into which the data has been organized. There is no requirement that the step locations be identical for the two CDFs being compared. This means that the bins defining the two CDFs need have no relation to one another. We do not impose consistency on the state space discretization. It is also possible to reconstruct piecewise linear or higher order interpolants for the CDFs, which also have no requirement of consistency between the Young measures being compared.

The result of the comparison for CDFs using an L_1 norm on the state space CFD differences defines a function on the supercells. For this purpose, we assume the supercell grids are identical. If one supercell is a strict refinement of the other, comparison is still possible, and with the two choices also faced in grid convergence studies, either to average the finer one onto the coarser, or to inject the coarser into the finer supercell grid.

2.1 *W* design and usage*

W* is a post processing tool written in Python. As Python is a well supported cross-platform scripting language, W* requires no compilation and is supported on any platform with Python V2.6 installed. The post processing is done in two parts: to generate PDF's, and to compare CDF's.

The first script parses an input file and a data file (or set of datafiles) to generate the PDF's on a coarse grained grid of supercells. The coarse graining factors and bin sizes are inputs at this stage. The PDF is stored as an array of bin frequencies as data is read in from the data files. As each datum is read, it is determined to which supercell it belongs, and then to which bin it belongs. The corresponding data counter is incremented for that cell and bin, and the process continues until all of the data is read in this way. At this stage the PDF is normalized, and the frequency counts are replaced with probability densities in a corresponding array. This data is then stored in an intermediate file that contains the coarse grained grid of PDF's.

Once intermediate PDF files have been created for separate grid level simulations, these files may be compared pairwise. A second script takes two of these PDF files as input. It integrates the PDF's into CDF's, and then does bin by bin differencing in each supercell. These differences are evaluated in the L_1 norm, and thus a number is assigned to each supercell. The result is output to a file.

2.2 *Technical capabilities*

W* supports reading in simulation data from the VTK data format, as well as our native FronTier output format. The file reading module is designed extensibly, and more file reading capabilities are planned for the future. Output of the final grid of L_1 norms is also in the VTK file format. W* also has the capability to compute means, variances,

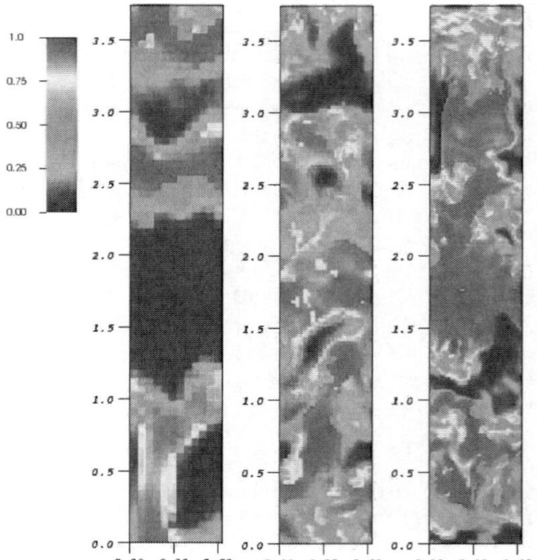

Figure 1. Plot of heavy fluid concentration at the midplane, $t = 59$. Coarse grid (left). Medium grid (middle). Fine grid (right).

and covariances of correlated variables in each supercell.

2.3 *Download and installation*

In order to run W*, Python version 2.6 or greater must be installed. (http://python.org/). W* can be downloaded from the web: http://www.ams.sunysb.edu/wstar.

The package comes with two scripts: wstar.py (the PDF generator) and wstar_compare.py (the comparison generator); several module files (grid.py, fgrid.py, arraynd.py, PDF.py), as well as some sample input files to each.

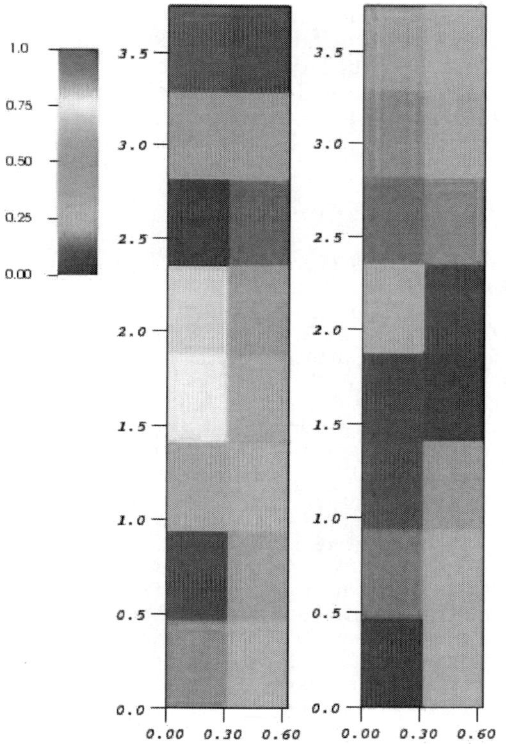

Figure 2. Spatial array of L_1 norms of CDF mesh differences for heavy fluid concentrations at the midplane. Coarse to fine (left). Medium to fine (right).

3 USE OF W* IN V&V STUDIES FOR LES OF TURBULENT MIXING

We use the W* tool to study mesh convergence of a supercell 8×2 array of CDFs for a Rayleigh-Taylor simulations. At the latest time completed for the fine grid, we compare the CDFs on the coarse to fine and medium to fine grids. The comparison is to compute the L_1 norm of the pairwise differences for each of the 8×2 CDFs. These differences yield an 8×2 array of norms, i.e. numbers, which is plotted in Figure 2.

4 USE OF W* IN V&V STUDIES OF TURBULENT COMBUSTION

A key advantage for the use of w* convergence of Young measures is illustrated in the case of turbulent combustion. Because the fluctuations (as well as mean values) converge, the solution can be used directly in a combustion calculation. This type of calculation is called finite rate chemistry. In contrast to common combustion algorithms, no assumption about flame structure is assumed.

We consider the flow in the engine of a scram jet, which is a $M = 7$ experimental aircraft. After a series of shock waves, the flow in the combustion chamber is reduced to $M = 2.4$, still strongly supersonic. Into this flow, H_2 fuel is injected through a nozzle.

In Figure 3, we show the H_2 concentration resulting from the combustion. For comparison of grid levels, we consider pure mixing (no combustion), with two levels of grid compared, and the difference analyzed by W*. Again we compare H_2 concentrations. Figure 4 shows the two grid levels (240 μ and 120 μ), while Figure 5 shows W* analysis of the norm difference between these two simulations. The norm difference has first an L_1 norm for the solution differences in the space of concentration CDFs and then an L_1 norm in space for the concentration CDF norms. The region of strong mismatch shows in blue, and it reflects an evident difference between the two grid levels obvious from the comparison in Figure 4. We can conclude that W* revealed what is evidentially correct: the coarse grid is not adequate for resolution of details of mixing. Returning to Figure 3, we anticipate that finer grids will be needed.

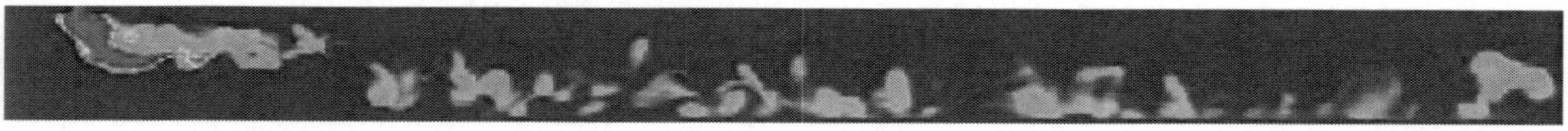

Figure 3. Combustion of H_2 as fuel injected into a $M = 2.4$ cross flow of pure O_2. Plot shows the H_2 levels resulting from combustion. The grid resolution is 240 μ. Preliminary computation.

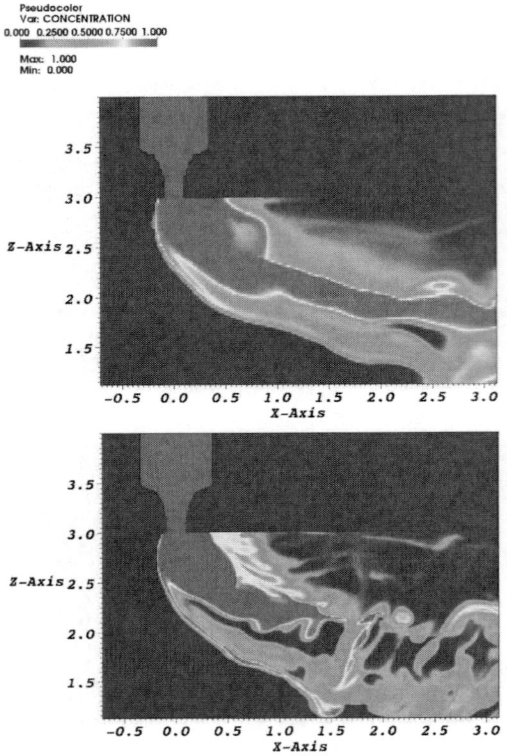

Figure 4. H_2 concentration for injection into a $M = 2.4$ cross flow in air, without combustion. Above: grid spacing 240 μ. Below: grid spacing 120 μ.

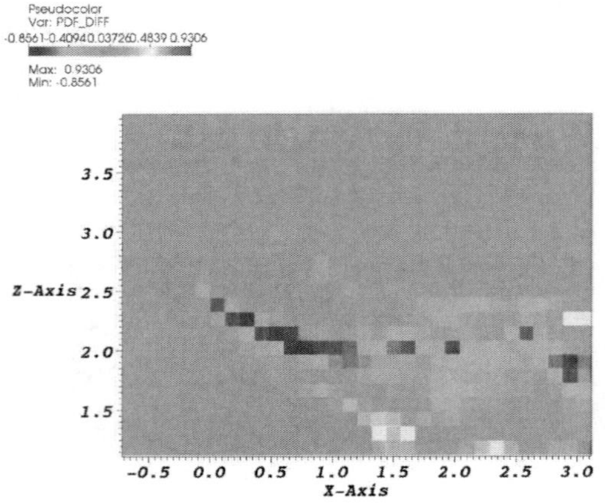

Figure 5. Differences of the H_2 concentration, compared with the difference of the CFDs, compared through an L_1 norm in concentration space, and the results indicated in each supercell to display the localization of the Young measure differences.

5 CONCLUSIONS

We see that w* convergence of Young measures is an intuitively appealing reinterpretation for the convergence of LES simulations of inertial range turbulence. The method will be of increased importance as computational power increases in an environment for which LES simulations become routine, while for many problems, Direct Numerical Simulations (DNS) remain out of reach. The method captures the fluctuations, important for modeling of many processes which depend on inertial range turbulence. With these observations as a backdrop, we have explained a software tool, W*, to implement these ideas. Fundamentally the ideas are of a post processing nature, and for this reason will be easy to adopt into existing simulation practice. W* deals with the comparison of Young measures, whether to assess grid convergence, comparison to experimental or observational data, or the effect of variation of parameters. The algorithm for comparison and its use is explained, and its use has been illustrated in two example problems.

REFERENCES

Chen, G.-Q. (1986). Convergence of the Lax-Friedrichs scheme for isentropic gas dynamics III. *Acta Mathematica Scienitia*, 75–120.

Chen, G.-Q. and J. Glimm (2010). Kolmogorov's theory of turbulence and inviscid limit of the Navier-Stokes equations in R^3. *Commun. Math. Phys.*. In Press.

Ding, X., G.-Q. Chen, and P. Luo (1985). Convergence of the Lax-Friedrichs scheme for isentropic gas dynamics I and II. *Acta Mathematica Scientia*, 415–432, 433–472.

Gangbo, W. and M. Westerdickenberg (2009). Optimal transport for the system of isentropic Euler equations. *Comm. PDEs*, 1041–1073.

Kaman, T., R. Kaufman, J. Glimm, and D.H. Sharp (2011). Uncertainty quantification for turbulent mixing flows: Rayleigh-taylor instability. In *IFIP Advances in Information and Communication Technology*. Springer. Submitted. Stony Brook University Preprint number SUNYSB-AMS-11-08.

Kaman, T., H. Lim, Y. Yu, D. Wang, Y. Hu, J.-D. Kim, Y. Li, L. Wu, J. Glimm, X. Jiao, X.-L. Li, and R. Samulyak (2011). A numerical method for the simulation of turbulent mixing and its basis in mathematical theory. In *Lecture Notes on Numerical Methods for Hyperbolic Equations: Theory and Applications: Short Course Book*, pp. 105–129. London: CRC/Balkema. Stony Brook University Preprint number SUNYSB-AMS-11-02.

Kaman, T., J. Melvin, P. Rao, R. Kaufman, H. Lim, Y. Yu, J. Glimm, and D.H. Sharp (2011). Recent progress in turbulent mixing. *Physica Scripta*. Submitted. Stony Brook University Preprint number SUNYSB-AMS-11-09. Los Alamos National Laboratory preprint LA-UR 11-06770.

Lim, H., T. Kaman, Y. Yu, V. Mahadeo, Y. Xu, H. Zhang, J. Glimm, S. Dutta, D. H. Sharp, and B. Plohr (2011). A mathematical theory for LES convergence. *Acta Mathematica Scientia*. Submitted for publication. Stony Brook Preprint SUNYSB-AMS-11–07 and Los Alamos National Laboratory preprint number LA-UR 11-05862.

Perna, R.D. (1983). Convergence of approximate solutions to conservation laws. *Arch. Rational Mech. Anal.*, 27–70.

Perna, R.D. (1985). Compensated compactness and general systems of conservation laws. *Trans. Amer. Math. Soc.*, 383–420.

Numerical Methods for Hyperbolic Equations – Vázquez-Cendón et al. (eds)
© 2013 Taylor & Francis Group, London, ISBN 978-0-415-62150-2

Numerical relativistic magnetohydrodynamics in dynamical spacetimes

José-María Ibáñez
Department of Astronomy and Astrophysics, University of Valencia, Spain

ABSTRACT: The most spectacular events in the sky detected by current astronomical observatories (ground or space-based) involve matter evolving in the presence of strong gravitational fields. In these astrophysical scenarios, matter can reach velocities near the speed of light, and the gravitational field (background or generated by the flow itself) can be so strong that a description in terms of Einstein's theory of General Relativity is necessary. Some examples of these scenarios are: the stellar core collapse (as a mechanism of hydrodynamical supernovae), the relativistic outflows released from collapsars or from the remnants of coalescing binary neutron stars (as a models of formation of gamma-ray bursts), the relativistic jets associated with active galactic nuclei or microquasars, etc. To improve our knowledge about the physical nature of these energetic phenomena, modern Astrophysics is demanding the development of robust algorithms and numerical multidimensional relativistic Magnetohydrodynamical (MHD) codes. These codes have to include the coupling to an Einstein solver. The main goal of this proceeding is to review the present status of the field of Numerical Relativity extended to the inclusion of matter fields, i.e., the Numerical Relativistic MHD in a dynamical space-time, and to give a general overview about some of the many astrophysical applications.

1 INTRODUCTION

Astrophysical scenarios involving relativistic flows have drawn the attention and efforts of many researchers since the pioneering studies of (May and White 1966), (May and White 1967) and (Wilson 1972). Some examples of systems in which the evolution of matter is described within the framework of the theory of relativity (special or general) are (see the reviews (Font 2008), (Fryer and New 2011), (Postnov and Yungelson 2006), (Sathyaprakash and Schutz 2009), (Shibata and Taniguchi 2011)): i) Relativistic jets in Active Galactic Nuclei (AGN) or associated to microquasars. ii) Accretion onto compact objects in X-ray binaries or in the inner regions of AGNs. The canonical formation of a relativistic jet involves an accretion torus around a Black Hole (BH). iii) Stellar core collapse: Hydrodynamical or core-collapse Supernovae (SNe) as a mechanism of SNIb/c and SNII types. iv) Tidal disruption by a supermassive BH (SMBH). v) Progenitors of Gamma-Ray Bursts (GRBs) of long duration (more than two seconds): Collapsing star or collapsars. vi) Progenitors of GRBs of short duration (less than two seconds): mergers of two Neutron Stars (NS). vii) Coalescing compact binaries: White Dwarfs (WD), NS, BH. Black Holes (BH). Recently, the merger of a NS and a Helium star has been proposed as a progenitor of GRB101225A (Thäne et al.). These are one of the most promising sources of gravitational radiation.

Twenty years ago authors in (Martí et al. 1991) paved the way for modern numerical Relativistic hydrodynamics (RHD) extending to this field the strategy of using High-Resolution Shock-Capturing (HRSC) methods. By exploiting the hyperbolic and conservative character of the RHD equations we derived the *full spectral decomposition* of the Jacobian matrices associated to the fluxes in a series of papers: (Martí et al. 1991), in 1D (One-Dimensional) in General Relativity (GRHD); (Font et al. 1994) and (Donat et al. 1998), in multidimensional Special Relativity (SRHD); (Banyuls 1997) and (Ibáñez et al. 2001), in multidimensional GRHD; (Antón et al. 2006), (Antón 2008) and (Antón et al. 2010), in multidimensional Relativistic magnetohydrodynamics (RMHD).

The use of Riemann solvers in computational RHD has proved to be successful in handling complex flows, with high Lorentz factors and strong shocks, superseding traditional methods which failed to describe ultrarelativistic flows (Norman and Winkler 1986). The task of developing robust, stable and accurate (special or general) relativistic hydrocodes is a challenge in the field of Relativistic Astrophysics. A general relativistic hydrocode is a useful research tool for studying flows which evolve in a background spacetime. Furthermore, when appropiately coupled with Einstein Equations (EE), such a general relativistic hydrocode is crucial to model the evolution of matter in a dynamical spacetime.

2 THE EQUATIONS OF GRHD

The evolution of a relativistic fluid is governed by a system of equations which summarize *local conservation laws*: the local conservation of baryon number, $\nabla \cdot \mathbf{J} = \mathbf{0}$, and the local conservation of energy-momentum, $\nabla \cdot \mathbf{T} = 0$ ($\nabla \cdot$ stands for the covariant divergence).

If $\{\partial_t, \partial_i\}$ define the coordinate basis of 4-vectors which are tangents to the corresponding coordinate curves, then, the *current of rest-mass*, $\mathbf{J}$, and the *energy-momentum tensor*, $\mathbf{T}$, for a perfect fluid, have the components: $J^\mu = \rho u^\mu$, and $T^{\mu\nu} = \rho h u^\mu u^\nu + p g^{\mu\nu}$, respectively, ρ being the rest-mass density, p the pressure and h the specific enthalpy, defined by $h = 1 + \varepsilon + p/\rho$, where ε is the specific internal energy. u^μ is the four-velocity of the fluid and $g_{\mu\nu}$ defines the metric of the spacetime M where the fluid evolves. As usual, Greek (Latin) indices run from 0 to 3 (1 to 3) – or, alternatively, they stand for the general coordinates $\{t,x,y,z\}$ ($\{x,y,z\}$) – and the system of units is the so-called geometrized ($c = G = 1$).

An equation of state $p = p(\rho,\varepsilon)$ closes, as usual, the system. Accordingly, the local sound velocity c_s satisfies: $hc_s^2 = \chi + \left(p/\rho^2\right)\kappa$, with $\chi = \partial p/\partial \rho|_\in$ and $\kappa = \partial p/\partial \in|_\rho$.

Following (Banyuls 1997), let M be a general spacetime described by the four dimensional metric tensor $g_{\mu\nu}$. According to the well-known $\{3 + 1\}$ formalism (see recent monographies as, e.g., (Alcubierre 2008), (Choquet-Bruhat 2008), (Gourgoulhon 2012) the metric is split into the objects α (*lapse*), β^i (*shift*) and γ_{ij}, keeping the line element in the form:

$$ds^2 = -\left(\alpha^2 - \beta_i\beta^i\right)dt^2 + 2\beta_i dx^i dt + \gamma_{ij}dx^i dx^j \qquad (1)$$

If $\mathbf{n}$ is a unit timelike vector field normal to the spacelike hypersurfaces Σ_t (t = const.), then, by definition of α and β^i is (see Fig. 1): $\partial_t = \alpha\mathbf{n} + \beta^i\partial_i$, with $\mathbf{n}\cdot\partial_i = 0$, $\forall_i$. Observers, O_E, at rest in the slice Σ_t, i.e., those having $\mathbf{n}$ as four-velocity (*Eulerian observers*), measure the following velocity of the fluid

$$v^i = \frac{u^i}{\alpha u^t} + \frac{\beta^i}{\alpha} \qquad (2)$$

where $W := -(\mathbf{u}\cdot\mathbf{n}) = \alpha u^t$, the Lorentz factor, satisfies $W = (1 - \mathrm{v}^2)^{-1/2}$ with $\mathrm{v}^2 = v_i v^i$ ($v_i = \gamma_{ij}v^j$).

Let me define a basis adapted to the observer O_E, $\mathbf{e}_{(\mu)} = \{\mathbf{n}, \partial_i\}$, and the following five four-vector fields $\{\mathbf{J}, \mathbf{T}\cdot\mathbf{n}, \mathbf{T}\cdot\partial_1, \mathbf{T}\cdot\partial_2, \mathbf{T}\cdot\partial_3\}$. Hence, the system of equations of GRHD can be written

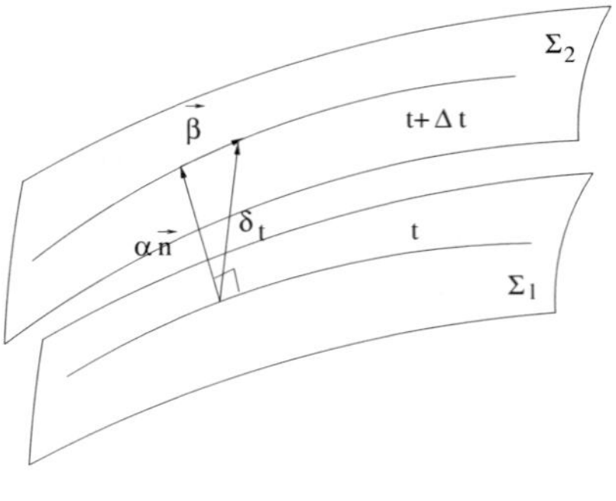

Figure 1. A time slicing of spacetime: Σ_1 and Σ_2 are two spacelike hypersurfaces defined, respectively, by $t = const$ and $t + \Delta t = const$.

$$\nabla \cdot \mathbf{A} = s, \qquad (3)$$

where $\mathbf{A}$ denotes any of the above 5 vector fields, and s is the corresponding source term.

The set of *conserved variables* gathers those quantities which are directly measured by O_E, i.e., the rest-mass density (D), the momentum density in the j-direction (S_j) and the total energy density (E). In terms of the *primitive variables* $\mathbf{w} = (\rho, v_i, \varepsilon)$ ($v_i = \gamma_{ij}v^j$) they are

$$D = \rho W, \quad S_j = \rho h W^2 v_j, \quad E = \rho h W^2 - p \qquad (4)$$

Taking all the above relations together, the fundamental first-order, flux-conservative system reads

$$\frac{1}{\sqrt{-g}}\left(\frac{\partial\sqrt{\gamma}\mathbf{F}^0(\mathbf{w})}{\partial x^0} + \frac{\partial\sqrt{-g}\mathbf{F}^i(\mathbf{w})}{\partial x^i}\right) = \mathbf{s}(\mathbf{w}) \qquad (5)$$

where the quantities $\mathbf{F}^\alpha(\mathbf{w})$ are

$$\mathbf{F}^0(\mathbf{w}) = (D, S_j, \tau) \qquad (6)$$

$$\mathbf{F}^i(\mathbf{w}) = \left(D\left(v^i - \frac{\beta^i}{\alpha}\right), \; S_j\left(v^i - \frac{\beta^i}{\alpha}\right) + p\delta^i_j, \; \tau\left(v^i - \frac{\beta^i}{\alpha}\right) + pv^i \right) \qquad (7)$$

and the corresponding sources $\mathbf{s}(\mathbf{w})$ are

$$\mathbf{s}(\mathbf{w}) = \left(0, T^{\mu\nu}\left(\frac{\partial g_{\nu j}}{\partial x^\mu} - \Gamma^\delta_{\nu\mu}g_{\delta j}\right), \; \alpha\left(T^{\mu 0}\frac{\partial \ln\alpha}{\partial x^\mu} - T^{\mu\nu}\Gamma^0_{\nu\mu}\right)\right) \qquad (8)$$

being $\tau := E\!-\!D$ (total energy density excluding the rest-mass one). $g = \det(g_{\mu\nu})$ is such that $\sqrt{-g} = \alpha\sqrt{\gamma}\left(\gamma := \det\left(\gamma_{ij}\right)\right)$

The three 5×5-Jacobian matrices $\mathcal{B}^i$ associated to system (5) are

$$\mathcal{B}^i = \alpha\frac{\partial \mathbf{F}^i}{\partial \mathbf{F}^0}. \tag{9}$$

The full spectral decomposition of the above three 5×5–Jacobian matrices $\mathcal{B}^i$ can be found in Ibáñez *et al.* (2001)[1].

In particular, the characteristic speeds (associated to one spatial direction, e.g.) are much more complex than the classical ones:

$\lambda_0 = \alpha v^x - \beta^x$ (triple)

which defines the *material waves*, and

$$\lambda_\pm = \frac{\alpha}{1 - v^2 c_s^2}\left\{v^x\left(1 - c_s^2\right) \pm c_s\sqrt{\left(1 - v^2\right)\left[\gamma^{xx}\left(1 - v^2 c_s^2\right) - v^x v^x\left(1 - c_s^2\right)\right]}\right\} - \beta^x$$

which are associated with the *acoustic waves*. The presence of the geometrical factors, and the strong coupling through the Lorentz factor with the tangential velocity, makes the GRHD much more rich than the classical one.

The SRHD limit (Minkowski spacetime) is recovered (in Cartesian coordinates) just doing: $\alpha = 1$, $\beta^i = 0$, $\gamma_{ij} = \delta_{ij}$.

2.1 *A note on GRHD convexity*

Let $\{\lambda_\alpha, \mathbf{r}_\alpha\}$ be the eigenvalues and righteigenvectors coming from the spectral decomposition of matrix $\mathcal{B}^x$ in (9). A characteristic field $\mathcal{C}_\alpha$ ($\alpha = 1, 2, \ldots, d$; $d = 5$ for system (5)) satisfying

$$\mathcal{C}_\alpha : \frac{dx}{dt} = \lambda_\alpha(\alpha = 1, 2, \ldots, d) \tag{10}$$

is said to be *genuinely nonlinear* or *linearly degenerate* if, respectively,

$$\mathcal{P}_\alpha := \vec{\nabla}_{\mathbf{u}}\lambda_\alpha \cdot \mathbf{r}_\alpha \neq 0, \tag{11}$$

$$\mathcal{P}_\alpha := \vec{\nabla}_{\mathbf{u}}\lambda_\alpha \cdot \mathbf{r}_\alpha = 0 \tag{12}$$

where the operator $\vec{\nabla}_{\mathbf{u}}$ acts on the space of conserved variables (see, e.g., (LeVeque 1992), (Toro 2009)).

[1]The right-eigenvectors published in (Banyuls 1997) corre-spond to spacetimes having a diagonal metric.

In a convex system, all the characteristic fields are either genuinely non-linear or linearly degenerate. Non-convexity is associated to those states for which the condition (11) is not fulfiled.

In (Ibáñez et al. 2011) we reported the results of applying the above analysis to the SRHD equations. The non-linear SR characteristic fields satisfy: $\mathcal{P}_\pm \sim \mathcal{T}(\rho, \epsilon)$ where

$$\mathcal{T} = \frac{\partial c_s}{\partial \rho} + \frac{p}{\rho^2}\frac{\partial c_s}{\partial \epsilon} + \frac{c_s}{\rho}\left(1 - c_s^2\right) \tag{13}$$

From the above relationship (13) it turns out that the loss of convexity is closely related with the thermodynamical properties of the flow; in particular, the behavior of the function $c_s(\rho, \epsilon)$ plays a fundamental role (it was alredy noticed by (Menikoff and Plohr 1989), for classical flows undergoing phase transitions). In (13) the last term $\left(1 - c_s^2\right)$ shows a purely relativistic correction of second order $\left(c_s^2\right)$.

The analysis of the characteristic fields associated to GRHD leads to similar conclusions. The quantity $\mathcal{P}_\pm$, in GRHD, can be given in terms of two factors, a purely kinematical one, κ, and the purely thermodynamical one, $\mathcal{T}$ in (13):

$$\begin{aligned} \mathcal{P}_\pm &= \pm\kappa_\pm\left(\gamma_{ij}, v_m\right)\mathcal{T}(\rho, \epsilon) \\ \kappa_\pm &= \left(v^x c_s \pm \Delta^{1/2}\right)^{-2}\left(\gamma^{xx} - v^x v^x\right)^2 W^2 \Delta^{-1/2} \\ \Delta &:= W^2\gamma^{xx}\left(1 - v^2 c_s^2\right) - v^x v^x\left(1 - c_s^2\right) \end{aligned} \tag{14}$$

The factor $\mathcal{T}$ can be rewritten

$$\mathcal{T} = \frac{c_s}{\rho}\left(\mathcal{G} - \frac{3}{2}c_s^2\right) \tag{15}$$

where $\mathcal{G}$ is the fundamental derivative defined in (Menikoff and Plohr 1989):

$$\begin{aligned} \mathcal{G} &:= -\frac{1}{2}V\frac{\left(\frac{\partial^2 p}{\partial V^2}\right)_s}{\left(\frac{\partial p}{\partial V}\right)_s} \\ &= 1 + \left(\frac{\partial \ln c_s}{\partial \ln \rho}\right)_s \end{aligned} \tag{16}$$

being $V := 1/\rho$ the specific volume.

Hence, unlike classical HD which is convex there where $\mathcal{G} > 0$, GRHD is convex there where the following relation is satisfied:

$$\mathcal{G} - \frac{3}{2}c_s^2 > 0$$

2.2 *Exact solution of the Riemann problem in SRHD*

In (Martí et Müller 1994) authors have derived the analytical solution of the Riemann problem in SRHD, for the case of zero tangential velocities (strictly 1D problem). The exact solution of the Riemann problem with non-zero tangential velocities were obtained in (Pons and Müller 2000). Authors provide two FOR-TRAN programs called RIEMANN and RIEMANN-VT, which allow one to compute the exact solution of an arbitrary special relativistic Riemann problem for an ideal gas EOS with constant adiabatic index, both with zero and non-zero tangential speeds. Some improvements of the above results can be found in (Rezzolla and Zanotti 2001) and (Rezzolla et al. 2003). The knowledge of the exact solution of the Riemann problem in SRHD is a milestone in the field of numerical relativistic hydrodynamics. Since then, any SR hydro-code has to be able to recover a battery of test-beds for which the exact solution is known.

The scientific literature on *Special Relativistic Riemann Solvers* (SRRS) has known a spectacular progress during the second half of 1990s. Although some of the SRRS proposed are a straigthforward extension of the corresponding in classical fluid dynamics, most of them have been specifically designed to handle the Riemann problem of the equations of SRHD (for perfect fluids). An up-to-dated list of SRRS can be found in (Martí et Müller 2003).

In (Pons et al. 1998) we have deloped a general procedure allowing any SRHD Riemann solver to be applied to GRHD flows. Our proposal consists in performing, at each numerical interface, a coordinate transformation to locally Minkowskian coordinates, and assuming that the solution of the Riemann problem is the one in special relativity and planar symmetry.

3 THE EQUATIONS OF RMHD

The equations of ideal RMHD correspond to the conservation of rest-mass and energy-momentum, and the Maxwell equations. In the following, the standard Einstein sum convention is assumed. We use units in which the $(4\pi)^{1/2}$ factor is absorbed in the definition of the magnetic field. Specializing for a flat space-time and Cartesian coordinates, these equations can be written as a system of conservation laws, which reads

$$\frac{\partial \mathbf{U}}{\partial t} + \frac{\partial \mathbf{F}^i}{\partial x^i} = 0, \tag{17}$$

where the state vector, $\mathbf{U}$ (vector of *conserved variables*), and the fluxes, $\mathbf{F}^i$ ($i = 1,2,3$ or $i = x,y,z$), are the following column vectors,

$$\mathbf{U} = \left(D, S^j, \tau, B^k\right)^T \tag{18}$$

$$\mathbf{F}^i = (Dv^i, S^j v^i + p^* \delta^{ij} - b^j B^i/W \,, \quad \tau v^i + p^* v^i - b^0 B^i/W, \quad v^i B^k - v^k B^i)^T. \tag{19}$$

where the superscript ^{T} stands for the transposition. In the preceding equations, D, S^j and E stand, respectively, for the rest-mass density, the momentum density of the magnetized fluid in the j-direction and its total energy density as measured in the laboratory (i.e., Eulerian) frame,

$$D = \rho W, \; S^j = \rho h^* W^2 v^j - b^0 b^j, \quad E = \rho h^* W^2 - p^* - \left(b^0\right)^2 \tag{20}$$

where ρ is the proper rest-mass density, $h^* = 1 + \epsilon + p/\rho + b^2/\rho$ is the specific enthalpy including the contribution from the magnetic field (b^2 stands for $b^\mu b_\mu$), ϵ is the specific internal energy, p the thermal pressure, and $p^* = p + b^2/2$ the total pressure. Like in GRHD, instead of E we use the conserved variable $\epsilon := E-D$ The four-vectors representing the fluid velocity, u^μ, and the magnetic field measured in the fluid rest frame, b^μ, and there is an equation of state relating the thermodynamic variables, p, ρ and ϵ, $p = p(\rho,\epsilon)$. All the discussion will be valid for a general equation of state but results will be shown for an ideal gas, for which $p = (\gamma - 1)\rho\epsilon$, where γ is the adiabatic exponent. Quantities v^i stand for the components of the fluid velocity trivector as measured in the laboratory frame; they are related with the components of the fluid four-velocity according to the following expression $u^\mu = W(1, v^x, v^y, v^z)$, where W is the flow Lorentz factor, $W^2 = 1/(1 - v^i v_i)$.

The following fundamental relations hold between the components of the magnetic field four-vector, in the comoving frame, and the three vector components B^i measured in the laboratory frame,

$$b^0 = W\, \mathbf{B} \cdot \mathbf{v}, \quad b^i = \frac{B^i}{W} + b^0 v^i \tag{21}$$

$\mathbf{v}$ and $\mathbf{B}$ being, respectively, the tri-vectors (v^x, v^y, v^z) and (B^x, B^y, B^z).

$$b^2 = \frac{\mathbf{B}^2}{W^2} + (\mathbf{B} \cdot \mathbf{v})^2 \tag{22}$$

The preceding system must be complemented with the usual divergence constraint

$$\frac{\partial B^i}{\partial x^i} = 0, \tag{23}$$

which should be fulfilled at all times. Fluxes $\mathbf{F}^i$ $(i = x, y, z)$ are functions of the conserved variables, $\mathbf{U}$, although for the RMHD this dependence, in general, can not be expressed explicitly. It is therefore necessary to introduce another set of variables, the so-called *primitive variables*, derived from the conserved ones, in terms of which the fluxes can be computed explicitly. We have used the following set of primitive variables

$$\mathbf{V} = \left(\rho, p, v^x, v^y, v^z, B^x, B^y, B^z\right)^T. \tag{24}$$

The hyperbolicity of the equations of RMHD including the derivation of wavespeeds and the corresponding eigenvectors, and the analysis of various degeneracies has been reviewed by (Anile 1089), in a covariant framework, using a set of variables of dimension 10, the so-called *covariant variables* (Anile's variables, in the next):

$$\tilde{\mathbf{U}} = \left(u^\mu, b^\mu, p, s\right)^T, \tag{25}$$

where s is the specific entropy.

In terms of variables $\tilde{\mathbf{U}}$, the system of RMHD equations can be written as a quasi-linear system of the form

$$\mathcal{A}^\mu \tilde{\mathbf{U}}_{;\mu} = 0, \tag{26}$$

where the subscript; μ stands for the covariant derivative, and four 10×10 Jacobian matrices $\mathcal{A}^\mu$ can be found in Anile's book. It is important to remark that the 10 covariant variables we have used to write the system of equations are not independent, since they are related by the constraints

$$u^\alpha u_\alpha = -1,\ b^\alpha u_\alpha = 0,\ \partial_\alpha\left(u^\alpha b^0 - u^0 b^\alpha\right) = 0 \tag{27}$$

The latter condition, is a covariant representation of the divergence constraint (Eq. 23). The system of (ideal) RMHD equations have the same seven wavespeeds as in classical MHD: the entropic, Alfvén, slow magnetosonic, and fast magnetosonic waves. They can be ordered as follows

$$\lambda_f^- \le \lambda_a^- \le \lambda_s^- \le \lambda_e \le \lambda_s^+ \le \lambda_a^+ \le \lambda_f^+, \tag{28}$$

where the subscripts e, a, s and f stand for *entropic*, *Alfvén*, *slow magnetosonic* and *fast magnetosonic* respectively, and the superscript – or + refer to the lower or higher value of each pair. Unlike classical MHD, it is however not possible, in general, to obtain simple expressions for the magnetosonic speeds since they are given by the solutions of a quartic equation.

As in the case of classical MHD, degeneracies are encountered for waves propagating perpendicular to the magnetic field direction (Type I) and for waves propagating along the magnetic field direction (Type II). Finally, a particular subcase of Type II degeneracy appears when the sound speed is equal to c_a. For Type I degeneracy, the two Alfvén waves, the entropic wave and the two slow magnetosonic waves propagate at the same speed $(\lambda_a^- = \lambda_s^- = \lambda_e = \lambda_s^+ = \lambda_a^+)$. For Type II degeneracy, an Alfvén wave and a magnetosonic wave (slow or fast) of the same class propagate at the same speed $(\lambda_f^- = \lambda_a^-$ or $\lambda_a^- = \lambda_s^-$ or $\lambda_s^+ = \lambda_a^+$ or $\lambda_a^+ = \lambda_f^+)$. In the special Type II' subcase, an Alfvén wave and both the slow and fast magnetosonic waves of the same class propagate at the same speed $(\lambda_f^- = \lambda_a^- = \lambda_s^-$ or $\lambda_s^+ = \lambda_a^+ = \lambda_f^+)$.

3.1 *Renormalized right eigenvectors*

As it is well known in classical MHD, Alfvén and magnetosonic eigenvectors have a pathological behaviour at degeneracies, since they become zero or linearly dependent and they do not form a basis. In (Antón et al. 2010), we have derived a new set of renormalized Alfvén and magnetosonic eigenvectors for RMHD. Our renormalized Alfvén right eigenvectors (following the methodology of (Brio and Wu 1988)) are a linear combination of the ones proposed by (Komissarov 1999), for the Type II degeneracy case. However, contrary to the Komissarov's choice, our expressions are free of pathologies not only in the Type II degeneracy but also in the Type I degeneracy case. Our derivation of the renormal-ized magnetosonic right eigenvectors is algebraically more cumbersome and reader interested is addressed to (Antón et al. 2010). The final result of this analysis allows to have a complete set of right eigenvectors linearly independent for all possible states. Following the same procedure we have used to renormalize the right eigenvectors, we have derived (Antón et al. 2010) left eigenvectors well behaved for degenerate states.

3.2 *A Full Wave Decomposition Riemann Solver in RMHD (FWD)*

Let me summarize the main steps in our derivation of a FWD Riemann Solver in RMHD: 1) Let $r_{\tilde{\mathbf{u}}}$ be a generic right eigenvector derived in terms of *Anile's variables* (25). 2) The corresponding eigenvector in terms of the *primitive variables* (24) is derived according to: $r_{\mathrm{v}} = \left(\partial_{\tilde{\mathrm{U}}}\mathbf{V}\right) r_{\tilde{\mathrm{U}}}$.3) Finally, the corresponding vector in terms of the *conserved variables* (18) is obtained from $\mathbf{R} := r_{\mathrm{U}} = (\partial_{\mathrm{V}}\mathbf{U})r_{\mathrm{v}}$. 4) Analogously, for the *left eigenvectors*. This procedure, which starts with renormalized eigenvectors, allows one to get the full spectral decomposition of the Jacobian matrices (associated to the fluxes), and free of pathologies in the degeneracies. 5) We use a *linearized (Roe's type, (Roe 1981)) Riemann solver*:

$$\hat{\mathbf{f}}_{j\pm\frac{1}{2}} = \frac{1}{2}\left(\mathbf{f}\left(\mathbf{u}^{L}_{j\pm\frac{1}{2}}\right) + \mathbf{f}\left(\mathbf{u}^{R}_{j\pm\frac{1}{2}}\right) - \sum_{\alpha=1}^{p}\left|\tilde{\lambda}_{\alpha}\right|\Delta\tilde{\omega}_{\alpha}\tilde{r}_{\alpha}\right)$$

where $\mathbf{u}^L$, $\mathbf{u}^R$, are the left and right reconstructed variables; $\Delta\omega$, is the jump of characteristic variables. 6) Our FWD linearized Riemann solver has been exhaustively tested in (Antón et al. 2010).

3.3 *A note on RMHD convexity*

Authors in (Brio and Wu 1988) noted that the equations of classical MHD are non-convex at the degenerate states. We have faced on the convexity properties of RMHD. Preliminar results allows one to conclude that, like classical MHD, the degenerate states are non-convex (magnetosonic waves change from genuinely non-linear to linearly degenerate). We refer the reader to (Antón 2008) (appendix G), where an analysis of the characteristic fields of RMHD in terms of Anile's variables is presented. Much more theoretical work is necessary in order to asses all the richness of other possible non-convex states in RMHD. The previous analysis in SRHD serves us as a road-map to the full characterization of RMHD pathological behaviours (the non-convex character of both the classical and relativistic MHD equations is source of several pathologies, as the development of the so-called compound waves).

4 EINSTEIN SOLVER

The field equations of the Theory of General Relativity (Einstein equations, EE) are

$$G^{\mu\nu} = 8\pi T^{\mu\nu}. \tag{29}$$

The tensor on the left hand side is a combination of the first and second derivatives of the components of the metric. The tensor on the right hand side, the energy-momentum tensor, is a combination of matter fields. The richness (and beauty) of these Partial Differential Equations (PDEs) have been summarized in (Misner et al. 1973.): *Geometry tells to matter how to move, and matter tells geometry how to curve.*

In the 3+1 formalism (see section 2) and following the ADM formulation (Arnowitt et al. 1962), the EE can be written as a system of first order PDEs governing the evolution of the following geometrical (dynamical) quantities

$$\partial_t \gamma_{ij} = -2\alpha K_{ij} + \nabla_i \beta_j + \nabla_j \beta_i \tag{30}$$

$$\begin{aligned}\partial_t K_{ij} = {} & \beta^m \nabla_m K_{ij} + K_{im}\nabla_j \beta^m + K_{jm}\nabla_i \beta^m \\ & + \alpha\left(R_{ij} + KK_{ij} - 2K_{im}K^m_j\right) \\ & + \nabla_i \nabla_{j\alpha} - 8\pi T_{ij}\end{aligned} \tag{31}$$

which have to satisfy the *four constraints:*

$$0 = R + K^2 - K_{ij}K^{ij} - 16\pi\alpha^2 T^{00} \tag{32}$$

$$0 = \nabla_i\left(K^{ij} - \gamma^{ij}K\right) - 8\pi S^j \tag{33}$$

where $R_{\mu\nu}$ is the *Ricci tensor*, R the *Riemann scalar*, K_{ij} the *extrinsic curvature* and $K = \gamma^{ij}K_{ij}$.

The mathematical analysis of the above formulation of EE (30, 31, 32, 33) lead—in the '60 s—to the conclusion that they were not suitable for applications. The field of *Numerical Relativity* was born, having as the main objective to build up a code, to solve EE, that simultaneously (Shapiro and Teukol-sky 1985): i) Avoids singularities. ii) Handles black holes. iii) Maintains high accuracy. iv) Runs forever. Many groups during many years have tried it. Only recently (since 2005) we can say, according to Pretorius in (Pretorius 2007), that the binary BH problem (the basic two body problem in GR) is just beginning to be fully revealed by recent breakthroughs in numerical relativity.

In connection with the formulation of EE (in the 3+1 formalism) two main strategies have been developed: (A) *Free evolution*, and (B) *Fully Constrained Formulation.*

4.1 *(A) Free evolution formulations*

The strategy consists in solving the constraint equations only to get the initial data.

The most popular one is the so-called *BSSN* (from Baumgarte, Shapiro, Shibata, Nakamura; see (Shi-bata and Nakamura 1995), (Baumgarte and Shapiro 1999)) Its main properties can be summarized: i) It shows a much more improved stability than the standard ADM formulation. ii) It has been subjected to a rigurous mathematical analysis (well-posedness, symmetric hyperbolic, …) mainly by Gundlach & Martín-García (see, e.g., (Gundlach and Martín-García 2006)). iii) The most successful computations in numerical relativity to date has been done in BSSN: the merger of binary compact objects (BH/BH, BH/NS, NS/NS), long-term evolution of NS, gravitational collapse of NSs to BHs, ... (see the review (Font 2008)).

EE in the BSSN formulation are described by the following auxiliary variables and system of PDEs:

1. A conformal metric: $\tilde{\gamma}_{ij} = e^{-4\phi}\gamma_{ij}$ such that the algebraic constraint: $\tilde{\gamma} := det\left(\tilde{\gamma}_{ij}\right) = 1$ (where $\gamma := det(\gamma_{ij})$) has to be satisfied. As a consequence: $\phi = (\ln\gamma)/12$,
2. A conformal rescaling of the traceless part of the extrinsic curvature:

$$A_{ij} := K_{ij}^{TF} \Rightarrow \tilde{A}_{ij} = e^{-4\phi}A_{ij}$$

3. An important choice of auxiliary variables was the following conformal connection functions:

$$\tilde{\Gamma}^i := \tilde{\gamma}^{jk}\tilde{\Gamma}^i_{jk} = -\tilde{\gamma}^{ij}_{,j}$$

4. The evolution equations are governed by a hyperbolic system (17 PDEs) of balance laws:

$$\partial_t\phi - \beta^k\partial_k\phi = s^{(1)}\left(\alpha, \partial_i\beta^i, K\right)$$

$$\partial_t K = s^{(2)}\left(\alpha, \beta^i, \gamma^{ij}, \tilde{A}_{ij}, K, \rho, S\right)$$

$$\partial_t\tilde{\gamma}_{ij} - \beta^k\partial_k\tilde{\gamma}_{ij} = s^{(3)}_{ij}\left(\alpha, \beta^i, \gamma^{ij}, \tilde{A}_{ij}\right)$$

$$\partial_t\tilde{A}_{ij} - \beta^k\partial_k\tilde{A}_{ij} = s^{(4)}_{ij}\left(\alpha, \beta^i, \gamma^{ij}, \phi, K, \tilde{A}_{ij}, R_{ij}, S_{ij}\right)$$

$$\partial_t\tilde{\Gamma}^i - \beta^k\partial_k\tilde{\Gamma}^i = s^{(5)i}\left(\alpha, \beta^i, \gamma^{ij}, \phi, \partial_j K, \tilde{A}_{ij}, \tilde{\Gamma}^i_{jk}, S_j\right)$$

Other First Order Hyperbolic formulations can be found in the review (Reula) and the book (Bona and Palenzuela-Luque 2005).

4.2 *(B) Fully constrained formulation*

The strategy, in this formulation (FCF), was designed by the Meudon group (Bonazzola et al. 2004), and consists in solving the four constraint equations at each time step.

The FCF is based on the following assumptions: i) Elliptic equations are much more stable than hyperbolic ones. ii) The constraint-violating modes do not exist by construction. iii) The equations describing stationary spacetimes are usually elliptic and are naturally recovered when taking the steady-state limit. iv) There are very efficient numerical techniques, based on spectral methods, to solve systems of elliptic PDEs. v) The five elliptic PDEs of the popular Isenberg-Wilson-Mathews (IWM) approximation to General Relativity (Isenberg 1978), (Isenberg and Nester 1980), (Wilson and Mathews 1989) are naturally recovered.

EE in the FCF formulation are described by the following auxiliary variables and system of PDEs (see (Cordero-Carrión et al. 2011) for an updated version of FCF):

1. A time independent fiducial flat metric f_{ij}, and conformal decomposition of the 3+1 fields:

$$\gamma_{ij} = \psi^4\tilde{\gamma}_{ij}, \quad \psi = \left(\gamma / f\right)^{1/12}. \tag{34}$$

2. The deviation of the conformal metric from the flat metric: $h^{ij} := \tilde{\gamma}^{ij} - f^{ij}$ contains the physics of the gravitational waves.
3. Gauge conditions: maximal slicing ($K = 0$) and the generalized Dirac gauge $\left(H^i := \mathcal{D}_k\tilde{\gamma}^{ki} = 0,\right)$
4. We introduce the conformal decomposition

$$\hat{A}^{ij} := \psi^{10}K^{ij}, \tag{35}$$

and its decomposition in longitudinal and transverse-traceless parts

$$\hat{A}^{ij} = (LX)^{ij} + \hat{A}^{ij}_{TT}, \tag{36}$$

where

$$(LX)^{ij} := \mathcal{D}^i X^j + \mathcal{D}^j X^i - \frac{2}{3}f^{ij}\mathcal{D}_k X^k \tag{37}$$

and $\mathcal{D}_i\hat{A}^{ij}_{TT} = 0$. These decompositions are motivated by the local uniqueness properties of elliptic equations shown in (Cordero-Carrión et al. 2009).

5. The Hyperbolic Sector: We define the auxil-iar quantities $w^{ij}_k := \mathcal{D}_k\tilde{\gamma}^{ij}$. The hyperbolic system for h^{ij} can be written as a first order evolution system for the tensors $\left(h^{ij}, \hat{A}^{ij}, w^{ij}_k\right)$,

$$\frac{\partial h^{ij}}{\partial t} = \mathsf{F}^{ij}_{(1)}(\alpha, \psi, \beta^i, \tilde{\gamma}_{ij}, \hat{A}^{ij}) \tag{38}$$

$$\frac{\partial \hat{A}^{ij}}{\partial t} = \mathsf{F}^{ij}_{(2)}(\alpha, \psi, \beta^i, \tilde{\gamma}_{ij}, \hat{A}^{ij}, h^{ij}, w^{ij}_k, S^{ij}) \tag{39}$$

$$\frac{\partial w^{ij}_k}{\partial t} = \mathsf{F}^{ij}_{(3)}(\alpha, \psi, \beta^i, \tilde{\gamma}_{ij}, \hat{A}^{ij}, w^{ij}_k) \tag{40}$$

being $S_{ij} := T_{\mu\nu}\gamma^\mu_i\gamma^\nu_j$ the stress tensor, and $S := \gamma^{ij}S_{ij}$ its trace. The system obeys the constraint of the Dirac gauge, $w^{ij}_i = 0$, and the determinant of the conformal metric: $\tilde{\gamma} = f$.

In (Cordero-Carrión et al. 2008) and (Jaramillo et al. 2008) we have studied the above hyperbolic system of 30 first order PDEs. The main conclusions are:

a. Characteristic velocities:

$$\lambda_0 = 0(multiplicity = 18),$$

$$\lambda_{\pm}^{(\zeta)} = -\beta^{\mu}\zeta_{\mu} + \frac{\alpha}{\psi^2}\left(\tilde{\gamma}^{\mu\nu}\zeta_{\mu}\zeta_{\nu}\right)^{1/2}$$

$$= -\beta^{\mu}\zeta_{\mu} + \alpha\left(\zeta^{\mu}\zeta_{\mu}\right)^{1/2}(multiplicity = 6+6)$$

b. Dirac's gauge is a sufficient condition for the hyperbolicity of the PDEs governing the evolution of h^{ij}. c) The (right-)eigenvectors define a complete system iff: i) the lapse α does not vanish, and ii) $(\beta^{\|})^2 \neq \alpha^2$. d) *Strong hyperbolicity* if t^{μ} is timelike, i.e. if $\alpha \neq 0$ and $\alpha^2 - \beta^i\beta_i > 0$. e) All the characteristic fields are *linearly degenerate*. f) For a coordinate system adapted to a spacelike inner worldtube H, where $\beta^{\perp} > \alpha$, no ingoing radiative modes flow into the integration domain Σ_t at the excision surface.

6. The Elliptic Sector (see (Cordero-Carrión et al. 2009) and (Cordero-Carrión et al. 2011))

$$\Delta\psi = \varepsilon_{(1)}\left(\psi, \tilde{\gamma}_{ij}, \hat{A}^{ij}, E\right)$$

$$\Delta(\alpha\psi) = \varepsilon_{(2)}\left(\alpha, \psi, \tilde{\gamma}_{ij}, \hat{A}^{ij}, E, S,\right)$$

$$\Delta\beta^i + \frac{1}{3}D^iD_j\beta^j = \varepsilon^i_{(3)}\left(\alpha, \psi, \beta^i, \tilde{\gamma}_{ij}, \hat{A}^{ij}, S^i\right)$$

where $\Delta := \tilde{\gamma}^{kl}\mathcal{D}_k\mathcal{D}_l$. $E := T_{\mu\nu}n^{\mu}n^{\nu}$ and $S_i := -\gamma^{\mu}_{i}T_{\mu\nu}n^{\nu}$ are, respectively, the energy density and the momentum density measured by the (Eulerian) observer of 4-velocity n^{μ}. The IWM approximation of EE (see below) is covered in the limit: $\tilde{\gamma}^{ij} \to f^{ij} \Leftrightarrow h^{ij} \to 0$

7. The elliptic-hyperbolic structure of EE in the FCF is fully consistent with causality. On one hand, the lapse function and shift vector just tell us how spacetime is foliated (they are *gauge fields*). On the other hand, the conformal factor is not a metric component; this factor multiplies the sum of the tensors $f_{ij} + h_{ij}$ to obtain the 3-metric components. The f_{ij} tensor represents the flat metric and it does not depend on the coordinate time, i.e., it does not change for different slices of the foliation. The evolution of the h_{ij} tensor is governed by a hyperbolic evolution system having eigenvalues that are bounded by the speed of light in the corresponding coordinates. In brief, the evolution of the metric components are governed by a hyperbolic system such that their characteristic speeds are bounded by the speed of light[2].

[2]Author appreciates discussions on this topic with I. Cordero-Carrión.

4.3 *Some numerical RHD-codes*

Nowadays there are several numerical codes built up to solve the evolution of matter fields in dynamical spacetimes (see the reviews (Martí et Müller 2003) and (Font 2008)). I will focus on some of the ones developed inside european collaborations:

1. Whisky code (see (Whisky 2005) and (Baiotti et al. 2005))
   In the framework of an EU-Training Network (funded during the period: 2000–2003) members of the Valencia group have contributed to the development of fully general relativistic three-dimensional codes such as the CACTUS code for Numerical Relativity (www.cactuscode.org) and the WHISKY code for Numerical Relativistic Hydrodynamics (Whisky 2005). The URL sites provide details about the codes. They are public domain and can be downloaded from these sites.
2. MR-Genesis code (see (Aloy et al. 1999), (Leismann et al. 2005) and (?))
   Current version of the former 3D SRHD-code Genesis. MRGenesis is a common framework for RHD, RMHD (relativistic ideal magnetohydrodynamics) and RRMHD (relativistic resistive magneto-hydrodynamics). It is a multidimensional (1D, 2D or 3D) parallel (MPI) code which allows one to study astrophysical scenarios governed by Special Relativistic magnetohydrodynamical processes. Main features of MRGenesis are: i) Finite Volume approach. ii) Method of lines: separate semi-discretization of space and time. iii) Time advance: TVD Runge Kutta methods of 2nd and 3rd order. iv) High-Resolution Shock Capturing methods (HRSC). v) Inter-cell reconstruction: Up to 3rd order using PPM algorithm. vi) RMHD (relativistic ideal magneto-hydrodynamics): constrained transport. vii) RRMHD (relativistic resistive magneto-hydrodynamics): Munz's method (Lagrange multipliers) to conserve $\vec{\nabla} \cdot \vec{B}$ and charge. viii) Several orthogonal coordinate systems: Cartesian, Cylindrical, Spherical. ix) SPEV (module for spectral evolution of particles) for problems where the non-thermal emission of R(M)HD models is sought.
3. CoCoNuT code (see (CoCoNuT code 2005) and (Dimmelmeier et al. 2005))
   CoCoNuT (Core Collapse Code with Nu Technologies) is a general-relativistic (magneto-) hydro code to evolve matter fields in a FCF dynamical spacetime. Its first version (CoCoA, see (Dim-melmeier et al. 2002b)) used HRSC methods for the Hydro Solver, coupled with standard methods for the Einstein solver (simplified to the IWM approximation). CoCoNuT was born as an improvement of CoCoA by using spectral methods (Dimmelmeier et al. 2005) in the Einstein solver. Current version of

CoCoNuT evolves matter (including magnetic fields) in the dynamical spacetime of FCF. Recently, we (Cordero-Carrión et al. 2011) have succeeded in getting the gravitational waveform from an oscillating NS in FCF.

5 ASTROPHYSICAL APPLICATIONS

In this part I am going to summarize the main results obtained mainly by the Valencia/Garching/Meudon collaboration in the study of astrophysical systems where flows evolve reaching velocities near the speed of light and/or in the presence of strong dynamical gravitational fields.

5.1 *Extragalactic relativistic jets*

Almost every galaxy hosts a Supermassive Black Hole (SMBH). A signature of the existence of a SMBH is its associated *jet*. About **1%** of these SMBHs accretes (from a surrounding torus) enough matter to become detectable. About 0.1% of the SMBHs is able to launch relativistic jets, i.e., higly collimated structures of plasma spanning linear scales up to Mpc scales (exceeding in some cases the size of the host galaxy), and where the bulk velocity reaches values > 0.995 the speed of light (Lorentz factor > 10). Although first discovered in radio, jets emit most of their radiation in the—ray band. After almost half a century since their discovery (in 1918) we still do not know what produces and collimates them. Key quantities that start to be known for several relativistic jets is the mass of the black hole and the mass accretion rate of their disks. In terms of the distance to the central SMBH powering the nuclear activity in radio loud AGN we can distinguish, in their associated relativistic jets, three main regions: *subparsec scale*, *parsec scale* and *kiloparsec scale*.

5.1.1 *Long term simulations of large scale jets*

At kiloparsec scales, the implications of relativistic flow speeds and/or relativistic internal energies in the morphology and dynamics of jets have been the subject of a detailed investigation (see (Martí et al. 1997), the review (Martí et Müller 2003), and references therein). Broadly speaking, relativistic jets give rise to two different morphologies according to the relevance of relativistic effects: i) Hot jets, i.e., those having internal energies of the order or larger than their rest-mass energies. They show little internal structure and smooth cocoons without backflow. ii) Cold jets, in which kinematic relativistic effects due to high Lorentz factors dominate (over thermal effects), display extended overpressured cocoons.

In long term simulations (Scheck et al. 2002) the evolution is dominated by a strong deceleration phase during which large lobes of jet material start to inflate around the jet's head. The numerical simulations reproduce some properties observed in powerful extragalactic radio jets (lobe inflation, hot spot advance speeds and pressures, deceleration of the beam flow along the jet) and can help to constrain the values of basic parameters (such as the particle density and the flow speed) in the jets of real sources.

5.1.2 *Hydrodynamical models of superluminal sources*

The development of multidimensional RHD-codes has allowed the simulation of parsec scale jets and superluminal radio components. The presence of emitting flows at almost the speed of light enhance the importance of relativistic effects (relativistic Doppler boosting, light aberration, light-travel time delays) in the appearance of these sources. Hence, models should use a combination of hydrodynamics and synchrotron radiation transfer to compare with observations. The equations of radiative transfer for the synchrotron radiation are integrated along the line of sight with retarded emission and absortion coefficients in order to account for light-travel time delays. In the hydrodynamical models discussed by (Gómez et al. 1997) (and references cited therein) moving radio components are obtained from perturbations in steady relativistic jets. These jets propagate through pressure decreasing atmospheres causing them to expand and accounting for the observed jet opening angles. Where pressure mismatches exist between the jet and the surrounding atmosphere reconfinement shocks are produced. The energy density enhancement produced downstream from these shocks can give rise to stationary radio knots as observed in many VLBI sources. Superluminal components are produced by triggering small perturbations in these steady jets which propagate at almost the jet flow speed.

In (Aloy et al. 2000a) authors have studied, for the first time, the radio emission properties of 3D relativistic hydrodynamic jet models paying particular attention on the observational signatures of the interaction between the relativistic jet and the surrounding medium, which gives rise to a stratification of the jet where a fast spine is surrounded by a slow high energy shear layer.

5.1.3 *Relativistic Kelvin-Helmholtz instabilities*

The linear stability analysis of relativistic flows against Kelvin-Helmholtz (KH) perturbations goes back to the '70 s. Very recently, Martí and coll. (in (Perucho et al. 2004a) and (Perucho et al. 2004a), (Perucho et al. 2005) and (Perucho et al. 2007)) have investigated in detail the process of transition of the KH instability modes from the linear to the non-linear regime in relativistic jets with the aim of characterising the relation between the long-term evolution of the instability and the initial jet parameters. Martí and coll. use the so-called temporal approach in which the time evolution of the instability is studied in a fixed slice of the jet allowing for a huge transversal spatial resolution (e.g., 400 zones per beam radius). These authors report about qualitatively new aspects of the stability of sheared relativistic jets in the linear and non-linear regimes related to the properties of very high-order reflection modes which have the largest growth rates. The corresponding linear problem was solved for more than 20 models with different specific internal energies of the jet, Lorentz factors and shear layer widths, fixing jet/ambient rest-mass density contrast (= 0.1).

5.1.4 *Magnetized relativistic jets*

In (Leismann et al. 2005) authors have performed a comprehensive parameter study of the morphology and dynamics of magnetized relativistic jets (see Fig. 2). The simulations have been performed with an upgraded version of the Genesis code. Starting from pure hydrodynamic models they consider the effect of a magnetic field of increasing strength (up to 3.3 times the equipartition value) and different topology (purely toroidal or poloidal). Two important parameters of the initial setup are: the beam Lorentz factor and the adiabatic index.

5.2 *Relativistic Bondi-Hoyle accretion*

Historically, the canonical astrophysical scenario in which matter is accreted, in a non-spherical way, by a compact object was suggested early in the '40 s by Hoyle, Lyttleton and Bondi. This will be referred to as the *Bondi-Hoyle-Lyttleton* (BHL) accretion onto a Schwarzschild black hole. Using Newtonian gravity these authors studied the accretion onto a gravitating point mass moving with constant velocity through a nonrelativistic gas which is at rest and has a uniform density at infinity. Since then, this pioneering analytical work has been numerically investigated, for a finite size accretor, by a great number of authors over the years. In a series of papers Font & Ibáñez have made an extensive numerical study of the relativistic extension of the BHL scenario (see, e.g., (Font 2008) for an updated reference list). In particular, in (Font et al. 1998) and (Font et al. 1999), a detailed analysis is made of the morphology and dynamics of the flow evolving in the equatorial plane of a Kerr black hole. The analysis made is novel in its use of *the Kerr-Schild (KS) coordinates system*, which is the simplest within the family of *horizon adapted coordinate systems*, where all fields (metric, fluid and electromagnetic fields) are free

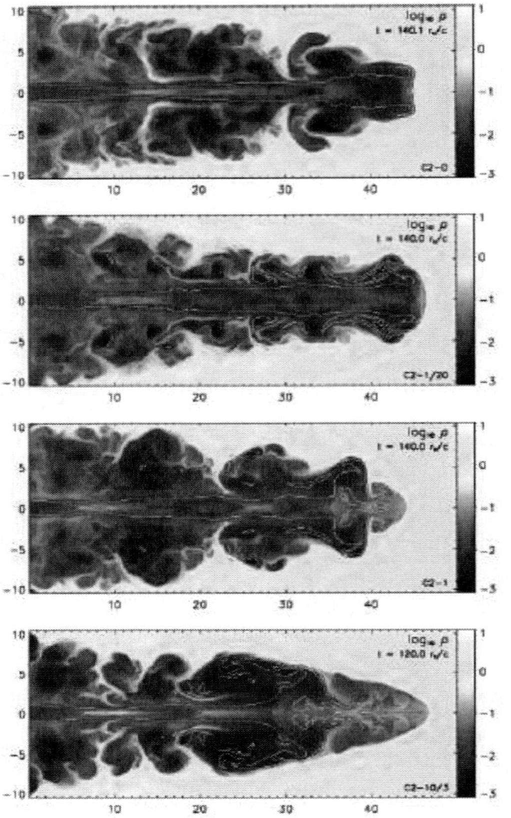

Figure 2. Snapshots showing the rest-mass density of a magnetized jet. The inclusion of the magnetic field leads to diverse effects which, contrary to Newtonian magnetohydrodynamics models, do not always scale linearly with the (relative) strength of the magnetic field. Relativistic jets with toroidal magnetic fields produce a cavity which consists of two parts: an inner one surrounding the beam which is compressed by magnetic forces, and an adjacent outer part which is inflated due to the action of the magnetic field. The outer border of the outer part of the cavity is given by the bow-shock where its interaction with the external medium takes place. Toroidal magnetic fields well below equipartition combined with a value of the adiabatic index of 4/3 yield extremely smooth jet cavities and stable beams. Prominent nose cones form when jets are confined by toroidal fields and carry a high Poynting flux. In contrast, none of the models possessing a poloidal field develops such a nose cone.

of coordinate singularities at the event horizon. This procedure allows to perform accurate numerical studies of relativistic accretion flows around black holes since it is possible to extend the computational grid inside the black hole horizon (see Fig. 3, left panel). In Boyer-Lindquist (BL) coordinates, for a near-extreme Kerr BH, the shock

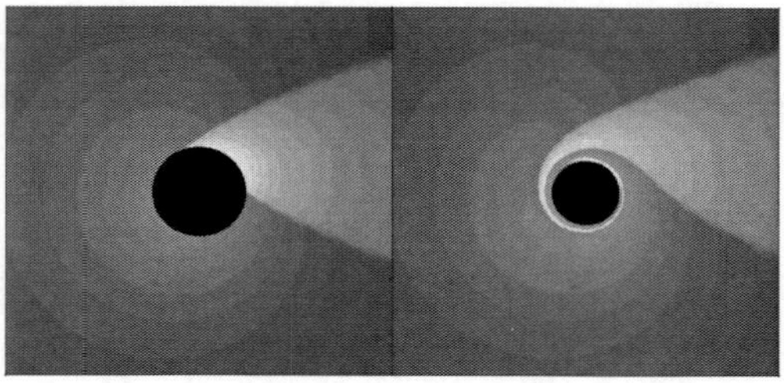

Figure 3. Relativistic wind accretion onto a rapidly-rotating Kerr black hole ($a = 0.999\ M$), the black-hole spin is counterclockwise) in Kerr-Schild coordinates coordinates (left panel). Isocontours of the logarithm of the density are plotted at the final stationary time t = 500 M. Brighter colors (yellow-white) indicate high density regions, while darker colors (blue) correspond to low density zones. The right panel shows how the flow solution looks when transformed to Boyer-Lindquist coordinates. The shock appears here totally wrapped around the horizon of the black hole. The box is 12 M units long. The simulation employed a (r,ϕ)-grid of 200×160 zones.

wraps many times before reaching the horizon due to coordinate effects (see Fig. 3, right panel). This is a pathology of the BL system associated to the collapse of the lapse function at the horizon. The wrapping in the shock wave has an important and immediate consequence: its computation in BL coordinates, although possible in principle, would be much more challenging than in KS coordinates, and the numerical difficulties would increase the closer to the horizon one would impose the boundary conditions in the BL framework.

5.3 *General-relativistic core collapse*

Stars (of Population I) more massive than ≈ 10–$12 M_\odot$ are the primary candidates for what is considered the standard formation scenario behind core-collapse Supernovae, or hydrodynamical SNe (type Ib/c and type II SNe). The so-called *prompt mechanism* proceeds in a hydrodynamical characteristic time: $\tau_{dyn} \approx (G\rho)^{-1/2} \approx 4\rho_{12}^{-1/2}\ msec$, where $\rho_{12} = \rho/(10^{12}\ g/cm^3)$. Several stages can be considered: i) Infall of the "iron" core formed at the center of these massive stars at the end of their evolution. ii) Formation of a strong shock around the sonic point (*bounce*). The energetics involved is $\approx(4$–$10) \times 10^{51}$ ergs. iii) Propagation of that hydrodynamical shock. This epoch is the most complex and still is still a matter of debate. The "numerical" success is strongly dependent of many details on the physics (convection, magnetic fields, neutrino transport, …). Reader interested on the main achievements, in this field, can address to the reviews (Font 2008) and (Fryer and New 2011).

5.3.1 *General-relativistic spherical collapse*

We have studied in (Romero et al. 1996) the general-relativistic spherically symmetric stellar core collapse, paying particular attention to the numerical treatment of the formation and propagation of strong shocks, and extending HRSC techniques to the general-relativistic hydrodynamic equations. The following main results merit to be pointed out: i) The formation and evolution of a shock is sharply solved in one or two zones and is free of spurious oscillations. ii) The conservative features of the code, consistent with the conservation laws of baryonic mass and gravitational mass (or binding energy). iii) As a by-product, authors in (Romero et al. 1996) have presented a battery of test-beds that any GRHD-code has to overcome.

5.3.2 *General-relativistic non-spherical collapse*

Thirty years ago E.Müller /shortciteEwald82 obtained the gravitational waveforms coming from the collapse of rotating stellar cores. The calculations were 2D (axisymmetric) and Newtonian gravity (using, as a post-processing, the quadrupole formula). It was the first numerical evidence of the low gravitational efficiency ($E < 10^{-6}\ M\ c^2$) of the core-collapse scenario. Twenty years later, authors in /shortciteDFM02a and /shortciteDFM02b undertook the first relativistic attempt to deepen into the knowledge of the dynamics of the collapse of rotating stellar cores. The calculations were 2D, in the IWM approximation of EE. The code developed (CoCoA) was the former version of the current CoCoNuT. Some of the goals of these studies were: i) To extend Newtonian simulations in order to determine whether GR effects make a difference in overcoming the angular momentum threshold. ii) To extract gravitational radiation from core collapse and to get more realistic waveforms. iii) To develop a versatile and extensible code for simulating highly relativistic rotating stars. iv) To make an exhaustive analysis (26 models) of the gravitational waveforms generated during axisym-metric relativistic rotational core collapse. The so-called DFM catalogue of wave-forms has been used by the experts involved in data analysis.

The work by Dimmelmeier, Font and Müller (see, e.g., /shortciteDFM02a and /shortciteD-FM02b) has been an outstanding step towards further studies of fully multidimensional general-relativistic stellar core collapse.

5.3.3 *Magnetorotational core collapse*

An increasing number of authors have recently performed axisymmetric magnetorotational core collapse simulations (within the ideal MHD limit) employing Newtonian treatments of magnetohydrodynamics, gravity, and microphysics (see, e.g., (Kotake et al. 2004), (Yamada and Sawai 2004),

(Obergaulinger et al. 2006a), (Obergaulinger et al. 2006b), (Burrows et al. 2007) and references therein). The weakest point of all existing magnetorotational core collapse simulations to date is the fact that the strength and distribution of the initial magnetic field in the core are unknown. If the magnetic field is initially weak, there exist several mechanisms which may amplify it to values, which can have an impact on the dynamics, among them differential rotation (O-dynamo) and the Magnetorotational Instability (MRI). The former transforms rotational energy into magnetic energy, winding up any seed poloidal field into a toroidal field, while the latter, which is present as long as the radial gradient of the angular velocity of the fluid is negative, leads to exponential growth of the field strength. Specific magnetorota-tional effects on the gravitational wave signature were first studied in detail by (Kotake et al. 2004), (Ya-mada and Sawai 2004), who found differences with purely hydrodynamic models only for very strong initial fields ($\geq 10^{12}$ G). The most exhaustive parameter study of magneto-rotational core collapse to date has been carried out by (Obergaulinger et al. 2006a) and (Obergaulinger et al. 2006b). Authors in these references, have studied a complete parameter set up (>50 models). They have employed rotating polytropes, Newtonian (and modified Newtonian) gravity and hydrodynamics, and ad-hoc initial poloidal magnetic field distributions (as no self-consistent solution is yet known). These authors have found that for weak initial fields ($\leq 10^{11}$ G, which is the most relevant case) there are no differences in the collapse dynamics nor in the resulting gravitational wave signal, when comparing with purely hydrodynamic simulations. However, strong initial fields ($\geq 10^{11}$ G) manage to slow down the core efficiently (leading even to retrograde rotation in the proto-NS), which causes qualitatively different dynamics and gravitational wave signals. For some models, they even find highly bipolar, jet-like outflows. The creation and propagation of MHD jets is also studied in the multigroup, radiation MHD simulations of supernova core collapse in the context of rapid rotation performed by (Burrows et al. 2007). These simulations suggest that for rotating cores a SN explosion might be followed by a secondary, weak MHD jet explosion, which might be generic in the collapsar model of GRBs. Finally, authors in (Obergaulinger et al. 2006a) and (Ober-gaulinger et al. 2006b) have found new types of gravitational waveforms, related with the presence of the magnetic field, which are the signatures of the production of jets.

The collapse of magnetized NS in full GR has also been numerically undertaken recently by (Shibata et al. 2006) and (Duez et al. 2006). Both works study the fate of a supramassive NS (see below) as it collapses to form a rotating black hole surrounded by a massive torus, a process which does not happen in prompt timescales due to the transport of angular momentum via magnetic braking and the MRI.

Black-hole formation When effects on the structure of NS coming from, e.g., strong magnetic fields, high temperatures, or rapid (differential) rotation are taken into account, the stability analysis of these relativistic equilibrium configurations leads to the existence of supramassive NSs, i.e., those having a mass (baryonic or gravitational) higher than the corresponding one when these effects are neglected. These supramassive NSs are the natural progenitors of stellar BHs.

In full 3D, hydrodynamic simulations of the collapse of supramassive uniformly-rotating NSs to rotating BHs were first presented in /short-citeShibata00. In (Baiotti et al. 2005) the gravitational collapse of uniformly-rotating NSs to Kerr BHs was studied in three dimensions, using the cactus/whisky code. Long-term simulations were possible by excising the region of the computational domain that includes the curvature singularity when the BH forms and lies within the apparent horizon. It was found that, for sufficiently high angular velocities, the collapse can lead to the formation of a differentially-rotating unstable disk.

5.4 *Self-gravitating accretion tori around a black hole*

Accretion tori around a compact object is a very common astrophysical scenario. The system BH-torus has received the attention of many researchers as a natural extension of the pioneering studies by Bondi-Hoyle and once the so-called Algol paradox was unveiled (in the '50 s) thanks to the paradigm of the mass transfer in close binary stars. The field reached its maturity in the '70 s (see, e.g., (Shakura 1972), (Sunyaev 1972), (Pringle and Rees 1972), (Novikov and Thorne) and (Bardeen and Petterson 1975)). Only very recently, the first fully general relativistic numerical simulations, in axisymmetry, of a system formed by a BH surrounded by a self-gravitating torus in equilibrium have been carried out by the authors in (Montero et al. 2010) and (Kiuchi et al. 2011). Self-gravitating tori orbiting BHs may form after the merger of binary systems formed by BH/NS or NS/NS, or as the result of the gravitational collapse of the rotating core of massive stars. Important parameters of the initial setup are: the torus-to-BH mass ratio and the angular momentum distribution orbiting in equilibrium around the BH. In these references the authors focus on the study of the conditions favoring the onset of the runaway instability. A by-product of these studies is the extraction of the gravitational wave signal coming from this scenario.

5.5 *GRBs of long duration (lGRB): Relativistic jets from collapsars*

Following the review by Piran (Piran 2005), Gamma-Ray Bursts (GRBs) are short and intense pulses of soft γ-rays, lasting from a fraction of a second up to two seconds (short GRBs, sGRB) to several hundred seconds (long GRBs, lGRB). The isotropic luminosities involved range from 10^{51}–10^{52} ergs/sec, making GRBs the most luminous objects in the sky.

According to the current view models of progenitors of GRBs require a common engine, namely a stellar mass BH which accretes up to several solar masses of matter. A fraction of the gravitational binding energy released by accretion is thought to power a pair fireball. If the baryon load of the fireball is not too large, the baryons are accelerated together with the $e^+ e^-$ pairs to Lorentz factors > 10^2.

Let me summarize the current *collapsar* model as a progenitor of lGRB. In (MacFadyen and Woosley 1999) the authors have explored the evolution of rotating helium stars ($M_\alpha \geq 10\ M_\odot$) whose iron core collapse does not produce a successful outgoing shock, but instead forms a BH surrounded by a compact accretion torus. Using a collapsar progenitor, provided by MacFadyen & Woosley, we have simulated (Aloy et al. 2000b) the propagation of an axisymmetric jet through a collapsing rotating massive star with the multi-dimensional RHD-code Genesis. The jet forms as a consequence of an assumed energy deposition in the range 10^{50}–10^{51} ergs s^{-1} within a 30° cone around the rotation axis. We have assumed a space-time corresponding to a Schwarzschild BH. Effects due to the self-gravity of the star on the dynamics are neglected. The equation of state includes the contributions of non-relativistic nucleons treated as a mixture of Boltzmann gases, radiation, and an approximate correction due to e^+e^-–pairs. We advect nine non-reacting nuclear species which are present in the initial model: C^{12}, O^{16}, Ne^{20}, Mg^{24}, Si^{28}, Ni^{56}, He^4, neutrons and protons. The jet flow is strongly beamed, spatially inhomogeneous, and time dependent. The jet reaches the surface of the stellar progenitor. At break-out, the maximum Lorentz factor of the jet flow is 33 (see Fig. 4). After breakout, the jet accelerates into the circumstellar medium, whose density is assumed to decrease exponentially and then become constant ($\approx 10^{-5}$ g cm^{-3}. Outside the star, the flow begins to expand laterally but the beam remains very well collimated. When the simulation ends, the Lorentz factor has increased to 44 (bulk velocity 0.99975).

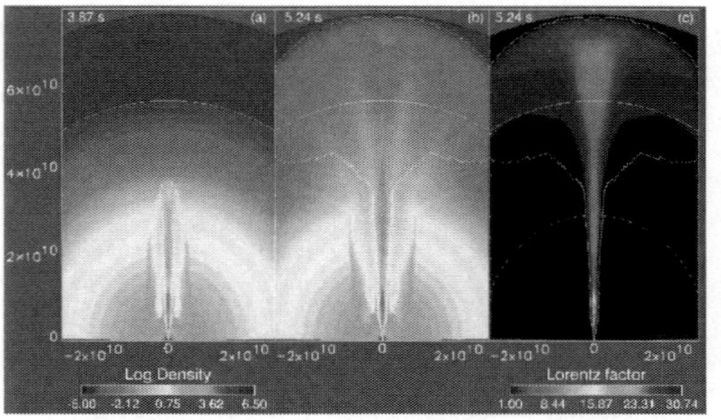

Figure 4. Contour maps of the logarithm of the rest–mass density after 3.87 s and 5.24 s (left two panels), and of the Lorentz factor (right panel) after 5.24 s. X and Y axis measure distance in centimeters. Dashed and solid arcs mark the stellar surface and the outer edge of the exponential atmosphere, respectively. The other solid line encloses matter whose radial velocity > $0.3c$, and whose specific internal energy density > 5×10^{19} erg g^{-1}.

5.6 *Evolution of magnetized and unmagnetized afterglows*

In (Mimica et al. 2009b) the authors have made a detailed analysis of the evolution of magnetized and un-magnetized afterglows coming from a GRB. Figure 5 shows several snapshots of the interaction between ejecta and the interstellar medium: a *forward shock* forms in both magnetized (lower panel) and unmag-netized ejecta (upper panel), paired with either a *"reverse" shock* (yellow arc in the third snapshot of the upper panel) or a *"reverse" rarefaction* (red-to-blue shades in the third snapshot of the lower panel), depending on the degree of magnetization.

The work carried out in (Mimica et al. 2009b) has allowed to put quantitative limits on the typical strength of the magnetic field in the ejecta (afterglow of a GRB), above which we do not expect to observe an optical flash. A new set of scaling laws has been derived allowing one to extrapolate the results of numerical models with moderate values of the initial bulk Lorentz factor (~15) of the ejecta to *equivalent* models with much larger Lorentz factors (~100).

Figure 6 displays a snapshot of the thin magnetized shell evolution taken after the RS has formed and before it has crossed the shell. Full and dashed black lines show the logarithms of the rest-mass density (normalized to the initial shell density ρ_0) and of the pressure (normalized to $\rho_0 c^2$). The red line shows the logarithm of the magnetization σ, while the blue line shows the fluid Lorentz factor γ in the linear scale. All quantities are shown as a function of radius r. Positions of the Forward Shock (*FS*), Contact Discontinuity (*CD*), Reverse Shock (*RS*), and left- (R_1) and right-going (R_2) rarefactions

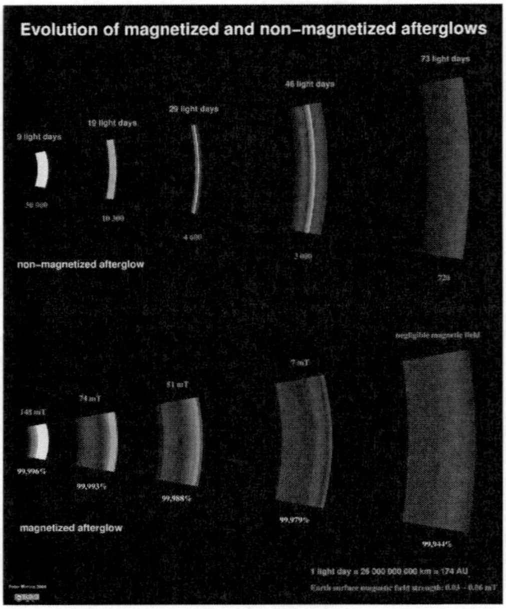

Figure 5. Several snapshots of the interaction between the interstellar medium and magnetized (lower panel) or unmagnetized ejecta (upper panel).

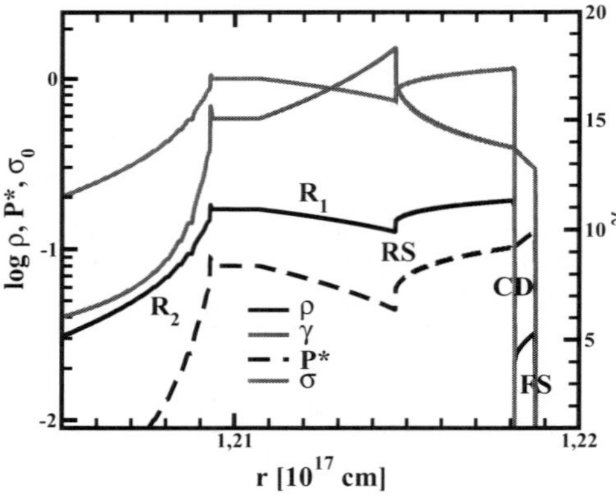

Figure 6. Snapshot of the thin magnetized shell evolution taken after the RS has formed and before it has crossed the shell. See the text for details.

have been indicated. There exist another rarefaction (moving backwards in the external medium) and another contact discontinuity that form at the rear edge of the shell, which are not shown here. Both structures are located to the right of R_2. The unshocked external medium is located in front of the FS and its density $\rho_e \approx 6 \times 10^{-4}\rho_0$ is smaller than the minimum density shown in the plot. The rarefaction R_2, the CD and the FS display a qualitatively similar profile in the non-magnetized case. The rarefaction R_1 and the late steepening of the conditions at its tail, resulting into the formation of RS, are specific of magnetized ejecta.

5.7 *Merger of binary neutron stars*

Short Gamma-Ray Bursts (sGRBs) are among the most luminous events in the universe, releasing in less than two seconds the energy emitted by our Galaxy over one year. Progenitors of sGRBs are the merger of compact binary objects (NS/NS, NS/BH). Double NS/BH systems result from the evolution of initially massive binaries. In particular, double NSs have been discovered because one of the components of the binary is observed as a radio pulsar (see, e.g., the reviews by Postnov and Yungelson (Postnov and Yungelson 2006) on the formation of these systems, and the one by Shibata and Taniguchi (Shibata and Taniguchi 2011) for an uptodated information about current simulations). One of the imprints of these mergers is the emission of gravitational radiation that can be detected in the Earth-based observatories of gravitational waves. In this section I will discuss two particular issues: the development of magnetized Kelvin-Helmholtz instabilities during the merger stage, and the recent results on the merger of two magnetized NS that fill up the gap—in the scientific literature—between the merger epoch and the formation of a sGRB.

5.7.1 *SGRBs: Magnetized Kelvin-Helmholtz instability in neutron-star mergers*

Authors in (Obergaulinger et al. 2010) have carried out global (classical) magnetohydrodynamic simulations showing the growth of Kelvin-Helmholtz instabilities at the contact surface of two merging NSs (see Fig. 7). That region has been identified as the site of efficient amplification of magnetic fields. The magnetic field may be amplified efficiently to very high field strengths, the maximum field energy reaching values of the order of the kinetic energy associated with the velocity components transverse to the interface between the two NSs. In 2D, the magnetic field is amplified on time scales of less than 0.01 ms until it reaches locally equipartition with the kinetic energy. In 3D, the hydromagnetic mechanism seen in 2D may be dominated by purely hydrodynamic instabilities leading to less filed amplification.

5.7.2 *sGRBs: Merger of magnetized BNS*

In/shortciteRezzolla11 authors have found the missing link between the merger of two magnetized NSs and the formation of jet-like structures that can power sGRBs. Considering a binary of

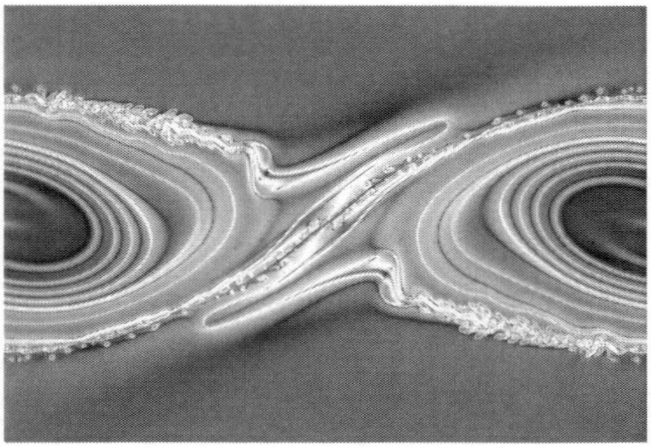

Figure 7. Snapshot of the structure of the model with an initial Mach number M = 1 and an Alfvén number A = 125 taken shortly afterwards the termination of the kinematic amplification phase. Reddish colors indicate regions where magnetic structures are larger than one computational zone, and blueish colors where they are smaller.

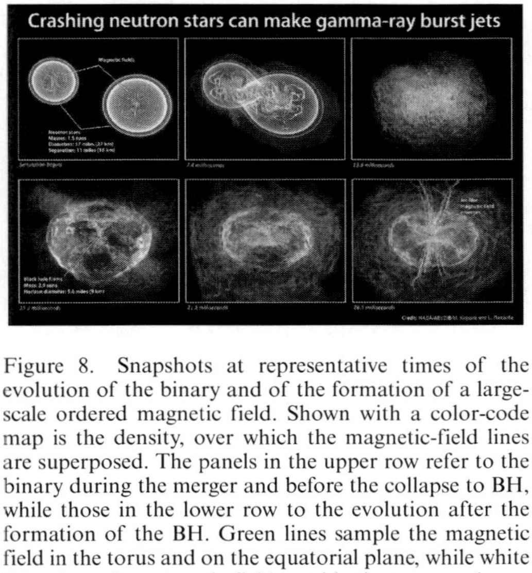

Figure 8. Snapshots at representative times of the evolution of the binary and of the formation of a large-scale ordered magnetic field. Shown with a color-code map is the density, over which the magnetic-field lines are superposed. The panels in the upper row refer to the binary during the merger and before the collapse to BH, while those in the lower row to the evolution after the formation of the BH. Green lines sample the magnetic field in the torus and on the equatorial plane, while white lines show the magnetic field outside the torus and near the BH spin axis. The inner/outer part of the torus has a size of ~90/170 km, while the horizon has a diameter of ≃ 9 km.

magnetized NSs and solving the EE, authors in /shortciteRez-zolla11 show that their merger results in a rapidly spinning black hole surrounded by a hot and highly magnetized torus (see Fig. 8). Magnetohydrodynamical instabilities amplify an initially turbulent magnetic field of $\sim 10^{12}$ G to produce an ordered poloidal field of $\sim 10^{15}$ G along the black-hole spin-axis, within a half-opening angle of ~30°, which may naturally launch a relativistic jet.

6 CONCLUSIONS

The present status regarding with the mathematical properties of the equations governing the evolution of matter fields (perfect fluid, and ideal MHD), in a background spacetime, in connection with the HRSC techniques, can be summarized: i) The theoretical foundations on the characteristic structure of the GRHD & GRMHD equations are available. ii) The exact solution of the Riemann problem (both in RHD, and RMHD) is known. iii) The full spectral decomposition of the flux vector Jacobians of the RHD and RMHD equations is known. A renormalized sets of right and left eigenvectors of the flux vector Jacobians of the RMHD equations, which are regular and span a complete basis in any physical state including degenerate ones have achieved. iv) As a consequence of the above statements, it is possible to develop robust and efficient linearized Riemann solvers based on the renormalized spectral decomposition of the RHD and RMHD equations. v) As a by-product, choosing the adequate restrictions on characteristic variables serves optimally the purpose of setting boundary conditions. The computation of characteristic variables needs of the knowledge of the complete set of left and right eigenvectors.

Regarding with the mathematical properties of the EE governing the dynamics of the spacetime itself, in connection with the numerical techniques to solve them, we can conclude: i) There are several stable formulations of EE which differ in the way they treat the constraints: the free evolution ones (hyperbolic:BSSN, Z4), and the full constrained ones (elliptic-hyperbolic: FCF) ii) The Holy Grail of Numerical Relativity such as it was defined in (Shapiro and Teukolsky 1985) has recently been achieved. In particular, the FCF formulation seems to be appropriate to face on long-term evolution studies.

In the short and middle term the perspectives are: a) To develop HRD-solvers to solve the equations beyond the ideal case: Non-perfect fluids, exotic equations of state, ... b) To extend these studies to relativistic radiation transport, in order to face on the suitable way to incorporate the coupling between matter and photons (or neutrinos). c) To extend these studies to non-ideal MHD (finite conductivity) in order to get an appropriate way of facing on problems such as reconnection. d) The first waveforms coming from an oscillating NS have been obtained in the framework of FCF (Cordero-Carrión et al. 2011), using the current version of CoCoNuT. It will be interesting to explore long term evolutions of astrophysical sources of gravitational radiation (e.g., lGRBs). To end, definitively, we are entering into the golden age of (General) Relativistic Astrophysics/Astrophysical Relativity.

ACKNOWLEDGEMENTS

This contribution relies on extensive analytical and numerical work done in collaboration with M.A. Aloy, L. Antón, P. Cerdá-Durán, I. Cordero-Carrión, H. Dimmelmeier, R. Donat, J.A. Font, E. Gourgoul-hon, J.L. Jaramillo, A. Marquina, J.Mª. Martí, P. Mimica, J.A. Miralles, E. Müller, J. Novak, J.A. Pons, V. Quilis, J.V. Romero and O. Zanotti. The work has been partially supported by the Spanish Ministery of Science (grant AYA2010-21097-C03-01).

REFERENCES

Alcubierre M. 2008. *Introduction to 3 + 1 Numerical Relativity,* Oxford University Press, Oxford.

Aloy M.A., J.Mª. Ibáñez, J.Mª. Martí E. Müller. 1999. Genesis: A High-Resolution Code for Three-dimensional Relativistic Hydrodynamics. *Astrophys. J. Suppl.,* **122**, 151.

Aloy M.A., J.L. Gómez, J.Mª. Ibáñez, J.Mª. Martí & E. Müller. 2000. Radio Emission from Three-dimensional Relativistic Hydrodynamic Jets: Observational Evidence of Jet Stratification. *Astrophys. J. Letters*, **528**, L85.

Aloy M.A., E. Müller, J.Mª. Ibáñez, J.Mª. Martí and A. MacFadyen. 2000. Relativistic Jets from Collapsars. Astrophys. J. Letters, **531**, L119.

Anile M.A. 1989. *Relativistic fluids and magneto-fluids* (Cambridge, England: Cambridge University Press).

Antón L., O. Zanotti, J.A. Miralles, J.Mª. Martí, J. Mª. Ibáñez, J.A. Font & J.A. Pons. 2006. Numerical 3+1 general relativistic magnetohydrodynamics: A local characteristic approach. *Astrophys. J.* **637**, 296.

Antón L. 2008. Magnetohidrodinámica Relativista Numérica: Aplicaciones en Relatividad Especial y General. *Ph.D. Thesis, University of Valencia, Valencia, Spain.*

Antón L., J.A. Miralles, J.Mª. Martí, J. Mª. Ibáñez, M.A. Aloy & P. Mimica. 2010 Relativistic Magnetohydrodynamics: Renormalized eigenvectors and full wave decomposition Riemann solver. *Astrophys. J. Suppl.*, **187**, 1.

Arnowitt R., S. Deser & C.W. Misner. 1962. The dynamics of general relativity, in Witten, L., ed., *Gravitation: An Introduction to Current Research*, 227, (Wiley, New York, USA).

Baiotti L., I. Hawke, P.J. Montero, F. Löffler, L. Rezzolla, N. Stergioulas, J.A. Font & E. Seidel. 2005. Three-dimensional relativistic simulations of rotating neutron-star collapse to a Kerr black hole. *Phys. Rev. D*, **71**, 024035.

Banyuls F., J.A. Font, J.Mª. Ibáñez, J. Mª. Martí & J.A. Miralles. 1997. Numerical 3+1 General Relativistic Hydrodynamics: a Local Characteristic Approach. *Astro-phys. J.* **476**, 221.

Bardeen J.M. & J.A. Petterson. 1975. The Lense-Thirring Effect and Accretion Disks around Kerr Black Holes. *Astrophysical J. Letters*, **195**, L65.

Baumgarte T.W. & S.L. Shapiro. On the numerical integration of Einstein's field equations. *Phys. Rev. D*, **59**, 1, 1999.

Bona C. & C. Palenzuela-Luque. 2005. *Elements of Numerical Relativity,*, Editor: C. Bona, C. Palenzuela-Luque, Lecture Notes in Physics, **673**, (Springer, Berlin).

Bonazzola S., E. Gourgoulhon, P. Grandclément & J. Novak. 2004. Constrained scheme for the Einstein equations based on the Dirac gauge and spherical coordinates. *Phys. Rev. D*, **70**, 104007.

Brio M. & C.C. Wu. 1988. An upwind differencing scheme for the equations of ideal magnetohydrodynamics. *J. Comput. Phys.* **75**, 400.

Burrows, A., L. Dessart, E. Livne, C.D. Ott, & J. Murphy. 2007. Simulations of Magnetically Driven Supernova and Hypernova Explosions in the Context of Rapid Rotation *Astrophys. J.*, **664**, 416.

Choquet-Bruhat Y. 2008. *General Relativity and the Ein-stein Equations,* Oxford University Press, Oxford.

Cordero-Carrión I., P. Cerdá-Durán, H. Dimmelmeier, J.L. Jaramillo, J. Novak & E. Gourgoulhon. 2009. Improved constrained scheme for the Einstein equations: An approach to the uniqueness issue. *Phys. Rev. D*, **79**, 024017.

Cordero-Carrión I., P. Cerdá-Durán & J.Mª. Ibáñnez. 2011. Gravitational waves in dynamical spacetimes with matter content in the Fully Constrained Formulation. *Phys. Rev. D, submitted.*

Cordero-Carrión, I., J.Mª. Ibáñez, E. Gourgoulhon, J.L. Jaramillo & J. Novak. 2008. Mathematical issues in a fully constrained formulation of the Einstein equations. *Phys. Rev. D*, **77**, 084007.

Dimmelmeier H., J.A. Font & E. Müller. 2002. Relativistic simulations of rotational core collapse. I. Methods, initial models, and code tests. *Astron. and Astrophys.*, **388**, 917.

Dimmelmeier H., J.A. Font & E. Müller. 2002. Relativistic simulations of rotational core collapse. II. Collapse dynamics and gravitational radiation. *Astron. and Astrophys.*, **393**, 523.

Dimmelmeier H., J. Novak, J.A. Font, J. Mª. Ibáñez & E. Müller. 2005. Combining spectral and shock-capturing methods: A new numerical approach for 3D relativistic core collapse simulations. *Phys. Rev. D*, **71**, 064023.

Donat R., J.A. Font, J.Mª. Ibáñez & A. Marquina. 1998. A Flux-Split Algorithm applied to Relativistic Flows. *J. Comput. Phys.*, **146**, 58, 1998.

Duez, M.D., Liu, Y.T., Shapiro, S.L., Shibata, M., and Stephens, B.C. 2006. Collapse of magnetized hypermassive neutron stars in general relativity. *Phys. Rev. Lett.*, **96**, 031101.

Font J.A. 2008. Numerical Hydrodynamics in General relativity. *Living Rev. Relativity, 11, lrr-2008-7.* http://www.livingreviews.org/.

Font J.A., J.Mª. Ibáñez, J. Mª. Martí & A. Marquina. 1994. Multidimensional relativistic hydrodynamics: characteristic fields and modern high-resolution shock-capturing schemes. *Astron. Astrophys.*, **282**, 304.

Font J.A., J.Mª. Ibáñez & P. Papadopoulos. 1998. A horizon-adapted approach to the study of relativistic accretion flows onto rotating black holes. *Astrophys. J. Lett.*, **507**, L67.

Font J.A., J.Mª. Ibáñez & P. Papadopoulos. 1999. Non-axisymmetric Relativistic Bondi-Hoyle Accretion onto a Kerr Black Hole. *Mon. Not. R. Astron. Soc.*, **305**, 920.

Fryer C.L. & K.C.B. New. 2011. Gravitational Waves from Gravitational Collapse. *Living Rev. Relativity, 14, lrr-2011-1.* http://www.livingreviews.org/.

Giacomazzo B. & L. Rezzolla. 2006. The exact solution of the Riemann problem in relativistic magnetohydrodynamics. *J. Fluid Mech.* **562**, 223.

Gómez J.L., J.Mª. Martí, A.P. Marscher, J.Mª. Ibáñez & A. Alberdi. 1997. Hydrodynamical Models of Superluminal Sources. *Astrophys. J. Letters,* **482**, L33.

Gourgoulhon E. 2012. *3 + 1 Formalism in General Relativity: Bases of Numerical Relativity,* Lecture Notes in Physics 846, (Springer). arXiv:gr-qc/0703035.

Gundlach C. & J.Mª. Martín-García. 2006. Well-posedness of formulations of the Einstein equations with dynamical lapse and shift conditions. *Phys. Rev. D*, **74**, 024016.

Ibáñez J.Mª., M.A. Aloy, J.A. Font, J.Mª. Martí, J.A. Mi-ralles and J.A. Pons. 2001. Riemann Solvers in General Relativistic Hydrodynamics. *Proc. from the International Conference Godunov Methods: Theory and Applications, Oxford, United Kingdom, October 18–22, 1999. Ed. E.F. Toro (Kluwer Academic/Plenum Publishers, New York),* 485.

Ibáñez J.Mª., M.A. Aloy, P. Mimica, L. Antón, J.A. Mi-ralles & J.Mª. Martí. 2011. A Roe-type Riemann solver based on the spectral decomposition of the equations of Relativistic Magnetohydrodynamics. *Proc. from the 6th International Conference on Numerical Modeling of Space Plasma Flows (Astronum 2010, San Diego, USA), in press.*

Isenberg J.A. 1978. Waveless approximation theories of gravity. *Preprint University of Maryland*, unpublished.

Isenberg J. & J. Nester. 1980. in *General Relativity and Gravitation*, edited by A. Held (Plenum, New York), Vol. 1.

Jaramillo J.L., E. Gourgoulhon, I. Cordero-Carrión & J.Mª. Ibáñez. 2008. Trapping horizons as inner boundary conditions for black hole spacetimes. *Phys. Rev. D*, **77**, 047501.

Kotake, K., Yamada, S., Sato, K., Sumiyoshi, K., Ono, H. & Suzuki, H. 2004. Gravitational radiation from rotational core collapse: Effects of magnetic fields and realistic equations of state. *Phys. Rev. D*, **69**, 124004.

Kiuchi K., M. Shibata, P.J. Montero & J.A. Font. 2011. Gravitational Waves from the Papaloizou-Pringle Instability in Black-Hole-Torus Systems. *Phys. Rev. Lett.*, **106**, 251102.

Komissarov S.S. 1999. A Godunov-Type Scheme for Relativistic Magnetohydrodynamics. *Mon. Not. R. Astron. Soc.*, **303**, 343.

Leismann T., L. Antón, M.A. Aloy, E. Müller, J.Mª. Martí, J.A. Miralles & J.Mª. Ibáñez. 2005. Relativistic MHD simulations of extragalactic jets. *Astronomy and Astrophysics*, **436**, 503.

LeVeque R.J. 1992. Numerical Methods for Conservation Laws. *Birkhaüser, Basel, Switzerland.*

MacFadyen A.I. and S.E. Woosley. 1999. Collapsars: Gamma-Ray Bursts and Explosions in “Failed Supernovae”. *Astrophys. J.*, **524**, 262.

Martí J.Mª., J.Mª. Ibáñez & J.A. Miralles. 1991. Numerical Relativistic Hydrdynamics: A local characteristic approach. *Phys. Rev.* **D43**, 3794.

Martí J.Mª. & E. Müller. 1994. The Analytical Solution of the Riemann Problem in Relativistic Hydrodynamics. *J. Fluid Mech.* **258**, 317.

Martí J.Mª. & E. Müller. 2003. Numerical Hydrodynamics in Special Relativity. *Living Rev. Relativity, 6, lrr-2003–7.* http://www.livingreviews.org/.

Martí J.Mª., E. Müller, J.A. Font, J.Mª. Ibáñez & A. Marquina. 1997. Morphology and Dynamics of Relativistic Jets. *Astrophys. J.*, **479**, 151.

Max Planck Institute for Gravitational Physics. 2005. Whisky: the EU Network GR Hydrodynamics code http://www.whiskycode.org.

Max Planck Institute for Astrophysics. CoCoNuT code http://www.mpa-garching.mpg.de/hydro/, 2005.

May M.M. & R.H. White. 1966. Hydrodynamic calculations of general-relativistic collapse. *Phys. Rev.*, **141**, 1232.

May M.M. & R.H. White. 1967. Stellar dynamics and gravitational collapse. *Methods Comput. Phys.*, **7**, 219.

Menikoff R. & B. Plohr. 1989. The Riemann problem for fluid flow of real materials. *Rev. Mod. Phys.*, **61**, 75.

Mimica P., M.A. Aloy, I. Agudo, J.Mª. Martí, J.L. Gómez & J.A. Miralles. 2009. Spectral Evolution of Superluminal Components in Parsec-Scale Jets. *Astrophys. J.*, **696**, 1142.

Mimica P., D. Giannios & M.A. Aloy. 2009. Deceleration of arbitrarily magnetized GRB ejecta: the complete evolution. *Astronomy and Astrophysics*, **494**, 879.

Misner C.W., K.S. Thorne & J.A. Wheeler. 1973. *Gravitation,* W.H. Freeman, San Francisco, U.S.A.

Montero P.J., J.A. Font & M. Shibata. 2010. Influence of Self-Gravity on the Runaway Instability of Black-Hole-Torus Systems. *Phys. Rev. Lett.*, **104**, 191101.

Müller E. 1982. Gravitational radiation from collapsing rotating stellar cores. *Astron. and Astrophys.*, **114**, 53.

Norman M.L. & K.-H.A. Winkler. 1986. Why Ultrarelativis-tic Numerical Hydrodynamics is Difficult. in Astrophysical Radiation Hydrodynamics, Winkler, K.-H.A., and Norman, M.L., eds., Proceedings of the NATO Advanced Research Workshop, Garching, Germany, 1982, NATO ASI Series C, **188**, 449 (Reidel, Dordrecht, Netherlands; Boston, U.S.A.

Novikov I.D. and K.S. Thorne. 1973. Astrophysics of black holes. *Black holes (Les astres occlus)*, pp 343–450.

Obergaulinger M., M.A. Aloy & E. Müller. 2006. Axisym-metric simulations of magneto-rotational core collapse: dynamics and gravitational wave signal *Astron. and Astrophys.*, **450**, 1107, 2006.

Obergaulinger M., M.A. Aloy & E. Müller. 2006. Axisymmetric simulations of magnetorotational core collapse: Approximate inclusion of general relativistic effects *Astron. and Astrophys.*, **457**, 209.

Obergaulinger M., M.A. Aloy & E. Müller. 2010. Local simulations of the magnetized Kelvin-Helmholtz instability in neutron-star mergers. *Astronomy and Astrophysics,* **515**, 300.

Perucho M., M. Hanasz, J.M$^{\underline{\Delta}}$ Martí & H. Sol. 2004. Stability of hydrodynamical relativistic planar jets. I. Linear evolution and saturation of Kelvin-Helmholtz modes. *Astron. Astrophys.*, **427**, 415.

Perucho M., J.Mª. Martí & M. Hanasz. 2004 Stability of hydrodynamical relativistic planar jets. II. Long-term nonlinear evolution. *Astron. Astrophys.*, **427**, 431.
Perucho M., J.Mª. Martí & M. Hanasz. 2005. Nonlinear stability of relativistic sheared planar jets. *Astron. Astrophys.*, **443**, 863.
Perucho M., M. Hanasz, J.Mª. Martí & J.A. Miralles. 2007. Resonant Kelvin-Helmholtz modes in sheared relativistic flows. *Phys. Rev. E*, **75**, 056312.
Piran T. 2005. The physics of gamma-ray bursts. *Rev. Mod. Phys.*, **76**, 1143, 2005.
Pons J.A., J.A. Font, J.Mª. Ibáñez, J.Mª. Martí & J.A. Miralles. 1998. General Relativistic Hydrodynamics with Special Relativistic Riemann Solvers. *Astronomy and Astrophysics*, **339**, 638.
Pons J.A., J.M. Martí & E. Müller. 2000. The exact solution of the Riemann problem with non-zero tangential velocities in relativistic hydrodynamics. *J. Fluid Mech.* **422**, 125.
Postnov K.A. & L.R. Yungelson. 2006. The Evolution of Compact Binary Star Systems. *Liv. Rev. Relativity, lrr-2006-6* http://www.livingreviews.org/. Pretorius F. 2007. Binary Black Hole Coalescence. http://arXiv.org/abs/0710.1338.
Pringle J.E. & M.J. Rees. 1972. Accretion Disc Models for Compact X-Ray Sources. *Astron. and Astrophys.*, **21**, 1.
Reula O.A. 1998. Hyperbolic Methods for Ein-stein's Equations. *Liv. Rev. Relativity, lrr-1998-3* http://www.livingreviews.org/.
Rezzolla L., B. Giacomazzo, L. Baiotti, J. Granot, Ch. Kou-veliotou & M.A. Aloy. 2001. The missing link: Merging neutron stars naturally produce jet-like structures and can power short Gamma-Ray Bursts. *Astrophys. J. Lett.* **732**, L6.
Rezzolla L. & O. Zanotti. 2001. An improved exact Riemann solver for relativistic hydrodynamics. *J. Fluid Mech.*, **449**, 395.
Rezzolla L., O. Zanotti & J.A. Pons. 2003. An improved exact Riemann solver for multidimensional relativistic flows. *J. Fluid Mech.*, **479**, 199.
Roe P.L. 1981. Approximate Riemann solvers, parameter vectors and difference schemes. *J. Comput. Phys.*, **43**, 357.
Romero J.V., J.Mª. Ibáñez, J.Mª. Martí, & J.A. Miralles. 1996. A new spherically symmetric general relativistic hydrodynamical code. *Astrophys. J.*, **462**, 839.
Romero R., J.Mª. Martí, J.A. Pons, J. Mª. Ibáñez & J.A. Miralles. 2005. The exact solution of the Riemann problem in relativistic magnetohydrodynamics with tangential magnetic fields. *J. Fluid Mech.*, **544**, 323.
Sathyaprakash B.S. & B.F. Schutz. 2009. Physics, Astrophysics and Cosmology with Gravitational Waves. *Liv. Rev. Relativity, 12, lrr-2009-2* http://www.livingreviews.org/.
Scheck L., L., M.A. Aloy, J.Mª. Martí, J.L. Gómez & E. Müller. 2002. Does the plasma composition affect the long-term evolution of relativistic jets? *Month. Not. Royal Astron. Soc.*, **331**, 615.
Shakura N.I.. 1972. Disk Model of Gas Accretion on a Relativistic Star in a Close Binary System. *Astronomich-eskii Zhurnal*, **49**, 921.
Shapiro S. & S. Teukolsky. 1985. Test-bed calculations in Numerical Relativity. in J.M. Centrella, ed.: Dynamical Spacetimes and Numerical Relativity, Proceedings of the Workshop held at Drexel University on October 7–11, 1985, pp. 326, Cambridge University Press (Cambridge, U.K.; New York, U.S.A.).
Shibata M., T.W. Baumgarte & S.L. Shapiro. 2000. Stability and collapse of rapidly rotating, supramassive neutron stars: 3D simulations in general relativity. *Phys. Rev. D*, **61**, 1.
Shibata M., Duez, M.D., Liu, Y.T., Shapiro, S.L., & Stephens, B.C. 2006. Magnetized Hypermassive Neutron Star Collapse: A Central Engine for Short Gamma-Ray Bursts *Phys. Rev. Lett.*, **96**, 031102.
Shibata M. & T. Nakamura. 1995. Evolution of three-dimensional gravitational waves: Harmonic slicing case. *Phys. Rev. D*, **52**, 5428, 1995.
Shibata M. & K. Taniguchi. 2011. Coalescence of Black Hole-Neutron Star Binaries. *Liv. Rev. Relativity, lrr-2011-6* http://www.livingreviews.org/.
Sunyaev R.A. 1972. Variability of X Rays from Black Holes with Accretion Disks. *Astronomicheskii Zhurnal*, **49**, 1153.
Thöne C.C. et al.. 2011. The unusual γ-ray burst GRB 101225 A from a helium star/neutron star merger at red-shift 0.33 *Nature*, **480**, 72.
Toro E.F. 2009. Riemann solvers and numerical methods for fluid dynamics: a practical introduction. *3rd. edition, Springer, Berlin, Germany*.
Wilson J.R. 1972. Numerical study of fluid flow in a Kerr space. *Astrophys. J.* **173**, 431.
Wilson J.R. & G.J. Mathews, 1989, in *Frontiers in numerical relativity*, edited by C.R. Evans, L.S. Finn and D.W. Hobill (Cambridge University Press, Cambridge, England).
Yamada, S., & H. Sawai. 2004. Numerical Study on the Rotational Collapse of Strongly Magnetized Cores of Massive Stars *Astrophys. J.*, **608**, 907.

Numerical Methods for Hyperbolic Equations – Vázquez-Cendón et al. (eds)
© 2013 Taylor & Francis Group, London, ISBN 978-0-415-62150-2

Initiation of reactive blast waves by external energy sources

A. Liñán
E.T.S.I. Aeronáuticos, Universidad Poliécnica de Madrid, Spain

V. Kurdyumov
Departamento de Energía, CIEMAT, Spain

A.L. Sánchez
Grupo de Mecánica de Fluidos, Universidad Carlos III de Madrid, Spain

ABSTRACT: This paper is devoted to the analysis of the direct initiation, by concentrated external energy sources, of self-sustained cylindrical or spherical detonation waves in gaseous reactive mixtures. The dynamics of the detonation front will be described in the fast reaction limit, when the thickness of the reaction layer that follows the shock front is very small compared with the shock radius. At early times, after the initiation of the external thermal energy deposition, the detonation front, which generates a strongly expanding flow, is overdriven; so that it is reached by expansion waves that decrease its velocity towards the Chapman-Jouguet value, for which the expansion waves can no longer reach the front. In cylindrical and spherical detonations, the transition to the constant Chapman-Jouguet velocity occurs at a finite value of the detonation radius, which is scaled by the radius for which the energy released by the external source equals the heat released by the chemical reaction. The variation with deposition time of the critical transition radius is computed. A detailed description of the near-front flow structure near the transition time is given here for the first time, along with the analysis of the evolution towards the Zel'dovich-Taylor cylindrical or spherical self-similar flow structure, which corresponds to a Chapman-Jouguet detonation front ideally initiated at the center without any external energy source. A brief discussion will be given on how the reaction may be quenched by the expansion waves if the initiating energy is smaller than a critical value, thus failing to generate a self-propagating detonation wave.

1 INTRODUCTION

We shall show below how an external source of thermal energy in a reactive gaseous medium, when spatially concentrated and with a small deposition time, initiates a detonation front that propagates with supersonic speeds in the medium. This is so because of the high temperature sensitivity of the chemical reactions, which are frozen at the initial low temperatures but become very fast when the reactive mixture is heated by the shock wave generated by the thermal energy source at early times. Thus the chemical reactions occur rapidly behind the shock front and are therefore completed in a thin layer, which, together with the much thinner shock, forms the detonation front. The heat release due to the reactions contribute to sustain the detonation, which will become self-sustained if the reaction layer remains sufficiently thin during the first overdriven stage.

The internal structure of the detonation front, whose thickness is much smaller than the shock radius, is quasi-planar and quasi-steady in the first approximation. Zel'dovich (1940), Neumann (1942), and Döring (1943) described independently in the early1940's the planar and steady internal detonation structure as a non-reactive shock wave followed by a non-viscous diffusionless reactive layer. This ZND analysis, which is used in the following, is based on the small value of the fraction of molecular collisions that lead to a chemical reaction.

The detonation front separates the chemically-frozen outer gas mixture, which is typically stagnant and spatially uniform, from the reacted gas, whose motion, initially started by the external heat release, is sustained at later stages by the release of heat due to the reaction. The velocities associated with the motion of the reactive gases are of the order of the local sound velocity behind the detonation front; so that the corresponding Reynolds numbers are very large and the flow behind the detonationis inviscid, outside a hot core of nearly uniform pressure created, as we shall see, by the external concentrated heat addition.

The infinitely fast reaction limit, in which the reactive front is treated as a discontinuity, is applicable to the description of reactive blast waves when the thickness of the reaction layer is small compared with the detonation radius $r'_d(t')$, an increasing function of the time t'. The flow is described by the Euler equations for the density ρ', pressure p' and radial velocity v'. The outer boundary conditions are given by the values of the flow variables, $\rho' = \rho'_d$, $p' = p'_d$ and $v' = v'_d$, immediately behind the detonation front, at $r' = r'_d$. These values are determined in terms of the ambient properties $\rho' = \rho'_o$, $p' = p'_o$ and $v' = 0$ by the Rankine-Hugoniot jump conditions of conservation of mass, momentum and energy $p'_o D' = \rho' (D' - v'_d)$, $p'_o + p'_o D'^2 = p'_d + \rho'_d (D' - v'_d)^2$ and $h'_o + q_R + D'^2/2 = \gamma/(\gamma - 1)(p'_d/p'_d) + (D' - v'_d)/2$. In writing the energy equationwe have assumed that the reacted gas behaves as a perfect gas with a constant value γ of the ratio of the specific heats. In the formulation, $D' = \mathrm{d}r'_d/\mathrm{d}t'$ is the detonation velocity, h'_o is the ambient thermal enthalpy, and q_R is the amount of heat released by the reaction per unit mass of gas mixture. ations. Typical detonations propagate at large Mach numbers, that is, with propagation velocities much larger than the ambient sound velocity. Then, the ambient pressure p'_o is found to be smaller than the momentum flux $\rho'_o D'^2$ by a factor of order of the square of the propagation Mach number, and can be consequently neglected in the momentum jump condition. Note that, in this strong-shock approximation, one could in principle neglect also the initial thermal enthalpy h'_o in the energy balance equation, but the errors in evaluating the temperature associated with this additional simplification are found to be more important, when computing the reaction rates, than those involved in neglecting the effect of the ambient pressure; so that for increased accuracy, when evaluating the temperature behind the shock, we choose to retain this relatively small term in the analysis.

Detonations can be initiated directly by releasing an amount of energy E at a concentrated location (e.g., explosive charge) or along a line (e.g., electric spark, laser beam), which generates a strong overdriven detonation bounding a region of rapidly expanding radial motion of the burnt gases. The expansion waves propagating from the central nearly empty region reach the detonation front and continuously weaken it. If the detonation propagation velocity decreases to reach the Chapman-Jouguet (CJ) value, defined by the condition that the gas velocity relative to the front, $u' = D' - v'$, behind the detonation is equal to the local value $c'_d = (\gamma p'_d/\rho'_d)^{1/2}$ of the sound velocity, then the expansion waves can no longer reach the front. Afterwards, the front propagation will be self-sustained.

The sonic condition $D_d - v'_d = c'd$ together with the Rankine-Hugoniot jump equations, written with the strong-shock approximation, leads to

$$\frac{c'_{\mathrm{CJ}}}{D'_{\mathrm{CJ}}} = \frac{D'_{\mathrm{CJ}} - v'_{\mathrm{CJ}}}{D'_{\mathrm{CJ}}} = \frac{\rho'_o}{\rho'_{\mathrm{CJ}}} = \frac{\gamma}{\gamma+1} \tag{1}$$

and

$$\frac{p'_{\mathrm{CJ}}}{\rho'_o (D'_{\mathrm{CJ}})^2} = \frac{1}{\gamma+1} \tag{2}$$

for the values c'_{CJ}, v'_{CJ}, ρ'_{CJ}, and ρ'_{CJ} behind the CJ detonation, along with the corresponding propagation velocity

$$(D'_{\mathrm{CJ}})^2/2 = (\gamma^2 - 1)(q_R + h'_o). \tag{3}$$

Zel'dovich (1942) and Taylor (1950) independently described the self-similar flow structure, with central symmetry, behind a constant-velocity CJ detonation front originating, ideally, at the center without an external initiating energy. In the present paper, after a brief description of the Zel'dovich-Taylor (ZT) structure, we shall show how it is reached for large times after the detonation is initiated by an external thermal-energy source; which is an essential mechanism for initiation of detonations in unconfined gaseous reactive mixtures.

The characteristic scales involved in the direct initiation process of CJ detonations can be based on the propagation velocity D'_{CJ}, together with the total initiating energy E, the deposition time t'_{d}, and the ambient density ρ'_o. In particular, for short deposition times the condition $E = \rho'_o (D'_{\mathrm{CJ}})^2 (r'_{\mathrm{E}})^{j+1}$ that the initiating energy E (or energy per unit length in line sources) be comparable to the energy released by chemical reaction of the gas inside the front defines a characteristic length

$$r'_{\mathrm{E}} = \left(\frac{E}{\rho'_o (D'_{\mathrm{CJ}})^2} \right)^{1/(j+1)} \tag{4}$$

and a corresponding characteristic time

$$t'_{\mathrm{E}} = r'_{\mathrm{E}}/D'_{\mathrm{CJ}}. \tag{5}$$

Here, the heat released per unit mass q_R is taken to be of order $(D'_{\mathrm{CJ}})^2$ in the above estimates, in agreement with (3). For long deposition times the transition to the CJ detonation is based on alternative length and time scales. For example, for cases in which the external heating rate q' is constant, the only case considered in this paper, the appropritate scales are given by

$$r'_q = \left(\frac{q'}{\rho'_o (D'_{\text{CJ}})^3} \right)^{1/j} \tag{6}$$

and

$$t'_q = r'_q / D'_{\text{CJ}}, \tag{7}$$

which can be used to give a universal description of the flow for times t' smaller than the deposition time t'_d.

We shall analyze below the initiation of self-sustained blast waves by localized energy sources of characteristic size $r'_s \ll r'_{\text{E}}$ with deposition times $t'_d \lesssim t'_{\text{E}}$. Concentrated sources were also considered in the early numerical work of Bishimov (1968), who addressed in particular the problem of instantaneous heat release corresponding to the limit $t'_d/t'_{\text{E}} \to 0$ (see Korobeinikov (1971) for additional entries into the early literature). The initial conditions for the problem were written for $t' \ll t'_{\text{E}}$, when the effect of the chemical heat release is negligible and the solution is that described, independently, by Sedov (1946) and Taylor (1950) for a strong point explosion. The expansion waves were found to progressively weaken the overdriven detonation causing its velocity D' to decay continuously. As anticipated by (Levin & Chernyi, 1967), for spherical and cylindrical detonations the CJ velocity is reached at a finite critical time t'_c when $r'_d = r'_c \sim r'_{\text{E}}$. As pointed out by Korobeinikov (1969), the computation *becomes very complex due to the presence of peculiar conditions on the shock wave front as it nears the C.J. point*. These complexities have prevented in the past the numerical computation of the flow for $t' \geq t'_c$. It is seen below that understanding the changes in the pattern of the characteristics occurring at $t' = t'_c$ enables the integration to be extended to larger times. Our calculations consider energy sources of finite deposition time, $t'_d \lesssim t'_c$, to determine the variation of $r_c = r'_c/r'_{\text{E}}$ and $t_c = t'_c/t'_{\text{E}}$ with $t_d = t'_d/t_c$, including the limit $t_d \to 0$ of instantaneous heat release analyzed previously in the literature.

The paper is organized as follows. The formulation of the problem is given in Section 2, including the nondimensional Euler equations and the outer boundary conditions at the detonation front. This is followed by a description of the self-similar flow structure of Zel'dovich and Taylor. Next we present the flow structure generated by external energy sources, of central symmetry, for source sizes r'_s small compared with r'_{E} and finite deposition times $t'_d \sim t'_{\text{E}}$. The flow field is found to include a hot core of radius r'_e, with very high temperatures, very small densities, and spatially near-uniform pressure, and an outer region filled with the reacted gases between two free boundaries, thecore surface and the detonation front. The flow in this outer region is governed by the inviscid, Euler equations, to be integrated subject to the Rankine-Hugoniot jump conditions behind the detonation front and the piston-type conditions at the interface with the hot core. The initial conditions are given in Section 2.3. Numerical solutions are presented next in Section 3, and the near-front solution close to the transition time, i.e., for $|t' - t'_c|/t'_{\text{E}} \ll 1$ and $(r'_d - r')/r'_{\text{E}} \ll$, will be described analytically in Section 4. Finally concluding remarks will be given in Section 5, including comments on the relevance of the present work for the computation of the minimum initiation energy.

2 FORMULATION

As previously mentioned, the flow of the reacted gases behind the detonation is effectively inviscid. In writing the Euler equations for the inviscid flow behind the detonation

$$\frac{\partial \rho}{\partial t} + \frac{1}{r^j} \frac{\partial}{\partial r}(r^j \rho v) = 0 \tag{8}$$

$$\frac{\partial v}{\partial t} + v \frac{\partial v}{\partial r} + \frac{1}{\rho} \frac{\partial p}{\partial r} = 0 \tag{9}$$

$$\frac{\partial}{\partial t}(p/\rho^\gamma) + v \frac{\partial}{\partial r}(p/\rho^\gamma) = 0, \tag{10}$$

the time and radial distance are scaled with t'_E and r'_E, respectively, giving the dimensionless coordinates $r = r'/r'_{\text{E}}$ and $t = t'/t'_{\text{E}}$ togetherwith the detonation propagation velocity $D = D'/D'_{\text{CJ}}$, whereas the density, pressure and velocity are scaled with p'_o, D'_{CJ} and $\rho'_o (D'_{\text{CJ}})^2$, to give the variables ρ, p and v. Cylindrical and spherical geometries correspond to $j = 1$ and $j = 2$, respectively. Note that, for computational purposes, it is often convenient to write the Euler equations (8)–(10) in their characteristic form

$$\frac{\text{d}p}{\text{d}t} \pm \rho c \frac{\text{d}v}{\text{d}t} + \frac{j \rho c^2 v}{r} = 0 \text{ on } C^{\pm} : \frac{\text{d}r^{\pm}}{\text{d}t} = v \pm c \tag{11}$$

$$\frac{\text{d}p}{\text{d}t} - c^2 \frac{\text{d}\rho}{\text{d}t} = 0 \text{ on } C^0 : \frac{\text{d}r^0}{\text{d}t} = v \tag{12}$$

where $c = (\gamma p/\rho)^{1/2}$ is the sound velocity. The first two characteristics, C^+ and C^-, correspond to the surfaces of acoustic wave propagation, whereas C^0 are the particle paths.

The Rankine-Hugoniot jump conditions determine the boundary conditions

$$r = r_d(t): \; v - v_d = \rho - \rho_d = p - p_d = 0, \tag{13}$$

where the boundary values can be expressed in the form

$$\frac{D - v_d}{D} = \frac{1}{\rho_d} = \frac{\gamma}{\gamma+1} - \frac{1}{\gamma+1}(1 - D^{-2})^{1/2} \tag{14}$$

and

$$p_d = \frac{D^2}{\gamma+1}[1 + (1 - D^{-2})^{1/2}] \tag{15}$$

in terms of the propagation velocity $D = \mathrm{d}r_d/\mathrm{d}t$. For $D = 1$, these expressions reduce to the CJ values

$$\rho_{CJ} = \frac{\gamma+1}{\gamma} \text{ and } p_{CJ} = v_{CJ} = \frac{1}{\gamma+1}, \tag{16}$$

with the corresponding sound velocity being $c_{CJ} = \gamma/(\gamma+1)$.

2.1 *Zel'dovich-Taylor solution*

As previously mentioned, the initially overdriven detonations generated by heat deposition decay eventually to the CJ solution. This was recognized early by Zel'dovich and Taylor, who addressed the fundamental problem of propagation of a CJ detonation instantaneously started at the origin at $t' = 0$, $r' = 0$. In the absence of geometrical scales, the flow is self-similar in terms of the variable $\xi = r'/(D'_{CJ}t') = r/t$. The solution includes an outer shell of outward moving gas anda core of stagnant gas with uniform properties, separated by a weak discontinuity at $\xi = \xi_o^+ = r_o^+/t$, where $v = 0$, thereby providing the boundary condition needed at the inner boundary to close the problem formulation. Along the particle paths the entropy conserves the value reached behind the CJ detonation front, which is constant; so that the pressure p and the density ρ can be evaluated in terms of the sound velocity c by the isentropic relationships

$$\left(\frac{p}{p_{CJ}}\right)^{(\gamma-1)/\gamma} = \left(\frac{\rho}{\rho_{CJ}}\right)^{\gamma-1} = \left(\frac{c}{c_{CJ}}\right)^2, \tag{17}$$

to be used in integrating the continuity and momentum equations

$$\frac{(v-\xi)}{\rho}\frac{\mathrm{d}\rho}{\mathrm{d}\xi} + \frac{\mathrm{d}v}{\mathrm{d}\xi} + \frac{jv}{\xi} = 0 \tag{18}$$

$$(v-\xi)\frac{\mathrm{d}v}{\mathrm{d}\xi} + \frac{1}{\rho}\frac{\mathrm{d}p}{\mathrm{d}\xi} = 0 \tag{19}$$

with the conditions $v = v_{CJ}$, $\rho = \rho_{CJ}$ and $p = p_{CJ}$ at $\xi = 1$. The solution for cylindrical ($j = 1$) and spherical ($j = 2$) detonations depends only on the specific heat ratio γ. Sample solutions are shown below in Figure 1.

Since the flow behind the detonation is homentropic, the characteristic equations (11) reduce in this case to

$$\frac{\mathrm{d}I^+}{\mathrm{d}t} = -\frac{jcv}{r} \text{ on } C^+ : \frac{\mathrm{d}r^+}{\mathrm{d}t} = v + c \tag{20}$$

and

$$\frac{\mathrm{d}I^-}{\mathrm{d}t} = +\frac{jcv}{r} \text{ on } C^- : \frac{\mathrm{d}r^-}{\mathrm{d}t} = v - c, \tag{21}$$

where $I^\pm = v \pm 2c/(\gamma-1)$. For later reference, Figure 2 shows a plot of the three characteristics of the problem C^+, C^-, and C^0. As can be seen, in the ZT flow all three characteristics depart from the detonation front. No information can therefore reach the front from the burnt side. The boundary radius r_o^+ separating the outward moving gas from the stagnant gas corresponds to the first C^+ characteristic that leaves the front at $t = 0$; note that all other characteristics C^+ are tangent to the detonation path $r_d(t)$, as expected from the sonic condition found behind the detonation, but are soon left behind the front.

2.2 *Flow induced by a heat source*

The self-similar ZT solution does not consider an initiation stage, in that it assumes that the CJ detonation departs from the origin at the initial instant. The direct initiation problem addressed here considers the localized deposition of a finite amount of energy E at the origin. For sources with deposition time t'_d much smaller than t'_E, the case considered in previous studies, the release is effectively instantaneous, with the initial solution for $t'_d \ll t \ll t'_E$ corresponding to the nonreactive Sedov-Taylor blast wave. For the case $t'_d \sim t'_E$ considered here the flow associated with the finite-rate heat source must be analyzed when describing the initiation of the self-sustained detonation. As shown in

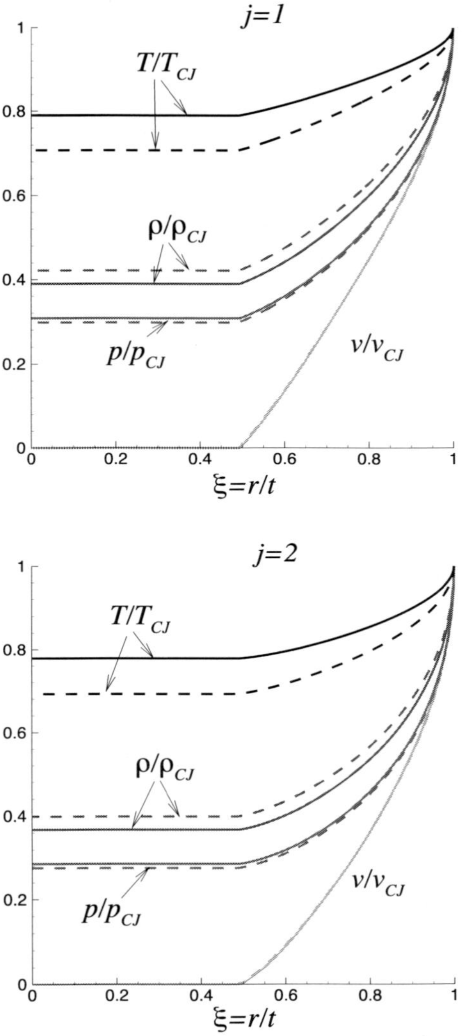

Figure 1. ZT solution for $\gamma = 1.4$ (dashed curves) and $\gamma = 1.25$ (solid curves), the latter being a value more typical of detonation products.

(Kurdyumov, Sánchez & Liñán, 2003) when analyzing the flow induced by concentrated external heat sources in non-reactive gases, the flow includes a neatly defined core, of radius $r'_e(t) \sim r'_d(t)$, of very hot expanding gas dominated by heat conduction from the source. Because of the resulting high temperatures, the local Mach number is very small, although it takes valuesof order unity outside. Since the velocities are of the same order inside and outside the hot core and the Mach number is of order unity outside, the pressure in this hot kernel is uniform in the first approximation, with a

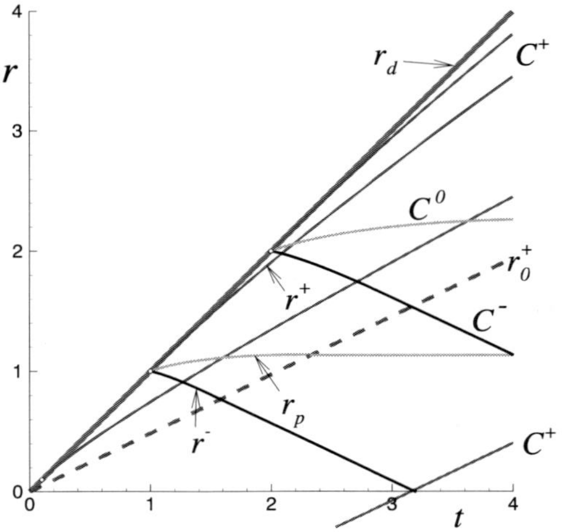

Figure 2. The characteristics for the ZT solution with $\gamma = 1.25$ and $j = 2$.

value $p'_e(t)$ comparable to p'_d, whereas the density is much lower. The edge of the resulting almost-empty region appears as a contact surface that acts as a piston for the outer flow, assisting the motion of the outward propagating detonation. Integrating with respect to r the low-Mach-number energy equation in the hot core $0 < r' \le r'_e$, dominated by heat conduction from the external energy source at the origin with heating rate $q'(t')$, yields an equation linking the temporal evolution of the contact-surface radius $r'_e(t')$ and the hot core pressure p'_e. For instance, for the spherical case, the equation can be written in dimensional form as

$$\frac{\mathrm{d}}{\mathrm{d}t'}\left(\frac{4}{3}\pi(r'_e)^3\frac{p'_e}{\gamma-1}\right)+4\pi(r'_e)^2 p'_e\frac{\mathrm{d}r'_e}{\mathrm{d}t'}=q'(t'), \quad (22)$$

stating that the heat added at the origin is partly employed to increase the internal energy of the hot conductive pocket, and partly transfered to the outer fluid when displacing it with the velocity $\mathrm{d}r'_e/\mathrm{d}t'$. This last condition, togetherwith the condition that the edge of the internal expanding hot core be a contact surface, can be expressed in dimensionless form to give the additional boundary conditions

$$\frac{2^j\pi}{j+1}\frac{\mathrm{d}}{\mathrm{d}t}\left(\frac{r_e^{j+1}p_e}{\gamma-1}\right)+2^j\pi r_e^j p_e\frac{\mathrm{d}r_e}{\mathrm{d}t}=q(t). \quad (23)$$

and

$$v=\frac{\mathrm{d}r_e}{\mathrm{d}t} \quad (24)$$

at $r = r_e(t)$ for the Euler equations (8)–(10). The heating rate is scaled according to $q = q'/(E/t'_E)$, so that the heating function $q(t)$ must satisfy $\int_0^{t_d} q\mathrm{d}t = 1$, where $t_d = t'_d/t'_E$.

As can be inferred from (23), sources with $q \propto t^j$ yield constant values of p_e and $\dot{r}_e = \mathrm{d}r_e/\mathrm{d}t$. In the inviscid region, the associated gas motion is therefore self-similar for $t < t_d$, and corresponds to that foundin the piston-supported detonations discussed by John Lee (2007); although unfortunately they cannot be used for the analysis of the initiation of the ZT detonation. As indicated before, for the analysis of the direct initiation we consider below, as an example, sources of constant heating rate, for which the function q becomes $q = 1/t_d$ for $t < t_d$ and $q = 0$ for $t > t_d$, with t_d entering as a parameter in the solution. We shall see below that, for the analysis of heat sources with deposition times t'_d larger than the critical time required to reach the CJ detonation t'_c, it is more convenient to use the scales r'_q and t'_q defined in (6) and (7) to nondimensionalize the problem. The corresponding alternative formulation involves the replacement of r and t with $\tilde{r} = r'/r'_q$ and $\tilde{t} = t'/t'_q$ in the Euler equations and boundary conditions; resulting in the value $q = 1$ for the nondimensional heating rate $q(t)$ appearing in (23).

2.3 *Initial conditions for the initiation problem*

The initial conditions are obtained by analyzing the solution at early times, $t \ll 1$, when the detonation is very strong, with $D \sim (t_d t)^{-1/(3+j)}$, so that behind the detonation the gas properties are approximately given by those of the Neumannspike values $\rho_d = \rho_N = (\gamma + 1)/(\gamma - 1)$, $p_d = p_N = 2D^2/(\gamma+1)$, and $v_d = v_N = 2D/(\gamma+1)$. The description, given in an appendix, employes the similarity variables $\zeta = r/(t^3/t_d)^{1/(3+j)}$, $\rho = R(\zeta)$, $p = (t_d t^j)^{-2/(3+j)} P(\zeta)$, and $v = (t_d t^j)^{-1/(3+j)} V(\zeta)$. Sample profiles are shown in Figure 3 for $\gamma = 1.4$ and $j = 1$. The solution determines in particular the initial evolution of the radii of the detonation front and of the hot core, $r_d = 0.671(t^3/t_d)^{1/4}$ and $r_e = 0.587(t^3/t_d)^{1/4}$, as well as the values of the density, pressure, and velocity behind the detonation front, given by $\rho_d = (\gamma+1)/(\gamma-1) = 6$, $p_d = 0.211(t_d t)^{-1/2}$, and $v_d = 0.419(t_d t)^{-1/4}$.

The results for the initial evolution of the flow can be directly related to the self-similar flows investigated by Rogers (1958), except that in his work he equated the external energy input to the energy transfered to the heated gases by the piston, neglecting the energy stored in the hot core, i.e., the first term in (22).

The solution approaches the case of instantaneous heat deposition $t_d = 0$ for values $t_d \ll 1$, at times $t_d \ll t \ll 1$, with a remaining hot core of small size. To treat exactly the case of vanishing deposition times, one should replace the boundaryconditions at $r = r_e$ with the condition $v = 0$ at $r = 0$ and the initial profiles with those corresponding to the Sedov-Taylor solution.

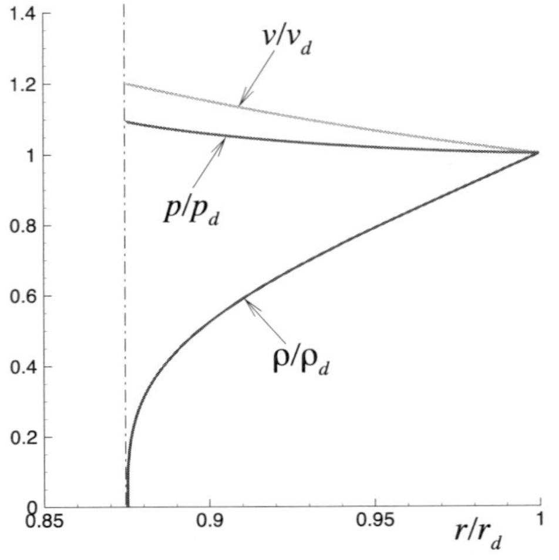

Figure 3. The radial variation of pressure, velocity, and density between the detonation and the hot core for $t \ll 1$ with $\gamma = 1$ and $j = 1$.

3 NUMERICAL RESULTS

The solution requires integration of (8)–(10) for $r_e(t) \le r \le r_d(t)$ with boundary conditions (13) at $r = r_d$ and (24) and (23) at $r = r_e$ and initial profiles $\rho = R(\zeta)$, $p = (t_d t^j)^{-2/(3+j)} P(\zeta)$, and $v = (t_d t^j)^{-1/(3+j)} V(\zeta)$ evaluated at $t \ll t_d$. The integration provides for given values of γ and t_d the evolution with time of the profiles of pressure, density and velocity for $r_e(t) \le r \le r_d(t)$ along with $r_e(t)$ and $r_d(t)$ and the corresponding front velocities $\dot{r}_e = \mathrm{d}r_e/\mathrm{d}t$ and $D = \mathrm{d}r_d/\mathrm{d}t$.

As an illustrative example, results are shown in Figures 4–6 for $t_d = 0.1$ and $\gamma = 1.4$. Figure 4 shows the variation with time of the two fronts r_e and r_d and their velocities $\dot{r}_e$ and D. As can be seen, the detonation velocity decreases continuously until it reaches the CJ velocity $D = 1$ at $t = t_c = 1.767$ when $r_d = r_c = 2.071$. The piston-like contact surface r_e moves outwards to reach a final radius for $t \gg 1$. The logarithmic scale used in the right-hand-side plot clearly shows how $\dot{r}_e$ changes abruptly when heat deposition ceases at $t = t_d = 0.1$, in agreement with (23). The sudden deceleration is felt at a later instant $t \simeq 0.2$ by the propagating detonation front, when reached by the expansion wave leaving the contact surface at $t = t_d$.

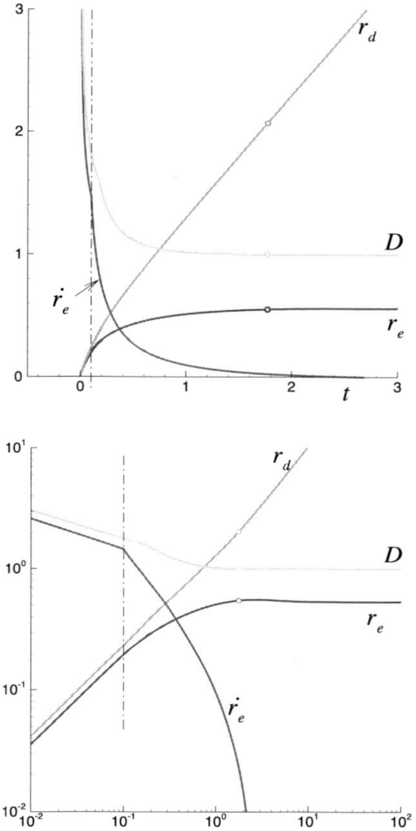

Figure 4. The evolution with time of r_d, r_e, D and $\dot{r}_e = \mathrm{d}r_e/\mathrm{d}t$ for $t_d = 0.1$, $\gamma = 1.4$ and $j = 1$.

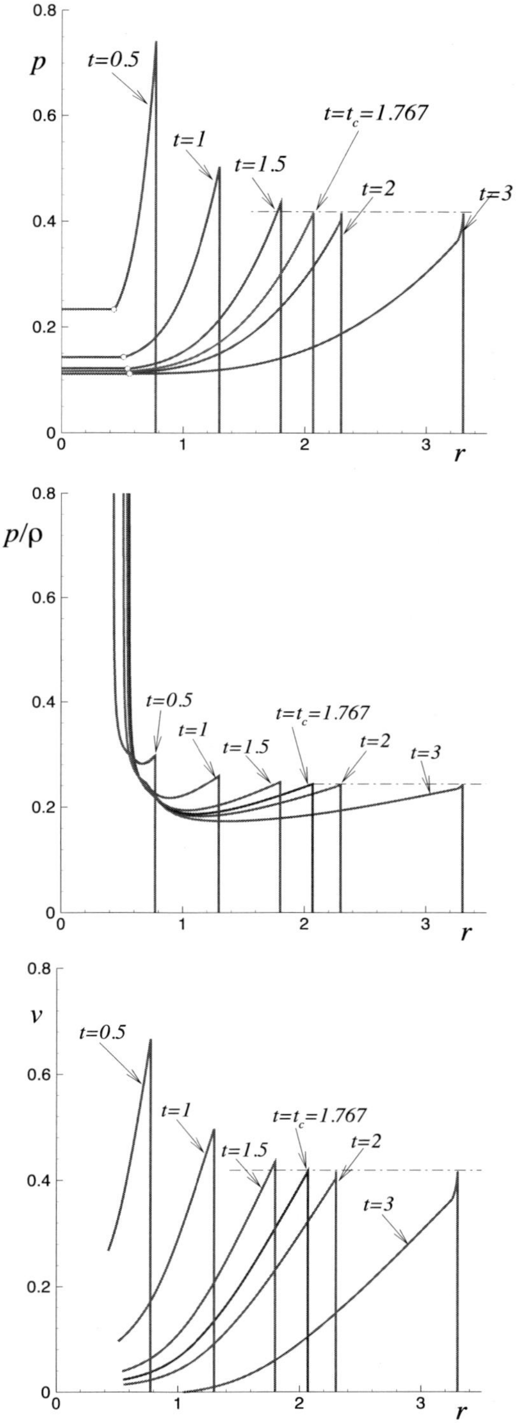

Figure 5. The evolution with time of the profiles of p, p/ρ and v for $t_d = 0.1$, $\gamma = 1.4$ and $j = 1$.

The evolution of the accompanying profiles of pressure, density and velocity are shown in Figure 5. Note in this figure the gradual variation of the gradients of the flow variables at the detonation front for $t < t_c$, while they become infinite for $t < t_c$. To investigate how the ZT solution is established for $t \gg t_c$ the profiles scaled with their CJ values are also shown in Figure 6 as a function of r/r_d. The presence of the high-temperature core associated with the finite deposition time is clearly visible in the profiles of density and temperature. Since r_e approaches a constant value, the corresponding value of r_e/r_d becomes proportional to the t^{-1} for large times. The plots also show that for $t < t_c$ there exists behind the detonation a shell of increasing width where the numerical solution is exactly that of a corresponding shell of the self-similar ZT solution initiated at the origin at the time $t_c - r_c$. This result is to be discussed further below based on the pattern of characteristics encountered once the CJ velocity is reached.

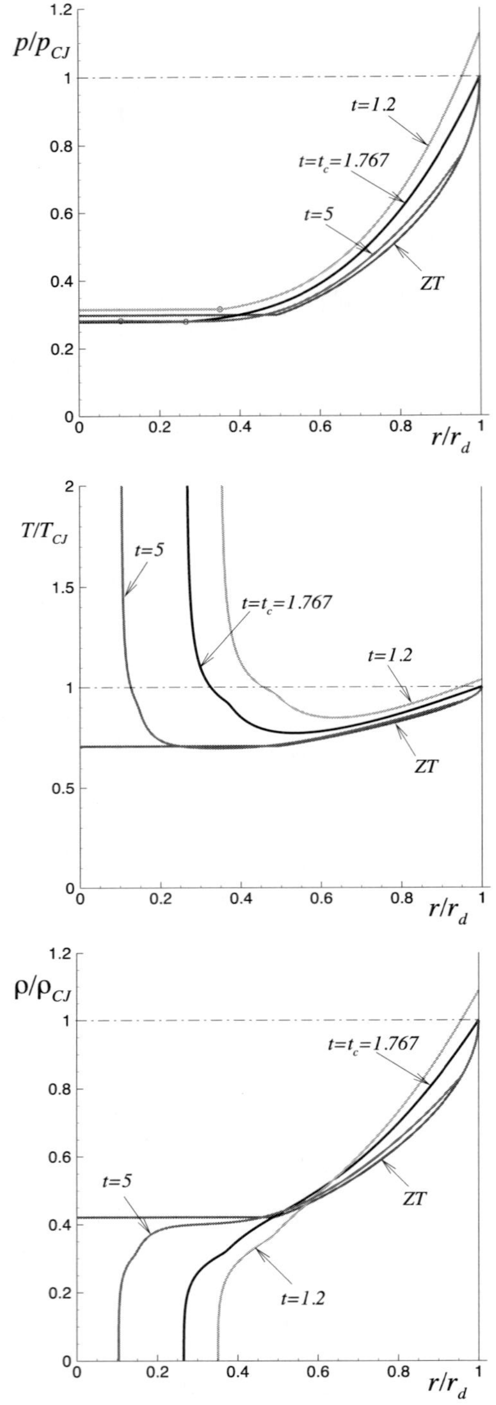

Figure 6. The evolution with time of the profiles of p, p/ρ and v for $t_d = 0.1$, $\gamma = 1.4$ and $j = 1$ compared with the ZT profiles.

As shown in Figure 7, of the three characteristics of the Euler equations for these flows with central symmetry, two, namely C^0 and C^-, leave always the detonation front toward the burnt gases, while the third, the acoustic wave C^+, moving with velocity $v + c > D$, can reach the detonation from inside when $t \le t_c$. These expansion waves are responsible for the continuous deceleration of the detonation. The relative slope $\mathrm{d}r_*^+/\mathrm{d}t - D$ with which C^+ intersects the front trajectory $r_d(t)$ decreases as the CJ point is approached, vanishing at $t = t_c$, so that the critical characteristic $C^+ = C_*^+$ reaching—and also leaving—the front at $t = t_c$ is tangent to the front

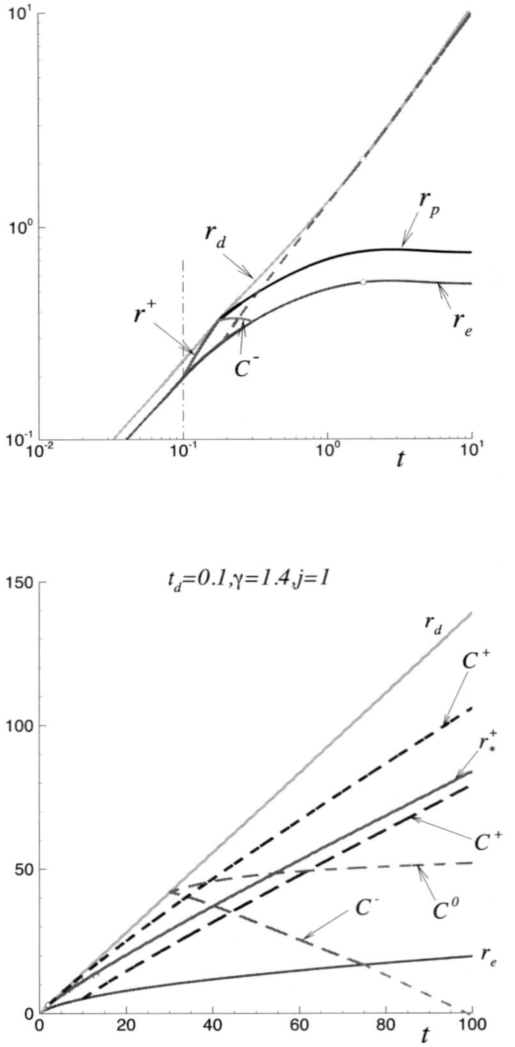

Figure 7. The pattern of characteristics emerging for $t_d = 0.1$, $\gamma = 1.4$ and $j = 1$.

trajectory $r_d(t)$, with $\mathrm{d}r_*^+/\mathrm{d}t = v + c = 1$ equal to the CJ detonation velocity $D = 1$. Since no C^+ characteristic can reach the front for $t > t_c$, the locus $r_*^+(t)$ of the critical characteristic C_*^+ defines for $t > t_c$ the inner boundary of a shell of gas $r_*^+ < r < r_d$ whose solution is independent of that found inside. As clearly shown in the plots of Figure 6, the solution in the shell corresponds, exactly, to the the ZT solution, and can be determined, for $t > t_c$, by integrating along the three characteristics leaving the detonation front $r_d = r_c + t - t_c$, where $\rho_d = \rho_{CJ} = (\gamma+1)/\gamma$, $p_d = p_{CJ} = 1/(\gamma+1)$, and $v_d = v_{CJ} = 1/(\gamma+1)$; with the family of C^+ characteristics being tangent to the detonation locus. These profiles will certainly coincide with those shown in Figure 1 for the ZT solution, with ξ replaced by $r/(r_c + t - t_c)$ as long as $r > r_*^+$, the location of a weak discontinuity.

This changing pattern of characteristics was accounted for in the numerical solution of the problem, which employed the characteristic form of the conservation equations (11) and (12) written in the transformed radial variable $(r - r_e)/(r_d - r_e)$. First-order finite differences were used in the integration, with up to 5000 cell points in the radial discretization. An implicit scheme was used, with time steps ranging from $\delta t = 10^{-6}$ for $t \ll 1$ to $\delta t = 10^{-5}$ for larger times and with the number of iterations at each time step being as large as 20 for small times. The characteristic equations C^- and C^0 were integrated from $r = r_d$. Implicit central differences were used for the characteristic equations C^+. This scheme enables the transition to CJ to be described, including the precise computation of r_c and t_c and the long-time evolution towards the ZT solution.

Critical values r_c and t_c are shown in Figure 8. As can be seen, a weak dependence on the deposition time t_d is observed below a given value $t_d = t_d^*$, whereas for larger values of $t_d > t_d^*$, both r_c and t_c undergo a sharp decrease. To understand this, note that, as can be seen in the upper plot of Figure 7, the limiting characteristic $C^+ = C_*^+$ reaching the detonation at $t = t_c$ leaves the contact surface at an earlier time, with the characteristics C^+ leaving the contact surface at later times never reaching the detonation. There exists a limiting case for which the C_*^+ characteristic leaves the contact surface at the exact instant when the heat deposition ends. The associated deposition time is precisely the transition value t_d^* in Figure 8. For $t_d > t_d^*$, the fraction of the energy released for $t_d^* < t < t_d$ is not useful in the creation of a self-propagating shell of detonated gas bounded outside by a CJ detonation front, which originates at $r = r_c$ at the time $t = t_c$; but affects the evolution of the gas inside the inner boundary $r = r_*^+(t)$ of the shell. Clearly, for an effective use of the energy released, detonation igniters should be designed with deposition times $t_d < t_d^*$.

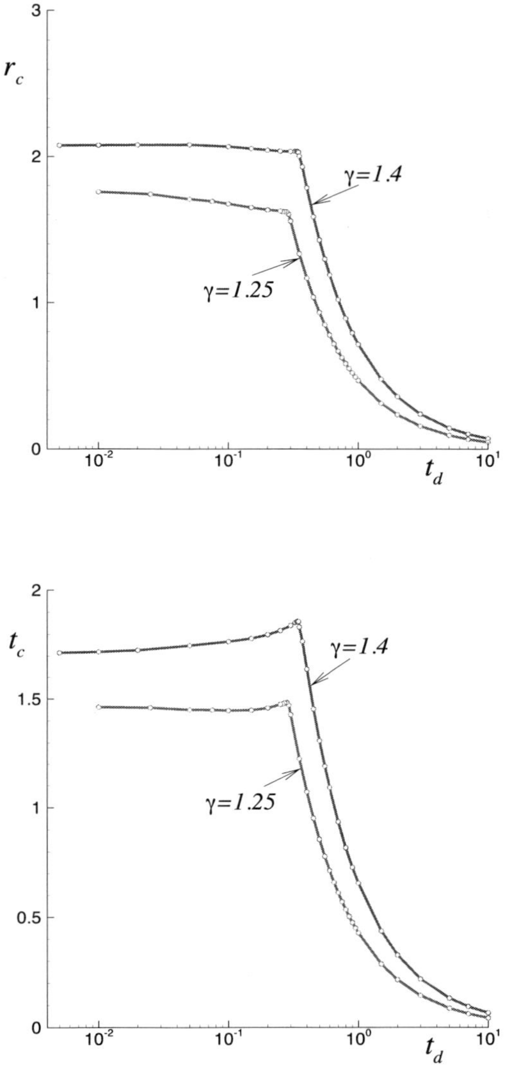

Figure 8. The variation of r_c and t_c with t_d for $\gamma = (1.25, 1.4)$ and $j = 1$ (lower plot) and $j = 2$ (upper plot).

To analyze initiation events with $t_d > t_d^*$, for which the energy source remains active until after the transition to CJ takes place, it is convenient to use the scales r'_q and t'_q introduced above in (6) and (7). The resulting problem, which depends only on j and γ, was integrated for $j = 1$ and $\gamma = 1.4$, giving the results shown in Figure 9. The computation provides in particular the critical values $\tilde{t}_c = t'_c/t'_q = 0.655$ and $\tilde{r}_c = r'_c/r'_q = 0.715$ at transition. These values can be used to determine the laws $r_c = \tilde{r}_c/t_d^{1/j}$ and $t_c = \tilde{t}_c/t_d^{1/j}$ for the variation of

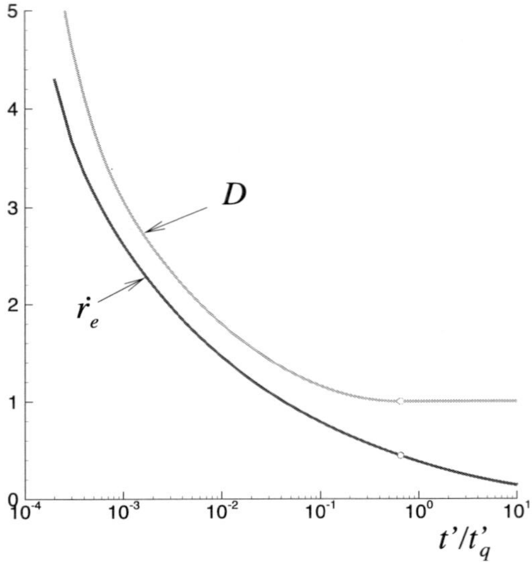

Figure 9. The variation with t'/t'_q of D and $\dot{r}_e$ for $j = 1$ and $\gamma = 1.4$.

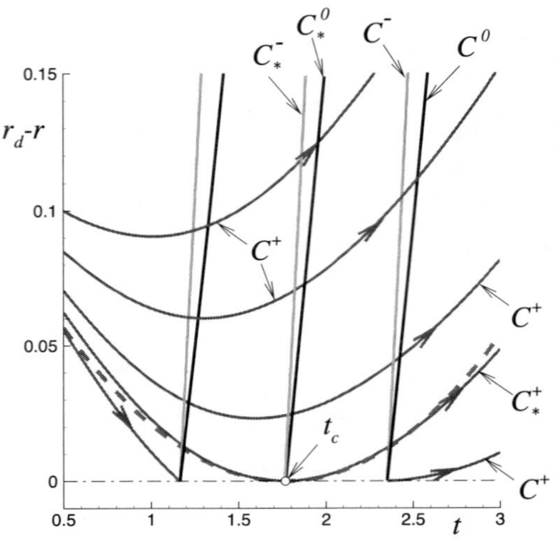

Figure 10. The pattern of characteristics emerging for $t_d = 0.1$, $\gamma = 1.4$ and $j = 1$.

r_c and t_c with t_d when $t_d > t_d^*$, as follows from the identities $r_c = \tilde{r}_c r'_q / r'_E$ and $t_c = \tilde{t}_c r'_q / r'_E$.

4 THE NEAR-FRONT SOLUTION AT TRANSITION TO CJ

Figure 10 shows a detailed plot of the near-front characteristics obtained numerically. The plot includes the critical characteristic C_*^+, which is the last C^+ characteristic that reaches the front, at $t = t_c$, and then leaves, also tangentially, to bound the ZT shell. Also shown in the plot is the characteristic $C^0 = C_*^0$ leaving the front at $t = t_c$, which defines the boundary of the region of homentropic flow; since the entropy of all fluid particles crossing the CJ detonation for $t > t_c$ is strictly constant, the equation $p/p_{CJ} = (\rho/\rho_{CJ})^\gamma$ holds everywhere in the flow field to the right of C_*^0 in Figure 10.

The flow field behind the front near the transition time t_c can be determined in terms of the local variables $\eta = (r_d - r)/r_c \ll 1$ and $\tau = (t - t_c)/r_c \ll 1$, which reduce the Euler equations (8)–(10) to

$$\frac{\partial \rho}{\partial \tau} + (D - v)\frac{\partial \rho}{\partial \eta} - \rho\frac{\partial v}{\partial \eta} + j\rho v = 0 \quad (25)$$

$$\frac{\partial v}{\partial \tau} + (D - v)\frac{\partial v}{\partial \eta} - \frac{1}{\rho}\frac{\partial p}{\partial \eta} = 0 \quad (26)$$

$$\frac{\partial}{\partial \tau}(p/\rho^\gamma) + (D - v)\frac{\partial}{\partial \eta}(p/\rho^\gamma) = 0. \quad (27)$$

For $0 < -\tau \ll 1$ the detonation velocity decays according to $D - 1 = b^2\tau^2/2$, so that $r_d/r_c = 1 + \tau + b^2\tau^3/6$. The corresponding profiles of density, pressure and velocity for $\eta \ll 1$ are linear functions of the form

$$\begin{aligned} \rho &= \rho_d(1 + A\eta) \\ p &= p_d(1 + B\eta) \\ v &= v_d(1 + C\eta), \end{aligned} \quad (28)$$

with

$$\rho_d = \frac{\gamma + 1}{\gamma}(1 - b\tau/\gamma) \quad (29)$$

and

$$p_d = v_d = \frac{1}{\gamma + 1}(1 - b\tau). \quad (30)$$

Introducing these expressions into (25)–(27) provides

$$b = \frac{\gamma j}{2(\gamma + 1)} \quad \text{and} \quad \gamma A = B = C - j/2, \quad (31)$$

with C expected to be negative. As can be seen in the first equation, the decay law of the detonation velocity as the CJ point is approached is independent of t_d when expressed in terms of τ. The second equation shows that the local profiles are functions of a single constant (for instance, the density gradient A), which must be obtained numerically by integration of the initiation problem in terms of γ and t_d.

The equation $\gamma A = B$ states that, because of the slow quadratic deceleration $D-1=b^2\tau^2/2$ towards CJ, the solution for $-\tau \ll 1$ is homentropic in the first approximation, with small errors of order τ^2. Equations (25) and (26) can be therefore written in characteristic form in terms of the Riemann variables $v-2c/(\gamma-1)$ and $v+2c/(\gamma-1)$. Integration of the C^- characteristic equation

$$\frac{d}{d\tau}\left(v-\frac{2c}{\gamma-1}\right)=\frac{j\gamma}{(\gamma+1)^2} \quad (32)$$

on

$$\frac{d\eta^-}{d\tau}=\frac{2\gamma}{\gamma+1}, \quad (33)$$

with the condition $v-2c/(\gamma-1)=-1/(\gamma-1)$ at the front $\eta=0$, yields a first relationship

$$v-\frac{2c}{\gamma-1}=-\frac{1}{\gamma-1}+\frac{j\eta}{2(\gamma+1)}, \quad (34)$$

which is uniformly valid near the transition with errors of order τ^2 for $\tau \sim \eta \ll 1$. Similarly, integration of the C^+ characteristic equation

$$\frac{d}{d\tau}\left(v+\frac{2c}{\gamma-1}\right)=-\frac{j\gamma}{(\gamma+1)^2} \quad (35)$$

on

$$\frac{d\eta^+}{d\tau}=D-v-c \quad (36)$$

with the condition

$$v+\frac{2c}{\gamma-1}=\frac{3\gamma-1}{(\gamma-1)(\gamma+1)}+\frac{\eta}{\gamma+1}(2\gamma A+j/2) \quad (37)$$

at $\tau = 0$, obtained with use made of the profiles given above in (28), yields

$$v+\frac{2c}{\gamma-1}=\frac{3\gamma-1}{(\gamma-1)(\gamma+1)}+\frac{(2\gamma A+j/2)}{\gamma+1}\eta-\frac{j\gamma}{2(\gamma+1)^2}\tau. \quad (38)$$

Along the separating characteristic $\eta_*^+(\tau)$ $D-v-c=j\gamma\tau/[4(\gamma+1)]+O(\tau^2)$, which can be used in integrating $d\eta_*^+/d\tau=D-v-c$ with $\eta_*^+=0$ at $\tau=0$ to give

$$\eta_*^+=\frac{j\gamma}{8(\gamma+1)}\tau^2, \quad (39)$$

shown as a dashed curve in Figure 10. For the C^+ characteristics leaving the detonation for $\tau>0$, integration of (35) with condition $v+2c/(\gamma-1)=(3\gamma-1)/(\gamma^2-1)+O(\tau^2)$ at $\eta=0$ yields

$$v+\frac{2c}{\gamma-1}=\frac{3\gamma-1}{(\gamma-1)(\gamma+1)}-\left(\frac{8j\gamma\eta}{(\gamma+1)^3}\right)^{1/2}, \quad (40)$$

to be used in place of (38) to describe in the first approximation the ZT shell emerging for $\tau>0$ and $0<\eta<\eta_*^+$.

In summarty, the solution found above is non-analytic, because of a weak discontinuity located at $\eta=\eta_*^+(\tau)$, given in (39). For $\tau<0$ and also for $\tau>0$ with $\eta>\eta_*^+$, the flow variables take the linear form

$$v-\frac{1}{\gamma+1}=-\frac{j\gamma}{2(\gamma+1)^2}\tau+\frac{(\gamma A+j/2)}{\gamma+1}\eta \quad (41)$$

$$c-\frac{\gamma}{\gamma+1}=-\frac{j\gamma(\gamma-1)}{4(\gamma+1)^2}\tau+\frac{\gamma(\gamma-1)A}{2(\gamma+1)}\eta \quad (42)$$

as can be obtained by combining (34) and (38). However, for $\tau>0$ with $\eta<\eta_*^+$, the solution is determined from (34) and (40) to give the ZT profiles

$$\frac{v-\frac{1}{\gamma+1}}{\tau}=\frac{2}{\gamma-1}\frac{c-\frac{\gamma}{\gamma+1}}{\tau}=-\frac{j\gamma}{2(\gamma+1)^2}\left(\frac{\eta}{\eta_*^+}\right)^{1/2}. \quad (43)$$

As can be seen, at the order computed here (i.e., the above expressions contain errors of order τ^2), the solution at $\eta=\eta_*^+(\tau)$ is continuous, with boundary values of v and c decreasing linearly in time as determined by evaluating (43) at $\eta=\eta_*^+$. The slope of the profiles of v and c is however discontinuous; for instance, the gradient $dv/d\eta$ takes the negative values $(\gamma A+j/2)/(\gamma+1)$ and $-2/[(\gamma+1)\tau]$ on the inner and outer sides of r_*^+, respectively.

5 DISCUSSION OF FINITE-RATE EFFECTS AND CONCLUDING REMARKS

The analysis presented above is based on the assumption of infinite rates of the chemical reactions, when the reaction layer together with the shock can be considered as an infinitesimally thin detonation front. The analysis should be complemented with an analysis of the structure of the thin reaction layer, which in the ZND model follows the shock front, and is determined by the finite-rate effects. These effects can be expected to be very relevant when influenced by the infinite gradients encountered inthe ZT structure immediately behind the detonation front; these gradients are smoothed by the finite-rate effects.

There are numerous numerical studies, incorporating finite-rate effects, of the transition to the self-propagating mode of detonation. They show that the transition involves oscillations of the shock front and its accompanying reaction layer, and that only for initiation energies above a critical value this transition is possible. However, these numerical studies do not provide a sufficiently clear physical understanding of why the critical initiation energy turns out to be several orders of magnitude larger than that predicted by the criterion put forward by Zel'dovich et al. (1956), namely, that the initiation energy should be large enough to ensure that the detonation radius at transition r'_c be larger than the thickness of the reaction layer. In 1994, He and Clavin carried out a quasi-steady analysis of the structure of the reaction layer behind the shock front, propagating with the CJ velocity, for a one-step irreversible reaction with Arrhenius kinetics with large activation energy. Their analysis incorporates the effects of the radius of curvature of the front, leading to a critical value

$$\frac{(r'_d)^*}{l_i} = \frac{8e\gamma^2 j}{\gamma^2 - 1}\beta \qquad (44)$$

below which no quasi-steady solution exists. Here, $\beta = E_a/(R^o T'_N)$ represents the nondimensional activation energy based on the Newmann temperature T'_N behind the shock wave. The thickness of the reaction layer is based on the induction length l_i, which increases rapidly with decreasing values of T'_N according to $l_i \propto \exp[E_a/(R^o T'_N)]$. This temperature decreases as the detonation front weakens, as can be seen in Figure 11, which shows the evolution with time of T'_N/T'_{CJ} and T'_d/T'_{CJ} for a detonation front with $t_d = 0.1$, $\gamma = 1.4$ and $j = 1$. The value of l_i, which is exponentially small during the overdriven stage, increases to reach its maximum value as the CJ detonation is reached.

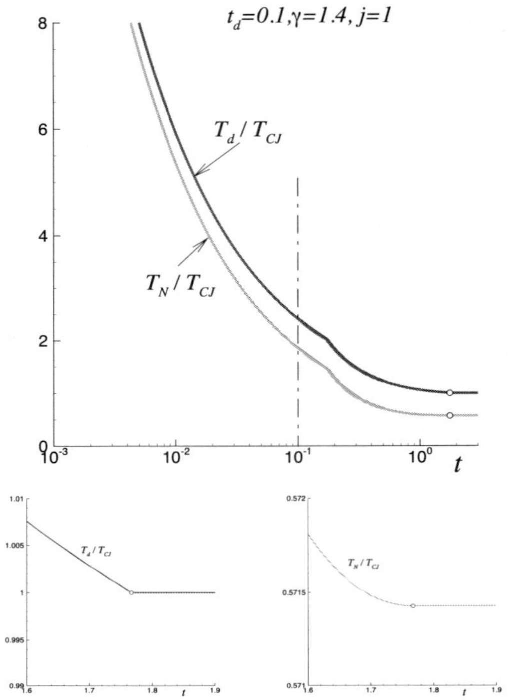

Figure 11. The evolution with time of T_d and T_N for $t_d = 0.1$, $\gamma = 1.4$ and $j = 1$.

According to the criterion developed by He and Clavin (1994), for a successful initiation, when the CJ solution is reached, the detonation-front radius, r'_c, must be larger than the critical value $(r'_d)^*$ given in (44). Substituting into this criticality condition $r'_c > (r'_d)^*$ the expression for $(r'_d)^*$ given in (44) and the equation $r'_c = r_c r'_E$, with r'_E defined in (4), yields the criterion

$$E > E_c = \rho'_o D'^2_{CJ} \left(\frac{\beta l'_i}{r_{CJ}}\right)^{1+j} \left(\frac{8ej\gamma^2}{\gamma^2 - 1}\right)^{1+j} \qquad (45)$$

for the minimum initiation energy, where the value of r_c is given as a function of t_d in Figure 8.

ACKNOWLEDGMENTS

We express here our appreciation for the outstanding work of Professor Toro in Numerical Gas Dynamics. This work was supported by the Spanish MCINN through the Program CONSOLIDER-Ingenio 2010 (project # CSD2010-00011).

APPENDIX A

Zel'dovich-Taylor problem for $\gamma - 1 \ll 1$.

For $\gamma - 1 \ll 1$, the boundary values given in (16) become $\rho_{CJ} = 2$ and $v_{CJ} = p_{CJ} = 1/2$, while the condition of isentropic flow (17) reduces in the first approximation to the condition of isothermal flow

$$p = \rho/4. \qquad \text{(A.46)}$$

Using this last equation to eliminate the pressure in (19) yields

$$\frac{1}{\rho}\frac{d\rho}{d\xi} = -4(v - \xi)\frac{dv}{d\xi}, \qquad \text{(A.47)}$$

which can be substituted into (18) to give the reduced problem

$$\left[4(v-\xi)^2 - 1\right]\frac{dv}{d\xi} = \frac{jv}{\xi},\ \xi = 1 : v = \frac{1}{2} \qquad \text{(A.48)}$$

for the velocity field. Note in particular that the coefficient in square brackets indicates that the weak discontinuity is located for $\gamma - 1 \ll 1$ at $\xi_o^+ = 1/2$ regardless of the value of j.

An exact solution exists for $j = 0$, when $v = \xi - 1/2$, whereas the solution for $j = 1$ and $j = 2$ requires in general numerical integration. The solution near the weak discontinuity is given by $v = 1/2(\xi - 1/2)$ for $j = 1$. For the spherical case $j = 2$, the structure of the discontinuity is more complex, as disccused by Landau and Lifshitz (see Eq. 130.7), with the velocity vanishing according to $Cv - v\ln v = \xi - 1/2$, where $C = -1.597$ is determined as part of the integration.

APPENDIX B

Self-similar flow induced by a concentrated source of constant heating rate at $t \ll 1$.

The corresponding self-similar solution is determined by integrating (8)–(10), written in terms of the similarity variables $\zeta = r/(t^3/t_d)^{1/(3+j)}$, $\rho = R(\zeta)$, $p = (t_d t^j)^{-2/(3+j)} P(\zeta)$, and $v = (t_d t^j)^{-1/(3+j)} V(\zeta)$ in the form,

$$-\frac{3}{3+j}\zeta\frac{dR}{d\zeta} + \frac{1}{\zeta^j}\frac{d}{d\zeta}(\zeta^j R V) = 0 \qquad \text{(B.49)}$$

$$-\frac{j}{3+j}V + \left(V - \frac{3}{3+j}\zeta\right)\frac{dV}{d\zeta} + \frac{1}{R}\frac{dP}{d\zeta} = 0 \qquad \text{(B.50)}$$

$$-\frac{2j}{3+j}\frac{P}{R^\gamma} + \left(V - \frac{3}{3+j}\zeta\right)\frac{d}{d\zeta}\left(\frac{P}{R^\gamma}\right) = 0 \qquad \text{(B.51)}$$

with boundary conditions at the contact surface

$$\zeta = \zeta_e : \begin{cases} V = \dfrac{3}{3+j}\zeta_e \\ 2^j\pi\left[\dfrac{1}{(1+j)(\gamma-1)} + \dfrac{3}{3+j}\right]\zeta_e^{1+j}P = 1 \end{cases} \qquad \text{(B.52)}$$

and at the leading shock

$$\zeta = \zeta_d : \begin{cases} R = \dfrac{\gamma+1}{\gamma-1} \\ V = \dfrac{2}{\gamma+1}\dfrac{3\zeta_d}{3+j} \\ P = \dfrac{2}{\gamma+1}\left(\dfrac{3\zeta_d}{3+j}\right)^2. \end{cases} \qquad \text{(B.53)}$$

Integration of (B.49)–(B.53) determines, for a given value of γ, the self-similar profiles $P(\zeta)$, $V(\zeta)$ and $R(\zeta)$, along with the values of ζ_d and ζ_e, the former giving the front velocity $D = [\zeta_d/(3+j)](t_d t)^{-1/(3+j)}$, with $\zeta_d = 0.671$ for $\gamma = 1.4$ and $j = 1$. When we take into account that the specific heat ratio γ is close to unity, the flow is seen to be confined to a relatively thin shell located behind the leading shock with $\zeta_d - \zeta_e \ll \zeta_d$, a characteristic of the flow that can be observed even for the value $\gamma = 1.4$ employed in the computation of Figure 3.

REFERENCES

Bishimov, E. 1968 Numerical solution of a problem of a strong point explosion in a detonating gas. *Differential Equations and Their Applications.* Nauka, Alma-Ata.

Döring, W. 1943 On the detonation process in gases. *Ann. Physik.* 43, 421–436.

He, L. & Clavin, P. 1994 On the direct initiation of gaseous detonations by an energy source. *J. Fluid Mech.* 277 227–248.

Korobeinikov, V.P. 1969 The problem of point explosion in a detonating gas. *Astronautica Acta* 14, 411–419.

Korobeinikov, V.P. 1971 Gas dynamics of explosions. *Ann. Rev. Fluid Mech.* 3, 317–346.

Kurdyumov, V., Sánchez, A.L. & Liñán, A. 2003 Heat propagation from an external energy source in a gas. *J. Fluid Mech.* 491, 379–410.

Landau, L.D. & Lifshitz, E.M. 1987 *Fluid Mechanics*, Second Edition: Volume 6 (Course of Theoretical Physics), Butterworth-Heinemann.

Lee, J.H.S. 2008 *The Detonation Phenomenon*, Cambridge University Press, New York.
Levin, V.A. & Chernyi, G.G. 1967 Asymptotic laws of behavior of detonation waves. *PMM* 31, 393–405.
von Neumann, J. 1942 Theory of detonation waves. *Prog. Rept. No.* 238 (April 1942) *O.S.R.D. Rept. No.* 549.
Rogers, M.H. 1958 Similarity flows behind strong shock waves. *Q.J. Mech. Appl. Maths* 11, 411–422.
Sedov, L.I. 1946 Propagation of strong shock waves. *Journal of Applied Mathematics and Mechanics*, 10, 241–250.
Taylor, G.I. 1950 The dynamics of the combustion products behind plane and spherical detonation fronts in explosives. *Proc. R. Soc. Lond. A* 200, 235–247.
Taylor, G.I. 1952 The formation of a blast wave by a very intense explosion. I. Theoretical discussion. II. The atomic explosion of 1945. *Proc. R. Soc. Lond. A* 201, 159–174, 175–186.
Zel'dovich, Ya.B. 1940 On the theory of the propagation of detonations in gaseous systems. *Zhur. Eksp. Teor. Fiz.* 10, 542–568.
Zel'dovich, Ya.B. 1942 On the distribution of pressure and velocity in products of detonation blasts, in particular for spherically propagating detonation waves. *Zhurn. Eksper. Teor. Fiz.* 12, 389–406.
Zel'dovich, Ya.B., Kogarko, S.M., & Simonov, N.N. 1956 An experimental investigation of spherical detonation of gases. *Sou. Phys. Tech. Phys.* 1, 1689–1731.

Numerical Methods for Hyperbolic Equations – Vázquez-Cendón et al. (eds)
© 2013 Taylor & Francis Group, London, ISBN 978-0-415-62150-2

Numerical performance of discontinuous and stabilized continuous Galerkin methods for convection–diffusion problems

Paola F. Antonietti
MOX, Dipartimento di Matematica, Politecnico di Milano, Italy

Alfio Quarteroni
MOX, Dipartimento di Matematica, Politecnico di Milano, Italy
CMCS, Ecole Polytechnique Federale de Lausanne (EPFL), Switzerland

ABSTRACT: We compare the performance of two classes of numerical methods for the approximation of linear steady–state convection–diffusion equations, namely, the Discontinuous Galerkin (DG) method and the continuous Streamline Upwind Petrov-Galerkin (SUPG) method. We present a fair comparison of such schemes considering both diffusion–dominated and convection–dominated regimes, and present numerical results obtained on a series of test problems including smooth solutions, and test cases with sharp internal and boundary layers.

1 INTRODUCTION

Convection–diffusion equations occur in the mathematical modeling of a wide range of phenomena as semiconductor devices modeling, magnetostatics and electrostatic flows, heat and mass-transfer, and flows in porous media related to oil and groundwater applications. The linear steady–state convection–diffusion equation with constant coefficients posed on a domain $\Omega \subseteq \mathbb{R}^d$, d = 2,3, is a boundary value problem of the form

$$\begin{aligned} -\varepsilon\Delta u + \operatorname{div}(\beta u) &= f \text{ in } \Omega \\ u &= 0 \text{ on } \partial\Omega, \end{aligned} \tag{1}$$

where $f \in L^2(\Omega)$ is a given real-valued function, $\varepsilon > 0$ is the diffusion parameter and $\beta \in \mathbb{R}^d$, $d = 2,3$, represents a constant (for simplicity) velocity field. In many physical phenomena the convection has much greater magnitude than the diffusion, i.e., $\|\beta\|/\varepsilon >> 1$. In such cases, problem (1) is an example of a singularly perturbed problem (with respect to ε): that is, the solution in the case $\varepsilon = 0$ (with boundary conditions prescribed not on the whole boundary $\partial\Omega$ but on $\partial\Omega_{in} = \{x \in \partial\Omega : \beta \cdot n(x) < 0\}$ with n(x) denoting the unit outward normal vector to $\partial\Omega$ at the point $x \in \partial\Omega$, see (3) below) is not equal, at all points, to the limit of the solutions as $\varepsilon \searrow 0^+$. From the numerical viewpoint, the design of robust numerical schemes for the solution of (1) represents a challenging problem, andindeed there has been an extensive development of discretizations methods for convection–diffusion equations that are robust in all the regimes, see (Roos et al. 1996), for example (also (Ferziger and Perić 1999), (LeVeque 2007), (LeVeque 2002)). Among the numerical methods developed so far, we are interest here in discontinuous and stabilized continuous Galerkin finite element methods which restore two opposite paradigms within the finite element approach, namely non-conforming (discontinuous) versus conforming (continuous) approximation spaces.

Stabilized conforming finite element methods have been extensively developed for hyperbolic and parabolic conservation laws. The original conforming stabilized method for advection-diffusion equations, the Streamline-Upwind Petrov-Galerkin (SUPG), was developed by (Brooks and Hughes 1982) and then analyzed by (Johnson et al. 1984) (see also (Burman and Smith 2010) for the analysis of the SUPG method for transient convection-diffusion equations). It provides an upwinding effect to standard finite element methods using the Petrov-Galerkin framework. It is well known that the SUPG method has the capability of improving numerical stability for convection–dominated flows, while satisfying a strong consistency property. Like many other conforming stabilized methods, the SUPG scheme contains an elementwise stabilization parameter τ that has to be tuned in practice, and, except for simplified situations, the "optimal" value is not known. More precisely, for linear finite element discretizations on a uniform decomposition of a one-dimensional problem with constant coefficients $\varepsilon \in \mathbb{R}^+$, $\beta \in \mathbb{R}$, and constant data, it can be shown that the choice

$$\tau = \frac{h}{2|\beta|}\left(\coth(\mathrm{Pe}_{\mathcal{T}}) - \frac{1}{\mathrm{Pe}_{\mathcal{T}}}\right), \tag{2}$$

where $\mathrm{Pe}_{\mathcal{T}}$ is the mesh Pclet number (cf. also (6), below), leads to a nodally exact approximate solution (Brooks and Hughes 1982), see also (Christie et al. 1976). In the multi-dimensional case, many different choices of the stabilization parameter τ have been proposed see (Franca et al. 2001), and the references therein. It is also known that the choice of the stabilization parameter τ also influences the behavior of the iterative solvers employed to solve the resulting linear system of equations. For example, GMRES converges more slowly for worse stabilization parameters, as pointed out in (Liesen and Strakos 2005).

On the other hand, Discontinuous Galerkin (DG) methods are parameters free since they do not rely on the addiction of *ad-hoc tuned* streamline stabilization terms. Since their introduction in 1970s (Reed and Hill 1973), (Douglas and Dupont 1976), (Arnold 1982), the discontinuous Galerkin method has become a topic of extensive research for the numerical solution of differential equations. DG methods employ piecewise polynomial spaces which may be discontinuous across elements boundaries. Therefore, DG can accommodate non-matching meshes as well as variable polynomial approximation orders. These features make them ideally suited for both geometrical and functional adaptive discretizations. Moreover, DG methods are locally mass conservative and can capture possible discontinuities in the solution thanks to the discontinuous approximation spaces. For such capabilities, DG methods have been extensively developed for convection–diffusion problems, for example (Cockburn 1999), (Baumann and Oden 1999), (Houston et al. 2002), (Buffa et al. 2006), (Antonietti and Houston 2008), (Ayuso de Dios and Marini 2009), and the references therein.

In this paper we aim at comparing DG and SUPG solutions, working, for the sake of simplicity, on a linear steady–state convection–diffusion equation with constant coefficients. A detailed assessment of the performance of such schemes (employing the lowest order elements) with respect to accuracy is presented in a number of test problems featuring smooth solutions, and other problems featuring sharp layers (either internal or boundary's).

The rest of the paper is organized as follows. In Section 2 we introduce some notation and recall the variational formulation of problem (1). In Section 3 and Section 1 we recall the DG and SUPG formulations and the corresponding error estimates, respectively. Section 1 contains an extensive set of numerical experiments to compare the DG and SUPG solutions. Finally, in Section 17 we draw some conclusions.

2 MODEL PROBLEM

We consider the linear steady–state convection–diffusion equation with constant coefficients (1) posed on Ω, a bounded convex open polygonal (resp. polyhedral) domain in $\mathbb{R}^2$ (resp. $\mathbb{R}^3$). The corresponding weak formulation reads as: Find $u \in H_0^1(\Omega)$ such that

$$\int_{\Omega} (\varepsilon \nabla u - \beta u) \cdot \nabla v \, dx = \int_{\Omega} f v \, dx \quad \forall v \in H_0^1(\Omega).$$

The existence of a unique solution of the above problem follows from the Lax-Milgram Theorem. For the analysis of more general convection–diffusion equations see, e.g., (Quarteroni and Valli 1994). For the design and analysis of DG methods in the case of a diffusion tensor whose entries are bounded, piecewise continuous real-valued functions defined on $\overline{\Omega}$ and a piecewise continuous real-valued velocity field, we refer to (Ayuso de Dios and Marini 2009). By $n \equiv n(x)$ we denote the unit outward normal vector to $\partial\Omega$ at the point $x \in \partial\Omega$, and define the sets $\partial\Omega_{in}$ and $\partial\Omega_{out}$ of inflow and outflow boundary, respectively, as

$$\begin{aligned} \partial\Omega_{in} &= \{x \in \Gamma : \beta \cdot n < 0\}, \\ \partial\Omega_{out} &= \{x \in \Gamma : \beta \cdot n \geq 0\}. \end{aligned} \tag{3}$$

3 DG AND SUPG DISCRETISATIONS

The aim of this section is to introduce the DG and SUPG discretisations.

Let $\mathcal{T}_h$ be a conforming shape-regular quasi-uniform partition of Ω into disjoint open triangles T where each $T \in \mathcal{T}_h$ is the affine image of the reference open unit d -simplex in $\mathbb{R}^d$, d = 2,3. Denoting by $h_{\mathcal{T}}$ the diameter of an element $T \in \mathcal{T}_h$, we define the mesh size $h = \max_{T \in \mathcal{T}_h} h_T$.

3.1 *The DG method*

We start by introducing suitable trace operators, defined in the usual way (Arnold, Brezzi, Cockburn, and Marini 2002). Let $\mathcal{F}_h^I$ and $\mathcal{F}_h^B$ be the sets of all the interior and boundary (d – 1)-dimensional (open) faces if d = 2, "face" means "edge"), respectively, we set $\mathcal{F}_h = \mathcal{F}_h^I \cup \mathcal{F}_h^B$. We also decompose $\mathcal{F}_h^B = \mathcal{F}_h^{out} \cup \mathcal{F}_h^{in}$, where $\mathcal{F}_h^{B_{out}} = \{F \in \mathcal{F}_h^B : F \subset \partial\Omega_{out}\}$ and $\mathcal{F}_h^{B_{in}} = \{F \in \mathcal{F}_h^B : F \subset \partial\Omega_{in}\}$. Implicit in these definitions is the assumption that T_h respects the decomposition of $\partial\Omega$ in the sense that each $F \in \mathcal{F}_h$ that lies on $\partial\Omega$ belongs to the interior of exactly one of $\partial\Omega_{out}$, $\partial\Omega_{in}$. Let $F \in \mathcal{F}_h^I$ be an interior face shared by the elements T^+ and T^- with outward unit normal vectors n^+ and n^-, respectively. For piecewise smooth vector–valued and scalar functions τ and z, respectively, let $\tau^\pm$ and $z^\pm$ be the traces of τ and z on $\partial T^\pm$ taken within the interior of $T^\pm$, respectively. We define the *jump* and the *average* across F by

$$[[\tau]] = \tau^+ \cdot n^+ + \tau^- \cdot n^-, \qquad [[z]] = z^+ n^+ + z^- n^-,$$
$$\{\{\tau\}\} = (\tau^+ + \tau^-)/2, \qquad \{\{z\}\} = \left(z^+ + z^-\right)/2.$$

On a boundary face $F \in \mathcal{F}_h^B$, we set, analogously,

$$[[z]] = z\,n, \quad \{\{\tau\}\} = \tau.$$

We do not need either $[[\tau]]$ or $\{\{v\}\}$ on boundary faces, and leave them undefined. On each interior face $F \in \mathcal{F}_h^I$, we introduce an *upwind* average $\{\{\cdot\}\}_\beta$ defined as

$$\{\{\beta u\}\}_\beta = \begin{cases} \beta u^+ & \text{if } \beta \cdot n^+ > 0 \\ \beta u^- & \text{if } \beta \cdot n^+ < 0 \\ \beta\{\{u\}\} & \text{if } \beta \cdot n^+ = 0 \end{cases}$$

On $F \in \mathcal{F}_h^B$, we set $\{\{\beta u\}\}_\beta = \beta u$, if $\beta \cdot n > 0$, $\{\{\beta u\}\}_\beta = 0$ otherwise. For a given approximation order $\ell \geq 1$, we introduce the DG finite element space V_h^{DG} as

$$V_h^{DG} = \{v \in L^2(\Omega) : v|_T \in \mathbb{P}^\ell(T) \quad \forall T \in T_h\},$$

where $\mathbb{P}^\ell(T)$ is the set of polynomials of total degree ℓ on T.

We define the bilinear form $\mathcal{A}_h^{DG}(\cdot,\cdot) : V_h^{DG} \times V_h^{DG} \to \mathbb{R}$ as

$$\begin{aligned} \mathcal{A}_h^{DG}(u,v) = & \sum_{T \in \mathcal{T}_h} \int_T (\varepsilon \nabla u - \beta u) \cdot \nabla v \, dx \\ & - \sum_{F \in F_h} \int_F \left(\{\{\varepsilon \nabla u\}\} \cdot [[v]] + [[u]] \cdot \{\{\varepsilon \nabla v\}\} \right) ds \\ & + \sum_{F \in F_h} \int_F \left(\sigma_F [[u]] \cdot [[v]] + \{\{\beta u\}\}_\beta \cdot [[v]] \right) ds \end{aligned}$$

with the convention that the application of the operator ∇ has to be intendend elementwise whenever the function to which is applied is elementwise discontinuous. Here σ_F is defined for all $F \in \mathcal{F}_h$ as

$$\sigma_F = \alpha \frac{\varepsilon \ell^2}{h_F} \quad \forall F \in F_h, \tag{4}$$

with α a positive real number at our disposal, and h_F the diameter of the face $F \in \mathcal{F}_h$. The DG approximation of problem (1) is defined as follows: find $u_h^{DG} \in V_h^{DG}$ such that

$$\mathcal{A}_h^{DG}(u_h^{DG}, v_h) = \int_\Omega f v_h \, dx \quad \forall v_h \in V_h^{DG}.$$

Remark 3.1. For the approximation of the convection and diffusion terms we have employed a DG method with *upwinded fluxes* (Brezzi et al. 2004) and the *Symmetric Interior Penalty* (SIPG) method (Arnold 1982), respectively. With minor changes, any other (possibly unsymmetric) DG approximation of the second order term could be considered.

We close this section observing that, denoting by n_F the normal to the face $F \in \mathcal{F}_h$, it holds

$$\begin{aligned} \sum_{F \in F_h} \int_F \{\{\beta u\}\}_\beta \cdot [[v]] ds & = \sum_{F \in F_h^I \cup F_h^{B_{out}}} \int_F \{\{\beta u\}\} \cdot [[v]] ds \\ & + \sum_{F \in F_h^I} \int_F \frac{|\beta \cdot n_F|}{2} [[u]] \cdot [[v]] ds, \end{aligned}$$

for all $u, v \in V_h^{DG}$. Therefore, the *upwind* value of βu, can be written as the sum of the usual (symmetric) average plus a jump penalty term. Such equivalent representation has several advantages. For example, notice that both terms on the right hand side are of the same kind as the ones already present in the treatment of the diffusive part of the operator. This can be favorably exploited in the implementation process.

3.2 *The SUPG method*

Let the SUPG discrete space be defined as the subspace of V_h^{DG} of continuous polynomials that also satisfy the boundary condition, i.e.,

$$V_h^{SUPG} = \{v \in C^0(\overline{\Omega}) : v|_T \in \mathbb{P}^\ell(T) \quad \forall T \in \mathcal{T}_h, \text{ and } v = 0 \text{ on } \partial\Omega\},$$

For piecewise positive parameters τ_T that we will specify later, we define the bilinear form $\mathcal{A}_h^{SUPG}(\cdot,\cdot) : V_h^{SUPG} \times V_h^{SUPG} \longrightarrow \mathbb{R}$ as

$$\begin{aligned} \mathcal{A}_h^{SUPG}(u,v) = & \int_\Omega (\varepsilon \nabla u - \beta u) \cdot \nabla v \, dx \\ & + \sum_{T \in \mathcal{T}_h} \int_T \tau_T (-\varepsilon \Delta u + \beta \cdot \nabla u) \beta \cdot \nabla v \, dx, \end{aligned}$$

and the functional $F_h^{SUPG}(\cdot) : V_h^{SUPG} \longrightarrow \mathbb{R}$

$$\mathcal{F}_h^{SUPG}(v) = \sum_{T\in\tau_h} \int_T f(v + \tau_T\beta \cdot \nabla v)dx.$$

Then, the SUPG formulation reads: find $u_h^{SUPG} \in V_h^{SUPG}$ such that

$$\mathcal{A}_h^{SUPG}(u_h^{SUPG}, v_h) = \mathcal{F}_h^{SUPG}(v_h) \quad \forall v_h \in V_h^{SUPG}.$$

It is well known that the choice of the stabilization parameters τ_T, $T \in \tau_h$, is a delicate matter since it may influence the performance of the method. For such a reason, the problem of finding the optimal stabilization parameter has been a subject of an extensive research in the last years, and many choices are available in the literature ((John and Knobloch 2007), (John and Knobloch 2008), for example). Here, we consider a generalization of (2) to the multi-dimensional case and set

$$\tau|_T \equiv \tau_T = \frac{h_T}{2\|\beta\|\ell}\,\xi(\mathrm{Pe}_T), \tag{5}$$

with Pe_T being the local Pclet number, i.e.,

$$\mathrm{Pe}_T = \frac{\|\beta\| h_T}{2\varepsilon\ell}, \tag{6}$$

and where $\|\beta\|$ is the Euclidean norm of β. The upwind function $\xi(\cdot)$ can be chosen, for instance, as

$$\xi(\theta) = \coth(\theta) - 1/\theta, \quad \theta > 0, \tag{7}$$

cf. Figure 1. Notice that $\xi(\theta) \to 1$ for $\theta \to \infty$, and $\xi(\theta)/\theta \to 1/3$ for $\theta \to 0^+$ (and the SUPG stabilization is not necessary for $\theta \to 0^+$). Moreover, if convection strongly dominates diffusion, i.e., $\mathrm{Pe}_T \gg 1$, we have $\tau_T \approx h_T/(2\|\beta\|\ell)$. As already pointed out, many other definitions of τ_T are possible. For example, in (5) and (6) h_T can be replaced by the element diameter in the direction of the convection β, or the upwind function $\xi(\cdot)$ can be approximated either by $\xi(\theta) \approx \max\{0, 1-1/\theta\}$ or by $\xi(\theta) \approx \min\{1, \theta/3\}$. We refer to (John and Knobloch 2007), (John and Knobloch 2008) (Brooks and Hughes 1982), for further details.

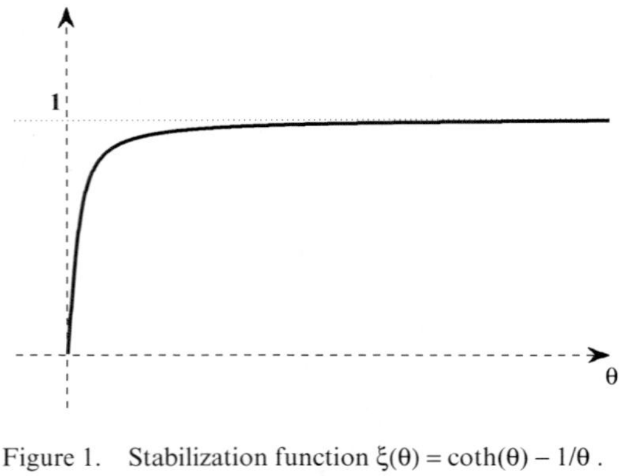

Figure 1. Stabilization function $\xi(\theta) = \coth(\theta) - 1/\theta$.

4 ERROR ESTIMATES

In this section we briefly recall the error estimates for the DG and SUPG formulations.

To deal with the convection–dominated regime, we first introduce a norm which controls the streamline derivative:

$$\|v\|_\beta = \left(\sum_{T\in T_h} \left\| \tau_T^{1/2}\beta \cdot \nabla v \right\|_{0,T}^2 \right)^{1/2} \quad \forall v \in H^1(\tau_h), \tag{8}$$

where τ is defined in (5), and where the piecewise broken Sobolev space $H^1(T_h)$ consists of functions v such that $v|_T \in H^1(T)$ for any $T \in \mathcal{T}_h$. We endow the DG space with the norm

$$\|v\|_{DG} = \left(\|v\|^2 + \|v\|_\beta^2 \right)^{1/2} \quad \forall v \in V_h^{DG}, \tag{9}$$

where $\|\cdot\|_\beta$ is defined in (8), and where

$$\|v\|^2 = \sum_{T\in\mathcal{T}_h} \left\|\varepsilon^{1/2}\nabla v\right\|_{0,T}^2 + \sum_{F\in\mathcal{F}_h} \left\|\alpha_F^{1/2}[\![v]\!]\right\|_{0,F}^2$$

for all $v \in V_h^{DG}$. Here σ_F is defined as

$$\alpha_F = \sigma_F + |\beta \cdot n| \quad \forall F \in F_h,$$

with σ_F given in (4).

Provided the exact solution u is regular enough and the stability parameter α in (4) is chosen sufficiently large, the following error estimates hold for the DG solution

$$\begin{aligned} \left\| u - u_h^{DG} \right\| &\lesssim h^\ell \left(\varepsilon^{1/2} + h^{1/2}\|\beta\|_\infty^{1/2} \right), \\ \left\| u - u_h^{DG} \right\|_{DG} &\lesssim h^\ell \left(\varepsilon^{1/2} + h^{1/2}\|\beta\|_\infty^{1/2} \right). \end{aligned} \tag{10}$$

We refer to (Houston et al. 2002) and (Ayuso de Dios and Marini 2009) for the proof (in a more general framework) and further details.

For the SUPG method, we define the norm

$$\|v\|_{SUPG} = \left(\left\| \varepsilon^{1/2}\nabla v \right\|_{L^2(\Omega)}^2 + \|v\|_\beta^2 \right)^{1/2} \tag{11}$$

for all $v \in V_h^{SUPG}$. Then, the following estimate holds:

$$\left\| u - u_h^{SUPG} \right\|_{SUPG} \lesssim h^{\ell} \left(\varepsilon^{1/2} + h^{1/2} \| \beta \|_{\infty}^{1/2} \right), \quad (12)$$

see (Quarteroni and Valli 1994), for example.

Note that the two error estimates (10) and (12) feature the same right hand side (apart from the hidden multiplicative constants).

Remark 4.1 In the error estimates (10) and (12) the hidden constants depend on the domain Ω, the shape regularity constant of the partition T_h, and the polynomial approximation degree ℓ, but are independentof the mesh size h and the physical parameters ε and β of the problem.

5 NUMERICAL RESULTS

We compare the numerical performance of the DG and SUPG methods on four test cases. In the first example, we choose a smooth exact solution and compare the approximation properties of the two schemes. In the second and third examples, the exact solutions exhibit an internal and an exponential boundary layer, respectively. Finally, in the last example the (unknown) analytical solution features sharp internal and boundary layers. Throughout this section we set $\Omega = (0,1) \times (0,1)$ and consider a sequence of uniform unstructured triangulations obtained by uniformly refining the initial mesh shown in Figure 2 (left). Denoting by h_0 the mesh size of this initial grid, after R = 1,2, ... refinements the corresponding mesh size is given by $h = h_0/2^R$ (Fig. 2 (right), for R = 2). Since the grids are quasi uniform, it holds $h \approx 1/\sqrt{N}$, N being the number of elements of the partition. Finally, throughout the section, we restrict ourselves to piecewise linear elements, i.e., $\ell = 1$.

5.1 *Example 1*

In this section, we aim at comparing the approximation properties of the DG and SUPG schemes on a smooth analytical solution. To this aim, we chose $\beta = (1,1)^T$, and the source term f such that the solution of problem (1) is $u(x,y) = \sin(2\pi x)(y - y^2)$, for any positive diffusion coefficient ε. Figure 3 (log-log scale) shows the computed errors in the energy norms (9) and (11) versus 1/h, for different choices of the diffusion coefficient ε, $\varepsilon = 10^{-k}$, k = 0,1,5,9. The expected convergence rates are clearly observed: O(h) and O($h^{3/2}$) convergence rates are attained in the diffusion– and convection–dominated regimes, respectively. We have also compared the errors computed in the $L^2(\Omega)$ and $L^\infty(\Omega)$ norms. The results are reported in Figure 4 (log-log scale) and in Figure 5 (log-log scale), respectively. For all the cases, a quadratic convergence rate is clearly observed, for any value of the diffusion coefficient ε. We have run the same set of experiments on a sequence of structured triangular

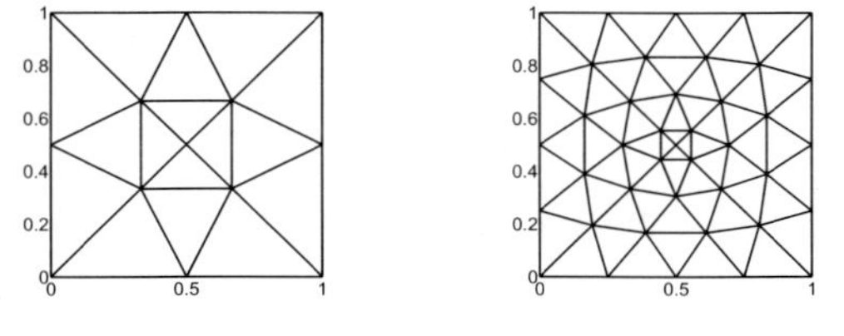

Figure 2. Left: Initial triangulations with mesh size h_0. Right: triangulation obtained after R = 2 levels of refinement (mesh size $h_0/4$).

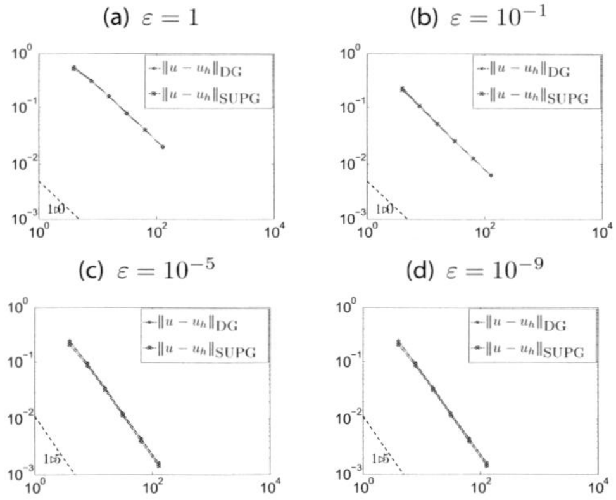

Figure 3. Example 1. Errors in the energy norms (9) and (11) for $\varepsilon = 10^{-K}$, K = 0,1,5,9.

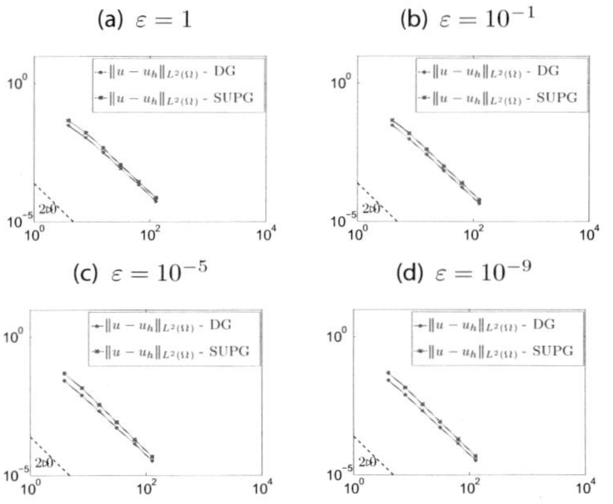

Figure 4. Example 1. Errors in the $L^2(\Omega)$ norm for $\varepsilon = 10^{-K}$, K = 0,1,5,9.

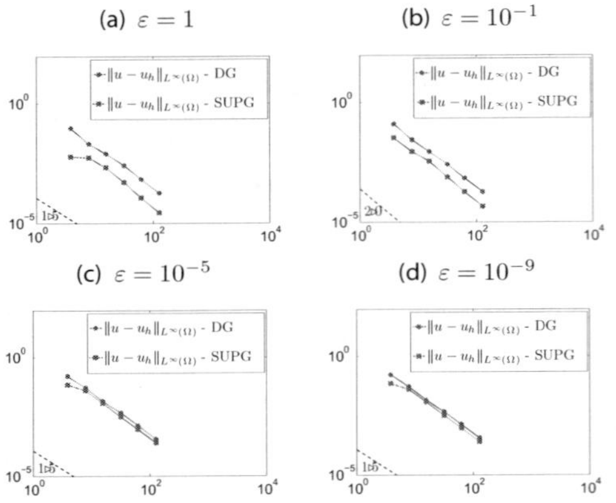

Figure 5. Example 1. Errors in the $L^\infty(\Omega)$ norm for $\varepsilon = 10^{-K}$, $K = 0,1,5,9$.

grids, the results (not reported here, for the sake of brevity) are completely analogous. From the results reported in Figures 3–5, we can infer that in the case of a smooth analytical solution, the DG and SUPG enjoy almost the same accuracy, for any value of the diffusion coefficient ε. Indeed, the error curves reported in Figures 3 show that the errors computed in the energy norms (9) and (11) are almost identical. Concerning the error in average (Fig. 4) the DG method seems to be slightly more accurate than the SUPG; whereas if we measure the error in the $L^\infty(\Omega)$ norm the SUPG method seems to be slightly better than the DG method (Fig. 5). Nevertheless, the difference between the DG and SUPG solutions are negligible, as highlighted by the results reported in Figure 6 where we have plotted, for the same choices of ε, the pointwise absolute value of the difference between the DG and SUPG numerical solutions, on a grid with mesh size $h = h_0/4$.

5.2 *Example 2*

In the second example, we set again $\beta = (1,1)$ and let ε vary, however now the exact solution reads

$$u(x,y) = -\mathrm{atan}\left(\frac{(x-1/2)^2 + (y-1/2)^2 - 1/16}{\sqrt{\varepsilon}}\right),$$

(the source term f and the non-homogeneous boundary conditions being set accordingly). We notice that, as $\varepsilon \searrow 0^+$, the solution exhibits a sharp internal layer, Figure 7 (left) for $\varepsilon = 10^{-3}$ and Figure 7 (right) for $\varepsilon = 10^{-5}$.

Figure 8 compare the DG and SUPG discrete solutions, respectively, obtained on an unstructured triangular grid of mesh size $h = h_0/4$, for

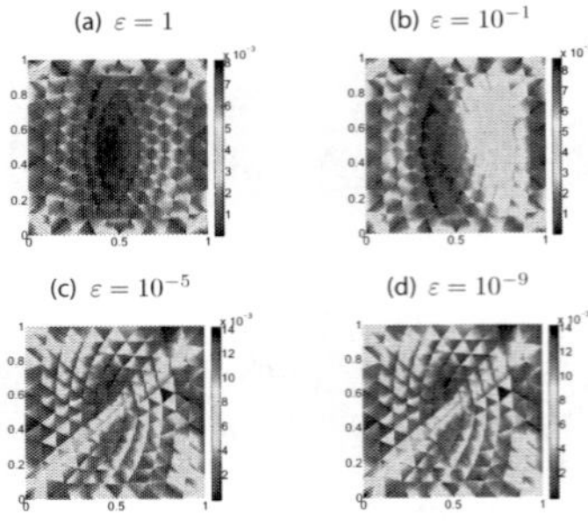

Figure 6. Example 1. Absolute value of the pointwise difference between the DG and SUPG approximate solutions for $\varepsilon = 10^{-k}$, $k = 0,1,5,9$. Grids with mesh sizes $h = h_0/4$.

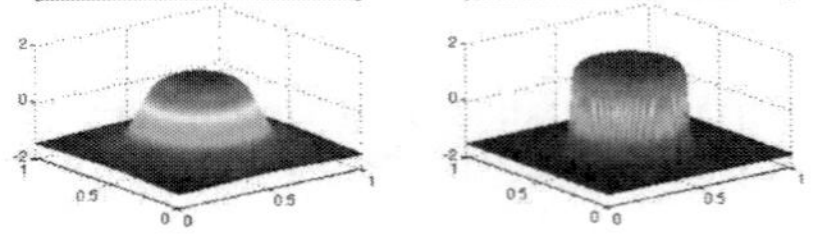

Figure 7. Example 2. Exact solutions for $\varepsilon = 10^{-3}$ (left) and $\varepsilon = 10^{-5}$ (right).

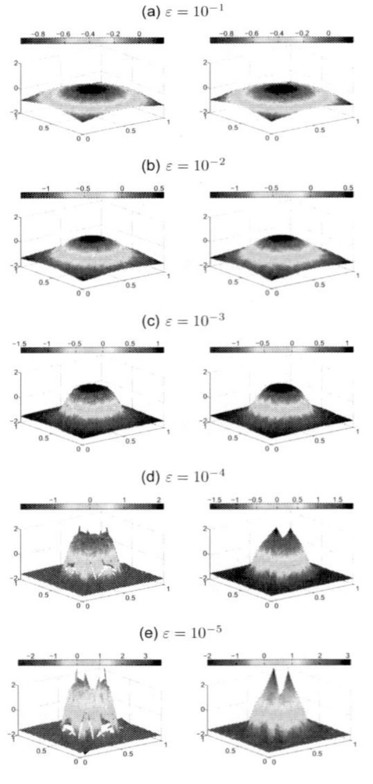

Figure 8. Example 2. DG (left) and SUPG (right) approximate solutions on an grid of mesh size $h = h_0/4$ for $\varepsilon = 10^{-K}$, $K = 1,2,3,4,5$.

$\varepsilon = 10^{-k}$, k = 1,2,3,4,5. We can observe that the discrete solutions approximate quite well the corresponding analytical solution. Nevertheless, in the convection–dominated regime both methods exhibit spurious oscillations near the internal layer.

We next compare the numerical solutions provided by the two methods. In Figure 9 the pointwise absolute error of the DG and SUPG approximate solutions is reported on a grid with mesh size $h = h_0/4$, for $\varepsilon = 10^{-k}$, $k = 1,2,3,4,5$. Notice that, in this case, the approximate solutions provided by the two methods are almost identical in the diffusion–dominated regime (cf. Fig. 9, top), whereas they differ substantially near the internal layer in the convection–dominated regime (cf. Fig. 9, bottom). Indeed, as ε becomes small, they both exhibit oscillations near the internal layer, but of different kind.

Finally, for $\varepsilon = 10^{-k}$, $k = 1,2,3,4$, Figure 10 (log-log scale) and Figure 11 (log-log scale) show the computed errors in the energy norms (9) and (11) and in the $L^2(\Omega)$ norm, respectively. The results reported in Figures 10 and 11 have been obtained on unstructured triangular grids as the ones shown in Figure 2; analogous results (not reported here) have been obtained on structured triangular grids and the same convergence behavior has been observed. The error in the energy norms goes to zero at the predicted rate, namely linearly in the diffusion–dominated case and at a rate of 3/2 in the convection–dominated regime. Nevertheless, notice that such a rate seems to be achieved only asymptotically for $\varepsilon = 10^{-4}$ by both the DG and SUPG methods.

Concerning the mean error (Fig. 11), we observe that both methods achieve a quadratic convergence rate, and the DG scheme seems to be

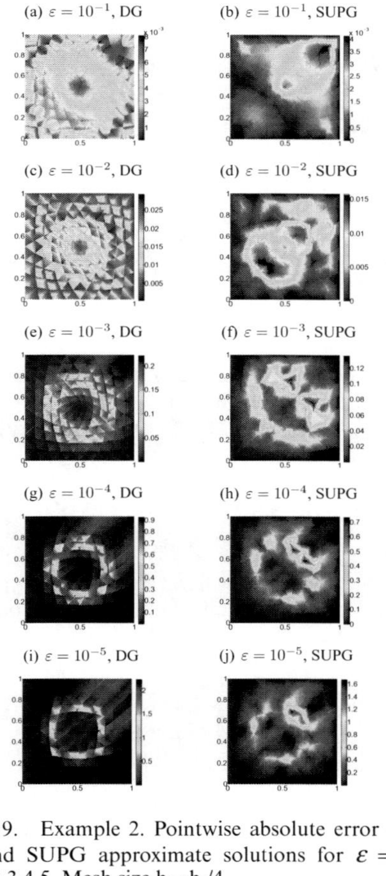

Figure 9. Example 2. Pointwise absolute error of the DG and SUPG approximate solutions for $\varepsilon = 10^{-K}$, K = 1,2,3,4,5. Mesh size $h = h_0/4$.

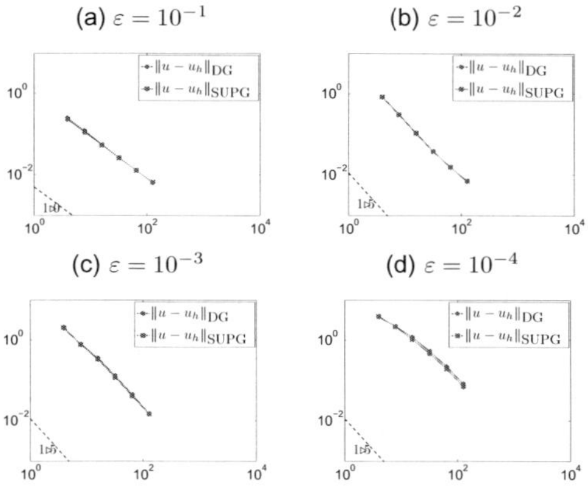

Figure 10. Example 2. Errors in the energy norms for $\varepsilon = 10^{-K}$, K = 1,2,3,4.

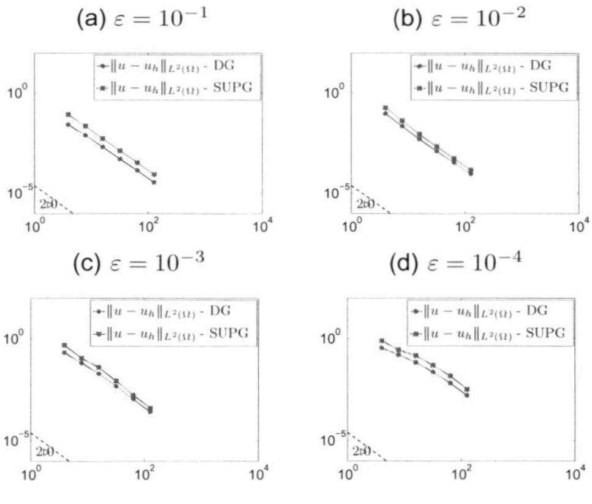

Figure 11. Example 2. Errors in the $L^2(\Omega)$ norm for $\varepsilon = 10^{-K}$, K = 1,2,3,4.

slightly more accurate than the SUPG method. If we measure the errors in the $L^\infty(\Omega)$ norm (cf. Fig. 12) we observe again a quadratic convergence rate but in this case the SUPG method seems to be slightly more accurate.

5.3 *Example 3*

In this example, we consider a test case with an exponential boundary layer taken from (Houston et al., 2002) and (Ayuso de Dios and Marini 2009), which is the multi-dimensional version of the one-dimensional test proposed by (Melenk and Schwab 1999). We set $\beta = (1,1)$, and vary the diffusion coefficient ε. The right hand side f and the non-homogeneous boundary condition are chosen in such a way that the exact solution is given by:

$$u(x,y) = x + y - xy + \frac{e^{-1/\varepsilon} - e^{-(1-x)(1-y)/\varepsilon}}{1 - e^{-1/\varepsilon}}.$$

Figure 13 shows the discrete solutions computed with the DG and SUPG methods for $\varepsilon = 10^{-k}$, $k = 0,1,2,9$ on a grid with mesh size $h = h_0/4$. We can observe that, for $\varepsilon = 1, 10^{-1}$ the two solutions are almost identical, whereas in the intermediate regime $\varepsilon = 10^{-2}$ the DG method exhibits some spurious oscillations near the boundary layer, while the SUPG solution seems to be definitely more accurate. Finally, we observe that in the convection–dominated regime, the DG solution is totally free of spurious oscillations but the boundary layer is not captured. Such a behavior, already observed in (Ayuso de Dios and Marini 2009), is due to the fact that the boundary conditions are imposed weakly. On the other hand, for $\varepsilon = 10^{-9}$ the SUPG method results in overshoot near the point (1,1). The above

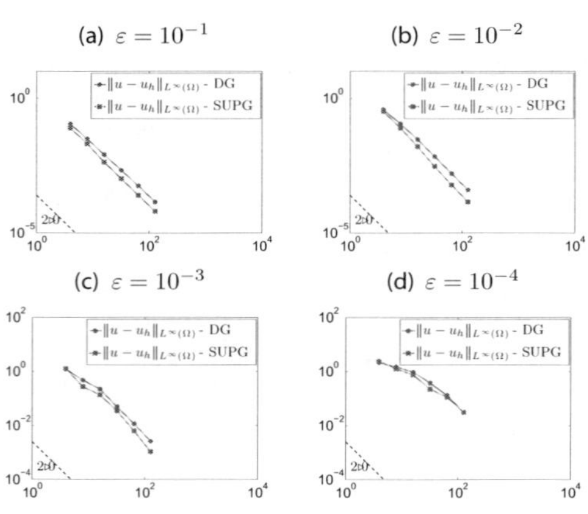

Figure 12. Example 2. Errors in the $L^\infty(\Omega)$ norm for $\varepsilon = 10^{-K}$, $K = 1,2,3,4$.

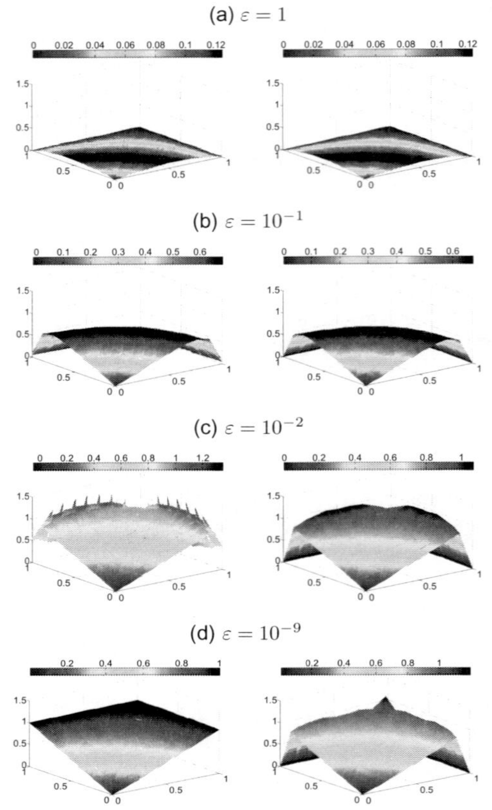

Figure 13. Example 3. DG (left) and SUPG (right) discrete solutions for $\varepsilon = 10^{-K}$, $K = 0,1,2,9$. Mesh size $h = h_0/4$.

considerations are also confirmed by the results reported in Figure 14 where the absolute value of the pointwise error is shown on grid with mesh size $h = h_0/4$.

Finally, Figure 15 shows the computed errors in the $L^2(\Omega)$ and $L^1(\Omega)$ norms for $\varepsilon = 10^{-k}$, $k = 0,1,9$. In the diffusion–dominated regime and for both the methods the errors go to zero quadratically as the mesh is refined, whereas for $\varepsilon = 10^{-9}$ the DG method is quadratically convergent in both the $L^2(\Omega)$ and $L^1(\Omega)$ norm while the SUPG method convergences at a rate of order $\mathcal{O}(h^{1/2})$ and $\mathcal{O}(h)$ in the $L^2(\Omega)$ and $L^1(\Omega)$ norms, respectively.

5.4 *Example 4*

In the last example, always with $\Omega = (0,1)^2$, we consider the following convection–diffusion equation:

$$-\varepsilon \Delta u + \beta \cdot \nabla u = 0 \quad \text{in } \Omega, \qquad u = g \quad \text{on } \partial\Omega.$$

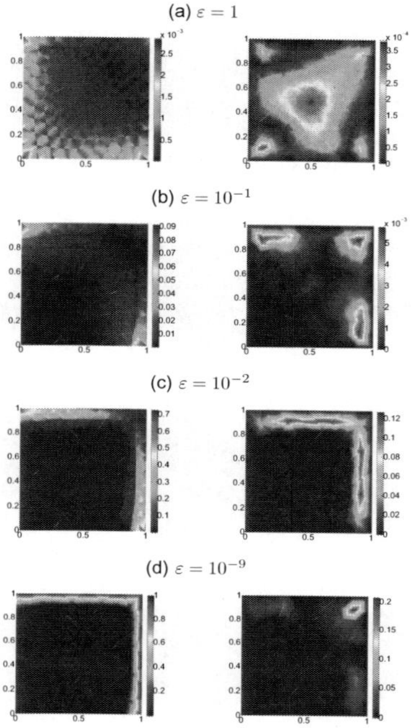

Figure 14. Example 3. Pointwise absolute error of the DG and SUPG approximate solutions for $\varepsilon = 10^{-K}$, K = 0,1,2,9. Mesh size $h = h_0/4$.

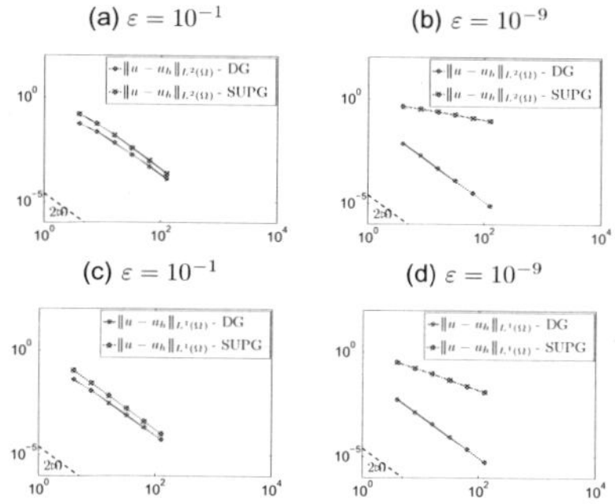

Figure 15. Example 3. Errors in the $L^2(\Omega)$ (top) and $L^1(\Omega)$ (bottom) norms for $\varepsilon = 10^{-K}$, K = 1,9.

The Dirichlet boundary conditions are set as follows:

$$g = \begin{cases} 1 & \text{if } y = 0 \\ 1 & \text{if } x = 0 \text{ and } y \leq 1/2, \\ 0 & \text{otherwise,} \end{cases}$$

the velocity β is chosen as $\beta = (\cos(\pi/3), \sin(\pi/3))$, and we let ε vary. Notice that in this case, a close expression for the exact solution is not available. We know however that, as ε tends to zero, the exact solution exhibits a strong internal and boundary layer.

We compare the discrete solutions computed by the DG and SUPG schemes for $\varepsilon = 10^{-k}$, k = 0,3,9. The results on the unstructured triangular mesh with mesh size $h = h_0/8$ are shown in Figure 16.

The following conclusions can be drawn:

i. in the diffusion–dominated regime, the DG and SUPG methods are substantially equivalent;
ii. in the intermediate regime, the DG method exhibits spurious oscillations, whereas the SUPG method seems to be more stable and only an overshooting effect is observed near the point (1,1);
iii. in the convection–dominated regime, DG method exhibits some spurious oscillations on the internal layer whereas no oscillations along boundary layer are present, however unfortunately the boundary layer is completely missed. On the other hand SUPG method exhibits again an overshooting phenomenon near the point (1,1) and also some spurious oscillations along the internal layer.

The above considerations are also confirmed by the results reported in Figure 17 where the absolute value of the pointwise difference between the DG and SUPG approximate solutions for $\varepsilon = 10^{-k}$,

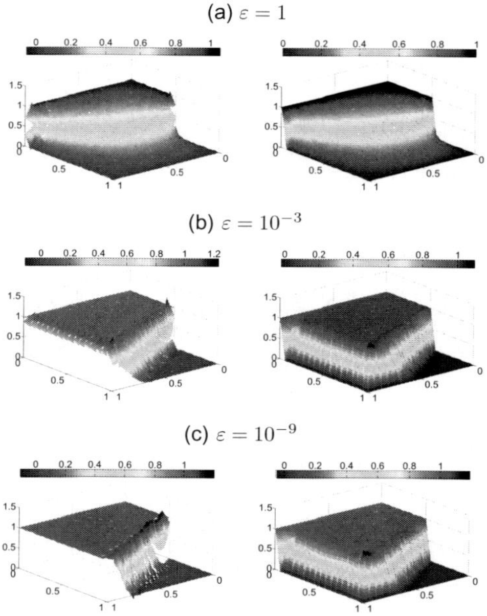

Figure 16. Example 4. DG (left) and SUPG (right) discrete solutions for $\varepsilon = 10^{-K}$, K = 0,3,9. Mesh size $h = h_0/8$.

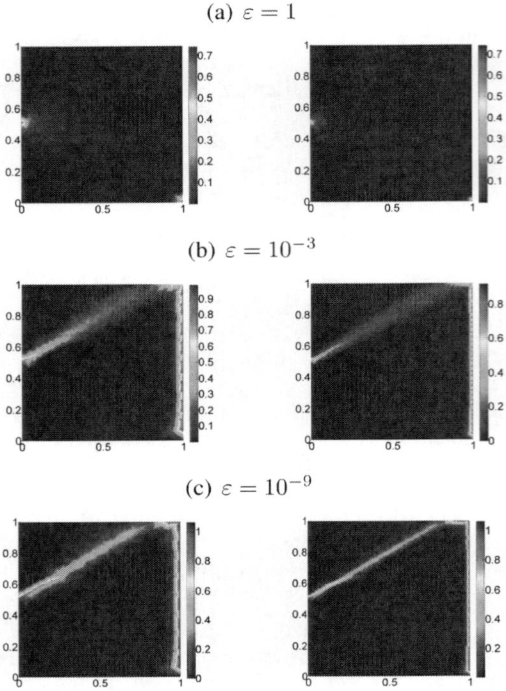

Figure 17. Example 4. Pointwise absolute difference between the DG and SUPG approximate solutions for $\varepsilon = 10^{-K}$, K = 0,3,9. Mesh size $h = h_0/8$ (left) and $h = h_0/16$ (right).

k = 0,3,9, are reported. Results shown in Figure 17 (left) have been obtained on a grid with mesh size $h = h_0/8$; the analogous ones obtained on a finer triangulation with mesh size $h = h_0/16$ are reported in Figure 17 (right).

6 CONCLUSIONS

In this paper we have recalled the formulation and principal properties of DG and SUPG approximations of linear steady–state convection–diffusion equations. The aim was to carry out a careful comparison of their performance (as of stability, accuracy and numerical robustness) in three different solution regimes: diffusion dominated, intermediate, and convection dominated. Our comparative analysis is made for piecewise linear elements on a number of significant test problems. As expected, DG and SUPG methods are substantially equivalent in the diffusion dominated regime. In the intermediate regime, SUPG looks more robust (even if overshooting may occur in the boundary layer). In the convection–dominated regime, SUPG still exhibits overshooting whereas DG is oscillation free, at least when the exact solution features sharp boundary layers. However, it is fair to notice that the DG method is not able to capture the boundary layers (indeed, the latter are completely missed because of the weak enforcement of boundary conditions). An in depth comparison between discontinuous and stabilized continuous Galerkin methods in the case of higher order polynomial approximations and transient convection diffusion problems is currently under investigation (Antonietti and Quarteroni 2011).

ACKNOWLEDGEMENTS

P.F.A. has been partially supported by Italian MIUR PRIN 2008 "Analisi e sviluppo di metodi numerici avanzati per EDP". A.Q. acknowledges the financial support of Italian MIUR PRIN 2009 "Modelli numerici per il calcolo scientifico e applicazioni avanzate".

REFERENCES

Antonietti, P.F. and Houston, P. (2008). A pre-processing moving mesh method for discontinuous Galerkin approximations of advection-diffusion-reaction problems. *Int. J. Numer. Anal. Model.* 5(4), 704–728.

Antonietti, P.F. and Quarteroni, A. (2011). In preparation.

Arnold, D.N. (1982). An interior penalty finite element method with discontinuous elements. *SIAM J. Numer. Anal.* 19(4), 742–760.

Arnold, D.N., Brezzi, F., Cockburn, B. and Marini, L.D. (2002). Unified analysis of discontinuous Galerkin methods for elliptic problems. *SIAM J. Numer. Anal.* 39(5), 1749–1779 (electronic).

Ayuso de Dios, B. and Marin, L. D. (2009). Discontinuous Galerkin methods for advection-diffusion-reaction problems. *SIAM J. Numer. Anal.* 47(2), 1391–1420.

Baumann, C.E. and Oden, J.T. (1999). A discontinuous *hp* finite element method for convection-diffusion problems. *Comput. Methods Appl. Mech. Engrg.* 175(3–4), 311–341.

Brezzi, F., Marini, L.D. and Süli, E. (2004). Discontinuous Galerkin methods for first-order hyperbolic problems. *Math. Models Methods Appl. Sci.* 14(12), 1893–1903.

Brooks, A.N. and Hughes, T.J.R. (1982). Streamline upwind/Petrov-Galerkin formulations for convection dominated flows with particular emphasis on the incompressible Navier-Stokes equations. *Comput. Methods Appl. Mech. Engrg.* 32(1–3), 199–259. FENOMECH '81, Part I (Stuttgart, 1981).

Buffa, A., Hughes, T. and Sangalli, G. (2006). Analysis of a multiscale discontinuous galerkin method for convection diffusion problems. *SIAM J. Numer. Anal.* 44(4), 1420–1440.

Burman, E. and Smith, G. (2010). Analysis of the space semi-discretized supg-method for transient convection–diffusion equations. Mathematical Models and Methods in Applied Sciences, DOI 10.1142/S0218202511005659.

Christie, I., Griffiths, D.F., Mitchell, A.R. and Zienkiewicz, O.C. (1976). Finite element methods for second order differential equations with significant first derivatives. *IJNME* 10(6), 1389–1396.

Cockburn, B. (1999). Discontinuous Galerkin methods for convection-dominated problems. In *High-order methods for computational physics*, Volume 9 of *Lect. Notes Comput. Sci. Eng.*, pp. 69–224. Berlin: Springer.

Douglas, Jr.J. and Dupont, T. (1976). Interior penalty procedures for elliptic and parabolic Galerkin methods. In *Computing methods in applied sciences (Second Internat. Sympos., Versailles, 1975)*, pp. 207–216. Lecture Notes in Phys., Vol. 58. Berlin: Springer.

Ferziger, J.H. and Perić, M. (1999). *Computational methods for fluid dynamics* (revised ed.). Berlin: Springer-Verlag.

Franca, L., Harari, I. and Olivera, S. (2001). Streamline design of stability parameters for advection diffusion problems. *J. Comput. Phys.*, 171, 115–131.

Houston, P., Schwab, C. and Süli, E. (2002). Discontinuous *hp*-finite element methods for advection-diffusion-reaction problems. *SIAM J. Numer. Anal.* 39(6), 2133–2163 (electronic).

John, V. and Knobloch, P. (2007). On spurious oscillations at layers diminishing (SOLD) methods for convection-diffusion equations. I. A review. *Comput. Methods Appl. Mech. Engrg.* 196(17–20), 2197–2215.

John, V. and Knobloch, P. (2008). On spurious oscillations at layers diminishing (SOLD) methods for convection-diffusion equations. II. Analysis for P_1 and Q_1 finite elements. *Comput. Methods Appl. Mech. Engrg.* 197(21–24), 1997–2014.

Johnson, C., Nävert, U. and Pitkäranta, J. (1984). Finite element methods for linear hyperbolic problems. *Comput. Methods Appl. Mech. Engrg.* 45(1–3), 285–312.

LeVeque, R.J. (2002). *Finite volume methods for hyperbolic problems*. Cambridge Texts in Applied Mathematics. Cambridge: Cambridge University Press.

LeVeque, R.J. (2007). *Finite difference methods for ordinary and partial differential equations*. Philadelphia, PA: Society for Industrial and Applied Mathematics (SIAM). Steady-state and time-dependent problems.

Liesen, J. and Strakos, Z. (2005). Gmres convergence analysis for a convection-diffusion model problem. *SIAM J. on Scientific Computing* 26,1989–2009.

Melenk, J.M. and Schwab, C. (1999). An hp finite element method for convection-diffusion problems in one dimension. *IMA Journal of Numerical Analysis* 19(3), 425–453.

Quarteroni, A. and Valli, A. (1994). Numerical approximation of partial differential equations, Volume 23 of Springer Series in Computational Mathematics. Berlin: Springer-Verlag.

Reed, W. and Hill, T. (1973). Triangular mesh methods for the neutron transport equation. Technical Report Tech. Report LA-UR-73-479, Los Alamos Scientific Laboratory.

Roos, H.-G., Stynes, M. and Tobiska, L. (1996). Numerical methods for singularly perturbed differential equations: convection-diffusion and flow problems. Berlin: Springer Verlag.

Numerical Methods for Hyperbolic Equations – Vázquez-Cendón et al. (eds)
© 2013 Taylor & Francis Group, London, ISBN 978-0-415-62150-2

The Riemann problem in computational science

Eleuterio F. Toro
Laboratory of Applied Mathematics, University of Trento, Italy

ABSTRACT: A review of the Riemann problem and its impact in computational science is given, starting from basic definitions and simple examples. I then concentrate on approximate Riemann solvers for use in numerical methods, such as finite volume and discontinuous Galerkin finite element methods. A brief overview of the most well known approximate methods is given. There follows a review of some specific approximate Riemann solvers in more detail; I have chosen examples that are connected to theauthor's own research and have included some recent developments. The paper ends with a discussion on generalisations of the Riemann problem, to account for source terms and piece-wise smooth initial data, and the construction of very-high order ADER methods based on the generalised Riemann problem.

1 INTRODUCTION

The Riemann problem for a system of hyperbolic conservation laws is a special initial-value problem in which the initial condition consists of just two constant states separated by a discontinuity at some position x_0. This mathematical problem, firststudied by Riemann in 1860 (Riemann 1860), is closely associated with the shock-tube problem, a well-known physical problem in classical gas dynamics. In the shock tube problem the initial velocities are identically zero. Pioneering work on hyperbolic equations is due to Lax and others, see for example (Lax 1954), (Lax 1972) and (Lax 1990). A detailed discussion on the shock-tube problem is found the classical book of Courant and Friedrichs (Courant and Friedrichs 1985). For shallow water flow with a free surface, the analogue of the Riemann problem is the dam-break problem. A more general problem is the Cauchy problem, in which the initial data is a piece-wise smooth function. The Cauchy problem is hard to solve, not only due to the non-linear character of the equations but also to the generality of the initial condition. For the Riemann problem the situation is simpler due to the special initial condition and often it is possible to solve the problem *exactly*. Since the seminal work of Godunov (Godunov 1959), there has been an explosion of the research activity centred around the Riemann problem in a large variety of disciplines associated with hyperbolic partial differential equations, or extensions of these. The Riemann problem can also be posed, and in some simple cases solved, for other classes of evolutionary partial differential equations, such as the diffusion or heat equation (Toro and Hidalgo 2009).

The exact solution of the Riemann problem is useful in a number of ways. First, it represents the solution to a system of hyperbolic equations subject to the simplest, non-trivial, initial conditions. In spite of the simplicity of the initial data the solution of the Riemann problem contains the essential mathematical and physical features of the relevant system. The exact Riemann problem solution is also an invaluable reference solution that is useful in assessing the performance of numerical methods and to check the correctness of programmes in the early stages of development. Generally, there is no exact, closed-form solution to the Riemann problem for a non-linear hyperbolic system. In gas dynamics, for example, the Riemann problem for the one-dimensional Euler equations cannot be solved in closed-form, not even for ideal gases. In fact, not even for much simpler models, such as the isentropic and isothermal equations. By exact solution one means a solution obtained by a method involving a numerical iterative scheme, at some stage. Godunov is credited with the first exact Riemann solver for the Euler equations (Godunov 1959). By today's standards, Godunov's first Riemann solver is regarded as cumbersome and computationally inefficient. Later, Godunov (Godunov (Ed.) 1976) proposed a second exact Riemann solver. Distinct features of this solver are: the equations used are simpler, the variables selected are more convenient from the computational point of view and the iterative procedure is rather sophisticated. Much of the work that followed contains the fundamental features of Godunov's second Riemann solver. Chorin independently (Chorin 1976), produced improvements to Godunov's first Riemann solver. In 1979, van Leer (van Leer 1979) produced another improvement to Godunov's first Riemann solver resulting in a scheme that is similar to Godunov's second solver. Smoller (Smoller 1994) proposed a rather different approach; later, Dutt (Dutt 1986) produced a practical implementation of the scheme. Gottlieb and Groth (Gottlieb and Groth 1988) presented another Riemann solver for ideal gases. Of the schemes they tested, theirs is shown to be the most efficient. Toro (Toro 1989) presented an exact Riemann solver for ideal and covolume gases of comparable efficiency to that of Gottlieb and Groth. Schleicher (Schleicher 1993) and Pike (Pike 1993) have also presented fast exact Riemann solvers. For gases obeying a general equation of state the reader is referred to the pioneering work of Colella and Glaz (Colella and Glaz 1985). The paper by Menikoff and Plohr (Menikoff and Plohr 1989) is highly recommended. Further information on the Riemann problem is found in Toro (Toro 2009) and the many references therein.

The solution of the Riemann problem has found many uses. First, even if special, it is the solution of a difficult problem and contains the basic mathematical features of the equations. The information given by this solution may be used as a starting point for the study of more general problems. From the physical point of view, the solution of the Riemann problem may also be of practical use. The shock tube problem and the dam-break problem are two prominent examples. In attempting to solve the more general Cauchy problem one can also use the Riemann problem *locally*, in this manner reducing the general problem to a sequence of solutions of local Riemann problems. A very prominent example is the work of Glimm (Glimm 1965), in which he usedlocal Riemann problems to develop a powerful existence and uniqueness theory for non-linear systems of hyperbolic equations.

The use of the Riemann problem in the designing of numerical methods for hyperbolic equations is prominent. In 1959 Godunov (Godunov 1959) proposed a numerical method in which the Riemann problem was used locally, giving rise to a class of finite volume methods, Godunov-type methods. These methods have become very prominent in the last five decades and are also known as upwind methods. Later, the Riemann-problem based technique developed by Glimm for theoretical studies was converted into a numerical method to compute approximate solutions (Chorin 1976). This method is also known as the Random Choice Method, or RCM. Two special methods, akin to RCM, that use the Riemann problem as a building block are the Front Tracking Method (Glimm et al. 1998), (Glimm et al. 1998), (Risebro 2002) and the Shock Fitting Method (Salas 2009). More recent methods, such as Finite Element Methods, have also been influenced by the Riemann problem. Discontinuous Galerkin finite element methods require a numerical flux, for which the solution of the Riemann problem is used locally, as in Godunov-type methods. The influence of the Riemann problem has also reached other numerical methodologies, the Smooth Particle Hydrodynamics method, or SHP, is a prominent example (Cha and Whitworth 2003). But even if one does not use the Riemann problem for constructing numerical methods, the application of boundary conditions will require, in one way or another, the Riemann problem. This is most obvious in the case of impermeable boundaries, fixed or moving. For example, assume a fixed impermeable boundary at $x = 0$ for the shallow water equations, such as a very high vertical concrete wall. One obvious condition to impose is zero (normal) velocity at the boundary, which naturally gives zero mass flux. However for the momentum flux such zero velocity condition is not sufficient to determine the flux, as one requires some pressure condition too. Posing a reflecting Riemann problem at the boundary resolves the problem in the most elegant way. Its solution gives exactly the zero velocity at the boundary and, as a bonus, the required pressure, consistent with the zero velocity condition. Application of these ideas to the Euler equations are found in Chapter 6 of (Toro 2009).

Once the Riemann problem enters the computational tools for solving hyperbolic equations, it becomes prominent in the study of many problems in science and engineering. Shock wave physics is a huge area with applications in aerodynamics and space technology. Hyperbolic equations supplemented by suitable source terms are used in the study of combustion problems, detonation waves and propulsion technology. More complex physics requires extensions of the hyperbolic models to include dissipative and dispersive effects. Still, the role of the Riemann problem remains crucial. In addition to the traditional applications inspired by compressible gas dynamics, the use of the Riemann problem has entered the fields of geophysics and environmental disciplines. Shallow water hydrodynamics is as prominent example. See (Toro 2001), (Toro and García-Navarro 2007) and references therein. Particular applications include seismic wave propagation and tsunami wave propagation. The Riemann problem and associated numerical techniques have also seen incursions in medical applications, see (Takayama and Saito 2004), (Toro and Siviglia 2011), for example.

Astrophysics is a huge and growing area in which the Riemann problem figures prominently. Common systems of equations used are the relativistic Euler equations and the equations of ideal and relativistic magnetohydrodynamics, or MHD equations. The exactsolution of the Riemann problem for relativistic gas dynamics was first reported by Martí and Müller (Martí and Müller 1994). See also the papers by Schneider et al. (Schneider et al. 1993); Falle and Komissarov (Falle and Komissarov 1996), (Falle and Komissarov 1997); Gurski (Gurski 2004); Mignone and Bodo (Mignone and Bodo 2006); Klingenberg et al. (Klingenberg et al. 2007); Bodo et al. (Bodo et al. 2008); Dumbser et al. (Dumbser et al. 2008). An important work is the paper by Giacomazzo and Rezzolla (Giacomazzo and Rezzolla 2006); they found the exact solution of the Riemann problem for relativistic magneto-hydrodynamics, which is very valuable for assessing numerical schemes. A very recent and important work to mention here, in which the Riemann problem is prominent, is the paper by Antón et al. (Antón et al. 2010). In this scientific context it is mandatory to mention the advances of numerical relativity, a growing area of application of numerical methods in which the Riemann problem plays an important role. See Lehner (Lehner 2001), for example.

In this paper I shall concentrate the attention on the Riemann problem as applied to finite volume numerical methods for solving the general initial-boundary value problem for hyperbolic equations. Efficiency is an important issue here and thus the need for approximate Riemann solvers. Then I shall also include a discussion on generalisations of the Riemann problem to include source terms and initial conditions other than two constant states. Mention is also made of a corresponding Godunov-type method that emerges naturally from this generalisation of the Riemann problem.

The remaining part of this paper includes definitions and examples in Section 2. Then, in sections 3 to 6 I review three particular Riemann solvers. In Section 7 we study the Generalised, as opposed to classical, Riemann problem and give details on its use to construct Godunov-type schemes of arbitrary order of accuracy in space and time. Conclusions are drawn in Section 8.

2 DEFINITIONS AND EXAMPLES

First we define the general initial-value problem, or Cauchy problem, for a system of m conservation laws, namely

$$\left.\begin{aligned} &\text{PDEs:} \quad \partial_t \mathbf{Q}(x,t) + \partial_x \mathbf{F}(\mathbf{Q}(x,t)) = 0\,, \\ &\text{ICs:} \quad \mathbf{Q}(x,0) = \mathbf{Q}_0(x)\,, \end{aligned}\right\} \tag{1}$$

for $-\infty < x < \infty$ and $t \geq 0$. Here we assume that the conservation laws are hyperbolic, that is the Jacobian matrix

$$\mathbf{A}(\mathbf{Q}) = \partial \mathbf{F}(\mathbf{Q})/\partial \mathbf{Q} \tag{2}$$

has m real eigenvalues $\lambda_i(\mathbf{Q})$ with associated m linearly independent right eigenvectors $R_i(\mathbf{Q})$.

The Riemann problem is defined as the special Cauchy problem

$$\left.\begin{aligned} &\text{PDEs:} \quad \partial_t \mathbf{Q}(x,t) + \partial_x \mathbf{F}(\mathbf{Q}(x,t)) = 0 \\ &\text{ICs:} \quad \mathbf{Q}(x,0) = \begin{cases} \mathbf{Q}_L & \text{if} \quad x < x_0\,, \\ \mathbf{Q}_R & \text{if} \quad x > x_0\,, \end{cases} \end{aligned}\right\} \tag{3}$$

in which the two data states $\mathbf{Q}_L$ and $\mathbf{Q}_R$ are both constant. This is also called the *classical* Riemann problem, to distinguish it from the Generalised Riemann problem, to be considered later.

In general, the Cauchy problem is difficult to solve exactly, which is partly due to the generality of the initial condition $Q(x,0) = \mathbf{Q}_0(x)$. For the simpler case in which the initial condition is piece-wise constant, as in the Riemann problem, part of the difficulty disappears. What remains however is the non-linear character of the system. For some well-known systems it is possible to find the solution of the Riemann problem. Invariably, however, the solution procedure involves, at some stage, a numerical iteration procedure. The general Cauchy problem along with boundary conditions can then be solved approximately by using solutions to the classical Riemann problem locally, in a Godunov-type numerical approach. In what follows we consider some examples.

Example 1: The linear advection equation. The simplest example is furnished by the Riemann problem for the linear advection equation with constant characteristic speed λ, namely

$$\left.\begin{aligned} &\text{PDE:} \quad \partial_t q + \lambda \partial_x q = 0\,, \\ &\text{IC:} \quad q(x,0) = h(x) = \begin{cases} q_L & \text{if} \quad x < 0\,, \\ q_R & \text{if} \quad x > 0\,. \end{cases} \end{aligned}\right\} \tag{4}$$

The exact solution, in terms of the characteristic line $x = \lambda t$ emanating from the origin, is

$$q(x,t) = \begin{cases} q_L & \text{if} \quad \dfrac{x}{t} < \lambda\,, \\ q_R & \text{if} \quad \dfrac{x}{t} > \lambda\,. \end{cases} \tag{5}$$

Figure 1 shows the solution of the Riemann problem for the linear advection equation; frame (a) illustrates the piece-wise constant initial condition with discontinuity at $x = 0$. Frame (b) illustrates the complete solution in the x–t plane, with two constant regions, region R_0 to the left of the line $x/t = \lambda$ and R_1 to the right

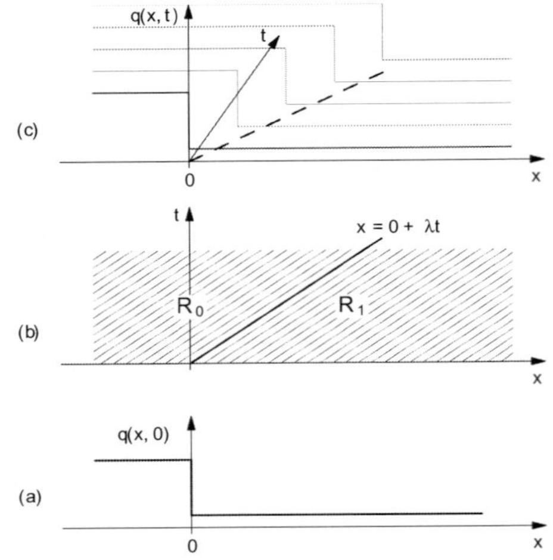

Figure 1. The Riemann problem for the linear advection equation for $\lambda > 0$. Frame (a): initial comdition; frame (b): solution in x–t plane, frame (c): solution as function of x and t.

of the line $x/t = \lambda$. Frame (c) shows the solution $q(x,t)$ as a function of x and t.

More generally, for the linear PDE in (4) the solution of the Cauchy problem with initial condition $q(x,0) = h(x)$ is

$$q(x,t) = h(x - \lambda t)\,, \tag{6}$$

from which the special case (5) follows. A distinguishing feature of the solution of the Cauchy problem for the linear advection equation with constant coefficient is that there are no new values generated as the solution evolves; the initial profile is simply translated in space as time evolves.

Example 2: isothermal gas dynamics. The (augmented) governing equations are

$$\partial_t \mathbf{Q} + \partial_x \mathbf{F}(\mathbf{Q}) = 0\,, \tag{7}$$

with

$$\left.\begin{array}{l} \mathbf{Q} = \begin{bmatrix} q_1 \\ q_2 \\ q_3 \end{bmatrix} = \begin{bmatrix} \rho \\ \rho u \\ \rho\psi \end{bmatrix}, \\ \mathbf{F}(\mathbf{Q}) = \begin{bmatrix} f_1 \\ f_2 \\ f_3 \end{bmatrix} = \begin{bmatrix} \rho u \\ \rho u^2 + a^2\rho \\ \rho\psi u \end{bmatrix}. \end{array}\right\} \tag{8}$$

Here ρ is density, u is velocity and a is the (constant) sound speed. The isothermal equations result from taking the very simple equation of state

$$p = p(\rho) = a^2\rho\,. \tag{9}$$

Here, the standard isothermal equations have been augmented by an extra transport equation for a passive scalar $\psi(x,t)$, to represent the concentration of some chemical species, for example.

The Jacobian matrix with entries $a_{ij} = \partial f_i / \partial q_j$ is computed to be

$$\mathbf{A}(\mathbf{Q}) = \begin{bmatrix} 0 & 1 & 0 \\ a^2 - u^2 & 2u & 0 \\ -u\psi & \psi & u \end{bmatrix}. \tag{10}$$

The eigenvalues are the roots of the *characteristic polynomial*

$$P(\lambda) = Det(\mathbf{A} - \lambda\mathbf{I}) = 0, \tag{11}$$

where $\mathbf{I}$ is the identity matrix and λ is a parameter. It is easily verified that

$$P(\lambda) = (u - \lambda)[\lambda(2u - \lambda) + a^2 - u^2] = 0, \tag{12}$$

a cubic equation, for which three solutions exist, namely

$$\lambda_1 = u - a\,, \quad \lambda_2 = u, \quad \lambda_3 = u + a\,. \tag{13}$$

Note that all three roots are real; they are also distinct if $a \neq 0$.

The right eigenvectors corresponding the the eigenvalues, suitably normalized, are

$$\mathbf{R}_1 = \begin{bmatrix} 1 \\ u - a \\ \psi \end{bmatrix}, \quad R_2 = \begin{bmatrix} 0 \\ 0 \\ 1 \end{bmatrix}, \quad R_3 = \begin{bmatrix} 1 \\ u + a \\ \psi \end{bmatrix}. \tag{14}$$

There are three characteristic fields. It can easily be shown that the $\lambda_1(\mathbf{Q})$ and $\lambda_3(\mathbf{Q})$ characteristic fields are genuinely non-linear and the the $\lambda_2(\mathbf{Q})$ characteristic field is linearly degenerate. The Riemann problem (3) for the isothermal equations (7)–(8) has structure as depicted in Figure 2. There are three wave families corresponding to the three characteristic fields. The non-linear wave families can either be shocks or rarefactions and one does not know in advance the particular wave type that might emerge from the solution. However, the linear (middle) wave is always a contact discontinuity. The solution includes four constant regions, as depicted in Figure 2, separated by three waves. In Figure 3 we show each type of wave separately. If all three waves are discontinuities then the four regions depicted in Figure 2 form the complete the solution. This situation is analogous to the linear

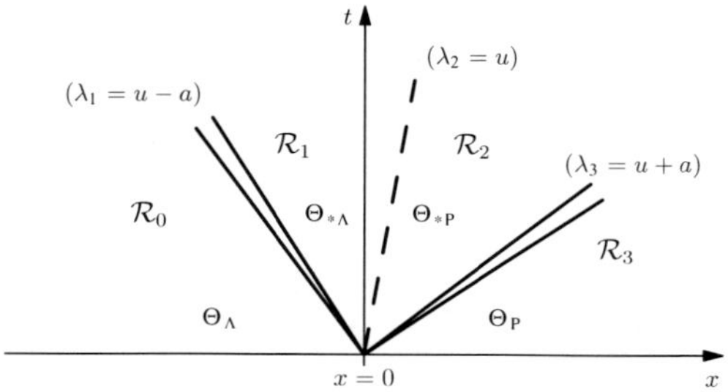

Figure 2. Structure of solution of the Riemann problem for the augmented isothermal equations.

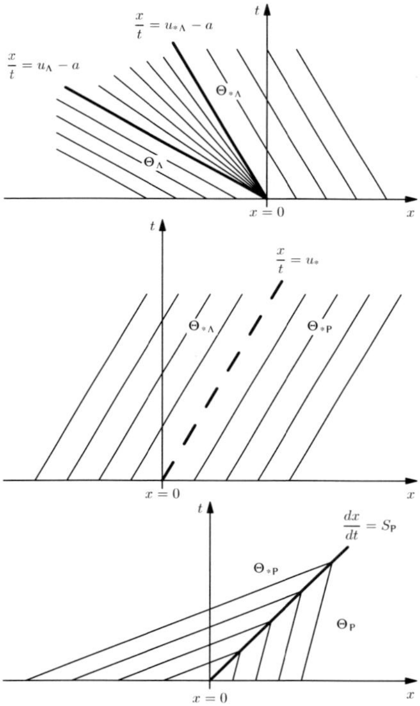

Figure 3. Elementary wave solutions of the Riemann problem for the augmented isothermal equations. Bottom frame: shock, middle frame: contact, top frame: rarefaction.

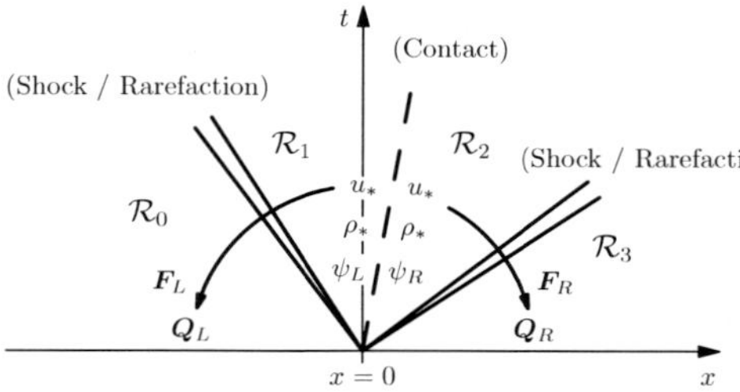

Figure 4. General wave pattern for the solution of the Riemann problem for the augmented isothermal equations. Star region connected to left and right data states via functions F_L and F_R respectively.

system case. If rarefactions waves are present then there will be additional wedge-like regions to account for rarefaction fans.

A solution strategy for the Riemann problem for (7) identifies two subproblems. The first, which we call the *Star Problem*, consists in determining the solution in the constant regions $\Re_1$ and $\Re_2$ in Figure 2, called the *Star Region*. Figure 4 illustrates the procedure to solve this problem. One exploits the fact that both density ρ and velocity u are constant throughout the *Star Region*, whose unknown state is connected to the left data via a function F_L and to the right data state via a function F_R. These functions discriminate between shocks and rarefactions. For shocks we apply Rankine-Hugoniot conditions and for rarefactions one uses generalised Riemann invariants. Then equality of velocity in the unknown region leads to a single non-linear algebraic equation for density in the entire Star Region. This equation is solved numerically by an iterative method to find ρ_*. Then u_* follows directly from exact relations involving F_L, or F_R or both. This completes the solution of the *Star Problem*. To complete the solution procedure in the entire x–t half-plane one must perform, for a given pair (x,t), a sampling procedure involving the determination of the wave types presentin the solution and finding their associated speeds. If rarefactions are present, see top frame of Figure 3, then one must determine the exact solution inside these wedge-like regions using generalised Riemann invariants.

This solution strategy applies almost without modification to some other hyperbolic systems. For the Euler equations for ideal gases see (Toro 2009) and for the shallow water equations see (Toro et al. 2001). For gases with general equationof state the procedure is more involved and requires the simultaneous solution of two non-linear equations. See Quartapelle et al. (Quartapelle et al. 2003), for example. The exact solution is also possible for even more complicated systems. For the Baer-Nunziato equations of two-phase compressible flow a (direct) Riemann solver is due to Schwendeman et al. (Schwendeman et al. 2006). For the Riemann problem in relativistic MHD see Giacomazzo and Rezzolla (Giacomazzo and Rezzolla 2006) and Antón et al. (Antón et al. 2010), for example.

In this paper I do not discuss closure conditions, such as equations of state for compressible gas dynamics, or tube laws for blood flow in highly

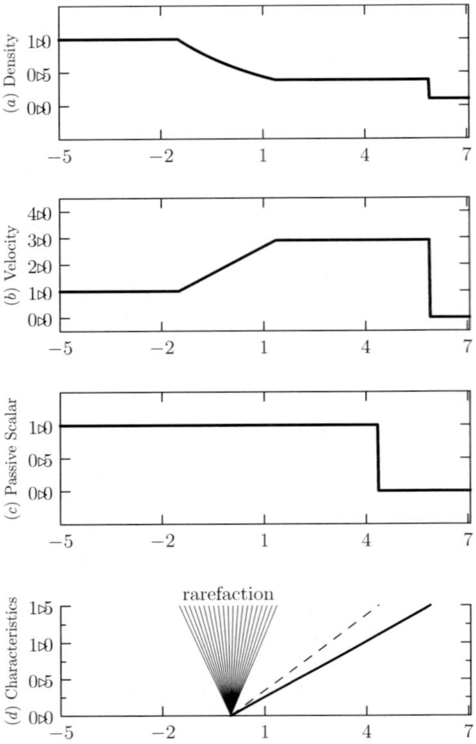

Figure 5. Complete solution of a Riemann problem for the augmented isothermal equations at the output time T_{out} =1.5. Solution profiles for (a) density ρ, (b) velocity u and (c) passive scalar ψ. Bottom frame (d) summarises the solution pattern as a function of space and time, in the x–t plane. There is right-propagating shock, a right-propagating contact discontinuity and a transonic left rarefaction wave.

compliant vessels. Closure laws have a profound influence on the behaviour of the equations, determining notjust their type but the fundamental question of well posedness, that is existence, uniqueness and continuous dependence on data. Another very important omission is hyperbolic systems that cannot be written in conservation-law form; on this I particularlyrecommend the seminal papers of LeFloch (LeFloch 1988) and Dalmaso et al. (Dalmaso et al. 1995).

In Figure 5 we show results for the following test at time t=1.5. We solve the augmented isothermal equations in the spatial domain $-5 \leq x \leq 5$, with initial conditions: for $x \leq 0.5$ $\rho(x,0) = 1$, $u(x,0) = 1$, $\psi(x,0) = 1$. For $x > 0.5$ $O(x,0) = 0.1$, $u(x,0) = 0$ and $\psi(x,0) = 0$. The sound speed is taken as $a = 2$.

In the next section I review some of the main approaches to obtain approximate solutions to the Riemann problem for numerical purposes.

3 APPROXIMATE RIEMANN SOLVERS

One of the main uses of the solution of the Riemann problem is in the approximation of solutions to initial-boundary values problems for hyperbolic conservation laws in one and multiple space dimensions. Finite volume and discontinuous Galerkin finite element methods are the main approaches nowadays. Consider a one-dimensional system of hyperbolic balance laws

$$\partial_t \mathbf{Q} + \partial_x \mathbf{F}(\mathbf{Q}) = \mathbf{S}(\mathbf{Q}), \qquad (15)$$

where $\mathbf{S}(\mathbf{Q})$ is the source term vector, containing no differential terms of the unknown vector $\mathbf{Q}$. A general framework for finite volume methods can be constructed from integrating the equations in space and time in a control volume $[x_{i-\frac{1}{2}}, x_{i+\frac{1}{2}}] \times [t^n, t^{n+1}]$. The resulting formula is

$$\mathbf{Q}_i^{n+1} = \mathbf{Q}_i^n - \frac{\Delta t}{\Delta x}\left(\mathbf{F}_{i+\frac{1}{2}} - \mathbf{F}_{i-\frac{1}{2}}\right) + \Delta t \mathbf{S}_i \qquad (16)$$

where

$$\left.\begin{aligned} \mathbf{Q}_i^n &= \frac{1}{\Delta x}\int_{x_{i-\frac{1}{2}}}^{x_{i+\frac{1}{2}}} \mathbf{Q}(x,t^n)dx, \\ \mathbf{F}_{i+\frac{1}{2}} &= \frac{1}{\Delta t}\int_{t^n}^{t^{n+1}} \mathbf{F}(\mathbf{Q}(x_{i+\frac{1}{2}},t))dt, \\ \mathbf{S}_i &= \frac{1}{\Delta t \Delta x}\int_{t^n}^{t^{n+1}}\int_{x_{i-\frac{1}{2}}}^{x_{i+\frac{1}{2}}} \mathbf{S}(\mathbf{Q}(x,t))dxdt. \end{aligned}\right\} \qquad (17)$$

To construct a specific finite volume method one needs to specify the *numerical flux* $\mathbf{F}_{i+\frac{1}{2}}$ and the *numerical source* S_i in (16), as approximations to the appropriate integrals in (17). In the rest ofthis paper, unless otherwise stated, I only consider the homogenous case, $\mathrm{S}(\mathbf{Q}) = 0$.

The original Godunov method finds a solution $\mathbf{Q}_{i+\frac{1}{2}}(x/t)$, approximate or exact, to a local Riemann problem. The similarity solution $\mathbf{Q}_{i+\frac{1}{2}}(x/t)$ is constant along the t- axis and thus the numerical flux becomes

$$\mathbf{F}_{i+\frac{1}{2}} = \mathbf{F}(\mathbf{Q}_{i+\frac{1}{2}}(0)). \qquad (18)$$

Godunov (Godunov 1959) utilised the exact solution to find the numerical flux, but later proposed a simple solver obtained from a linearisation of the equations written in characteristic form, see Chapter 9 of (Toro 2009). Unfortunately Godunov's simple linearization is not robust enough and is not entropy-satisfying.

Twenty years later, Roe (Roe 1981) proposed a linearised Riemann solver to find an approximate flux directly. For details see the original paper or Chapter 11 of (Toro 2009). Roe's solver is more robust that the linearised solver of Godunov but is not entropy satisfying. A variety of approaches exist to correct this defect. Roe's solver has been applied to solve many practical problems associated with hyperbolic systems and is probably the most popular of all Riemann solvers to date. There are however systems for which the so called Roe averages have not been derived, due to the algebraic complexity involved. An extension of the Roe approach that makes it possible to deal with any hyperbolic system is included in the paper by Dumbserand and Toro (Dumbserand et al. 2010). An entropy fix is still needed. This is an aspect currently being studied by the author and collaborators.

In 1982, Osher and Solomon published their approximate Riemann solver (Osher and Solomon 1982). This is an unusual solver. Its strong points: non-linear, robust and entropy satisfying. However, its derivation is exceedingly complex. It requires analytical integration in phase space along pre-selected integration paths. Moreover, in order to define tractable paths one requires the path-intersection points in phase space. This in turn makes use of the two-rarefaction Riemann solver. It is interesting to remark that this two-rarefaction solver is a very accurate Riemann solver and can be used by itself without the need for integration points, integration paths and integration procedures. Complete details on the Osher-Solomon solver are found in the original paper and in Chapter 12 of (Toro 2009). A generalisation of the Osher-Solomon solver is possible by adopting a numerical approach to the phase-space integration (Dumbser et al. 2010). This extension will be reviewed in Section 6.

Up to this point, all Riemann solvers are *complete*, that is the structure of the approximate solution contains all the characteristic fields present in the exact Riemann problem. In other words the *wave model* of the approximate solver has the same number of characteristic fields as the exact solver.

In 1961 Rusanov (Rusanov 1961) published an approximate Riemann solver whose wave model contains just one wave, regardless the number of waves present in the exact solver. This solver is indeed very simple and very robust. It is widely used today and is sometimes termed the *Local Lax-Friedichs* flux, but note that it is distinct from the original Lax-Friedrichs flux.

Twenty two years later Harten, Lax ad van Leer (Harten 1983) proposed an approach, the HLL approach, to construct approximate Riemann solvers to generate a numerical flux. The resulting fluxes may be seen as a generalisation of the Rusanov one-wave model. The HLL approach offers a two-wave model. The flux results from assuming prescribed lower S_L and upper S_R bounds for the wave speeds emerging from the Riemann problem solution. Then, integration of the conservation laws in appropriate volumes in x^{-t} space gives a very simple expression for the flux. The HLL approach assumes that S_L and S_R are known. Harten and collaborators did not propose practical algorithms to estimate S_L and S_R. This was later done independently by Davis (Davis 1988) and Einfeldt (Einfeldt 1988). A complete description of the HLL approach is found in Chapter 10 of (Toro 2009). A shortcoming of HLL is its two-wave model. It results in excessive dissipationof intermediate characteristic fields not included in the two-wave model. In 1992, a correction of this defect was proposed by Toro and collaborators (Toro et al. 1992), (Toro et al. 1994), (Toro and Chakraborty 1994), whereby a three-wave model is adopted. In section 4 we review an updated version of this Riemann solver.

A radically different approach to deriving numerical fluxes results from simply ignoring the explicit use of wave propagation information and effectively adopting a zero-wave model. An example of this approach is the FORCE flux proposed by Toro (Toro 1996), (Toro and Billett 1996), (Toro and Billett 2000). A review of this is found in Section 5.

4 THE HLLC RIEMANN SOLVER

The HLLC Riemann solver is a modified version of the HLL solver of Harten, Lax and van Leer to account for the presence of intermediate waves, such as the contact discontinuity. The original HLLC Riemann solver was first put forward by Toro et al. in 1992 in the Cranfield Technical Report (Toro et al. 1992), later published in 1994 in (Toro et al. 1994), see also Toro and Chakraborty (Toro and Chakraborty 1994). Subsequent developments and applications are reported in (Batten et al. 1997), (Batten et al. 1997) amongst others.

Here we review the current version of HLLC as applied to the three-dimensional Euler equations along with m species equations

$$\partial_t \mathbf{Q} + \partial_x \mathbf{F}(\mathbf{Q}) + \partial_y \mathbf{G}(\mathbf{Q}) + \partial_z \mathbf{H}(\mathbf{Q}) = 0 \,, \qquad (19)$$

with equation of state

$$p = p(\rho, e) \,. \qquad (20)$$

The vector $\mathbf{Q}$ of conserved variables and the flux in the x-direction are

$$\mathbf{Q} = \begin{bmatrix} \rho \\ \rho u \\ \rho v \\ \rho w \\ E \\ \rho q_1 \\ \dots \\ \rho q_l \\ \dots \\ \rho q_m \end{bmatrix}, \mathbf{F}(\mathbf{Q}) = \begin{bmatrix} \rho u \\ \rho u^2 + p \\ \rho uv \\ \rho uw \\ u(E+p) \\ \rho u q_1 \\ \dots \\ \rho u q_l \\ \dots \\ \rho u q_m \end{bmatrix}. \quad (21)$$

For the purpose of determining a numerical flux for the three-dimensional Euler equations that is normal to a cell interface, by virtue of the rotational invariance of the equations, it is sufficient to consider the augmented one-dimensional Euler equations aligned in that normal direction. Here, without loss of generality we assume the normal direction to be the x-direction. First we assume wave speed estimates S_L, S_* and S_R for the three waves depicted in Figure 6 are available. See (Toro 2009) for details. Then, by integrating in appropriate control volumes around the waves of speeds S_L and S_R we obtain averaged Rankine-Hugoniot conditions

$$\left.\begin{aligned} \mathbf{F}_{*L} &= \mathbf{F}_L + S_L(\mathbf{Q}_{*L} - \mathbf{Q}_L), \\ \mathbf{F}_{*R} &= \mathbf{F}_R + S_R(\mathbf{Q}_{*R} - \mathbf{Q}_R). \end{aligned}\right\} \quad (22)$$

This gives rise to a large algebraic system with more unknowns than equations. One way to resolve this difficulty is by introducing a number of assumptions, all consistent with the exact solution, leading to the following expressions for the intermediate state vectors

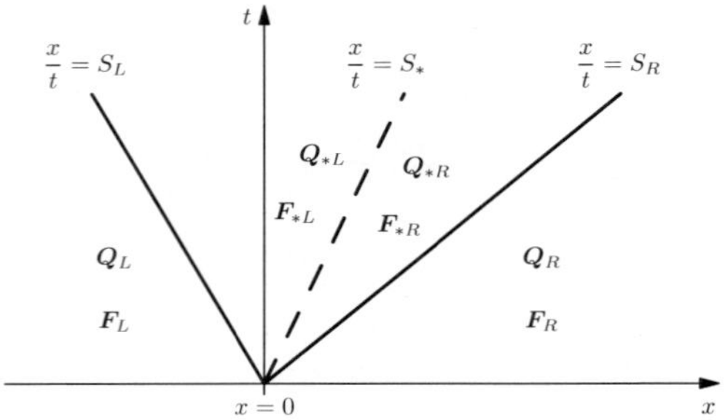

Figure 6. Wave model for the HLLC flux for the 3D Euler equations in normal direction.

$$\mathbf{Q}_{*K} = \Omega \begin{bmatrix} 1 \\ S_* \\ v_K \\ w_K \\ \frac{E_K}{\rho_K} + (S_* - u_K)\left[S_* + \frac{p_K}{\rho_K(S_K - u_K)}\right] \\ (q_1)_K \\ \dots \\ (q_l)_K \\ \dots \\ (q_m)_K \end{bmatrix}, \quad (23)$$

with

$$\Omega = \rho_K \left(\frac{S_K - u_K}{S_K - S_*} \right), \quad (24)$$

for $K = L$ and $K = R$. Then the intermediate fluxes F_{*L} and F_{*R} are completely determined and the numerical flux is given as

$$\mathbf{F}^{hllc}_{i+\frac{1}{2}} = \begin{cases} \mathbf{F}_L & \text{if } 0 \le S_L, \\ \mathbf{F}_{*L} & \text{if } S_L \le 0 \le S_*, \\ \mathbf{F}_{*R} & \text{if } S_* \le 0 \le S_R \\ \mathbf{F}_R & \text{if } 0 \ge S_R. \end{cases} \quad (25)$$

Test 0: isolated stationary contact discontinuity. We solve the one-dimensional Euler equations in the domain $0 \le x \le 1$ with initial conditions for density $\rho(x,0) = 1.4$ for $x \le 0.5$ and $\rho(x,0) = 1.4$ for $x > 0.5$. Velocity $u(x,0) = 0$ and pressure $p(x,0) = 1$ $\forall x$.

Figure 7 shows numerical results from the HLL and HLLC solvers, both in first-order mode, for Test 0. Results are obtained with M = 100 cells, CFL number $CFL = 0.9$ at output time $t = 5.0$.

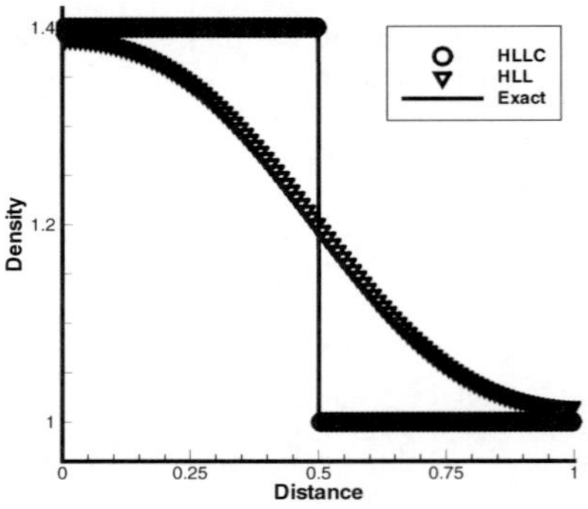

Figure 7. Test 0: isolated stationary contact in the 1D Euler equations. HLLC (circles) and HLL (triangles) numerical solutions compared to exact solution (line) at time $t = 5$.

There are many applications and extensions of the HLLC solver in the literature. Examples include the work of Mignone and collaborators (Mignone and Bodo 2005), (Mignone and Bodo 2006), (Mignone et al. 2009).

5 THE FORCE FLUX

The aim is to construct a numerical flux for use in (16) without making explicit use of wave propagation information, as contained in the Riemann problem solution, for example.

5.1 *Review of FORCE in 1D*

The inspiration for the construction of the FORCE flux is found in Glimm's method (or Random Choice Method) (Glimm 1965) on a staggered grid as illustrated in Figure 8. The FORCE scheme was first proposed by Toro in (Toro 1996) and by Toro and Billett in (Toro and Billett 1996), and (Toro and Billett 2000).

In Glimm's method, the solution $\mathbf{Q}_i^n$ in i at time level n is advanced to $\mathbf{Q}_i^{n+1}$ at time level n+1 in two steps. In the first step one solves two Riemann problems yielding solutions $\hat{\mathbf{Q}}_{i-\frac{1}{2}}^{n+\frac{1}{2}}(x,t)$ and $\hat{\mathbf{Q}}_{i+\frac{1}{2}}^{n+\frac{1}{2}}(x,t)$. This is followed by a random sampling procedure of each of these solutions at the time $t = t_n + \frac{1}{2}\Delta t$, producing randomly sampled values $\mathbf{Q}_{i-1/2}^{n+1/2}$ and $\mathbf{Q}_{i-1/2}^{n+1/2}$ at each interface. In the second step one solves a Riemann problem using these states as initial data. Then the emerging solution $\hat{\mathbf{Q}}_i^{n+1}(x,t)$ is sampled at the complete time yielding the sought solution $\mathbf{Q}_i^{n+1}$.

The FORCE scheme replaces sampled values in RCM by integral averages as follows

$$\mathbf{Q}_{i-\frac{1}{2}}^{n+\frac{1}{2}} = \frac{1}{\Delta x}\int_{-\frac{1}{2}\Delta x}^{\frac{1}{2}\Delta x} \hat{\mathbf{Q}}_{i-\frac{1}{2}}^{n+\frac{1}{2}}\left(x, \frac{\Delta t}{2}\right)dx \tag{26}$$

and

$$\mathbf{Q}_{i+\frac{1}{2}}^{n+\frac{1}{2}} = \frac{1}{\Delta x}\int_{-\frac{1}{2}\Delta x}^{\frac{1}{2}\Delta x} \hat{\mathbf{Q}}_{i+\frac{1}{2}}\left(x, \frac{\Delta t}{2}\right)dx. \tag{27}$$

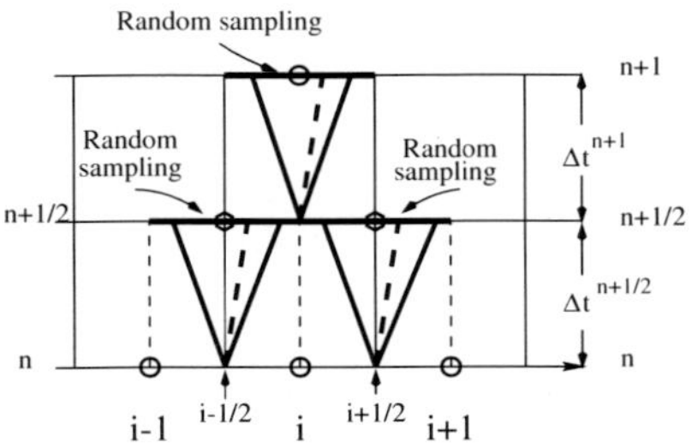

Figure 8. Glimm's method on a staggered mesh. The solution is updated in two stages of half-time step each.

As a matter of fact one can do away with the computation of the Riemann problem solutions under each of the above integrals by resorting to the integral form of the conservation laws in (26) and (27). This operation yields

$$\mathbf{Q}_{i-\frac{1}{2}}^{n+\frac{1}{2}} = \frac{1}{2}(\mathbf{Q}_{i-1}^n + \mathbf{Q}_i^n) + \frac{\Delta t}{2\Delta x}(\mathbf{F}_{i-1}^n - \mathbf{F}_i^n) \tag{28}$$

and

$$\mathbf{Q}_{i+\frac{1}{2}}^{n+\frac{1}{2}} = \frac{1}{2}(\mathbf{Q}_i^n + \mathbf{Q}_{i+1}^n) + \frac{\Delta t}{2\Delta x}(\mathbf{F}_i^n - \mathbf{F}_{i+1}^n). \tag{29}$$

At the complete time level n+1 after a timestep Δt the solution is

$$\mathbf{Q}_i^{n+1} = \frac{1}{\Delta x}\int_{-\frac{1}{2}\Delta x}^{\frac{1}{2}\Delta x} \hat{\mathbf{Q}}_i(x, \frac{1}{2}\Delta t)dx. \tag{30}$$

But again the integral form of the conservation laws gives

$$\mathbf{Q}_i^{n+1} = \frac{1}{2}\left[\mathbf{Q}_{i-\frac{1}{2}}^{n+\frac{1}{2}} + \mathbf{Q}_{i+\frac{1}{2}}^{n+\frac{1}{2}}\right] + \frac{\Delta t}{2\Delta x}\left[\mathbf{F}_{i-\frac{1}{2}}^{n+\frac{1}{2}} - \mathbf{F}_{i+\frac{1}{2}}^{n+\frac{1}{2}}\right], \tag{31}$$

where

$$\mathbf{F}_{i+\frac{1}{2}}^{n+\frac{1}{2}} = \mathbf{F}\left(\mathbf{Q}_{i+\frac{1}{2}}^{n+\frac{1}{2}}\right). \tag{32}$$

The update formula (31) can be written as a one-step, non-staggered conservative scheme

$$\mathbf{Q}_i^{n+1} = \mathbf{Q}_i^n - \frac{\Delta t}{\Delta x}\left(\mathbf{F}_{i+\frac{1}{2}}^{\text{force}} - \mathbf{F}_{i-\frac{1}{2}}^{\text{force}}\right), \tag{33}$$

with numerical flux

$$\mathbf{F}_{i+\frac{1}{2}}^{\text{force}} = \frac{1}{2}\left(\mathbf{F}_{i+\frac{1}{2}}^{\text{LW}} + \mathbf{F}_{i+\frac{1}{2}}^{\text{LF}}\right). \tag{34}$$

It turns out that $\mathbf{F}_{i+\frac{1}{2}}^{\text{LW}}$ and $\mathbf{F}_{i+\frac{1}{2}}^{\text{LF}}$ are the Lax-Wendroff and the Lax-Friedrichs fluxes, given respectively by

$$\mathbf{F}_{i+\frac{1}{2}}^{\text{LW}} = \mathbf{F}\left(\mathbf{Q}_{i+\frac{1}{2}}^{n+\frac{1}{2}}\right) \tag{35}$$

and

$$\mathbf{F}_{i+\frac{1}{2}}^{\text{LW}} = \frac{1}{2}(\mathbf{F}(\mathbf{Q}_i^n) + \mathbf{F}(\mathbf{Q}_{i+1}^n)) - \frac{1}{2}\frac{\Delta x}{\Delta t}(\mathbf{Q}_{i+1}^n - \mathbf{Q}_i^n). \tag{36}$$

The FORCE scheme has the following properties:

- The scheme is a one-step (unstaggered) method in conservative form.
- For the linear advection equation FORCE is monotone and linearly stable under the CFL condition

$$0 \leq |c = \frac{\Delta t \lambda}{\Delta x}| \leq 1. \quad (37)$$

- For the linear advection equation FORCE has modified equation

$$\left.\begin{aligned} & q_t + \lambda q_x = \alpha_{\text{fo}} q_{xx}, \\ & \alpha_{\text{fo}} = \frac{1}{4}\lambda\Delta x\left(\frac{1-c^2}{c}\right) = \frac{1}{2}\alpha_{\text{lf}}, \end{aligned}\right\} \quad (38)$$

  where α_{lf} is the numerical viscosity of the Lax-Friedrichs scheme.
- For the isentropic gas dynamics and shallow water equations the FORCE scheme has been shown to be convergent (Chen and Toro 2004).

5.2 *FORCE in multiple space dimensions*

The extension of FORCE to multiple space dimensions is quite recent and is due to Toro and collaborators (Toro et al. 2009). Recall that a conservative scheme in 3D general unstructured meshes has the form

$$\mathbf{Q}_i^{n+1} = \mathbf{Q}_i^n - \frac{\Delta t}{|\mathbf{C}_i|}\sum_{j=1}^{n_i} |S_{i,j}^0| \, \bar{\mathbf{F}}_{i,j}^{force\alpha} \cdot \mathbf{n}_j. \quad (39)$$

Here index j visits all the neighbours of element i being updated, $|\mathbf{C}_i|$ is its volume and n_i is the number of neighbours of element i. $|S_{i,j}^0|$ is the cell interface area in the 3D case and the interface length in the 2D case. $\bar{\mathbf{F}} = (\mathbf{F}, \mathbf{G}, \mathbf{H})$ is the tensor of fluxes and $\mathbf{n}_j$ is the outward unit normal vector to the interface j between element i and element j. Δt is the time step. The FORCE flux for the interface (i,j) is

$$\bar{\mathbf{F}}_{i,j}^{force\alpha} = \frac{1}{2}\left(\bar{\mathbf{F}}_{i,j}^{lw\alpha}(\mathbf{Q}_i^n, \mathbf{Q}_j^n) + \bar{\mathbf{F}}_{i,j}^{lf\alpha}(\mathbf{Q}_i^n, \mathbf{Q}_j^n)\right). \quad (40)$$

The FORCE flux is, again as in the 1D case, the arithmetic average of a Lax-Wendroff and a Lax-Friedrichs type flux, which as far as I am aware had not been published before 2009.

The Lax-Wendroff type flux is

$$\bar{\mathbf{F}}_{i,j}^{lw\alpha}\left(\mathbf{Q}_i^n, \mathbf{Q}_j^n\right) = \bar{\mathbf{F}}(\mathbf{Q}_{i,j}^{n+\frac{1}{2}}), \quad (41)$$

where

$$\left.\begin{aligned} \mathbf{Q}_{i,j}^{n+\frac{1}{2}} = & \frac{\alpha_L \mathbf{Q}_i^n + \alpha_R \mathbf{Q}_j^n}{\alpha_L + \alpha_R} \\ & - \frac{1}{2}\frac{\Delta t\,|S_{i,j}^{(0)}|}{\alpha_L + \alpha_R}\left(\bar{\mathbf{F}}(\mathbf{Q}_j^n) - \bar{\mathbf{F}}(\mathbf{Q}_i^n)\right)\cdot \mathbf{n}_j, \end{aligned}\right\} \quad (42)$$

with

$$\alpha_L = |S_{i,j}^{(-)}|, \quad \alpha_R = |S_{i,j}^{(+)}|. \quad (43)$$

Here for each interface (i, j) we consider an interface volume decomposed into three disjoint sets: the set of points right on the interface with volume $|S_{i,j}^{(0)}|$; a set of points inside cell i of volume $\alpha_L = |S_{i,j}^{(-)}|$ and a setof points inside neighbour j, of volume $\alpha_R = |S_{i,j}^{(+)}|$.

The Lax-Friedrichs type flux is

$$\begin{aligned} \bar{\mathbf{F}}_{i,j}^{lf\alpha} = & \frac{\alpha_R \bar{\mathbf{F}}(\mathbf{Q}_i^n) + \alpha_L \bar{\mathbf{F}}(\mathbf{Q}_j^n)}{\alpha_L + \alpha_R} \\ & - \frac{2}{\Delta t\,|S_{i,j}^{(0)}|}\frac{\alpha_L \alpha_R}{\alpha_L + \alpha_R}\left(\mathbf{Q}_j^n - \mathbf{Q}_i^n\right)\cdot \mathbf{n}_j. \end{aligned} \quad (44)$$

In the special case of a 3D Cartesian regular mesh the FORCE scheme becomes

$$\left.\begin{aligned} \mathbf{Q}_{i,j,k}^{n+1} = & \mathbf{Q}_{i,j,k}^n - \frac{\Delta t}{\Delta x}[\mathbf{F}_{i+\frac{1}{2},j,k} - \mathbf{F}_{i-\frac{1}{2},j,k}] \\ & - \frac{\Delta t}{\Delta y}\left[\mathbf{G}_{i,j+\frac{1}{2},k} - \mathbf{G}_{i,j-\frac{1}{2},k}\right] \\ & - \frac{\Delta t}{\Delta z}\left[\mathbf{H}_{i,j,k+\frac{1}{2}} - \mathbf{H}_{i,j,k-\frac{1}{2}}\right], \end{aligned}\right\} \quad (45)$$

with the obvious notation for the numerical fluxes and the mesh parameters. The FORCE flux has the form

$$\mathbf{F}_{i+\frac{1}{2},j,k}^{force\alpha} = \frac{1}{2}\left(\mathbf{F}_{i+\frac{1}{2},j,k}^{lw\alpha} + \mathbf{F}_{i+\frac{1}{2},j,k}^{lf\alpha}\right). \quad (46)$$

The Lax-Wendroff type flux is now

$$\mathbf{F}_{i+\frac{1}{2},j,k}^{lw\alpha} = \mathbf{F}(\mathbf{Q}_{i+\frac{1}{2},j,k}^{lw\alpha}), \quad (47)$$

with

$$\left.\begin{aligned} \mathbf{Q}_{i+\frac{1}{2},j,k}^{lw\alpha} = & \frac{1}{2}\left(\mathbf{Q}_{i,j,k}^n + \mathbf{Q}_{i+1,j,k}^n\right) \\ & - \frac{1}{2}\frac{\alpha\Delta t}{\Delta x}\left[\mathbf{F}(\mathbf{Q}_{i+1,j,k}^n) - \mathbf{F}(\mathbf{Q}_{i,j,k}^n)\right]. \end{aligned}\right\} \quad (48)$$

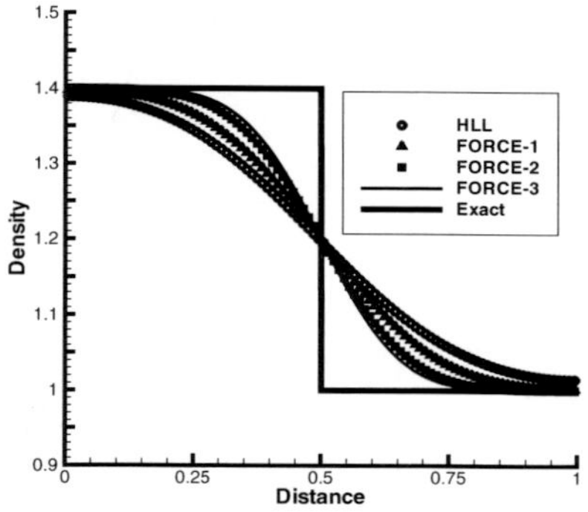

Figure 9. Test 0: isolated stationary contact in the 1D Euler equations. FORCE schemes compared to HLL and exact solutions. Output time $t = 5$.

The Lax-Friedrichs type flux is

$$\left.\begin{aligned} \mathbf{F}^{lf\alpha}_{i+\frac{1}{2},j,k} &= \frac{1}{2}\left[\mathbf{F}\left(\mathbf{Q}^{n}_{i,j,k}\right) + \mathbf{F}\left(\mathbf{Q}^{n}_{i+1,j,k}\right)\right] \\ &- \frac{1}{2}\frac{\Delta x}{\alpha \Delta t}\left(\mathbf{Q}^{n}_{i+1,j,k} - \mathbf{Q}^{n}_{i,j,k}\right). \end{aligned}\right\} \quad (49)$$

The above FORCE scheme is valid in 3D when $\alpha = 3$, valid in 2D when $\alpha = 2$ and valid in 1D when $\alpha = 1$. That is, α is a parameter representing the number of space dimensions. One could even conceive FORCE schemes in α space dimensions, with $\alpha > 3$. Such schemes appear to have some interesting properties yet to be properly investigated.

Figure 9 shows the performance of the FORCE schemes for **Test 0**, the stationary contact test problem for the 1D Euler equations, at time $t = 5$. Compare with results of Figure 7. Of course, the FORCE scheme being a centred scheme, will smear the contact discontinuity. However it is worth noting that the FORCE scheme ($\alpha = 1$) give results that are better than those of the HLL Riemann solver, at least for large CFL numbers. This is somehow surprising. On the other hand we know that the HLL Riemann solver is not complete, thereby resulting in smearing of intermediate characteristic fields. One conclusion is that Riemann solvers, if not complete, give results that are comparable to those of centred methods. I also remark that if the problem is solved in 2D ($\alpha = 2$) and in 3D ($\alpha = 3$), the corresponding results are better that those from the 1D sheme ($\alpha = 1$).

6 A NUMERICAL FLUX OF THE OSHER-TYPE

In this section I briefly review a very recent generalization of the Osher-Solomon Riemann solver which, unlike the original scheme, is applicable to any hyperbolic system (Dumbser amd Toro 2011), (Toro and Dumbser 2011).

Recall that the Osher-Solomon (Osher and Solomon 1982) numerical flux is constructed from

$$\mathbf{F}_{i+\frac{1}{2}} = \frac{1}{2}\left(\mathbf{F}(\mathbf{Q}_0) + \mathbf{F}(\mathbf{Q}_1)\right) - \frac{1}{2}\int_{\mathbf{Q}_0}^{\mathbf{Q}_1} |\mathbf{A}(\mathbf{Q})|\, d\mathbf{Q}\,. \quad (50)$$

Here we keep their original notation for the the data states, namely $\mathbf{Q}_0 = \mathbf{Q}_L$, $\mathbf{Q}_1 = \mathbf{Q}_R$. Evaluation of the integral in phase space is required. Full details of the Osher-Solomon scheme are given in Chapter 12 of (Toro 2009). There is no unique way of calculating these integrals. Osher and Solomon choose a way so as to make the procedure tractable, relying on integration along paths in phase space. Osher and Solomon considered two ways of ordering the integrations paths, namely the *Physical Ordering*, or P-ordering and the original *Osher Ordering*, or O-ordering.

Dumbser and Toro (Dumbser and Toro 2011), (Toro and Dumbser 2011) proposed to use the simple path

$$\psi(s, \mathbf{Q}_0, \mathbf{Q}_1) = \mathbf{Q}_0 + s(\mathbf{Q}_1 - \mathbf{Q}_0)\,,\ s \in [0,1]\,. \quad (51)$$

Then, under a change of variables the flux becomes

$$\left.\begin{aligned} \mathbf{F}_{i+\frac{1}{2}} &= \frac{1}{2}\left(\mathbf{F}(\mathbf{Q}_0) + \mathbf{F}(\mathbf{Q}_1)\right) \\ &- \frac{1}{2}\left(\int_0^1 \left|\mathbf{A}\left(\psi(s;\mathbf{Q}_0,\mathbf{Q}_1)\right)\right| ds\right)\Delta\mathbf{Q}\,, \end{aligned}\right\} \quad (52)$$

with $\Delta\mathbf{Q} = \mathbf{Q}_1 - \mathbf{Q}_0$ denoting the jump. The integration is performed numerically. Using a Gauss-Legendre quadrature rule with G points s_j and associated weights ω_j in the unit interval $I = [0,1]$ we have

$$\left.\begin{aligned} \mathbf{F}_{i+\frac{1}{2}} &= \frac{1}{2}\left(\mathbf{F}(\mathbf{Q}_0) + \mathbf{F}(\mathbf{Q}_1)\right) \\ &- \frac{1}{2}\left(\sum_{j=1}^{G} \omega_j \left|\mathbf{A}\left(\psi(s_j,\mathbf{Q}_0,\mathbf{Q}_1)\right)\right|\right)\Delta\mathbf{Q}\,. \end{aligned}\right\} \quad (53)$$

For each s_j, $|\mathbf{A}\,\psi(s_j, \mathbf{Q}_0, \mathbf{Q}_1))|$ is decomposed using the standard characteristic decomposition

$$|\mathbf{A}(\mathbf{Q})| = \mathbf{R}(\mathbf{Q})\,|\Lambda(\mathbf{Q})|\,\mathbf{R}^{-1}(\mathbf{Q})\,. \quad (54)$$

Some of the features of the new Osher-type Riemann solver include the following: the flux is indeed very simple to calculate and is applicable to any hyperbolic system provided the complete eigenstructure of the system is available, either analytically ornumerically. The solver is non-linear and complete, that is the wave model contains all characteristic fields. Numerical experiments suggest also that the scheme is entropy satisfying. The resulting numerical scheme is very robust requiring no special fixes or tuning of parameters. In oder to illustrate some features of the new solver we show some results for two test problems for the Euler equations.

Test A: Non-isolated stationary contact discontinuity. We solve the one-dimensional Euler equations in the domain $0 \le x \le 1.0$, $\gamma = 1.4$, mesh $M = 100$ cells, CFL number $CFL = 0.9$, output time $t = 0.012$ and initial conditions given by

$$\left.\begin{array}{l} \rho(x,0) = 1\,,\, u(x,0) = -19.59745\,,\, 0 \le x \le 1 \\ p(x,0) = \begin{cases} 1000 & \text{if} \quad 0 \le x \le 0.8\,, \\ 0.01 & \text{if} \quad 0.8 < x \le 1.0\,. \end{cases} \end{array}\right\} \tag{55}$$

Test B: Shock-wave collision. We solve the one-dimensional Euler equations in the domain $0 \le x \le 1.0$, $\gamma = 1.4$, mesh $M = 100$ cells, CFL number $CFL = 0.9$, output time $t = 0.8$ and initial conditions given by

$$\left.\begin{array}{l} \rho(x,0) = 1\,,\, p(x,0) = 0.1\,,\, 0 \le x \le 1 \\ u(x,0) = \begin{cases} 2.0 & \text{if} \quad 0.0 \le x \le 0.5\,, \\ -2.0 & \text{if} \quad 0.5 < x \le 1.0\,. \end{cases} \end{array}\right\} \tag{56}$$

Figure 10 shows results for Test A: non-stationary contact discontinuity. Results from the numerical Osher scheme are shown in symbols, while those of the exact solution are depicted by the full line. Figure 11 shows the corresponding results

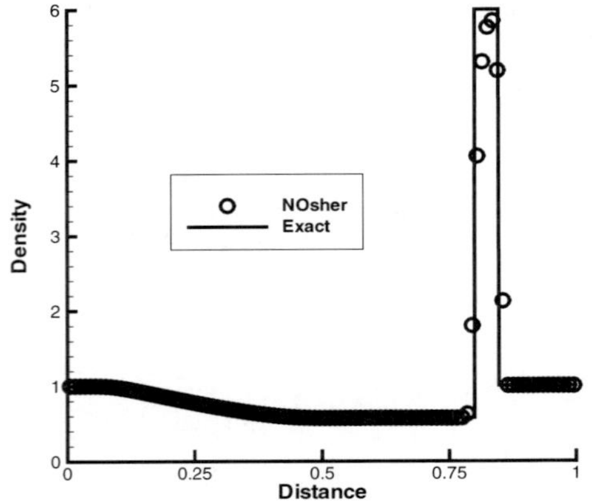

Figure 10. Test A: Non-stationary contact. Numerical Osher (symbol) and exact (line) solutions at time $t = 0.012$.

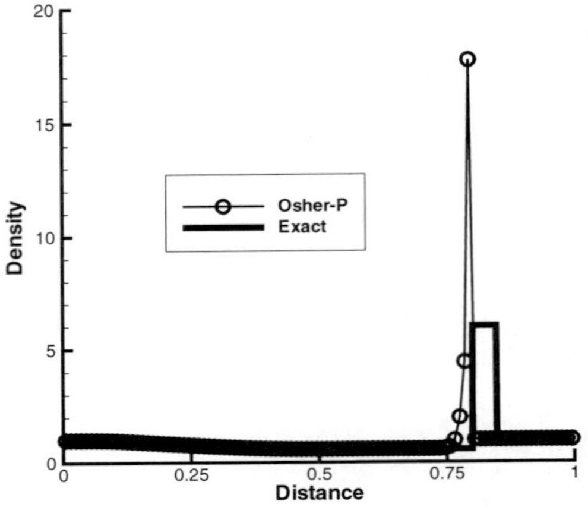

Figure 11. Test A: Non-stationary contact. Osher-Solomon scheme with P-ordering (symbol) and exact (line) solutions at time $t = 0.012$.

from the conventional (analytical) Osher-Solomn scheme with P-ordering of integration paths. The new Osher-type scheme gives very satisfactory results, even though the (stationary) contact discontinuity is somewhat smeared. The classical Osher-Solomon scheme fails, being unable to recognised the shock wave and producing a huge overshoot in density. Also, we note that the Osher-Solomon scheme with the O-ordering crushes for this test problem and thus results for this version of the scheme are not shown.

Figures 12 and 13 show results for Test B, a shock collision problem. Numerical solutions are compared with the exact solution. The results from the new Osher-type scheme depicted in Figure 12 are very satisfactory and we note that the usual oscillations behind these slowly-moving shocks are quite small. The dip in density around the trivial contact discontinuity is visible and is similar to that of other schemes for this class of test problems. In the literature, this difficulty is known as the *wall-heating problem* and is notoriously difficult to cure. Figure 13 shows the result from the classical Osher-Solomon scheme with P-ordering. The result is comparable to that of the new Osher-type scheme shown in Figure 12. A small difference is observed inthe amplitude of the spurious post-shock oscillations, the new scheme giving better results, but the shock is slightly sharper in the classical Osher-Solomon scheme. We note that for this test problem the classical Osher-Solomon scheme with O-ordering fails and thus results are not shown.

In conclusion, we could state that the performance of the new Osher-type scheme is superior to that of the classical Osher scheme. The latter crashes for some types of test problems. But perhaps the most

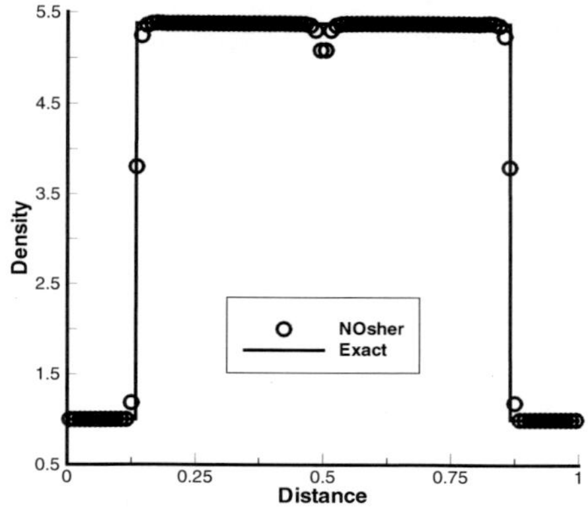

Figure 12. Test B: Shock Collision. Numerical Osher (symbol) and exact (line) solutions at time $t = 0.8$.

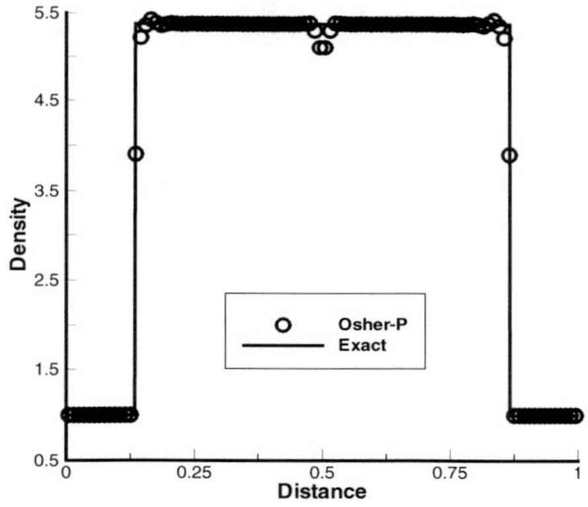

Figure 13. Test B: Shock Collision. Osher-Solomon scheme with P-ordering (symbol) and exact (line) solutions at time $t = 0.8$.

important feature of the new Osher-type scheme is that it is applicable to any hyperbolic system, unlike the classical Osher-Solomon scheme that remains applicable to just a few simple hyperbolic systems. The new scheme uses the simple linear path for all cases. Recent experience suggests that for some particular applications one may need to consider more sophisticated choices for the integration path. This and other issues are currently the subject of investigations by the author and collaborators.

7 THE GENERALISED RIEMANN PROBLEM

The generalised Riemann problem (or GRP) for hyperbolic systems of balance laws is the special Cauchy problem

$$\left.\begin{array}{l} \text{PDEs}: \ \partial_t \mathbf{Q} + \partial_x \mathbf{F}(\mathbf{Q}) = \mathbf{S}(\mathbf{Q}), x \in R, t > 0 \\ \text{ICs}: \ \mathbf{Q}(x,0) = \begin{cases} \mathbf{Q}_L(x) & \text{if} \quad x < 0, \\ \mathbf{Q}_R(x) & \text{if} \quad x > 0. \end{cases} \end{array}\right\} \quad (57)$$

Figures 14 and 15 compare the form of the initial condition and of the solution structure for the classical and the generalised Riemann problems. One of the earliest reported generalisations of the Riemann problem is due to Glimm et al. (Glimm et al. 1984), who studied the GRP for improving the performance of the Random Choice Method, or Glimm's method, for one-dimensional flow with area variation. The challenge was the numerical treatment of the source term due to area variation present in the equations.

The same year, Ben-Artzi and Falcovitz (Ben-Artzi and Falcovitz 1984) posed the GRP for the homogeneous Euler equations (no source terms) with piece-wise linear initial condition, a special case of the GRP (57). They used the solution of the GRP problem for constructing a fully discrete scheme of second order of accuracy in space and time. Then, in a less obvious way, Harten and collaborators (Harten et al. 1987) posed and solved the homogenous GRP for more general initial conditions than those of Ben-Artzi and Falcovitz. Their resulting schemes are fully discrete and of high order of accuracy in space and time. The way Harten and collaborators constructed their scheme was more akin to the MUSCL-Hancock scheme.

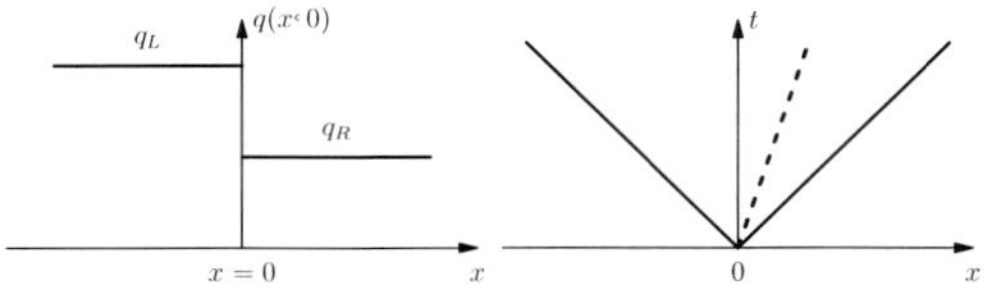

Figure 14. Initial condition for single component of vector of unknowns (left) and full solution structure (right) of classical Riemann problem with piece-wise constant data, for typical 3×3 hyperbolic system.

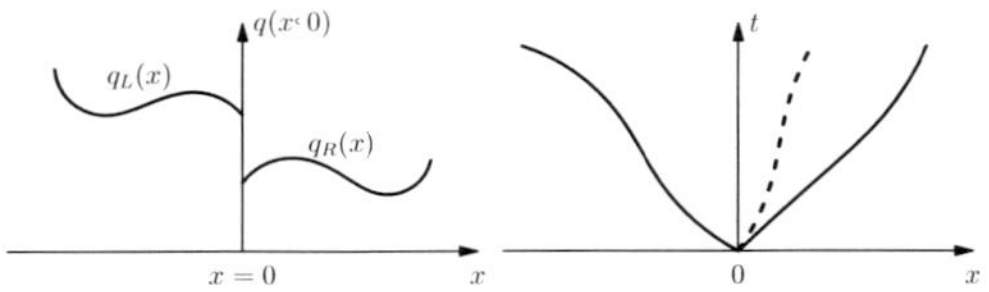

Figure 15. Initial condition for single component of vector of unknowns (left) and full solution structure (right) of generalized Riemann problem with piece-wise constant data, for typical 3×3 hyperbolic system.

Hence one may regard their scheme as a higher order version of the MUSCL-Hancock second order scheme. Later, LeFloch and Raviart (Lefloch and Raviart 1988) posed the general, though homogeneous, GRP problem (57) and developed an existence and uniqueness theory. Men'shov (Men'shov 1990) solved exactly the second-order GRP for the ideal Euler equations in the full x-t semi-plane, constructing, as a result, a second-order Godunov method.

Toro and Titarev (Toro and Titarev 2002) solved the complete problem (57) for any order of accuracy, generalising the solver proposed by Toro et al. (Toro et al. 2001) for linear systems. It also generalises the Ben-Artzi and Falcovitz (Ben-Artzi and Falcovitz 1984) second-order method and that of Harten and collaborators (Harten et al. 1987), which did not include source terms. Alternative ways of solving the GRP have been presented by Dumbser et al. (Dumbser et al. 2008) and by Toro and Castro (Castro and Toro 2008), who also discuss the relationship between various solvers for the GRP.

In what follows we describe, in algorithmic form, the Toro-Titarev solver for the GRP. A solution $\mathbf{Q}_{LR}(\tau)$ at $x = 0$ is sought as as function of time, in terms of the power series expansion

$$\mathbf{Q}_{LR}(\tau) = \mathbf{Q}(0,0_+) + \sum_{k=1}^{K} [\partial_t^{(k)}\mathbf{Q}(0,0_+)]\frac{\tau^k}{k!}, \quad (58)$$

where the leading term of the expansion is

$$\mathbf{Q}(0,0_+) = \lim_{t\to 0_+} \mathbf{Q}(0,t). \quad (59)$$

In what follows we describe three steps to compute the various terms of the expansion.

Step A: The leading term. Solve classical, homogeneous Riemann problem

$$\left.\begin{array}{ll} \text{PDEs}: & \partial_t\mathbf{Q} + \partial_x\mathbf{F}(\mathbf{Q}) = 0, \\ \text{ICs}: & \mathbf{Q}(x,0) = \begin{cases} \mathbf{Q}_L(0) & \text{if} \quad x < 0, \\ \mathbf{Q}_R(0) & \text{if} \quad x > 0 \end{cases} \end{array}\right\} \quad (60)$$

and denote the similarity solution by $\mathbf{D}^{(0)}(x/t)$. Then the sought leading term is

$$\mathbf{Q}(0,0_+) = \mathbf{D}^{(0)}(0). \quad (61)$$

Step B: Higher order terms.

- Cauchy-Kowalewski procedure:

$$\partial_t^{(k)}\mathbf{Q} = \mathbf{P}^{(k)}(\partial_x^{(0)}\mathbf{Q}, \partial_x^{(1)}\mathbf{Q}, \ldots, \partial_x^{(k)}\mathbf{Q}). \quad (62)$$

- Evolution equations for spatial derivatives:

$$\partial_t(\partial_x^{(k)}\mathbf{Q}) + \mathbf{A}(\mathbf{Q})\partial_x(\partial_x^{(k)}\mathbf{Q}) = \mathbf{H}^{(k)}. \quad (63)$$

- Riemann problems for spatial derivatives:

$$\left.\begin{array}{ll} \text{PDEs}: & \partial_t(\partial_x^{(k)}\mathbf{Q}) + \mathbf{A}_{LR}^{(0)}\partial_x(\partial_x^{(k)}\mathbf{Q}) = 0 \\ \text{ICs}: & \partial_x^{(k)}\mathbf{Q}(x,0) = \begin{cases} \mathbf{Q}_L^{(k)}(0) & \text{if } x < 0 \\ \mathbf{Q}_R^{(k)}(0) & \text{if } x > 0 \end{cases} \end{array}\right\} \quad (64)$$

The non-linear inhomogeneous systems (63) have been linearised and the source terms have been neglected. Empirical convergence rate studies consistently suggest that such simplifications do not deteriorate the high accuracy, for smooth solutions. A related theoretical study is due to Goetz and Iske (Goetz and Iske 2012), in this book.

Denote the solution of the linear system (64) by $\mathbf{D}^{(k)}(x/t)$ and set

$$\partial_x^{(k)}\mathbf{Q}(0,0_+) = \mathbf{D}^{(k)}(0). \quad (65)$$

Then, the k-th time derivative in (62) is known and hence the sought solution (58) is determined, completing the solution procedure.

Based on this solution of the K-th order generalized Riemann problem, including source terms, Toro and Titarev (Toro and Billett 2000), (Titarev and Toro 2002) constructed (K+1)-th order fully discrete, one-step finite volume schemes of the form

$$\mathbf{Q}_i^{n+1} = \mathbf{Q}_i^n - \frac{\Delta t}{\Delta x}(\mathbf{F}_{i+\frac{1}{2}} - \mathbf{F}_{i-\frac{1}{2}}) + \Delta t\mathbf{S}_i. \quad (66)$$

The *numerical flux* is computed by evaluating the time integral in (17) numerically, in which $\mathbf{Q}(x_{i+\frac{1}{2}},t)$ is approximated by the GRP solution (58). The computation of the *numerical source* $\mathbf{S}_i$ is obtained from the space-time integral in (17), where an analogous procedure to that of the GRP solution is applied inside the space-time control volume. For details see the original paper (Toro and Titarev 2002) and Chapters 19 and 20 of (Toro 2009). Note that the complete scheme has the same form as the first-order Godunov method, of which is a very natural extension. It is also an extension of the Ben-Artzi second order method.

8 CONCLUSIONS

A brief review of the Riemann problem and its impact in computational science has been presented. Some basic definitions and simple examples of Riemann problems have been given. Then I have given an overview of the most well known methods to find approximate solutions to the Riemann problem to be used in finite volume and discontinuous Galerkin finite element methods. I then reviewed some specific Riemann solvers in more detail, with the choice made on grounds of personal contributions and on recent developments. The final part of the paper discusses generalisations of the Riemann problem, the GRP, and reviews one way of solving the GRP for any order of accuracy, and including source terms.

REFERENCES

Antón, L., J. A. Miralles, J. M. Martí, J. M. Ibáñez, and M. A. Aloy (2010). Relativitic Magnetohydrodynamics: Renormalized Eigenvectors and Full Decomposition Riemann Solver. *The Astrophysical Journal Supplement Series* 188, 1–31.

Batten, P., N. Clarke, C. Lambert, and D. Causon (1997). On the Choice of Wave Speeds for the HLLC Riemann Solver. *SIAM J. Sci. and Stat. Comp.* 18, 1553–1570.

Batten, P., M. A. Leschziner, and U. C. Goldberg (1997). Average–State Jacobians and Implicit Methods for Compressible Viscous and Turbulent Flows Flows. *J. Comput. Phys.* 137, 38–78.

Ben-Artzi, M. and J. Falcovitz (1984). A Second Order Godunov–Type Scheme for Compressible Fluid Dynamics. *J. Comput. Phys.* 55, 1–32.

Bodo, G., S. Masaglia, A. Mignone, and P. Rossi (Editors) (2008). Jets from Young Stars III. Numerical MHD and Instabilities. Lecture Notes in Physics 754. Springer.

Castro, C. E. and E. F. Toro (2008). Solvers for the High–Order Riemann Problem for Hyperbolic Balance Laws. *J. Comput. Phys.* 227, 2481–2513.

Cha, S.-H. and A. P. Whitworth (2003). Implementations and tests of Godunov-type particle hydrodynamics. *Monthly Notices of the Royal Astronomical Society* 340, 73–90.

Chen, G. Q. and E. F. Toro (2004). Centred Schemes for Non–Linear Hyperbolic Equations. *Journal of Hyperbolic Differential Equations* 1(1), 531–566.

Chorin, A. J. (1976). Random Choice Solutions of Hyperbolic Systems. *J. Comput. Phys.* 22, 517–533.

Colella, P. and H. H. Glaz (1985). Efficient Solution Algorithms for the Riemann Problem for Real Gases. *J. Comput. Phys.* 59, 264–289.

Courant, R. and K. O. Friedrichs (1985). *Supersonic Flow and Shock Waves*. Springer-Verlag.

Dal Maso, G., P. G. LeFloch, and F. Murat (1995). Definition and Weak Stability of Nonconservative Products. *J. Math. Pure Appl.* 74, 483–548.

Davis, S. F. (1988). Simplified Second–Order Godunov–Type Methods. *SIAM J. Sci. Stat. Comput.* 9, 445–473.

Dumbser, M., D. Balsara, E. F. Toro, and C. D. Munz (2008). A Unified Framework for the Construction of One–Step Finite–Volume and Discontinuous Galerkin Schemes. *J. Comput. Phys.* 227, 8209–8253.

Dumbser, M., C. Enaux, and E. F. Toro (2008). Finite Volume Schemes of Very High Order of Accuracy for Stiff Hyperbolic Balance Laws. *J. Comput. Phys.* 227(8), 3971–4001.

Dumbser, M., A. Hidalgo, M. Castro, C. Parés, and E. F. Toro (2010). FORCE Schemes on Unstructured Meshes II: Non-conservative Hyperbolic Systems. *Comput. Methods Appl. Mech. Engrg.* 199, 625–647.

Dumbser, M. and E. F. Toro (2011). On universal Osher–type schemes for general nonlinear hyperbolic conservation laws. *Communications in Computational Physics* 10, 635–671.

Dutt, P. (1986). A Riemann Solver Based on a Global Existence Proof for the Riemann Problem. Technical Report ICASE 86-3, NASA Langley Research Center, USA.

Einfeldt, B. (1988). On Godunov–Type Methods for Gas Dynamics. *SIAM J. Numer. Anal.* 25(2), 294–318.

Falle, S. A. E. G. and S. S. Komissarov (1996). An Upwind Numerical Scheme for Relativistic Hydrodynamics with a General Equation of State. *Monthly Notices of the Royal Astronomical Society* 278, 586–602.

Falle, S. A. E. G. and S. S. Komissarov (1997). A Multidimensional Upwind Scheme for Magnetohydrodynamics. *Monthly Notices of the Royal Astronomical Society* 297, 265–277.

Giacomazzo, B. and L. Rezzolla (2006). The Exact Solution of the Riemann Problem in Relativistic Magnetohydrodynamics. *Journal of Fluid Mechanics* 562, 223–259.

Glimm, J. (1965). Solution in the Large for Nonlinear Hyperbolic Systems of Equations. *Comm. Pure. Appl. Math.* 18, 697–715.

Glimm, J., M. J. Graham, J. W. Grove, X.-L. Li, T. M. Smith, D. Tan, F. Tangerman, and Q. Zhang (1998). Front Tracking in Two and Three Dimensions. *Computers Math. Applic.* 35, 1–11.

Glimm, J., J. W. Grove, X.-L. Li, K. M. Shyue, Q. Zhang, and Y. Zeng (1998). Three Dimensional Front Tracking. *SIAM J. Sci. Comp.* 19, 703–727.

Glimm, J., G. Marshall, and B. Plohr (1984). A Generalized Riemann Problem for Quasi–One–Dimensional Gas Flows. *Advances in Applied Mathematics* 5, 1–30.

Godunov, S. K. (1959). A Finite Difference Method for the Computation of Discontinuous Solutions of the Equations of Fluid Dynamics. *Mat. Sb.* 47, 357–393.

Godunov (Ed.), S. K. (1976). Numerical Solution of Multi–Dimensional Problems in Gas Dynamics. Nauka Press, Moscow.

Goetz, C. R. and A. Iske (2012). Approximate Solutions of Generalized Riemann Problems: the Toro-Titarev Solver and the LeFloch-Raviart Expansion. In –. –.

Gottlieb, J. J. and C. P. T. Groth (1988). Assessment of Riemann Solvers for Unsteady One–Dimensional Inviscid Flows of Perfect Gases. *J. Comput. Phys.* 78, 437–458.

Gurski, K. F. (2004). An HLLC–Type Approximate Riemann Solver for Ideal Magnetohydrodynamics. *SIAM J. Sci. Comput.* 25(6), 2165–2187.

Harten, A. (1983). High Resolution Schemes for Hyperbolic Conservation Laws. *J. Comput. Phys.* 49, 357–393.

Harten, A., B. Engquist, S. Osher, and S. R. Chakravarthy (1987). Uniformly High Order Accuracy Essentially Non–oscillatory Schemes III. *J. Comput. Phys.* 71, 231–303.

Klingenberg, C., W. Schmidt, and K. Waagan (2007). Numerical Comparison of Riemann Solvers for Astrophysical Hydrodynamics. *J. Comput. Phys.* 227(1), 12–35.

Lax, P. D. (1954). Weak Solutions of Nonlinear Hyperbolic Equations and Their Numerical Computation. *Comm. Pure. Appl. Math. VII*, 159–193.

Lax, P. D. (1972). The Formation and Decay of Shock Waves. *American Mathematical Monthly* 79, 227–241.

Lax, P. D. (1990). *Hyperbolic Systems of Conservation Laws and the Mathematical Theory of Shock Waves.* Society for Industrial and Applied Mathematics, Philadelphia.

LeFloch, P. (1988). Entropy Weak Solutions to Nonlinear Hyperbolic Systems under Nonconservative Form. *Commun. Part. Diff. Eqn.* 13(6), 669–727.

LeFloch, P. and P. A. Raviart (1988). An Asymptotic Expansion for the Solution of the Generalized Riemann Problem. Part 1: General Theory. *Ann. Inst. Henri Poincaré. Analyse non Lineáre* 5(2), 179–207.

Lehner, L. (2001). Numerical Relativity: a Review. *Class. Quantum Grav.* 18, R25—R86.

Martí, J. M. and E. Müller (1994). The Analytical Solution of the Riemann Problem for Relativistic Hydrodynamics. *Journal of Fluid Mechanics* 258, 317–333.

Menikoff, R. and B. J. Plohr (1989). The Riemann Problem for Fluid Flow of Real Materials. *Reviews of Modern Physics* 61, 75–130.

Men'shov, I. S. (1990). Increasing the Order of Approximation of Godunov's Scheme Using the Generalized Riemann Problem. *USSR Comput. Math. Phys.* 30(5), 54–65.

Mignone, A. and G. Bodo (2005). An HLLC Riemann solver for relativistic flows Ð I. Hydrodynamics. *Monthly Notices of the Royal Astronomical Society* 364, 126–136.

Mignone, A. and G. Bodo (2006). An HLLC Riemann solver for relativistic flows Ð II. Magnetohydrodynamics. *Monthly Notices of the Royal Astronomical Society* 368, 1040–1054.

Mignone, A., M. Ugliano, and G. Bodo (2009). A five-wave HartenÐLaxÐvan Leer Riemann solver for relativistic magnetohydrodynamics. *Monthly Notices of the Royal Astronomical Society* 393, 1141–1156.

Osher, S. and F. Solomon (1982). Upwind Difference Schemes for Hyperbolic Conservation Laws. *Math. Comp.* 38,158, 339–374.

Pike, J. (1993). Riemann Solvers for Perfect and Near–Perfect Gases. *AIAA Journal* 31(10), 1801–1808.

Quartapelle, L., L. Castelletti, A. Guardone, and G. Quaranta (2003). Solution of the Riemann Problem of Classical Gasdynamics. *J. Comput. Phys.* 190, 118–140.

Riemann, B. (1860). Über die Fortpflanzung ebener Luftwellen von endlicher Schwingungsweite. *Abhandlungen der Königlichen Gesellschaft der Wissenschaften zu Göttingen* 8, 43–65.

Risebro, N. H. (2002). Front Tracking for Hyperbolic Conservation Laws. Springer.

Roe, P. L. (1981). Approximate Riemann Solvers, Parameter Vectors, and Difference Schemes. *J. Comput. Phys.* 43, 357–372.

Rusanov, V. V. (1961). Calculation of Interaction of Non–Steady Shock Waves with Obstacles. *J. Comput. Math. Phys. USSR* 1, 267–279.

Salas, M. (2009). *A Shock-Fitting Primer*. CRC Press.

Schleicher, M. (1993). Ein Einfaches und Effizientes Verfahren zur Loesung des Riemann–Problems. *Z. Flugwiss. Weltraumforsch.* 17, 265–269.

Schneider, V., U. Katscher, D. H. Rischke, B. Waldhauser, J. A. Maruhn, and C. D. Munz (1993). New Algorithms for Ultra–Relativistic Numerical Hydrodynamics. *J. Comput. Phys.* 105, 92–.

Schwendeman, D. W., C. W. Wahle, and A. K. Kapila (2006). The Riemann Problem and a High–Resolution Godunov Method for a Model of Compressible Two–Phase Flow. *J. Comput. Phys.* 212, 490–526.

Smoller, J. (1994). Shock Waves and Reaction–Diffusion Equations. Springer–Verlag.

Takayama, K. and T. Saito (2004). Shock Wave/Geophysical and Medical Applications. *Annual Review of Fluid Mechanics* 36, 347–379.

Titarev, V. A. and E. F. Toro (2002). ADER: Arbitrary High Order Godunov Approach. *J. Scientific Computing* 17, 609–618.

Toro, E. F. (1989). A Fast Riemann Solver with Constant Covolume Applied to the Random Choice Method. *Int. J. Numer. Meth. Fluids* 9, 1145–1164.

Toro, E. F. (1996). On Glimm–Related Schemes for Conservation Laws. Technical Report MMU-9602, Department of Mathematics and Physics, Manchester Metropolitan University, UK.

Toro, E. F. (2001). Shock–Capturing Methods for Free–Surface Shallow Flows. Wiley and Sons Ltd.

Toro, E. F. (2009). Riemann Solvers and Numerical Methods for Fluid Dynamics, Third Edition. Springer–Verlag.

Toro, E. F. and S. J. Billett (1996). Centred TVD Schemes for Hyperbolic Conservation Laws. Technical Report MMU–9603, Department of Mathematics and Physics, Manchester Metropolitan University, UK.

Toro, E. F. and S. J. Billett (2000). Centred TVD Schemes for Hyperbolic Conservation Laws. *IMA J. Numerical Analysis* 20, 47–79.

Toro, E. F. and A. Chakraborty (1994). Development of an Approximate Riemann Solver for the Steady Supersonic Euler Equations. *The Aeronautical Journal* 98, 325–339.

Toro, E. F. and M. Dumbser (2011). Reformulated Osher–type Riemann solver. In *Computational Fluid Dynamics 2010. Alexander Kuzmin (editor)*, pp. 131–136. Springer-Verlag.

Toro, E. F. and P. García-Navarro (2007). Godunov–Type Methods for Free–Surface Shallow Flows: A Review. *J. Hydraulic Research* 45(6), 736–751.

Toro, E. F. and A. Hidalgo (2009). ADER finite volume schemes for non-linear reaction-diffusion equations. *Applied Numerical Mathematics* 59, 73–100.

Toro, E. F., A. Hidalgo, and M. Dumbser (2009). FORCE Schemes on Unstructured Meshes I: Conservative Hyperbolic Systems. *J. Comput. Phys.* 228, 3368–3389.

Toro, E. F., R. C. Millington, and L. A. M. Nejad (2001). Towards Very High–Order Godunov Schemes. In *Godunov Methods: Theory and Applications. Edited Review, E. F. Toro (Editor)*, pp. 905–937. Kluwer Academic/Plenum Publishers.

Toro, E. F. and A. Siviglia (2011). Simplified blood flow model with discontinuous vessel properties: analysis and exact solutions. In *Modelling Physiological Flows Series: Modelling, Simulation and Applications. Editors: D. Ambrosi, A. Quarteroni and G. Rozza*, pp. 19–39. Springer–Verlag, ISBN 978-88-470-1934-8.

Toro, E. F., M. Spruce, and W. Speares (1992). Restoration of the Contact Surface in the HLL–Riemann Solver. Technical Report CoA–9204, Department of Aerospace Science, Collegue of Aeronautics, Cranfield Institute of Technology, UK.

Toro, E. F., M. Spruce, and W. Speares (1994). Restoration of the Contact Surface in the HLL–Riemann Solver. *Shock Waves* 4, 25–34.

Toro, E. F. and V. A. Titarev (2002). Solution of the Generalised Riemann Problem for Advection–Reaction Equations. *Proc. Roy. Soc. London A* 458, 271–281.

van Leer, B. (1979). Towards the Ultimate Conservative Difference Scheme V. A Second Order Sequel to Godunov's Method. *J. Comput. Phys.* 32, 101–136.

II Recent advances in the numerical computation of environmental conservation laws with source terms

Numerical Methods for Hyperbolic Equations – Vázquez-Cendón et al. (eds)
© 2013 Taylor & Francis Group, London, ISBN 978-0-415-62150-2

Finite volume discretisation of depth averaged scalar transport equations coupled to shallow water models

L. Cea
Environmental and Water Engineering Group, University of A Coruña, Spain

M.E. Vázquez-Cendón
Department of Applied Mathematics, University of Santiago de Compostela, Spain

ABSTRACT: In this paper we analyse the relation between the numerical scheme used to solve the 2D shallow water equations and the scheme used to solve a scalar transport model linked to the shallow water equations. It is shown that the numerical scheme used to solve the scalar transport equation must take into consideration the scheme used to solve the hydrodynamic equations. The most important implication is that a well balanced and conservative scheme for the scalar transport equation cannot be formulated just from the water depth and velocity fields, but has to consider also the way in which the hydrodynamic equations have been solved.

1 INTRODUCTION

Shallow water models provide a useful and efficient tool in order to study environmental free surface flows. In this field, two-dimensional depth averaged models have become very popular not only to compute the hydrodynamics, but also to compute the dispersion of contaminants, sediment transport, or any other physical process which depends on the water depth and velocity fields.

The numerical discretisation of the convective flux in the shallow water equations has been extensively studied in many previous works (LeVeque 2002; Toro 2001). These works have shown that it is important for the numerical discretisation of the bathymetry to be related to the discretisation of the convective flux in order to have a well-balanced scheme. The same applies for the discretisation of any scalar transport equation linked to the hydrodynamic equations (Murillo et al. 2008; Latorre et al. 2009).The scalar transport is strongly determined by the velocity and water depth fields. Therefore, in order to define a well-balanced conservative scheme, the numerical discretisation of the hydrodynamic and the scalar transport equations cannot be done independently of each other. This means that it is not only important to consider the flow field, but it is also necessary to take into account the numerical scheme used to discretise the hydrodynamic equations, in particular the mass conservation equation, otherwise the resulting scheme for the scalar transport equation might not be conservative. This is specially true in the presence of unsteady wet-dry fronts, because in those cases the numerical implementation of the wetting and drying algorithm defines how the water front moves and therefore, how the mass of the scalar is transported through the domain.

In this paper we analyse the previous issues using a finite volume solver for the hydrodynamic equations. An alternative discretisation for computing the convective transport of a scalar is analysed, which guarantees the mass conservation of the scalar inthe presence of unsteady wet-dry fronts over irregular bathymetries.

2 NUMERICAL MODEL

The numerical model used in this work solves the two-dimensional St. Venant equations, written in conservative form as:

$$\frac{\partial h}{\partial t} + \frac{\partial q_x}{\partial x} + \frac{\partial q_y}{\partial y} = 0 \tag{1}$$

$$\frac{\partial q_x}{\partial t} + \frac{\partial}{\partial x}\left(\frac{q_x^2}{h} + \frac{gh^2}{2}\right) + \frac{\partial}{\partial y}\left(\frac{q_x q_y}{h}\right) = -gh\frac{\partial z_b}{\partial x} - \frac{\tau_{b,x}}{\rho} \tag{2}$$

$$\frac{\partial q_y}{\partial t} + \frac{\partial}{\partial x}\left(\frac{q_x q_y}{h}\right) + \frac{\partial}{\partial y}\left(\frac{q_y^2}{h} + \frac{gh^2}{2}\right) = -gh\frac{\partial z_b}{\partial y} - \frac{\tau_{b,y}}{\rho} \tag{3}$$

where z_b is the bed elevation, $q_x = hU_x, q_y = hU_y$ are the two components of the unit discharge, U_x, U_y are the two horizontal components of the depth averaged velocity, h is the water depth, $\tau_{b,x}, \tau_{b_y}$ are the two horizontal components of the bed friction, ρ is the water density, and g is the gravity acceleration. In addition, the following depth averaged scalar transport equation written in conservative form is solved coupled to the hydrodynamic equations:

$$\frac{\partial hc}{\partial t} + \frac{\partial q_x c}{\partial x} + \frac{\partial q_y c}{\partial y} = \frac{\partial}{\partial x}\left(\Gamma_e h \frac{\partial c}{\partial x}\right) + \frac{\partial}{\partial x}\left(\Gamma_e h \frac{\partial c}{\partial y}\right) \tag{4}$$

where c is the concentration of any passive scalar, and Γ_e is the effective diffusivity. In this work we will focus on the discretisation of the convective terms in the scalar transport equation and therefore, the diffusion terms will be ommited for the sake of simplicity in the notation. With this approximation, a usual finite volume discretisation of the scalar transport equation is given by:

$$\frac{w_{c,i}^{n+1} - w_{c,i}^{n}}{\Delta t} A_i + \sum_{j \in K_i} \Phi_{c,ij} \,|\, \mathbf{n}_{ij} \,| = 0 \tag{5}$$

where c is the concentration of the scalar, $w_c = hc$ is the conservative variable, and the numerical flux Φ_c is an approximation to the convective flux normal to the face between the control volumes i and j $(Z_c = (U_x \tilde{n}_x + U_y \tilde{n}_y) w_c = U_n w_c = q_n c)$.

A usual approach used to compute the numerical flux Φ_c is given by the splitting of the flux in a convective part and a convected part as (Versteeg and Malalasekera 1995):

$$\Phi_{c,ij} = q_{n,ij} c_{ij}^{*} \tag{6}$$

where the convective part is now given by the unit discharge normal to the control volume face ($q_{n,ij} = h_{ij} U_{n,ij}$), and the convected part is the scalar concentration itself. Using a first order upwind discretisation, c_{ij}^{*} is defined as:

$$c_{ij}^{*} = \frac{c_i + c_j}{2} - \frac{1}{2}\operatorname{sgn}(q_{n,ij})(c_j - c_i) \tag{7}$$

As it will be shown in the following, the discretisation defined in Equation (6) does not preserve a solution given by a spatially constant concentration, regardless of the discretisation used for c_{ij}^{*}. This is because the convective part in Equation (5) does not include any information about the numerical scheme used to compute the hydrodynamics of the flow ($q_{n,ij}$).

In the particular case in which the initial concentration is constant in the whole domain and equal to $c(x,t) = C_0$, in the absence of any source, according to Equation (3) the concentration must remain constant in space and time, andequal to C_0. In that case the diffusion term vanishes, and the convective numerical flux given by Equation (5) is equal to:

$$\Phi_{c,ij} = q_{n,ij} C_0 \tag{8}$$

The total contribution of the convective flux in Equation (4) is then computed as:

$$\begin{aligned} \sum_{j \in K_i} \Phi_{c,ij} \,|\, \mathbf{n}_{ij} \,| &= \sum_{j \in K_i} q_{n,ij} C_0 \,|\, \mathbf{n}_{ij} \,| \\ &= C_0 \sum_{j \in K_i} q_{n,ij} \,|\, \mathbf{n}_{ij} \,| = C_0 \sum_{j \in K_i} \Phi_{ij}^{C} \\ &= C_0 \Phi_i^{C} \end{aligned} \tag{9}$$

where Φ_i^C represents the centred contribution of the convective flux discretisation in the hydrodynamic mass conservation equation. Using Equation (8) in Equation (4), assuming that the initial concentration is equal to C_0 at every control volume ($c_i^n = C_0$), gives:

$$c_i^{n+1} h_i^{n+1} = C_0 h_i^n - \frac{\Delta t}{A_i} C_0 \Phi_i^C \tag{10}$$

The problem with Equation (9) is that it does not guarantee that the concentration remains constant and equal to C_0. The reason is that the discretisation of the shallow water equations includes not only the centred contribution of the convective flux, but also the upwind contributions of the convective flux, the bed slope and the bed friction source terms. In fact, the computation of h_i^{n+1} from the hydrodynamic mass conservation equation reads:

$$\begin{aligned} h_i^{n+1} &= h_i^n + \frac{\Delta t}{A_i} \sum_{j \in K_i} \left(S_{ij}^U - \Phi_{ij}^C - \Phi_{ij}^U \right) \\ &= h_i^n + \frac{\Delta t}{A_i} \left(S_i^U - \Phi_i^C - \Phi_i^U \right) \end{aligned} \tag{11}$$

where S_{ij}^U is the upwind contribution of the bed slope term. Substitution of h_i^n from Equation (10) into Equation (9) gives after some simple mathematical manipulation:

$$c_i^{n+1} = C_0 \left(1 + \frac{1}{h_i^{n+1}} \frac{\Delta t}{A_i} \left(\Phi_i^U - S_i^U \right) \right) \tag{12}$$

Hence, with this numerical scheme the concentration remains constant only in the particular cases in which $\Phi_i^U = S_i^U$. In fact, the term $\Phi_i^U - S_i^U$ is the error introduced in the mass conservation equation due to the numerical diffusion of the scheme and therefore, it will be reduced if the shallow water equations are solved with a high order scheme. On the other hand, if the hydrodynamic equations are solved with a first order scheme, the numerical diffusion will be much larger.

In order to improve the accuracy of the numerical scheme used for the discretisation of the scalar transport equation, we notice that in Equation (10) a new numerical flux can be defined at each cell face as:

$$q_{n,ij}^* := \left(\Phi_{ij}^C + \Phi_{ij}^U - S_{ij}^U\right) \tag{13}$$

which allows us to write Equation (10) as:

$$h_i^{n+1} = h_i^n - \frac{\Delta t}{A_i} \sum_{j \in K_i} q_{n,ij}^* \tag{14}$$

Considering the similarity between the conservation equations for the mass of water and for the mass of a scalar, we propose to replace Equation (5) to compute the convective numerical flux at each cell face with the following:

$$\Phi_{c,ij} = q_{n,ij}^* c_{ij}^* \tag{15}$$

with $q_{n,ij}^*$ given by Equation (12). Using Equation (14) to compute the convective flux, the discretisation of the scalar transport equation reads (omitting the diffusion term for the sakeof clarity):

$$\begin{aligned} c_i^{n+1} h_i^{n+1} &= c_i^n h_i^n - \frac{\Delta t}{A_i} \sum_{j \in K_i} q_{n,ij}^* c_{ij}^* \\ &= c_i^n h_i^n - \frac{\Delta t}{A_i} \sum_{j \in K_i} c_{ij}^* \left(\Phi_{ij}^C + \Phi_{ij}^U - S_{ij}^U\right) \end{aligned} \tag{16}$$

It should be noticed the similarity between the discretisation given by Equation (15) for the scalar transport equation, and the discretisation of the mass conservation equation (Equations (10) and (13)).

The scheme given by Equation (15) is the one analysed in this paper for a scalar transport equation linked to the shallow water equations. As it will be shown in the examples, this scheme preserves the constant concentration solution under any flow conditions, and it is more stable and accurate than the classic scheme given by Equation (5). Assuming that the initial concentration is equal to C_0 at every control volume ($c_i^n = C_0$), the concentration at the time step t^{n+1} computed from Equation (15) is equal to:

$$c_i^{n+1} h_i^{n+1} = C_0 h_i^n - \frac{\Delta t}{A_i} C_0 \left(\Phi_i^C + \Phi_i^U - S_i^U\right) \tag{17}$$

Equation (16), obtained with the definition for the convective flux given by Equation (14), should be compared with Equation (9), which was obtained using the classic convective flux definition given by Equation (5). Combining Equations (10) and (16) gives:

$$c_i^{n+1} h_i^{n+1} = C_0 h_i^n - C_0 \left(h_i^n - h_i^{n+1}\right) = C_0 h_i^{n+1} \tag{18}$$

and therefore $c_i^{n+1} = C_0$, and the concentration remains constant in the whole spatial domain, regardless the diffusion introduced by the numerical scheme in the hydrodynamic equations.

Considering the splitting of the scalar convective flux given by Equations (5) and (6), a high order discretisation of $\Phi_{c,ij}$ can be easily defined by using a linear reconstruction of the scalar concentration. Several alternatives might be used for this purpose. In this work we have tested two different high-order upwind discretisations, which are described in the following.

The first one is based on the high order scheme presented in (Cea and Vázquez-Cendón 2010), and computes c_{ij}^* as:

$$c_{ij}^* = \frac{c_i + c_j}{2} - \frac{1}{2} \operatorname{sgn}(q_{n,ij})(c_L - c_R) \tag{19}$$

where $c_L$and c_R are the values of the concentration extrapolated at the left and right side of the face between the control volumes *i* and *j*. Notice that in Equation (18) the reconstruction of the concentration at the control volume faces (c_L and c_R) is just used in the upwind part of c_{ij}^*, which represents the numerical diffusion introduced to stabilise the scheme. A centred discretisation of c_{ij}^* reads:

$$c_{ij}^* = \frac{c_i + c_j}{2} \tag{20}$$

The centred discretisation is of second order accuracy, but unstable. The numerical diffusion stabilises the scheme but reduces its accuracy. The idea behind the discretisation given by Equation (18) is to improve the accuracy of the scheme by reducing the numerical diffusion within acceptable limits so that the scheme remains stable.

The second high-order upwind discretisation for c_{ij}^* is given by the Gamma scheme (Jasak et al.1999). The gamma scheme uses second order central differencing wherever this scheme satisfies the boundedness criterion in the normalised variable diagram (Leonard 1988; Gaskell and Lau 1988), and first order upwind differencing otherwise. A blending factor is introduced in order to establish a smooth transition between the first order upwind and the second order centred schemes. The value of c_{ij}^* at the control volume face is computed as:

$$c_{ij}^* = \left[1-\gamma_{ij}(1-f_{ij})\right]c_i + \gamma_{ij}(1-f_{ij})c_j \tag{21}$$

where f_{ij} is the linear interpolation coefficient for the control volume face, and γ_{ij} is the blending factor. For $\gamma_{ij}=0$ the first order upwind scheme is recovered, while $\gamma_{ij}=1$ gives the second order centred scheme. In order to obtain a stable scheme the blending factor γ_{ij} is computed at each control volume face attending to the following cases (Jasak et al. 1999):

$$\begin{aligned} \gamma_{ij} &= 0 && \text{if} \quad 1 \le \hat{c}_i \\ \gamma_{ij} &= 1 && \text{if} \quad \beta_m \le \hat{c}_i < 1 \\ \gamma_{ij} &= \hat{c}_i/\beta_m && \text{if} \quad 0 < \hat{c}_i < \beta_m \\ \gamma_{ij} &= 0 && \text{if} \quad \hat{c}_i \le 0 \end{aligned} \tag{22}$$

where the normalised variable $\hat{c}_i$ is computed as:

$$\hat{c}_i = 1 - \frac{c_j - c_i}{2(\nabla \mathbf{c})_i \mathbf{r}_{i,j}} \tag{23}$$

where the gradient $(\nabla \mathbf{c})_i$ is computed applying Gauss theorem to the control volume, and the distance vector $\mathbf{r}_{i,j}$ joins the nodes N_i and N_j. In Equation (22) the fluid flows through the control volume face from the node N_i to the node N_j. The coefficient β_m in Equation (21) controls the amount of blending between the first order upwind scheme and the second order central differencing scheme. Small values of β_m introduce less blending and therefore produce more accurate but also more unstable solutions, while large values of β_m give a more stable but less accurate scheme. In this work we have used $\beta_m = 0.1$. In (Jasak et al. 1999) a value of $beta_m$ between 0.1–0.5 is recommended.

3 EXAMPLES

Table 1 specifies the numerical discretisations which will be compared in the test cases presented in this section.

Table 1. Schemes used in the discretisation of the convective flux in the scalar transport equation.

	Scalar transport
Flux A	Equations (5) and (6)
Flux B2	Equations (12), (14) and (6)
Flux B3	Equations (12), (14) and (18)
Flux B4	Equations (12), (14) and (20)
Flux B5	Equations (12), (14) and (19)

3.1 *Dam break with scalar transport*

The dam-break test case is used to compare the discretisation described in section 2 for the convective flux in the transport equation of a passive scalar with the classic implementation. In order to do so, a dam-break problem with scalar transport has been solved in which the initial concentration of the scalar is different at both sides of the dam.

Figure 1 shows some water depth and concentration profiles computed from different initial conditions. In all the cases the water depth jump across the dam is $\Delta h = 0.9\ m$, but the jump in concentration varies from one case to another, taking the following values $\Delta c = 0.9, 0.5, 0.2, 0.0$. When the difference in concentration between both masses of water is large ($\Delta c = 0.9$), both the conservative and non-conservative discretisations of the convective flux give very similar results. However, as the difference in concentration diminishes, the classic implementation (Flux A) generates non-physical oscillations in the concentration near the shock front. The spurious oscillation is larger as Δc tends to zero. On the other hand, the implementation described in section 2 (Flux B) gives a smooth concentration profile, provided that it is used with an upwind discretisation for c_{ij}^*. In the case that a centred discretisation is used for c_{ij}^*, strong oscillations of the concentration develop in the solution. It is also interesting to notice the different numerical diffusion associated to the definition of c_{ij}^*, being the less diffusive the Gamma scheme (Flux B4).

3.2 *Tidal flow in a shallow estuary*

In this test case we compute the concentration of a passive scalar (salinity) in the inner part of a shallow estuary with a rather complex geometry and bathymetry (Fig. 2). The numerical model covers a surface of approximately 20 Km^2, with an unstructured mesh of 14287 control volumes. Under average tidal conditions there are extensive intertidal regions which dry and flood with every tidal cycle. In order to analyse the accuracy of the numerical discretisation of the scalar transport equation under different hydrodynamic conditions two different situations are considered. In the first case the mean level of the tidal wave is increased artificially in order to assure that the whole estuary remains wet at low tide and therefore, no wet-dry fronts appear in the solution. In this way the effect of the numerical discretisation

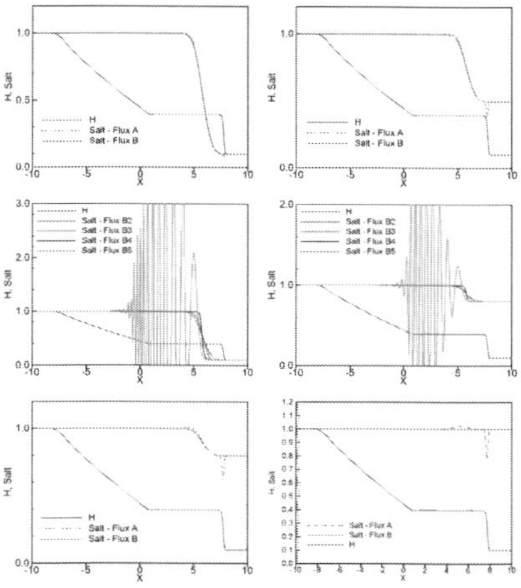

Figure 1. Dam break with scalar transport with different jumps in water depth (Δh) and concentration (Δc). Profiles of water depth and concentration. Flux B includes Fluxes B2 to B5 as detailed in Table 1.

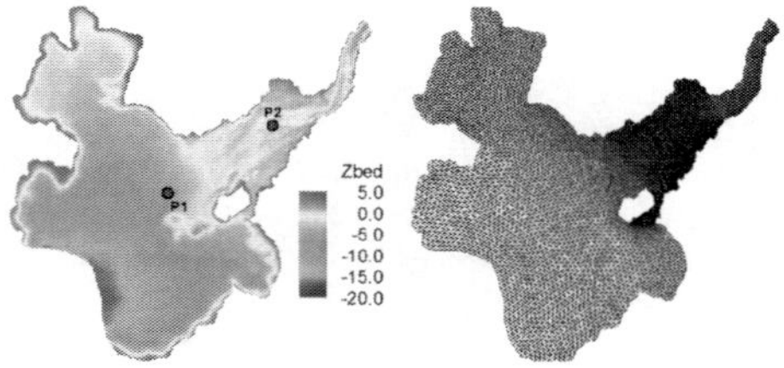

Figure 2. Tidal flow in a shallow estuary. Bathymetry and computational mesh.

can be analysed without taking into account the way in which the wetting and drying algorithm is implemented in the model. In the second case a real tidal waveis used in the computations and thus, certain regions of the estuary dry at low tide. In both situations a discharge of 120 m^3/s is imposed at the river boundary located at the head of the estuary, while at the open sea boundary, a semidiurnal harmonictidal wave with an amplitude equal to the mean tidal range ($\Delta Z_s = 2.52\ m$) is imposed.

Regarding the boundary condition for the concentration of salt, two cases are considered. In the first case a constant concentration of $c_{sea} = c_{river} = 30$ is imposed at both, the sea and the river boundaries. This is obviously not the real boundary condition since the water of the river is fresh water, but it allows us to verify if the numerical scheme is able to mantain a constant salt concentration over the tidal cycle or if it generates spurious oscillations. In the second case, a concentration of $c_{sea} = 30$ is imposed at the open sea boundary, while a fresh water condition $c_{river} = 0$ is used at the river boundary. In whole, considering the flow and boundary conditions, 4 different cases have been modelled, which are summarized in Table 2.

In the test cases with a constant salinity concentration (T2a and T2c), the classic discretisation for the convective flux produces unphysical oscillations of the salinity even when there are no dry regions in the estuary (Figure 3). These oscillations are more relevant near the boundaries and in regions where the bathymetry is more irregular. The value of the oscillations is low, specially considering the salinity variations in real applications, but it reveals that thenumerical scheme is not well-balanced. On the other hand, the discretisation described in section 2 is well-balanced and therefore, it gives a constant concentration in the whole estuary.

Tables 3 and 4 show the results for test cases T2c and T2d regarding mass conservation. The scheme described in section 2 to compute the convective flux is mass conservative and therefore, the error in mass conservation is virtually zero. On the other hand, the classic scheme does not guarantee mass conservation, although the relative error in the total mass of salt in the estuary is of the order of 2%.

The amount of numerical diffusion of the several discretisation schemes defined in section 2 is clearly shown in Figures 4 and 5. The less diffusive is the centred scheme, although it generates some spurious oscillations, specially near the river boundary. The Gamma scheme is the less diffusive bounded scheme. The first order schemes (Flux B1 and B2) introduce a similar level of numerical diffusion on the solution. Visual observation of the results does

Table 2. Tidal flow in a shallow estuary. Flow conditions.

Test	Wet-dry	Boundary conditions
T2a	No dry regions	$c_{sea} = 30$, $c_{river} = 30$
T2b	Wet-dry fronts	$c_{sea} = 30$, $c_{river} = 30$
T2c	No dry regions	$c_{sea} = 30$, $c_{river} = 0$
T2d	Wet-dry fronts	$c_{sea} = 30$, $c_{river} = 0$

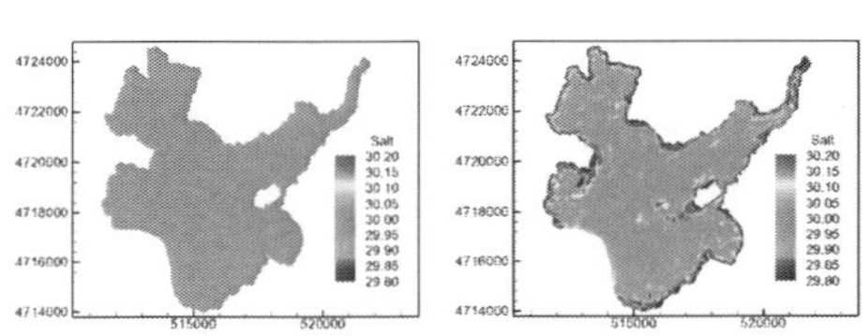

Figure 3. Test case T2a: tidal flow in a shallow estuary with no dry regions and constant concentration. Salinity concentration field at mean ebb tide. Flux A (left) and Flux B (right). Flux B includes Fluxes B2 to B5 as detailed in Table 1.

Table 3. Tidal flow in a shallow estuary with no dry regions and variable concentration. Mass of fresh water in the estuary.

Time	Analytical	Flux A	Flux B
0	0	0	0
43200	1.56E+08	1.53E+08	1.56E+08
86400	3.11E+08	3.05E+08	3.11E+08
172800	6.22E+08	6.10E+08	6.22E+08
378000	1.36E+09	1.33E+09	1.36E+09

Table 4. Tidal flow in a shallow estuary with wet-dry fronts and variable concentration. Mass of fresh water in the estuary.

Time	Analytical	Flux A	Flux B
0	0	0	0
43200	1.56E+08	1.54E+08	1.56E+08
86400	3.11E+08	3.07E+08	3.11E+08
172800	6.22E+08	6.14E+08	6.22E+08
378000	1.36E+09	1.34E+09	1.35E+09

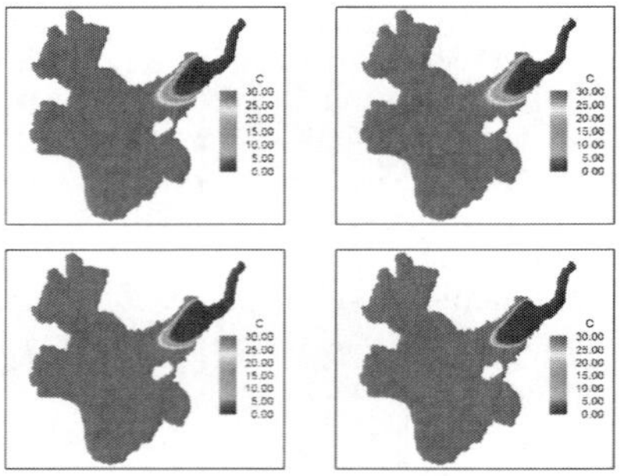

Figure 4. Tidal flow in a shallow estuary with no dry regions and variable concentration. Salinity concentration field at mean ebb tide computed with several schemes.

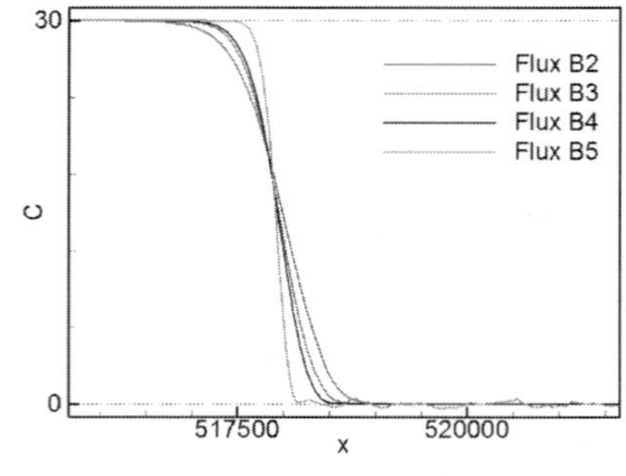

Figure 5. Tidal flow in a shallow estuary with no dry regions and variable concentration. Longitudinal profiles of salinity along the main channel axis, computed with different schemes.

not reveal any significant differences in the salinity fields computed although, as it is shown in Tables 3 and 4, the classic scheme (Flux A) does not conserve the total mass of fresh water in the whole estuary.

4 CONCLUSIONS

The discretisation analysed for the convective flux of a passive scalar is linked to the numerical scheme used to solve the hydrodynamic equations. It has been shown that this is a necessary requirement in order to define a conservative scheme for a depth-averaged scalar transport equation. The most important implication is that a well balanced and conservative scheme for the scalar transport equation cannot be formulated just from the water depth and velocity fields, but has to consider also the way in which the hydrodynamic equations have been solved. Several first and high order conservative discretisation alternatives have been analysed and tested in simple and complex geometries including unsteady wet-dry fronts.

REFERENCES

Cea, L. and M.E. Vázquez-Cendón (2010). Unstructured finite volume discretisation of two-dimensional depth averaged shallow water equations with porosity. *Int. J. Numer. Meth. Fluids 63*(8), 903–930.

Gaskell, P.H. and A.K.C. Lau (1988). Curvature-compensated convective transport: SMART, a new boundedness-preserving transport algorithm. *Int. J. Numer. Meth. Fluids 8*, 617–641.

Jasak, H., H.G. Weller, and A.D. Gosman (1999). High resolution NVD differencing scheme for arbitrarily unstructured meshes. *Int. J. Numer. Meth. Fluids 31*, 431–449.

Latorre, B., P.G. a Navarro, J. Murillo, and J. Burguete (2009). An efficient and conservative model for solute transport in unsteady shallow water flow. In P.I.-R.G. L.-P. F.J. M.-S. G.P.-S. P.A. Lopez-Jimenez, V.S. Fuertes-Miquel (Ed.), *Proceedings of the International Workshop on Environmental Hydraulics,*, pp. 115–117. CRC Press.

Leonard, B.P. (1988). Simple high-accuracy resolution program for convective modelling of discontinuities. *Int. J. Numer. Meth. Fluids 8*, 1291–1318.

LeVeque, R.J. (2002). Finite Volume Methods for Hyperbolic Problems, Volume 31 of Cambridge Texts in Applied Mathematics. Cambridge University Press.

Murillo, J., J. Burguete, and P. Garcia-Navarro (2008). Analysis of a second-order upwind method for the simulation of solute transport in 2D shallow water flow. *International Journal of Numerical Methods in Fluids 56*, 661–686.

Toro, E.F. (2001). *Shock-capturing Methods for Free-Surface Shallow Flows*. Chichester, West Sussex PO19 1UD, England: Wiley.

Versteeg, H.K. and W. Malalasekera (1995). *An introduction to computational fluid dynamics. The finite volume method.* Essex, UK: Addison Wesley Longman Ltd.

Numerical Methods for Hyperbolic Equations – Vázquez-Cendón et al. (eds)
© 2013 Taylor & Francis Group, London, ISBN 978-0-415-62150-2

Modeling of shallow curved flows with 3D unsteady RANS

B. Fraga, E. Peña & L. Cea
GEAMA (Water and Environmental Engineering Group), A Coruña University, Spain

ABSTRACT: Complex flows around in-stream structures or curved channels exhibit three dimensional features that 2-D or 1-D models cannot realistically represent and non-hydrostatic phenomena that Shallow Waters Equations (SWE) models ignore. Flow in meandering or curved channels is characterized by circulation cells generated by centrifugal forces around the meander bend that move in opposite directions from one meander apex to the next. This results in a highly complex configuration, even for shallow flows.

In this work, a 3-D layer-structured finite volume model for free surface flows was applied to a 270° bend curved channel. There is a strong separation between horizontal and vertical scales, exhibiting a shallow flow behavior. 3-D unsteady momentum and mass conservation equations were solved using a Finite Volume code developed by the authors. 3-D grid is collocated in the vertical coordinate adding several horizontal layers to a 2D horizontal unstructured mesh. 29 horizontal layers were calculated in simulations to get a good resolution of the shear stress at the bottom. 3-D Unsteady Reynolds Averaged Navier-Stokes equations were solved. $k - \varepsilon$ model of turbulence are included in the equations.

The model was tested in some study cases which exhibit important non-hydrostatic 3D flow behavior. Results in the curved channel mentioned above were used preliminarily to validate model's accuracy. Comparisons of simulated primary and secondary velocities with experimental results taken from the literature were made. Good agreement was found between numerical velocities and measured ones. Transversal flow patterns were simulated and analyzed. Influence of turbulence modelling, spatial discretization and bed friction is determinant for a good description of the flow.

1 INTRODUCTION

Numerical models based on the Shallow Waters Equations (specially 2D but also 3D approaches) have been widely used in river engineering applications with success. These models can accurately predict important features of the flow like water elevation, primary velocities and wet-dry fronts, among others. However, the common feature of these models is the hydrostatic assumption, which is not compatible with computation of some phenomena of practical interest.

Non-hydrostatic models describe the 3D flow equations with no assumption about pressure distribution. Their use has been somehow restricted in practical hydrodynamic applications because of computational requirements. They have been more popular in aerodynamics, where hydrostatic assumption is not feasible at all. Most of three-dimensional non-hydrostatic models were adapted to the features of compressible flow applications, in which turbulence modeling and/or solving in small length scales is the key issue and the computational cost is extremely high. Many of them have little practical application to river engineering, due to the extent and complexity of the spatial domains and the variety of length scales of the flow in real rivers.

Nevertheless, non-hydrostatic phenomena are of great practical importance for some hydrodynamic problems, such as flows around in-stream structures like piers or bridge's columns. But also in curved or meandering reaches. Meanders are formed by helical recirculation cells which transport high momentum water from the outer bank to the inner one, contributing to erode the outer bank and and depositing sediments at the inner side. This is a very important process both from the research in hydromorphology and hydrodynamics point of view but also for practical design of protection structures against floods or regeneration strategies. Shear stresses calculation is a key issue for engineers because of its influence in erosive processes. It is very important to characterize the balance of energy in this cases for accurately computing of sediment transport.

Simulation of hydrodynamics in natural rivers is a challenge for CFD models. Patterns of flow are very complex and a lot of processes are involved. Turbulence is important, but also curvature has a big effect on velocities and shear stresses distribution, as it generates flow separation and transverse pressure gradients. Other parameters like irregular bed shape, heterogeneous roughness or interaction between in-bank and out-bank flow (when floods) contribute to become velocity and vorticity patterns even more complicated.

Curved open channels give both experimental and numerical researchers the chance to study some of the previously described phenomena separately. Even reducing the sources of complexity, modeling turbulent flow in open non-straight laboratory flumes is a challenging case of study. Only a few researchers have investigated 3D structures and secondary velocities in curved channels, and many of these works are focused on detailed turbulent modeling and coherent structures catching (Stoesser et al. 2010).

Flow in bends is characterized by a strong transversal recirculation cell from the outer bank to the inner one. Curvature generates an unbalance of centripetal forces and, therefore, a span-wise pressure gradient. On the other hand, difference between boundary conditions on bed (impervious rough wall) and top (free surface) generates also a non-hydrostatic pressure gradient in the vertical direction. High momentum water hits the outer bank (this creates high shear stress and would generate erosion with mobile bed/banks) and moves to the bed towards the inner bank. Once there, water has lost momentum (this would create deposition of sediments on the inner part) and moves up to the surface. All these processes are driven by the non-hydrostatic part of pressure, so only mathematical and numerical models that take this into account are able to reproduce it.

Many models whose scope is in detailed solving of turbulent structures could not be applied to real engineering problems, as consequent computational cost is too high. In this work, a Finite Volume code to solve 3D Unsteady Reynolds Averaged Navier-Stokes equation (RANS) is used. Simulations in an open, turbulent channel flow with high curvature and shallowness was made, obtaining a good compromise between description of secondary velocities and three-dimensional flow features and a good computational performance for practical engineering work. In order to validate the parameters, experimental results obtained from () were used to compare with the numerical ones.

2 MATHEMATICAL MODEL AND NUMERICAL FRAMEWORK

In this section a non-hydrostatic 3D CFD model is presented.

2.1 *Governing equations*

3D Unsteady RANS equations for an incompressible Newtonian fluid can be expressed in scalar form as

$$\frac{\partial U_j}{\partial x_j} = 0 \tag{1}$$

$$\frac{\partial U_i}{\partial t} + \frac{\partial U_i U_j}{\partial x_j} = -\frac{1}{\rho}\frac{\partial P}{\partial x_i} + \frac{\partial}{\partial x_j}\left[\frac{T_{ij}^v}{\rho} - \overline{u_i u_j}\right] - \delta_{i3g}, \quad i = 1,3 \tag{2}$$

where (1) and (2) are the mass and momentum conservation equations, respectively. Regarding dependent variables, U_i and P are the velocity and total pressure averaged over time, respectively. Pressure can be divided into a dynamic part plus a hydrostatic contribution $P = P_d + P_h \cdot$ ρ is the fluid density, g is the gravity acceleration, μ is the fluid dynamic viscosity, T_{ij}^v is the viscous stress tensor (described in equation (3)) and $\rho\overline{u_i u_j}$ are the Reynolds stresses.

$$T_{ij}^v = \mu\left(\frac{\partial U_i}{\partial x_j} + \frac{\partial U_j}{\partial x_i}\right) \tag{3}$$

The hydrostatic pressure balances the gravity force in the vertical coordinate $x_3 = z$, so we can express (2) in terms of dynamic pressure:

$$\frac{\partial U_i}{\partial t} + \frac{\partial U_i U_j}{\partial x_j} = -g\frac{\partial z_s}{\partial x_i} - \frac{1}{\rho}\frac{\partial P_d}{\partial x_i} + \frac{\partial}{\partial x_j}\left[\frac{T_{ij}^v}{\rho} - \overline{u_i u_j}\right], \quad i = 1,2 \tag{4}$$

$$\frac{\partial U_3}{\partial t} + \frac{\partial U_3 U_j}{\partial x_j} = -\frac{1}{\rho}\frac{\partial P_d}{\partial x_3} + \frac{\partial}{\partial x_j}\left[\frac{T_{3j}^v}{\rho} - \overline{u_3 u_j}\right] \tag{5}$$

2.2 *Turbulence model*

As in RANS model turbulent fluctuations are not solved, Reynolds stress term is unknown and needs to be modelled to enclose the equation system. In this case, the standard $k - \varepsilon$ is used. This is an isotropic two-equation model based on the Boussinesq approximation, which establishes a relationship between the turbulent stresses and the averaged velocity gradients:

$$-\overline{u_i u_j} = \nu_t\left(\frac{\partial U_i}{\partial x_j} + \frac{\partial U_j}{\partial x_i}\right) - \frac{2}{3}k\delta_{ij} \tag{6}$$

This relationship is regulated by a variable called turbulent or eddy viscosity ν_t. In the case of $k - \varepsilon$ model, this "viscosity" is determined considering the dynamics of turbulence through the turbulent kinetic energy k and its dissipation rate ε:

$$\nu_t = C_\mu \frac{k^2}{\varepsilon} \tag{7}$$

As a result, two more transport equations and two more variables are added to the system formed by (1), (4) and (5).

$$\frac{\partial k}{\partial t} + U_j \frac{\partial k}{\partial x_j} = \frac{\partial}{\partial x_j}\left[\left(\nu + \frac{\nu_t}{\sigma_t}\right)\frac{\partial k}{\partial x_j}\right] + P_k - \varepsilon \tag{8}$$

$$\frac{\partial \varepsilon}{\partial t} + U_j \frac{\partial k}{\partial x_j} = \frac{\partial}{\partial x_j}\left[\left(\nu + \frac{\nu_t}{\sigma_\varepsilon}\right)\frac{\partial \varepsilon}{\partial x_j}\right] - C_{\varepsilon 1}\frac{\varepsilon}{k}P_k - C_{\varepsilon 2}\frac{\varepsilon^2}{k} \tag{9}$$

where P_k is the production of turbulent kinetic energy term, defined as:

$$P_k = \nu_t \frac{\partial U_i}{\partial x_j}\left(\frac{\partial U_i}{\partial x_j} + \frac{\partial U_j}{\partial x_i}\right) \tag{10}$$

The parameters $C_\mu = 0.09$, $C_{\varepsilon 1} = 1.44$, $C_{\varepsilon 2} = 1.92$, $\sigma_\varepsilon = 1.3$ and $\sigma_k = 1.0$ are constants. There is a general agreement in the literature about their standard values ().

2.3. *Initial and boundary conditions*

Equations (1), (4), (5), (8) and (9) are unsteady. In order to solve them, initial value of all the dependent variables must be specified:

$$\begin{aligned} &U_i(x, y, z, t=0) = U_i^0, \quad i = 1, 3 \\ &P_d(x, y, z, t=0) = P_d^0 \\ &z_s(x, y, z, t=0) = z_s^0 \\ &k(x, y, z, t=0) = k^0 \\ &\varepsilon(x, y, z, t=0) = \varepsilon^0 \end{aligned} \tag{11}$$

where z_s is the free surface elevation.

Regarding boundary conditions, four types are considered: open boundaries (inlet and outlet), wall and free surface.

At inlet boundary the three components of velocity and derivative of the dynamic pressure in the normal direction (x_n) are defined:

$$\begin{aligned} &U_i(\Gamma_{in}, t) = U_{in}, \quad i = 1,3 \\ &\frac{\partial P_d}{\partial x_n}(\Gamma_{in}, t) = 0 \end{aligned} \tag{12}$$

where Γ_{in} represents the nodes of the inlet boundary and U_{in} is the value imposed for each velocity component. Inlet conditions for k and ε consist in a given value which can be evaluated in several ways. In this case:

$$\begin{aligned} &k(\Gamma_{in}, t) \approx \frac{3}{2}(T_u U_i)^2, \quad i = 1, 3 \\ &\varepsilon(\Gamma_{in}, t) \approx C_\mu \frac{k_{in^2}}{\nu_{t,in}} \end{aligned} \tag{13}$$

where T_u is the turbulence intensity and $\nu_{t,in}$ the eddy viscosity at the inlet. Both can be approximated from experimental values or previous works.

At the outlet boundary water surface elevation and derivative of the dynamic pressure in the normal direction (x_n) are prescribed:

$$\begin{aligned} &z_s(\Gamma_{out}, t) = z_{out} \\ &\frac{\partial P_d}{\partial x_n}(\Gamma_{out}, t) = 0 \\ &\frac{\partial k}{\partial x_n}(\Gamma_{out}, t) = 0 \\ &\frac{\partial \varepsilon}{\partial x_n}(\Gamma_{out}, t) = 0 \end{aligned} \tag{14}$$

where Γ_{out} represents the points of the outlet boundary and z_{out} is the value imposed for water surface elevation at the outlet boundary. This is the most restrictive condition of all in terms of physical meaning. Null normal velocity gradient $\left(\partial U_n / \partial x_n = 0\right)$ could also be imposed. This means flow is fully developed and outlet is conveniently far away from our main area o study. But this is not necessary as long as a non-centered numerical discretization scheme is used, as it will be shown in the following section. No velocity should be imposed, even far away from the area of interest, in order to ensure mass conservation.

Wall boundaries include lateral solid walls and bottom bed. They are impervious and therefore the normal velocity component (U_n) is zero. Transverse velocities (U_t) are affected by the wall friction, generating shear stresses. Although the code allows to use different strategies to simulate this phenomenon, in this case wall function approach was used in all wall boundaries. Additionally, normal derivative of pressure is also set to zero.

$$\begin{aligned} &U_n(\Gamma_{wall}, t) = 0 \\ &U_t(\Gamma_{wall}, t) = \frac{u_*}{\kappa} ln\left(\frac{y u_* E}{\nu}\right) \\ &\frac{\partial P_d}{\partial x_n}(\Gamma_{wall}, t) = 0 \end{aligned} \tag{15}$$

where Γ_{wall} is the wall boundary, u_* is the friction velocity, k is the von Karman's constant ($k = 0.41$), ν is the cinematic viscosity and E is a parameter dependent of the roughness. To determine its value the non-dimensional roughness K_s^+ should be known first:

$$K_s^+ = \frac{k_s u_*}{\nu} \tag{16}$$

For the simulations which are going to be presented in this job, $k_s = 1.3\ mm$, so K_s^+ value is between 5 and 70. According to (), value of E in this range is defined as:

$$E = \frac{1}{0.11 + 0.033 K_s^+} \tag{17}$$

Finally, k and ε boundary condition with wall function are:

$$\begin{aligned} k(\Gamma_{wall}, t) &= \frac{u_*^2}{\sqrt{C_\mu}} \\ \varepsilon(\Gamma_{wall}, t) &= \frac{u_*^3}{\kappa \delta_{wall}} \end{aligned} \tag{18}$$

where δ_{wall} is the distance from the node to the wall.

Although the code is able to compute position and time evolution of free surface, this feature was not used in this case. This is because the focus was on secondary flow characterization in stationary state, and free surface evolution has very little impact on results once the stationary is achieved. Experimental data and numerical results indicate that this approach is very close to reality once the flow is stabilized. In consequence, computational cost is lower.

Therefore a fixed lid parallel to the bottom was imposed. This is a common approach in CFD. Symmetry condition was set in this boundary:

$$\begin{aligned} \frac{\partial U_n}{\partial x_n}(\Gamma_{surf}, t) &= 0 \\ P_d(\Gamma_{surf}, t) &= 0 \end{aligned} \tag{19}$$

where Γ_{surf} are the points of the free surface boundary. U_t value is completely free, as zero shear stress is considered (no wind). Dynamic pressure is zero at the surface due to atmospheric pressure is taken to be the reference.

2.4 *Finite volume solver*

The former equations are solved with a layer-structured Finite Volume solver. Spatial discretization is done in a semi-structured cell-centered collocated grid. 3D mesh is built as a 2D unstructured horizontal mesh with several horizontal layers of variable thickness. Views of the mesh built for this particular test case can be seen on Figures 2 and 3. Variables are stored at the nodes of the cells, which can have any shape.

The 3D momentum equations are solved with an unsteady semi-implicit Finite Volume solver in a collocated grid. Equations can be discretized in time following an implicit ($t^{n+\theta}$) or explicit (t^n) schemes. Under-relaxation factors for velocity and pressure can be set by the user, resulting in higher stability of the code.

Convective fluxes are computed through an upwind discretization scheme, whereas pressure gradient term is discretized using a centered scheme. As a collocated grid is used for the numerical discretization, it is necessary to use some kind of stabilization technique in order to avoid unrealistic results in the case of a highly irregular (checkerboard) pressure field. In the present code, Rie-Chow interpolation () has been used to stabilize the dynamic pressure.

Mass continuity is achieved by an iterative pressure-velocity correction using the SIMPLE algorithm ().

During the iterative procedure described in this section, six systems of equations need to be solved: three for the momentum equations, one for the pressure correction (that ensures mass conservation) and two more for the turbulence model. There are as many equations on each system as volumes in the mesh. User can solve each system iteratively by horizontal layers or vertical columns. In each layer or column a sparse linear system needs to be solved. The method implemented in the code is a preconditioned GMRES (). For further details regarding discretization method or algorithm and solver implementation, see ().

3 CASE OF STUDY AND SETUP

3.1 *Experimental data-set*

This non-hydrostatic 3D numerical model has been validated using measurements found in the literature. Following results correspond to the experiment of Steffler in 1984, later published by Gahmry and Steffler () and Tritthart and Gutknecht (). This experiment consist of a fully turbulent, stationary and shallow flow in a 270 degrees bend of high curvature. The flume is 32.4 m in total, consisting of a straight inlet of 6.13 m in which flow is supposed to achieve a full development and one straight outlet of 2.53 m. Both are connected by a bend with 3.66 m radius, as seen in Figure 1. Width is 1.07 m and bed slope is 0.083%.

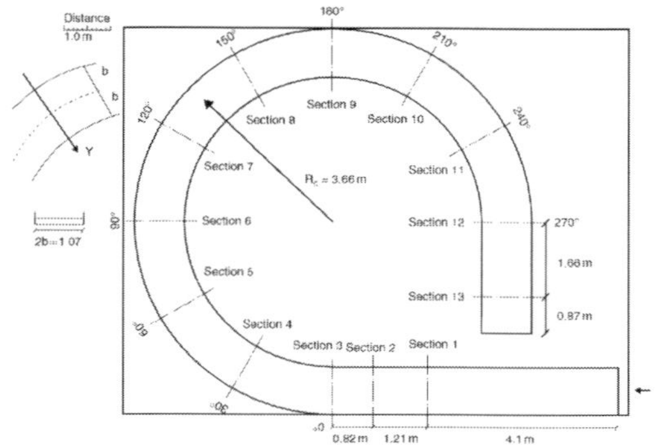

Figure 1. Layout of Steffler's experimental flume.

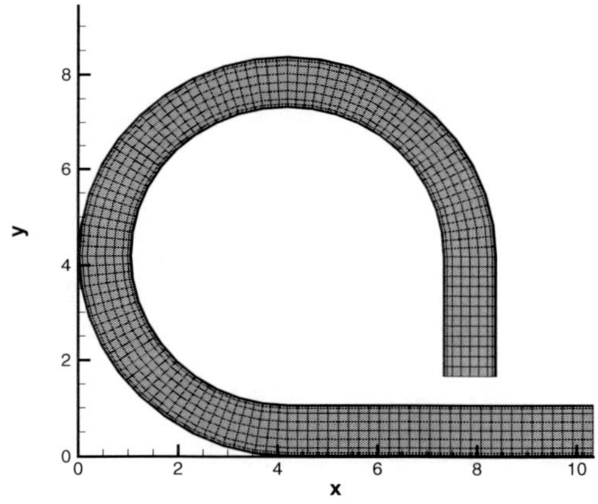

Figure 2. Plant view of the numerical mesh. Resolution is higher in span-wise direction, specially at the walls. This is possible due to RANS codes performance.

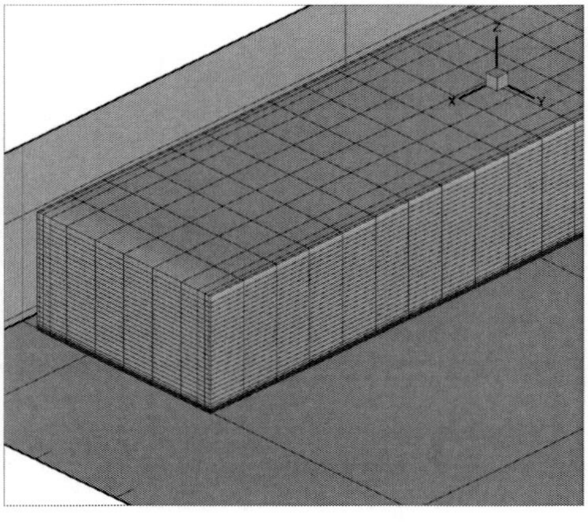

Figure 3. Detail of the 3D mesh at the inlet. 29 vertical layers can be observed with higher resolution at the bottom.

Experiment data:

- Reynolds Number: 3.852E+5 (length scale: width of the channel) and 2.196E+4 (length scale: water depth).
- Total discharge: 0.0235 m^3/s.
- Water depth at outlet: 6.1 *cm*.
- Bulk velocity: 0.36 *m/s*.
- Equivalent sand height: $k_s = 1.3$ *mm*.

Two values for Re were included because length scales are strongly different in span-wise and vertical direction. A different behaviour of turbulence in each of these scales is expected.

3.2 *Simulation setup*

Simulations were made in the collocated grid shown in Figures 2 and 3. These are the main characteristics of the mesh:

- Number of cells: 45936.
- Number or vertical layers: 29.
- Distance to first node: $y^+ \approx 130$ (lateral walls) and $y^+ \approx 70$ (bottom).

Spatial discretization scheme used is first-order Upwind scheme.

The concrete values for boundary conditions were taken from the experiment data.

- Inlet velocity: $U_{in} = 0.36$ m/s.
- Water surface elevation at the outlet: z_{out}=0.061 *m*. This is also the height of the top boundary.

4 RESULTS AND DISCUSSION

4.1 *Non-hydrostatic effects*

The focus of this work lies on simulating the 3D non-hydrostatic effects that a usual 2D Shallow Waters model is not able to represent. As commented on Introduction, the main feature of flow in bend is a strong secondary current which drives high momentum fluid from the outer bank to the inner one along the channel bed. To ensure continuity, low momentum fluid will flow back to the outer bank near the surface, generating a recirculation cell. This phenomena was correctly captured by the model, as seen in Figures 4 and 5.

4.2 *Primary flow*

Figure 6 compares the calculated primary velocity with data from Steffler's experiment. Velocity profiles were measured in different points at 0°, 90°, 180° and 270° cross-sections of the bend. –0.8

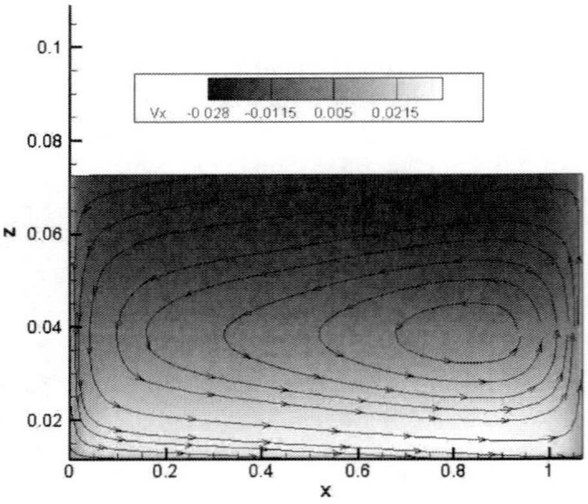

Figure 4. Slice at 90° of the bend. Z coordinate is 10% distorted. Inner bank is in the right side.

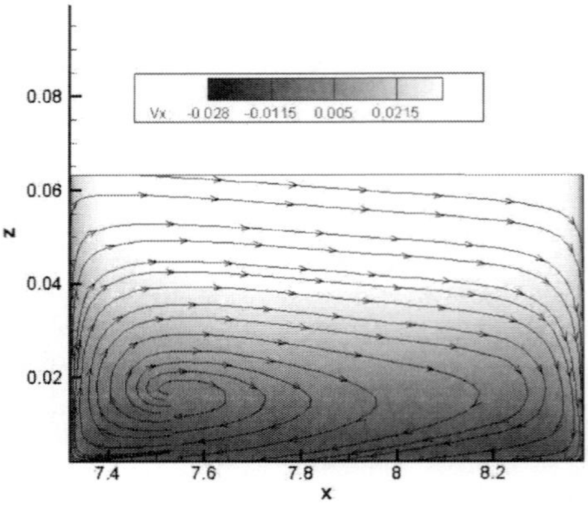

Figure 5. Slice at 270° of the bend. Z coordinate is 10% distorted. Inner bank is in the left side.

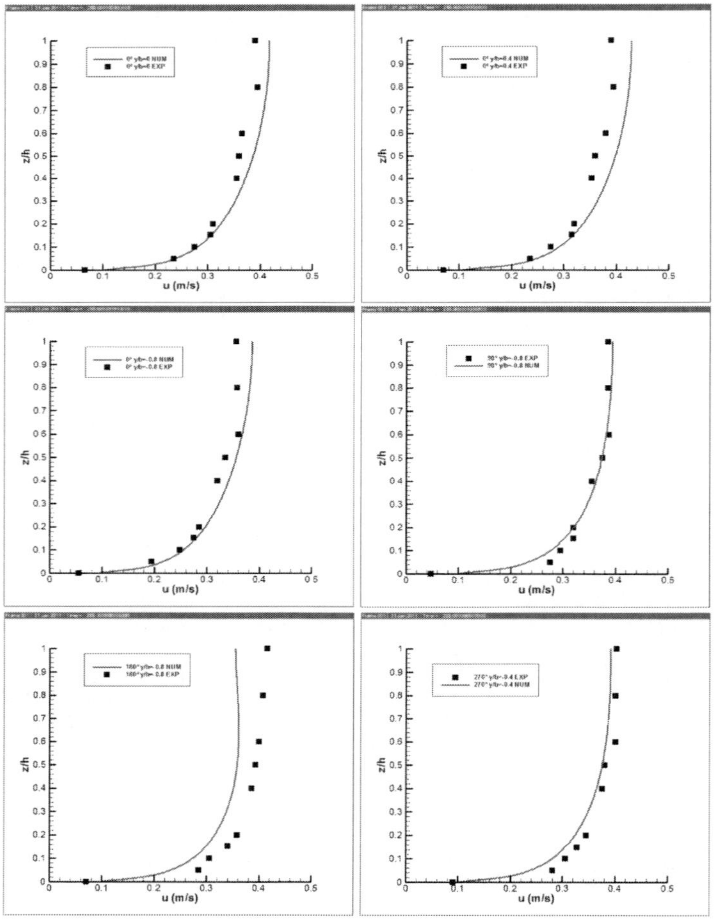

Figure 6. Primary velocities for several profiles in different sections. Square points represent experimental measurements while lines represent numerical results.

and 0.8 y/b are close to the wall, while 0 y/b corresponds to the middle line of each section.

The agreement of calculated with measured longitudinal velocities is very satisfying, both in the distribution of the velocity field along the vertical profile and the estimated values.

4.3 *Secondary flow*

Figure 7 shows two profiles of transversal velocities for the 90°, 180° and 270° sections. There are no profiles for 0° section because secondary currents are generated by the centripetal forces due to the curvature, so the effect is only remarkable once into the bend.

The agreement between numerical and experimental results is quite good. Simulations are able to capture the shape of the transversal velocity field, which means a good quantitative description of the recirculation cell characteristic of bends. This phenomenon is far more difficult to describe than primary flow because absolute velocity values are much lower (about one order of magnitude), so relative error gets higher. But its main complexity relies on the dependence on the wall friction modeling (boundary conditions), on the turbulence phenomena (turbulence model), the discretization scheme (high diffusive schemes may hide some effects) and on mesh resolution.

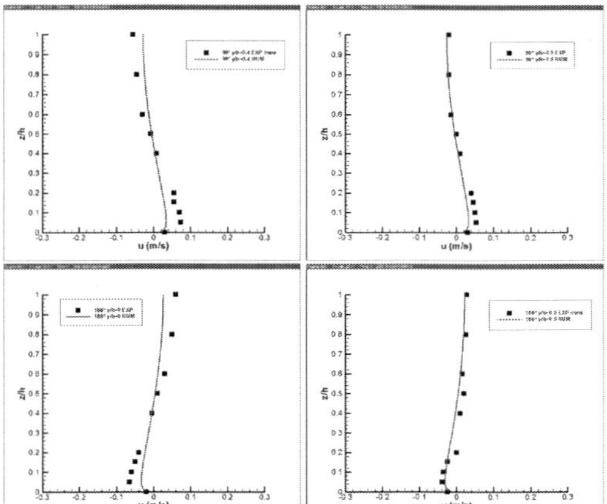

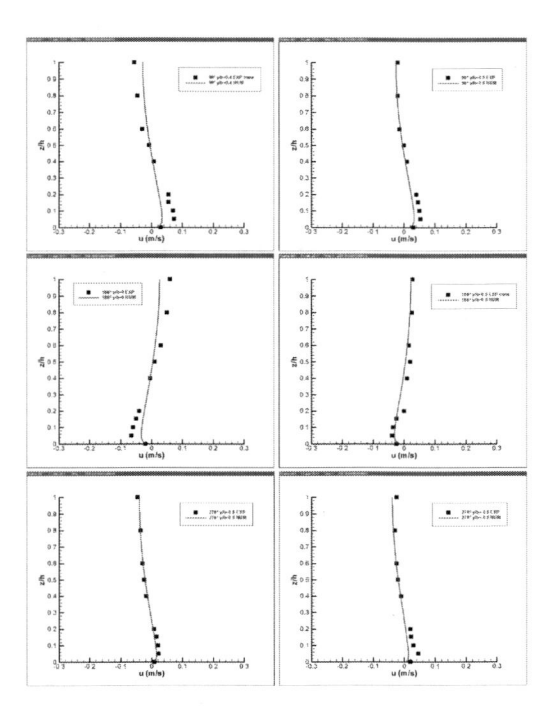

Figure 7. Secondary velocities for several profiles in different sections. Square points represent experimental measurements while lines represent numerical results.

For the present case, a very good agreement was found for the profiles closest to the wall ($y/b = \pm 0.8$), as it can be observed on the graphics of the right column. For profiles closer to the center of the flume, numerical results tend to sightly minimize the rotation effect observed in the experiments.

5 CONCLUSIONS AND FUTURE RESEARCH

Agreement in primary velocities of the flow is very high when comparing with experimental results. Transfer os mass and momentum is accurately characterized in the stream-wise direction.

Simulation of secondary flow is very satisfactory. Main non-hydrostatic structures of the flow were captured and good agreement was found with experimental profiles in different sections.

However, while secondary velocities show a very good agreement with vertical profiles near the wall, the more to the center of the flume, the less the accuracy. Results are still good, but part of flow vorticity is being somehow diffused. These disagreements may be generated by:

- Turbulence model: isotropic $k-\varepsilon$ has some limitations, specially when flow is clearly anisotropic, like in this case, and near the walls.
- Boundary conditions: law of the wall is always an approximation to reality. But it works really well in RANS models. Apparently this shouldn't be the main problem, as long as the agreement between experimental an numerical estimations near the wall is very good.
- Computational grid: mesh resolution is coarser in the center of the flume.
- Discretization scheme of advective terms: simulations were performed using a first-order upwind scheme. It is very stable, but adds an important numerical diffusion to the flow.

According to the former analysis, present and future lines of analysis and work in this field are:

- Turbulence modelling and solving: simulations with PANS model () and () are being carried out for the same test cases previously modeled with RANS. Implementation of anisotropic features on turbulence models may also enhance secondary flow characterization.
- Computational grid: a sensitivity analysis for different mesh resolutions is being done.
- Discretization scheme of advective terms: implementation of second-order schemes is being tested.

Some CFD models can handle 3D non-hydrostatic simulations with high accuracy. As long as turbulence is solved instead of modelled (LES, DNS, DES, PANS...) results achieve a high level of detail. But this enhancement comes together with a high computational cost and some constraints regarding computational domain and mesh definitions. In this work, an Unsteady RANS code is applied with very reasonable computational requirements (low resolution mesh, 1st order schemes, conventional turbulence models...) and very good computational performance. Results show very good both qualitative and quantitative agreement with experiments, and therefore a good validation for the code. A good preliminary compromise between accuracy in solving complex phenomena and applicability of the code for real large scale river engineering applications is obtained.

ACKNOWLEDGEMENTS

This work is part of a research project supported by the Spanish Science and Innovation Ministry (CGL2008-03319). PhD research of the first author is funded by the IN.CI.TE. program of the regional Government of Galicia (Spain).

REFERENCES

Cea L., Puertas J. and Vázquez-Cendón M.E. 2007. Depth Averaged Modelling of Turbulent Shallow Water Flow with Wet-Dry Fronts. *Archives of Computational Methods in Engineering* 14(3), 303–341.

Cea L., Stelling G., Zijlema M. 2009. Non-hydrostatic 3D free surface layer-structured finite volume model for short wave propagation. *International Journal for Numerical Methods in Fluids* 81, 382–410.

Duan J.G. 2004. Simulation of flow and mass dispersion in meandering channels. *Journal of Hydraulic Engineering* 130(10), 964–976.

Gahmry H. K., Steffler P. M. 2005. Two-dimensional depth-averaged modeling of flow in curved open channels. *Journal of Hydraulic Research* 43(1), 44–55.

Girimaji S. S., Jeong E., Srinivasan R. 2006. Partially Averaged Navier-Stokes Method for Turbulence: Fixed Point Analysis and Comparison with Unsteady Reynolds Averaged Navier-Stokes. *Journal of Applied Mechanics* 73, 422–429.

Ma J.M., Peng S.-H., Davidson L., Wang F.J. 2011. A low Reynolds number variant of partially-averaged Navier-Stokes model for turbulence. *International Journal of Heat and Fluid Flow* 32(3). 652–669.

Patankar S.V. 1980. Numerical Heat Transfer and Fluid Flow. *McGraw-Hill, New York*.

Rhie C.M., Chow W.L. 1983. Numerical study of the turbulent flow past an airfoil with trailing edge separation. *AIAA Journal* 21(11), 1525–1532.

Rodi W. 1980. Turbulence models and their application in hydraulics. *IAHR Monograph Series, Rotterdam: Balkema*.

Shiono, K., Spooner, J., Chan, T., Rameshwaran, J., and Chandler, J. 2008. Flow Characteristics in Meandering Channels with Non-Mobile and Mobile Beds for Overbank Flows. *Journal of Hydraulic Research* 46(1), 595–609.

Stoesser T., Ruether N. and Olsen N.R.B. 2010. Calculation of primary and secondary flow and boundary shear stresses in a meandering channel. *Advances in Water Resources* 33(2), 158–170.

Tritthart M., Gutknecht D. 2007. Three-dimensional simulation of free-surface flows using polyhedral finite volumes. *Engineering Applications of Computational Fluid Mechanics* 1, 1–14.

Versteeg H.K., Malalasekera W. 1995. An Introduction to Computational Fluid Dynamics. *Pearson. Prentice Hall, Essex*.

Vuik C., van der Vorst H.A. 1992. A comparison of some GMRES-like methods. *Linear Algebra and its Applications* 160, 131–162.

Wormleaton P.R., Ewunetu M. 2006. Three-dimensional k—e numerical modelling of overbank flow in a mobile bed meandering channel with floodplains of different depth, roughness and platform. *Journal of Hydraulic Research* 44(1), 18–32.

Numerical Methods for Hyperbolic Equations – Vázquez-Cendón et al. (eds)
© 2013 Taylor & Francis Group, London, ISBN 978-0-415-62150-2

A finite volume/duality method for Bingham viscoplastic flow

José M. Gallardo
Department of Análisis Matemático, Facultad de Ciencias, Universidad de Málaga, Spain

Enrique D. Fernández-Nieto
Department of Matemática Aplicada I, ETS Arquitectura, Universidad de Sevilla, Spain

Paul Vigneaux
Unité de Math. Pures et Appliquées, ENS de Lyon, France

ABSTRACT: This work focuses on the numerical analysis of a Bingham viscoplastic model. In particular, the numerical scheme shows a coupling between a well-balanced finite volume solver for the mass conservation law and a duality method for solving the variational inequality related to the conservation of momentum. We investigate the optimal choice of parameters in the duality method in order to achieve the greatest rate of convergence. A number of numerical experiments are proposed to test the performances of the proposed coupled algorithm.

1 INTRODUCTION

In recent years there has been increasing interest in the understanding and numerical simulation of Bingham fluid flows: see (Dean, E.J. & Glowinski 2007), (Oswald 2009). Bingham fluids behave rigidly below a certain strees yield and flow as an incompressible viscous fluid above that yield. Recently, the authors in (Bresch, D. & Fernández-Nieto, E.D. & Ionescu 2010) have deduced a viscous shallow water formulation for a Bingham fluid in the particular case of a plane slope bottom. The model has been applied to the simulation of avalanches in mountainous regions (on this topic, see the review (Ancey 2007)).

Our work focuses on the numerical analysis of the shallow water model in (Bresch, D. & Fernández-Nieto, E.D. & Ionescu 2010), which consists of a mass conservation law coupled with a variational inequality related to the conservation of momentum, which takes into account the viscoplastic behaviour of the fluid.

The conservation law is solved be means of a well-balanced finite volume method, built along the guidelines in (Chacón, T. & Castro, M.J. & Fernández-Nieto 2007). In particular, the method is able to preserve stationary solutions representing material at rest, and also stationary solutions of constant height over an inclined plane.

On the other hand, in (Bresch, D. & Fernández-Nieto, E.D. & Ionescu 2010) the variational inequality is solved by means of an Augmented Lagrangian method, whose main drawback is that its speed of convergence relies on a good choice of a scalar parameter. Usually, this choice is done by a trial-and-error approach.

In the present work, the solution of the variational inequality is approximated by using a duality method. In particular, we have adapted to this case the Bermúdez-Moreno duality algorithm (Bermúdez 1981), which is based on the properties of the Yosida regularization of maximal monotone operators. An important issue in the numerical model, also present in the Augmented Lagrangian approach, is the coupling of the finite volume solver with the duality method, that must be performed maintaining the well-balancing properties.

The convergence of the Bermúdez-Moreno algorithm strongly depends on the choice of two constant parameters. To overcome this difficulty and improve the rate of convergence, we have adapted the algorithms studied in (Gallardo, J.M. & Parés 2005) and the references there in, for which an automatic selection of the parameters is carried out, which gives the greatest rate of convergence.

The paper ends with several experiments showing the performances of the proposed coupled method.

2 MATHEMATICAL MODEL

We are interested in the flow of an inhomogeneous Bingham fluid in a domain $\mathcal{D}(t) \subset \mathbb{R}^3$ with smooth boundary $\partial\mathcal{D}(t)$, in the time interval $t \in (0,T)$. Denote by u the velocity field and σ the Cauchy stress tensor; $p = -\text{trace}(\sigma)/3$ is the pressure and $\sigma' = \sigma + pI$ is the stress deviator tensor. The momentum balance law reads as

$$\rho\left(\mathrm{St}\frac{\partial u}{\partial t}+(u\cdot\nabla)u\right)-\mathrm{div}\,\sigma'+\nabla p=g\rho f \quad \text{in } \mathcal{D}(t),$$

where $\rho > 0$ is the density, f denotes body forces, St is the Strouhal number and g is gravity. The constitutive equations for a Bingham fluid (Bingham 1922) has the following form:

$$\begin{cases} \sigma' = \dfrac{2}{\mathrm{Re}}\eta_1 D(u) + \eta_2 B\dfrac{D(u)}{|D(u)|} & \text{if } |D(u)| \neq 0, \\ |\sigma'| \leq \eta_2 \mathrm{B} & \text{if } |D(u)| = 0, \end{cases}$$

where $\eta_1 \equiv \eta_1(\rho)$ and $\eta_2 \equiv \eta_2(\rho)$ are, respectively, the viscosity and the yield limit, Re and B are the Reynolds and Bingham numbers, and

$$\mathrm{D(u)} = \frac{1}{2}(\nabla u + \nabla^T u)$$

is the rate deformation tensor.

On the other hand, the conservation of mass reads as

$$\mathrm{St}\frac{\partial \rho}{\partial t} + u\cdot\nabla\rho = 0 \quad \text{in } \mathcal{D}(t), \tag{1}$$

while

$$\mathrm{div}\,u = 0 \quad \text{in } \mathcal{D}(t)$$

stands for the incompressibility condition. The fact that the fluid is advected by the flow is written as

$$\mathrm{St}\frac{\partial 1_{\mathcal{D}(t)}}{\partial t} + u\cdot\nabla 1_{\mathcal{D}(t)} = 0,$$

where $1_{\mathcal{D}(t)}$ is the characteristic function of $\mathcal{D}(t)$.

Assume that $\partial\mathcal{D}(t) = \Gamma_b(t)\cup\Gamma_s(t)$, where Γ_b represents the bottom and Γ_s is the free surface. Then, the following boundary conditions are considered:

- On Γ_b, a Navier condition with friction coefficient β ($\sigma_t = -\beta u_t$) and a no-penetration condition ($u\cdot n = 0$).
- On Γ_s, a no-stress condition: $\sigma n = 0$.

In the above expressions, the index t denotes the tangential part and n is the outward unit normal on $\partial\mathcal{D}(t)$.

Finally, initial conditions of the form

$$u|_{t=0} = u_0, \quad \rho|_{t=0} = \rho_0 \tag{2}$$

are considered.

Following the guidelines in (Duvaut 1972), it is possible to construct a variational inequality related to the velocity field and pressure. In particular, if the test space

$$\mathcal{V}(t) = \{\Phi \in H^1(\mathcal{D}(t))^3 : \Phi = 0,\ \Phi\cdot n = 0 \text{on } \Gamma_b(t)\}$$

is introduced, we consider the following problem: Find the velocity field u and the pressure p such that

$$\begin{aligned} &\int_{\mathcal{D}(t)} \rho\left(\mathrm{St}\frac{\partial u}{\partial t} + (u\cdot\nabla)u\right)\cdot(\Phi - u) \\ &-\int_{\mathcal{D}(t)} p(\mathrm{div}\,\Phi - \mathrm{div}\,u) \\ &+\frac{1}{\mathrm{Re}}\int_{\mathcal{D}(t)} 2\eta_1(\rho)D(u):(D(\Phi)-D(u)) \\ &+\mathrm{B}\int_{\mathcal{D}(t)} \eta_2(\rho)(|D(\Phi)| - |D(u)|) \\ &+\int_{\Gamma_b(t)} \beta u_t\cdot(\Phi_t - u_t) \geq g\int_{\mathcal{D}(t)} \rho f\cdot(\Phi - u), \end{aligned} \tag{3}$$

and

$$\int_{\mathcal{D}(t)} q\,\mathrm{div}\,u = 0, \tag{4}$$

hold for all $\Phi \in \mathcal{V}(t)$ and $q \in L^2(\mathcal{D}(t))$ (see (Bresch, D. & Fernández-Nieto, E.D. & Ionescu 2010)).

In conclusion, the problem to consider is: Find the velocity field u, the pressure p, and the mass density field ρ satisfying (1), (2), (3), and (4).

3 THE PLANE SLOPE CASE

We focus now on the plane slope case, for which the domain has the form

$$D(t) = \{(x,z)\in\mathbb{R}^3 x : \in \Omega, 0 < z < h(t,x)\},$$

where $\Omega \subset \mathbb{R}^2$ is a fixed bounded domain and $h(t,x)$ denotes the thickness of the fluid layer. For the sake of simplicity, we assume that the density ρ is constant, so they are the viscosity η_1 and the yield limit η_2. After a standard procedure of vertical averaging, rescaling and asymptotic analysis of the equations (see (Bresch, D. & Fernández-Nieto, E.D. & Ionescu 2010) for the details), the model can be reduced to the relation

$$\mathrm{St}\frac{\partial H}{\partial t} + \mathrm{div}(HV) = 0,$$

and the variational inequality

$$\int_\Omega H\rho\big((\mathrm{St}\,\partial_t V\cdot(\Psi - V) + V\cdot\nabla V(\Psi - V)\big)dx$$
$$+\int_\Omega \beta V\cdot(\Psi - V)dx$$
$$+\int_\Omega \frac{2}{\mathrm{Re}}\eta_1 H D(V) : D(\Psi - V)dx$$
$$+\int_\Omega \frac{2}{\mathrm{Re}}\eta_1 H\,\mathrm{div}\,V(\mathrm{div}\,\Psi - \mathrm{div}\,V)dx$$
$$+\int_\Omega B\eta_2 H\Big(\sqrt{|D(\Psi)|^2 + (\mathrm{div}\,\Psi)^2}$$
$$-\sqrt{|D(V)|^2 + (\mathrm{div}\,V)^2}\Big)d^X$$
$$\geq \int_\Omega gH\rho f_H\cdot(\Psi - V)dx$$
$$-\int_\Omega gH^2 f_V(\mathrm{div}\,\Psi - \mathrm{div}\,V)dx.$$

Here, H and V are the height and the velocity in the rescaled variables, Ψ stands for the corresponding test functions, while f_V and f_H represent the vertical and horizontal components of the averaged body forces. As observed in (Bresch, D. & Fernández-Nieto, E.D. & Ionescu 2010), the above expressions give a viscous shallow water formulation of Bingham type. In fact, when $\eta_2 = 0$ the viscous shallow water model proposed by Gerbeau and Perthame (Gerbeau 2001) is recovered, as we have the same boundary condition at the bottom.

4 THE 1D PROBLEM

We shall focus throughout the rest of the paper on the one-dimensional version of the Bingham model, as most of the numerical difficulties involving its numerical solution appear in this model. Moreover, for the sake of clarity, we will consider the case of constant density ρ.

Thus, the one-dimensional Bingham model reads as:

$$\mathrm{St}\frac{\partial H}{\partial t} + \frac{\partial(HV)}{\partial x} = 0,$$
$$\int_0^L H\rho(\mathrm{St}\,\partial_t V(\Psi - V) + \frac{1}{2}\partial_x(V^2)(\Psi - V))dx$$
$$+\int_0^L \beta V(\Psi - V)dx$$
$$+\int_0^L \frac{4}{\mathrm{Re}}\eta_1 H\partial_x V\,\partial_x(\Psi - V)dx$$
$$+\int_0^L \sqrt{2}B\eta_2 H(|\partial_x\Psi| - |\partial_x V|)dx$$
$$\geq -\int_0^L gH\rho\sin\theta(\Psi - V)dx$$
$$+\int_0^L \frac{g}{2}\cos\theta H^2\rho(\partial_x\Psi - \partial_x V)dx,$$

where H and V depend on $(x,t)\in[0,L]\times[0,T]$.

We are also interested in preserving two types of stationary solutions associated to the system (*well-balanced* property):

- Horizontal free surface:

  $$x\sin\theta + H\cos\theta = \text{constant}.$$

- Constant height: $H = \text{constant}$, which is a stationary solution under the condition (see (Bresch, D. & Fernández-Nieto, E.D. & Ionescu 2010))

  $$g\rho H\sin\theta \leq 2\sqrt{2}\eta_2 B.$$

5 NUMERICAL SCHEME

The semidiscretization in time of the model has the following form:

$$\frac{H^{n+1} - H^n}{\Delta t} + \frac{\partial(H^n V^n)}{\partial x} = 0, \tag{5}$$

$$\begin{aligned}
&\int_0^L H^n\Big(\frac{V^{n+1} - V^n}{\Delta t}(\Psi - V^{n+1}) \\
&\quad + \frac{1}{2}\partial_x[(V^n)^2](\Psi - V^{n+1})\Big)dx \\
&+\int_0^L \beta V^{n+1}(\Psi - V^{n+1})dx \\
&+\int_0^L 4\eta_1 H^n\partial_x V^{n+1}\partial_x(\Psi - V^{n+1})dx \\
&+\int_0^L \sqrt{2}\eta_2 H^n(|\partial_x\Psi| - |\partial_x V^{n+1}|)dx \\
&\geq -\int_0^L g\sin\theta H^n(\Psi - V^{n+1})dx \\
&+\int_0^L \frac{g}{2}\cos\theta(H^n)^2(\partial_x\Psi - \partial_x V^{n+1})dx,
\end{aligned} \tag{6}$$

where, for the sake of clarity, we have set $\rho = \mathrm{St} = \mathrm{Re} = B = 1$. Thus, we have to solve two coupled problems associated, respectively, to the height and to the velocity.

5.1 *Treatment of the variational inequality*

The variational inequality (6) can be expressed in the form

$$\langle AV^{n+1}, \Psi - V^{n+1}\rangle + j(\Psi) - j(V^{n+1})$$
$$\geq \langle L, \Psi - V^{n+1}\rangle, \quad \forall\Psi\in\mathcal{V},$$

where $V = H_0^1([0,L])$ with duality pairing $\langle\cdot,\cdot\rangle$; $AV \to V'$ is the coercive linear operator

$$A : V^{n+1} = \left(\frac{1}{\Delta t}H^n + \beta\right)V^{n+1};$$

$j : \mathcal{V} \to \mathbb{R}$ is the convex functional

$$j(V) = \int_0^L \Phi(BV)dx,$$

where $\Phi : \mathcal{H} \to \mathbb{R} \cup \{\infty\}$ is given by

$$\Phi(q) = \eta_2 H^n \, |q|,$$

with $\mathcal{H} = L^2([0,L]); B : \mathcal{V} \to \mathcal{H}$ is the derivative operator

$$BV = \partial_x V;$$

finally, $L \in \mathcal{V}'$ is a functional depending on H^n and V^n.

In turn, (6) can be expressed as a fixed-point problem of the form:

$$\begin{cases} AV + \omega B^* BV + B^* \theta = L, \\ \theta = G_\lambda^\omega (BV + \lambda\theta), \end{cases}$$

where $V = V^{n+1}$ and G_λ^ω is the *Yosida regularization* of the operator $\partial\Phi - \omega I$ where $\partial\Phi$ is the subdifferential of Φ (see (Bermúdez 1981)). Notice that G_λ^ω depends on two constant parameters λ and ω. In our case, G_λ^ω can be computed explicitely:

$$G_\lambda^\omega(Z)(x) = \begin{cases} \dfrac{-\omega Z(x) - \sqrt{2}\eta_2 H^n(x)}{1-\lambda\omega} & \text{if } Z(x) < -\alpha, \\ \dfrac{Z(x)}{\lambda} & \text{if } Z(x) \in [-\alpha, \alpha] \\ \dfrac{-\omega Z(x) - \sqrt{2}\eta_2 H^n(x)}{1-\lambda\omega} & \text{if } Z(x) > \alpha, \end{cases}$$

where $\alpha = \sqrt{2}\eta_2 \lambda H^n(x)$ and $Z \in \mathbb{R}$ is arbitrary.

Under this abstract form, the variational inequality can be solved by means of the Bermúdez-Moreno (BM) algorithm (see (Bermúdez 1981) for details), which has the following form:

- For $m \geq 0$, assume that Θ^m is known.
- Compute the approximate velocity V^m by solving the linear equation

$$AV^m + \omega B^* BV^m + B^* \Theta^m = L. \quad (7)$$

- Update the multiplier by

$$\Theta^{m+1} = G_\lambda^\omega (BV^m + \lambda\Theta^m).$$

- Loop until a stop condition on Θ^m is achieved.

The convergence of the sequence V^m to the solution V can be proved if we choose λ and ω verifying

$$\lambda\omega = \frac{1}{2}.$$

We will assume this condition throughout the rest of the paper.

As it has been documented in the literature, the rate of convergence of the BM algorithm strongly depends on the choice of parameters λ and ω. Several efforts have been made to overcome this problem, allowing the choice of appropriate parameters in different functional frameworks (see (Gallardo, J.M. & Parés 2005), (Parés, C. & Macías 2001), (Parés, C. & Castro 2002)).

Following the guidelines in (Parés, C. & Castro 2002), the main idea being to minimize the contractivity constant of the sequence $\Theta^m - \Theta$, we have found that the optimal choice of the parameter ω would be

$$\omega_{\text{opt}} = \left(\frac{H^n_{\max}}{\Delta t} + \beta \right) \frac{1}{\gamma_1 \gamma_2} + 4\eta_1 H^n_{\max},$$

where $H^n_{\max = \|H^n\|_\infty}$ and γ_1, γ_2 are Poincaré constants verifying

$$\gamma_1 |\Psi| \leq |\partial_x \Psi| \leq \gamma_2 |\Psi|, \quad \forall \Psi \in \mathcal{V}.$$

To approximate these constants, let $\mathcal{V}_h$ be the space of standard conforming P_1 finite elements and consider the following spectral problem: Find $v \in \mathcal{V}_h$ such that

$$(v', \psi') = \mu(v, \psi), \quad \forall \psi \in \mathcal{V}_h.$$

It is well-known (Boffi 2010) that, for uniform mesh size h, there exists an orthonormal basis of $\mathcal{V}_h$ composed by eigenvectors $\{\varphi_h^{(1)}, \ldots, \varphi_h^{(N)}\}$ associated to the eigenvalues $0 < \mu_h^{(1)} \leq \ldots \leq \mu_h^{(N)}$ of the spectral problem. Indeed, these eigenvalues have the following form:

$$\mu_h^{(j)} = \frac{6}{h^2} \frac{1 - \cos(j\pi h/L)}{2 + \cos(j\pi h/L)}, \quad j = 1, \ldots, N,$$

where N denotes the number of internal nodes. For an arbitrary $v = \sum_{j=1}^N v_j \varphi_h^{(j)}$ we have:

$$|\partial_x v|^2 = \sum_{j=1}^N v_j^2 (\partial_x \varphi_h^{(j)}, \partial_x \varphi_h^{(j)}) = \sum_{j=1}^N v_j^2 \mu_j (\varphi_h^{(j)}, \varphi_h^{(j)}),$$

so

$$\mu_1^{1/2} |v| \leq |\partial_x v| \leq \mu_N^{1/2} |v|.$$

As noticed in (Boffi 2010), the following optimal estimate holds, as $h \to 0$:

$$|\mu^{(j)} - \mu_h^{(j)}| = \mathcal{O}(h^2),$$

where $\mu^{(j)} = (j\pi/L)^2$. These considerations lead us to the idea of approximating the Poincaré constants as $\gamma_1 = \sqrt{\mu^{(1)}} = \pi/L$ and $\gamma_N = \sqrt{\mu^{(N)}} = N\pi/L$.

Summarizing, the optimal choice of the parameter ω_{opt} will be given by

$$\omega_{\text{opt}} = \left(\frac{H^n_{\max}}{\Delta t} + \beta\right)\frac{L^2}{N\pi^2} + 4\eta H^n_{\max}.$$

5.2 *Spatial discretization*

In this section we describe the discretization in space of (5) and (7), that is done in a coupled way in order to obtain a well-balanced scheme able to preserve stationary solutions corresponding to a horizontal free surface or a constant height.

Thus, we consider the system

$$\begin{cases} \dfrac{H^{n+1} - H^n}{\Delta t} + \dfrac{\partial(H^n V^n)}{\partial x} = 0, \\ H^n\left(\dfrac{V^{m+1} - V^n}{\Delta t} + \partial_x\left(\dfrac{1}{2}(V^n)^2 + g\cos\theta H^n\right)\right) \\ = -\beta V^{m+1} - g\sin\theta H^n \\ \quad + \partial_x((\omega + 4\eta_1 H^n)\partial_x V^{m+1}) + \partial_x \Theta^m. \end{cases}$$

Following the guidelines in (Chacón, T. & Castro, M.J. & Fernández-Nieto 2007), a finite volume discretization of the form

$$W_i^{n+1} = W_i^n - \frac{\Delta t}{\Delta x}(\phi_{i+1/2} - \phi_{i+1/2})$$

is proposed (see (Bresch, D. & Fernández-Nieto, E.D. & Ionescu 2010) for the details), where $W = (H, V)^T$ and the numerical flux $\phi_{i+1/2}$ is defined as

$$\begin{aligned}\phi_{i+1/2} &= \frac{1}{2}(F(W_i) + F(W_{i+1})) \\ &\quad - \frac{1}{2}D_{i+1/2}(W_{i+1} - W_i - A^{-1}_{i+1/2}G_{i+1/2})\end{aligned}$$

In the above expression, the flux is given by

$$F(W) = (HV, \frac{1}{2}V^2 + g\cos\theta H)^T,$$

with associated Jacobian $A_{i+1/2}$, while the viscosity matrix is $D_{i+1/2} = \text{CFL}\Delta x/\Delta t\, I$ (modified Lax-Friedrichs). The source term discretization consists of two parts, one of them related to the topography and the other one to the BM multiplier:

$$G_{i+1/2} = G_{top,i+1/2} + G_{\Theta,i+1/2},$$

with

$$G_{top,i+1/2} = (0, -g\sin\theta)^T$$

and

$$G_{\Theta,i+1/2} = (0, \Theta^k_{i+1} - \omega V^k_{i+1}) - (\Theta^k_i - \omega V^k_i))^T.$$

It is worth noticing that this well-balanced approach induces a coupling between the height and the velocity problems, in which the BM multiplier Θ^m is interpreted as an additional source term.

6 NUMERICAL TESTS

6.1 *Well-balancing test*

The test in this section shows the well-balancing property of the scheme in the case of a horizontal free surface (see Fig. 1). We have considered a domain with length $L = 10$, $N = 300$ partitions, $\text{CFL} = 0.8$, slope angle $\theta = 5\pi/180$, viscosity $\eta_1 = 0.01$, yield limit $\eta_2 = 5$, friction $\beta = 0$, gravity $g = 9.81$ and final time $T = 2$.

The stationary solution with constant free surface equals to 1 and velocity $V \equiv 0$ is preserved up to the precision determined by the fixed-point BM algorithm. To see this, in Table 1 are given the L^∞ errors obtained at the final time for several choices of the BM tolerance.

Similar results are obtained for the case of a stationary solution with constant height.

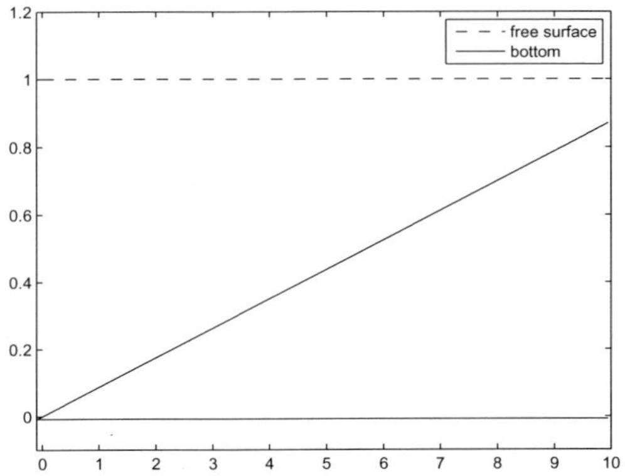

Figure 1. Test 6.1: Stationary solution.

Table 1. Test 6.1: L^{∞} errors for H and V.

Tol.	Error H	Error V
1E-03	3.115E-09	2.296E-06
1E-05	9.229E-11	2.290E-08
1E-07	5.273E-12	2.223E-10
1E-09	2.082E-13	1.886E-12
1E-11	3.358E-15	4.435E-16
1E-13	3.365E-16	1.493E-17

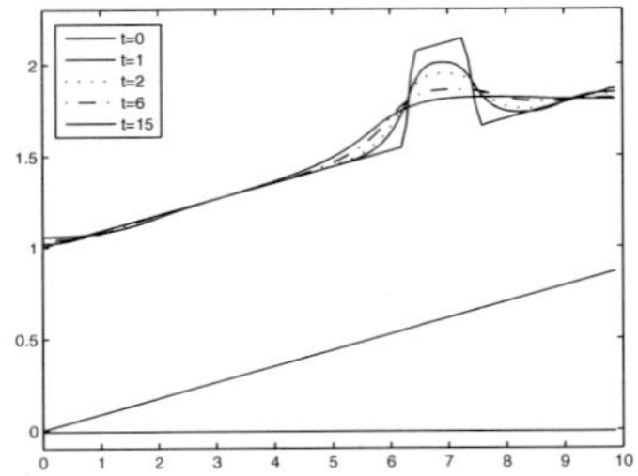

Figure 2. Evolution of the free surfaces in test 6.2.

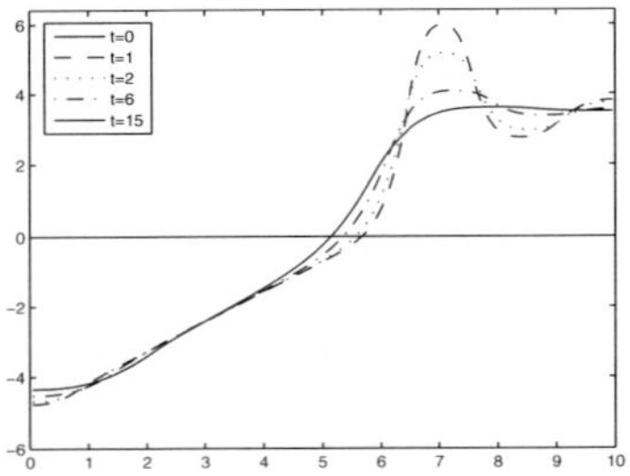

Figure 3. Evolution of the multipliers Θ^m in test 6.2.

6.2 *Convergence towards a stationary state*

We consider the same data as in the preceding example with $N = 100$, and take as initial conditions:

$$H(0,x) = \begin{cases} 1.5 & \text{if } 6.5 \le x \le 7.5, \\ 1 & \text{otherwise} \end{cases}$$

$$V(0,x) = 0.$$

The purpose of this experiment is to test the robustness of the scheme when converging to a non-trivial stationary state. The initial free surface consists of a bump that deforms until reaching a non-trivial stationary

state with null velocity. In Figure 2 are shown the free surfaces obtained at several times, ending at time $T = 15$, when the stationary solution has been reached. The evolution of the BM multipliers, departing from $\Theta^0 \equiv 0$, is shown in Figure 3.

7 CONCLUSIONS

We have presented a numerical scheme for solving the viscous shallow water model of Bingham type introduced in (Bresch, D. & Fernández-Nieto, E.D. & Ionescu 2010). This scheme couples a well-balanced finite volume method with the BM duality algorithm, which relies on the properties of maximal monotone operators (the subdifferential, in particular). An study on the optimal choice of parameters in the BM algorithm has been done, which leads to the greatest speed of convergence. We have focused on the one-dimensional version of the model, as most of the numerical difficulties related to the duality algorithm appear in this case.

REFERENCES

Ancey, C. (2007). Plasticity and geophysical flows: A review. *Journal of Non-Newtonian Fluid Mechanics 142*, 4–35.

Bermúdez, A. & Moreno, C. (1981). Duality methods for solving variational inequalities. *Computers & Mathematics with Applications 7*(1), 43–58.

Bingham, E.C. (1922). *Fluidity and plasticity*. Mc Graw-Hill.

Boffi, D. (2010). Finite element approximation of eigenvalue problems. *Acta Numerica 19*, 1–120.

Bresch, D. & Fernández-Nieto, E.D. & Ionescu, I.R. & Vigneaux, P. (2010). Augmented lagrangian method and compressible visco-plastic flows: Applications to shallow dense avalanches. In G. P. Galdi et al. (Ed.), *New Directions in Mathematical Fluid Mechanics*, Advances in Mathematical Fluid Mechanics, pp. 57–89. Birkhäuser Basel.

Chacón, T. & Castro, M.J. & Fernández-Nieto, E.D. & Parés, C. (2007). On well-balanced finite volume methods for non-conservative non-homogeneous hyperbolic systems. *SIAM J. Sci. Comput. 29(3)*, 1093–1126.

Dean, E.J. & Glowinski, R. & Guidoboni, G. (2007). On the numerical simulation of Bingham visco-plastic flow: Old and new results. *Journal of Non Newtonian Fluid Mechanics 142*, 36–62.

Duvaut, G. & Lions, J.-L. (1972). *Les inéquations en Mécanique et en Physique*. Dunod, Paris.

Gallardo, J.M. & Parés, C. & Castro, M. (2005). A generalized duality method for solving variational inequalities. applications to some nonlinear dirichlet problems. *Numer. Math. 100*(2), 259–291.

Gerbeau, J.-F. & Perthame, B. (2001). Derivation of viscous saint-venant system for laminar shallow water; numerical validation. *Discrete Contin. Dyn. Syst., Ser. B 1*(1), 89–102.

Oswald, P. (2009). *Rheophysics. The Deformation and Flow of Matter*. Cambridge University Press.

Parés, C. & Castro, M. & Macías, J. (2002). On the convergence of the bermúdez-moreno algorithm with constant parameters. *Numer. Math. 92*(1), 113–128.

Parés, C. & Macías, J. & Castro, M. (2001). Duality methods with an automatic choice of parameters. Application to shallow water equations in conservative form. *Numer. Math. 89*(1), 161–189.

Numerical Methods for Hyperbolic Equations – Vázquez-Cendón et al. (eds)
© 2013 Taylor & Francis Group, London, ISBN 978-0-415-62150-2

Accounting for building effects in shallow waters: Porosity approach, resolved approach and Manning coefficient approach. Comparison and experimental validation

M. Garrido, L. Cea & J. Puertas
University of A Coruña, Spain

M.E. Vázquez-Cendón
University of Santiago de Compostela, Spain

ABSTRACT: Urbanized areas have important effects on flow development and runoff quantity, so their influence on floods and rainfall-runoff development is crucial. For an accurate prediction of runoff quantity, velocities and depths in each single street of an urban area, a detailed 2D model is generally desirable, but this approach is costly in data requirements and computational time. In recent years some efforts have been devoted to study how to account for the effects of urbanized areas in flow fields in a macroscopic manner, taking into account water storage and obstruction effects without including a detailed geometry of the urban area. The present paper presents a comparison between three different approaches in the representation of urban areas in a two-dimensional model, one of them taking into account the detailed geometry of the urban area (resolved approach) and another two being large scale approaches that compute for the effects of urban geometry in a macroscopic manner (porosity approach and Manning coefficient approach).

1 INTRODUCTION

The two-dimensional modelling of floodplain dynamics and rainfall–runoff processes involving urbanized areas are hampered by the strong geometrical variability of the urban tissue, inducing an important hydraulic variability. Taking into account the detailed geometry of an urban area in a two-dimensional model is extremely difficult, sometimes if not impossible, because of the number of computational cells (or points) needed to represent the details of the flow. As a result, practising engineers often prefer to use rough models with coarse grids, the results of which may be interpolated to provide boundary conditions for nested, refined flow models within the urban areas.

In such coarse-grids models, urban areas are generally represented as areas with a higher friction, to account for the increased resistance induced by the presence of buildings in the floodplains.

The spatial variation of the friction coefficient can be estimated from airborne scanning laser altimetry (LiDAR) data for example, as was done by Cobby *et al.* (2003) for vegetated floodplains. Another type of simplified models is the raster-based model developed by Bates & De Roo (2000). This method was compared to a finite-element simulation to determine the extent of an inundated area (Horritt & Bates, 2001). The key advantage of raster-based models is the much lower computational cost, but still, urban areas are not accounted for specifically. Soares-Frazao et al. (2008) proposed and intermediate approach that also account for the effect of urbanized areas in flow development in a macroscopic way, but try to represent three essential features of the flow that could be misrepresent using a coarse grid over the urban area, namely (i) the reduced volume available for water storage within the urbanized area; (ii) the reduction in the section available to the flow due to the presence of buildings, structures, etc.; and (iii) the extra energy loss induced by the buildings that act as obstacles in the case of severe transient floods (Soares-Frazão *et al.*, 2004).

They proposed an interesting path for the representation of urban areas in large-scale shallow-water flow models that consists in characterizing the urban area by a porosity that acts on both the storage and the fluxes. A first formulation of the shallow-water equations with porosity was proposed by Soares-Frazão *et al.* (2008) and later its discretization was studied by Cea & Vázquez-Cendón (2010).

The present paper presents a comparison between three different approaches in the representation of urban areas in a two-dimensional model:

(i) porosity approach; (ii) resolved approach (detailed geometry of urban area); and (iii) Manning coefficient approach (coarse grid with higher friction over the urban area). The numerical model used in this paper is presented in Section 2. Section 3 is devoted to the description of the validation test cases and the analysis of the performance of the three approaches. Section 4 provides concluding remarks.

2 NUMERICAL MODEL

The numerical model used in this paper is based on the depth averaged shallow water equations (2D-SWE), also known as 2D St. Venant equations. A complete description of the numerical model (Turbillon) can be found in Cea (2010) and the porosity implementation in this kind of models in Cea & Vázquez-Cendón (2010) The terms of the equations that define the porosity approach are the ones proposed by Soares-Frazao et al. (2008). This section presents just a brief description of the equations and numerical schemes solved by the numerical model.

The numerical model used in this paper solves the Two-dimensional shallow water equations with porosity, which are given by

$$\frac{\partial \phi h}{\partial t}+\frac{\partial}{\partial x}(\phi h U_x)+\frac{\partial}{\partial y}(\phi h U_y)=R \tag{1}$$

$$\frac{\partial}{\partial t}(\phi h U_x)+\frac{\partial}{\partial x}\left(\phi h U_x^2+\phi\frac{gh^2}{2}\right)+\frac{\partial}{\partial y}(\phi h U_x U_y)$$
$$=\frac{1}{2}gh^2\frac{\partial\phi}{\partial x}-gh\phi\frac{\partial z_b}{\partial x}-\phi\frac{\tau_{b,x}}{\rho}+\frac{\partial\phi h\,\tau_{xx}^e}{\partial x}$$
$$+\frac{\partial\phi h\,\tau_{xy}^e}{\partial y}-\frac{\tau_{d,x}}{\rho}+R\,V_x$$

$$\frac{\partial}{\partial t}(\phi h U_y)+\frac{\partial}{\partial x}(\phi h U_x U_y)+\frac{\partial}{\partial x}\left(\phi h U_y^2+\phi\frac{gh^2}{2}\right)$$
$$=\frac{1}{2}gh^2\frac{\partial\phi}{\partial y}-gh\phi\frac{\partial z_b}{\partial y}-\phi\frac{\tau_{b,y}}{\rho}+\frac{\partial\phi h\,\tau_{xy}^e}{\partial x}$$
$$+\frac{\partial\phi h\,\tau_{yy}^e}{\partial y}-\frac{\tau_{d,y}}{\rho}+R\,V_y$$

Where ϕ is the porosity, which is equal to one in the resolved and Manning coefficient approaches, h is the water depth, U_x, U_y are the depth averaged horizontal components of the water velocity, g is the gravity acceleration, z_b is the bed elevation, τ_b is the bed friction, ρ the water density, τ^e_{xx}, τ^e_{xy}, τ^e_{yy} are the horizontal turbulent shear stresses, τ_d is the additional drag due to non-resolved obstructions in the urban region, R is the rainfall intensity, and V_x, V_y are the 2 horizontal components of the rain velocity. The equations are solved in a 2D unstructured mesh with a finite volume solver.

For the present applications, the effects of bed friction, bed slope and precipitation are considered in the model. The fact of neglecting the turbulent horizontal stresses is justified in this case because the turbulent vertical shear caused by bed friction is much larger than the horizontal turbulent shear. In the 2D-SWE the turbulent vertical shear is introduced in the bed friction term, which in this case is computed with Manning's formula.

Other mass losses as evapotranspiration, interception, infiltration and retention can also be considered in the model. However, in the applications presented in this paper these losses do not exist or are not significant, and do not need to be considered in the computations.

The only parameters of the model which need calibration are the bed friction coefficient and the infiltration properties of the soil. The effects of small scale microtopography which is not resolved by the model must be included via the bed friction coefficient, in the same way as the effects of ripples and dunes in rivers are included in the bed friction stress. By microtopography we mean the bed surface features with a length scale smaller than the mesh size used in the numerical discretization.

An unstructured finite-volume solver with a first order explicit time discretization is used to solve the mean flow equations. The convective flux is discretized with a second order scheme, upwind Godunov's schemes based on Roe's average (Toro, 2001).

In order to avoid spurious oscillations of the free surface when the bathymetry is irregular, an upwind discretization of the bed slope source term is used. This has proved to be more stable and accurate than a centered scheme (Bermudez, 1998). Bed friction, rainfall and infiltration are discretized with a centered semi-implicit scheme at the cell nodes. The numerical scheme is explicit in time, so the CFL restriction applies over the time step.

3 EXPERIMENTAL VALIDATION

3.1 *Experimental setup and test cases*

An experimental campaign was carried out to validate and compare the numerical results of the three approaches: porosity approach, resolved approach and Manning coefficient approach. During the experimental campaign a uniform rainfall was generated over different simplified urban configurations on 2D laboratory geometry. The rainfall simulator used in the experiments was a rectangular basin with dimensions 2 × 2.5 m, made of three planes of stainless-steel, each of them with an approximate slope of 0.05 (Fig. 1). Several idealized urban configurations were built on the basin including aligned, staggered and random patterns for different building densities and scales. Rainfall is simulated with a grid of 100 nozzles distributed evenly over the basin. The only variable measured in the experiments campaign was the discharge hydrograph generated at the outlet of the basin, which was used to calibrate de model and validate the numerical results.

The global experimental uncertainty in the outlet discharge data was estimated as an 11%. This is due to the uncertainty in the measurement technique of the outlet discharge and to the uncertainty in the imposed rainfall intensity.

An urban laboratory configuration, with 182 square buildings of 9 × 9 cm, was built in the laboratory and computed with the numerical model. The urban configuration was tested with six different hyetographs, which are defined by a constant and spatially uniform rainfall discharge of 15 and 25 l/min, with a duration of 20, 40 and 60 s (Table 1).

In the resolved approach buildings are represented as holes in the numerical mesh (Fig. 3) and a slip wall boundary condition is imposed at the buildings boundary in contraposition to methods where buildings shapes are included in the mesh,

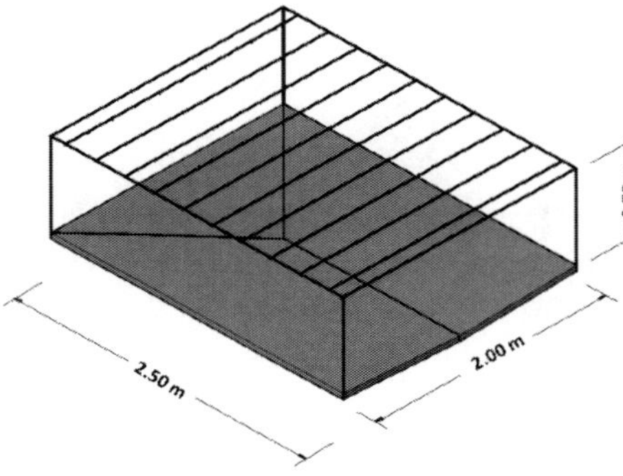

Figure 1. 3D sketch of the rainfall simulator over a stainless-steel basin used for experimental validation.

Table 1. Hyetographs used in the experiments.

Hyetograph	Rainfall discharge	Rainfall intensity	Rainfall duration
Q15T20	15 l/min	180 mm/h	20s
Q15T40	15 l/min	180 mm/h	40s
Q15T60	15 l/min	180 mm/h	60s
Q25T20	25 l/min	300 mm/h	20s
Q25T20	25 l/min	300 mm/h	40s
Q25T20	25 l/min	300 mm/h	60s

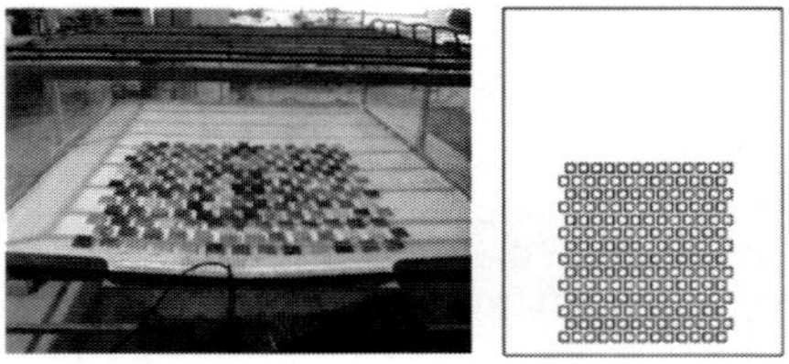

Figure 2. Experimental setup: photograph of laboratory catchment (left) and sketch of urban geometry (right).

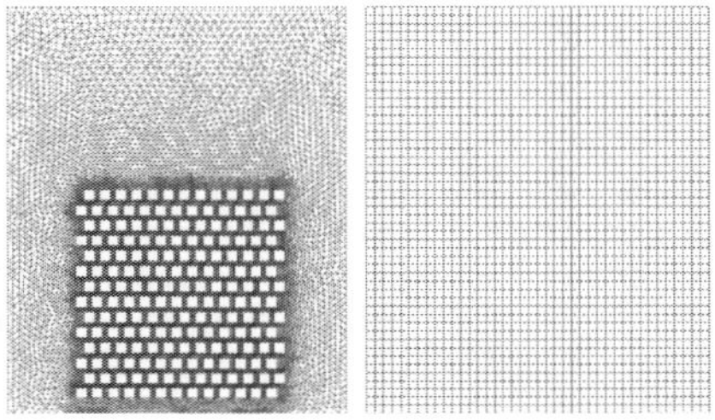

Figure 3. Numerical meshes: resolved approach (left) and porosity (right).

and the bed elevation is raised according to their height. Both approaches have also been used by Schubert et al. (2008) to model urban flood inundation. The selected approach can represent the buildings shape accurately with rather coarse meshes, making flood inundation modelling more efficient from a numerical point of view. Both ways of building representation were tested for these cases obtaining almost identical results.

Numerical meshes have 6395 elements for the resolved approach and 1044 for both large scale approaches (Fig. 3).

3.2 *Calibration and validation against experimental data*

Bed friction was modeled with Manning's formula for turbulent flow. In all the cases presented in this paper, using Manning's formulation for the whole experiment gave good results and therefore, it was not necessary to use more sophisticated bed friction formulations. A free slip wall boundary condition was imposed at all the closed boundaries. At the outlet boundary a critical depth boundary condition was used. The wet-dry tolerance parameter was set to 10^{-6} m.

The calibration of the bed friction coefficient was done in the resolved approach and fixed to the same value ($n = 0.02\ s \cdot m^{-1/3}$) for all the experiments except for the urban area en in Manning coefficient approach since in this case the friction coefficient also encompasses the effect of the buildings.

Table 2 shows the calibration parameters for the three approaches.

Figure 4 shows the numerical-experimental comparison of the outlet hydrograph for the three approaches for hyetograph Q25T40. In all the cases the differences in the outlet hydrographs computed with the three approaches are not significant considering the uncertainty of the experimental data. The differences between the three approaches in water depths and velocity fields inside the catchment are compared in next section.

3.3 *Comparison between approaches in urban representation*

Once the three models were calibrated and the best fit of outlet hydrograph was achieved, the numerical results obtained inside and outside the urban area were compared for the three cases. So three different sections in x and y directions of the urban area were defined to compare the numerical results of water depths and velocities with each approach (Fig. 5). Velocity and water depths fields were also compared (Fig. 6). The sections defined are: AA' in $x = 0$ m; BB' in $y = 0.8$ m; and CC' in $y = 1.4$ m.

Table 2. Calibration parameters and numerical scheme.

	Urban approach		
Parameter	Resolved approach	Porosity approach	Manning approach
Porosity	1	0.5	1
n* urban area	0.02	0.02	0.17
n* non-urban area	0.02	0.02	0.02
Head loss coefficient	–	25	–
Numerical scheme	Roe 2**	Roe 2**	Roe 2**

* Manning coefficient.
** Roe Second Order.

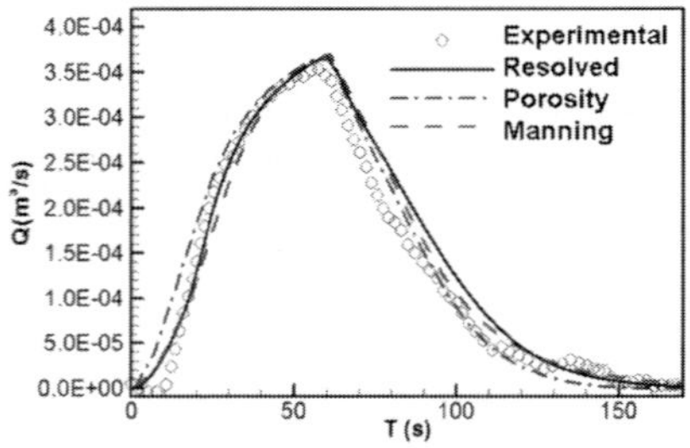

Figure 4. Numerical and experimental adjustment for hyetograph Q25T40.

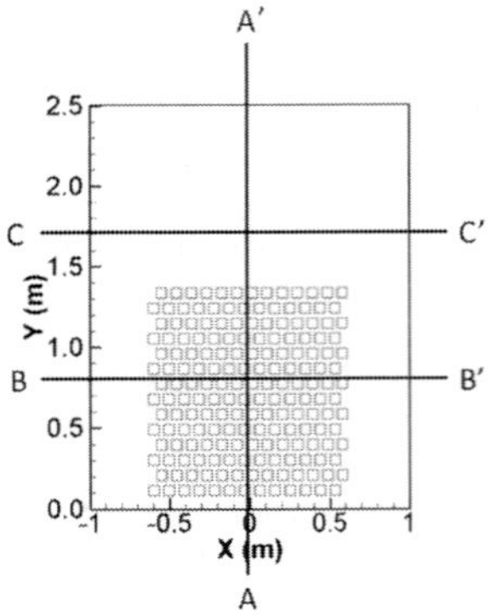

Figure 5. Cross sections defined to compare flow fields in the three approaches.

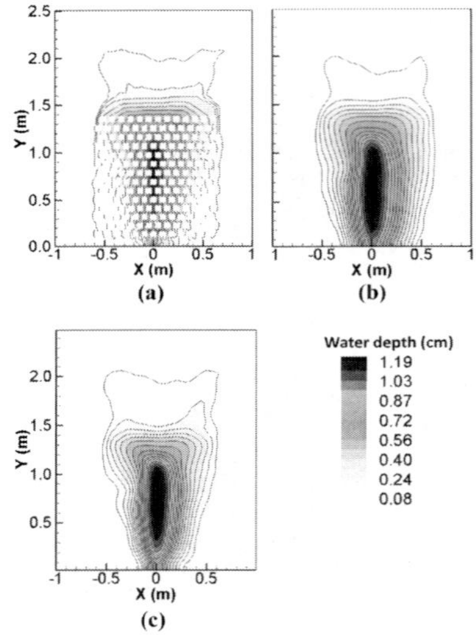

Figure 6. Water depths for hytograph Q25T60 and t = 60 s: (a) resolved approach; (b) porosity approach; (c) Manning coefficient approach.

When solving classical 2D equations the numerical model used in this paper has been widely validated, so after calibration it is reasonable to assume that resolved approach has the most reliable results and that those are the very close to reality. With this consideration in mind it is assumed that the other approaches are more accurate the more they fit the resolved approach results.

For section BB' (Fig. 5), inside the urban area, the porosity approach has better predictions than manning approach in terms of water depths and velocities. Studying flow fields upstream the urban area in section CC' both large scale approaches

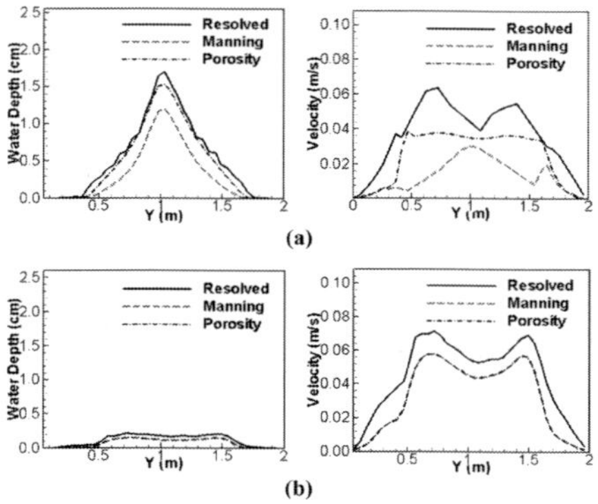

Figure 7. Numerical water depths and velocities in sections BB' (a) and CC' (b).

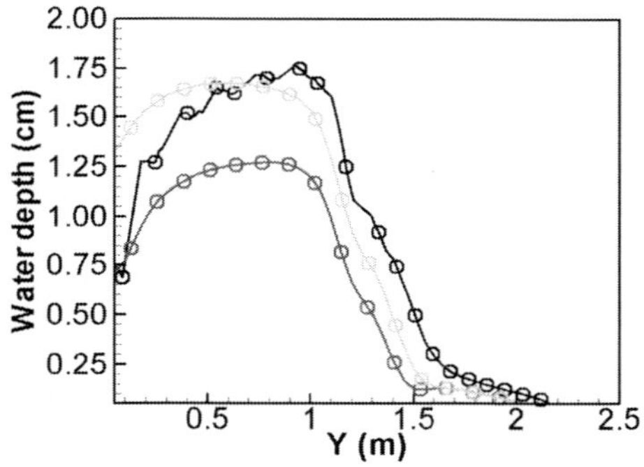

Figure 8. Numerical water depths in section AA'.

have very similar results being in both cases their accuracy less than that provided by the resolved approach (Fig. 7).

When studying section AA' (Fig. 8) we can see that there is region upstream of $y = 0.7$ m in which both porosity and Manning coefficient approaches underestimate water depths but the porosity approach is most accurate of them, that's the case of section BB'. Though, when getting close to the outlet of the catchment the porosity approach star overestimating water depths while Manning coefficient approach falls short, and in the outlet, specifically from $y = 0.15$ to $y = 0$ Manning coefficient approach becomes more precise than porosity approach in terms of water depths and velocities.

4 CONCLUSIONS

The discharge hydrograph, the water depth and the velocity fields in a laboratory urban catchment, were computed for six different hyetographs. The outlet hydrograph was measured in laboratory and compared with the numerical results obtained with three different approaches to account for buildings effect on flow development. With a proper calibration of the drag coefficient and friction coefficients the porosity approach and the Manning coefficient approach are able to reproduce the resolved approach results in the outlet hydrograph with a much lower computational effort (Fig. 4).

When comparing flow fields inside the catchment, obviously both large scale approaches do not reproduce flow details inside the urban region but in most cases the porosity approach yields more accurate results. However, when getting closer to the outlet Manning approach was more precise than porosity approach. In both cases global velocity and water depth fields are well represented.

The numerical results were also compared with the experimental hydrographs obtaining a good fit not only with the resolved approach but also with both large scale approaches. In all cases it has been compared the peak discharge and the shape of the outlet hydrograph. It has been also compared the computational time and numerical stability of each approach.

After applying the technique to scale models, its applicability to real scales should be investigated.

REFERENCES

Bermúdez, A., Dervieux, A., Desideri, J.A., and Vázquez-Cendón, M.E. (1998). "Upwind schemes for the two-dimensional shallow water equations with variable depth using unstructured meshes". *Comput. Methods Appl. Mech. Eng.* 155, 49–72.

Bates, P.D., Horrit, M.S., and Wilson, M.D. (2007). "Simple spatially-distributed models for predicting flood inundation: A review". *Geomorphology* 90, 208–225.

Cea, L., Garrido, M., Puertas, J. (2010). Experimental validation of two-dimensional depth averaged models for forecasting rainfall-runoff from precipitation data in urban areas. *Journal of Hydrology*, 382, 88–102.

Cea, L., Vázquez-Cendón, M.E. (2010) "Unstructured finite volume discretisation of two-dimensional depth averaged shallow water equations with porosity". *International Journal for Numerical Methods in Fluids*, Vol.63 (8) pp. 903–930, 2010, ISSN 0271-2091. Editor: Wiley, Reino Unido.

Soares-Frazão, S., Noël, B. and Zech,Y. (2004). "Experiments of Dam-Break Flow in the Presence of Obstacles." *Proceedings of the River Flow 2004 Conference*, Naples, Italy, 23–25 June 2004, Vol. 2, 911–918.

Soares-Frazao, S., Lhomme, J., Guinot, V., Zech, Y. (2008). "Two-dimensional shallow-water model with porosity for urban flood modelling". *Journal of Hydraulic Research* **46** (1).

Schubert, J.E., Sanders, B.F., Smith, M.J. and Wright, N. (2008). "Unstructured mesh generation and landcover-based resistance for hydrodynamic modelling of urban flooding". *Advances in Water Resources*.

Numerical Methods for Hyperbolic Equations – Vázquez-Cendón et al. (eds)
 ISBN 978-0-415-62150-2

A numerical approximation for a climatological model with dynamic and diffusive boundary condition

A. Hidalgo
Departamento Matemática Aplicada y Métodos Informáticos, ETSI Minas, Universidad Politécnica de Madrid, Spain

L. Tello
Departamento Matemática Aplicada a la Edificación, al Medio Ambiente y al Urbanismo, ETS Arquitectura, Universidad Politécnica de Madrid, Spain

ABSTRACT: In this work we compute the numerical solution of a global climate model including deep ocean effect. The model is based on that proposed by Watts and Morantine in (Watts & Morantine 1990) but including a coalbedo temperature dependent and nonlinear diffusion at the boundary. One of the main features of the model, which makes the problem particularly interesting, is the dynamic and diffusive boundary condition that represents the coupling between ocean and atmosphere. The numerical method used is a finite volume scheme with high order WENO reconstruction in space and third order Runge-Kutta TVD for time discretization.

1 INTRODUCTION

Climate system is very complex and involves many components and complicate mechanisms. Different climate models consider only a few of these components to understand only some of the mechanisms. We are concerned with a diagnostic model trying to understand the evolution of the global climate for a long time scale. This kind of models has been used in the study of Milankovitch theory of the ice-ages. The model includes the coalbedo feedback mechanism and the deep ocean effect. The main difficulty in the mathematical treatment is that the boundary condition is dynamic and diffusive (nonlinear diffusion). In the next section we describe a mathematical model which represents the coupling surface/deep ocean based on that proposed by Watts-Morantine in (Watts & Morantine 1990).

2 THE MODEL

The model represents the evolution of the temperature inside an ocean of depth H. The spatial domain considered is $\Omega = (-1,1) \times (-H,0)$ where the spatial variables (x,z) represent $x = sin(latitude)$ and $-z = depth$. The boundary of Ω is denoted by $\Gamma_H \cup \Gamma_0 \cup \Gamma_1 \cup \Gamma_{-1}$, where $\Gamma_H = \{(x,z) \in \bar{\Omega} : z = -H\}$, $\Gamma_0 = \{(x,z) \in \bar{\Omega} : z = 0\}$, $\Gamma_1 = \{(x,z) \in \bar{\Omega} : x = 1\}$, $\Gamma_{-1} = \{(x,z) \in \bar{\Omega} : x = -1\}$ and $\bar{\Omega}$ denote the closure of Ω.

The governing equation for the ocean interior is given by

$$U_t - \left(\frac{K_H}{R^2}(1-x^2)U_x \right)_x - K_V U_{zz} + \omega U_z = 0, \quad \text{in } \Omega \times (0,T) \tag{1}$$

where the subscript x, z or t denotes partial derivation with respect to the variable x, z or t, $U = U(x,z,t)$ represents the temperature, K_H and K_V are thermal conductivities in x and z direction respectively, w is the velocity, assumed vertical, R is the radius of the Earth and T is the final simulation time. In this model the temperature is assumed to be constant over each parallel, therefore it only depends on latitude and depth. In the rest of the paper we shall denominate the equation (1) as DOM (Deep Ocean Model).

Concerning the boundary conditions for the ocean bottom, Γ_H, we have

$$\omega x \frac{\partial U}{\partial x} + K_V \frac{\partial U}{\partial z} = 0 \qquad \text{on } \Gamma_H \times (0,T).$$

Budyko (Budyko 1969) and Sellers (Sellers 1969) formulated one layer thermodynamic models of the Earth's zonally averaged mean annual surface temperature field as a balance. Both models include one important nonlinear mechanism: ice albedo feedback. The boundary condition in Γ_0 is based in such a balance:

$$Du_t - \frac{DK_{H_0}}{R^2}\left((1-x^2)^{\frac{p}{2}}|u_x|^{p-2}u_x\right)_x + Bu + C + K_V\frac{\partial U}{\partial n} + \omega x u_x \in \frac{1}{\rho c}QS(x)\beta(u). \quad (2)$$

According to Budyko's model, $Bu+C$ represents the emitted energy by cooling (that is, the Newton's cooling law) with B and C representing cooling parameters (assumed constant in time and space), D is the depth of the mixed layer, K_{H_0} is the horizontal thermal diffusivity in the mixed layer, Q is a solar parameter, ρ is the density, c is the specific heat, $S(x)$ is an insolation function and β is the co-albedo. The coalbedo represents the fraction of the incoming radiation flux which is absorbed by the surface. Coalbedo is eventually discontinuous in u and it is introduced in the equation as a continuous graph in order to get a well posed problem.

In the rest of the paper we shall denominate the equation (2) as EBM (Energy Balance Model). Let us remark that upper-case letter U is used for the ocean temperature while lower-case $u := U(x,0,t)$ represents surface temperature, this means that $U|_{\Gamma_0} = u$.

Regarding the diffusive part in the equation (2) we have followed the ideas proposed in (Stone 1972) where nonlinear diffusion is considered and it is given by the term $\mathrm{div}(|\nabla u|\nabla u)$, which after being changed into spherical coordinates and assuming u is constant over each parallel give raise to $\left((1-x^2)^{3/2}|u_x|u_x\right)_x$, where x is the sine of the latitude. We notice that this is a particular case of the p-laplacian operator, $\Delta_p u := \mathrm{div}(|\nabla u|^{p-2}\nabla u)$, with $p=3$.

2.1 *Structural hypotheses*

(H_1) β is a bounded maximal monotone graph, i.e. $|v| \le M \ \forall v \in \beta(s),\ s \in R$.

(H_2) $S \in L^\infty(\Gamma_0)$ and $s_1 \ge S(x) \ge s_0 > 0$ a.e. $x \in \Gamma_0$.

(H_3) $w \in C^1(\Omega)$.

(H_4) The constants B, C, R, Q, K_V, K_H and K_{H_0} are positive.

The existence of solutions of this problem under the previous hypotheses is proved in (Díaz & Tello 2001) by fixed point arguments and extended to higher dimension in (Díaz & Tello 2007). One of the model's main features is its high sensitivity front variation of parameters. Multiplicity of steady states depending on the parameter Q was studied in (Díaz & Tello 2008). We have studied the numerical model in the works (Díaz et al. 2012; Hidalgo & Tello 2011).

Remark. The term $K_V\,\partial U/\partial n$ stands for the coupling atmosphere-ocean in the sense of analyzing the influence of the ocean temperature in the atmosphere. We shall show results with and without this term.

Many works are dedicated to the mathematical treatment of global climate energy balance models (one layer), among them, we mention (Díaz 1996) and the references there in, (Hetzer 1990), (North 1990), (Xu 1991). In (Bermejo et al. 2009) a finite element approach is given to a 2D climate energy balance model without deep ocean effect.

3 NUMERICAL APPROXIMATION

We are interested in computing a numerical solution for the problem described in the previous section. To start with we rewrite this problem as advection-reaction-diffusion equations, both for the upper boundary EBM and for the DOM. In particular, for the EBM we have

$$u_t - \left(f(x,u,u_x)\right)_x = \sigma\left(x,u,\frac{\partial U}{\partial n}(x,0,t)\right), \quad (3)$$

where

$$f(x,u,u_x) := \frac{K_{H_0}}{R^2}(1-x^2)^{3/2}|u_x(x,t)|u_x - \frac{w}{D}xu,$$
$$\sigma\left(x,u,\frac{\partial U}{\partial n}\right) := \frac{1}{D}\left(-C + \frac{Q}{\rho c}S(x)\beta(u) + (\omega + x\omega_x - B)u - K_V\frac{\partial U}{\partial n}\right). \quad (4)$$

and the source DOM reads

$$U_t - (F(x,U_x))_x - (G(U,U_z))_z = \gamma(x,U), \quad (5)$$

where

$$F(x,U_x) := \frac{K_H}{R^2}(1-x^2)U_x,$$
$$G(U,U_z) := K_V U_z - wU,$$
$$\gamma(x,U(x,z,t)) := \omega_z U.$$

In this work we have obtained a numerical solution of this problem using the finite volume method with Weighted Essentially Non-Oscillatory (WENO) reconstruction in space and third-order Runge-Kutta TVD for time integration. For each time step, a numerical solution of the EBM model equation, (2), is computed and then used as a Dirichlet boundary condition for the DOM, given by (1).

3.1 *The finite volume framework*

We construct a semi-discrete finite-volume scheme for both the 1D part of the problem as formulated in equation (3), and the 2D part, which is formulated in (5). To start with, we consider the upper boundary condition, equation (3). Let us discretize the 1D domain [−1,1] in N_x intervals, denominated as control volumes. Then, we integrate equation (3) over each control volume and divide by its length. That is, given one general control volume $S_i = \left[x_{i-\frac{1}{2}}, x_{i+\frac{1}{2}}\right]$ of dimension $\Delta x_i = x_{i+1/2} - x_{i-1/2}$ we integrate equation (3) in S_i and divide it by its length Δx_i to obtain the following Ordinary Differential Equation (ODE)

$$\frac{du_i(t)}{dt} = \frac{1}{\Delta x_i}\left(f_{i+\frac{1}{2}} - f_{i-\frac{1}{2}}\right) + \sigma_i(t) \equiv l_i(u(t)), \tag{6}$$

where

$$u_i(t) = \frac{1}{\Delta x_i}\int_{x_{i-\frac{1}{2}}}^{x_{i+\frac{1}{2}}} u(x,t)dx, \tag{7}$$

is the spatial cell average of the solution $u(x,t)$ in the control volume S_i at time t,

$$f_{i+\frac{1}{2}} = f\left(x_{i+\frac{1}{2}}, u\left(x_{i+\frac{1}{2}},t\right), u_x\left(x_{i+\frac{1}{2}},t\right)\right), \tag{8}$$

is the right interface numerical flux at time t, and

$$\sigma_i(t) = \frac{1}{\Delta x_i}\int_{x_{i-\frac{1}{2}}}^{x_{i+\frac{1}{2}}} \sigma\left(x,u,\frac{\partial U}{\partial n}\right)dx, \tag{9}$$

is the spatial average of the source term $\sigma(u(x,t))$ in the control volume S_i at time t.

In a similar way we apply a finite volume scheme for the DOM. We discretize the 2D domain $[-1,1]\times[0,-H]$ in $N_x \cdot N_z$ rectangular cells, also denominated as control volumes. Let $V_{i,j}$ be one of these 2D control volumes of dimensions $\Delta x_i \times \Delta z_j$ where $\Delta x_i = x_{i+1/2} - x_{i-1/2}$ and $\Delta z_j = z_{j+1/2} - z_{j-1/2}$.

We integrate the equation in this control volume to yield

$$\begin{aligned}\frac{dU_{i,j}(t)}{dt} &= \frac{1}{\Delta x_i}\left(F_{i+\frac{1}{2},j} - F_{i-\frac{1}{2},j}\right)\\ &+ \frac{1}{\Delta z_j}\left(G_{i,j+\frac{1}{2}} - G_{i,j-\frac{1}{2}}\right) + \Gamma_{ij} \equiv L_{ij}(t),\end{aligned} \tag{10}$$

where

$$U_{i,j}(t) = \frac{1}{\Delta x_i \Delta z_j}\int_{x_{i-\frac{1}{2}}}^{x_{i+\frac{1}{2}}}\int_{z_{j-\frac{1}{2}}}^{z_{j+\frac{1}{2}}} U(x,z,t)dzdx, \tag{11}$$

is the cell average of the unknown inside the cell V_{ij}, while the value $F_{i+\frac{1}{2},j}$ is the right intercell numerical flux in $x-$direction and $G_{i,j+1/2}$ is the upper intercell numerical flux in $z-$direction at time t, and

$$F_{i+\frac{1}{2},j} = \frac{1}{\Delta z_j}\int_{z_{j-\frac{1}{2}}}^{z_{j+\frac{1}{2}}} F(x_{i+\frac{1}{2}}, U_x(x_{i+\frac{1}{2}},z,t))dz, \tag{12}$$

$$G_{i,j+\frac{1}{2}} = \frac{1}{\Delta x_i}\int_{x_{i-\frac{1}{2}}}^{x_{i+\frac{1}{2}}} G(U(x,z_{j+\frac{1}{2}},t), U_z(x,z_{j+\frac{1}{2}},t))dx, \tag{13}$$

are the spatial average of physical fluxes over cell faces at time t and

$$\Gamma_{ij} = \frac{1}{\Delta x_i \Delta z_j}\int_{x_{i-\frac{1}{2}}}^{x_{i+\frac{1}{2}}}\int_{z_{j-\frac{1}{2}}}^{z_{j+\frac{1}{2}}} \gamma(x,U(x,z,t))dzdx, \tag{14}$$

is the spatial average of the source term $\gamma(x,U(x,z,t))$ over the control volume V_{ij}.

The numerical solution of the EBM given by (6) and the DOM given by (10) may be advanced in time by means of, for instance, a TVD Runge-Kutta method. The one we have used in this paper is the third-order method, as described in (Shu 1988; Titarev & Toro 2004), whose expressions are

$$\begin{aligned}\eta^{k,1} &= \eta^n + \Delta t \Lambda(\eta^n),\\ \eta^{k,2} &= \frac{3}{4}\eta^n + \frac{1}{4}\eta^{k,1} + \frac{1}{4}\Delta t \Lambda(\eta^{k,1}),\\ \eta^{k+1} &= \frac{1}{3}\eta^n + \frac{2}{3}\eta^{k,2} + \frac{2}{3}\Delta t \Lambda(\eta^{k,2}),\end{aligned} \tag{15}$$

where $\eta^n := u(x_{i+1/2},t^n)$ for the EBM and $\eta^n := U(x_{i+1/2},t^n)$ for the DOM. Moreover the operator $\Lambda(\cdot)$ is the operator $l(\cdot)$ for the EBM and the operator $L(\cdot)$ for the DOM part.

Appart from the Runge-Kutta TVD scheme other possible options can be chosen for time integration, such as the ADER scheme developed by Toro and collaborators, see for instance (?; Toro & Hidalgo 2009; Toro 2009).

The process we have applied to solve the problem can be summarized as follows:

1. Compute the initial cell averages of the solution, both for the EBM and the DOM. Therefore we have the values

$$u_i^0 = \frac{1}{\Delta x_i}\int_{x_{i-\frac{1}{2}}}^{x_{i+\frac{1}{2}}} u(x,0)dx,$$

$$U_i^0 = \frac{1}{\Delta x_i \Delta z_j}\int_{x_{i-\frac{1}{2}}}^{x_{i+\frac{1}{2}}}\int_{z_{j-\frac{1}{2}}}^{z_{j+\frac{1}{2}}} U(x,0)dzdx.$$

2. For each time step:
   a. Solve the EBM, according to the following steps:
      i. Compute the intercell numerical fluxes using WENO technique, which will be briefly explained in the next subsection.
      ii. Use the intercell numerical fluxes (8) to obtain the operator $l(\cdot)$ for the EBM using (6).
      iii. Solve the ODE (6) using third-order TVD Runge-Kutta scheme (15) to obtain the cell averages of the numerical solution of the EBM, u_i^n.
   b. Solve the DOM using the solution of the EBM, u_i^n, as Dirichlet boundary condition at the upper boundary. The steps to follow are:
      i. Compute the intercell numerical fluxes using WENO technique.
      ii. Use the numerical fluxes (12) to obtain the operator $L(\cdot)$ for the DOM using (10).
      iii. Solve the ODE (10) using third-order TVD Runge-Kutta scheme (15) to obtain the cell averages of the numerical solution of the DOM, U_i^n.

It can be seen that we need to know intercell values and intercell spatial derivatives from cell averages of the solution in order to be able to calculate the numerical fluxes to be used in (6) and (10). This is achieved via the procedure usually called *reconstruction*. The one used in this paper is described in the next subsection. Details about WENO procedure can be found in many references such as (Balsara & Shu 2000; Casper & Atkins 1992; Titarev & Toro 2004; Titarev & Toro 2005).

3.2 *WENO reconstruction*

We distinguish two different parts: reconstruction for the Energy Balance Model and reconstruction for the Deep Ocean Model. In the case of the Energy Balance Model, the numerical solution of (6) requires the computation of the intercell numerical fluxes $f_{i\pm\frac{1}{2}}$. If we consider the expression (8) we notice that we need to obtain the solution and the spatial derivative of the solution at each cell interface from the spatial cell averages that are given by

$$u_i(t) = \frac{1}{\Delta x_i}\int_{x_{i-\frac{1}{2}}}^{x_{i+\frac{1}{2}}} u(x,t)dx. \tag{16}$$

Since we are working in one space dimension, for an order of accuracy r we have r candidate stencils each one of them with r cells. We can denote the r stencils as $\{S_{i-r+1}, S_{i-r+2}, \ldots, S_i\}$, $\{S_{i-r+2}, S_{i-r+3}, \ldots, S_{i+1}\}, \ldots,$ $\{S_i, S_{i+1}, \ldots, S_{i+r-1}\}$. For each one of those stencils we can consider a $(r-1)-th$ degree interpolating polynomial $p_l(x)$, $l = 0, \cdots, r-1$ The WENO procedure defines the reconstructed values: $u(x_{i+1/2}, t)$, $u_x(x_{i+1/2}, t)$ as a convex combination of the rth- order accurate values of all polynomials taken with positive nonlinear weights.

Each one of the polynomials considered must be conservative, in the sense that the integral average of the polynomial is equal to the integral average of the solution within each cell in the stencil

$$\frac{1}{\Delta x_k}\int_{S_k} p_l(x)dx = u_k(t), 0 \le (l,k) \le r-1, \tag{17}$$

where S_k is each one of the r cells in the stencil used to construct the polynomial p_l.

Let us now denote as $u_{i+1/2}^{(k,0)}$, $(1 \le k \le r)$ the r values taken by the r polynomials at the cell interface $x_{i+1/2}$ and we denote as $u_{i+1/2}^{(k,1)}$, $(1 \le k \le r)$ the r values taken by the first spatial derivative of the r polynomials at $x_{i+1/2}$. We define the values $u(x_{i+1/2}, t)$ and $u_x(x_{i+1/2}, t)$ as

$$u(x_{i+\frac{1}{2}}, t) = \sum_{k=0}^{r-1} \omega_k u_{i+\frac{1}{2}}^{(k,0)}, \quad u_x(x_{i+\frac{1}{2}}, t) = \sum_{k=0}^{r-1} \omega_k u_{i+\frac{1}{2}}^{(k,1)} \tag{18}$$

where ω_r are the so-called nonlinear weights. They are calculated using

$$\omega_j = \frac{\alpha_j}{\sum_{k=0}^{r-1}\alpha_k} \quad \text{where} \quad \alpha_j = \frac{d_j}{(\varepsilon + \beta_j)^p} \tag{19}$$
$$(0 \le j \le r-1),$$

where we use $p = 2$ and $\varepsilon = 10^{-6}$ which is introduced to avoid division by zero. In the expression (19) we have used the so-called smoothness indicators that are obtained from

$$\beta_k = \sum_{m=0}^{r-1}\int_{x_{i-\frac{1}{2}}}^{x_{i+\frac{1}{2}}} \frac{d^m}{dx^m}(p_k(x))^2 \Delta x^{2m-1}dx \tag{20}$$
$$(0 \le k \le r-1).$$

Apart from the numerical fluxes we also need to compute the source term integral given in (9). One possible option is to use a Gaussian quadrature formula, as the following two-point one, which for the reference interval $\hat{S} = [-1,1]$ reads

$$\int_{-1}^{1} \phi(\xi)d\xi \approx \phi\left(-\frac{1}{\sqrt{3}}\right) + \phi\left(\frac{1}{\sqrt{3}}\right). \tag{21}$$

We can easily map the interval $S_i = [x_{i-1/2}, x_{i+1/2}]$ onto the reference interval $\hat{S}$ by means of a linear transformation and obtain the gaussian quadrature points, denoted as $x_{\alpha,i}$ and $x_{\beta,i}$, and weights, denoted as $w_{\alpha,i}$ and $w_{\beta,i}$, for S_i which are expressed as $x_{\alpha,i} = x_i - \Delta x_i/2\sqrt{3}$ and $x_{\beta,i} = x_i + \Delta x_i/2\sqrt{3}$, where $x_i = (x_{i+1/2} + x_{i-1/2})/2$. The weights are $w_{\alpha,i} = w_{\beta,i} = \Delta x_i$.

Therefore we can approximate the integral of the source term as

$$\sigma_i(t) = \frac{1}{\Delta x_i}\int_{xi-\frac{1}{2}}^{xi+\frac{1}{2}} \sigma(u(x,t))dx \approx \frac{1}{2}\Big(\sigma\big(u\big(x_{\alpha,i},t\big)\big) + \sigma\big(u\big(x_{\beta,i},t\big)\big)\Big). \quad (22)$$

The expression (22) requires to compute the values $u\left(x_{\alpha,i},t\right)$ and $u\left(x_{\beta,i},t\right)$, which can be achieved using the WENO polynomials, p_l, previously used. We obtain the value of the unknown for each Gaussian point as

$$u\left(x_{\alpha,i},t\right) = \sum_{k=0}^{r-1} \Omega_k u^{(k,\alpha)},\quad u\left(x_{\beta,i},t\right) = \sum_{k=0}^{r-1} \Omega_k u^{(k,\beta)}, \quad (23)$$

where $u^{(k,\alpha)}$ and $u^{(k,\beta)}$ are the values of each polynomial at the gaussian points x_α, i, x_β, i and Ω_k are the nonlinear weights for WENO interpolation.

In this work we have used piecewise-cubic reconstruction which means that we have taken the value $r = 4$. Therefore, we use the four following stencils:

$$\{S_{i-3}, S_{i-2}, S_{i-1}, S_i\},\quad \{S_{i-2}, S_{i-1}, S_i, S_{i+1}\}$$

$$\{S_{i-1}, S_i, S_{i+1}, S_{i+2}\},\quad \{S_i, S_{i+1}, S_{i+2}, S_{i+3}\}.$$

The extrapolated values of the solution for $r = 4$ at cell interface $x_{i+1/2}$ are given by the general expression:

$$u^{(k,0)}_{i+\frac{1}{2}} = \sum_{j=-3}^{3} C_j u_{i+j}(t), \quad (24)$$

where the coefficients C_j are given in Table 1.

The extrapolated derivatives of the solution for $r = 4$ at cell interface $x_{i+1/2}$ are given by the general expression:

$$u^{(k,1)}_{i+\frac{1}{2}} = \frac{1}{\Delta x_i}\sum_{j=-3}^{3} D_j u_{i+j}(t), \quad (25)$$

where the coefficients D_j are given in Table 2.

As optimal weights for intercell values computation we have followed the idea first proposed in (?) in which a much higher weight is assigned to the central stencils. The values of this optimal weights we propose are: $d_0 = d_3 = 1$, $d_1 = d_2 = 10^{10}$. The smoothness indicators, optimal weights in Gaussian calculation and extrapolated values at Gaussian points can be found in (Balsara & Shu 2000; Titarev & Toro 2004).

In order to obtain $u^{(k,\alpha)}$ and $u^{(k,\beta)}$ we consider the general expressions:

$$u^{(k,\alpha)} = \sum_{j=-3}^{3} E_j u_{i+j}(t), \qquad u^{(k,\beta)} = \sum_{j=-3}^{3} F_j u_{i+j}(t),$$

where the coefficients E_j, F_j are given in Tables 3 and 4.

In formula (4) we have a term depending on the normal derivative, $\frac{K_V}{D}\frac{\partial U}{\partial n}$, which represents the influence of the temperature in the deep ocean on the interface with the atmosphere. In order to consider thisterm in the formulation we first approximate the normal derivative and integrate this term in the term in the 1D control volume $[x_{i-1/2}, x_{i+1/2}]$ to yield: $\frac{\partial U}{\partial n} \approx \frac{u_i(t) - U_{i,N_z}(t)}{\Delta z/2}$ where we have considered integral averages of the solution of the EBM and DOM, as defined in (7) and (11). In the numerical example the effect of the consideration of this term will be studied. In the 2D part of our problem, that is the DOM, we need to obtain the numerical solution of the scheme (10) which requires to compute the intercell numerical fluxes $F_{i\pm1/2,j}$ and $G_{i,j\pm1/2}$. Following the ideas put forward in the previous subsection we shall produce a reconstruction polynomial which allows us to obtain point-wise values and spatial derivatives wherever they are needed (in particular at cell interfaces) from cell averages of the solution calculated in the previous time step

$$U_{ij}(t) = \frac{1}{\Delta x_i \Delta z_j}\int_{x_{i-\frac{1}{2}}}^{x_{i+\frac{1}{2}}}\int_{z_{j-\frac{1}{2}}}^{z_{j+\frac{1}{2}}} U(x,z,t)dzdx.$$

Following a similar procedure to the one used in the EBM case we apply WENO reconstruction.

Table 1. WENO coefficients for intercell extrapolated values.

k	C_{-3}	C_{-2}	C_{-1}	C_0	C_1	C_2	C_3
0	0	0	0	1/4	13/12	−5/12	1/12
1	0	0	−1/12	7/12	7/12	−1/12	0
2	0	1/12	−5/12	13/12	1/4	0	0
3	−1/4	13/12	−23/12	25/12	0	0	0

Table 2. WENO coefficients for intercell derivative extrapolated values.

k	D_{-3}	D_{-2}	D_{-1}	D_0	D_1	D_2	D_3
0	0	0	0	−11/12	9/12	3/12	−1/12
1	0	0	1/12	−15/12	15/12	−1/12	0
2	0	1/12	−3/12	11/12	9/12	0	0
3	−11/12	45/12	−69/12	35/12	0	0	0

Table 3. WENO coefficients for extrapolated values at first Gaussian point. ($\vartheta = \sqrt{3}$)

k	E_{-3}	E_{-2}	E_{-1}	E_0	E_1	E_2	E_3
0	0	0	0	$1 + 65\vartheta/216$	$-35\vartheta/72$	$17\vartheta/72$	$-11\vartheta/216$
1	0	0	$11\vartheta/216$	$1 + 7\vartheta/72$	$-13\vartheta/72$	$7\vartheta/216$	0
2	0	$-7\vartheta/216$	$13\vartheta/72$	$1 - 7\vartheta/72$	$-11\vartheta/216$	0	0
3	$11\vartheta/216$	$-17\vartheta/72$	$35\vartheta/72$	$1 - 65\vartheta/216$	0	0	0

Table 4. WENO coefficients for extrapolated values at second Gaussian point. ($\vartheta = \sqrt{3}$)

k	F_{-3}	F_{-2}	F_{-1}	F_0	F_1	F_2	F_3
0	0	0	0	$1 - 65\vartheta/216$	$35\vartheta/72$	$-17\vartheta/72$	$11\vartheta/216$
1	0	0	$-11\vartheta/216$	$1 - 7\vartheta/72$	$13\vartheta/72$	$-7\vartheta/216$	0
2	0	$7\vartheta/216$	$-13\vartheta/72$	$1 + 7\vartheta/72$	$11\vartheta/216$	0	0
3	$-11\vartheta/216$	$17\vartheta/72$	$-35\vartheta/72$	$1 + 65\vartheta/216$	0	0	0

In order to proceed we can choose between using a fully 2D reconstruction polynomial or a dimension-by-dimension reconstruction, which consists of obtaining two 1D polynomial for each cartesian direction. The dimension-by-dimension option is the one we have chosen in this work, as it is easier to implement, it is less computationally expensive and it gives good results. Furthermore it can be extended straightforward to the three-dimensional case. See for example (Casper & Atkins 1992; Titarev & Toro 2004; Titarev & Toro 2005) for details on applications of this kind of reconstruction. Therefore we are using two 1D reconstructions for each 2D control volume and applying WENO procedure for both of them. This means that, for each reconstruction we can use the same process as described previously in this section.

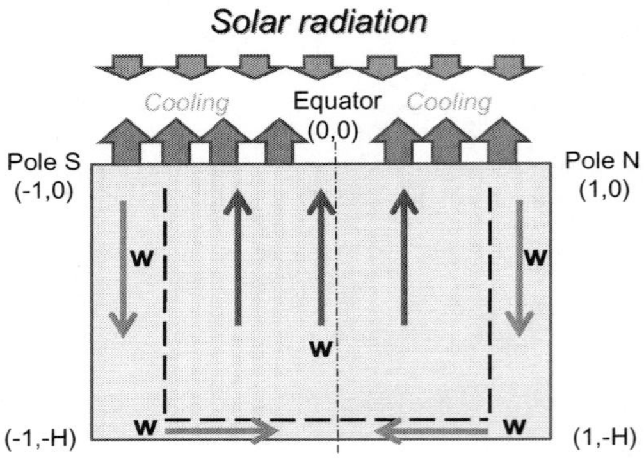

Figure 1. Physical process involved in the numerical example.

4 NUMERICAL EXAMPLE

We consider the system deep ocean-atmosphere. The main processes involved are incoming solar radiation onto the ocean surface, cooling in the interface ocean-atmosphere and sinking cold water owing to ice melting in the poles. We consider that melting water sinks all the way down to the bottom, spreads throughout the bottom of the ocean and, in certain latitudes, rises up towards the surface.

The mathematical model is the one proposed in this work using the initial condition $U(x,z,0) = (80e^{-x^2} - 60)e^{6z} + 18e^{-x^2-z^2}$ whose restriction to the upper boundary is $u(x,0) = U(x,0,0) = 98e^{-x^2} - 60$. The physical data used in this example are $K_H = 0.049$, $K_{H_0} = 0.555 \times 10^{-3}$, $K_V = 0.0125$, $C = 190$, $B = 2$, $c = 1$, $\rho = 1$, $Q = 340$ and $D = 60$. We also need to define, to be used in (2), the function $S(x)$. It represents how the incoming solar radiation is distributed throughout the surface of the ocean such that most of the heat goes to the Equator and a little amount of heat goes to the poles. Moreover this function must be non-negative everywhere. To simulate this effect we give the value $S(x) = 1 - \frac{1}{2}P_2(x)$ where $P_2(x) = \frac{1}{2}(3x^2 - 1)$ is the second Legendre polynomial in the interval $[-1,1]$. The coalbedo $\beta(u)$ is given by

$$\beta(u) = \begin{cases} m \text{ if } u < -10, \\ [m, M] \text{ if } u = -10, \\ M \text{ if } u > -10, \text{ with } 0 < m < M, \end{cases} \tag{26}$$

where $m = 0.4$ and $M = 0.69$. The considered velocity depends only on x and is defined as

$$\omega(x,z)=\frac{10(x+0.75)(x-0.75)}{(0.1+10|x+0.75|)(0.1+10|x-0.75|)}. \quad (27)$$

The spatial domain is the rectangle $[-1,1]\times[0,-1]$. We have discretized the domain using 40 cells in $x-$direction and also 40 cells in $z-$direction. Regarding the discretization in time we have taken the time step

$$\Delta t = min\left(\frac{\alpha\Delta x^2}{K_H},\frac{\alpha\Delta z^2}{K_H},\alpha\Delta x^2\left(K_{H_0}\left|\frac{du}{dx}\right|\right)^{-1}\right),$$

where $\alpha = 0.3$ is a diffusion parameter which controls the stability of the numerical scheme. Numerical experiments have shown that higher values of the parameter α yield an unstable numerical solution. In Figure 2 we display the contour plot of the temperature inside the deep ocean showing that, if we consider the effect of deep ocean, the temperature in the atmosphere has less variation, which agrees with the thermostatic behaviour of the ocean.

More precisely, if we call $\tilde{v}(x,t)$ the approximate temperature at the upper boundary without the deep ocean effect (without the term $K_V\frac{\partial U}{\partial n}$) and if we call $v(x,t)$ the approximate temperature at the upper boundary with the deep ocean effect (obtained by this numerical aproximation) then we get that, for every t,

$$\min_{\Gamma_0}(\tilde{v}(x,t)) \le \min_{\Gamma_0}(v(x,t)) \le \max_{\Gamma_0}(v(x,t)) \le \max_{\Gamma_0}(\tilde{v}(x,t)).$$

In Figure 3 it is displayed the temperature distribution on the surface of the ocean, assumed constant on each parallel for an output time t = 3.

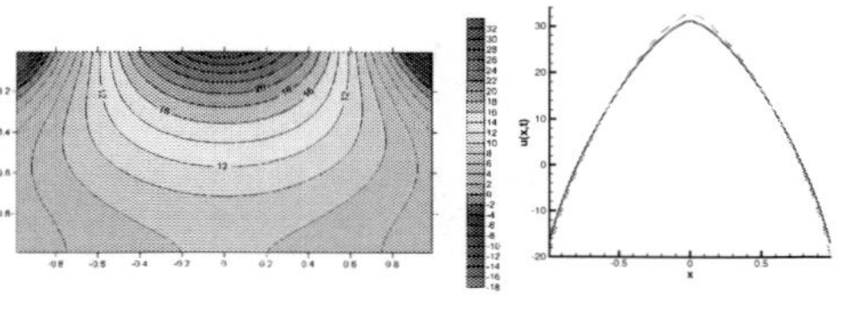

Figure 2. Solution in the deep ocean for t = 2. Left: Contour plot for DOM. Right: Solution of the EBM: Full line is with effect of deep ocean, dashed line is without effect of deep ocean.

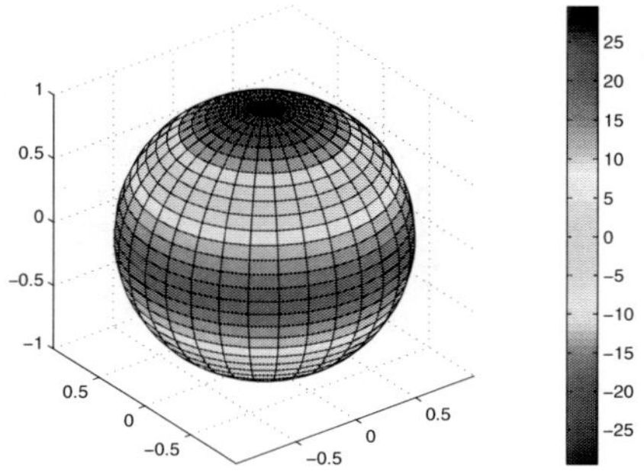

Figure 3. Temperature distribution on the surface of the ocean: assumed constant on each parallel. for t = 3.

5 VALIDATION OF THE NUMERICAL SCHEME

In order to carry out the validation of our numerical scheme we consider the function $U(x,z,t)=\frac{1}{1+t}(x^2-1)^2(1+z)^2$ and construct the auxiliary test problem

$$\begin{aligned}
&U_t-\left(\frac{K_H}{R^2}(1-x^2)U_x\right)_x - K_V U_{zz} + wU_z = \Phi(x,z,t)\\
&\text{in}(0,T)\times\Omega,\\
&wxU_x + K_V U_z = 0 \quad \text{in}(0,T)\times\Gamma_H,\\
&Du_t-\frac{DK_{H_0}}{R^2}\left((1-x^2)^{3/2}|u_x|u_x\right)_x + K_V\frac{\partial U}{\partial n}\\
&+wxU_x + C + Bu = \frac{1}{\rho c}QS(x)\beta(x,U)+\psi(x,t)\\
&\text{in}(0,T)\times\Gamma_0,\\
&(1-x^2)^{3/2}|U_x|U_x = 0 \quad \text{in}(0,T)\times\Gamma_{-1}\cup(0,T)\times\Gamma_1,\\
&U(x,z,0)=(1+x^2)(1+z^2) \quad \text{in}\,\Omega,\\
&u(x,0)=1+x^2 \quad \text{in}\,\Gamma_0,
\end{aligned} \quad (28)$$

where the source terms $\Phi(x,z,t)$ and $\psi(x,t)$ have been added to the original model due to the exact solution considered. Their expressions are not dis-

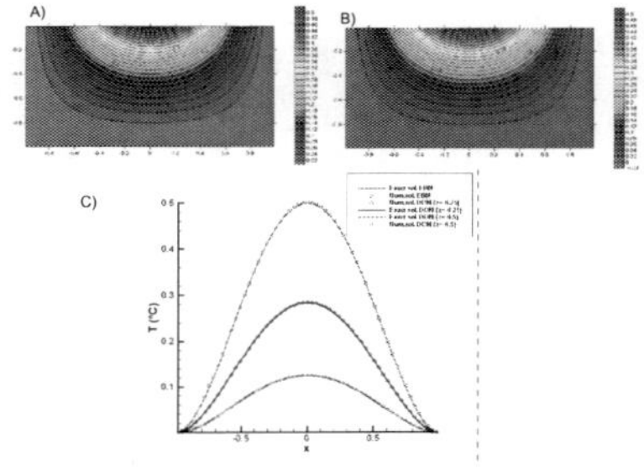

Figure 4. A) Contour plot of numerical solution of (28), B) Contour plot of exact solution of (28), C) Comparison exact solution (full line) versus numerical solution (symbols) for: Upper boundary (1D EBM) and DOM, along the lines z = −0.25 and z = −0.5 (with t = 1).

played here but they can be easily obtained using a symbolic computation tool. The physical and discretization parameters are the same as those used in the previous example. In Figure 4 *A*) and *B*) they are shown the contour plot of the numerical solution and the exact one, respectively, while *C*) represents the comparison between the numerical solution and the exact one in the upper boundary, which means the 1D EBM, and in the DOM along the lines $z = -0.25$ and $z = -0.5$. The results show a good agreement between the numerical solution and the exact one.

6 CONCLUSIONS

We have obtained a numerical solution of the coupled model atmosphere-deep ocean, by means of a finite volume approach with WENO-7 reconstruction and third order TVD Runge-Kutta for time discretization. The evolution of the temperature in the deep ocean is due to the combined effect of water going down from the Earth poles and heating-cooling processes taking place in the interface atmosphere-ocean. We have checked the numerical solution using a test problem in which we have added a source term in order to have an analytical solution, obtaining good results. The results obtained also allow us to compare the mean surface temperature with and without the deep ocean effect. The variation of the surface temperature is lower when the deep ocean effect is considered, showing the thermostatic effect of the ocean.

REFERENCES

Balsara D.S. & Shu Ch.W. 2000. Monotonicity preserving weighted essentially non-oscillatory schemes with increasingly high order of accuracy. *J. Comp. Phys.* 160:405–452.

Bermejo R., Carpio J., Díaz J.I. & Tello L. 2009. Mathematical and Numerical Analysis of a Nonlinear Diffusive Climate Energy Balance Model. *Mathematical and Computer Modelling* 49:1180–1210.

Budyko MI (1969) The effects of solar radiation variations on the climate of the Earth, *Tellus*, 21: 611–619.

Casper J. & Atkins H. 1993. A finite volume high order ENO scheme for two dimensional hyperbolic systems. *J. Comp. Phys.* 106:62–76.

Díaz J.I. (Edit.) 1996. The Mathematics of Models in Climatology and Environment. ASI NATO Global Change Series I, n.48. Springer-Verlag, Heidelberg.

Díaz J.I., Hidalgo A., Tello L. 2012. Multiple solutions and numerical analysis to the dynamic and stationary models coupling a delayed energy balance model involving latent heat and discontinuous albedo with a deep ocean (submitted).

Díaz J.I., Tello L. 2001. Sobre un modelo climático de balance de energía superficial acoplado con un océano profundo. Actas XVII CEDYA/VI CMA. Univ. Salamanca.

Díaz J.I. & Tello L. 2007. A 2D climate energy balance model coupled with a 3D deep ocean model. *Electronic Journal of Differential Equations*, Conf. 16:129–135.

Díaz J.I. & Tello L. 2008. On a climate model with a dynamic nonlinear diffusivity boundary condition. *Discrete and Continuous Dynamical systems*, Series S 1(2):253–262.

Dumbser M., Enaux C., Toro E.F. 2008. Finite volume schemes of very high order of accuracy for stiff hyperbolic balance laws. *J. Comp. Phys.* 227:3971–4001.

Hetzer G. 1990. The structure of the principal component for semilinear diffusion equations from energy balance climate models. *Houston J. of Math.* 16:203–216.

Hidalgo A., Tello L. 2011. A finite volume scheme for simulating the coupling between deep ocean and an atmospheric balance model. Modern Mathematical Tools and techniques in capturing complexity. Springer Series Understanding Complex Systems. Springer:239–255.

North G.R. 1990. Multiple solutions in energy balance climate models. *Paleogeography, Paleoclimatology, Paleoecology* 82:225–235.

Sellers W.D. 1969. A global climatic model based on the energy balance of the earth-atmosphere system. *J. Appl. Meteorol.* 8:392–400.

Shu Ch.W. 1988. Total variation diminishing time discretizations. *SIAM J. Sci. Stat. Comput.* 9:1073–1084.

Stone P.H. 1972. A simplified radiative—dynamical model for the static stability of rotating atmospheres. *J. of the Atmospheric Sciences* 29(3):405–418.

Titarev V.A. & Toro E.F. 2004. Finite volume WENO schemes for three-dimensional conservation laws. *J. Comp. Phys.* 201:238–260.

Titarev V.A. & Toro E.F. 2005. ADER schemes for three-dimensional non-linear hyperbolic systems. *J. Comp. Phys.* 204(2):715–736.

Toro E.F. & Hidalgo A. 2009. ADER finite volume schemes for nonlinear reaction–diffusion equations. *Appl. Num. Math.* 59(1):73–100.

Toro E.F. 2009. Riemann solvers and numerical methods for fluid dynamics (3*rd* ed). Springer-Verlag Berlin Heidelberg.

Watts R.G. & Morantine M. 1990. Rapid climatic change and the deep ocean. *Climatic Change* 16:83–97.

Xu X. 1991. Existence and Regularity Theorems for a Free Boundary Problem Governing a Simple Climate Model. *Applicable Anal.* 42: 33–59.

Numerical Methods for Hyperbolic Equations – Vázquez-Cendón et al. (eds)
 ISBN 978-0-415-62150-2

A large time step upwind scheme for the shallow water equations with source terms

M. Morales-Hernandez, J. Murillo & P. García-Navarro
Fluid Mechanics, Universidad Zaragoza, Spain

J. Burguete
Soil and Water, EEAD, CSIC, Spain

ABSTRACT: It is possible to relax the condition over the time step size when using explicit schemes. A generalization of the first order explicit upwind and Roe method, modified to allow large time steps, was explored by Leveque. It becomes stable for CFL's larger than 1 and provides an accurate and correct solution of shocks. The technique is devised to cope well with flow transients and even discontinuities, being able to give a resolution of the shocks even sharper for CFL > 1. When rarefactions are present it needs some adjustments and the preferable procedure is not clear. A way to overcome that situation will be proposed and explored. The extension of the explicit method considers situations with or without source terms. In the first case, they are discretized according to the upwind formulation. The performance of the above scheme will be evaluated to solve the shallow water equations.

1 INTRODUCTION

Upwind methods have proved very successful in Computational Fluid Dynamics, mainly in connection with aerodynamics where they have gained widespread acceptance. They are, perhaps, the most widely researched algorithms in connection with flow simulation codes. Finite difference schemes for time dependent equations are traditionally divided in two main groups, according to the way of discretization used for the time derivative, as explicit and implicit. Implicit schemes offer unconditional numerical stability at the extra cost of having to deal with the resolution of an algebraic, and often nonlinear, system with as many unknowns as grid points at every time step. On the other hand, conceptual simplicity is the most valuable characteristic of explicit schemes in which the variables at a future time can be independently evaluated at every single point. The allowable time step size is nevertheless restricted by stability reasons to fulfil the CFL condition.

However, it is possible to relax the condition over the time step (CFL condition) when using explicit schemes. This method was proposed by Leveque (Leveque 1981; Leveque 1985) first for the linear and non-linear scalar case and then adapted to systems of equations. It provides accurate and correct solutions of shocks so that rarefactions need some adjustments. The way to do this will be explored. The source terms that can appear in conservation laws introduce an extra difficulty in practical applications. The way to deal with source terms is incorporated into the proposed procedure according to the upwind formulation (Murillo and Garcia-Navarro 2010). The outline is as follows: the discretization is described first, for 1D scalar equations with and without source terms. Then, the scheme is extended to solve the shallow water equations and bed slope and friction source terms are incorporated into the proposed procedure. Steady and unsteady open channel flow problems with analytical solutions where bed slope and/or friction source terms play an important role are used as validation test cases.

2 ONE DIMENSIONAL SCALAR EQUATION

2.1 *Linear scalar equation*

The easiest way to understand a numerical method can be illustrated by application to the linear scalar equation:

$$\frac{\partial u}{\partial t} + \frac{\partial f(u)}{\partial x} = 0 \qquad (1)$$

where u is the conserved variable and $f(u)$ is a linear function, $f(u) = \lambda u$, $\lambda = constant$.

A first order cell-centred upwind finite volume method for the equation (1) is based on a piecewise constant approximation of u so that the updating at cell i for a time step Δt on a uniform grid can written as follows:

$$u_i^{n+1} = u_i^n - \frac{\Delta t}{\Delta x}(f_{i+1/2} - f_{i-1/2}) \tag{2}$$

where the numerical flux $f_{i+1/2}$ can be determined using an approximate solver.

In terms of Roe's approach (Roe 1986), an average discrete velocity can be defined at the interface $(i, i+1)$ as:

$$\lambda_{i+1/2} = \frac{f(u_{i+1}) - f(u_i)}{u_{i+1} - u_i} = \lambda \tag{3}$$

Following the upwind philosophy, to discriminate the sense of propagation according to the sign of the advection velocity, the quantities

$$\lambda_{i+1/2}^{\pm} = \frac{\lambda_{i+1/2} \pm |\lambda_{i+1/2}|}{2} \tag{4}$$

let us express the numerical fluxes in (2) as:

$$f_{i+1/2} = f_i + \lambda^- \delta u_{i+1/2} \qquad f_{i-1/2} = f_i - \lambda^+ \delta u_{i-1/2} \tag{5}$$

Therefore, the cell updating in Δt can be reformulated as resulting from the sum of two signals instead of the difference of two numerical fluxes (Fig. 1):

$$u_i^{n+1} = u_i^n - \frac{\Delta t}{\Delta x}(\delta f_{i-1/2}^{+} + \delta f_{i+1/2}^{-}) \tag{6}$$

This is a finite volume point of view centered at the cells which accumulates the arriving signals to update the value of the function at every cell. There is another way to consider this situation by looking where the signals go from each interface (Leveque 1981). For example, at interface $(i, i+1)$ the quantity $\nu\, \delta u_{i+1/2}$, where $\nu = \Delta t / \Delta x\, \lambda$ can be defined and it is sent according to the sign of λ followingthe algorithm:

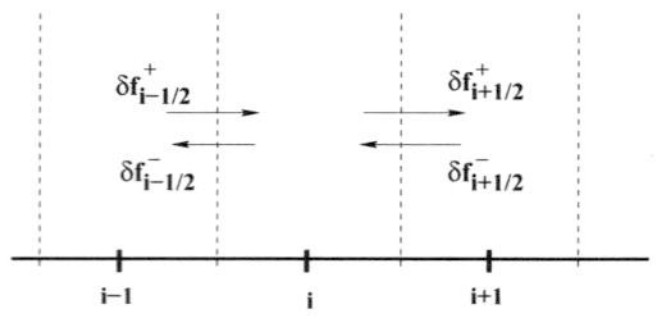

Figure 1. Contributions from left and right to cell i.

$$\begin{aligned} &\text{if } \lambda > 0 \Rightarrow \nu\, \delta u_{i+1/2} \quad \text{updates } i+1 \\ &\text{if } \lambda < 0 \Rightarrow |\nu|\, \delta u_{i+1/2} \quad \text{updates } i \end{aligned} \tag{7}$$

Both versions of the scheme are equivalent if

$$CFL = \frac{\Delta t}{\Delta x}\lambda \leq 1 \tag{8}$$

The second approach is nevertheless preferable to extend the scheme to $CFL > 1$. As described by Leveque (Leveque 1981), the extension of the scheme to larger time steps is achieved by allowing each wave or signal to propagate independently of all others waves according to the following algorithm:

$$\begin{aligned} &\text{If } \lambda > 0 \\ &\quad \delta u_{i+1/2} \text{ updates } i+1, \ldots, i+\mu_{i+1/2} \\ &\quad (\nu - \mu)\, \delta u_{i+1/2} \text{ updates } i+\mu_{i+1/2}+1 \end{aligned} \tag{9}$$

$$\begin{aligned} &\text{If } \lambda < 0 \\ &\quad \delta u_{i+1/2} \text{ updates } i, \ldots, i+\mu_{i+1/2}+1 \\ &\quad |\nu - \mu|\, \delta u_{i+1/2} \text{ updates } i+\mu_{i+1/2} \end{aligned} \tag{10}$$

where $\mu = int(\nu)$

The proposed scheme is explicit and remains conservative in the sense of Roe. It will be called LTS scheme from now on in this work. It is important to remark that if $CFL \leq 1$ the scheme recovers the original first order explicit upwind scheme. Figure 2 shows how the information is sent from interface $(i, i+1)$ to the involved cells when $\lambda > 0$ (a) and when $\lambda < 0$ (b).

2.2 *Non-linear scalar equation*

Consider now the conservation law:

$$\frac{\partial u}{\partial t} + \frac{\partial f(u)}{\partial x} = 0 \tag{11}$$

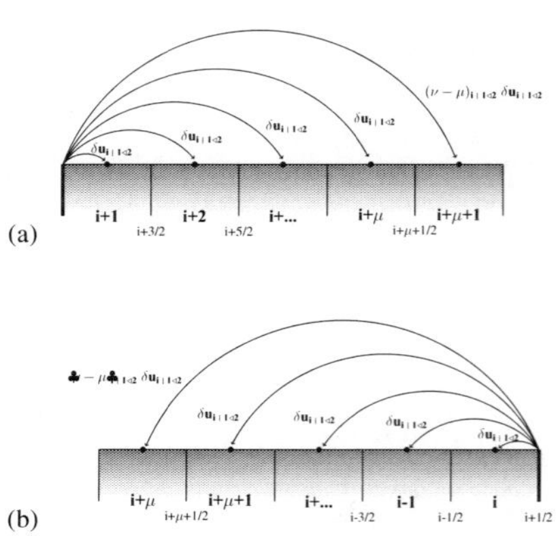

Figure 2. Scheme of the contributions from intercell i+1/2 for $\lambda > 0$ (a) and for $\lambda < 0$ (b).

where $f(u)$ is a convex non-linear function of u. It is possible to find an advection or transport velocity λ:

$$\lambda = \frac{df}{du} \quad \lambda = \lambda(u). \tag{12}$$

which is no longer constant. The LTS scheme, when applied to (11), requires the definition of an approximate advection celerity at the intercell as follows:

$$\tilde{\lambda}_{i+1/2} = \frac{f(u_{i+1}) - f(u_i)}{u_{i+1} - u_i} \tag{13}$$

Certain new elements appear in this case that are going to be explored using the Burgers equation as an example.

2.2.1 *Burgers equation and the Riemann problem*
The inviscid Burgers equation is a particular case of scalar conservation law of the type (11) with $f(u) = \frac{1}{2}u^2$. This equation can be written as

$$\frac{\partial u}{\partial t} + \frac{\partial}{\partial x}\left(\frac{u^2}{2}\right) = 0 \quad \frac{\partial u}{\partial t} + u\frac{\partial u}{\partial x} = 0 \tag{14}$$

Considering the following initial value problem (also called Riemann Problem, RP) for the inviscid Burgers equation

$$\frac{\partial u}{\partial t} + \frac{\partial}{\partial x}\left(\frac{u^2}{2}\right) = 0 \quad u(x,0) = \begin{cases} u_L & if \quad x < 0 \\ u_R & if \quad x > 0 \end{cases} \tag{15}$$

two different situations appear depending on u_L and u_R. When $u_L > u_R$ a right moving shock develops (see Fig. 3) with a speed of the discontinuity $\tilde{\lambda} = \frac{1}{2}(u_L + u_R)$.

When $u_L < u_R$ the solution consists of a smooth rarefaction wave connecting the two constant states u_L an u_R which is plotted in Figure 4.

As described in (Leveque 1981), the LTS scheme can be used to provide an accurate and correct solution of shocks provided (Leveque 1985). In presence of a rarefaction, the explicit upwind scheme replaces several characteristic lines with one line and a unique intermediate state u^* is defined

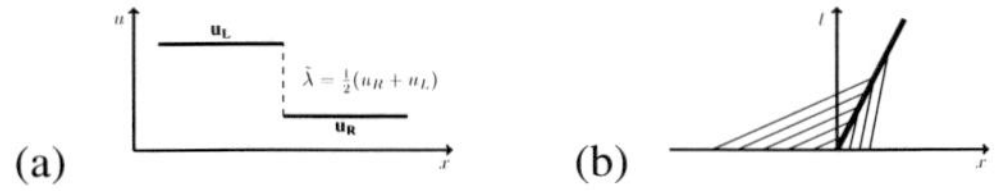

Figure 3. (a) Initial data of a shock; (b) Map of characteristic lines of a shock.

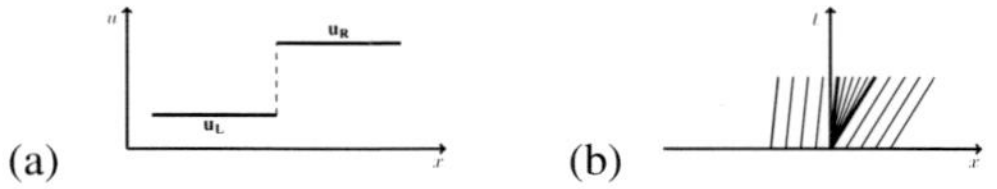

Figure 4. (a) Initial data of a rarefaction; (b) Map of characteristic lines of a rarefaction.

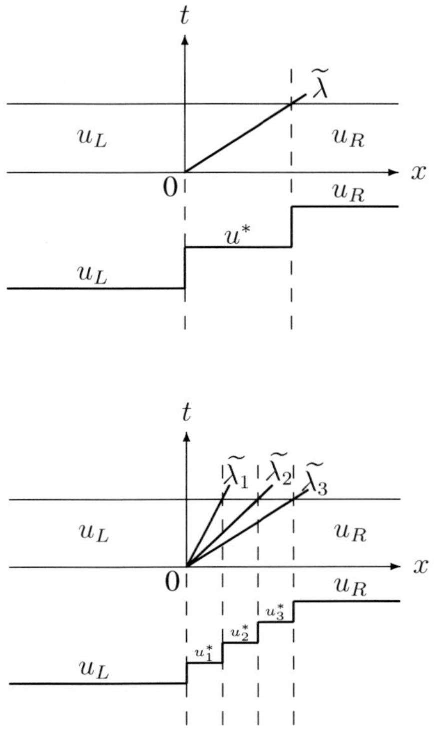

(a)

(b)

Figure 5. (a) Classical treatment of rarefaction waves in the upwind scheme; (b) Splitting treatment of rarefaction waves in the LTS scheme.

(see Fig. 5). In the method proposed in the present work, several intermediate states are defined corresponding to several discontinuities travelling at different speeds (Fig. 5). The required number of discontinuities is related with the strength of the wave. A good approximation could be the integer part of $\delta u \Delta t / \delta x$ where $\delta u = u_R - u_L$. The proposed way of handling rarefaction waves is always conservative in the sense of Roe.

In order to illustrate the performance of LTS in presence of a rarefaction, consider the case with the initial data:

$$u(x,0) = \begin{cases} 1.0 & if \quad x < 50.0 \\ 4.0 & if \quad x > 50.0 \end{cases} \tag{16}$$

The exact solution for this case is:

$$u(x,t)=\begin{cases}1.0 & if \quad \dfrac{x}{t}<1.0\\ \dfrac{x}{t} & if \quad 1.0<\dfrac{x}{t}<4.0\\ 4.0 & if \quad \dfrac{x}{t}>4.0\end{cases} \tag{17}$$

Figure 6(a) shows the exact solution at $t=5s$ and the numerical results obtained with the LTS scheme on a regular mesh of $\Delta x=1.0$. The discretization of the rarefaction in a single wave has been used and different CFL values are associated to different number of Time Steps (TS) as summarized in Table 1. It can be concluded that only in the case of $CFL=1.0$ an accurate solution is achieved although using 20 TS.

Figures 6(b) and (c) show again the exact solution at $t=5s$ and the numerical results obtained with the LTS scheme on a regular mesh of $\Delta x=1.0$, but now supplied with the splitting wave treatment. Different CFL values have been used and they have been summarized in Table 1. The number of time steps used to compute the numerical solution and the number of pieces that the discontinuity has been split into (the integer part of $\delta u\Delta t/\Delta x$) are also indicated.

The larger the CFL value is, the more accurate the numerical solution is. Moreover, there is no upper bound in the choice of the CFL value, only one time step provides the exact solution.

(a)

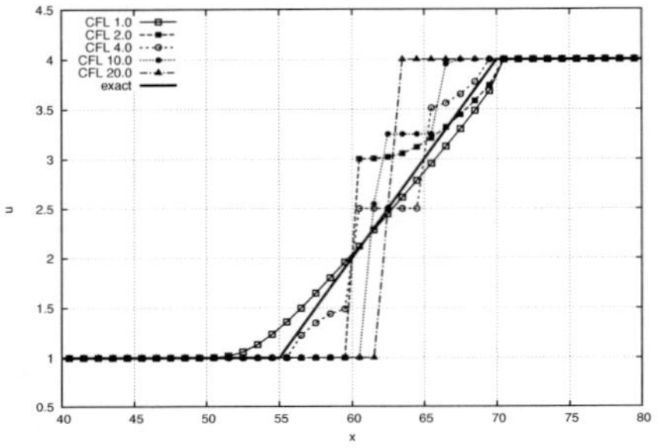

(b)

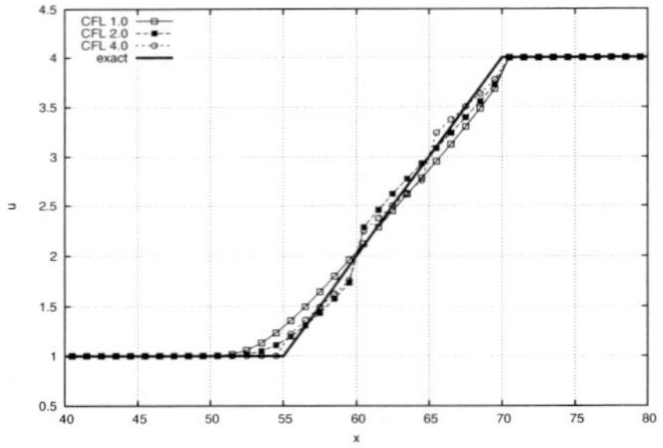

(c)

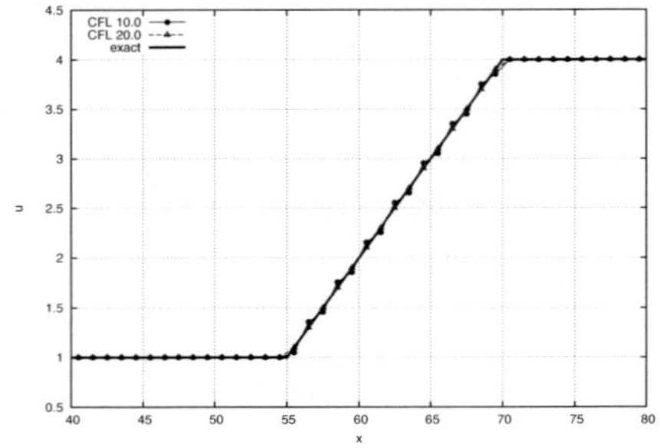

Figure 6. Exact and numerical solution of (14) with the initial data (16) using LTS scheme (a) No splitting rarefaction wave; (b),(c) Splitting rarefaction wave.

2.3 *Source term dicretization*

Consider now the nonlinear scalar equation with source term:

$$\frac{\partial u}{\partial t}+\frac{\partial f(u)}{\partial x}=s \tag{18}$$

and the local Riemann Problem:

$$u(x,0)=\begin{cases}u_L=u_i & if \quad x<0\\ u_R=u_{i+1} & if \quad x>0\end{cases} \tag{19}$$

According to Roe's approach, the solution of the RP is achieved from an approximate solution $\hat{u}(x,t)$ of the locally linearized problem that must fulfil the Consistency Condition (Leveque 2002).

Table 1. Summary of numerical solutions.

	CFL value	Time steps (TS)	Number of pieces
Without splitting waves	1.0	20	–
	2.0	10	–
	4.0	5	–
	10.0	2	–
	20.0	1	–
By splitting waves	1.0	20	1
	2.0	10	2
	4.0	5	3
	10.0	2	7
	20.0	1	15

Let Δt be the time step. Now, integrating over a suitable control volume $\left[-\frac{\Delta x}{2}, \frac{\Delta x}{2}\right]$

$$\int_{-\frac{\Delta x}{2}}^{\frac{\Delta x}{2}} \hat{u}(x,t^{n+1})dx = \Delta x\,(u_{i+1} + u_i) - (f(u_{i+1}) - f(u_i)) + \int_0^{\Delta t}\int_{-\frac{\Delta x}{2}}^{\frac{\Delta x}{2}} s\,dx\,dt \tag{20}$$

For the last integral involving the source term s, the following linearization is assumed

$$s_{i+1/2} = \int_{-\frac{\Delta x}{2}}^{\frac{\Delta x}{2}} s(x,0)dx \tag{21}$$

Therefore, a new RP from Roe's approximate solution can be defined:

$$\frac{\partial \hat{u}}{\partial t} + \lambda^*(u_{i+1},u_i)\frac{\partial \hat{u}}{\partial x} = 0$$
$$\hat{u}(x,0) = \begin{cases} u_i & if \quad x<0 \\ u_{i+1} & if \quad x>0 \end{cases} \tag{22}$$

where $\lambda^*(u_{i+1},u_i)$ is a constant. Now, integrating over the same control volume as in (20)

$$\int_{-\frac{\Delta x}{2}}^{\frac{\Delta x}{2}} \hat{u}(x,1)dx = \Delta x\,(u_{i+1} + u_i) - \lambda^*_{i+1/2}(u_{i+1} - u_i) \tag{23}$$

Since the consistency condition (20) must be satisfied,

$$(\delta f - s)_{i+1/2} = f(u_{i+1}) - f(u_i) - s_{i+1/2} = \lambda^*_{i+1/2}(u_{i+1} - u_i) \tag{24}$$

Following (Murillo and Garcia-Navarro 2010), a new quantity $\tilde{\theta}_{i+1/2}$ can be introduced in order to compact the notation. This measures the relative influence of the source and flux terms. Therefore, $\lambda^*_{i+1/2}$ can be expressed

$$\lambda^*_{i+1/2} = \tilde{\lambda}_{i+1/2}\tilde{\theta}_{i+1/2} \tag{25}$$

where

$$\tilde{\theta}_{i+1/2} = 1 - \frac{s_{i+1/2}}{f(u_{i+1}) - f(u_i)} \tag{26}$$

and $\tilde{\lambda}$ is the advection velocity when the source term s is not present.

Therefore, the LTS scheme could be written as follows:

If $\tilde{\lambda}_{i+1/2} > 0$
$(\tilde{\theta}\delta u)_{i+1/2}$ updates $i+1,\ldots,i+\mu_{i+1/2}$ (27)
$(\nu - \mu)_{i+1/2}\,(\tilde{\theta}\delta u)_{i+1/2}$ updates $i + \mu_{i+1/2} + 1$

If $\tilde{\lambda}_{i+1/2} < 0$
$(\tilde{\theta}\delta u)_{i+1/2}$ updates $i,\ldots,i+\mu_{i+1/2}+1$ (28)
$|\nu - \mu|_{i+1/2}\,(\tilde{\theta}\delta u)_{i+1/2}$ updates $i + \mu_{i+1/2}$

where $\nu_{i+1/2} = \tilde{\lambda}\Delta t/\Delta x$ and $\mu_{i+1/2} = int(\nu_{i+1/2})$

In several situations with large source terms that could have influence in the convective term, using the LTS proposed scheme, this linearization could lead us to wrong solutions due to an overestimation or underestimation of this value. The alternative way proposed consists of estimating the adequate wave celerity using directly the information provided by the analytical solution, constructed by means of the appropriate Rankine-Hugoniot conditions.

2.3.1 *Application to Burgers's equation with source terms*

Consider Burgers's equation with source terms as proposed by Murillo (Murillo and Garcia-Navarro 2010).

$$\frac{\partial u}{\partial t} + \frac{1}{2}\frac{\partial u^2}{\partial x} = -u\frac{\partial z}{\partial x} \tag{29}$$

with the initial data

$$u(x,0) = \begin{cases} u_L & if\ x<0 \\ u_R & if\ x>0 \end{cases} \qquad z(x) = \begin{cases} z_L & if\ x<0 \\ z_R & if\ x>0 \end{cases} \tag{30}$$

The numerical solutions are computed using the scheme proposed by Murillo with CFL = 1 (–▵–) and using the proposed LTS scheme with CFL = 30 (–•–). They are compared with the exact solution (—) at $t = 15s$, computing them with $\Delta x = 1$ and the source term approximation:

$$s_{i+1/2} = -\frac{1}{2}(u_{i+1} + u_i)(z_{i+1} - z_i) \tag{31}$$

The source term is represented in dashed line and all the cases are summarized in Table 2. More information about the nature and the exact solution of each test case can be found in (Murillo and Garcia-Navarro 2010). The results can be observed in Figures 7–10.

Table 2. Summary of test cases.

Test case	u_L	u_R	z_L	z_R
1	2.0	1.0	0.0	0.5
2	2.0	1.0	0.0	–0.5
3	1.0	2.0	0.5	0.0
4	1.0	2.0	0.0	0.5

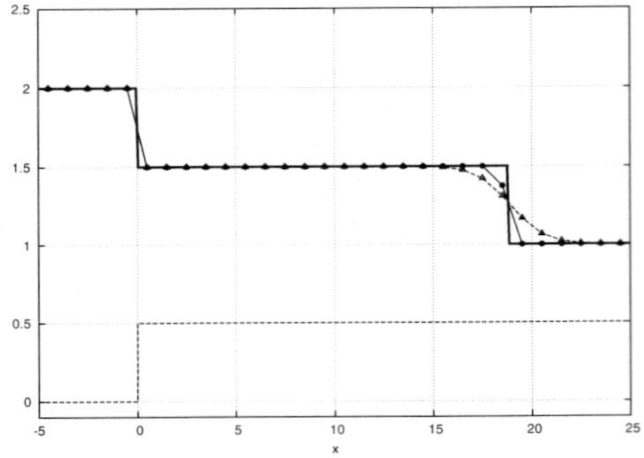

Figure 7. Exact and computed solutions at $t = 15$ for test case 1.

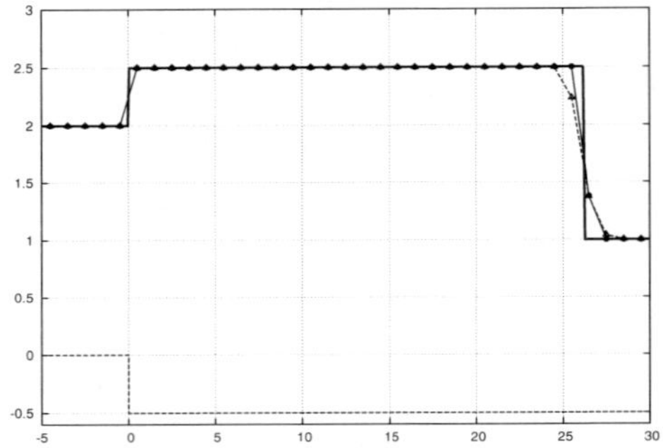

Figure 8. Exact and computed solutions at $t = 15$ for test case 2.

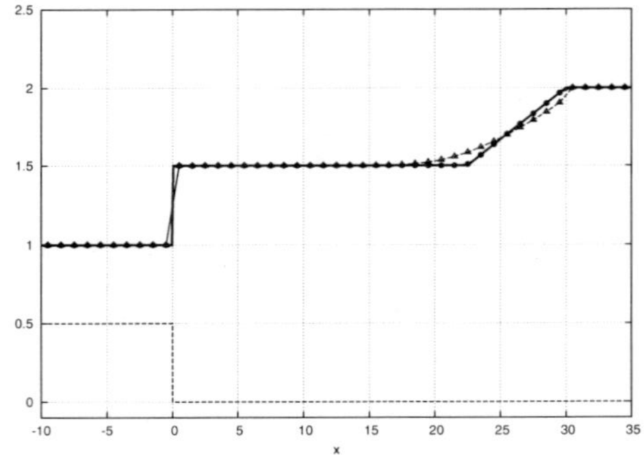

Figure 9. Exact and computed solutions at $t = 15$ for test case 3.

The main conclusion is that the LTS scheme, including a good source term treatment is less diffusive than the scheme provided by Murillo (Murillo and Garcia-Navarro 2010) and it can be able to reproduce the exact solution.

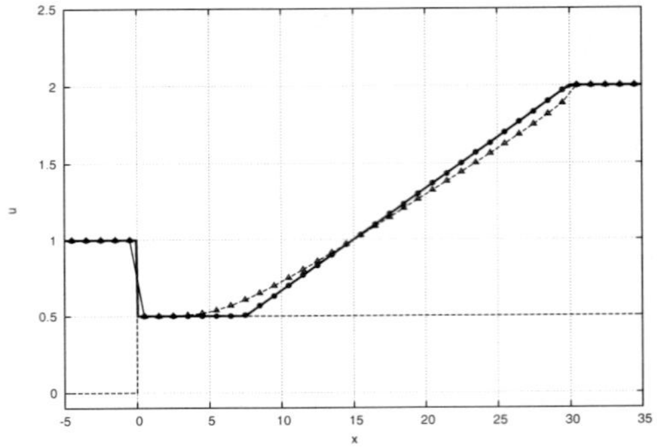

Figure 10. Exact and computed solutions at $t = 15$ for test case 4.

3 ID SYSTEMS OF CONSERVATION LAWS WITH SOURCE TERMS

The extension of this proposed LTS scheme to systems of equations with source terms is discussed in this section. A hyperbolic nonlinear system of equations can be expressed of the form

$$\frac{\partial \mathbf{U}}{\partial t} + \frac{\partial \mathbf{F}}{\partial x} = \mathbf{S} \tag{32}$$

where U are the conserved variables, F the fluxes of these conserved variables and S represents the source term. A Jacobian matrix J can be defined

$$\mathbf{J} = \frac{d\mathbf{F}}{d\mathbf{U}} \tag{33}$$

Applying Roe's linearization (Roe 1986) and extending the ideas from the scalar case:

$$(\delta \mathbf{U})_{i+1/2} = (\tilde{\alpha}^1 \tilde{e}^1 + \tilde{\alpha}^2 \tilde{e}^2)_{i+1/2} \tag{34}$$

$$(\delta \mathbf{F})_{i+1/2} = (\tilde{\lambda}^1 \tilde{\alpha}^1 \tilde{e}^1 + \tilde{\lambda}^2 \tilde{\alpha}^2 \tilde{e}^2)_{i+1/2} \tag{35}$$

$$(\mathbf{S})_{i+1/2} = (\tilde{\beta}^1 \tilde{e}^1 + \tilde{\beta}^2 \tilde{e}^2)_{i+1/2} \tag{36}$$

where $\lambda^1_{i+1/2}, \lambda^2_{i+1/2}$ and $\mathbf{e}^1_{i+1/2}, \mathbf{e}^2_{i+1/2}$ are the eigenvalues and the eigenvectors of the Jacobian matrix respectively.

A new quantity

$$\tilde{\gamma}^m_{i+1/2} = \left(\tilde{\alpha} - \frac{\tilde{\beta}}{\tilde{\lambda}} \right)^m_{i+1/2} \tag{37}$$

can be defined allowing us to express the LTS scheme as follows:

If $\tilde{\lambda}_{i+1/2} > 0$

$$\begin{aligned}&\left(\gamma\tilde{\mathbf{e}}\right)^m_{i+1/2} \text{ updates } i+1,\ldots,i+\mu_{i+1/2}\\&(\nu-\mu)^m_{i+1/2}\left(\gamma\tilde{\mathbf{e}}\right)^m_{i+1/2} \text{updates } i+\mu_{i+1/2}+1\end{aligned} \tag{38}$$

If $\tilde{\lambda}_{i+1/2} < 0$

$$\begin{aligned}&\left(\gamma\tilde{\mathbf{e}}\right)^m_{i+1/2} \text{ updates } i,\ldots,i+\mu_{i+1/2}+1\\&|\,\nu-\mu\,|^m_{i+1/2}\left(\gamma\tilde{\mathbf{e}}\right)^m_{i+1/2} \text{ updates } i+\mu_{i+1/2}\end{aligned} \tag{39}$$

where $\nu^m_{i+1/2} = \dfrac{\Delta t}{\Delta x}\tilde{\lambda}^m_{i+1/2}$ and $\mu^m_{i+1/2} = int(\nu^m_{i+1/2})$

4 APPLICATION TO THE 1D SHALLOW WATER EQUATIONS

4.1 *Equations*

The 1D shallow water mass and momentum system can be written:

$$\mathbf{U} = \begin{pmatrix} A \\ Q \end{pmatrix},\quad \mathbf{F} = \begin{pmatrix} Q \\ \frac{Q^2}{A} + gI_1 \end{pmatrix}$$
$$\mathbf{S} = \begin{pmatrix} 0 \\ g\left[I_2 + A\left(S_0 - S_f\right)\right] \end{pmatrix} \tag{40}$$

where Q is the discharge, A is the wetted cross section, S_0 is the bed slope, S_f is the friction slope represented by the empirical Manning law:

$$S_f = \frac{Q^2}{n^2 A^2 R^{4/3}} \tag{41}$$

and I_1 and I_2 represent hydrostatic pressure force integrals.

The approximate Jacobian $\tilde{\mathbf{J}}$ is

$$\tilde{\mathbf{J}}_{i+1/2} = \begin{pmatrix} 0 & 1 \\ \tilde{c}^2 - \tilde{u}^2 & 2\tilde{u} \end{pmatrix}_{i+1/2} \tag{42}$$

with

$$\tilde{c} = \sqrt{g\frac{(A/b)_i + (A/b)_{i+1}}{2}} \qquad \tilde{u} = \frac{Q_{i+1}\sqrt{A_{i+1}} + Q_i\sqrt{A_i}}{\sqrt{A_{i+1}} + \sqrt{A_i}} \tag{43}$$

where b is the cross section width and the resulting set of approximate eigenvalues and eigenvectors are

$$\tilde{\lambda}^1 = \tilde{u} - \tilde{c} \qquad \tilde{\lambda}^2 = \tilde{u} + \tilde{c}$$
$$\tilde{\mathbf{e}}^1 = \begin{pmatrix} 1 \\ \tilde{u} - \tilde{c} \end{pmatrix} \qquad \tilde{\mathbf{e}}^2 = \begin{pmatrix} 1 \\ \tilde{u} + \tilde{c} \end{pmatrix} \tag{44}$$

4.2 *Application to steady flow with source terms*

MacDonald (MacDonald 1996) supplied a set of realistic open channel flow test cases with analytical solution very well suited to validate the numerical schemes. Two examples from (MacDonald et al. 1997) are used here. They both apply a Manning friction coefficient $n = 0.03$, have been simulated with $\Delta x = 1.0$ and the inlet discharge is 20 m^3/s. In test case 1 the flow is subcritical all along the 150 m length and the 10 m wide rectangular channel. The downstream boundary condition is a fixed height. The steady water depth is:

$$h(x) = 0.8 + 0.25\, exp\left(33.75\left(\frac{x}{150} - 1/2\right)^2\right) \tag{45}$$

Testcase 2 corresponds to a trapezoidal channel with 10 m bottom width and 200 m length. The side slope of the channel is 2, and there is not downstream boundary condition. Hence, a smooth transition between subcritical flow upstream (at the first half of the reach) and supercritical flow downstream (at the second half) takes place.

Here, the steady water depth is expressed as follows:

$$h(x) = 0.706033 - 0.25\, tanh\left(\frac{x - 100}{50}\right) \tag{46}$$

The results for these test cases can be observed in Figures 11 and 12 where the numerical solution using CFL 60.0 (– • –) is compared with the exact solution (——). Also the bed level is represented in dashed line.

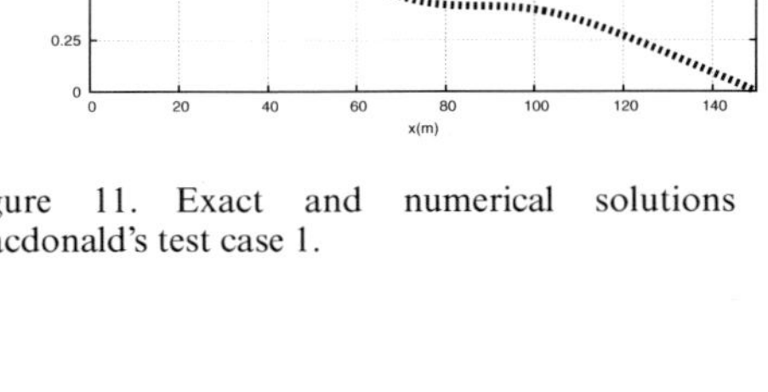

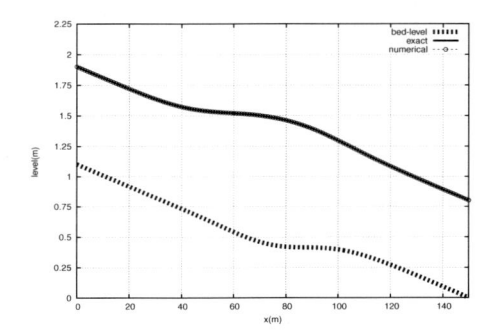

Figure 11. Exact and numerical solutions for Macdonald's test case 1.

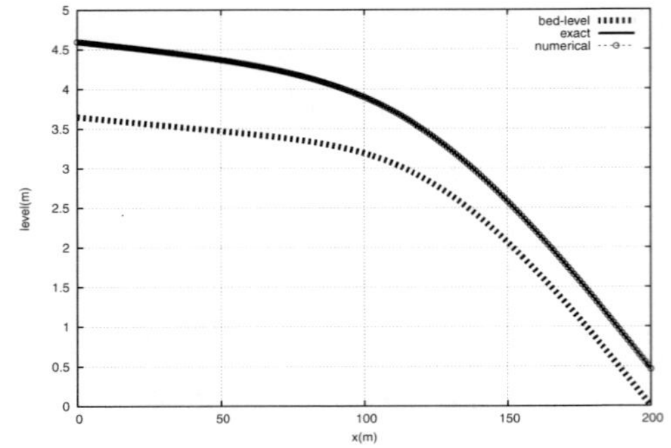

Figure 12. Exact and numerical solutions for Macdonald's test case 2.

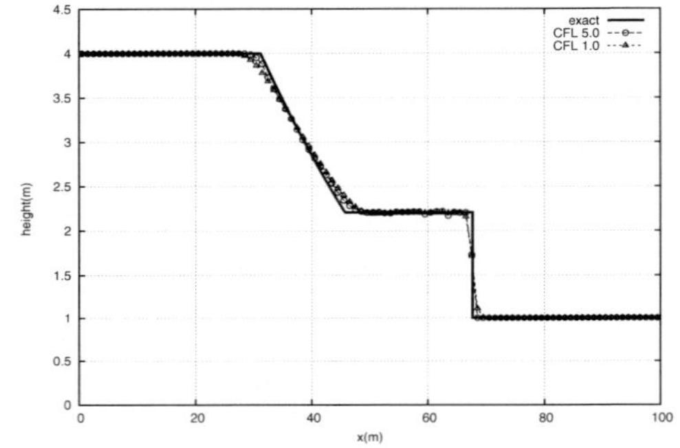

Figure 13. Exact and numerical solutions at $t = 3s$ for the dambreak problem.

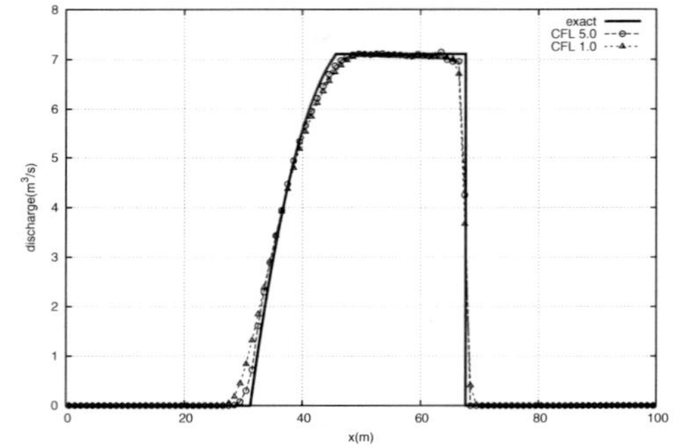

Figure 14. Exact and numerical solutions at $t = 3s$ for the dambreak problem.

4.3 *Application to unsteady flow*

The unsteady flow induced by and ideal dambreak is the most widely used test case for numerical schemes of the kind considered here. For this test case, a rectangular flat and frictionless channel is considered. The initial conditions defining this non-linear problem are as follows:

$$h(x,0) = \begin{cases} 4.0 & if\ x < 50.0 \\ 1.0 & if\ x > 50.0 \end{cases} \tag{47}$$

The results at $t = 3s$ can be observed in Figure 13 for the water depth and in Figure 14 for the discharge. The numerical solutions computed with CFL 1.0 (–▵–) and CFL 5.0 (– • –) are compared with the exact solution (——). Although the LTS scheme produces some oscillations near the zone of shock wave propagation, the solution is less diffusive than the classical upwind explicit scheme with CFL = 1.0.

5 CONCLUSIONS

The proposed LTS scheme is an explicit scheme, and the advantages related with this kind of schemes are conserved. Moreover, the CFL condition is relaxed and larger time steps can be used, so that a computational gain can be achieved. The scheme is able to reproduce faithfully not only one dimensional scalar equations or systems of conservation laws, but also several classical problems including source terms and it is valid for computing steady states and also for unsteady flows. Finally it has been proved less diffusive than the conventional upwind scheme.

REFERENCES

Leveque, R. (1981). *Large Time Step Shock-Capturing Techniques for Scalar Conservation Laws*. Standford Univertsity, Standford: Numerical Analysis Project Manuscript NA-81-13.

Leveque, R. (1985). A large time step generalization of godunov's method for system of conservation laws. *SIAM J. Numer. Anal. 22*, 1051–1073.

Leveque, R. (2002). *Finite Volume Methods for Hyperbolic Problems*. New York: Cambridge Univeristy Press.

MacDonald, I. (1996). *Analysis and computation of steady open channel flow*. University of Reading: PhD Thesis.

MacDonald, I., M. Baines, N. Nichols, and P. Samuels (1997). Analytical benchmark solutions for open-channel flows. *ASCE Journal of Hydraulic Engineering 123 (11)*, 1041–1045.

Murillo, J. and P. Garcia-Navarro (2010). Weak solutions for partial differential equations with source terms: Application to the shallow water equations. *Journal of Computational Physics 229*(11), 4327–4368.

Roe, P. (1986). A basis for upwind differencing of the two-dimensional unsteady euler equations. *Numerical Methods in Fluid Dynamics II*, 55–80.

Numerical Methods for Hyperbolic Equations – Vázquez-Cendón et al. (eds)
© 2013 Taylor & Francis Group, London, ISBN 978-0-415-62150-2

Augmented Roe's approaches for Riemann problems including source terms: Definition of stability region with application to the shallow water equations with rigid and deformable bed

Javier Murillo & P. García-Navarro
Fluid Mechanics, Universidad de Zaragoza

ABSTRACT: Approximate solutions of problems with m equations with source terms can be defined using an augmented Riemann solver with $m+1$ states instead of increasing the number of involved equations. These weak solutions use propagating jump discontinuities connecting the $m+1$ states to approximate the Riemann solution, and are of great interest when applied to the shallow water equations in complex scenarios. The average of the propagated waves in the computational cell leads to a reinterpretation of the Roe's approach depending on the type of the source term involved. From the analysis of the approximate solutions it is possible to establish the water depth positivity requirement, providing correct rules to control the global stability of the method.

1 INTRODUCTION

When managing source terms in a given specific finite volume scheme (Roe's scheme is used here) the main focus has been to establish a discrete balance between flux and source terms, leading to the notion of well-balanced schemes, (Bermudez and Vazquez 1994), (Vazquez-Cendon 1999), (Hubbard and Garcia-Navarro 2000). It can be argued that the presence of source terms warrants the construction of new weak solutions appropriate to the nature of the equations, rather than the use of those constructed for the simple, homogeneous case (Toro 1997),(Toro 2001). Even ensuring the discrete equilibrium formulated in well-balanced schemes, the direct application of the conclusions derived for the homogeneous case to cases with source terms leads to important difficulties as wet/dry fronts (Murillo et al. 2007). Recently, a few authors have been developed augmented approximate Riemann solvers involving source terms (George 2008), (Rosatti et al. 2008), (Murillo and Garcia-Navarro 2010). In this work, the fundamentals of the Roe solver are revisited to define weak solutions first in one dimension and then in two dimensions, putting especial emphasis in explaining the requirements that arise when manipulating numerical fluxes that involve source terms. The importance of the source term discretization is revealed using different representations, which even being well balanced, provide different results in cases with flow in movement.

2 MATHEMATICAL MODEL

Hyperbolic nonlinear systems of equations with source terms in 1D, can be expressed in integral formulation as

$$\frac{\partial}{\partial t}\int_{x_1}^{x_2} \mathbf{U}dx + \mathbf{F}\big|_{x_2} - \mathbf{F}\big|_{x1} - \int_{x_1}^{x_2} \mathbf{S}dx = 0 \qquad (1)$$

where x_1, x_2 are the limits of a generic control volume, with

$$\mathbf{U} = \begin{pmatrix} h \\ hu \end{pmatrix} \quad \mathrm{F} = \begin{pmatrix} hu \\ hu^2 + \frac{1}{2}gh^2 \end{pmatrix} \qquad (2)$$

and

$$\mathbf{S} = \left(0, \ \frac{p_b}{\rho_w} - \frac{\tau_b}{\rho_w}\right) \qquad (3)$$

where h represents the water depth, u the depth averaged component of the velocity vector and g is the acceleration of the gravity. The source term of the system is split in two kind of terms. The terms p_b and τ_b are the pressure force along the bottom and the shear stress in the x direction respectively, with ρ_w the density of water.

The differential formulation is obtained assuming smooth variation of the variables and an infinitesimal width of the control volume

$$\frac{\partial \mathbf{U}}{\partial t} + \frac{\partial \mathbf{F}}{\partial x} = \mathbf{S} \tag{4}$$

where the source term becomes

$$\mathbf{S} = \left(0, -gh\frac{\partial z}{\partial x} - ghS_o, 0\right)^T \tag{5}$$

with S_o the friction slope losses. From the differential formulation it is possible to define a Jacobian matrix for the convective part $\mathbf{J}$

$$\mathbf{J} = \frac{d\mathbf{F}}{d\mathbf{U}} \tag{6}$$

Assuming that the convective part in (1) is strictly hyperbolic with two real eigenvalues λ^1, λ^2 and eigenvectors $\mathbf{e}^1, \mathbf{e}^2$, it is possible define two matrices $\mathbf{P} = (\mathbf{e}^1, \mathbf{e}^2)$ and $\mathbf{P}^{-1}$ with the property that they diagonalize the Jacobian $\mathbf{J}$

$$\mathbf{J} = \mathbf{P}\Lambda\mathbf{P}^{-1} \tag{7}$$

3 NONLINEAR 1D SYSTEM OF EQUATIONS WITHOUT SOURCE TERMS

In the homogeneous case the system of equations takes the form

$$\frac{\partial \mathbf{U}}{\partial t} + \frac{\partial \mathbf{F}}{\partial x} = 0 \tag{8}$$

Using the information stored in $\mathbf{J}$ the solution of the classical Riemann problem for a typical 2×2 non-linear homogeneous system, can be solved. Figure 3 displays the initial problem at time $t = 0$, defined by a left initial state $\mathbf{U}_L$ and by a left initial state $\mathbf{U}_R$. Figure 3 shows how for the homogeneous case a constant value $\mathbf{U}^{\downarrow}$ appears in the position $x = 0$, and remains constant in time.

On the other hand, the system of equations can be discretized using a piecewise constant approximation

$$\mathbf{U}_i^{n+1} = \mathbf{U}_i^n - \frac{\Delta t}{\Delta x}\left(\mathbf{F}_{i+1/2}^n - \mathbf{F}_{i-1/2}^n\right) \tag{9}$$

This first order in time and space scheme requires the evaluation of the intermediate fluxes $\mathbf{F}_{i+1/2}$ at each intercell edge, as depicted in Figure 3.

According to Godunov's method (Godunov 1959), these fluxes have to be computed using the solution of each Riemann problem in each intercell edge and the numerical scheme has to be written as:

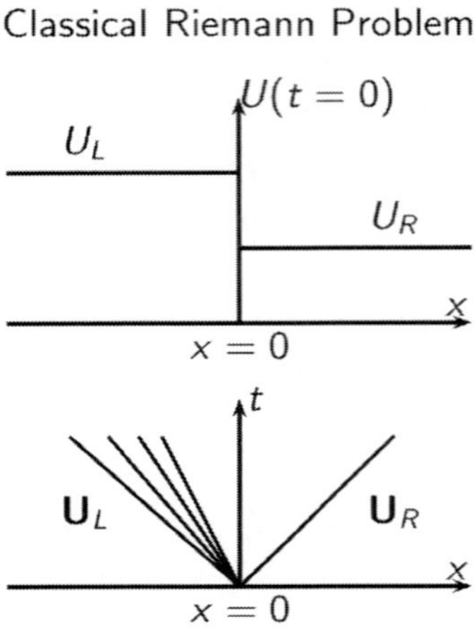

Figure 1. The classical Riemann problem for a typical 2×2 non-linear homogeneous system. Top frame: initial condition at $t = 0$ for a single component U of the vector of unknowns $\mathbf{U}$. Bottom frame: structure of the solution in the $x - t$ plane.

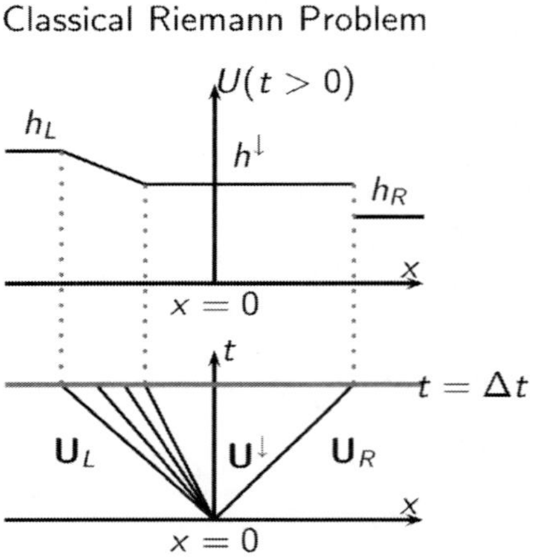

Figure 2. The classical Riemann problem for a typical 2×2 non-linear homogeneous system. Top frame: solution at $t = \Delta t$ for a single component U of the vector of unknowns $\mathbf{U}$. Bottom frame: structure of the solution in the $x - t$ plane.

$$\mathbf{U}_i^{n+1} = \mathbf{U}_i^n - \frac{\Delta t}{\Delta x}\left(\mathbf{F}(\mathbf{U}_{i+1/2}^{\downarrow}) - \mathbf{F}(\mathbf{U}_{i-1/2}^{\downarrow})\right) \tag{10}$$

Godunov's method tell us that $\mathbf{F}_{i+1/2} = \mathbf{F}(\mathbf{U}_{i+1/2}^{\downarrow})$ but do not offers the exact solution $\mathbf{U}_{i+1/2}^{\downarrow}$. Both values are difficult and costly to compute even if

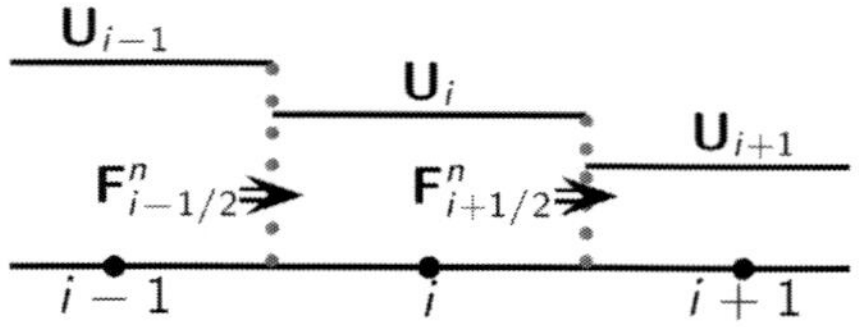

Figure 3. Flux components of the first order numerical scheme.

the exact solution can be defined. An approximate solution $\hat{\mathbf{U}}(x,t)$ can be used instead of the exact solution $\mathbf{U}(x,t)$, provided that the approximate solution $\hat{\mathbf{U}}(x,t)$ and the exact solution $\mathbf{U}(x,t)$ are equivalent. One commonly used approximate solution is the Roe solver. In the Roe formulation, the RP problem becomes linear

$$\frac{\partial \hat{\mathbf{U}}}{\partial t} + \tilde{\mathbf{J}}\frac{\partial \hat{\mathbf{U}}}{\partial x} = 0$$

$$\hat{\mathbf{U}}(x,0) = \begin{cases} \mathbf{U}_i & \text{if} \quad x < 0 \\ \mathbf{U}_{i+1} & \text{if} \quad x > 0 \end{cases} \tag{11}$$

by means of the definition of an approximate Jacobian, $\tilde{\mathbf{J}}$.

From $\tilde{\mathbf{J}}$ it is possible to extract a complete set of real eigenvalues and eigenvectors (Roe 1987). With them, it is possible to generate the following approximate matrix

$$\widetilde{\mathbf{P}} = (\tilde{\mathbf{e}}^1, \tilde{\mathbf{e}}^2) \tag{12}$$

that allows the linearization of the vector difference $\delta\mathbf{U}_{i+1/2} = \mathbf{U}_{i+1} - \mathbf{U}_i$

$$\delta\mathbf{U}_{i+1/2} = \widetilde{\mathbf{P}}_{i+1/2}\mathbf{A}_{i+1/2} = \sum_{m=1}^{N_\lambda}\left(\alpha\tilde{\mathbf{e}}\right)^m_{i+1/2} \tag{13}$$

with $\mathbf{A}_{i+1/2} = (\alpha^1\,\alpha^2)^T_{i+1/2}$. The set of coefficients in $\mathbf{A}_{i+1/2}$ allows the definition of an approximate intermediate state $\mathbf{U}^*_{i+1/2}$, that plays the role of the exact solution $\mathbf{U}^{\downarrow} = \mathbf{U}^*$. In the subcritical case, can be defined departing from both the left and right states

$$\begin{aligned} \mathbf{U}^*_{i+1/2}(\mathbf{U}_{i+1},\mathbf{U}_i) &= \mathbf{U}^n_i + \left(\alpha\tilde{\mathbf{e}}\right)^1_{i+1/2} \\ \mathbf{U}^*_{i+1/2}(\mathbf{U}_{i+1},\mathbf{U}_i) &= \mathbf{U}^n_{i+1} - \left(\alpha\tilde{\mathbf{e}}\right)^2_{i+1/2} \end{aligned} \tag{14}$$

as depicted in Figure 3. Also the flux in the star region is given by

$$\begin{aligned} \mathbf{F}^*_{i+1/2}(\mathbf{U}_{i+1},\mathbf{U}_i) &= \mathbf{F}^n_i + \left(\tilde{\lambda}\alpha\tilde{\mathbf{e}}\right)^1_{i+1/2} \\ \mathbf{F}^*_{i+1/2}(\mathbf{U}_{i+1},\mathbf{U}_i) &= \mathbf{F}^n_{i+1} - \left(\tilde{\lambda}\alpha\tilde{\mathbf{e}}\right)^2_{i+1/2} \end{aligned} \tag{15}$$

or in general case

$$\begin{aligned} \mathbf{F}^*_{i+1/2} &= \mathbf{F}^n_i + \sum_{m,\tilde{\lambda}>0}\left(\tilde{\lambda}\alpha\tilde{\mathbf{e}}\right)^m_{i+1/2} \\ \mathbf{F}^*_{i+1/2} &= \mathbf{F}^n_{i+1} - \sum_{m,\tilde{\lambda}<0}\left(\tilde{\lambda}\alpha\tilde{\mathbf{e}}\right)^m_{i+1/2} \end{aligned} \tag{16}$$

Then, first order for Godunov's can be written as setting $\mathbf{F}^{\downarrow} = \mathbf{F}^*$. Algebraic manipulations provide flux function values for the equivalent numerical flux-based finite volume scheme

$$\mathbf{F}^{\downarrow}_{i+1/2} = \frac{1}{2}(\mathbf{F}_{i+1} + \mathbf{F}_i) - \frac{1}{2}\left(\mathbf{P}\,|\,\widetilde{\Lambda}\,|\,\mathbf{P}^{-1}\delta\mathbf{U}\right)_{i+1/2} \tag{17}$$

Also, first order for Godunov's can be written in terms of flux difference in each cell edge, considering the left and right incoming waves to cell i, by simply combining numerical fluxes with $\mathbf{F}_i$

$$\mathbf{U}^{n+1}_i = \mathbf{U}^n_i - \left((\delta\mathbf{F})^+_{i-1/2} + (\delta\mathbf{F})^-_{i+1/2}\right)\frac{\Delta t}{\Delta x} \tag{18}$$

with

$$(\delta\mathbf{F})^{\pm}_{i+1/2} = \sum_{m=1}^{N_\lambda}\left(\tilde{\lambda}^{\pm}\alpha\tilde{\mathbf{e}}\right)^m_{i+1/2} \tag{19}$$

and

$$\tilde{\lambda}^{\pm,m}_{i+1/2} = \frac{1}{2}(\tilde{\lambda} \pm |\tilde{\lambda}|) \tag{20}$$

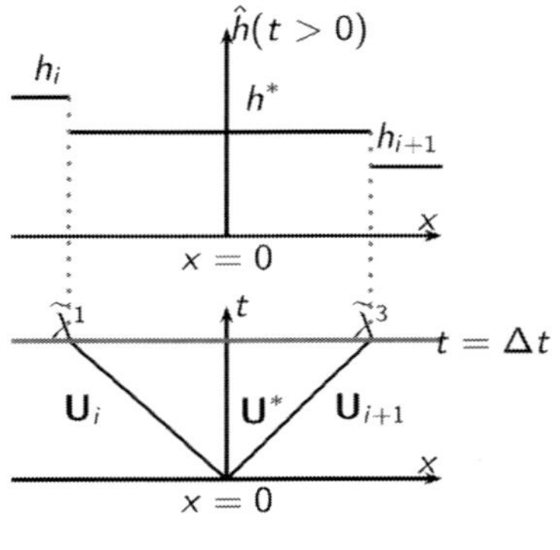

Figure 4. Top frame: approximate solution at $t = \Delta t$ for h. Bottom frame: structure of the approximate solution $\hat{\mathbf{U}}$ in the $x-t$ plane for the subcritical case, $\mathbf{U}^{\downarrow} = \mathbf{U}^*$.

4 NONLINEAR 1D SYSTEM OF EQUATIONS WITH SOURCE TERMS OVER RIGID BED

Although it is yet possible to retain a similarity solution for the evolution of the conserved variables in the associated RP, the presence of source terms lead to a discontinuity in the evolution of the solution when defining it in the right and the left side of the plane solution (Alcrudo and Benkhaldoun 2001), (Bernetti et al. 2008). Therefore the Godunov intercell flux has not a unique solution in the initial discontinuity and the general initial value problem is solved using the explicit conservative formula

$$\mathbf{U}_i^{n+1} = \mathbf{U}_i^n - \frac{\Delta t}{\Delta x}[\mathbf{F}_{i+1/2}^- - \mathbf{F}_{i-1/2}^+] \tag{21}$$

The first step to construct an approximate solution including source terms, is based in the linearization of the source term

$$\mathbf{S}_{i+1/2} = \widetilde{\mathbf{P}}_{i+1/2}\mathbf{B}_{i+1/2} = \sum_{m=1}^{N_\lambda}\left(\beta\tilde{\mathbf{e}}\right)_{i+1/2}^m \tag{22}$$

where in $\mathbf{S}_{i+1/2}$ source terms have been already integrated and are decomposed using the basis of the approximate eigenvalues for the homogeneous case, with $\mathbf{B}_{i+1/2} = (\beta^1 \beta^2)_{i+1/2}^T$. The β coefficients can be grouped with the rest of scalar coefficients to define a new variable

$$\theta_{i+1/2}^m = \left(1 - \frac{\beta}{\tilde{\lambda}\alpha}\right)_{i+1/2}^m \tag{23}$$

and the following diagonal matrix

$$\Theta_{i+1/2} = \begin{pmatrix} \theta^1 & 0 \\ 0 & \theta^2 \end{pmatrix}_{i+1/2} \tag{24}$$

Again, an approximate solution $\hat{\mathbf{U}}(x,t)$ can be used instead of the exact solution $\mathbf{U}(x,t)$. The linearized solution $\hat{\mathbf{U}}$ can be considered the solution of the following constant coefficient linear problem

$$\frac{\partial \hat{\mathbf{U}}}{\partial t} + \mathbf{L}_{i+1/2}\frac{\partial \hat{\mathbf{U}}}{\partial x} = 0 \tag{25}$$

where L includes source terms, with

$$\mathbf{L}_{i+1/2} = \left(\widetilde{\mathbf{P}}\widetilde{\Lambda}\Theta\widetilde{\mathbf{P}}^{-1}\right)_{i+1/2} \tag{26}$$

The new linealization provides the definition of an approximate solution. For the subcritical case this results in two different solutions, defined separately if departing from the left and right states

$$\begin{aligned} \mathbf{U}_i^*(\mathbf{U}_{i+1},\mathbf{U}_i,\mathbf{S}_{i+1/2}) &= \mathbf{U}_i^n + \left(\theta\alpha\tilde{\mathbf{e}}\right)_{i+1/2}^1 \\ \mathbf{U}_{i+1}^{**}(\mathbf{U}_{i+1},\mathbf{U}_i,\mathbf{S}_{i+1/2}) &= \mathbf{U}_{i+1}^n - \left(\theta\alpha\tilde{\mathbf{e}}\right)_{i+1/2}^2 \end{aligned} \tag{27}$$

leading to a discontinuous solution for the approximate solution in $x = 0$. Figure 5 shows how in presence of source terms the constant value $\mathbf{F}^{\downarrow}$ can not be defined. Following the same linealization, fluxes are defined as

$$\begin{aligned} \mathbf{F}_{i+1/2}^+ &= \mathbf{F}_i^n + \left(\theta\tilde{\lambda}\alpha\tilde{\mathbf{e}}\right)_{i+1/2}^1 \\ \mathbf{F}_{i+1/2}^- &= \mathbf{F}_i^n - \left(\theta\tilde{\lambda}\alpha\tilde{\mathbf{e}}\right)_{i+1/2}^2 \end{aligned} \tag{28}$$

or in general case

$$\begin{aligned} \mathbf{F}_{i+1/2}^+ &= \mathbf{F}_i^n + \sum_{m,\tilde{\lambda}>0}\left(\tilde{\lambda}\theta\alpha\tilde{\mathbf{e}}\right)_{i+1/2}^m \\ \mathbf{F}_{i+1/2}^- &= \mathbf{F}_{i+1}^n - \sum_{m,\tilde{\lambda}<0}\left(\tilde{\lambda}\theta\alpha\tilde{\mathbf{e}}\right)_{i+1/2}^m \end{aligned} \tag{29}$$

First order for Godunov's can also be written in terms of flux difference and sources in each cell egde:

$$\mathbf{U}_i^{n+1} = \mathbf{U}_i^n - \left((\delta\mathbf{F})_{i-1/2}^+ + (\delta\mathbf{F})_{i+1/2}^-\right)\frac{\Delta t}{\Delta x} \tag{30}$$

with

$$(\delta\mathbf{F})_{i+1/2}^{\pm} = \sum_{m=1}^{N_\lambda}\left(\tilde{\lambda}^{\pm}\theta\alpha\tilde{\mathbf{e}}\right)_{i+1/2}^m \tag{31}$$

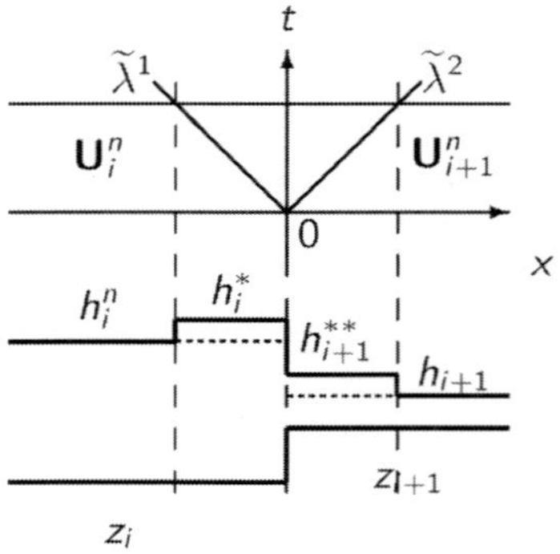

Figure 5. Values of the solution $\mathbf{U}$ in each wedge of the (x, t) plane for the subcritical case, $\tilde{u} > 0$, $\mathbf{U}^{\downarrow}$ does not exist!

5 NONLINEAR 1D SYSTEM OF EQUATIONS WITH SOURCE TERMS FOR DEFORMBLE BED

In case of deformable bed, the system of conserved variables must include a physical conservation law concerning the variation of the bed level z. The mathematical expression of such law is known as the Exner equation

$$\frac{\partial z}{\partial t} + \xi \frac{\partial q_s}{\partial x} = 0 \tag{32}$$

where $\xi = \frac{1}{1-p}$, p is the material porosity and q_s denotes the solid transport discharge influenced by the water depth h and the depth averaged velocities u. The formulation of the bed load discharge q_s can be based on deterministic laws or in probabilistic methods, always supported by experimentation.

If order to define a similarity solution for the evolution of the conserved variables in the associated RP, the presence of bed variation can not be defined as fixed discontinuity in the approximate solution. The dynamics must represent the influence involving an extra wave different from zero in this case. In order to define this extra wave, once the set of equations variables is enlarged,

$$\mathbf{U} = \begin{pmatrix} h \\ hu \\ z \end{pmatrix} \quad \mathbf{F} = \begin{pmatrix} hu \\ hu^2 + \frac{1}{2}gh^2 \\ \xi q_s \end{pmatrix} \tag{33}$$

and

$$\mathbf{S} = \left(0, \ \frac{p_b}{\rho_w} - \frac{\tau_b}{\rho_w}, \ 0\right) \tag{34}$$

two different matrices are defined, $\mathbf{M}$ and $\mathbf{H}$. $\mathbf{M}$ is defined using the Jacobian of the fluxes

$$\mathbf{M} = \begin{pmatrix} 0 & 1 & 0 \\ \left(gh - u^2\right) & 2u & 0 \\ \xi\left(\frac{\partial \mathbf{q}_s}{\partial h}\right) & \xi\left(\frac{\partial \mathbf{q}_s}{\partial q_x}\right) & 0 \end{pmatrix} \tag{35}$$

requiring a differential formulation of the solid transport discharge. Matrix $\mathbf{H}$ is constructed assuming that the bed slope source in integral form can be related to the vector of conserved variables ensuring $\mathbf{S}_b = \mathbf{H}\delta\mathbf{U}$ with

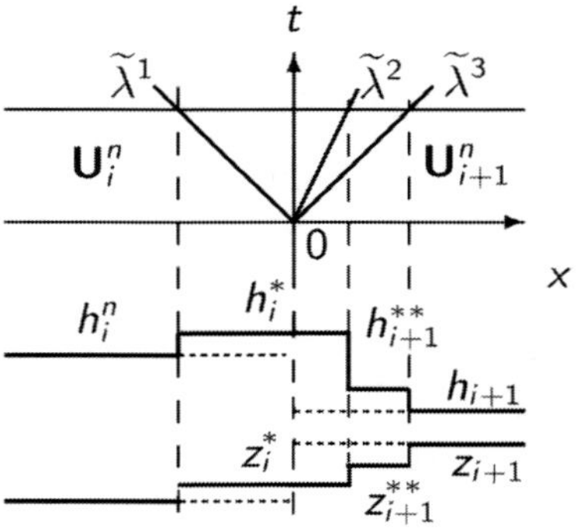

Figure 6. Values of the solution $\mathbf{U}$ in each wedge of the (x, t) plane for deformable bed.

$$\mathbf{H_n} = \begin{pmatrix} 0 & 0 & 0 \\ 0 & 0 & -\frac{1}{\delta z}\int \frac{p_b}{\delta z}\partial x \\ 0 & 0 & 0 \end{pmatrix} \tag{36}$$

The previous definitions lead to the following Jacobian matrix, $\mathbf{J_n}$

$$\mathbf{J_n} = \mathbf{M_n} - \mathbf{H_n} \tag{37}$$

that will allow us to define the whole system as belonging to the family of hyperbolic systems, making possible to recover the numerical formulation for the homogeneous case.

The new linealization provides the definition of an approximate solution, as shown in Figure 6, leading to two different solutions, for water depth, unit discharge and bed level. for instance

$$\begin{aligned} z_i^* &= z_i^n + \left(\alpha\tilde{e}\right)^1_{3,i+1/2} \\ z_{i+1}^{**} &= z_{i+1}^n - \left(\alpha\tilde{e}\right)^2_{3,i+1/2} \end{aligned} \tag{38}$$

where the wave strengths α can be determined using the same procedure described for the homogeneus case.

6 STABILITY REGION

The Roe method presented in this work provides linearised Riemann solutions that consist of discontinuous jumps only, and in practical computation transonic rarefactions encounter difficulties. As a result unphysical, discontinuous waves appear in such cases. Attending to the approximate solution given to the conserved variables, in situations where data that is near the vacuum state or situations where there is a strong expansion, Roe average gives a non-physical solution such as negative depth. To avoid unphysical results derived from the linearization, the version of the Harten-Hyman entropy fix (Harten and Hyman 1983) can be sucesfully applied.

In case of quiescent equilibrium over variable bed level any problem may be detected when applying the approximate solvers defined in this work, as integration of the source term can be done exactly. When moving to stationary cases with non-zero velocity, the integral of the source term is done setting its value constant in time and equal to the initial value. Considering that in the exact solution the inner states of the RP change in time and therefore the source terms also do, for both rigid and deformable bed, this time linearization may have negative consequences on the solution. The most dramatic is the the appearance of negative values of water depth when updating the solution. Unphysical solutions may be avoided in some cases reducing the *CFL* condition, but this option results in an excessive computational cost, making useless the Courant condition.

In the same way that the linearized Roe solver (Roe 1987) may fail in some situations, requiring an entropy fix to ensure positively conservative solutions, the augmented numerical solvers presented in this work require a source fix to recover the Courant condition. In case that the source term integration step leads to positive solutions of the inner states

$$h_i^* \geq 0 \qquad h_{i+1}^{**} \geq 0 \tag{39}$$

the average solution will remain positive. In any other case the value of the source term can be reduced to ensure (39) before compute the final intercell flux. Considering that the approximate solution is constructed using a completely linear algorithm the correction can be done straightforward or iteratively. If the inner solution becomes negative in wet/dry edges the flow will not evolve in both regions, and the solution behaves as a solid wall.

Once the approximate solution provides correct predictions for the intermediate states, the stability region is given by the Courant condition

$$\Delta t \leq CFL \frac{\Delta x}{\max(|\tilde{\lambda}|)} \tag{40}$$

with $CFL < 1$.

7 CONCLUSIONS

In this paper approximate Riemann solvers problems for equations and systems with source terms are been presented by adding one extra wave associated to the source term. The approximate solution is assumed discontinuous. The method is specifically designed to satisfy the integral formulation. This is the starting point to explore different approximations to the integral source terms in the search for the best properties in each case.

REFERENCES

Alcrudo, F. and F. Benkhaldoun (2001). Exact solutions to the Riemann problem of the shallow water equations with a bottom step. *Comput. Fluids 30*(6), 643–671.

Bermudez, A. and E. Vazquez (1994). Upwind Methods For Hyperbolic Conservation-Laws With Source Terms. *Comput. Fluids 23*(8), 1049–1071.

Bernetti, R., V. A. Titarev, and E. F. Toro (2008). Exact solution of the Riemann problem for the shallow water equations with discontinuous bottom geometry. *J. Comput. Phys. 227*(6), 3212–3243.

George, D. L. (2008). Augmented Riemann solvers for the shallow water equations over variable topography with steady states and inundation. *J. Comput. Phys. 227*(6), 3089–3113.

Godunov, S. (1959). A Finite Difference Method for the Computation of Discontinuous Solutions of the Equations of Fluid Dynamics. *Mat. Sb. 47,* 357–393.

Harten, A. and J. Hyman (1983). Self-Adjusting Grid Methods For One-Dimensional Hyperbolic Conservation-Laws. *J. Comput. Phys. 50*(2), 235–269.

Hubbard, M. and P. Garcia-Navarro (2000). Flux difference splitting and the balancing of source terms and flux gradients. *J. Comput. Phys. 165*(1), 89–125.

Murillo, J. and P. Garcia-Navarro (2010). Weak solutions for partial differential equations with source terms: Application to the shallow water equations. *J. Comput. Phys. 229*(11), 4327–4368.

Murillo, J., P. Garcia-Navarro, J. Burguete, and R. Brufau (2007). The influence of source terms on stability, accuracy and conservation in two-dimensional shallow flow simulation using triangular finite volumes. *Int. J. Num. Methods in Fluids 54*(5), 543–590.

Roe, P. (1987). Upwind Differencing Schemes For Hyperbolic Conservation-Laws With Source Terms. *Lecture Notes In Mathematics 1270*, 41–51.

Rosatti, G., J. Murillo, and L. Fraccarollo (2008). Generalized Roe schemes for 1 D two-phase, free-surface flows over a mobile bed. *J. Comput. Phys. 227*(24), 10058–10077.

Toro, E. (1997). *Riemann solvers and numerical methods for fluid dynamics.* Springer, Berlin.

Toro, E. (2001). *Shock-Capturing Methods for Free-Surface Shallow Flows.* Wiley, New York.

Vazquez-Cendon, M. (1999). Improved treatment of source terms in upwind schemes for the shallow water equations in channels with irregular geometry. *J. Comput. Phys. 148*(2), 497–526.

III Multiphase flow and porous media

Numerical Methods for Hyperbolic Equations – Vázquez-Cendón et al. (eds)
© 2013 Taylor & Francis Group, London, ISBN 978-0-415-62150-2

Dynamics of submerged gravitational granular flows

A. Armanini, M. Dumbser, E. Nucci & M. Larcher
Department of Civil and Environmental Engineering, University of Trento, Italy

ABSTRACT: Debris flows are complex, natural phenomena, characterized by a mixture of poorly sorted sediments and water driven by gravity. To investigate the basic physics of debris flows, it is very useful to analyze the flow of a mixture of identical, spherical particles saturated by water and driven by gravity down a steep channel in steady flow condition.

The flow presents three regions: an external one, near to the free surface, dominated by nearly instantaneous contacts among the particles (*collisional regime*), an internal region dominated by prolonged contacts among the particles (*frictional regime*) and a *static bed* in which the particles are immobile. Armanini *et al.* (2009) analysed the stratification of rheological mechanisms inside the flow, focusing on the coexistence of frictional and collisional regimes, on the stress transmission inside the flow and on particles kinematics. In particular, it was observed that debris flows may show locally a typical intermittence of the flow regime, switching alternatively from frictional to collisional. In general, the tensor of the granular phase can be assumed to be the composition of two tensors: one, T_{ij}^{g-coll}, representative of the stresses exchanged with a collisional mechanism and one, T_{ij}^{g-fric}, representative of the stresses expressed by a frictional mechanism. While the rheology of the collisional layers is well described by the dense gas analogy (*kinetic theory*), a persuasive theoretical description of the frictional regime does not yet exist. A Coulombian scheme is often assumed, but this hypothesis is rather limitative because it requires a constant concentration or a distribution of particles concentration known a priori. An interesting scheme of this kind was recently proposed by GDR-Midi (2004), but this model does not contain a suitable formulation for the granular pressure (equation of state of the mixture). Following Armanini (2010), we propose a reinterpretation of the model, as weighted average of a pure Coulombian stress (dependent on the static friction angle at the static bed level) and of a dynamic stress, represented by a dynamic friction angle. Besides, a state relation is introduced for the granular pressure and the dynamic friction angle is derived from the kinetic theory. The proposed relations are finally compared with the experimental data.

1 INTRODUCTION

Debris flows and mud flows are phenomena of very intense sediment transport, affecting the mountain streams in particular hydrological conditions. They are often sudden phenomena, that have caused lots of catastrophic damages with human losses during last decades, to such an extent that the research is turning to a better understanding of the mechanism of these hyper-concentrated flows (Armanini, Fraccarollo et al. 2005), (Takahashi 1981), (Takahashi 1991). New rational approaches, more convincing from a physical point of view, are necessary in order to predict such events and to properly design efficient protection measures to a smarter defence strategy of such kind of geophysical phenomena and to map the areas at risk. The first empirical approaches were more or less qualitative descriptions and classifications without any physical basis. In fact there is a great number of classifications that require a priori knowledge of the flow features, but this kind of information are generally known only a posteriori. Tamotsu Takahashi in his book *Debris flow* (Takahashi 1991) identifies four types of debris flows and he gives a specific rheology for each one of them. It is urgent and necessary to reduce the subjectivity and to make a conceptual simplification and to base their description and simulation on physical bases. A proper and realistic approach to the problem consists in making the hypothesis that debris flows are hyper-concentrated flows formed by two fluids. One is the interstitial fluid, that follows the fluid mechanic laws with an appropriate rheology, and the other one is the solid phase, which is composed by a granular fluid provided also with a specific rheology. The fluid phase is usually water, that can be treated as a Newtonian fluid and described by the Navier-Stokes equations. The smaller particles of the solid phase (smaller than 30 μm) are commonly supposed dispersed in the liquid phase, forming with the water a homogenous fluid. In this case, the interstitial fluid exhibits a non-Newtonian behavior. If we do not consider the presence of cohesive (clayish) particles in the flow, the solid phase is composed by particles with size bigger than fine sand and can be treated as a granular fluid. Under the hypothesis that the particles dimensions are much smaller than the control volume and that this volume is infinitesimal, the particles could be liken to a fluid with an own rheological law, which describes the interaction mechanism among the grains. Bagnold (1954) proposed the first theory for granular fluids. He distinguished two flow regimes, on the base of the relative importance of stresses due to inertia and stresses due to viscosity. His pioneering work inspired most of the formulations utilized until now and was refined by the introduction of kinetic theory proposed by Savage & Jeffrey (1981) and Jenkins & Savage (1983). This approach derived from the analogy with the kinetic theory of gases, pursuing the idea of deriving a set of continuum equations (mass, momentum and energy conservation) entirely from microscopic models of individual particle interactions. This theory can be employed to obtain a solution for the flow in terms of distribution of concentration, velocity and velocity fluctuations (granular temperature). However it was derived on the basis of some simplified hypothesis, which have to be improved and generalized in order to successfully apply kinetic theory to debris flow.

1.1 *The equations of liquid-granular two-phases fluids*

If there is no mass exchange between the two phases, the equations of mass and momentum conservation of both phases (m = *f*, *g* where *f* is fluid and *g* is granular) is (Truesdell 1984), (Iverson 1997):

$$\begin{cases} \dfrac{\partial \rho_m}{\partial t} + \dfrac{\partial\left(\rho_m u_i^m\right)}{\partial x_i} = 0 \\ \dfrac{\partial \rho_m u_i^m}{\partial t} + \dfrac{\partial\left(\rho_m u_i^m u_j^m\right)}{\partial x_j} = \rho_m g_i^m + \dfrac{\partial T_{ij}^m}{\partial x_j} + F_i^m \end{cases}$$

In these equations ρ_m is the density of the phases, that is: for the fluid phase $\rho_f = (1-c)\rho_w$, where c is the particle concentration and ρ_w is the density of the interstitial liquid, and $\rho_g = c\rho_s$, where ρ_s is the material density of the particle; u_i^m are the components of the velocity of both phases; F_i^m are the components of the force per unit volume that represents the interaction between solid and liquid phase; T_{ij}^m are the components of the stress tensor of phases; g_i^m are the components of the force of mass per unit volume that acts on each phase. Because the flow is governed by gravity, this force coincides with the gravity acceleration and so $g_i^g = g_i^g = -g\partial z/\partial x_i$, where z represents the vertical upward direction.

1.1.1 *2D uniform flow of liquid-granular mixture*

In the following, we will consider a uniform flow in the longitudinal direction x_1. By summing term by term the momentum equations of the two phases, the interaction forces, F_i^m, will be eliminated. Moreover, if the particles concentration is high enough, it is possible to neglect the term relative to the stresses internal to the fluid phase, τ^f, and in the end to write:

$$\frac{\partial \tau_{12}^g}{\partial x_2} = (1 + c\Delta)\rho_w g \frac{\partial z}{\partial x_1} \tag{1}$$

$$\frac{\partial p^g}{\partial x_2} = c\Delta\rho_w g \frac{\partial z}{\partial x_2} \tag{2}$$

where $\Delta = (\rho_s - \rho_w)/\rho_w$ is the submerged relative density of the particles.

2 CONCEPTUAL MODEL

2.1 *Rheology of the granular phase*

The rheology of the granular phase can be outlined by two modalities of interaction between particles: almost instantaneous contacts and long lasting contacts. Two regimes correspond to this two interactions, which are termed respectively *collisional regime*, represented by the stresses tensor T_{ij}^{g-coll}, and *frictional regime*, represented by T_{ij}^{g-fric} (Jenkins & Savage 1983), (Campbell 1990), (Goldhircsh 2003), (Forterre & Pouliquen 2008). Generally it is assumed that the stresses of both regimes can be added, but it is generally assumed (Jenkins & Savage 1983), (Johnson & Jackson 1987), (Meruane et al. 2010) that the two regimes are stratified and so physically separated: the collisional regime in the upper layer of the flow and the frictional regime in the lower layer, where the particle concentration is bigger. On the contrary, recent experimental investigations (Armanini et al. 2009) show that the two regimes are alternated in space and time through a intermittent mechanism, similar to the one that exists between the viscous sub-layer and the turbulent sub-layer in a wall boundary layer. Generally under both hypotheses, it is possible to assume that:

$$T_{ij}^g = T_{ij}^{g-coll} + T_{ij}^{g-fric} \tag{3}$$

The collisional part of the granular tensor is well described by the kinetic theory of granular flows (Jenkins & Savage 1983), (Haff 1983), (Lun et al. 1984), (Jenkins & Hanes 1998), (Lun & Savage 1988), derived by analogy with the kinetic theory of gases, according to which the flow of the particles is similar to the flow of the molecules of an ideal gas. The temperature is replaced by the granular temperature: $\Theta = \langle u'^{p}_{i} u'^{p}_{i} \rangle / 3$, in which the symbol $\langle\rangle$ represents the average done on all the particles that are in a control volume, small enough compared to the dimensions of the boundary and large enough compared to the particle size. According to this theory, the rheological law represents the collisional regime is expressed as (Chapman & Cowling 1971):

$$T_{ij}^{g-coll} = -f_1 \rho_s \Theta \delta_{ij} + \mu^{g-coll} \left(\frac{\partial u_i^g}{\partial x_j} + \frac{\partial u_j^g}{\partial x_i} \right) \tag{4}$$

where T^{g-coll} is the collisional component of the stresses tensor of the granular phase; δ_{ij} Kronecker's delta. The granular temperature represents the kinetic energy of the collisional flow of the particles, and its behavior is represented by particle kinetic energy balance:

$$\rho_s \left(\frac{\partial \Theta}{\partial t} + u_j^g \frac{\partial \Theta}{\partial x_j} \right) = \frac{\partial}{\partial x_j} \left(\kappa_\Theta \frac{\partial \Theta}{\partial x_j} \right) + \mu_{g-coll} \left(\frac{\partial u_j^g}{\partial x_i} + \frac{\partial u_i^g}{\partial x_j} \right)^2 - f_5 \rho_s \frac{\Theta^{1.5}}{d_p} \tag{5}$$

The system of equations (4)–(5) describes the behavior of the collisional scheme through a fairly convincing theoretical approach, which gave good results in many experimental situations. The expressions of the coefficients are reported in the footnote.[1]

On the contrary, the problem of the rheology of the frictional regime is still open, and validated general formulations do not exist yet. In this condition the collisions among particles are not instantaneous, but they become long lasting and they could involve more particles at the same time. In granular flows of heavy materials governed by gravity, under the material that is moving, if the boundary conditions allow it, it is possible to find an immobile layer, because the frictional forces among grains are so high that do not permit any flow. In uniform flow, this condition is identified as an *equilibrium condition* between the granular flow and the immobile bed (Armanini, Capart et al. 2005), because there is no net exchange of material between the flow and the bed. Also experiments have shown that the system becomes increasingly frictional (Armanini et al. 2009) while approaching the immobile bed, and on this frontier a Coulombian condition is established. Recently the GDR-MiDi group (Da Cruz et al. 2005), (Forterre & Pouliquen 2008), (GDR MiDi 2004) proposed a rheological model that combines the Coulombian, rate independent scheme with a rate dependent model. This approach was originally formulated for granular dry 2D flows. It is based on the observation, derived by molecular dynamics simulations, that the ratio between the shear stress and the pressure is a function of a single dimensionless parameter I, termed *inertial number* by Da Cruz *et al.* (Da Cruz et al. 2005), defined as:

$$I = \dot{\gamma} d / \sqrt{p^g / \rho_s} \tag{16}$$

where $\dot{\gamma}$ is the strain rate of the granular flow. The inertial number represents the rate between two temporal scales (Da Cruz et al. 2005): a micro scale $d_p / \sqrt{p^g / \rho_s}$ that is the time in which a particle falls in an empty space with dimension of the particle d_p by the action of a pressure p^g; and a macro scale proportional to the local strain rate $\dot{\gamma}$. The GDR-MiDi rheological model can be written as:

$$\frac{T_{ij}^g}{p^g} = \tan \varphi^{fric} + (\tan \varphi^{coll} + \tan \varphi^{fric}) \frac{I}{I + I_0} \tag{17}$$

[1]Coefficients of the system (4)–(5) according to Lun and Savage (Springer-Verlag 1988).

$$\mu^{g-coll} = f_2 \rho_s \sqrt{\Theta} d_p \tag{6}$$

$$k_\Theta = f_4 \rho_s \sqrt{\Theta} d_p \tag{7}$$

$$f_1 = (1 + 4c\,\eta_p\, g_o)c \tag{8}$$

$$f_2 = \frac{5\sqrt{\pi}}{96\eta_p(2-\eta_p)} \left(1 + \frac{8}{5} \eta_p\, c\, g_0 \right) \left(\frac{1}{g_0} + \frac{8}{5} \eta_p (3\eta_p - 2)c \right) + \frac{8/5}{\sqrt{\pi}} \eta_p c^2 g_0 \tag{9}$$

$$f_4 = \frac{25\sqrt{\pi}}{16\eta_p(41-33\eta_p)} \left(1 + \frac{12}{5} \eta_p c\, g_0 \right) \left(\frac{1}{g_0} + \frac{12}{5} \eta_p^2 (4\eta_p - 3)c \right) + \frac{4}{\sqrt{\pi}} \eta_s c^2 g_0 \tag{10}$$

$$f_5 = \frac{12}{\sqrt{\pi}} c^2 g_0 (1 - e^2) \tag{11}$$

$$g_0(c) = (1 - c/c*)^{-1/3} \tag{12}$$

$$e = 0.9 - 2.85 St^{-0.5} \tag{13}$$

$$St = \frac{\rho_s d_p \Theta^{0.5}}{18\mu_w} \tag{14}$$

$$\eta = (1 + e_p)/2 \tag{15}$$

I_0 is an experimental constant less then 1. One of the main limit of the Midi formulation is that the state equation was originally not provided. Later a linear relationship between particle concentration c and inertial parameter was suggested (GDR MiDi 2004).

In this paper we have tried to overcome this limit. The first suggestion (Armanini 2010) consists in the introduction of the kinetic theory expression in the rheological relationship (17). It is possible to observe in fact that eq (17), considering the granular shear stress τ_{ij}^g as a linear combination of a Coulombian stress $p^g \tan\varphi^{fric}$, with constant friction angle typical of the frictional regime, and a analogous Coulombian stress, in which the friction angle is depending on the shear rate trough the inertial number that in this case is used as weighting factor:

$$(I + I_o)\tau_{ij}^g = p^g \tan\varphi^{fric} I_o + p^g \tan\varphi^{coll} I \qquad (18)$$

In the above equation the second term, divided by $(I_o + I)$ is replaced by the corresponding expression derived by the kinetic theory:

$$\tau_{12}^g = p^g \frac{I_0}{I_0 + I} \tan\varphi^{fric} + \mu^{g-coll} \frac{\partial u_i^g}{\partial x_j} \qquad (19)$$

In analogy with this interpretation of the shear stress, it is possible (Armanini 2011) to derive an expression for the pressure (equation of state):

$$p^g = p_0^g \frac{I_0}{I_0 + I} + \rho_s f_1 \Theta \qquad (20)$$

It should be noted that when the MiDi inertial number I tends to infinite (pure collisional regime), the granular pressure and the shear stress tend to the kinetic one. While the inertial number tends to 0, the shear stress tend to become purely Coulombian and the kinetic component of the pressure vanish.

The system formed by eqs. 1, 2, 5, 19, 20 is well posed and can be numerically integrated provided that the proper boundary conditions are assigned.

2.2 *Numerical method*

The governing equations of the proposed model can be written under the general form of a nonlinear system of Differential Algebraic Equations (DAE) as follows:

$$\frac{d}{dt}\mathbf{E}(\mathbf{Q}(t)) = \mathbf{f}(\mathbf{Q}(t),t), \quad \mathbf{Q}(0) = \mathbf{Q}_0, \qquad (21)$$

where $\mathbf{Q} = \mathbf{Q}(t) = (q_1(t), q_2(t), \ldots q_n(t)) \in \mathbb{R}^n$ is the unknown state vector, and $\mathbf{E}(\mathbf{Q}) \in \mathbb{R}^n$ and $\mathbf{f}(\mathbf{Q},t) \in \mathbb{R}^n$ are two nonlinear functions of the state vector Q and the independent variable t. Q_0 is the known initial condition of the initial value problem (21). For its numerical solution we use a Galerkin method, based on the following expression for the unknown solution vector:

$$\mathbf{Q_h}(t) = \sum_{l=0}^{N} \theta_l(t)\hat{\mathbf{Q}}_l := \theta_l \hat{\mathbf{Q}}_l, \qquad (22)$$

where $\theta_l(t)$ represent piecewise polynomial basis functions of maximum degree N and $\hat{\mathbf{Q}}_l$ are the unknown coefficients of the numerical solution. In the above relation we have used classical tensor notation with the Einstein summation convention over two equal indices. Equation (22) is valid for one timestep $\Delta t = t^{n+1} - t^n$, where t^n is the current solution time. To obtain the unknown coefficients $\hat{\mathbf{Q}}_l$, the DAE is multiplied with test functions $\theta_k(t)$ that are identical with the basis functions (classical Galerkin approach), and is subsequently integrated over a time step to obtain the following weak formulation of the DAE:

$$\int_{t^n}^{t^{n+1}} \theta_k(t)\left(\frac{d}{dt}\mathbf{E}(\mathbf{Q_h}(t)) - \mathbf{f}(\mathbf{Q_h}(t),t)\right)dt. \qquad (23)$$

For the test and basis functions $\theta_k(t)$ we choose the Lagrange interpolation polynomials that pass through the $N+1$ equidistant Newton-Cotes quadrature points, $t_l^n = t^n + (l-1)/(N-1)\Delta t$, hence we use a *nodal basis*. Therefore, the numerical approximations of the nonlinear functions E and f are simply given by

$$\mathbf{E_h}(t) = \theta_l \hat{\mathbf{E}}_l, \quad \text{and} \quad \mathbf{f_h}(t) = \theta_l \hat{\mathbf{f}}_l, \qquad (24)$$

with

$$\hat{\mathbf{E}}_l = \mathbf{E}(\hat{\mathbf{Q}}_l), \quad \text{and} \quad \hat{\mathbf{f}}_l = \mathbf{f}(\hat{\mathbf{Q}}_l, \mathbf{t}_l^n) \qquad (25)$$

due to the choice of the nodal basis. The weak formulation (23) for the unknowns $\hat{\mathbf{Q}}_l$ therefore becomes

$$\left(\int_{t^n}^{t^{n+1}} \theta_k(t)\frac{d}{dt}\theta_l(t)dt\right)\hat{\mathbf{E}}_l = \left(\int_{t^n}^{t^{n+1}} \theta_k(t)\theta_l(t)dt\right)\hat{\mathbf{f}}_l, \qquad (26)$$

or, in a more compact matrix-vector notation:

$$\mathbf{K}_{kl}\mathbf{E}(\hat{\mathbf{Q}}_l) - \mathbf{M}_{kl}\mathbf{f}(\hat{\mathbf{Q}}_l, t_l^n) = 0, \qquad (27)$$

with $\hat{\mathbf{Q}}_0 = \mathbf{Q}(t^n)$, and the mass matrix $\mathbf{M}_{kl}$ and the stiffness matrix $\mathbf{K}_{kl}$, which can both be precomputed once and for all. The resulting nonlinear algebraic equation system (27) of dimension $n(N+1)$ is solved by a standard Newton method for systems with a line-search-type globalization strategy. The initial guess is provided using a second order Crank-Nicholson-type scheme for the DAE (21) to initialize the nodal values $\hat{\mathbf{Q}}_l$ at all time levels t_l^n:

$$\frac{\mathbf{E}(\hat{\mathbf{Q}}_{l+1})-\mathbf{E}(\hat{\mathbf{Q}}_l)}{t_{l+1}^n-t_l^n}=\frac{1}{2}\left(\mathbf{f}(\hat{\mathbf{Q}}_{l+1},t_{l+1}^n)+\mathbf{f}(\hat{\mathbf{Q}}_l,t_l^n)\right). \quad (28)$$

Equation (28) is again a nonlinear algebraic equation system, however, of smaller dimension n, which is again solved by a globally convergent Newton method. The proposed Galerkin-type method (27) is theoretically of arbitrary order of accuracy in the independent variable t and can be used inside a classical *shooting method* for solving DAE boundary value problems of the type

$$\frac{d}{dt}\mathbf{E}(\mathbf{Q}(t))=\mathbf{f}(\mathbf{Q}(t),t), \quad \mathcal{Q}(t_0)=\mathcal{Q}_0, \mathcal{Q}(t_1)=\mathcal{Q}_1, \quad (29)$$

where $\mathcal{Q}_0$ and $\mathcal{Q}_1$ are the known boundary values of the Boundary Value Problem (BVP) (29).

3 RESULTS AND DISCUSSION

In the next figures, we have reported a comparison between the prediction of the model and some experimental results obtained by Armanini *et al.* (Armanini et al. 2009). The figures represent the profiles in direction normal to bed of the most important physical variables.

It should be stressed, however, that the only parameter of the model that needs to be calibrated is I_o, whose value does not have much influence on results. We have assumed $I_o = 0.13$.

All the variable are made properly dimensionless, and in particular the distance from the static bed is expressed by $\eta = x_2/h$, where h is the flow depth. The system is solved according the following boundary conditions assigned at the boundary of the static bed $\eta = 0$: $u^g = 0.0000001$, $c = 0.999c^*$, $\Theta = 0.00001$, $du^g/d\eta = 0.07$, and $d\Theta/d\eta = 0.0000001$. The experimental conditions are: $\alpha = 8°$, $\tan\phi^{fric} = 0.353$, $\rho_s/\rho_w = 2.21$, $h = 6.17\,\text{cm}$ and $D = 6.0\,\text{mm}$.

In the Figures 1–3 we have reported the distributions of the dimensionless granular velocity, granular concentration and temperature respectively. The model captures just the tendency of the experimental results, and the results are reasonably comparable to them just in the proximity of the static bed, that is where the frictional contact are dominant. Near the free surface the results of the model tend to deviate systematically respect to experimental data.

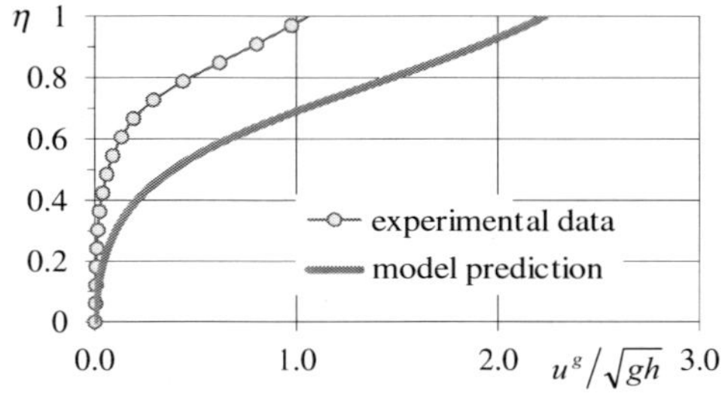

Figure 1. Velocity profile, comparison between results of the numerical simulation and experimental data.

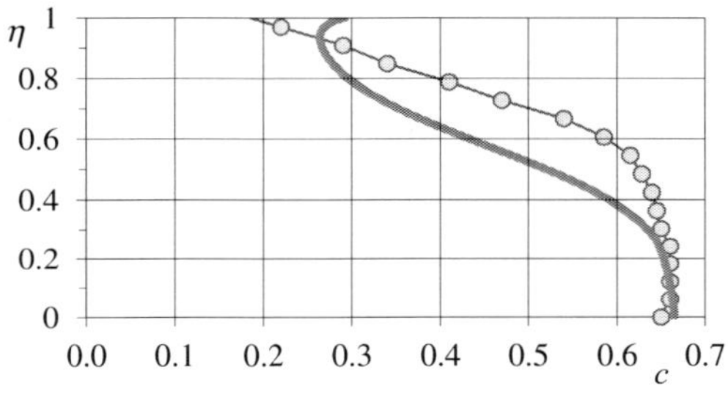

Figure 2. Particle concentration profile, comparison between results of the numerical simulation and experimental data.

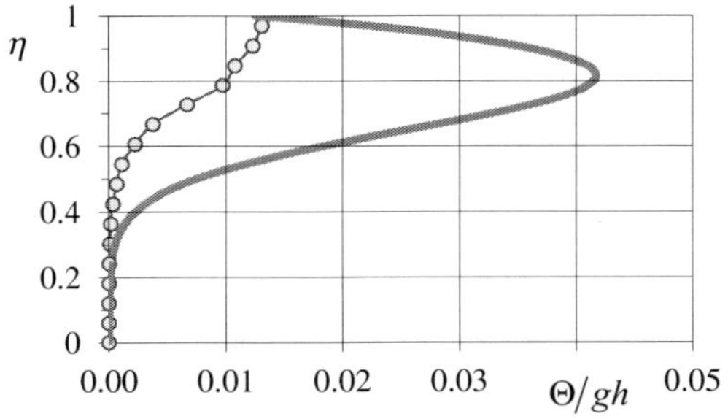

Figure 3. Granular temperature profile, comparison between results of the numerical simulation and experimental data.

Similar arguments can be made regarding the distributions of the shear stress and of the pressure (Figures 4 and 5). For these parameters the model seems to catch better the experiments, but this is just an apparent agreement, because the relative error on the contrary is definitely bigger.

It must be stressed, however, than we cannot conclude that the frictional model is correct, while

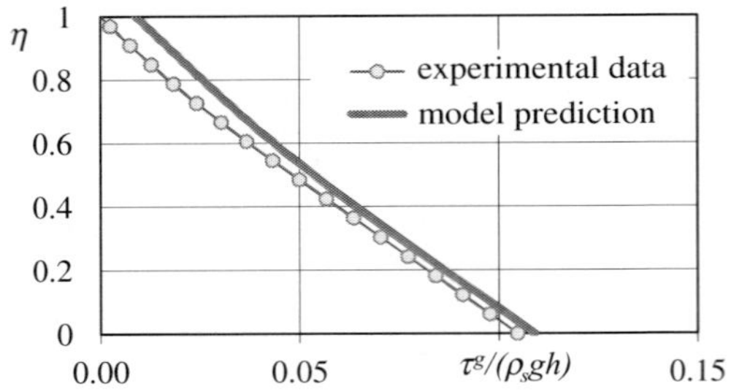

Figure 4. Granular shear stress profile, comparison between results of the numerical simulation and experimental data.

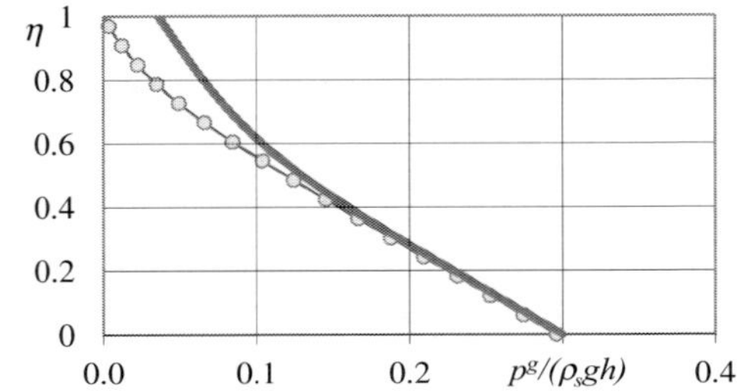

Figure 5. Granular pressure profile, comparison between results of the numerical simulation and experimental data.

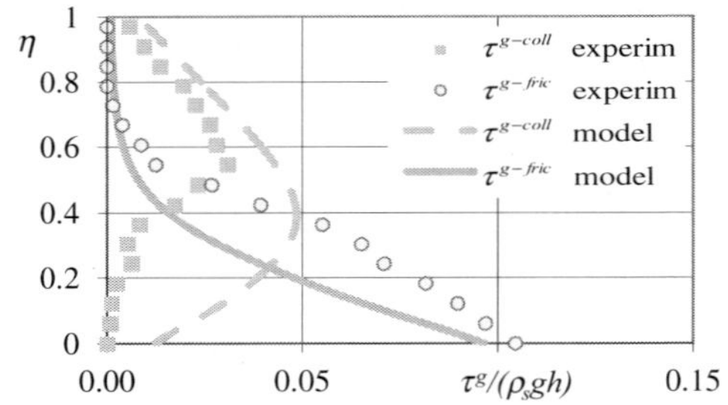

Figure 6. Distributions of the collisional and frictional components of the granular shear stress. Comparison between results of the numerical simulation and experimental data.

the collisional one is not, because the collisional model is strongly non linear, so it is very sensitive to boundary conditions, which are instead determined by the frictional regime. Some preliminary results suggest that we have to reconsider the frictional model, as it is possible to argue also from the Figures 6 and 7 in which we have reported the distribution of the frictional and collisional components of the shear stress and of the pressure.

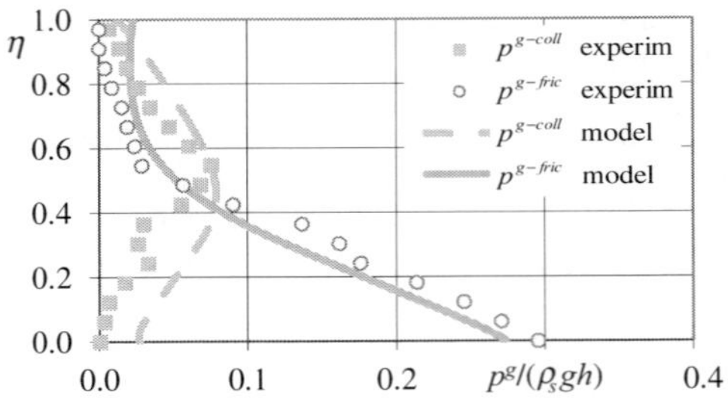

Figure 7. Distributions of the collisional and frictional components of the granular pressure. Comparison between results of the numerical simulation and experimental data.

The maximum deviations are in the area next to the static bed, where the frictional regime is dominant, specially for the shear stresses. A similar behavior, but less pronounced, is evident in the distribution of the components of the granular pressure.

4 CONCLUSIONS

In the paper we have presented a rheological model relative to a granular, submerged flow driven by gravity. Respect to the previous schemes, the model considers the simultaneous presence in the flow of the frictional and of the collisional regimes. The model can be considered as an evolution of the GDR-MiDi (GDR MiDi 2004) model. The results of the model catch in a reasonable way the trend of the experimental data, even if the quantitative comparison cannot be considered satisfactory.

A reconsideration of the GDR-MiDi for the collisional regime will be the next step to do.

REFERENCES

Armanini A., 2010. Modellazione idraulica dei fluidi bifasici solido liquidi, Palermo, XXXII Convegno Nazionale di Idraulica e Costruzioni Idrauliche (in Italian).

Armanini A., Larcher M. & L. Fraccarollo., 2009. Intermittency of rheological regimes in uniform liquid-granular flows, *Ph. Rev. E*, **79.051306.**

Armanini A., Capart H., Fraccarollo L. & Larcher M., 2005. Rheological Stratification in experimental free-surface of granular-liquid mixture *J. Fluid Mech.*, vol.532, pp. 269–319.

Armanini, A., Fraccarollo, L. & Larcher, M., 2005. Debris Flow. In *Encyclopedia of Hydrological Sciences* (Anderson, M.G., Ed.) Chap. 142, Vol. 4(12), pp. 2173–2186, John Wiley.

Bagnold R.A., 1954. Experiment on a gravity-free dispersion of large solid spheres in a Newtonian fluid under shear, Proceeding of the Royal Society, London, Deries A, vol. 225, pp. 49–63.
Campbell C.S., 1990. Rapid granular flows, *Annual Review of Fluid Mechanics*, vol. 22, pp. 57–92.
Chapman, S. & Cowling T.G., 1971. The mathematical theory of non-uniform gases. Third edition, Cambridge University Press.
Da Cruz F., Emam S., Prochnow M., Roux J.N. & Chevoi F., 2005. Rheophysics of dense granular materials: Discrete simulation of plane shear flows, *Physical Review E*.
Forterre Y. & Poulinque O., 2008. Flows of Dense Granular Media, *Annul Review of Fliud Mechanics*, vol. 40, pp. 1–24.
GDR MiDi, 2004. On dense granular flows, *Eur. Phys. J. E*, vol. 14, pp. 341–365.
Gidaspow D. & Hullin L., 1998. Equation of State and Radial Distribution of FCC Particles in a CBF, *AIChE Journal*, pp. 279–294, vol. 44, no. 2.
Goldhirsch I., 2003. Rapid granular flows. *Ann. Rev. Fluid Mech. 35*. pp. 267–93.
Haff, P.K., 1983. Grain flow as a fluid-mechanical phenomenon, *J. Fluid Mech.*, 134, pp. 401–430.
Jenkins J.T. & Hanes D.M., 1998. Collisional sheet-flow of sediment driven by a turbulent fluid, *J. Fluid Mech.*, vol. 370, pp. 29–52.
Jenkins, J.T. & Savage S.B., 1983. A theory for rapid flow of identical, smooth, nearly elastic, spherical particles, *J. Fluid Mech.*, Vol. 130, pp. 186–202.
Johnson, P.C. & R. Jackson, 1987. Frictional-collisional constitutive relations for granular materials, with application to plane shearing. *J. Fluid Mech.*, Vol. 176, pp. 67–93.
Iverson R.M., 1997. The physics of debris flows, *Review of Geophysics*, 35(3). pp. 245–296.
Lun C.K.K., Savage S.B., Jeffrey D.J. & Chepurniy N., 1984. Kinetic theories for granular flow: inelastic particles in Couette flow and slightly inelastic particles in a general flow field, *J. Fluid Mech.*, vol. 140, pp. 223–256.
Lun, C.K.K. & Savage, S.B., 1988. Particle size segregation in inclined chute flow of dry cohesionless granular solids, J. Fluid Mech., Vol. 189, 311–335.
Meruane C., Tamburrino A. & Roche O., 2010. On the role of the ambient fluid on gravitational granular flow dynamics, *J. Fluid Mech.*, vol. 648, pp. 381–404.
Savage, S.B. & D.J. Jeffrey, 1981. The stress tensor in a granular flow at high shear rates, *J. Fluid Mech*, vol. 110, pp. 255–272.
Takahashi T., 1981. Debris FLow, *Annual Review of Fluid Mechanichs*, vol. 13, pp. 57–77.
Takahashi T., 1991. Debris FLow, IAHR Monograph Series, A.A. Balkema, Rotterdam-Brookfild.
Truesdell C., 1984. Rational thermodynamics. Springer-Verlag.

Numerical Methods for Hyperbolic Equations – Vázquez-Cendón et al. (eds)
© 2013 Taylor & Francis Group, London, ISBN 978-0-415-62150-2

Three-phase Eulerian mixture formulation for the collapse of dense-suspension columns in ambient fluid

Patricio Bohorquez
Área de Mecánica de Fluidos, Departamento de Ingeniería Mecánica y Minera, Universidad de Jaén, Spain

ABSTRACT: We study numerically the collapse of a dense suspension made of liquid and unimodal sediment particles in an ambient fluid. To account for the inertia of each phase we adopt a full Eulerian model that is written in terms of mixture variables and the slip velocities between phases. The proposed formulation includes the continuity and momentum balance equations of the three-phase mixture, the scalar transport equations of sediment-particle and ambient-fluid volumetric concentrations, and a partial differential equation for the slip velocity of the sediment with respect to the liquid. To solve numerically the system of balance laws we employ a finite volume method in collocated meshes that captures with accuracy the presence of sharp interfaces and ensures global conservation in the numerical solution. We present preliminary supporting numerical simulations for the collapse of silted-up reservoirs that are compared successfully with dam-break experimental results at the initial instants of time, when the dense suspension exhibits a viscous regime and the pressure field is not hydrostatic.

1 INTRODUCTION

This study is motivated by a wish to develop a numerical tool able to simulate sedimentation, erosion and transport of sediment particles in free-surface laminar flows avoiding intrinsic limitations in classical viscous shallow-water models. As starting point we compare in this work numerical results from a multiphase model with experiments on the dam-break wave for silted-up reservoirs over a fixed horizontal boundary.

Sediment transport in laminar free-surface flows can be understood as a particular case of a multiphase system (Huppert 1986) made of a liquid phase transporting solid grains (i.e. sediments) trough an ambient fluid (i.e. atmospheric gas). The sediment-water mixture can also be seen as a suspension containing negatively buoyant particles moving, in general, at a different velocity than the liquid because of sedimentation, Fickian diffusion and viscous resuspension, among other transport phenomena (Leighton & Acrivos 1987; Nott et al. 2011). Therefore we adopt a multiphase Eulerian model accounting for the continuity and momentum balance equations of each involved phase that are well established under the axiom of continuum (Drew & Passman 1999; Ishii & Hibiki 2006), including a universal constitutive law recently proposed for dense suspensions (Bonnoit el al. 2010; Boyer et al. 2011). Numerical difficulties usually arise when solving the momentum balance equations in the limits of vanishing volumetric concentration or close to the maximum packing factor (Passalacqua & Fox 2011). So the author has recently proposed a re-formulation of the Eulerian model for the problem at hand in terms of mixture unknowns and slip velocities. The complete details of the model derivation, numerical method and results for shallow particle-laden flows on steep slopes can be found in a recent work by the author (Bohorquez 2011).

To the author's knowledge, experiments and numerical results on the collapse of dense-suspension columns in ambient fluid (e.g. dam-break wave for silted-up reservoirs) are very scarce. Experiments have been performed only by hydraulic research groups (Spinewine & Zech 2007; Duarte et al. 2011) and numerical models have been developed in the context of multi-layer depth averaged equations. Recent numerical results on the granular column collapse (Lagrée et al. 2011) have verified that dry sediments flowing trough air behave as a continuum with a well defined viscosity law (Forterre & Pouliquen 2008). So the formulation of a continuum model and numerical method for the collapse of dense-suspension columns in ambient fluid is an outstanding problem that really motivates this research.

We start summarising the mixture model for three phase flow and a Finite Volume Method (FVM) appropriated for the physical scenario at hand in § 2 and § 3, respectively. We devote § 4 to present preliminary supporting numerical simulations for the collapse of silted-up reservoirs (Duarte et al. 2011). A summary and some concluding comments are drawn in § 5.

2 THREE-PHASE MIXTURE FORMULATION

We are interested in modelling particle-laden flows moving in an ambient fluid with three involved phases: liquid (l), solid particles (p) and gas (g). To study such flow we adopt the full Eulerian model. The complete details of the model derivation, algebra and justification of the full Eulerian approach versus the algebraic drift-flux (Leighton & Acrivos 1987) and suspension balance (Nott et al. 2011) models can be found in a previous work (Bohorquez 2011). Below we summarise the most important points and some details of this formulation.

Under the axiom of continuum (Ishii & Hibiki 2006), the continuity equations of each phase are given by:

$$\frac{\partial \alpha_l \rho_l}{\partial t} + \nabla \cdot (\alpha_l \rho_l \vec{v}_l) = 0, \tag{1}$$

$$\frac{\partial \alpha_p \rho_p}{\partial t} + \nabla \cdot (\alpha_p \rho_p \vec{v}_p) = 0, \tag{2}$$

$$\frac{\partial \alpha_g \rho_g}{\partial t} + \nabla \cdot (\alpha_g \rho_g \vec{v}_g) = 0. \tag{3}$$

In these equations t is time, whilst ρ_k, α_k and $\vec{v}_k$ represent the density, volumetric concentration and conditionally averaged velocity of the kth phase, respectively ($k = l, p, g$). Because we shall consider that the solid phase cannot cross from the liquid to the gas phase, the phase volumetric concentrations $\alpha_{g,p}$ will be written as function of the volume fraction occupied by the solid-liquid mixture within a small control volume, say $\gamma \equiv 1 - \alpha_g$, and of the volumetric concentration of solid particles in the three phases mixture, denoted by $\beta \equiv \alpha_p$. The axiom of continuity, i.e. $\alpha_l + \alpha_p + \alpha_g = 1$, implies that the liquid volumetric concentration is given by $\alpha_l = \gamma - \beta$.

The three-phase mixture density ρ and the velocity of the centre of mass $\vec{v}$ are given, respectively, by

$$\rho \equiv \rho_p \beta + \rho_l (\gamma - \beta) + \rho_g (1 - \gamma), \tag{4}$$

$$\vec{v} \equiv \frac{1}{\rho} \{ \rho_p \beta \vec{v}_p + \rho_l (\gamma - \beta) \vec{v}_l + \rho_g (1 - \gamma) \vec{v}_g \}. \tag{5}$$

Furthermore, to simplify the notation, we introduce variables characterising the liquid-particle suspension, which are denoted with the subscript m. So we define the density (ρ_m), the conditionally averaged velocity of the centre of mass ($\vec{v}_m$) and volume ($\vec{u}_m$) in the liquid-sediment suspension as

$$\rho_m \equiv \rho_p \frac{\beta}{\gamma} + \rho_l \left(1 - \frac{\beta}{\gamma} \right), \tag{6}$$

$$\vec{v}_m \equiv \frac{1}{\rho_m} \left[\frac{\beta}{\gamma} \rho_p \vec{v}_p + \left(1 - \frac{\beta}{\gamma} \right) \rho_l \vec{v}_l \right], \tag{7}$$

$$\vec{u}_m \equiv \frac{\beta}{\gamma} \vec{v}_p + \left(1 - \frac{\beta}{\gamma} \right) \vec{v}_l. \tag{8}$$

Dividing the kth phase continuity equation (1)–(3) by the kth phase density and adding the resulting equations, yields

$$\nabla \cdot \vec{u} = 0 \quad \text{with} \quad \vec{u} \equiv \gamma \vec{u}_m + (1 - \gamma) \vec{v}_g. \tag{9}$$

This shows that the volumetric velocity $\vec{u}$ is a divergence-free vector.

Now we introduce the relative velocity of the solid particles with respect to the liquid,

$$\vec{v}_r \equiv \vec{v}_p - \vec{v}_l, \tag{10}$$

and the relative velocity of the dense suspension with respect to the gas phase,

$$\vec{u}_{r\gamma} \equiv \vec{u}_m - \vec{v}_g. \tag{11}$$

Taking into account (7)–(11), it is an easy exercise to write $\vec{v}$ (5) in terms of $\vec{u}$ (9), $\vec{v}_r$ (10) and $\vec{u}_{r\gamma}$ (11) as

$$\vec{v} = \vec{u} + \gamma(1 - \gamma) \frac{\rho_m - \rho_g}{\rho} \vec{u}_{r\gamma} + \beta s \frac{\gamma - \beta}{\gamma + \beta s} \frac{\rho_m}{\rho} \vec{v}_r, \tag{12}$$

with $s \equiv \rho_p / \rho_l - 1$.

In order to formulate a conservative numerical scheme to solve for (1)-(3), it is convenient to rewrite these equations in terms of the divergence-free velocity field $\vec{u}$, instead of $\vec{v}_k$. Dividing (1) and (2) by ρ_l and ρ_p, respectively, and adding the ensuing equations, yields

$$\frac{\partial \gamma}{\partial t} + \nabla \cdot (\gamma \vec{u}) + \nabla \cdot [\gamma (1 - \gamma) \vec{u}_{r\gamma}] = 0. \tag{13}$$

The equation for β requires some additional algebra. Eliminating $\vec{v}_l$ from (8) and (10), and $\vec{v}_g$ from (8) and (11), we can recast the velocity $\vec{v}_p$ as function of $\{\beta, \gamma, \vec{u}, \vec{u}_{r\gamma}, \vec{v}_r\}$,

$$\vec{v}_p = \vec{u} + (1-\gamma)\vec{u}_{r\gamma} + \left(1 - \frac{\beta}{\gamma}\right)\vec{v}_r. \quad (14)$$

Now substituting (14) into (2), one finds

$$\frac{\partial \beta}{\partial t} + \nabla \cdot (\beta \vec{u}) + \nabla \cdot [\beta(1-\beta)\vec{u}_{r\beta}] = 0, \quad (15)$$

in which

$$\vec{u}_{r\beta} \equiv \frac{\gamma(1-\gamma)\vec{u}_{r\gamma} + (\gamma - \beta)\vec{v}_r}{\gamma(1-\beta)}. \quad (16)$$

So the analogy between the scalar transport equation of the indicator function γ(13) and of the volumetric sediment concentration β (15) becomes evident and, therefore, one can apply exactly the same numerical scheme to solve them.

Analogous, the axiom of continuum allows us to write the momentum balance equations of each phase (Ishii & Hibiki 2006), given by:

$$\frac{\partial \alpha_l \rho_l \vec{v}_l}{\partial t} + \nabla \cdot (\alpha_l \rho_l \vec{v}_l \vec{v}_l) = \alpha_l \rho_l \vec{g} - (1-\alpha_p)\nabla p_l - K(\vec{v}_l - \vec{v}_p) + \nabla \cdot (\alpha_l \bar{\bar{\tau}}'_l), \quad (17)$$

$$\frac{\partial \alpha_p \rho_p \vec{v}_p}{\partial t} + \nabla \cdot (\alpha_p \rho_p \vec{v}_p \vec{v}_p) = \alpha_p \rho_p \vec{g} - \alpha_p \nabla p_l + K(\vec{v}_l - \vec{v}_p) - \nabla p_p + \nabla \cdot (\alpha_p \bar{\bar{\tau}}'_p), \quad (18)$$

$$\frac{\partial \alpha_g \rho_g \vec{v}_g}{\partial t} + \nabla \cdot (\alpha_g \rho_g \vec{v}_g \vec{v}_g) = \alpha_g \rho_g \vec{g} - \nabla p_g + \nabla \cdot (\alpha_g \bar{\bar{\tau}}'_g). \quad (19)$$

In (17)–(18), we have adopted the classical decomposition for the liquid (p_l) and solid (p_p) pressures, and the average interfacial momentum source is modelled as the drag function K (Meruane et al. 2010). In addition, weinclude the gas pressure p_g and the stresses $\bar{\bar{\tau}}'_k$ of each phase, given for the fluid phases (l, g) by

$$\bar{\bar{\tau}}'_k \equiv \mu_k \left[\nabla \vec{v}_k + (\nabla \vec{v}_k)^T - \frac{2}{3}(\nabla \cdot \vec{v}_k)\bar{\bar{I}} \right], \quad (20)$$

in which μ_l and μ_g are the liquid and gas dynamic viscosities, respectively. The solid stress term $\bar{\bar{\tau}}'_p$ is derived below. Notice that we have neglected the momentum source for the gas phase, which is consistent with vanishing surface tension and the presence of a thin (or sharp) interface between the liquid-particle suspension and the atmosphere. Finally, the solid pressure p_p is given by (Boyer et al. 2011):

$$p_p = \mu_l \dot{\gamma} \left(\frac{\beta}{\beta_M} \right)^2 \left(1 - \frac{\beta}{\beta_M} \right)^{-2}. \quad (21)$$

To obtain the momentum balance equation of the mixture, we add (17)–(19). Taking into account that $\alpha_g = 1-\gamma$, $\alpha_p = \beta$, $\alpha_l = \gamma - \beta$, and (5), the Left Hand Side (LHS) of the mixture momentum equation $\vec{C}$ reads

$$\vec{C} \equiv \frac{\partial \rho \vec{v}}{\partial t} + \nabla \cdot \bar{\bar{J}}, \quad (22)$$

where $\bar{\bar{J}}$ is the resultant (total) momentum flux associated with the averaged phase velocities,

$$\bar{\bar{J}} \equiv \beta \rho_p \vec{v}_p \vec{v}_p + (\gamma - \beta)\rho_l \vec{v}_l \vec{v}_l + (1-\gamma)\rho_g \vec{v}_g \vec{v}_g. \quad (23)$$

By using (5)–(11), $\bar{\bar{J}}$ can be recast into the form

$$\bar{\bar{J}} = \rho \vec{v}\vec{v} + \beta\left(1 - \frac{\beta}{\gamma}\right)\frac{\rho_l \rho_p}{\rho_m}\vec{v}_r \vec{v}_r + \gamma(1-\gamma)\frac{\rho_m \rho_g}{\rho}\vec{v}_{r\gamma}\vec{v}_{r\gamma}, \quad (24)$$

in which $\vec{v}_{r\gamma}$ denotes the slip velocity between the dense suspension and the air,

$$\vec{v}_{r\gamma} \equiv \vec{v}_m - \vec{v}_g = \vec{u}_{r\gamma} + \frac{\beta}{\gamma} s \frac{\gamma - \beta}{\gamma + \beta s}\vec{v}_r.$$

Subsequently, substituting (24) into (22), the LHS of the momentum balance equation is rewritten as

$$\vec{C} = \frac{\partial \rho \vec{v}}{\partial t} + \nabla \cdot (\rho \vec{v}\vec{v}) - \nabla \cdot \bar{\tau}'' - \nabla \cdot \bar{\bar{\tau}}''', \quad (25)$$

with

$$\bar{\bar{\tau}}'' \equiv -\beta\left(1 - \frac{\beta}{\gamma}\right)\frac{\rho_l \rho_p}{\rho_m}\vec{v}_r \vec{v}_r, \quad (26)$$

$$\bar{\bar{\tau}}''' \equiv -\gamma(1-\gamma)\frac{\rho_m \rho_g}{\rho}\vec{v}_{r\gamma}\vec{v}_{r\gamma}. \quad (27)$$

The tensors $\bar{\tau}''$ and $\bar{\bar{\tau}}'''$ represent the momentum diffusion due to the relative motion of sediment with respect to liquid and of the sediment-liquid mixture with respect to air, respectively. Therefore, they are referred hereafter to as *diffusion stress terms*. So, the mixture momentum equation can be expressed as

$$\frac{\partial \rho \vec{v}}{\partial t} + \nabla \cdot (\rho \vec{v}\vec{v}) = \rho \vec{g} - \nabla p + \nabla \cdot (\bar{\bar{\tau}}' + \bar{\bar{\tau}}'' + \bar{\bar{\tau}}'''), \quad (28)$$

where the mixture pressure is given by

$$p \equiv p_l + p_g + p_p. \tag{29}$$

For a thin interface between the liquid-particle suspension and the gas phase, the bulk viscosity of the three-phase mixture may be assumed as the arithmetic average of the laminar viscosity of the dense suspension μ_m and of the gas phase μ_g (Sethian & Smereka 2003):

$$\mu = \gamma\mu_m + (1-\gamma)\mu_g. \tag{30}$$

The harmonic mean of the viscosities is alternatively employed for low-disparity viscosity ratios (Ferziger 2003), i.e. $\mu_m/\mu_g \sim O(1)$. In (30) the dense suspension viscosity can be computed using a Krieger-Dougherty model in the viscous regime (Ovarlez et al. 2006; Bonnoit el al. 2010):

$$\mu_m(\beta) = \mu_l\left(1-\frac{\beta}{\beta_M}\right)^n, \tag{31}$$

in which β_M is the lowest stable packing of particles and $n = -2$. However, a more involved expression based on the dimensionless viscous number $I_v = \mu_l\,\dot{\gamma}/p_p$ is necessary to represent the transition to jamming (Boyer et al. 2011). The shear rate $\dot{\gamma}$ and the bulk laminar stress tensor $\overline{\overline{\tau}}'$ are given by

$$\begin{aligned}
&\dot{\gamma} \equiv \sqrt{2}\,|\overline{\overline{\gamma}}| \quad \text{with} \quad |\overline{\overline{\gamma}}| := (\overline{\overline{\gamma}} : \overline{\overline{\gamma}})^{1/2},\\
&\overline{\overline{\gamma}} = \frac{1}{2}\left[\nabla\vec{v} + (\nabla\vec{v})^T - \frac{2}{3}(\nabla\cdot\vec{v})\overline{\overline{I}}\right],
\end{aligned} \tag{32}$$

$$\overline{\overline{\tau}}' \equiv \sum_{k=\{l,p,g\}} \alpha_k\,\overline{\overline{\tau}}_k' = 2\mu\overline{\overline{\gamma}}. \tag{33}$$

Now we need to derive an additional partial differential equation for $\vec{v}_r$ that closes the mixture model. We apply the change rule to the LHS of (17)–(18), and taking into account (1)–(2) yields

$$\begin{aligned}
\frac{\partial\vec{v}_l}{\partial t} + (\vec{v}_l\cdot\nabla)\vec{v}_l = \frac{1}{\alpha_l\rho_l}\Big[&\alpha_l\rho_l\vec{g} - (1-\alpha_p)\nabla p_l\\
&-K(\vec{v}_l - \vec{v}_p) + \nabla\cdot(\alpha_l\overline{\overline{\tau}}_l')\Big],
\end{aligned} \tag{34}$$

$$\begin{aligned}
\frac{\partial\vec{v}_p}{\partial t} + (\vec{v}_p\cdot\nabla)\vec{v}_p = \frac{1}{\alpha_p\rho_p}\Big[&\alpha_p\rho_p\vec{g} - \alpha_p\nabla p_l\\
&+K(\vec{v}_l - \vec{v}_p) - \nabla p_p + \nabla\cdot(\alpha_p\overline{\overline{\tau}}_p')\Big].
\end{aligned} \tag{35}$$

From (20), (31) and (33) we get in the suspension

$$\alpha_p\overline{\overline{\tau}}_p' = 2\mu_m\overline{\overline{\gamma}} - \alpha_l\overline{\overline{\tau}}_l'. \tag{36}$$

Then, we simplify (34)–(36) recalling that inside the dense suspension $\gamma = 1$, i.e. $\alpha_l = 1-\alpha_p$ with $\alpha_p = \beta$, and $p_l = p - p_p$ (29). Also, by definition (10) we know that $\vec{v}_p = \vec{v}_l + \vec{v}_r$. Substituting this relation into (35) and using (34) one has

$$\begin{aligned}
&\frac{\partial\vec{v}_r}{\partial t} + \nabla\cdot(\vec{v}\vec{v}_r) + \nabla\cdot[w\vec{v}_r\vec{v}_r] - [\nabla\cdot(\vec{v} + 2\vec{v}_r)]\vec{v}_r\\
&+K\left(\frac{1}{\beta\rho_p} + \frac{1}{(1-\beta)\rho_l}\right)\vec{v}_r = -\left(\frac{1}{\rho_p} - \frac{1}{\rho_l}\right)\nabla p\\
&-\left(\frac{1}{\beta\rho_p} - \frac{1}{\rho_p} + \frac{1}{\rho_l}\right)\nabla p_p + \frac{1}{\beta\rho_p}\nabla\cdot\left(2\mu_m\overline{\overline{\gamma}}\right)\\
&-\left[\frac{1}{(1-\beta)\rho_l} + \frac{1}{\beta\rho_p}\right]\nabla\cdot[(1-\beta)\overline{\overline{\tau}}_l],
\end{aligned} \tag{37}$$

where $w \equiv 2 - \beta(1+s)/(1+\beta s)$.

3 FINITE VOLUME METHOD

In this section we present a brief summary of the segregated FVM employed to solve the mixture continuity equation (9), the hyperbolic scalar transport equations of γ (13) and β (15), the mixture momentum balance equation (28) and the partial differential equation of $\vec{v}_r$ (37). Full details of the numerical scheme can be found in Bohorquez (2011). All these equations are said to be in *strong conservation form* due to the fact that all terms have the form of the divergence of a vector or tensor (Patankar 1980; Jasak 1996; Ferziger & Perić 2002). The use of the strong conservation form together with the FVM in a fixed direction coordinate system automatically insures global conservation. For these reasons, we have paid special attention in § 2 to the mathematical formulation of the problem under consideration and adopted the Cartesian components as coordinate system.

Unknown dependent variables to be computed are stored at the cell centroid $\vec{x}_P$, using thus the so-called collocated arrangement, $\int_{V_P}(\vec{x} - \vec{x}_P)dV = 0$, where cell volume is denoted by V_P. To avoid the occurrence of oscillations in the pressure, an oscillation-free pressure-velocity coupling is adopted herein (Rhie & Chow 1983). Each cell has a neighbouring cell across each of its faces, of centroid denoted by N, and positional vector relative to $\vec{x}_P$ for the face f defined as $\vec{d}_f = \vec{x}_N - \vec{x}_P$. Similarly, using the centroid rule, the face centre, $\vec{x}_f$, is given by $\int_{S_f}(\vec{x} - \vec{x}_f)dS = 0$. Next, the face area vector $\vec{s}_f$ is a surface normal vector whose magnitude is equal to the area of the face.

The face area is calculated from the integrals $\vec{s}_f = \int_{S_f} \vec{n}\, dS$. When operating on a single cell, it is assumed that all face area vectors $\vec{s}_f$ point outwards of cell P. Hereafter superscript t denotes time, i.e. $\gamma^t \equiv \gamma(\vec{x}, t)$, and subscripts P and f correspond with cell centre and cell face magnitudes, respectively.

Numerical discretisation of differential operators were implemented up to second-order accuracy in space and time. For a detailed description of the numerical discretisation of the differential operator in collocated meshes (rate of change term, gradient operator, convection operator, diffusion operators as well as source and sink terms) we refer the reader to the work by Patankar (1980), Jasak (1996), and Ferziger & Perić (2002). A brief sketch of the numerical strategy, the segregated explicit FVM, and the extension of the PISO (Pressure Implicit with Splitting of Operators) algorithm are outlined below:

1. Update the normal flux of γ at the cell faces, $\phi_\gamma \equiv \left[\gamma\vec{u} + \gamma(1-\gamma)\vec{u}_{r\gamma}\right]_f^t \cdot \vec{s}_f$, and compute $\gamma^{t+\Delta t}$ by solving explicitly (Zalesak 1979) for the scalar transport equation (13).
2. Update the normal flux of β at the cell faces, $\phi_\beta \equiv \left[\beta\vec{u} + \beta(1-\beta)\vec{u}_{r\beta}\right]_f^t \cdot \vec{s}_f$, and compute $\beta^{t+\Delta t}$ by solving explicitly (Zalesak 1979) for the scalar transport equation (15).
3. Update the normal flux of mixture mass from $(\rho\vec{v})_f^t \cdot \vec{s}_f = \left[(\rho_m - \rho_g) - \beta/\gamma(\rho_p - \rho_l)\right]_f^t \phi_\gamma^t + \rho_g \phi_u^t + (\rho_p - \rho_l)\phi_\beta^t$, where $\phi_u^t = \vec{u}_f^t \cdot \vec{s}_f$.
4. Compute explicitly the mixture density from $\rho^{t+\Delta t} = \rho_l(\gamma^{t+\Delta t} - \beta^{t+\Delta t}) + \rho_g(1-\gamma^{t+\Delta t}) + \rho_p\beta^{t+\Delta t}$.
5. Discretise the momentum balance equation (28) and assemble the algebraic system of equations

$$\mathcal{A}(\vec{v}^{t+\Delta t}; \vec{v}^t) = -\vec{g}\cdot\vec{x}\nabla\rho^{t+\Delta t} - \nabla\hat{p}^{t+\Delta t}, \tag{38}$$

   where $\hat{p} \equiv p - \rho\vec{g}\cdot\vec{x}$ is the reduced pressure.
6. Then decompose $\mathcal{A}$ into a matrix containing its diagonal coefficients, $\mathcal{A}_D$, plus the off-diagonal coefficients, $\mathcal{A}_N$, and the source vector $\mathcal{A}_S$, i.e.

$$\mathcal{A} \equiv \mathcal{A}_D \vec{v}_P^{t+\Delta t} + \mathcal{A}_N \vec{v}_{nb}^{t+\Delta t} - \mathcal{A}_S. \tag{39}$$

   This classical decomposition is useful to formulate a Jacobi iteration scheme (Ferziger & Perić 2002), and facilitates the solution of the semidiscretised form of the momentum equation (38), as described below.
7. Predict the velocity field $\vec{v}^*$ by solving (38)–(39) with $\hat{p}^{t+\Delta t} \approx \hat{p}^t$.
8. Solve (38) subject to the incompressible constraint (9) employing Schur complement:
   (a) Predict convective flux $\phi^* = \vec{v}^* \cdot \vec{s}_f$ by using

$$\phi^* = \left(\frac{\mathcal{A}_S - \mathcal{A}_N \vec{v}^*_{nb}}{\mathcal{A}_D}\right)_{ff} \cdot \vec{s}_{ff} - \left(\frac{1}{\mathcal{A}_D}\right) |\vec{s}\;| \nabla_f^{\perp}\hat{p}^* - \left(\frac{1}{\mathcal{A}_D}\right)_f (\vec{g}\cdot\vec{x})_{ff} |\vec{s}_f| \nabla^{\perp}\rho^{t+\Delta t}$$

   Here the face normal gradient is denoted by $\nabla_f^{\perp}$ and is computed as the inner product of the face gradient ∇_f and unit normal to the face $\vec{n}_f$, e.g. $\nabla_f^{\perp}\rho \equiv \vec{n}_f \cdot \nabla_f \rho$.

   (b) Construct the pressure equation and solve for the reduced pressure $\hat{p}^*$,

$$\sum_{ff} \vec{s}_f \cdot \left[\left(\frac{1}{\mathcal{A}_D}\right)_f (\nabla\hat{p}^*)_f\right] = \sum \phi^* - \vec{s}_f \cdot \left[\gamma(1-\gamma)\frac{\rho_m - \rho_g}{\rho}\vec{u}_{r\gamma} - \beta s\frac{\gamma-\beta}{\gamma+\beta s}\frac{\rho_m}{\rho}\vec{v}_r\right]_f^{t+\Delta t}$$

   (c) Repeat (a)–(b) until convergence.
9. Update $\hat{p}^{t+\Delta t} = \hat{p}^*$, and obtain the velocity field $\vec{v}^{t+\Delta t}$ from (38).
10. Compute $\vec{v}_r^{t+\Delta t}$ from (37) and repeat 7–9 up to convergence.
11. Update flow properties.
12. Continue with the new time step if necessary.

The explicit approach imposes a Courant-Friedrichs-Lewy (CFL) stability restriction on the time step Δt, which is calculated and adjusted during the numerical simulation. The largest Courant number is associated with the effective velocity $\vec{u} + (1+\beta)\vec{u}_{r\beta}$, and is given by

$$\mathrm{CFL}_\beta = \frac{|[\vec{u} + (1+\beta)\vec{u}_{r\beta}]_f \cdot \vec{s}_f|}{\vec{d}_f \cdot \vec{s}_f}\Delta t. \tag{40}$$

So the time step is adjusted according to the following relation:

$$\Delta t \leftarrow \min\left\{\frac{\mathrm{CFL}_{\max}}{\mathrm{CFL}_o}\Delta t, \left(1 + \lambda_1\frac{\mathrm{CFL}_{\max}}{\mathrm{CFL}_o}\right)\Delta t, \lambda_2\Delta t\right\}, \tag{41}$$

in which $\mathrm{CFL}_{\max}$ is the desired Courant number, $\lambda_1 = 0.1$ and $\lambda_2 = 1.2$ are damping factors employed in order to avoid numerical instabilities arising from time step oscillations.

4 RESULTS

We have analysed numerically the flow after the sudden release of a finite volume of sand

and water on a solid horizontal boundary. For the sake of comparison with previous experimental works we considered the same parameter values as (Duarte et al. 2011): $L = 1.5\,m$, $0.25 \le h_s \le 0.39\,m$ and $h_w = 0.41\,m$, where h_s and h_w denote the sediment and water depths at $t = 0$ and L is the reservoir length upstream of the gate initially filled with sand and water. In addition we set the same sediment properties of constant density $\rho_p \approx 1414\,kg/m^3$ and mean diameter $d = 1\,mm$, water properties of $\rho_l = 1000\,kg/m^3$ and $\mu_l = 10^{-3}\,kg/(ms)$, and ambient fluid with $\rho_g = 1.225\,kg/m^3$ and $\mu_g = 18.5 \times 10^{-6}\,kg/(ms)$.

Figure 1 shows the computational domain and a sketch of the initial conditions for reservoir silting degrees of $h_s/h_w = 0.94$, 0.79 and 0.6. We have performed direct numerical simulations in a two-dimensional mesh of 1440×240 cells distributed uniformly along the streamwise and vertical directions, of length 3.6×0.6 meters. Unfortunately, the initial volumetric concentration of sediment β in the reservoir was not quantified in the experiments and, consequently, there is a large uncertainty in the modelling (Rondon et al. 2011): if the sediment is not fully packed (i.e. $\beta \lesssim \beta_M$), the sand-water suspension may flow as a Newtonian fluid (Ovarlez et al. 2006; Bonnoit el al. 2010); conversely, close to the lowest stable packing of particles (i.e. $\beta \lesssim \beta_M$), the dense suspension may exhibit jamming (Boyer et al. 2011).

We performed a first set of numerical simulations setting $h_s/h_w = 0.94$ so that the upper clear-water layer is much thinner than the lower sediment-water layer. Hence we expect the ensuing flow to be dominated by the dense suspension. We considered an initial homogeneous mixture, $\beta(t = 0) = \beta_o$, and simulated a wide range of possible concentration values within the interval of $0.5 \le \beta_o \le \beta_M = 0.635$. The optimal value of β_o was determined by comparing the temporal evolution of thewetting front position with the experimental values, which yield $\beta_o = 0.99\,\beta_M$. This calibration of the initial concentration of sediment was maintained also in the rest of simulations described below. Notice that the viscosity of the dense suspension μ_m (31) is 10^4 times larger than the water viscosity μ_l, indicating that the sediment-water mixture is extremely viscous. So we expect that the relative motion of sediment particles is hindered at the first stage of the flowand, consequently, we set to zero the slip velocity of the sediments, i.e. $\vec{v}_r = 0$. The effects of including the slip velocity $\vec{v}_r$ (37) in the numerical model might be relevant in the stopping phase of sediment motion and will be explored elsewhere.

Figures 2 and 3 depict the numerical solution at two instants of time ($t = 0.375$ and $t = 0.925\,s$, respectively). Panel (a) corresponds with the sediment volumetric concentration that remained constant inside the lower layerduring most part of the flood. It is found a clearly distinguished layer of water above the saturated sediments. The upper clear water layer flew above the dense suspension

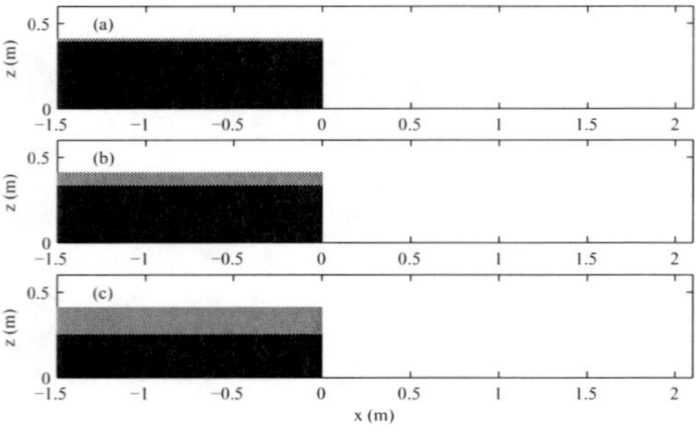

Figure 1. Initial conditions for $h_s/h_w = 0.94$ (a), 0.79 (b) and 0.6 (c). We also show the computational domain. Black, light grey and white colours correspond with sediment-water mixture, water and air, respectively.

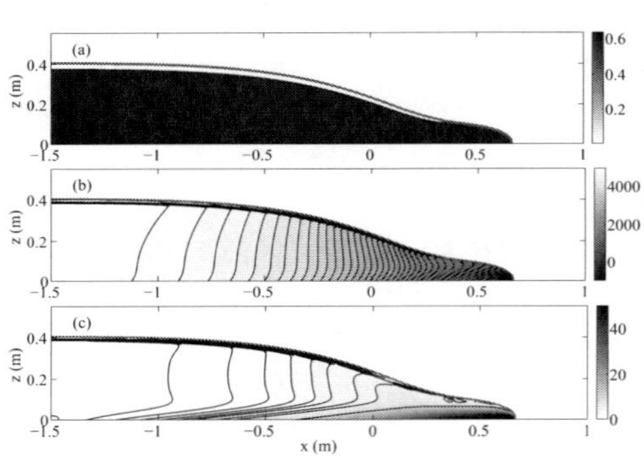

Figure 2. Numerical solution at $t = 0.375\,s$ for $h_s/h_w = 0.94$: (a) sediment volumetric concentration β, (b) reduced pressure $\hat{p}$ $[Pa]$ and (c) shear rate $\dot{\gamma}$ $[s^{-1}]$. Thin black solid lines in (c) correspond with shear rate values of 0.1, 0.5, 1, 1.5, 2, 2.5, 3, 4, 5 and 10 s^{-1}. Thick blue line highlights the free surface interface $\gamma = 0.5$.

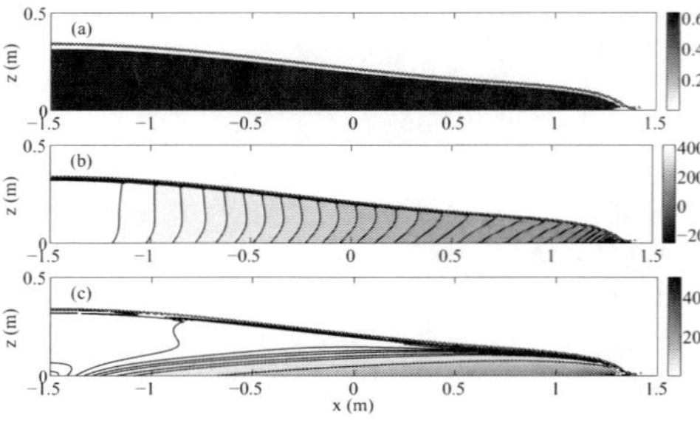

Figure 3. Numerical solution at $t = 0.925\,s$ as for Figure 2.

and reached the wetting front approximately at $t \approx 0.9\,s$. During this stagethe shape of the tip region changed according to experimental results. The levels of constant reduced pressure $\hat{p} = p - \rho \vec{g} \cdot \vec{x}$ shown in panels 2(b) and 3(b) indicate that the pressure field was not hydrostatic at any time because they are misaligned with respect to the vertical. Only at late time ($t > 0.9\,s$) and far enough from the tip region ($x < -1\,m$) the flow could be assumed hydrostatic. Conversely, in the rest of the domain the pressure distribution is very complex due to the curvature of the free surface owing non-vanishing vertical velocities and a two-dimensional flow. This phenomenon is much more severe closer to the wetting front, justifying the need of our formulation (see § 2) and numerical scheme (see § 3). In addition the numerical simulation allows us to evaluate the shear rate $\dot{\gamma}$ at every instant of time, which is difficult to be measured experimentally. Figures 2(c) and 3(c) illustrate the shear rate field and its temporal evolution. The shear rate is maximum in the tip region, decreases upstream and reaches minimum values close to the upstream vertical wall. Within the bore front the shear rate is very high, up to 50 s^{-1} at $t = 0.375\,s$, and lines of constant values are nearly parallel to the channel bottom. This fact is better observed at later time, $t = 0.925\,s$, though the shear rate decreases because of the lower velocities and developing uniform conditions.

Subsequently, we performed additional numerical simulations for different reservoir silting degrees. In the case $h_s/h_w = 0.79$ (Fig. 4) there were no marked up differences with respect to the previous simulation ($h_s/h_w = 0.94$). For instance, the pressure and shear rate fields were very similar in the sediment layer. However we found that the upper layer was thicker at late time because more water was put in motion, see Figure 4(a), and that the flow front was fluidisedsooner. Near the interface separating the clear water layer and sediment-water mixture we observe a jump in the reduced pressure $\hat{p}$, see Figure 4(b), caused by the different

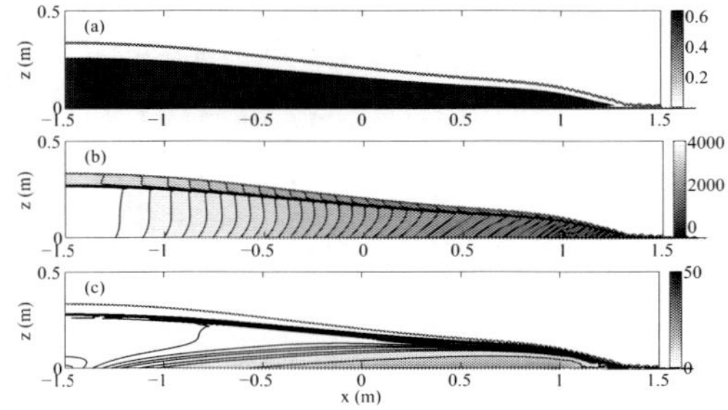

Figure 4. Numerical solution at $t = 0.925\,s$ for $h_s/h_w = 0.79$.

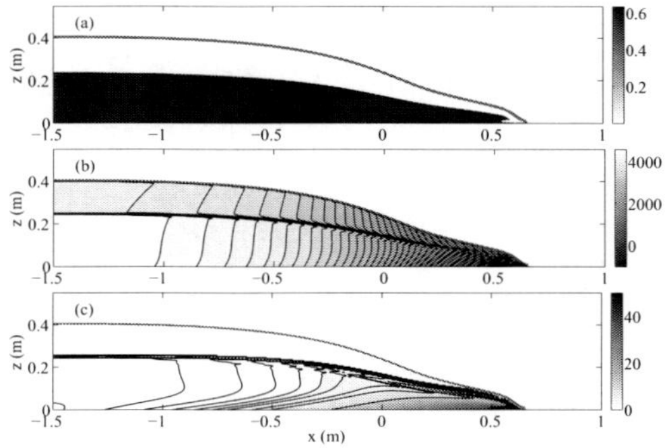

Figure 5. Numerical solution at $t = 0.375\,s$ for $h_s/h_w = 0.6$.

densities in each layer. Also, the shear rate was negligible in the water, indicating quasi-uniform flow conditions. However, in the dense suspension the shear rate was found as high as in the previous case.

These phenomena are better observed for the reservoir silting degree value of $h_s/h_w = 0.6$. In this case the clear water layer reached the sediment front at very early time, $t \approx 0.375\,s$, as shown in Figure 5(a). The shape of the tip region was found different with respect to the silting degree of 0.94, recall Figure 2(a), and agrees with the experimental result, see Figure 3(d) in (Duarte et al. 2011). The pressure jump across the sediment interface is also evident in the present case, as shown in Figure 5(b). Once again there is a misalignment of the lines of constant reduced pressure with respect to the vertical, indicating that the mixture pressure p is not hydrostatic. Finally, the shear rate was analogous to the previous simulations, being much higher in the sediment-water mixture than in the clear water layer.

5 CONCLUSION

In this work we have employed a three-phase Eulerian model written in terms of mixture variables (§ 2) and a finite volume method (§ 3) to study the collapse of hyperconcentrated mixtures of sediment and water in air (§ 4). The proposed formulation is valid for arbitrary values of the shallowness parameter (or reservoir aspect ratio) and does not impose any restriction on the mixture pressure p, centre of mass velocity $\vec{v}$, sediment-to-water slip velocity $\vec{v}_r$ and sediment volumetric concentration β fields that are solved locally at every time step during the numerical simulation.

We have performed direct numerical simulations for the same experimental conditions as described by (Duarte et al. 2011). According to experimental results, our numerical simulations show that the dense suspension behaves as a highly viscous fluid at the initial stage of the flow when we found a good matching between numerics and experiments.

The applications of the physical model and numerical tool developed in this work are numerous and include the study of roll waves in debris flows on steep slopes (Bohorquez & Fernandez-Feria 2008; Iverson et al. 2010; Bohorquez 2010). Also this tool could be valuable to better understand the dynamics of debris and hyperconcentrated flows as well as to check the physical assumptions made in the derivation of flow depth averaged models suitable for risk mitigation (Denlinger & Iverson 2004) and paleoflood hydraulic reconstruction (Bohorquez & Darby 2008).

ACKNOWLEDGMENTS

This work has been partially supported by the *Spanish MICINN* Project # DPI2008-06624-C03-02 and *Junta de Andalucia* Project # P07-TEP-02693.

REFERENCES

Bohorquez P. 2010. Competition between kinematic and dynamic waves in floods on steep slopes. *J. Fluid Mech. 675*, 375–409.

Bohorquez P. 2011. Non-hydrostatic numerical simulation of shallow particle-laden flows on steep slopes. (submitted).

Bohorquez P. & Darby S.E. 2008. The use of one- and two-dimensional hydraulic modelling to reconstruct a glacial outburst flood in a steep Alpine valley. *J. Hydrol. 361*, 240–261.

Bohorquez P. & Fernandez-Feria R. 2008. Transport of suspended sediment under the dam-break flow on an inclined plane bed of arbitrary slope. *Hydrol. Process. 22*, 2615–2633.

Bonnoit C., Darnige T., Clement E. & Lindner A. 2010. Inclined plane rheometry of a dense granular suspension. *J. Rheol. 54*, 65–79.

Boyer F., Guazzelli E. & Pouliquen O. 2011. Unifying suspension and granular rheology. *Phys. Rev. Lett. 107*, 188301.

Denlinger R.P. & Iverson R.W. 2004. Granular avalanches across irregular three-dimensional terrain: 1. Theory and computation. *J. Geophys. Res. 109*, F01014.

Drew, D.A. & Passman S.L. 1999. *Theory of Multicomponent Fluids*. New York: Springer.

Duarte R., Ribeiro J., Boillat J.L. & Schleiss A. 2011. Experimental study on dam-break waves for silted-up reservoirs. *J. Hydr. Engrg.*, doi:10.1061/(ASCE)HY.1943-7900.0000444.

Ferziger J.H. 2003. Interfacial transfer in Tryggvason's method. *Int. J. Numer. Meth. Fluids 41*, 551–560.

Ferziger J.H. and Perić M. 2002. *Computational Methods for Fluid Mechanics*. Springer.

Forterre Y. & Pouliquen O. 2008. Flows of dense granular media. *Annu. Rev. Fluid Mech. 40*, 1–24.

Huppert H.E. 1986. Intrusion of fluid mechanics into geology. *J. Fluid Mech. 173*, 557–594.

Ishii M. and Hibiki T. 2006. *Thermo-fluid dynamics of two-phase flow*. Springer.

Iverson R.M., Logan R., LaHusen R.G. & Berti, M. 2010. The perfect debris flow? Aggregated results from 28 large-scale experiments. *J. Geoph. Res. 115*, F03005.

Jasak H. 1996. *Error analysis and estimation in the Finite Volume method with applications to fluid flows*. Ph.D. thesis, Imperial College, University of London.

Lagrée P.-Y., Staron L. & Popinet S. 2011. The granular column collapse as a continuum: validity of a two-dimensional Navier–Stokes model with a $\mu(I)$-rheology. *J. Fluid Mech. 686*, 378–408.

Leighton D. and Acrivos, A. 1987. The shear-induced migration of particles in concentrated suspensions. *J. Fluid Mech. 181*, 415–439.

Meruane C., Tamburrino A. & Roche O. 2010. On the role of the ambient fluid on gravitational granular flow dynamics. *J. Fluid Mech. 648*, 381–404.

Nott P.R., Guazzelli E. & Pouliquen O. 2011. The suspension balance model revisited. *Phys. Fluids 23*(4), 043304.

Ovarlez G., Bertrand F. & Rodts S. 2006. Local determination of the constitutive law of a dense suspension of noncolloidal particles through magnetic resonance imaging. *J. Rheol. 50*, 259–292.

Passalacqua A. & Fox R.O. 2011. Implementation of an iterative solution procedure for multi-fluid gas-particle flow models on unstructured grids. *Powder Technol. 213*, 174–187.

Patankar S. V. 1980. *Numerical Heat Transfer and Fluid Flow*. Taylor & Francis.

Rhie C.M. & Chow W.L. 1983. Numerical study of the turbulent flow past an airfoil with trailing edge separation. *AIAA Journal 21*(11), 1525–1532.

Rondon L., Pouliquen O. & Aussillous P. 2011. Granular collapse in a fluid: Role of the initial volume fraction. *Phys. Fluids 23*, 073301.

Sethian J.A. and Smereka P. (2003). Level set methods for fluid interfaces. *Annu. Rev. Fluid Mech. 35*, 341–372.

Spinewine B. & Zech Y. 2007. Small-scale laboratory dam-break waves on movable beds. *45*(SPEC. ISS.), 73–86.

Zalesak S.T. 1979. Fully multidimensional flux-corrected transport algorithms for fluids. *J. Comput. Phys. 31*, 335–362.

Numerical Methods for Hyperbolic Equations – Vázquez-Cendón et al. (eds)
 ISBN 978-0-415-62150-2

Numerical simulation of three-dimensional transient variably saturated flow

D. Caviedes-Voullième, P. García-Navarro & J. Murillo
Fluid Mechanics Group, Department of Materials and Fluids Science and Technology, University of Zaragoza, Aragon, Spain

ABSTRACT: Subsurface flow is an important part of hydrological processes, from infiltration during rain events to surface-subsurface flow interactions in floodplains and hillslopes. Since subsurface flow may occur in unsaturated and saturated conditions, models which are able to represent such conditions in a continous fashion are necessary. Furthermore, adequate mass conservation in hydrological processes must account for the exchange of surface and subsurface water both in steady and unsteady cases. Hence, a model which allows to approximate the solution of transient variably saturated flow problems, aiming towards surface-subsurface interactions, has been developed.

Flow in variably saturated soils is described by Richards equation. This highly non-linear PDE requires in most cases the use of numerical methods for its solution. The proposed numerical model allows for the solution of the water content and pressure distribution in three dimensional domains by means of an implicit finite volume scheme which adequately conserves mass and represents wetting fronts both in time and space. The model has been developed progresively from one to three dimensions, testing it against an analytical solution of an infiltration problem and by means of test cases. In one dimensional form, several numerical schemes were tested to determine the best numerical approach and determine the response of the model to different configurationsand parameters. Validation against the analytical solution shows that the model is able to correctly solve Richards' equation in one, two and three dimensions. Test case solutions show good behavior of the model in all dimensions against particular geometries and different initial and boundary conditions.

1 INTRODUCTION

Groundwater flow has been classified conceptually into saturated and unsaturated flows. Saturated flows can be treated, and often are in hydrological models, by solving the Darcy equation using the Dupuit assumptions. The unsaturated region is treated by solving the Richards equation, a highly non-linear partial differential equation. In many applications accurate solution of the Richards equation may be unnecesary, as when solving large scale aquifer dynamics, but when solving infilitration and drainage processes, as well as complex surface-groundwater interactions it is essential. Furthermore, because the Richards equation degenerates to the Darcy equation under saturated conditions, it is possible in fact to use the Richards equation to solve for the entire groundwater domain.

Appropriate numerical techinques are necessary to solve the Richards equation while ensuring mass conservation and keeping low numerical diffusion. It is also of primary interest for the numerical model to be able to exploit the generality of the Richardsequation, and that it is able to solve de entire groundwater domain, regardless of its saturation condition. Finally, computational efficiency is a key factor, as hydrological modelling is often aimed towards solving large domains and time lapses.

2 MATHEMATICAL MODEL

The three-dimensional Richards equation in the mixed form (containing both water content θ and matric or suction pressure h) is

$$\frac{\partial \theta(h)}{\partial t} = \nabla\left[\mathbf{K}(h)\nabla(h+z)\right] \tag{1}$$

where $\theta\,[L^3/L^3]$ is the water content in the soil, $z\,[L]$ is the vertical coordinate in reference to a certain datum, and is considered positive upwards, $h\,[L]$ is the water head in the soil (which should be noted to be negative when unsaturated and zero or positive when saturated) and $\mathbf{K}\,[L/T]$ is the hydraulic conductivity tensor

$$\mathbf{K} = \begin{pmatrix} K_{xx} & 0 & 0 \\ 0 & K_{yy} & 0 \\ 0 & 0 & K_{zz} \end{pmatrix}$$

Note that this definition assumes that the principal directions of hydraulic conductivity are aligned with the cartesian coordinates that have been selected.

From this point onwards, the term *variably* saturated flow is prefered over *unsaturated* flow, as saturation can also be modeled by the equation in this form. It should be noted that, although Richards' equation is often written in water content (θ-based) and matric (h-based) forms, the first one is not continous into the saturated region, and the second one shows poor mass balance. The advantage of the mixed form is that it allows for a general treatment both of unsaturated and saturated regimes and that it combines the positive aspects of both forms (Celia et al. 1990; Rathfelder and Abriola 1994).

Furthermore, equation (1) can be written in terms of hydraulic capacity $C = \partial\theta/\partial h$ (L^3/L^4). Hence, equation (2) is obtained. This equation corresponds to the h-based or matric potential form of Richards' equation. This expression will not be used directly, as discretization of the C term from this equation introduces mass balance issues, however, it is useful to keep in mind that such relations exists between C, θ and h.

$$C(h)\frac{\partial h}{\partial t} = \nabla\left[\mathbf{K}(h)\nabla(h+z)\right] \tag{2}$$

It is usefull to recall that equation (1) implicitly carries the definition of flux, as it relates the change in water content to the divergence of the flux, which is written in equation (3) from Darcy's Law.

$$\mathbf{J} = -\mathbf{K}\nabla(h+z) \tag{3}$$

To complete the mathematical model, a soil model is necessary. In this work, the Modified Mualem-van Genuchten model is used (Schaap and van Genuchten 2006; Vogel et al. 2000), which allows for better treatment of the conductivity function near saturation.

$$\theta_m = (\theta_s - \theta_r)(1 + (\alpha \mid h_s \mid)^{\hat{\eta}})^{\mu} + \theta_r \tag{4}$$

$$S_e = \frac{\theta - \theta_r}{\theta_s - \theta_r} \tag{5}$$

$$\mu = 1 - \frac{1}{\hat{\eta}} \tag{6}$$

$$\theta = \begin{cases} \dfrac{\theta_m - \theta_r}{\left(1 + (\alpha \mid h \mid)\hat{\eta}\right)^{\mu}} + \theta_r & \text{if } h \le h_s \\ \theta_s & \text{if } h > h_s \end{cases} \tag{7}$$

The conductivity function when saturated is $K = K_s$, when unsaturated

$$K = K_s S_e^{1/2}\left[\frac{1 - \left[1 - \left(\dfrac{\theta_s - \theta_r}{\theta_m - \theta_r}S_e\right)^{\frac{1}{\mu}}\right]^{\mu}}{1 - \left[1 - \left(\dfrac{\theta_s - \theta_r}{\theta_m - \theta_r}S_e\right)^{\frac{1}{\mu}}\right]^{\mu}}\right]^2 \tag{8}$$

when saturated, the capacity function is $C = 0$, while unsaturated it is

$$C = \frac{\partial\theta}{\partial h} = -\mu\hat{\eta}\alpha(\theta_m - \theta_r)(1 + \alpha^{\hat{\eta}} \mid h \mid^{\hat{\eta}})^{-\mu-1} h \tag{9}$$

where $\theta_r[L^3/L^3]$ is the residual water content, $K_s[L/T]$ is the saturated hydraulic conductivity, $\hat{\eta}$ is a parameter which measures pore-size distribution and $\alpha[L^{-1}]$ is a parameter related to the inverse of the air-entry pressure. Note that the hydraulic capacity function $C(h)$ is obtained by analytical differentiation from the water content function. It is possible to approximate C in a discrete way, but it has been shown (Rathfelder and Abriola 1994) that the analytical approach is accurate and efficient. It is important to note that water content and conductivity have a maximum value at saturation, and that the hydraulic capacity is zero at saturation and has a maximum value for a certain suction pressure.

3 NUMERICAL MODEL

Richards equation contains differential terms both in time and space, and has no general analytical solution. Thus it is necessary to utilize a numerical scheme in order to approximate the solution. In this model, a cell-centered finite volume approach isimplemented to solve the equation. The scheme was developed to solve for unstructured tetrahedrical meshes which easily degenerates to unstructured triangular meshes in 2D. Although the Finite Element tmethod has been used extensively for this problem, the finite volume method can be well suited because of its straightforward physical interpretation, mass conservation and steep front aproximations, specially in multidimensional problems (Manzini and Ferraris 2004).

Consider each cell i as a control volume Ω which does not change in time. Time integration is done with an implicit Euler approach. Furthermore, integrating equation (1) over Ω and invoking Gauss' theroem for the right side yields

$$\frac{\theta^{n+1} - \theta^{n}}{\Delta t}\Omega = \int_{\partial\Omega} \left[\mathbf{K}^{n+1}\mathbf{n}\nabla(h^{n+1} + z)\right]\partial\Omega \tag{10}$$

where $\mathbf{n}$ is the outer-pointing normal vector of the surface $\partial\Omega$. Dependence of C and $\mathbf{K}$ on h has been dropped in notation for simplicity but it should be remembered that such dependence still exists. Superindices n denote the current time step.

By discretizing within a finite volume framework,

$$\frac{\theta_i^{n+1} - \theta_i^{n}}{\Delta t}V_i = \sum_{\omega=1}^{N_\omega}\left[\mathbf{K}_\omega^{n+1}\mathbf{n}_\omega \frac{h_\omega^{n+1} - h_i^{n+1} + z_\omega - z_i}{d_\omega}\right]A_\omega \tag{11}$$

where V_i is the volume of cell i, with N_ω polygonal faces of area A_ω, each with an outer-pointing normal vector $\mathbf{n}_\omega$ and distance between centroids beteween cells i and ω defined as d_ω. The distance between centers has been computed during testing in two ways.

At this point, it is important to note the following properties of equation (11). If cell i is saturated in time n, and remains saturated in time $n + 1$, the left side of the equation is zero, regardless of the fact that steady state flow exists or not, as hydrostatic pressure could still change in the same cell. Hydraulic conductivity $\mathbf{K}$ is a non-linear function of h, and both are evaluated in time $n + 1$. Hence linearization of this term needs to be done in order to solve the equation system, which can be done by Picard iteration. Furthermore, because θ is a function of h it is possible to express the time derivative term in terms of h, in a similar way as in equation (2). However, correct discretization of the hydraulic capacity term is critical for mass balance. Following Celia et al. (Celia et al. 1990), the term is expanded by a Taylor polynomial from the definition $C = \partial\theta/\partial h$ instead of directly evaluating C in time $n + 1$. It should also be noted that C is a non-linear function of h. The first order Taylor polynomial applied to the time derivative of θ, around $h^{n+1,m}$, where m is an intermediate step for linealization of C and K, is

$$\theta^{n+1,m+1} = \theta^{n+1,m} + C^{n+1,m}(h^{n+1,m+1} - h^{n+1,m}) \tag{12}$$

For simplicity of notation $n + 1$, m will be reduced to m, and $n + 1$, $m + 1$ time will be indicated as $m + 1$. Time n will still be noted as time n, hence whenever m or $m + 1$ superindices are shown, it refers to time $n + 1$ in the correspondent iteration.

Substituting the Taylor polynomial into equation (11), evaluating K in iteration m and the unknowns h in time $n + 1$ yields,

$$\theta_i^{m} + C_i^{m}(h_i^{m+1} - h_i^{m}) - \theta_i^{n} = \frac{\Delta t}{V_i}\sum_{\omega=1}^{N_\omega}\left[\mathbf{K}_\omega^{m}\mathbf{n}_\omega \frac{h_\omega^{m+1} - h_i^{m+1} + z_\omega - z_i}{d_\omega}\right]A_\omega \tag{13}$$

For simplicity, let

$$\Gamma_\omega = \frac{A_w}{d_w}\mathbf{n}_\omega \tag{14}$$

Rewriting the equation in a compact form, practical for matrix operations

$$a_i h_i^{m+1} - \sum_{\omega=1}^{N_\omega} b_\omega h_\omega^{m+1} = c_i \tag{15}$$

$$a_i = C_i^{m} + \frac{\Delta t}{V_i}\sum_{\omega=1}^{N_\omega}\Gamma_\omega \mathbf{K}_\omega^{m} \tag{16}$$

$$b_\omega = \frac{\Delta t}{V_i}\Gamma_\omega \mathbf{K}_\omega^{m} \tag{17}$$

$$c_i = C_i^{m}h_i^{m} - \theta_i^{m} + \theta_i^{n} + \frac{\Delta t}{V_i}\sum_{\omega=1}^{N_\omega}\Gamma_\omega \mathbf{K}_\omega^{m}(z_\omega - z_i) \tag{18}$$

The unknowns in this system are the pressure heads in every cell, however C^m, $\mathbf{K}^m$ and θ^m are dependent on such values, which need to be calculated and updated from the previous iteration, within the iterative solution of the matrix. In this model, a straightforward Jacobi solver has been used to solve the system.

Note that the $\mathbf{K}_\omega$ is a vector, not a second rank tensor as $\mathbf{K}$. But furthermore, it is evalauted at the cell interfaces, not the cell themselves. Hence, let $\mathbf{K}_{i\to\omega} = \mathbf{K}_i\mathbf{n}_\omega$ represent the conductivity vector from cell i into cell ω, then

$$\mathbf{K}_\omega^{m} = \Xi_K(\mathbf{K}_{i\to\omega}^{m}, -\mathbf{K}_{\omega\to i}^{m}) \tag{19}$$

where Ξ_K is an operator which represents the selected weighting technique to find conductivity at the cell interface. The negative sign of $\mathbf{K}_{\omega\to i}^{m}$ allows for the resulting weighted intercell conductivity to be an outer-pointing vector from i towards ω. The formal interpretation however, because hydraulic conductivity is a scalar, is that the weighted intercell conductivity corresponds to that of an outgoing flow from cell i through intercell face ω.

Many weighting techniques have been proposed in the literature (Baker 2006). In the cases presented in this paper an arithmetic mean was used for testing purposes. Some testing was performed with other techniques, obtaining results which are consistent with those reported in the literature. These differences are not large and are not the primary interest of this paper, hence they are ommited for brevity.

3.1 *Boundary conditions*

Let face ψ be a boundary face of cell i upon which an external flux J_ψ is imposed. Then, equation (15) can be rewritten to separate the boundary face ψ from all other faces $\omega \neq \psi$. Coefficient b_i remains unchanged.

$$a_i h_i^{m+1} - \sum_{\substack{\omega=1 \\ \omega\neq\psi}}^{N_\omega} b_\omega h_\omega^{m+1} = c_i \tag{20}$$

$$a_i = C_i^m + \frac{\Delta t}{V_i} \sum_{\substack{\omega=1 \\ \omega\neq\psi}}^{N_\omega} \Gamma_\omega \mathbf{K}_\omega^m \tag{21}$$

$$c_i = C_i^m h_i^m - \theta_i^m + \theta_i^n + \frac{\Delta t}{V_i} \sum_{\substack{\omega=1 \\ \omega\neq\psi}}^{N_\omega} \Gamma_\omega \mathbf{K}_\omega^m (z_\omega - z_i) - \frac{\Delta t}{V_i} \mathbf{J}_\psi^{m+1} \mathbf{n}_\psi A_\psi \tag{22}$$

Particular cases can be modeled with this boundary condition, such as an inflowing hydrograph: $J_i^{m+1} = J_i^{n+1}(t)$ as well as impervious conditions $J_i^{m+1} = 0$.

Dirichlet type boundary conditions require imposing a known pressure on boundary cells, which also implies that the boundary cell is independent from the contigous interior cells. This boundary condition can be implemented simply by imposing such value inthe matrix equation (15). The coefficients for Dirichlet boundary cells β become:

$$a_\beta = 1;\; b_\omega = 0;\; c_\beta = h_p \tag{23}$$

In such manner, the solution for the boundary cell i is trivial. Consider now an interior cell which shares a face with a boundary cell. One of the faces ω of the interior cell i corresponds to a boundary cell β. Such cell β has a known pressure and must not be included as an unknown in the equation of cell i. Hence, the term associated to cell β must be taken out of the coefficient matrix and moved into the free vector. Hence, the equation and coefficients are

$$a_i h_i^{m+1} - \sum_{\substack{\omega=1 \\ \omega\neq\beta}}^{N_\omega} b_\omega h_\omega^{m+1} = c_i \tag{24}$$

$$a_i = C_i^m + \frac{\Delta t}{V_i} \sum_{\omega=1}^{N_\omega} \Gamma_\omega \mathbf{K}_\omega^m \tag{25}$$

$$b_\omega = \frac{\Delta t}{V_i} \Gamma_\omega \mathbf{K}_\omega^m \tag{26}$$

$$c_i = C_i^m h_i^m - \theta_i^m + \theta_i^n + \frac{\Delta t}{V_i} \sum_{\omega=1}^{N_\omega} \Gamma_\omega \mathbf{K}_\omega^m (z_\omega - z_i) + \frac{\Delta t}{V_i} \Gamma_\beta \mathbf{K}_\beta^m h_\beta^{n+1} \tag{27}$$

This boundary condition allow to model water table variations at the boundaries, surface ponding and semi infinite soil stratum draining conditions, this last one by imposing $h_p = h_i$ in equation (27).

4 VERIFICATION

Verification of the numerical model was performed by comparing against an analytical solution for an infiltration problem with a semi-infinite stratum (Warrick et al. 1985). The domain is a 100 cm deep soil column, with a uniform initial state of $h = -800\,cm$. The soil parameters are $\theta_s = 0.363$, $\theta_r = 0.186$, $\alpha = 0.01\,cm^{-1}$, $\eta = 1.53$, $K_s = 0.0001\,cm/s$ and $h_s = 0\,cm$.

The case was simulated with the 1D, 2D and 3D versions of the model, although the process is one dimensional. Figure 1 shows results of the 1D model, discretized with 100 cells, with several time steps. Results are shown for 4 times (11700, 23400, 35100 and 46800 seconds), as are the values of the analytical solution (WAS). The figure shows that the position of the wetting front is, in general, well described even with very large time steps.

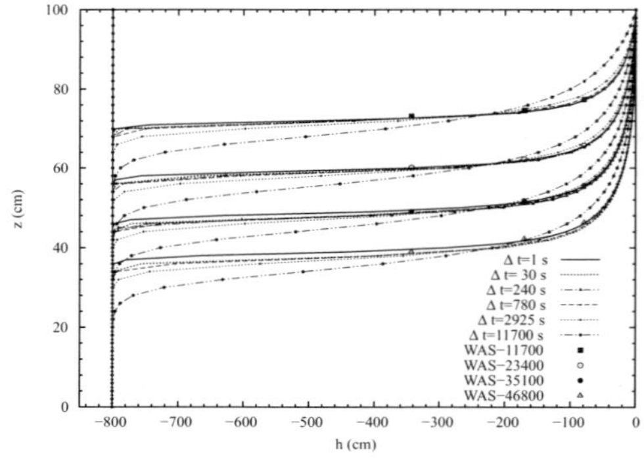

Figure 1. 1D verification results.

As the time step grows, the wetting front rotatess-lightly and is somewhat more diffused. CPU time decreases exponentially as the time step grows, e.g., with $\Delta t = 11700\,s$, CPU time was 0.03% of CPU time with $\Delta t = 1\,s$ (from around 5 hours to around 6 minutes). Figure 2a shows a comparison of results for the best 1D solution (400 cells and $\Delta t = 1\,s$) against a 2D solution with 1553 cells and $\Delta t = 780\,s$ and the analytical solution. The figure shows that the 2D solution is sufficiently accurate, The accelerated advance of the front is due to the larger time step. Further testing shows that coarser meshes result in numerical diffusion of the wetting front, with noticeable acceleration in the dryest region. The wet region is approximated farily well despite coarsening. Numerical diffusion is more significant in the dry region due to the small magnitude and the asymptotic behavior of $K(h)$ in such region. $K(h)$ can be interpreted as a nonlinear *diffusion coefficient* in the Richardsequation. When conductivity is very small the effects of numerical diffusion can be larger than those of

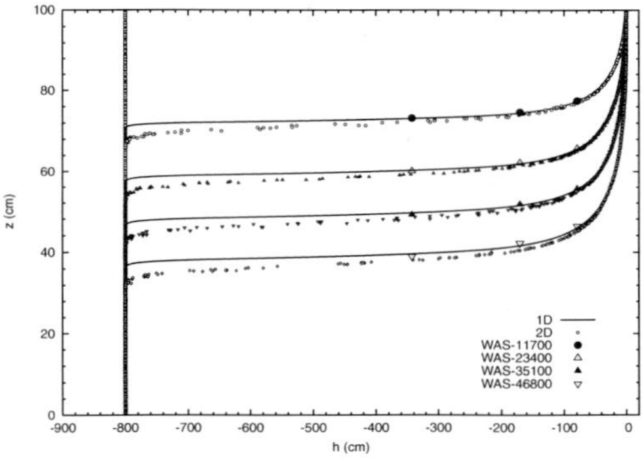

(a) 2D

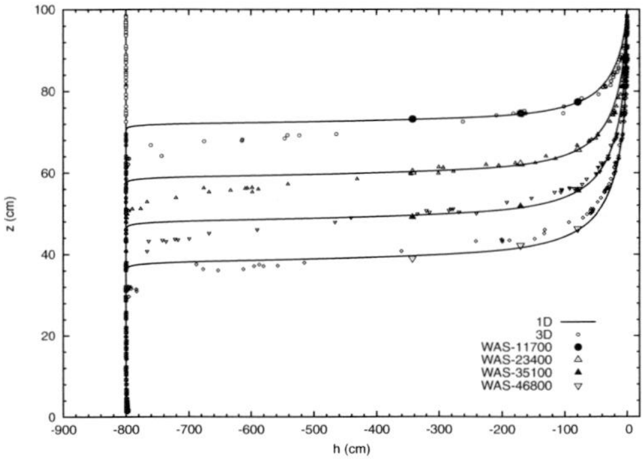

(b) 3D

Figure 2. 1D, 2D and 3D verification.

real diffusion (in fact, hydraulic conductivity). The asymptotic behavior can further amplify the effects of numerical diffusion when studying pressure, since large variations of h result in almost no variation in K in the dry region. Figure 2b shows a comparison of results for the best 1D solution (400 cells and $\Delta t = 1\,s$) against a 3D solution with 479 cells and $\Delta t = 780\,s$ and the analytical solution. Results show that the 3D solution is sufficiently accurate. It shows less tendency to accelerate with large time steps than the 2D solution. In turn, the wetting front rotates, as it does when using larger time steps in the 1D solution. This may be due to the different spatial resolutions of each solution.

Some additional testing was done using this case. The results are not shown for brevity, but a brief comment on the conclusions is in order. Some tests were run using other weighting techniques (see equation 19) with little differences. Because spatial resolution is rather high, this is expected. The way in which the distance d_ω is computed was also tested. Two methods were tested. First, the actual distance between centroids, which has an arbitrary direction depending only on the centroids themselves. Second, the projection of that distance over the normal vector of the wall between both cells, which in theory, is the only direction in which flow actually occurs. Both methods yield very similar results. It is likely that because the meshes were generated to achieve triangles and tetrahedrons approximately regular, or at least with appropriate aspect ratios, the difference between both distances and hence in the results, are small.

5 2D TEST CASE

The initial saturation condition was 0.877 for the entire domain. External and internal boundary conditions were set as impervious, hence all changes are due only to gravity. This test case is interesting as it forces 2D flows with wetting and drying, anda smooth saturation process while simulating only the internal dynamics, without any interference from boundary conditions.

Figure 3 shows the results of the saturation field. The darkest regions are dryer. The black line is a contour which corresponds to a saturation of 0.85. Note that the top dries as the bottom becomes wetter. The most interesting results are the wetting of the regions above the obstacles and the drying of the regions under the obstacles. In particular, the saturation of the small deposit in the lower obstacle and the drying of the triangle of the upper obstacle. The fact that total water flow is larger on the left side than the right side of the domain, which is clearly shown by the contour, is also noteworthy. The velocity field shows

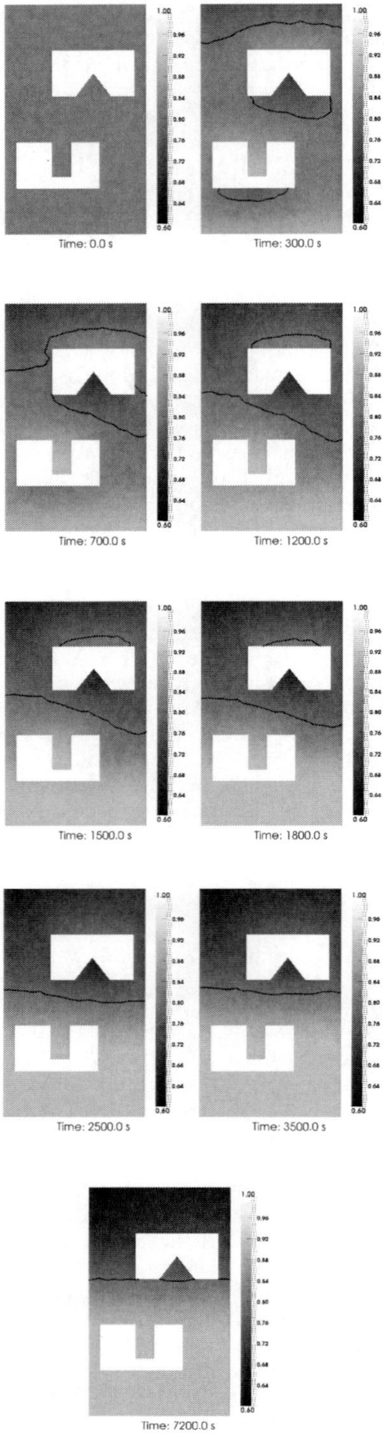

Figure 3. Saturation results for 2D test case.

the same behavior, though not shown here for brevity. Higher velocites are obtained in the narrow space to the left of the lower obstacle. The contour clearly shows a tendency for a large volume of water to flow downwards at the left of the upper obstacle (at 700 *s*). Furthermore, although it flows from that point to the right, a high speed flow is generated to the left of the lower obstacle, allowing for that region to saturate quickly, which is why the contour is tilted as can be seen clearly in times 1200 and 1500. The end of the process is a hydrostatic column with a linear pressure profile, reached after 7200 seconds.

REFERENCES

Baker, D. L. (2006, NOV-DEC). General validity of conductivity means in unsaturated flow models. *Journal of Hydrologic Engineering 11*(6), 526–538.

Celia, M., E. Bouloutas, and R. Zarba (1990, JUL). A General Mass-Conservative Numerical-Solution for the Unsaturated Flow Equation. *Water Resources Research 26*(7), 1483–1496.

Manzini, G. and S. Ferraris (2004). Mass-conservative finite volume methods on 2-d unstructured grids for the Richards' equation. *Advances in Water Resources 27*(12), 1199–1215.

Rathfelder, K. and L. Abriola (1994, SEP). Mass conservative numerical-solutions of the head-based Richards equation. *Water Resources Research 30*(9), 2579–2586.

Schaap, M. and M. van Genuchten (2006, FEB). A modified Mualem-van Genuchten formulation for improved description of the hydraulic conductivity near saturation. *Vadose Zone Journal 5*(1), 27–34.

Vogel, T., M. van Genuchten, and M. Cislerova (2000, NOV). Effect of the shape of the soil hydraulic functions near saturation on variably-saturated flow predictions. *Advances in Water Resources 24*(2), 133–144.

Warrick, A., D. Lomen, and S. Yates (1985). A generalized solution to infiltration. *Soil Science Society of America Journal 49*(1), 34–38.

Numerical Methods for Hyperbolic Equations – Vázquez-Cendón et al. (eds)
© 2013 Taylor & Francis Group, London, ISBN 978-0-415-62150-2

Modelling microbial chemo-tactic waves in saturated porous media using AMR

Samuel A.E.G. Falle
Department of Applied Mathematics, University of Leeds, UK

Nick F. Dudley-Ward
Otago Computational Modelling Group, New Zealand

Mira S. Olson
Department of Civil, Architectural and Environmental Engineering, Drexel University, Philadelphia, USA

ABSTRACT: Microbial transport in groundwater can be strongly influenced by chemo-taxis induced by chemical gradients and this can have a significant effect on the rate of degradation of groundwater contaminants. The main aim of this paper is to describe a method of simulating the propagation of a travelling microbial wave in a contaminated region and the resulting degradation of the contaminant. The presence of the chemo-tactic term and the relatively small bacterial diffusion means that the wave contains a very sharp wavefront (of the order of millimetres). We therefore use an upwind conservative numerical scheme to obtain accurate and numerically stable solutions. The accuracy of the method is verified by comparisons with an exact one dimensional solution of a simplified problem. The method is then used to simulate the propagation of a realistic chemo-tactic wave in one dimension. We then use Adaptive Mesh Refinement (AMR) to compute the propagation of chemo-tactic waves in two dimensions using the simplified model calibrated to give the wave speed as the full model. This technique makes it possible to study the propagation of such waves on scales of the order of kilometres, which typical for contaminated aquifers.

1 INTRODUCTION

Ground-water contaminants often persist in low-permeability regions, in which the flow rate is small. Fortunately, many soil microbes are able to degrade common pollutants and are chemo-tactic, meaning that they can migrate toward chemical concentrations that they find desirable, (Tros et al. 1996; Armitage 1999; Parales et al. 2000; Pandy et al. 2002; Singh & Olson 2008). In fact they can be highly motile: typical microbe swimming speeds are of order 10^{-3}–10^{-2} cm s^{-1}, whereas groundwater flow speeds are in the range 10^{-1}–10^{-4} cm s^{-1}.

Chemo-taxis may increase degradation of ground-water contaminants both by drawing microbes into contaminated regions with low ground-water flow rates, and by enabling them to follow contaminant gradients caused by their own consumption of pollutants (Witt et al. 1999; Olson et al. 2006; Long & Hilpert 2008; Ford & Harvey 2007).

In this work, we consider the movement and growth of microbes in a region in which ground-water flow can be neglected. Specifically, we consider a system of equations that describe the fate and transport of microbes induced by a contaminant gradient (chemo-taxis) and diffusion, and in which their growth and the degradation of the contaminant is described by a Monod kinetics model.

2 GOVERNING EQUATIONS

The system is described by the following equations

$$\frac{\partial c}{\partial t} = D_c \nabla^2 c - \frac{f(b,c)}{Y} \tag{1}$$

$$\frac{\partial b}{\partial t} = -\nabla \cdot (b\mathbf{v}) + D_b \nabla^2 b + f(b,c) - d_b b. \tag{2}$$

Here c is the contaminant concentration, b is the microbe concentration, D_c is the contaminant diffusion coefficient, D_b is the microbe diffusion coefficient, Y is the yield coefficient.

The microbe growth rate is given by Monod kinetics:

$$f(b,c) = \frac{\mu cb}{K_c + c}, \tag{3}$$

where μ is the growth coefficient and K_c is the saturation coefficient. d_b is the microbe death rate.

The chemo-tactic velocity is given by

$$\mathbf{v} = \frac{\chi K_v}{(K_v + c)^2}\nabla c, \tag{4}$$

where χ is the chemo-tactic sensitivity and K_v is the chemo-tactic saturation constant. Note that $\mathbf{v}$ is of the same form as $\nabla(f/b)$.

It is convenient to non-dimensionalise these equations by setting

$$x' = \frac{x}{L}, y' = \frac{y}{L}, t' = \frac{t\chi}{L^2}, c' = \frac{c}{c_0}, b' = \frac{b}{b_0}. \tag{5}$$

Here L is size of region and c_0, b_0 are the initial concentrations.

The equations then become

$$\frac{\partial c'}{\partial t'} = \frac{1}{P_c}\nabla'^2 c' - \frac{D_g}{Y'} f' \tag{6}$$

$$\frac{\partial b'}{\partial t'} = -\nabla'(b'\mathbf{v}') + \frac{1}{P_b}\nabla'^2 b' + D_g f' - D_d b' \tag{7}$$

where

$$f' = \frac{c'b'}{(K_c/c_0 + c')}, \quad Y' = \frac{c_0 Y}{b_0}, \quad \mathbf{v}' = \frac{L}{\chi}\mathbf{v}. \tag{8}$$

The dimensionless parameters are

$$\text{Contaminant Peclet No } P_c = \frac{\chi}{D_c}, \tag{9}$$

$$\text{Microbe Peclet No } P_b = \frac{\chi}{D_b}, \tag{10}$$

$$\text{Growth Damkohler No } D_g = \frac{\mu L^2}{\chi}, \tag{11}$$

$$\text{Death Damkohler No } D_g = \frac{d_b L^2}{\chi}. \tag{12}$$

For $L = 100$ cm with Pseudomomas Putida consuming naphthalene in a porous medium with beads of size 250–300 μm we get

$P_c = 0.89$, $P_b = 8.9\times10^3$
$D_g = 1.3\times10^6$, $D_d = 2.1\times10^4$

This tells us that microbe diffusion is negligible compared to contaminant diffusion and microbe death is much less important than microbe growth.

3 SHOCKS

In one dimension, equations (6) and (7) reduce to

$$\frac{\partial c}{\partial t} = -\frac{D_g}{Y} f, \tag{13}$$

$$\frac{\partial b}{\partial t} = -\frac{\partial(v_x b)}{\partial x} + D_g f \tag{14}$$

if we neglect diffusion (note that we have suppressed the primes).

Differentiate (13) to get

$$\frac{\partial}{\partial x}\frac{\partial c}{\partial t} = \frac{\partial c_x}{\partial t} = -\frac{D_g}{Y}\frac{\partial f}{\partial x} \quad \left(c_x \equiv \frac{\partial c}{\partial x}\right). \tag{15}$$

(14) (without the source term) and (15) constitute a non-linear hyperbolic system whose wave speeds are

$$\lambda_\pm = \frac{1}{2}[v_x \pm \sqrt{(v_x^2 + 4v_x f/c_x)}] \tag{16}$$

The shock relations are

$$s(b_l - b_r) = v_{xl}b_l - v_{xr}b_r, \tag{17}$$

$$s(c_{xl} - c_{xr}) = \frac{D_g}{Y}(f_l - f_r). \tag{18}$$

It can be seen that b and c_x can become discontinuous, whereas c must remain continuous.

4 SIMPLE SOLUTION FOR A CHEMO-TACTIC WAVE

Equations (13) and (14) admit chemo-tactic waves, which are analogous to detonation waves. On dimensional grounds we must have

Speed of wave $\propto D_g^{1/2}$.
Thickness of wave $\propto D_g^{-1/2}$.

Since D_g is large, this means that the wave are thin.

If we assume that

$$f = bc, \quad v_x = \frac{\partial c}{\partial x}, \tag{19}$$

then we can find a simple solution for steady travelling chemo-tactic wave.

Let $\xi = x - st$, where s is the speed of the wave. The solution consists of a shock travelling with speed s with $b = 0$ for $\xi > 0$, $b = Y$ for $\xi < 0$. c is given by

$$c = 1 \quad \text{for } \xi > 0,$$

$$c = \exp\left(\frac{D_g}{s}\xi\right) \qquad \text{for } \xi < 0.$$

The speed of the wave is $s = D_g^{1/2}$ and its thickness is $1/D_g^{1/2}$, as expected.

5 NUMERICAL SCHEME

Despite the apparent simplicity of the system, the Riemann problem is quite complicated. However, since there are only two waves, it is eminently suited to an HLL scheme. The diffusive fluxes can be calculated with a with central difference, as usual. However, there remains the problem of calculating ∇c for the chemo-tactic term. There are two obvious possibilities:

Method 1: Get ∇c by numerical difference of c. This works well enough if the waves are well resolved, but it is not properly upwind. However, it is simple and introduces no extra variables.

Method 2: Take the gradient of the c equation to get

$$\frac{\partial \nabla c}{\partial t} = \frac{1}{P_c}\nabla^2 \nabla c - \frac{D_g}{Y}\nabla f$$

Now define new variables ∇c and solve this equation. Although this gives better shock capturing, it has more variables and the numerical solutions for c and ∇c are not automatically compatible.

We therefore opt for method 1.

Figure 1 shows a comparison between the numerical and exact solutions for $D_g = 1.0$. It is clear that the agreement is very good except for a slight lack of monotonicity at the shock, which is a consequence of opting for method 1. The numerical wave speed is 0.994, which is is very close to the exact wave speed of 1.

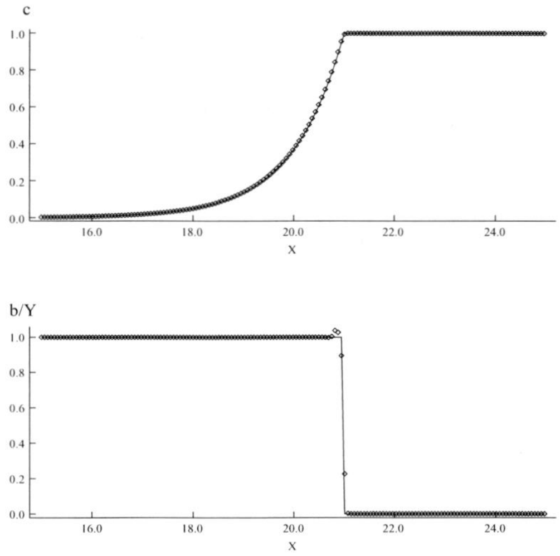

Figure 1. Simple one dimensional chemo-tactic wave. The line is the exact solution, the markers from the numerical calculation.

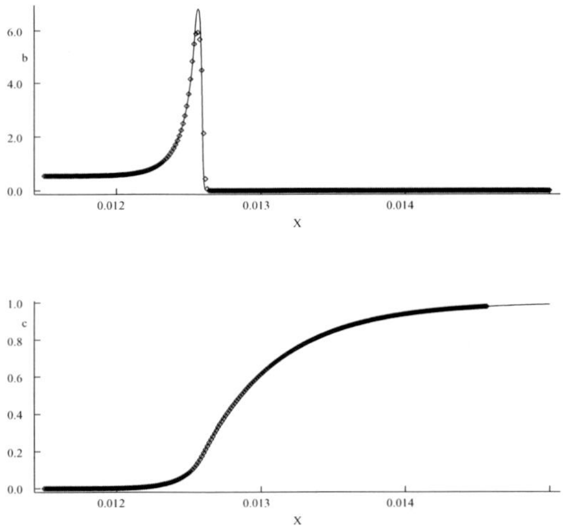

Figure 2. High (line) and low (markers) numerical solution for a realistic one dimensional chemo-tactic wave.

6 REALISTIC ONE DIMENSIONAL CHEMO-TACTIC WAVE

Figure 2 shows the numerical solution for a wave with the parameters discussed in section 2, except that microbial death is neglected since it has no effect on the wave. Note that $Y = 0.538$. The lower resolution run (markers) had $\Delta x = 12.8 \times 10^{-6}$ and gave a wave speed of 2.0728×10^3, whereas the higher resolution run (line) had $\Delta x = 6.4 \times 10^{-6}$

and a wave speed of 2.0732×10^3. These were computed using a cell-by-cell AMR code (Falle 2005) with 4 grid levels for the lower resolution calculation and 5 levels for the higher one. As one would expect, the large Peclet number for microbial diffusion means that the shock structure is very thin and is only just resolved in the higher resolution calculation. The entire wave is also very thin because of the large Damköhler number for microbial growth.

If one translates this back into physical units, then the wave thickness is $\simeq 0.05$ cm and its speed is 2.76×10^{-5} cm s^{-1}. Note that the speed is significantly smaller than the microbe swimming speed of $= 4.8 \times 10^{-3}$ cm s^{-1}.

7 REALISTIC WAVE IN A PETRI DISH

Having established that the numerical method works well in one dimension, we now consider a circular wave, such as the ones that are observed in laboratory experiments in Petri dishes. The initial conditions were $c = 1$, $b = Y$ in $0 \le r \le 0.001$, 0 elsewhere. The size of the domain is appropriate for a 3 inch Petri dish. This was an AMR calculation with 6 grid levels giving with a minimum mesh spacing of 12.8×10^{-6}.

Figure 3 shows the solution when the wave has almost reached the boundaries of the domain. It is clear that the wave is very close to circular and very thin, so thin in fact that the finest resolution is smaller than the pixel size. Figure 4 shows that the finest grid is confined to the wave itself, which

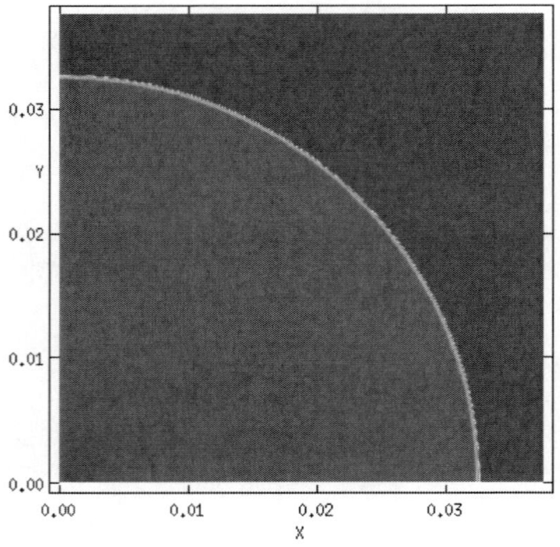

Figure 3. Microbial cell density for a realistic circular chemo-tactic wave.

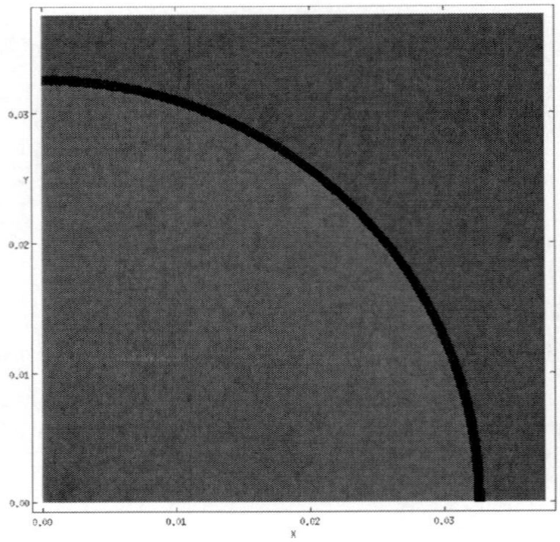

Figure 4. Finest grid for the circular chemo-tactic wave.

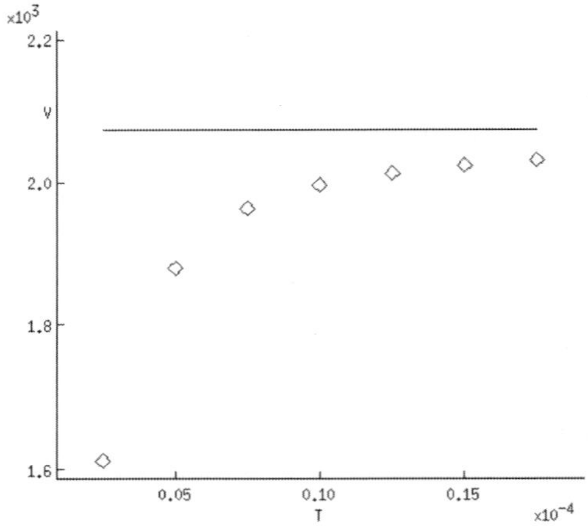

Figure 5. Numerical wave speed for the circular chemo-tactic wave (markers). The line shows the one dimensional speed.

means that the calculation is extremely efficient. It would clearly be extremely expensive to carry out such a calculation with a uniform grid.

Figure 5 shows the wave speed as a function of time. It is clear that the wave takes a while to accelerate and even at later times, there is a curvature effect. Nevertheless, the final speed is quite close to the one dimensional value.

8 GROUNDWATER POLLUTION

It can be seen that it is quite difficult to model a wave even on the scale of a Petri dish. In groundwater problems, the polluted region may have a scale of of many kilometres, which means that it is quite impossible to model the real wave, even with AMR.

Since the wave is one-dimensional, one could use a one dimensional calculation to get the wave speed and then use a level set method. The alternative is to use the simple model with an increased wave thickness, combined with AMR.

To do this, all we have to do is to multiply the chemo-tactic term by a factor, α. We then have

$$\text{Wave speed} = (\alpha D_g)^{1/2}, \tag{20}$$

$$\text{Wave thickness} = (\alpha / D_g)^{1/2}. \tag{21}$$

Now choose α, D_g to get desired thickness and correct wave speed. It is convenient to kill the microbes when $c < 0.001$, since this make no difference to the wave speed, but avoids refinement in regions that are of no interest.

Figure 6 shows how this works for a circular contaminated region. α and D_g were chosen to give the correct wave speed and a wave thickness of 0.02 i.e about 40 times larger than the real thickness. The initial conditions were

$$b = 0,\ c = 1 \text{ for } 0 \le r \le 1.0,$$
$$b = 1,\ c = 0 \text{ for } 1.0 < r \le 1.2.$$

This produces a circular wave that moves inwards, starting from $r = 1$. It can be seen from the figure that at this stage the wave is somewhat distorted from a circular shape, which indicates that this resolution is indequate once the radius of the wave has become sufficiently small. Figure 7 shows that one can improve on this by increasing the resolution. Figure 8 shows that one also needs more resolution to get the wave speed correct.

In an actual application to groundwater, one would want to increase the thickness of the wave by a much larger factor than 40. The only restriction is that the thickness of the wave should be

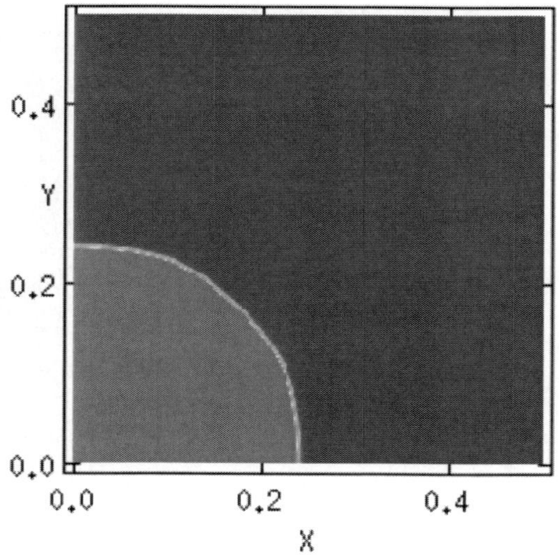

Figure 6. Contaminant concentration for the modified simple model with ≈8 cells in a wave thickness.

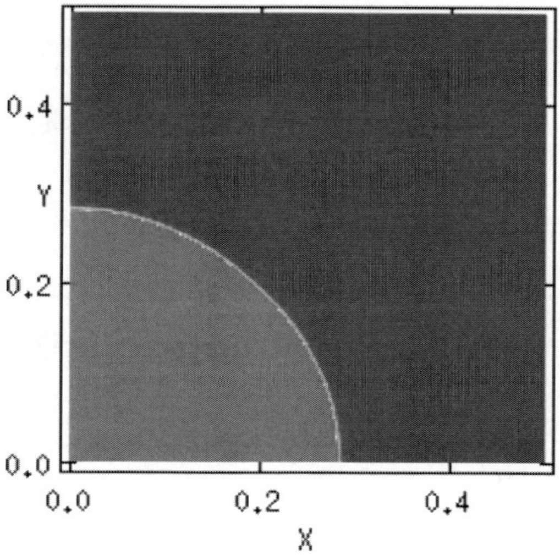

Figure 7. Contaminant concentration for the modified simple model with ≈32 cells in a wave thickness.

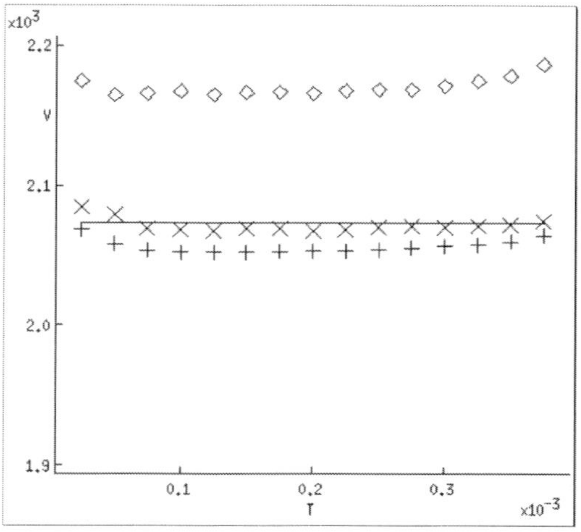

Figure 8. Wave speed as a function of time varous resolutions for the modified simple model. ◇8 cells, +16 cells, ×32 cells.

sufficiently small compared to the scale of the polluted region for curvature effects to be neglible.

9 CONCLUSIONS

We have shown that it is possible to construct an efficient upwind scheme to model chemo-taxis, but that even chemo-tactic waves on the scale of Petri dishes require very high resolution. AMR is an effective way of doing this, but even so, it is not possible to model the real wave in polluted groundwater. However, it is possible to construct a simple model that propagates the wave at the correct speed. In this simple model the wave thickness is artificially increased, but the use of AMR makes it possible to keep its thickness small compared to the overall scale of the polluted region.

REFERENCES

Armitage, J.P., 1999. Bacterial tactic responses. Advances in Microbial Physiology. 41, 229–289.

Falle, S.A.E.G., 2005. AMR applied to non-linear elastodynamics, in: Plewa T; Linde T., Weirs V.G.J. (Eds), Adaptive Mesh Refinement—Theory and Applications, Lecture Notes in Computational Science and Engineering, Springer: pp 235–253.

Ford, R.M. & Harvey, R.W., 2007. Role of chemotaxis in the transport of bacteria through saturated porous media, Advances in Water Resources 30 (6–7), 1608–1617.

Long, W., Hilpert, M., 2008. Lattice-Boltzmann modeling of contaminant degradation by chemotactic bacteria: exploring the formation and movement of bacterial bands. Water Resources Research. 44, doi:10.1029/2007/WR006129.

Olson, M.S., Ford, R.M., Smith, J.A., Fernandez, E.J., 2006. Mathematical modeling of chemotactic bacterial transport through a two-dimensional heterogeneous porous medium, Bioremediation Journal. 10, 13–23.

Pandy, G. & Jain, R.K., 2002. Bacterial chemotaxis toward Environmental Pollutants: Role in Bioremediation, Appl. Environ. Microbiol. 68(12), 5789–5795.

Parales, R.E., Ditty, J.L. & Harwood, C.S., 2000. Toluene-degrading bacteria are chemotactic towards the environmental pollutants benzene, toluene, and trichloroethylene, Applied and Environmental Microbiology. 66(9), 4098–4104.

Singh, R. & Olson, M.S., 2008. Application of bacterial swimming and chemotaxis for enhanced bioremediation, Shah V. (Ed), Emerging Environmental Technologies, Springer. pp. 149–172.

Tros, M.E., Schraa, G. & Zhender, A.J.B., 1996. Transformation of low concentrations of 3-chlorobenzoate by Pseudomonas sp. strain B13: kinetics and residual concentrations, Applied and Environmental Microbiology. 62(2), 437–442.

Witt, M.E., Dybas, M.J., Worden, R.M. and Criddle, C.S., 1999. Motility-Enhanced Bioremediation of Carbon Tetrachloride-Contaminated Aquifer Sediments, Environ. Sci. Technol. 33(17), 2958–2964.

Numerical Methods for Hyperbolic Equations – Vázquez-Cendón et al. (eds)
© 2013 Taylor & Francis Group, London, ISBN 978-0-415-62150-2

Mathematical modeling of heat transfer during quenching process

Diego N. Passarella & Fernando Varas
Departamento de Matemática Aplicada II, E.T.S. de Ing. de Telecomunicación, Universidad de Vigo, España

Elena B. Martín
Área de Mecánica de Fluidos, E.T.S. de Ing. Industriales, Universidad de Vigo, España

ABSTRACT: In quenching by submerging the workpiece is cooled due to vaporization, convective flow and interaction of both mechanisms. These phenomena are very complex and the corresponding heat fluxes are strongly dependent on local flow variables such as velocity and vapor fraction. In order to obtain an accurate description of cooling during quenching, a heat transfer model that takes into account the thermo-fluid dynamics of the quenching bath is presented. The model is based on the drift-flux mixture-model for multiphase flows, and also includes specific boundary conditions for the heat transfer mechanisms. To test the model, a simple flow condition is analyzed. Generation and collapse of vapor blanket, as well as dependence on flow and vapor variables is observed. These results highlight the necessity of an appropriate description of the process and its dynamics in order to reproduce the complex heat transfer rates that occur during the it.

1 INTRODUCTION

Quenching is a technological process that involves cooling of a heated workpiece in order to improve its mechanical properties. Undesired effects can also be developed during it, such as geometrical distortions and/or generation of residual stresses. Inorder to optimize the outcome of the process it is necessary to tailor the cooling rates that are produced.

As any other process of technological interest, several efforts were made in order to describe it through mathematical models. In this area it is important to remark that quenching of steels can be divided in at least three sub-problems. One major problem is the modeling of metallurgical transformations. This part is well known and the available models are robust and reliable (Porter 1992). On another side we have the generation of residual stresses and geometrical distortions. This complex phenomena are also well covered and refined models can be found in specialized literature (Denis 1992, Kang 2007, Sugianto 2009 & Vorster 2009). Finally, these two sub-problems strongly rely on an accurate description of the temperature evolution inside the piece. Several attempts to put together a thorough description of the quenching process can be found in the cited literature. In general, metallurgical and mechanical problems are solved using the most adequate model but, the thermal problem is usually solved taken a very rough description of heat exchange with the fluid through the boundaries conditions (Kang 2007, Sugianto 2009 & Vorster 2009). It is important to remark that all the complexity and accuracy of each sub-model is lost if the thermal problem is over-simplified. In this work the analysis of heat transfer in the workpiece-fluid interface is tackled. The microstructure and residual stresses problems will be coupled in further works.

2 MODEL

The analysis of multiphase flows presented here is based on the general methodology presented in (Clift 1978, Crowe 1998, Ishii 2006). Before the definition of the particular multiphase model it is necessary to compare characteristic times, coupling parameters between phases and finally assess the characteristic non-dimensional numbers of the specific problem. This methodology is presented in following subsections.

Some definitions along the text are: v, l, m and b sub-indexes refer to vapor, liquid, mixture and bubble respectively. α_i is the volume (or time) fraction of the ith-phase, while $c_i = \alpha_i\rho_i/\rho_m$ is the mass fraction. The mixture density is weighted by volume fraction ($\rho_m = \alpha_v\rho_v + \alpha_l\rho_l$), while mixture velocity and enthalpy are mass weighted. Properties with $=$ and ^ symbols correspond to phase-averaged and density-averaged values respectively. This nomenclature and averaging procedures are in accordance to (Ishii 2006).

2.1 *Characteristic times and phase coupling*

The idea behind the study of the characteristic times is to identify if there is some quantity that can be considered in equilibrium compared to the characteristic evolution time of the flow. In Table 1 the characteristic times and their range values are presented. The mass transfer process is the latent heat of a bubble divided by the heat transfer rate through its surface, while the energy transfer considers the sensible heat as it was developed in (Passarella 2010). The characteristic moment transfer time was taken from (Crowe 1998) and considers the drag coefficient of a bubble (C_D) and its corresponding Reynolds number ($Re_b = \rho_l D_b U_0/\mu_l$). In Table 1 $L_0 \in (0.1,1)$ m and $U_0 \in (0.1,1)$ m/s, these ranges were taken to try to cover industrial applications as well as laboratory experiments. h_b is the bubble heat transfer coefficient based on the Ranz-Marshall correlation for the Nusselt number. In this analysis it is clear that mass variations occurs in a time scale similar to the one of the flow, while moment and energy variations are much more faster.

The analysis of coupling parameters made in Table 2 show us that vapor and liquid phases are coupled by the transfer of mass and latent energy released, while the effects of transfer of momentum and sensible heat are negligible.

The results of Tables 1 & 2 dictate that for the characteristic time of the problem, both phases are in mechanical equilibrium and are coupled through the exchange of mass and its corresponding transformation latent heat. The mechanical equilibrium dictates that it is not necessary to solve the momentum equations for both phases, just one momentum equation for the whole mixture and a kinematic relationship of both phase velocities are enough. Based on these conclusions, our physical problem can be modeled as a mixture with mass transfer between phases plus an equation for conservation of energy. Regarding the thermal modeling of the problem, in Table 1 it is observed that the estimated time to release sensible heat of the vapor phase is very short, this effect can be assumed as a fast equalization of vapor temperature to its saturation temperature. Due to this result, we assume that the vapor phase is in saturation condition during the process, therefore just the energy equation of the liquid needs to be solved. This approach is commonly adopted in the resolution of similar thermal-multiphase problems (Koncar 2008, Koncar 2010a, b & Krepper 2007).

Table 1. Characteristic times [s].

Flow	
$\tau_{flow} = L_0/U_0$	$(10^{-1}, 10)$
Mass transfer	
$\tau_m = Q_L/\dot{Q} = (V_b\, \rho_v\, h_{lv})/(A_b\, h_b\, \Delta T)$	$(10^{-2}, 1)$
Moment transfer	
$\tau_v = (4\rho_v D_b)/(3\, \mu_l C_D Re_b)$	$(10^{-5}, 10^{-3})$
Energy transfer	
$\tau_T = Q_S/\dot{Q} = (V_b\, \rho_v\, c_{p,v}\, \Delta T)/(A_b\, h_b\, \Delta T)$	$(10^{-4}, 10^{-3})$

Table 2. Coupling parameters.

Mass	$\Pi_m = c_v \tau_{flow}/\tau_m$	$(10^{-1}, 10)$
Moment	$\Pi_v = c_v/(1 + \tau_v/\tau_{flow})$	$(10^{-5}, 10^{-1})$
Energy	$\Pi_T = c_v/(1 + \tau_T/\tau_{flow})$	$(10^{-5}, 10^{-1})$
Energy (latent)	$\Pi_L = \Pi_m h_{lv}/(c_{p,l} T_l)$	$(10^{-2}, 10)$

2.2 *Non-dimensional equations and characteristic numbers*

The analysis of the characteristic numbers from the non-dimensional conservation equations is necessary in order to assess the source terms that have to be retained. The non-dimensional system of partial differential equations (pde's) is:

$$\frac{\partial \rho_m^*}{\partial t^*} + \nabla^* \cdot \left(\boldsymbol{v}_m^* \rho_m^* \right) = 0 \tag{1}$$

$$\frac{\partial \alpha_v \rho_v^*}{\partial t^*} + \nabla^* \cdot (\alpha_v \rho_v^* \boldsymbol{v}_m^*) = N_{pch} \Gamma_v^* - N_D \nabla^* \cdot \left(\alpha_v \frac{\rho_l^* \rho_v^*}{\rho_m^*} \boldsymbol{V}_{vj}^* \right) \tag{2}$$

$$\frac{\partial \rho_m^* \boldsymbol{v}_m^*}{\partial t^*} + \nabla^* \cdot \left(\rho_m^* \boldsymbol{v}_m^* \boldsymbol{v}_m^* \right) = -\nabla^* p_m^* + \frac{1}{Re} \nabla^* \cdot (\overline{T} + T^T)^* + \frac{1}{Fr} \rho_m^* \frac{\boldsymbol{g}}{|\boldsymbol{g}|} - N_\rho N_D^2 \nabla^* \cdot \left(\frac{\alpha_v}{1-\alpha_v} \frac{\rho_l^* \rho_v^*}{\rho_m^*} \boldsymbol{V}_{vj}^* \boldsymbol{V}_{vj}^* \right) + \boldsymbol{\Phi}_{int} \tag{3}$$

$$\frac{\partial \rho_m^* h_m^*}{\partial t^*} + \nabla^* \cdot (\rho_m^* h_m^* \boldsymbol{v}_m^*) = -\frac{1}{Pe} \nabla^* \cdot (\overline{\boldsymbol{q}} + \boldsymbol{q}^T)^* - \frac{N_\rho}{(1-\alpha_v) + \alpha_v N_\rho} N_D \nabla^* \cdot \left[\alpha_v \frac{\rho_l^* \rho_v^*}{\rho_m^*} \boldsymbol{V}_{vj}^* h_{lv}^* \right] + Ec \left[\frac{D p_m^*}{D t^*} + \boldsymbol{\Phi}_{int} \right] \tag{4}$$

where * are non-dimensional quantities. The relevance of each source term depends on the relative value of the factor that multiplies it. Characteristic values and the definition of each number is presented in Table 3. It is observed that terms multiplied by $1/Pe$ and Ec can be neglected while

Table 3. Characteristic numbers.

Phase change: $N_{pch} \equiv (\Gamma_{v0} L_0)/(\rho_{v0} U_0)$	$(1, 10^2)$
Drift: $N_D \equiv (\rho_{l0} \mathbf{V}_{vj0})/(\rho_{m0} U_0)$	$(10^{-1}, 1)$
Density: $N_\rho \equiv (\rho_{v0})/(\rho_{l0})$	(10^{-3})
Reynolds: $Re \equiv (\rho_{m0} U_0 L_0)/(\mu_{m0})$	$(10^2, 10^4)$
Froude: $Fr \equiv (U_0^2)/(\|g\| L_0)$	$(10^{-3}, 1)$
Peclet: $Pe \equiv (\rho_{m0} U_0 h_{lv0} L_0)/(k_{m0} \Delta T_0)$	$(10^3, 10^5)$
Eckert: $Ec \equiv (U_0^2)/(h_{lv0})$	$(10^{-9}, 10^{-7})$

drift (i.e. the drag effect caused by the difference of velocity between phases, denoted by $\mathbf{V}_{vj}$, and whose relevance is given by the non-dimensional number N_D) and phase change are relevant. The $\mathbf{\Phi}_{int}$ term corresponds to interfacial terms and in the momentum equation can be neglected due to the relative mass of the interface compared to the bulk. Finally, the chosen kinematic relationship between phase velocities is the so called drift-velocity, whose general form is $\mathbf{V}_{vj} = \alpha_l(\hat{\mathbf{v}}_v - \hat{\mathbf{v}}_l)$ (Ishii 2006).

2.3 *Conservation equations*

Based on the results of previous section we define the pde's (in dimensional form) that describe our problem as:

$$\frac{\partial \rho_m}{\partial t} + \nabla \cdot (\mathbf{v}_m \rho_m) = 0 \tag{5}$$

$$\frac{\partial \alpha_v \overline{\overline{\rho_v}}}{\partial t} + \nabla \cdot (\alpha_v \overline{\rho_v} v_m) = \Gamma_v - \nabla \cdot \left(\alpha_v \frac{\overline{\overline{\rho_l \rho_v}}}{\rho_m} \mathbf{V}_{vj} \right) \tag{6}$$

$$\frac{\partial \rho_m v_m}{\partial t} + \nabla \cdot (\rho_m v_m \mathbf{v}_m) = -\nabla p_m + \rho_m \mathbf{g} + \nabla \cdot \left[\overline{T} + T^T - \nabla \cdot \left(\frac{\alpha_v}{\alpha_l} \frac{\overline{\overline{\rho_l \rho_v}}}{\rho_m} \mathbf{V}_{vj} \mathbf{V}_{vj} \right) \right] \tag{7}$$

$$\frac{\partial \alpha_l \overline{\overline{\rho_l}} c_{p,l} T_l}{\partial t} + \nabla \cdot (\alpha_l \overline{\overline{\rho_l}} c_{p,l} T_l \hat{\mathbf{v}}_l) = -\nabla \cdot \alpha_l (\overline{\overline{\mathbf{q}_l}} + \mathbf{q}_l^T) - \Gamma_v c_{p,l} (T_{sat} - T_l) \tag{8}$$

This set is the so called drift-flux mixture model. It is based on the description of the multiphase flow through the mixture properties and assumes a drift velocity between phases ($\mathbf{V}_{vj}$). The model is completed with a turbulence model, in our case standard $k - \in$ was chosen for T^T, closure functions for $\mathbf{V}_{vj}$ and Γ_v, and appropriate boundary conditions (bc's). It is important to note that in Equation 8 we use $h_l = c_{p,l} T_l$ and the hyperbolicity of Equation 6 is moderated through the introduction of a diffusive term in the formulation of drift velocity ($\mathbf{V}_{vj} = \sqrt{2}(\sigma g \Delta\rho / \rho_l^2)^{1/4} \alpha_l^2 \mathbf{g}/|\mathbf{g}| - D_d(\nabla \alpha_v)/\alpha_v$) based on (Ishii 2006 & Manninen 1996). This mechanism is predominant for vapor injection along a boundary oriented parallel to gravity.

Cooling of the workpiece is modeled using an extra conservation equation for the thermal energy of the solid. The multiphase model of the quenching bath and this extra equation are coupled through the boundary conditions.

2.4 *Boundary conditions*

For the solid-mixture interface, a shear stress bc is applied to the momentum equation (Equation 9). In this equation, a rough wall due to the presence of bubbles attached to the wall is considered by taking a $u^+_{Rough}(y^+, \alpha_v)$ law in a similar way to (Koncar 2010a, b).

$$\tau_w = (\mu_m + \mu_m^T) \frac{\partial v_m}{\partial n} = \rho_m \left(\frac{v_m}{u^+_{Rough}(y^+, \alpha_v)} \right)^2 \tag{9}$$

For the heating of the liquid, a heat flux bc is applied (Equation 10). The heat flux is divided in two parts, a convective term and a radiative one. The conductive law (Equation 11) also includes the effect of roughness, while the radiative term is similar to the one used in (Bromley 1953).

$$q_w^l = -\alpha_l k_l \frac{\partial T_l}{\partial n} = q_{conv} + q_{rad} \tag{10}$$

$$q_{conv} = \frac{\Delta T \rho_m Cp_m \left(v_m / u^+_{Rough}(y^+, \alpha_v) \right)}{T^+_{Rough}(y^+, \alpha_v)} \tag{11}$$

The cooling of the workpiece is also modeled by a heat flux condition (Equation 12). This heat flux contains two big contributions, one part devoted to heating the liquid and another responsible for the evaporation of it. The evaporation part can be divided into two terms, one in charge of the proper phase change ($\tilde{q}_{ev}$) and other necessary to heat the liquid to be evaporated up to saturation temperature (q_{q*}). This approach is inspired by the heat partition model used in (Kocar 2009, Koncar 2008, Koncar 2010a, b & Krepper 2007), but has some relevant differences. In particular, the definition of convective heat flux where local effects of fluid flow are taken into account and the definition of q_{q*} which, up to our knowledge, has not been proposed before.

$$q_w^s = -k_s \frac{\partial T_s}{\partial n} = -(q_{conv} + q_{rad} + \tilde{q}_{ev} + q_{q*}) \tag{12}$$

$$\widetilde{q_{ev}} = q_{ev}(l,v,T_w) \cdot f_L(\alpha_l) \quad (13)$$

$$q_{q*} = \dot{m}_v Cp_l \rho_l \Delta T \quad (14)$$

The evaporation term in Equation 13 contains a phase change function that is mainly dependent on the quenching bath properties and wall temperature ($q_{ev} = f_b N_a \pi/6 D_d^3 \rho_v h_{lv} = f_1(l,v) \cdot (T_w - T_{sat})^{f_2(l,v)}$) which is the same functional form if D_d, f_b and N_a are taken from (Kocamustafaogullari 1995, Krepper 2007, Li 2009) and a damping function (f_L) that reduces the evaporation process if the piece is covered by vapor ($f_L \to 1$ as $\alpha_v \to 0$, and $f_L \to 0$ as $\alpha_v \to 1$) with a steep change at $\alpha_v \approx \pi/4$ if a wall covered by semi-spherical bubbles is considered, function similar to the one presented in (Koncar 2008).

Finally, the vapor injection is controlled by the mass flux of vapor that results from the corresponding phase change function (Equation 15). It is important to note that the cooling and mass injection due to evaporation is moderated by the presence of vapor through the f_L function, and the vapor fraction depends on the resolution of the edp system presented in Section 2.3. By these means, the proposed heat transfer model is fully coupled with the resolution of the multiphase problem.

$$\alpha_v v_v \cdot n = \frac{\dot{m}_v}{\rho_v} = \frac{\widetilde{q_{ev}}}{h_{lv}\rho_v} \quad (15)$$

Other boundary conditions such as inlets, outlets, symmetries, etc. are taken in the typical way, therefore they are not detailed here.

3 RESULTS

3.1 *Case studied*

The cooling of an Inconel cylinder is proposed to test the model. The configuration of a 2-D axisymmetric cylinder supported by a thin test probe and immersed in a forced convection quenching bath is presented in Figure 1. The symmetry axis is at the left hand side, the inlet is from the bottom of the domain and the outlet from the top. This type of configuration is adopted because it resembles the typical standard test to measure the cooling velocity of quenching media (Houghton 2006).

The properties of the liquid, vapor and workpiece are summarized in Table 4. In addition to these data, the following parameters are considered: T_{sat} = 490 [K], σ = 3 × 10^{-2} [N/m], D_b = 10^{-3} [m] & $h_{lv} = 10^6$ [J/kg]. It is important to note that we assume that the oil behaves similar to a pure liquid, with properties such as saturation temperature taking one specific value, while this is not true since there is a range of temperatures where phase change occurs. It is expected that this source of discrepancy between the model and the physical reality do not play a mayor role.

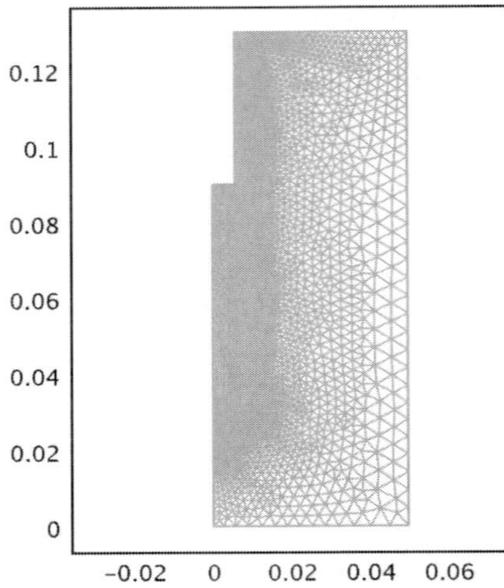

Figure 1. Domain and mesh.

Table 4. Properties of quenching liquid, its vapor and workpiece.

	Liquid		Vapor	Piece	
ρ	870		3	8470	
μ	0.094	(293 K)	2.3 × 10^{-5}	–	
	0.0235	(313 K)			
	0.0043	(373 K)			
k	0.14		0.014	14.8	(294 K)
				22.8	(811 K)
				28.8	(1144 K)
c_p	1800	(290 K)	3000	445	(294 K)
	2300	(420 K)		555	(811 K)
				625	(1144 K)

The numerical resolution was made using the Finite Element Method (FEM) and implemented in *Comsol Multiphysics* ®software. The mesh consists of 10853 triangular elements and 102385 degrees of freedom. Momentum, turbulence, transport of vapor, and both thermal problems were segregated. Lagrange P_2/P_1 elements were used for Equation (5) & (7), P_1 for Equation (6), P_2 for Equation (8), turbulence and energy of solid problems. The advective terms were numerically stabilized using the Streamline Upwind Petrov Galerkin (SUPG) scheme. The evolution of cooling during 60 s was solved using a Backward Differentiate Formula (BDF) of order 1 with adaptive time step. The resolution time took approximately 2 h in a 4-core processor at 2.66 GHz with 8 Gb of memory. Most of this time was consumed in the resolution of the generation and the subsequent collapse of the vapor blanket.

3.2 *Vapor and temperature fields*

Snapshots of workpiece temperature and vapor fraction distributions are presented in Figures 2 to 4. It can be observed that a blanket of high vapor fraction covers the workpiece and its test probe in very short time (0.1 s).

At 2.5 s (Fig. 3) it can be seen that the test probe and leading edge of the cylinder are getting cold and at the same time the vapor blanket start to collapse at these points.

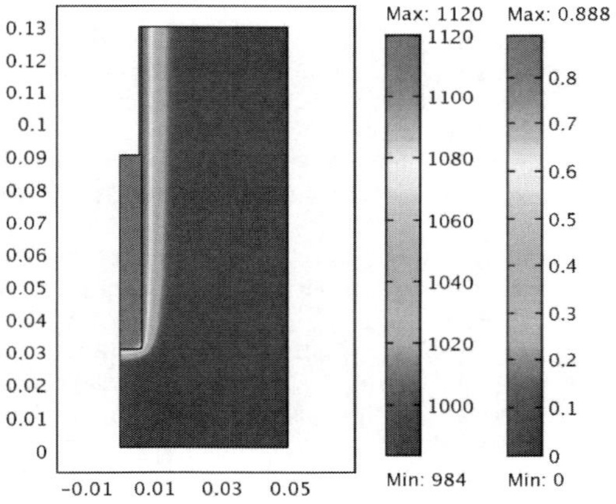

Figure 2. Workpiece temperature and vapor fraction at t = 0.1 s.

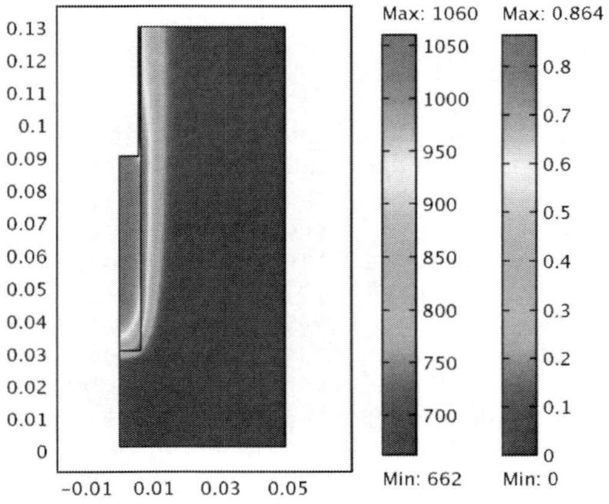

Figure 3. Workpiece temperature and vapor fraction at 2.5 s.

Finally, at 6.0 s (Fig. 4) a clear spatial correlation between the presence of the vapor blanket and the hottest zones of the workpiece's surface is observed.

3.3 *Heat flux evolution, its partition and cooling rates*

In order to study the effect of vapor blanket on heat transfer, the damping function f_L along the workpiece's surface (left: stagnation point, right: test probe) and at different times is plotted in Figure 5. It is observed that the damping parameter starts fairly constant at 0.1 s and then evolves as the vapor blanket collapses.

The combination of this effect with the transition between different heat transfer mechanisms can be seen in the spatial and temporal variation

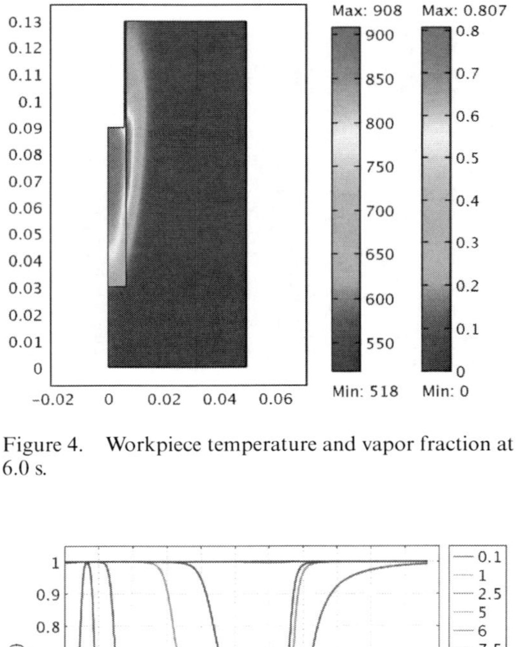

Figure 4. Workpiece temperature and vapor fraction at 6.0 s.

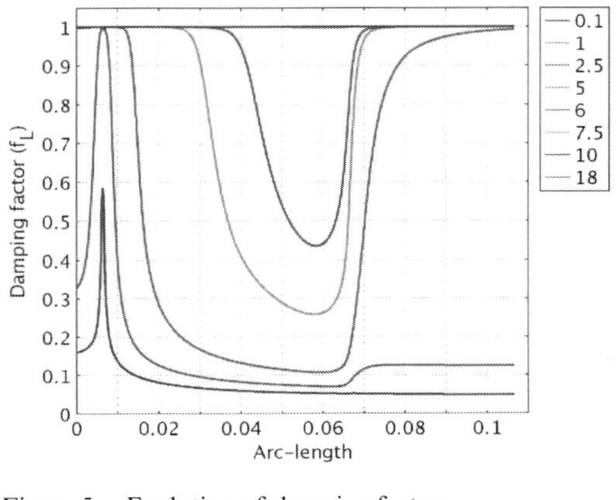

Figure 5. Evolution of damping factor.

that suffers the so called heat transfer coefficient $\left(h = q_w^s/(T_w - T_l)\right)$ seen from the solid side (Fig. 6). This variation is certainly hard to be captured by a regular correlation based on the Nusselt number, as it is usually performed in the literature, and this is the core advantage of this type of modeling. It is expected that for more complex geometries the benefits obtained from this approach would help to overcome problems in steel quenching such as generation of residual stresses, distribution of metallurgical phases, etc. and can also be applied to other engineering problems related to heat transfer to an evaporable liquid.

A detail of heat flux partition is presented in Figure 7. There it can be seen how the four contributions are combined to give the total heat flux that the solid releases during cooling. From 0 to approximately 6 s the damped evaporation heat flux is dominant, the damping function is disabled from then on and the evaporation keeps dominant up to 15 s. A mixture mode combining evaporation and convection runs up to 35 s approximately and from then on, single phase convection continues. It is also relevant to note that radiative heat transfer is negligible.

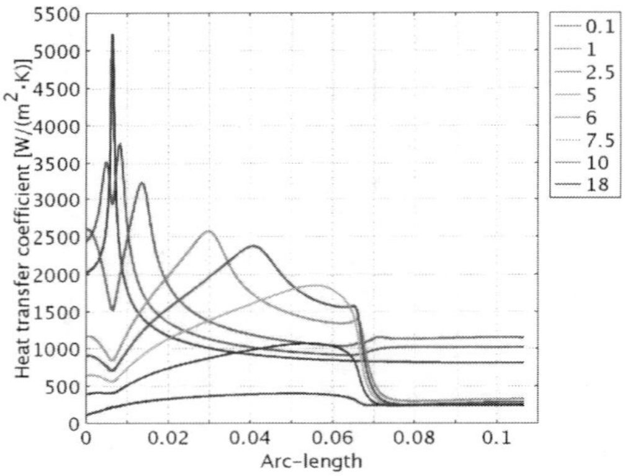

Figure 6. Heat transfer coefficient along workpiece's surface.

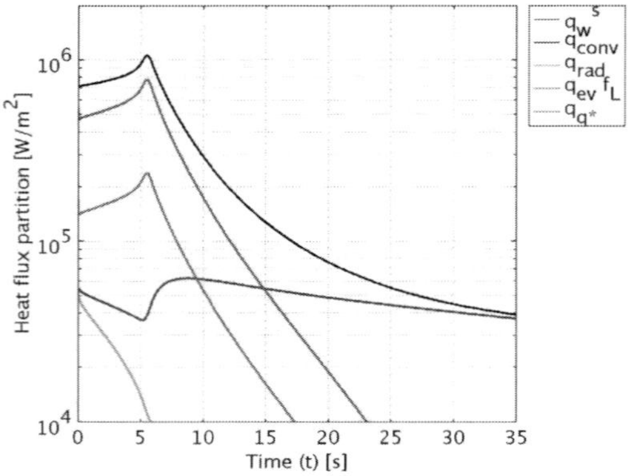

Figure 7. Heat flux partition at middle height of the cylinder.

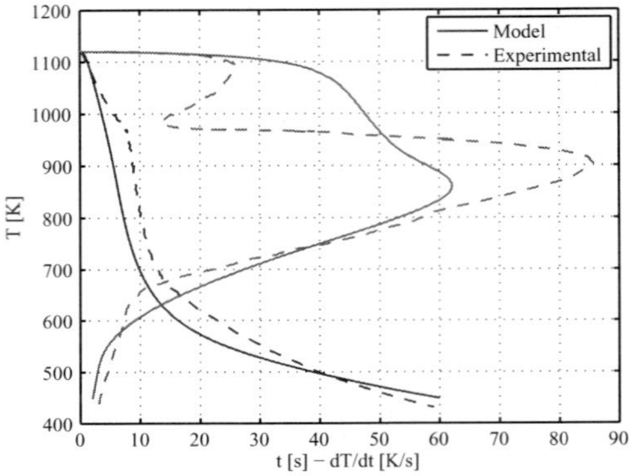

Figure 8. T-t and T-dT/dt curves at the center of the workpiece. Experimental data from (Houghton 2006).

Finally, a comparison of cooling curves from the numerical case against experimental data from an actual quenching oil (Houghton 2006) is made in Figure 8. It is observed that the final temperature is rather well recovered, but the cooling process present some discrepancies. In particular the first stages of cooling are overestimated, producing a decrease in temperature earlier than expected. It is believed that other mechanism than damped evaporation describes the heat transfer at high temperatures. The stages from 10 s and on are rather well recovered, with minor adjustments needed for evaporation-convection and convective processes.

4 CONCLUSIONS

The description of heat transfer during quenching process should include the evolution of the quenching bath and the corresponding boiling phenomena. This should be done through an appropriate multiphase flow model and adequate local description of heat fluxes exchanged with the treated workpiece. The reduction of the specific model from a general frame depends on characteristic times and numbers of the physical problem.

A model that couples multiphase dynamics, energy conservation of the liquid phase and heat transferred from the quenched piece was presented. Our model is based on the drift-flux mixture-model and includes specific heat flux functions to model the heat transfer.

The effect of vapor phase on the cooling dynamics of our tested case is clearly pointed out. Local analysis shows that cooling velocities are strongly correlated with the presence of vapor, and the corresponding heat transfer coefficients are quite complex to be described by regular heat transfer correlations.

The model presented here adequately captures the order of magnitude of the physical problem. Heat transfer processes at high temperature need further review in order to match the experimental behavior, and nucleate boiling and single phase convection stages need small tweaks in order to match experimental trends. These features are part of our ongoing work.

ACKNOWLEDGEMENTS

This work was partially funded by project MTM2010-21135-C02-02 of the Spanish MEC and FEDER (EU), and projects IN845B-2010/046 & 09DPI189E of Xunta de Galicia & CIE Galfor S.A. (Spain). DNP acknowledges his PhD scholarship from MAEC-AECID (Spain).

REFERENCES

Bromley, L.A. *et al.* 1953. Heat transfer in forced convection film boiling. *Industrial and Engineering Chemistry* 45(12): 2639–2646.

Clift, R. Grace, J.R. & Weber, M.E. 1978. Bubbles, *drops and particles*. New York: Dover Publications Inc.

Crowe, C. Sommerfeld, M. & Tsuji, Y. 1998. *Multiphase flows with droplets and particles*. Florida: CRC Press.

Denis, S. et al. 1992. Mathematical model coupling phase transformations and temperature evolutions in steels. Iron and Steel Institute of Japan 32: 316–25.

Houghton Ibérica S.A. 2006. Datasheet of Houghto-Quench 125 oil (http://www.houghtonintl.com/index.php).

Ishii, M. & Hibiki, T. 2006. *Thermo-fluid dynamics of two-phase flow*. New York: Springer.

Kang, S.-H. & Im, Y.-T. 2007. Thermo-elasto-plastic finite element analysis of quenching process of carbon steel. *Journal of Materials Processing Technology* 192(3): 381–90.

Kocamustafaogullari, G. & Ishii, M. 1995. Foundation of the interfacial area transport equations and its closure relations. *Int. J. Heat Mass Transfer* 38: 481–93.

Kocar, C. & Sokmen, C.M. 2009. A CFD model of two-phase flow in BWRs with results of sensitivity and uncertainty analysis, *Nuclear Engineering and Design* 239: 1839–46.

Koncar, B. & Mavko, B. 2008. Simulation of boiling flow experiments close to CHF with the Neptune_CFD code, *Science and Technology of Nuclear Installations*. Article ID 732158.

Koncar, B. & Tiselj, I. 2010. Influence of near-wall modelling on boiling flow simulation, *Nuclear Engineering and Design* 240: 275–83.

Koncar, B. & Borut, M. 2010. Wall function approach for boiling two-phase flows, *Nuclear Engineering and Design* 240: 3910–3918

Krepper, E. *et al.* 2007. CFD modelling of subcooled boiling—Concept, validation and application to the fuel assembly design, *Nuclear Engineering and Design* 237: 716–31.

Li, X. *et al.* 2009. Numerical and experimental investigation of heat transfer on heating surface during subcooled boiling flow of liquid nitrogen, *Int. J. Heat Mass Transfer* 52: 1510–16.

Manninen, M. *et al.* 1996. On the mixture model for multiphase flow, *VTT Publications, VTT Energy, Nuclear Energy*, Technical Research Center of Finland.

Passarella, D.N. *et al.* 2010. Heat transfer model for quenching by submerging, *J. Phys.*: Conf. Ser. 296 012004.

Porter, D.A. & Easterling, K.E. 1992. *Phase Transformations in Metals and Alloys, 2nd Ed.* London: Chapman & Hall.

Sugianto, A. *et al.* 2009. Failure analysis and prevention of quench crack of eccentric holed disk by experimental study and computer simulation, *Engineering Failure Analysis* 16: 70–84.

Vorster, W.J.J. *et al.* 2009. Influence of quenchant hydrodynamics and boiling phase incipient temperature shifts on residual stress formation, *Heat Transfer Engineering* 30(7): 564–573.

Numerical Methods for Hyperbolic Equations – Vázquez-Cendón et al. (eds)
 ISBN 978-0-415-62150-2

Conservative formulation for compressible fluid flow through elastic porous media

E. Romenski
Sobolev Institute of Mathematics, Novosibirsk, Russia

ABSTRACT: The model of compressible fluid flow in elastic porous media is derived on the basis of thermodynamically compatible system theory. Governing equations of the model are hyperbolic and all equations can be cast in a conservation form. The system of equations for small amplitude wave propagation in the unstressed immovable medium is derived and some numerical test problems for the one-dimensional wave propagation are presented.

1 INTRODUCTION

The development of advanced computational modelling for fluid flow in elastic porous media is of interest in a number of scientific and engineering disciplines including geophysical applications. In recent decades a theories for multiphase flows and associated numerical algorithms have been developed and exploited for the needs of industry and environmental sciences. But up to now there is no conventional form of the model and its governing equations even for compressible two-phase flows. Several approaches exist also for the modelling of fluid flow in elastic media. The Biot equations (Biot 1956) for small amplitude wave propagation in the saturated porous medium are widely used in geophysics but they are not applicable to the case of finite deformations. The phenomenological model based on the irreversible thermodynamics principles has been developed in (Blokhin & Dorovsky 1995) taking into account finite deformations, but its governing equations are not fully conservative and in addition have some disadvantages for the case of vanishing phase.

The main challenge in the multiphase flow modelling is associated with the formulation of a mathematical model that satisfies such important properties as hyperbolicity, fully conservative form of the governing equations and compatibility and consistency of the mathematical model with thermodynamic laws. These properties provide a solid mathematical framework in the study of initial-boundary value problems, and allow application of accurate numerical methods.

Recently a hierarchy of conservative models for compressible two-phase flow and high-accuracy numerical methods for solving the corresponding governing equations has been proposed in (Romenski et al. 2007; Romenski et al. 2010). The development of these models is based on the theory of thermodynamically compatible hyperbolic systems of conservation laws (Godunov & Romenski 2003).

Here we present an application of this theory to conservative formulation of isentropic flow of compressible fluid in the elastic porous media. Its core aspect is a phenomenological modelling of continuous media, where by using thermodynamic laws we determine a structure of the governing balance laws. In this context the mixture is supposed to be a continuum in which the multiphase character of flow is taken into account. The resulting system of differential equations is hyperbolic and all balance equations can be cast in conservation form.

As an alternative to the Biot theory, the governing equations for small amplitude wave propagation are derived from the general system of conservation laws. The dependence of the fast and slow pressure waves and shear wave on the porosity is studied. Some test problems are solved numerically.

2 EQUATIONS IN CONSERVATION FORM FOR COMPRESSIBLE FLUID FLOW IN SATURATED ELASTIC POROUS MEDIA

A phenomenological continuum mechanics approach to model fluid flow in porous elastic medium allows us to consider such a medium as a mixture, in which the two-phase character of flow is taken into account. Thus the set of parameters of state for the mixture comprises not only generally used density, momentum and temperature, but some additional parameters, such as phase mass and volume concentrations, and the relative velocity of phase motion. Simultaneously, the new governing equations should be formulated for these new parameters of state beyond the classical conservation laws of mass, momentum and energy.

In this Section a derivation of conservation form governing equations is presented, based on the thermodynamically compatible system theory (Godunov & Romenski 2003; Romenski 2001). Let us consider an isentropic flow of compressible fluid in elastic porous media neglecting inelastic deformations and thermal effects. Assume that physical parameters of state fully describing such a medium are: α_1—the volume fraction of elastic solid, ρ—the mass density of the mixture, c_1—the mass fraction of elastic solid, u_i—the velocity of the mixture, w_i—the relative velocity of constituents, F_{ij}—the deformation gradient of the element of the mixture. Note that the volume fraction of fluid (called also porocity) and mass fraction of fluid can be computed as $\alpha_2 = 1 - \alpha_1$ and $c_2 = 1 - c_1$ respectively.

Physical variables listed above are connected with the phase densities and velocities $\rho_1, \rho_2, u_i^1, u_i^2$, by the relationships:

$$\rho = \alpha_1\rho_1 + \alpha_2\rho_2,\ c_1 = \frac{\alpha_1\rho_1}{\rho},\ c_2 = \frac{\alpha_2\rho_2}{\rho},$$
$$u_i = c_1 u_i^1 + c_2 u_i^2,\ w_i = u_i^1 - u_i^2.$$

The derivation of conservation form equations with the thermodynamically compatible system theory is based on the few interrelated steps. If the physical variables characterizing medium are defined, then it is necessary to define a subsystem of the general thermodynamically compatible system of conservation laws for these variables (Godunov & Romenski 2003; Romenski 2001). This dissipation-free system of conservation laws is formulated in terms of generating potential and variables. The generating potential is connected with the generalized internal energy of the medium, while the generating variables can be defined with the use of the conservative variables. The next step consists in the introducing a source terms describing a phase interaction and dissipation into the equations. And finally, closing relations such as an equation of state for the mixture and functional dependence of source terms on the parameters of state should be defined.

Assume that the generalized internal energy E is a function of $\alpha_1, \rho, c_1, w_i, F_{ij}$:

$$E = E(\alpha_1, \rho, c_1, w_i, F_{ij}).$$

Then the thermodinamically compatible dissipation-free system for the fluid flow in the elastic porous medium reads as

$$\begin{aligned}
&\frac{\partial}{\partial t}(\rho\alpha_1) + \frac{\partial}{\partial x_k}(\rho\alpha_1 u_k) = 0,\\
&\frac{\partial}{\partial t}(\rho) + \frac{\partial}{\partial x_k}(\rho u_k) = 0,\\
&\frac{\partial}{\partial t}(\rho c_1) + \frac{\partial}{\partial x_k}(\rho c_1 u_k + \rho E_{w_k}) = 0,\\
&\frac{\partial}{\partial t}(w_k) + \frac{\partial}{\partial x_k}(u_\alpha w_\alpha + E_{c_1}) = 0,\\
&\frac{\partial}{\partial t}(\rho u_i) + \frac{\partial}{\partial x_k}(\rho u_i u_k + \rho^2 E_\rho \delta_{ik}\\
&\quad + \rho w_i E_{w_k} - \rho F_{k\alpha} E_{F_{i\alpha}}) = 0,\\
&\frac{\partial}{\partial t}(\rho F_{ij}) + \frac{\partial}{\partial x_k}(\rho F_{ij} u_k - \rho F_{kj} u_i) = 0.
\end{aligned} \tag{1}$$

Presented above differential equations correspond to conservation of the elastic phase volume fraction, mixture density, elastic phase mass fraction, phase relative velocity and deformation gradient of the medium respectively. Solution to system (1) satisfies the steady compatibility constraints:

$$\begin{aligned}
&\frac{\partial w_k}{\partial x_i} - \frac{\partial w_i}{\partial x_k} = 0, \quad i,k = 1,2,3;\ i \neq k,\\
&\frac{\partial \rho F_{kj}}{\partial x_k} = 0, \quad j = 1,2,3.
\end{aligned} \tag{2}$$

These constraints remain valid for $t > 0$, if they valid for the initial data at $t = 0$. To prove this one can use appropriately combined equations for w_k and F_{ij} differentiated with respect to spacial coordinates (Godunov & Romenski 2003).

An additional energy conservation laws holds for system (1):

$$\begin{aligned}
&\frac{\partial}{\partial t}\rho\left(E + \frac{u_i u_i}{2}\right) +\\
&\frac{\partial}{\partial x_k}\left(\rho u_k\left(E + \frac{u_i u_i}{2} + \frac{p}{\rho}\right)\right.\\
&\left. + \rho(u_\alpha w_\alpha + E_{c_1})E_{w_k} - \rho u_i F_{kj} E_{F_{ij}}\right) = 0,
\end{aligned} \tag{3}$$

where $p = \rho^2 E_\rho$.

Thus, equations (1) augmented with constraints (2) form the thermodynamically compatible system. Such kind of system can be transformed to the symmetric form and is called symmetric hyperbolic if the generating potential is a convex function (Godunov & Romenski 2003; Romenski 2001).

An interphase interaction can be taken into account by adding a source terms into the equations of system (1). Here only the phase pressure relaxation and interfacial friction are considered and an appropriate source terms appear in the equations for α_1 and w_k. The resulting system reads as

$$
\begin{aligned}
&\frac{\partial}{\partial t}(\rho\alpha_1)+\frac{\partial}{\partial x_k}(\rho\alpha_1 u_k)=-\lambda\rho E_{\alpha_1},\\
&\frac{\partial}{\partial t}(\rho)+\frac{\partial}{\partial x_k}(\rho u_k)=0,\\
&\frac{\partial}{\partial t}(\rho c_1)+\frac{\partial}{\partial x_k}(\rho c_1 u_k+\rho E_{w_k})=0,\\
&\frac{\partial}{\partial t}(w_k)+\frac{\partial}{\partial x_k}(u_\alpha w_\alpha+E_{c_1})=e_{klj}u_l\omega_j-\chi E_{w_k},\\
&\frac{\partial}{\partial t}(\rho u_i)+\frac{\partial}{\partial x_k}(\rho u_i u_k+\rho^2 E_\rho\delta_{ik}\\
&\quad+\rho w_i E_{w_k}-\rho F_{k\alpha}E_{F_{i\alpha}})=0,\\
&\frac{\partial}{\partial t}(\rho F_{ij})+\frac{\partial}{\partial x_k}(\rho F_{ij}u_k-\rho F_{kj}u_i)=0.
\end{aligned}
\tag{4}
$$

Here e_{klj} is the completely anti-symmetric Levy-Chivita tensor. The source term in the equation for volume fraction α_1 is responsible for the phase pressure relaxation, the rate of which is characterized by λ, while the source term in the equation for w_k represents an interfacial friction force with the friction coefficient χ. Note, that λ and χ can depend on the parameters of state. An artificial variable ω_j is introduced in order to write the equation in divergent form and satisfies the following compatibility conditions:

$$\frac{\partial w_k}{\partial x_i}-\frac{\partial w_i}{\partial x_k}=-e_{ki\alpha}\omega_\alpha$$

$$\frac{\partial \omega_k}{\partial t}+\frac{\partial(u_l\omega_k-u_k\omega_l+e_{kl\alpha}\chi E_{w_\alpha})}{\partial x_l}=0.$$

The latter gives a possibility to treat $e_{klj}u_l\omega_j$ in the equation for w_k as a lower-order term.

The introduced source terms change the energy conservation to the energy dissipation:

$$
\begin{aligned}
&\frac{\partial}{\partial t}\rho\left(E+\frac{u_i u_i}{2}\right)\\
&+\frac{\partial}{\partial x_k}\left(\rho u_k\left(E+\frac{u_i u_i}{2}+\frac{p}{\rho}\right)+\rho(u_\alpha w_\alpha+E_{c_1})E_{w_k}\right.\\
&\left.-\rho u_i F_{kj}E_{F_{ij}}\right)=-\lambda\rho E_{\alpha_1}^2-\chi\rho E_{w_k}E_{w_k}\le 0.
\end{aligned}
$$

If $\lambda\ge 0,\chi\ge 0$, then the right-hand side is negative and the total mixture energy dissipates owing to the phase interaction.

The generalized internal energy E can be defined in terms of known phase specific internal energies. Its simplest choice is based on the assumption that E is a mass averaged phase internal energy supplemented by the kinetic energy of relative motion. This choice does not take into account an influence of interface energy effects, but the proposed model can be considered as a basis for more general models describing effects of interface phenomena. The finite deformations of the element of the mixture can be characterized by the strain tensor constructed with the deformation gradient F. Here the Finger strain tensor $G=f^T f$ is used, where $f=F^{-1}$. In general it is impossible to factorize strain tensor to the strains of pure elastic skeleton and pure liquid. That is why the liquid density, elastic phase density and deviator of strain tensor of elastic skeleton can be taken as the parameters characterizing the deformation of the mixture.

Thus, supposing that the phase internal energies e^1, e^2 are known as

$$e^1=e^1(\rho_1,F_{11},F_{12},\ldots,F_{33}),\quad e^2=e^2(\rho_2),$$

the generalized energy $E(\alpha_1,\rho,c_1,F_{ij},w_k)$ is defined as

$$E=c_1e^1(\rho_1,F_{11},F_{12},\ldots,F_{33})+c_2e^2(\rho_2)+\frac{1}{2}w_iw_i.$$

Note that here it is assumed that the dependence e^1 on F is given by means of deviatoric part of the Finger strain tensor. Using relationships between $\rho,c_1,c_2,\rho_1,\rho_2$

$$\rho=\alpha_1\rho_1+\alpha_2\rho_2, c_1=\frac{\alpha_1\rho_1}{\rho}, c_2=1-c_1=\frac{\alpha_2\rho_2}{\rho},$$

one can derive expressions for the derivatives of the generalized energy:

$$
\begin{aligned}
&E_{\alpha_1}=\frac{1}{\rho}(\rho_2^2e^2_{\rho_2}-\rho_1^2e^1_{\rho_1})=\frac{1}{\rho}(p_2-p_1),\\
&E_{c_1}=e^1+\frac{p_1}{\rho_1}-e^2-\frac{p_2}{\rho_2}+(1-2c_1)\frac{w_iw_i}{2},\\
&E_\rho=\frac{1}{\rho^2}(\alpha_1\rho_1^2e^1_{\rho_1}+\alpha_2\rho_2^2e^2_{\rho_2})=\frac{1}{\rho^2}(\alpha_1p_1+\alpha_2p_2),\\
&E_{F_{ij}}=c_1e^1_{F_{ij}},\quad E_{w_i}=c_1(1-c_1)w_i.
\end{aligned}
$$

The following formulae are also useful for the final formulation of equations

$$
\begin{aligned}
&\rho F_{k\alpha}E_{F_{i\alpha}}=\alpha_1\rho_1F_{k\alpha}e^1_{F_{i\alpha}},\\
&u_\alpha w_\alpha+(1-2c_1)\frac{w_\alpha w_\alpha}{2}=\frac{u^1_\alpha u^1_\alpha}{2}-\frac{u^2_\alpha u^2_\alpha}{2}.
\end{aligned}
$$

With the use of above formulae equations (4) can be rewritten in terms of phase parameters as

$$
\begin{aligned}
&\frac{\partial \rho\alpha_1}{\partial t}+\frac{\partial \rho\alpha_1 u_k}{\partial x_k}=\lambda(p_1-p_2),\\
&\frac{\partial \rho}{\partial t}+\frac{\partial \rho u_k}{\partial x_k}=0,\\
&\frac{\partial \rho_1\alpha_1}{\partial t}+\frac{\partial \rho_1\alpha_1 u_k^1}{\partial x_k}=0,\\
&\frac{\partial (u_k^1-u_k^2)}{\partial t}+\frac{\partial}{\partial x_k}\left(\frac{u_i^1u_i^1}{2}-\frac{u_i^2u_i^2}{2}+e^1\right.\\
&\left.+\frac{p_1}{\rho_1}-e^2-\frac{p_2}{\rho_2}\right)=-e_{klj}u_l\omega_j-\chi E_{w_k},\\
&\frac{\partial(\alpha_1\rho_1u_i^1+\alpha_2\rho_2u_i^2)}{\partial t}+\frac{\partial}{\partial x_k}\big(\alpha_1\rho_1u_i^1u_k^1\\
&+\alpha_2\rho_2u_i^2u_k^2+\alpha_1p_1+\alpha_2p_2-\alpha_1\tilde{\sigma}_{ik}\big)=0,\\
&\frac{\partial \rho F_{ij}}{\partial t}+\frac{\partial(\rho F_{ij}u_k-\rho F_{kj}u_i)}{\partial x_k}=0.
\end{aligned}
\tag{5}
$$

Here $\rho=\alpha_1\rho_1+\alpha_2\rho_2$ is the mixture density, $u_k=c_1u_k^1+c_2u_k^2$ is the mixture velocity, $p_1=\rho_1^2e^1_{\rho_1}, p_2=\rho_2^2e^2_{\rho_2}$ are the pressures of elastic and liquid constituents, $\tilde{\sigma}_{ik}=\rho_1F_{kj}e^1_{F_{ij}}$ is the shear stress tensor of elastic phase. A set of compatibility constraints reads as follows:

$$
\begin{aligned}
&\frac{\partial w_k}{\partial x_i}-\frac{\partial w_i}{\partial x_k}=-e_{ki\alpha}\omega_\alpha,\\
&\frac{\partial \omega_k}{\partial t}+\frac{\partial(u_l\omega_k-u_k\omega_l+e_{kl\alpha}\chi E_{w_\alpha})}{\partial x_l}=0,\\
&\frac{\partial \rho F_{kj}}{\partial x_k}=0.
\end{aligned}
$$

Thus the system of governing equations for isentropic deformations of porous medium saturated with compressible liquid is formulated. These equations, in particular, can be used for the analysis of small amplitude wave propagation in porous media.

3 EQUATIONS FOR SMALL AMPLITUDE WAVE PROPAGATION IN SATURATED POROUS MEDIUM

In this Section the governing equations for small amplitude wave propagation in the stress free immovable medium are derived. For their derivation another, equivalent form of (5) is used, in which equations are transformed to the equations for phase velocities and densities:

$$
\begin{aligned}
&\frac{\partial u_i^1}{\partial t}+u_k^1\frac{\partial u_i^1}{\partial x_k}+\frac{1}{\rho_1}\frac{\partial p_1}{\partial x_i}-\frac{1}{\rho}\frac{\partial(\alpha_1\tilde{\sigma}_{ik})}{\partial x_k}\\
&+\frac{p_1-p_2}{\rho}\frac{\partial\alpha_1}{\partial x_i}+c_1u_k^1\left(\frac{\partial u_i^1}{\partial x_k}-\frac{\partial u_k^1}{\partial x_i}\right)\\
&+c_2u_k^2\left(\frac{\partial u_i^2}{\partial x_k}-\frac{\partial u_k^2}{\partial x_i}\right)=-\frac{\alpha_2\rho_2}{\rho}\chi(u_i^1-u_i^2),\\
&\frac{\partial u_i^2}{\partial t}+u_k^2\frac{\partial u_i^2}{\partial x_k}+\frac{1}{\rho_2}\frac{\partial p_2}{\partial x_i}-\frac{1}{\rho}\frac{\partial(\alpha_1\tilde{\sigma}_{ik})}{\partial x_k}\\
&+\frac{p_1-p_2}{\rho}\frac{\partial\alpha_1}{\partial x_i}+c_1u_k^1\left(\frac{\partial u_i^1}{\partial x_k}-\frac{\partial u_k^1}{\partial x_i}\right)\\
&+c_2u_k^2\left(\frac{\partial u_i^2}{\partial x_k}-\frac{\partial u_k^2}{\partial x_i}\right)=\frac{\alpha_1\rho_1}{\rho}\chi(u_i^1-u_i^2),\\
&\frac{\partial \rho_1}{\partial t}+u_k^1\frac{\partial \rho_1}{\partial x_k}+\rho_1\frac{\partial u_1^k}{\partial x_k}\\
&+\frac{\rho_1c_2}{\alpha_1}(u_k^1-u_k^2)\frac{\partial\alpha_1}{\partial x_k}=\frac{\rho_1}{\alpha_1\rho}\lambda(p_2-p_1),\\
&\frac{\partial \rho_2}{\partial t}+u_k^2\frac{\partial \rho_2}{\partial x_k}+\rho_2\frac{\partial u_2^k}{\partial x_k}\\
&+\frac{\rho_2c_1}{\alpha_2}(u_k^1-u_k^2)\frac{\partial\alpha_1}{\partial x_k}=-\frac{\rho_2}{\alpha_2\rho}\lambda(p_2-p_1),\\
&\frac{\partial F_{ij}}{\partial t}+u_k\frac{\partial F_{ij}}{\partial x_k}-F_{kj}\frac{\partial u_i}{\partial x_k}=0,\\
&\frac{\partial\alpha_1}{\partial t}+u_k\frac{\partial\alpha_1}{\partial x_k}=\lambda\frac{p_1-p_2}{\rho}.
\end{aligned}
\tag{6}
$$

Recall that in the above equations $u_k=c_1u_k^1+c_2u_k^2$, $\rho=\alpha_1\rho_1+\alpha_2\rho_2$, $p_1=\rho_1^2e^1_{\rho_1}$, $p_2=\rho_2^2e^2_{\rho_2}$, $e^1=e(\rho_1,F_{ij})$, $e^2=e^2(\rho_2)$, $\tilde{\sigma}_{ik}=\rho_1F_{k\alpha}e^1_{F_{i\alpha}}$.

The wave of small amplitude propagating in the unstressed stationary medium is a solution of the equations (6) for small perturbation of steady solution. Thus our goal is to find a solution to this system in the form

$$
u_i^k=(u_i^k)^0+\Delta u_i^k,\quad \rho_i^k=(\rho_i^k)^0+\Delta\rho_i^k,
$$

$$
F_{ij}=(F_{ij})^0+\Delta F_{ij},\quad \alpha_1=(\alpha_1)^0+\Delta\alpha_1,
$$

Where $(u_i^k)^0=0, (\rho_i^k)^0$ correspond to the unstressed state (phase pressures are zero), $(F_{ij})^0=\delta_{ij}$, and $(\alpha_1)^0$ is a given value connected with the initial porosity as $\Phi=(\alpha_2)^0=1-(\alpha_1)^0$.

Substituting above formulae into (6), setting the linear strain tensor as a new variable $\varepsilon_{ij}=(\Delta F_{ij}+\Delta F_{ji})/2$ and omitting Δ in the notation of perturbation, one can obtain the system for small perturbations of state parameters:

$$\frac{\partial u_i^1}{\partial t} + \frac{1}{\rho_1^0}\frac{\partial p_1}{\partial x_i} - \frac{\alpha_1^0}{\rho^0}\frac{\partial \tilde{\sigma}_{ik}}{\partial x_k} = -\frac{\alpha_2 \rho_2^0}{\rho^0}\chi(u_i^1 - u_i^2),$$
$$\frac{\partial u_i^2}{\partial t} + \frac{1}{\rho_2^0}\frac{\partial p_2}{\partial x_i} - \frac{\alpha_1^0}{\rho^0}\frac{\partial \tilde{\sigma}_{ik}}{\partial x_k} = \frac{\alpha_1 \rho_1^0}{\rho^0}\chi(u_i^1 - u_i^2),$$
$$\frac{\partial \rho_1}{\partial t} + \rho_1^0\frac{\partial u_k^1}{\partial x_k} = \lambda\frac{\rho_1^0}{\alpha_1^0\rho^0}(p_2 - p_1),$$
$$\frac{\partial \rho_2}{\partial t} + \rho_2^0\frac{\partial u_k^2}{\partial x_k} = -\lambda\frac{\rho_2^0}{\alpha_1^0\rho^0}(p_2 - p_1), \qquad (7)$$
$$\frac{\partial \varepsilon_{ij}}{\partial t} - \frac{c_1^0}{2}\left(\frac{\partial u_i^1}{\partial x_j} + \frac{\partial u_j^1}{\partial x_i}\right) - \frac{c_2^0}{2}\left(\frac{\partial u_i^2}{\partial x_j} + \frac{\partial u_j^2}{\partial x_i}\right) = 0,$$
$$\frac{\partial \alpha_1}{\partial t} = \lambda\frac{p_1 - p_2}{\rho^0}.$$

In the above system the spatial derivatives of phase pressures p_1, p_2 and shear stress tensor $\tilde{\sigma}_{ik}$ are connected with phase densities and small strain tensor by relations

$$\frac{\partial p_1}{\partial x_i} = K_1\frac{\partial \rho_1}{\partial x_i}, \quad \frac{\partial p_2}{\partial x_i} = K_2\frac{\partial \rho_2}{\partial x_i},$$
$$\frac{\partial \tilde{\sigma}_{ik}}{\partial x_k} = 2\mu\frac{\partial \varepsilon_{ik}}{\partial x_k} - \frac{2}{3}\mu\delta_{ik}\frac{\partial(\varepsilon_{11} + \varepsilon_{22} + \varepsilon_{33})}{\partial x_k},$$

Where K_1, K_2 are phase bulk modules, μ shear module of elastic medium.

Thus we derived linear equations system for small amplitude wave propagation in the unstressed immovable medium. In the equation for the volume fraction a phase pressure relaxation term is presented. Actually the rate of pressure relaxation in real porous media can be very high that is why it is reasonable to derive a simplified model for media with equal phase pressure. To do this it is necessary to replace in (7) the equation for α_1 by relations $p_1 = p_2 = p$. After some transformations we come to

$$\rho_1^0\frac{\partial u_i^1}{\partial t} + \frac{\partial p}{\partial x_i} - c_1^0 2\mu\frac{\partial \tilde{\varepsilon}_{ik}}{\partial x_k} = -\rho_1^0 c_2^0\chi(u_i^1 - u_i^2),$$
$$\rho_2^0\frac{\partial u_i^2}{\partial t} + \frac{\partial p}{\partial x_i} - c_1^0\frac{\rho_2^0}{\rho_1^0}2\mu\frac{\partial \tilde{\varepsilon}_{ik}}{\partial x_k} = \rho_2^0 c_1^0\chi(u_i^1 - u_i^2),$$
$$\frac{\partial p}{\partial t} + \alpha_1^0 < K > \frac{\partial u_k^1}{\partial x_k} + \alpha_2^0 < K > \frac{\partial u_k^2}{\partial x_k} = 0, \qquad (8)$$
$$\frac{\partial \varepsilon_{ij}}{\partial t} - \frac{c_1^0}{2}\left(\frac{\partial u_i^1}{\partial x_j} + \frac{\partial u_j^1}{\partial x_i}\right) - \frac{c_2^0}{2}\left(\frac{\partial u_i^2}{\partial x_j} + \frac{\partial u_j^2}{\partial x_i}\right) = 0,$$

where $< K >= \left(\alpha_1^0/(\rho_1^0 K_1) + \alpha_2^0/(\rho_2^0 K_2)\right)^{-1}$, $\tilde{\varepsilon}_{ik} = \varepsilon_{ik} - \delta_{ik}(\varepsilon_{11} + \varepsilon_{22} + \varepsilon_3)/3$.

The volume fraction α_1 can be recovered from the following equation

$$\frac{\partial \alpha_1}{\partial t} + \frac{\alpha_1^0\alpha_2^0}{\rho_2^0 K_2} < K > \frac{\partial u_k^1}{\partial x_k} - \frac{\alpha_1^0\alpha_2^0}{\rho_1^0 K_1} < K > \frac{\partial u_k^2}{\partial x_k} = 0.$$

Equations (8) with single pressure are designed for the modelling of small amplitude wave propagation in the unstressed immovable saturated porous medium and it would be interesting to find the sound velocities in such a medium. We study this problem only for the one-dimensional case that can be useful in the development of numerical methods for solving the full system. If the solution of one-dimensional version of (8) has the form of harmonic wave $\exp(\omega t - kx)$, then one can obtain two dispersion equations for longitudinal and shear waves. The fourth order dispersion equation for longitudinal wave reads as

$$\omega^4 + \chi\omega^3 - \left(\frac{(\alpha_1^0\rho_2^0 + \alpha_2^0\rho_1^0)K_1K_2}{\alpha_2^0\rho_1^0K_1 + \alpha_1^0\rho_2^0K_2} + \frac{4}{3}\frac{\mu}{\rho_1^0}c_1^0\right)k^2\omega^2$$
$$-\chi\omega\left(\frac{\rho_1^0\rho_2^0K_1K_2}{\rho^0(\alpha_2^0\rho_1^0K_1 + \alpha_1^0\rho_2^0K_2)} + \frac{4}{3}\frac{\mu}{\rho_1^0}c_1^0\right)$$
$$+\frac{4}{3}\frac{\mu}{\rho_1^0}c_1^0\frac{\alpha_1^0\alpha_2^0}{\rho^0}\frac{(\rho_1^0 - \rho_2^0)^2K_1K_2}{\alpha_2^0\rho_1^0K_1 + \alpha_1^0\rho_2^0K_2}k^4 = 0,$$

and the second order equation for shear wave reads as

$$\omega^2 - c_1^0\frac{\mu}{\rho_1^0}k^2 = 0.$$

One can see that the shear wave velocity does not depend on the interfacial friction coefficient χ and equal to

$$V_s = \sqrt{c_1^0\mu/\rho_1^0}.$$

There are two longitudinal waves (fast and slow) and their velocity depend on χ. For $\chi = 0$ (there is no interfacial friction) speeds of sounds can be found with the use of solution $\omega(k)$ of equation

$$\omega^4 - \left(\frac{(\alpha_1^0\rho_2^0 + \alpha_2^0\rho_1^0)K_1K_2}{\alpha_2^0\rho_1^0K_1 + \alpha_1^0\rho_2^0K_2} + \frac{4}{3}\frac{\mu}{\rho_1^0}c_1^0\right)k^2\omega^2$$
$$+\frac{4}{3}\frac{\mu}{\rho_1^0}c_1^0\frac{\alpha_1^0\alpha_2^0}{\rho^0}\frac{(\rho_1^0 - \rho_2^0)^2K_1K_2}{\alpha_2^0\rho_1^0K_1 + \alpha_1^0\rho_2^0K_2}k^4 = 0.$$

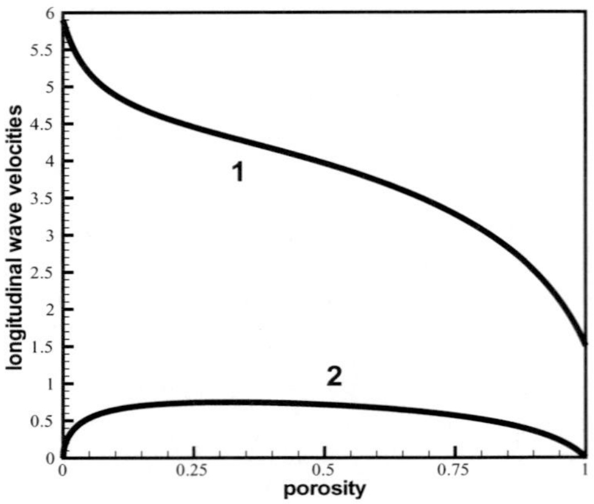

Figure 1. Fast (curve 1) and slow (curve 2) longitudinal sound wave velocities as a function of porosity.

For $\chi = \infty$ (there is no phase relative motion) only one longitudinal wave exists, which can be found from the equation

$$\omega^2 - \left(\frac{\rho_1^0 \rho_2^0 K_1 K_2}{\rho^0 (\alpha_2^0 \rho_1^0 K_1 + \alpha_1^0 \rho_2^0 K_2)} + \frac{4}{3} \frac{\mu}{\rho_1^0} c_1^0 \right) k^2 = 0.$$

With the use of these formulas one can predict sound velocities in the liquid saturated porous medium if sound velocities in liquid and elastic constituents are known. As an example a dependence on the porosity $\Phi = 1 - \alpha_1^0$ of fast and slow longitudinal waves for water saturated quartz sandstone is presented on Figure 1 in the case of interfacial friction absence ($\chi = 0$). The material parameters for water are: $\rho_2^0 = 1.0\ g/cm^3$—density, $C_2 = 1500\ m/s$—sound velocity, and for elastic skeleton are: $\rho_1^0 = 2.65\ g/cm^3$—density, $C_p = 5900\ m/s$—longitudinal wave velocity, $C_s = 3700\ m/s$—shearwavevelocity. Moduli K_1, K_2, μ are computed as $K_1 = C_p^2$, $K_2 = C_2^2$, $\mu = \rho_0^1 C_s^2$.

One can see that the fast speed of sound decreases monotonically if the porosity varies from 0 to 1. The slow speed of sound is equal to zero if the porosity is equal to 1 or 0 and has a maximum for some intermediate value.

4 NUMERICAL EXAMPLES

The numerical modelling of small amplitude wave propagation in saturated porous media for real problems, such as an acoustic logging, is very challenging problem. The point is that it is necessary to study waves of very high frequency and the wave run can be of order a hundred of its length. We have developed the Runge-Kutta WENO method to solve one-dimensional wave problems in order to study the capabilities of the model. It was found that the application of the fourth order Runge-Kutta integration in time method coupled with the finite difference method WENO reconstruction of the fifth order gives a very good agreement with exact solution for the single longitudinal sound wave. In this Section the results of computations for the problem of pressure wave propagation generated by the boundary velocity impulse are presented.

The system of equations to be solved is the one-dimensional simplification of (8), and it is obtained neglecting tangential velocities and reads as:

$$\rho_1^0 \frac{\partial u_1^1}{\partial t} + \frac{\partial p}{\partial x_1} - c_1^0 \frac{4\mu}{3} \frac{\partial \varepsilon_{11}}{\partial x_1} = -\rho_1^0 c_2^0 \chi (u_1^1 - u_1^2),$$

$$\rho_2^0 \frac{\partial u_1^2}{\partial t} + \frac{\partial p}{\partial x_1} - c_1^0 \frac{\rho_2^0}{\rho_1^0} \frac{4\mu}{3} \frac{\partial \varepsilon_{11}}{\partial x_1} = \rho_2^0 c_1^0 \chi (u_i^1 - u_i^2),$$

$$\frac{\partial p}{\partial t} + \alpha_1^0 < K > \frac{\partial u_1^1}{\partial x_1} + \alpha_2^0 < K > \frac{\partial u_1^2}{\partial x_1} = 0,$$

$$\frac{\partial \varepsilon_{11}}{\partial t} - c_1^0 \frac{\partial u_i^1}{\partial x_1} - c_2^0 \frac{\partial u_i^2}{\partial x_1} = 0.$$

The problem statement is as follows. Computational region is the right half-axis $x \geq 0$. The initial data are: velocities $u_1^1 = 0, u_1^2 = 0$, densities $\rho_1 = 0$, $\rho_2 = 0$ and deformation $\varepsilon_{11} = 0$. The porosity $\Phi = 1 - \alpha_1^0$ is equal to 0.2. On the left boundary $x = 0$ the phase velocities are given as a function of time: $u_1^1 = u_1^2 = U \sin(\Omega t)$ with $U = 100\ m/s, \Omega = 10^5 s^{-1}$. Note that the above velocity impulse is very sharp and it was chosen in order to test numerical method in extremal situation. The medium under consideration is the water saturated sandstone with phase material parameters given in the previous Section.

On Figure 2 a phase velocity profiles are shown at $t = 0.1810^{-3} s$. The interfacial friction coefficient χ is equal to zero. On this figure and all subsequent figures the solid line corresponds to the velocity of elastic skeleton whereas the dash-dotted line corresponds to the liquid phase. On Figure 2 one can see two waves (fast and slow) propagating to the right and each phase has its own mass velocity. This picture is in accordance with the theory predicting the two type of pressure waves. On Figure 3 the phase velocity profiles are presented at the same moment in time, but for $\chi = 10^7 s^{-1}$. One can see that the phase velocities are very close due to the big value of interfacial friction. On Figures 3 and 4 the two intermediate variants with $\chi = 10^3 s^{-1}$ and $\chi = 10^5 s^{-1}$ are presented. One can see that the relative velocity relaxation zone exists between the fast and slow waves and the slow

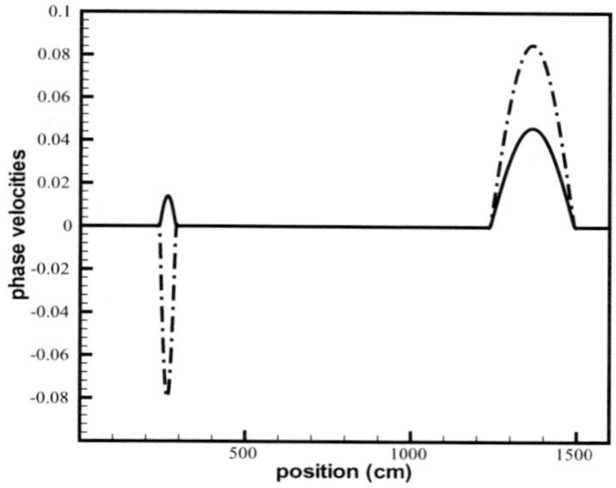

Figure 2. Two-wave configuration of phase velocites for $\chi = 0$. Elastic and liquid phase velocities are drawn by the solid and dash-dotted line respectively.

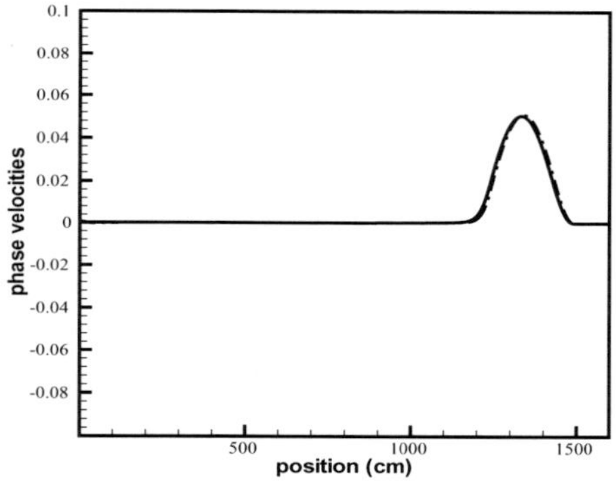

Figure 3. Phase velocity profiles for $\chi = 10^7 s^{-1}$. Elastic and liquid phase velocities are drawn by the solid and dash-dotted line respectively.

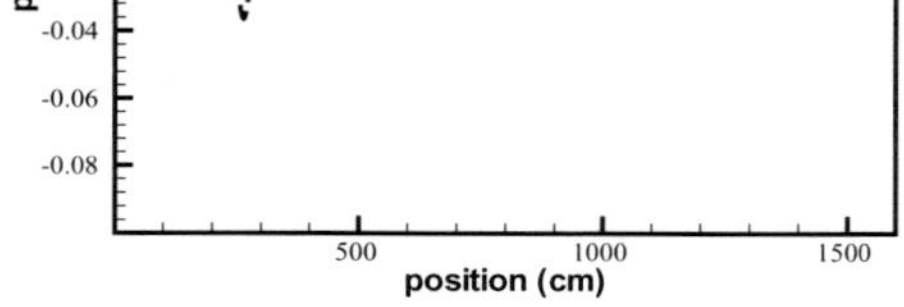

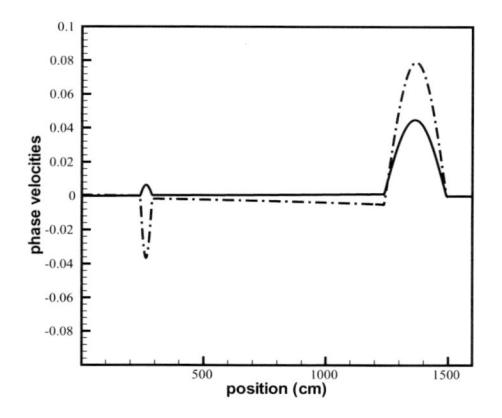

Figure 4. Phase velocity profiles for $\chi = 10^3 s^{-1}$. Elastic and liquid phase velocities are drawn by the solid and dash-dotted line respectively.

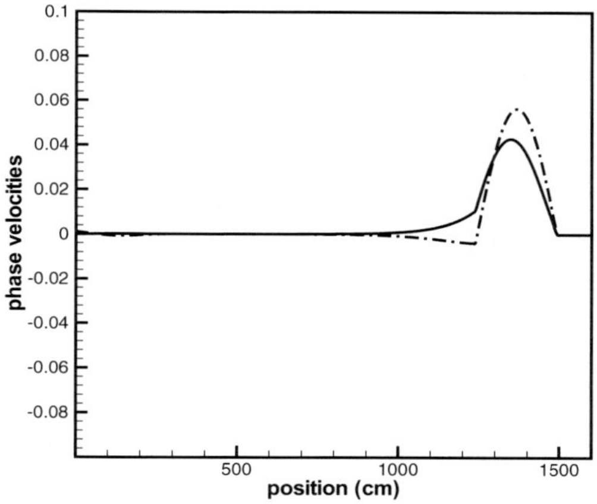

Figure 5. Phase velocity profiles for $\chi = 10^5 s^{-1}$. Elastic and liquid phase velocities are drawn by the solid and dash-dotted line respectively.

wave velocity amplitude decreases if the interfacial friction coefficient increases.

5 CONCLUSIONS

The new model for compressible fluid flow through elastic porous medium is proposed. The model is derived on the base of thermodynamically compatible systems theory. The governing equations of the model are designed for the case of isentropic finite deformations, they can be written in a conservation form and form a hyperbolic system. The advantage of this approach is that the advanced numerical methods can be directly applied to solving these equations. On the base of the formulated model a linear equations for small amplitude wave propagation in saturated porous media are derived. It is proved that there are two types of pressure wave (fast and slow), that is in accordance with existing experiments and another theories. An application of high accuracy Runge-Kutta WENO method to the one-dimensional longitudinal wave propagation demonstrates a good capability of the model.

ACKNOWLEDGEMENTS

The financial support of the Russian Foundation for Basic Research (grants 10-05-00233, 11-01-00147, 11-05-12022) and of the Presidium of Russian Academy of Sciences (Programme of Fundamental Research No 2 (project 121)) is greatly acknowledged. The author also would like to thank A. Romankov for providing results of numerical simulations for the one-dimensional wave propagation.

REFERENCES

Biot M.A. 1956. Theory of propagation of elastic waves in a fluid-saturated porous solid. I. Low frequency range. *J.Acoust. Soc. Am.* **28**(2), 168–178.

Blokhin A.M. & Dorovsky V.N. 1995. Mathematical modeling in the theory of multivelocity continuum. Nova Science. New York.

Godunov S.K. & Romenski E.I. 2003. Elements of continuum mechanics and conservation laws. Kluwer Academic/Plenum Publishers. New York.

Romenski E. 2001. Thermodynamics and hyperbolic systems of balance laws in continuum mechanics, in Godunov Methods: Theory and Applications, edited by E.F. Toro. Kluwer Academic/Plenum Publishers.

Romenski E., Resnyansky A.D. & Toro E.F. 2007. Conservative hyperbolic formulation for compressible two-phase flow with different phase pressures and temperatures. Q. Appl. Math. 65(2), 259–279.

Romenski E., Drikakis D. & Toro E.F. 2010. Conservative models and numerical methods for compressible two-phase flow *J. Sci. Comput.* **42**(1), 68–95.

Numerical Methods for Hyperbolic Equations – Vázquez-Cendón et al. (eds)
 ISBN 978-0-415-62150-2

On the coupling of compressible and incompressible fluids

Veronika Schleper
Institute of Applied Analysis and Numerical Simulation, University of Stuttgart, Germany

ABSTRACT: This article deals with the coupling of compressible and incompressible Euler equations to model e.g. the gas and liquid phases in two phase flows. We shortly review recently obtained analytical results on the one dimensional model and give a detailed description of the numerical treatment of the compressible–incompressible interface. We conclude the presentation with two illustrative numerical examples in one space dimension.

1 INTRODUCTION

In some applications in multiphase flow, compressible effects cannot be neglected. In this case, one assumes in general, that the liquid phase as well as the gaseous phase are modelled by compressible equations such as the Euler equations

$$\begin{aligned} &\partial_t \rho + \nabla_{\mathbf{x}} \cdot (\rho \mathbf{v}) = 0 \\ &\partial_t (\rho \mathbf{v}) + \nabla_{\mathbf{x}} \cdot (\mathbf{v} \otimes (\rho \mathbf{v})) + \nabla_{\mathbf{x}} p_i = 0 \\ &\partial_t E + \nabla_{\mathbf{x}} \cdot (\mathbf{v}(E + p_i)) = 0, \end{aligned} \tag{1}$$

where as usual ρ denotes the density, $\mathbf{v}$ the velocity, $E = \rho e + \frac{1}{2}\rho |\mathbf{v}|^2$ the total energy and e the internal energy. For any thermodynamically correct equation of state p_i, the above system is strictly hyperbolic as long as no vacuum states occur (i.e. for $\rho > 0$).

To model the different behaviours of the phases, each phase is described by a different equation of state $p_i(\rho, e)$, which accounts especially for the different compressibilities. In the literature (see e.g. (Ferrari et al. 2010; Liu et al. 2005)), it is for instance common to use the Tait equation of state

$$p_{Tait}(\rho) = k_0 \left(\left(\frac{\rho}{\rho_0} \right)^{\gamma} - 1 \right) + p_0 \tag{2}$$

to model the behaviour of very weakly compressible fluids such as liquids, where the constants are chosen to $\gamma \approx 7$ and $k_0 \approx 3000$. This leads to a very stiff pressure law and to a very large speed of sound in the liquid regions. Numerically, this results in severe time step restrictions due to the CFL-condition.

To circumvent this high restriction on the time step size, we propose a coupling of compressible and incompressible flows, where the flow in the liquid phase is given by the incompressible Euler equations

$$\begin{aligned} &\nabla_{\mathbf{x}} \cdot (\mathbf{v}) = 0 \\ &\partial_t \mathbf{v} + \nabla_{\mathbf{x}} \cdot (\mathbf{v} \otimes \mathbf{v}) + \frac{1}{\rho} \nabla_{\mathbf{x}} p_i = 0. \end{aligned} \tag{3}$$

We show that the interface conditions for the coupling of isothermal (compressible) and incompressible Euler equations in one space dimension are given by an ordinary differential equation and discuss how the interface conditions can be solved numerically.

Unlike the purely compressible case, where a one-dimensional Riemann solver is sufficient to solve numerically also problems in higher space dimension, this holds not true for the coupling of compressible and incompressible flows. We show in section 3 why this is the case and discuss the numerical treatment of the interface conditions in section 3 in one space dimension as well as in higher space dimensions. Section 4 illustrates the performance of the interface coupling in two one dimensional examples.

2 ANALYTICAL RESULTS

The starting point of the following considerations is a fully compressible model, where a *weakly* compressible liquid droplet is surrounded by a compressible gas phase. We assume that the interface is given by a sharp representation and does not contain any mass. Furthermore, liquid and gas phase cannot exchange mass and any phase transition from the liquid to gas or reverse is excluded. This assumption results in the fact that we have conservation of mass in the liquid and gas phase separately. To simplify the analysis, we assume further that an isothermal description of the fluids is sufficient. All together the fully compressible model is given by

$$\begin{cases} \partial_t\rho + \partial_x(\rho v) = 0, & \text{in the gas phase} \\ \partial_t(\rho v) + \partial_x\left(\rho v^2 + p_g(\rho)\right) = 0, & x \in \mathbb{R}\setminus[a(t),b(t)] \end{cases} \tag{4a}$$

$$\begin{cases} \partial_t\rho + \partial_x(\rho v) = 0, & \text{in the liquid phase} \\ \partial_t(\rho v) + \partial_x\left(\rho v^2 + p_l(\rho)\right) = 0, & x \in [a(t),b(t)] \end{cases} \tag{4b}$$

$$\begin{cases} v_g - v_l = 0, & \text{at the interfaces} \\ p_l(\rho_l) - p_g(\rho_g) = 0, & a(t) \text{ and } b(t) \end{cases} \tag{4c}$$

$$\begin{cases} \dot{a}(t) = v_g(t,a(t)-), & \text{kinetic relations} \\ \dot{b}(t) = v_g(t,b(t)+), & \text{(no mass transfer)} \end{cases} \tag{4d}$$

Hereby, p_g and p_l are thermodynamically correct equations of state and therefore fulfil $p_i(\rho) > 0$ for all $\rho > 0$.

In (Colombo and Schleper 2012), the well-posedness of the above model (4) as well as the existence of entropy solutions for initial data with small total variation is shown. This result also includes the continuous dependence of the solution on the initial data and on the equation of state. Furthermore, it is shown that the resulting interfaces $a(t)$ and $b(t)$ fulfil $\dot{a}(t), \dot{b}(t) \in \mathrm{BV}(\mathbb{R}^+, \mathbb{R})$ and the traces of the solution at $a(t)$ and $b(t)$ are well-defined.

For the mixed model, we substitute the weakly compressible description of the liquid phase by an incompressible version. In one space dimension, the incompressible Euler equations are given by

$$\begin{aligned} &\partial_x v = 0 \\ &\partial_t v + \frac{1}{\rho}\partial_x p = 0 \end{aligned} \tag{5}$$

Note that, since $\partial_x v = 0$, we have no spacial variation of the velocity in the incompressible phase. The one dimensional droplet therefore acts like a solid body. Integrating (5) with respect to the space variable $x \in [a(t),b(t)]$, we obtain therefore

$$\dot{v}(t) = \frac{p(t,a(t)) - p(t,b(t))}{\rho L}$$

and the mixed model in one space dimension is given by

$$\begin{cases} \partial_t\rho + \partial_x(\rho v) = 0, & \text{in the gas phase} \\ \partial_t(\rho v) + \partial_x\left(\rho v^2 + p_g(\rho)\right) = 0, & x \in \mathbb{R}\setminus[a(t),b(t)] \end{cases} \tag{6a}$$

$$\begin{cases} \rho = \rho_l, & \text{in the liquid phase} \\ \dot{v}_l(t) = \dfrac{p(t,a(t)-) - p(t,b(t)+)}{(b(0)-a(0))\rho_l}, & x \in [a(t),b(t)] \end{cases} \tag{6b}$$

$$\begin{cases} v_g - v_l = 0, & \text{at the interfaces} \\ \dot{a}(t) = v_l(t), & a(t) \text{ and } b(t) \\ \dot{b}(t) = v_l(t), & \text{(no mass transfer)} \end{cases} \tag{6c}$$

In the one dimensional case, we have therefore a coupling of a partial differential equation with an ordinary differential equation at a free boundary. This model fits into the framework of (Borsche et al. 2010), where the well-posedness and existence of solutions to mixed pde-ode models is proved.

We review now some important properties of the Riemann problem in the bulk phase as well as at the interface. To this end, denote by $c_i(\rho) := \sqrt{p_i(\rho)}$ the speed of sound in the liquid $(i = l)$ and gas phase $(i = g)$ respectively. The eigenvalues and eigenvectors of the isothermal Euler equations (4a) and (4b) are given by

$$\lambda_1^i = v - c_i(\rho), \qquad r_1^i = \begin{pmatrix} -1 \\ -v + c_i(\rho) \end{pmatrix}$$

$$\lambda_2^i = v + c_i(\rho), \qquad r_2^i = \begin{pmatrix} 1 \\ v + c_i(\rho) \end{pmatrix}$$

The corresponding (forward) 1-wave curve through a given left state $u_0^i := (\rho_0^i, v_0^i)$ is given by

$$\mathcal{L}_1^i(\sigma; u_0^i) = \begin{pmatrix} \rho_1^i(\sigma) \\ v_1^i(\sigma) \end{pmatrix} = \begin{cases} \begin{pmatrix} (1-\sigma)\rho_0^i \\ v_0 - \int_{\rho_0^i}^{(1-\sigma)\rho_0^i} \frac{c_i(r)}{r}\,dr \end{pmatrix}, & \sigma \geq 0 \\ \begin{pmatrix} (1-\sigma)\rho_0^i \\ v_0 - \sqrt{\left(\frac{1}{\rho_0^i} - \frac{1}{\rho_1^i(\sigma)}\right)\left(p(\rho_1^i(\sigma)) - p(\rho_0^i)\right)} \end{pmatrix}, & \sigma < 0 \end{cases} \tag{7a}$$

Analogously, the (backward) 2-wave curve through a given right state $u_0^i := (\rho_0^i, v_0^i)$ is given by

$$\mathcal{L}_2^i(\sigma; u_0^i) = \begin{pmatrix} \rho_2^i(\sigma) \\ v_2^i(\sigma) \end{pmatrix} = \begin{cases} \begin{pmatrix} \dfrac{\rho_0^i}{(1+\sigma)} \\ v_0 - \int_{\rho_0^i}^{\rho_2^i(\sigma)} \frac{c_i(r)}{r}\,dr \end{pmatrix}, & \sigma \geq 0 \\ \begin{pmatrix} \dfrac{\rho_0^i}{(1+\sigma)} \\ v_0 + \sqrt{\left(\frac{1}{\rho_0^i} - \frac{1}{\rho(\sigma)}\right)\left(p(\rho(\sigma() - p(\rho_0^i)\right)} \end{pmatrix}, & \sigma < 0 \end{cases} \tag{7b}$$

The solution of a standard Riemann problem in the gas or liquid phase with initial conditions

$$u_0^i = \begin{cases} u_{left}^i & x < 0 \\ u_{right}^i & x > 0 \end{cases}$$

consists now in finding σ_1 and σ_2 defined in (7a) and (7b) such that

$$v_1^i(\sigma_1) = v_2^i(\sigma_2)$$
$$p_i(\rho_2(\sigma_1)) = p_i(\rho_2(\sigma_2))$$

for $i = l, g$ respectively. The unique solvability of this Riemann problem at least locally is well-known in the literature. Analogously and using the interface conditions given in (4c) and (4d) we obtain a solution to the Riemann problem at the interface by solving the system

$$v_1^g(\sigma_1) = v_2^l(\sigma_2) \tag{8a}$$

$$p_g(\rho_2(\sigma_1)) = p_l(\rho_2(\sigma_2)) \tag{8b}$$

for the compressible system. Hereby, we assume without loss of generality that the interface separates a gaseous state to the left from a liquid state to the right of the interface.

Now, to establish a more formal link between the two models (4) and (6), we consider the Riemann problem at the interface in more detail, see also Figure 1. Without loss of generality, we assume that the interface separates a gaseous state to the left from a liquid state to the right. Denote by $u_g = (\rho_g, v_g)$ and $u_l = (\rho_l, v_l)$ the two constant initial states in the gas and in the liquid and assume that u_g and u_l satisfy the interface conditions $p_g(\rho_g) = p_l(\rho_l)$ and $v_g = v_l = \Lambda_-$, where Λ_- is the

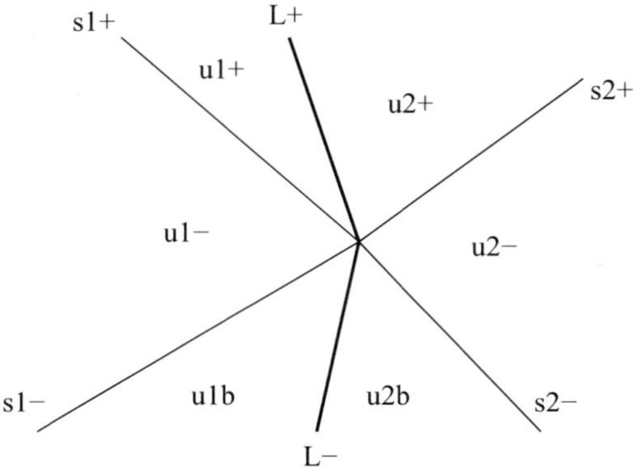

Figure 1. Illustration of the notation for the Riemann problem at the interface.

initial speed of the interface. Now, assume that two waves of size σ_1^- and σ_2^- hit the interface at time t_0, leading to a new Riemann problem with initial data u_g^- and u_l^- to the left and right of the interface respectively. As shown above, the solution of this Riemann problem is now given by exactly two waves: one wave of the first family in the gas phase and one wave of the second family in the liquid phase. We denote by σ_1^+ and σ_2^+ the size of the new waves and by $u_g^+ := \mathcal{L}_1^g(\sigma_1^+; u_g^-)$ and $u_l^+ := \mathcal{L}_2^l(\sigma_2^+; u_l^-)$ the new states to the left and right of the interface, satisfying $p_g(\rho_g^+) = p_l(\rho_l^+)$ and $v_g^+ = v_l^+ = \Lambda_+$. Using first order approximations, we can now estimate the size σ_1^+ and σ_2^+ of the new waves in terms of the incoming waves σ_1^- and σ_2^- as well as the initial data u_g and u_l. This yields the relations

$$\sigma_1^+ \simeq \frac{2}{\eta(1+\alpha\eta)}\sigma_1^- + \frac{1-\alpha\eta}{1+\alpha\eta}\sigma_2^- \tag{9a}$$

$$\sigma_2^+ \simeq -\frac{1-\alpha\eta}{1+\alpha\eta}\sigma_1^- + \frac{2\alpha\eta^2}{1+\alpha\eta}\sigma_2^- \tag{9b}$$

$$\Lambda^+ - \Lambda^- \simeq \frac{2c_g}{\eta(1+\alpha\eta)}\sigma_1^- - \frac{2c_g\alpha\eta}{1+\alpha\eta}\sigma_2^-, \tag{9c}$$

where $c_i = \sqrt{p_i(\rho_i)}$ for $i = g, l$, $\alpha = \frac{\rho_g}{\rho_l}$ and $\eta = \frac{c_g}{c_l}$. For more details see (Colombo and Schleper 2012).

In the zero Mach limit (i.e. $c_l \to \infty$), the liquid phase becomes incompressible. However, using the above relations, it is easy to see that in the limit $c_l \to \infty$ and thus $\eta \to 0$, any incoming wave σ_2^- is completely reflected by the compressible-incompressible interface, since (9a) reads $\sigma_1^+ = \sigma_2^-$ in the limit case. Note that in the limit case the wave σ_1^- is meaningless, since there cannot be any propagating wave in the incompressible phase. We can therefore assume that $\sigma_1^- = 0$. (9b) emphasises that no wave is transmittet to the liquid phase, as we get $\sigma_2^+ = 0$. Furthermore, (9c) shows that in this case, also the interface velocity doesnot change ($\Lambda^+ - \Lambda^- = 0$ in (9c)). All together, we can state that the liquid phase turns out to act like a solid wall in the Riemann solution at the interface in the zero Mach limit. This result however contradicts the expected behaviour of an incompressible droplet that should start to move due to any pressure difference between the left and the right interface and shows that it cannot be sufficient to solve a standard Riemann problem at the interface in a numerical simulation of mixed compressible-incompressible models.

However, in (Colombo and Schleper 2012) it is shown—for a special class of initial conditions—that the cumulative effect of the wave interactions with the interface in the fully compressible model yields an acceleration of the droplet that persists also in the limit case and yields the correct acceleration of the droplet. More precisely, consider the following special initial data for the fully compressible model (4):

$$\begin{cases} (p_g(\rho),v)^T(0,x)=(\bar{p},v_0)^T, & \text{for } x\in[-\infty,a(t)] \\ (p_l(\rho),v)^T(0,x)=(p_0,v_0)^T, & \text{for } x\in[a(t),b(t)], \\ (p_g(\rho),v)^T(0,x)=(p_0,v_0)^T, & \text{for } x\in[b(t),\infty], \end{cases} \tag{10}$$

where $\bar{p}$, p_0 and v_0 are constant states such that $(\bar{p},v_0)$ and (p_0,v_0) can be connected by a single Lax-shock wave of the second family in th gas phase. Note that the equations of state $p_i(\rho)$ are different in the liquid and gas regions, such that the above initial data results in a (possibly large) jump in the densities across the interface. For this special family of initial conditions, it is shown in (Colombo and Schleper 2012) that the movement of the interface in the zero Mach limit is given by the ordinary differential equation

$$\dot{\Lambda}(t)=\frac{p_g\big(t,a(t)-\big)-p_g\big(t,b(t)+\big)}{\big(a(0)-b(0)\big)\rho_l}.$$

We recover thus the model (6). This suggests that (6) is the right incompressible limit of the fully compressible model (4).

3 NUMERICAL TREATMENT OF THE INTERFACE

We are now faced with the problem to compute solutions to the coupled compressible-incompressible model. Unfortunately we cannot use the standard techniques from compressible simulations, since the Riemann problem at the interface will not yield the right solution to the mixed problem, as shown above. We have thus to deal with the full coupled model. In this section, we show in detail, how an efficient implementation of the coupling can be obtained. The resulting method allows to use existing compressible and incompressible codes that need to be supplemented by a simple routine for the interface treatment. In this sense the described method is *non invasive*, and can be obtained easily by code coupling. To fix the ideas, we start with the one dimensional case.

3.1 *One space dimension*

In the one dimensional case, the coupled model consists of a partial differential equation in the bulk phase and an ordinary differential equation at the interface. Note first that v_l is constant in x so that the interface velocities $\dot{a}(t)$ and $\dot{b}(t)$ are equal. It is therefore sufficient to compute only $a(t)$ and set $b(t)=a(t)+L$, where $L=b(0)-a(0)$ is the constant droplet size.

We have thus to find the states $a(t)$, $\rho(t,a(t)-)$, $\rho(t,b(t)+)$ and $v(t,a(t)-)$ at the interface such that $\big(\rho(t,a(t)-),v(t,a(t)-)\big)^T$ and $\big(\rho(t,b(t)+),v(t,a(t)-)\big)^T$ are valid boundary conditions for the system of conservation laws describing the gas phase and such that $a(t)$ fulfils

$$\dot{a}(t)=v(t,a(t)-)$$
$$\ddot{a}(t)=\frac{p\big(t,a(t)-\big)-p\big(t,b(t)+\big)}{\big(b(0)-a(0)\big)\rho_l}.$$

At first glance this looks like a boundary control problem, which is computationally very expensive to solve.

However, if we discretize the ordinary differential equation, we obtain a piecewise constant function defined through

$$v_l^{n+1}=v_l^n+F(\rho_-^n,\rho_+^n) \qquad \text{explicit} \tag{11a}$$
$$v_l^{n+1}=v_l^n+F(\rho_-^{n+1},\rho_+^{n+1}) \qquad \text{implicit} \tag{11b}$$
$$v_l^{n+1}=v_l^n+F(\rho_-^n,\rho_+^n,\rho_-^{n+1},\rho_+^{n+1}) \qquad \text{mixed,} \tag{11c}$$

where $\rho_-=\rho(t,a(t)-)$ and $\rho_+=\rho(t,b(t)+)$ are the states in the gas phase to the left and right of the droplet and F stands for a suitable discretization of the right hand side in (6b). To go from one time step n to $(n+1)$ in the coupled pde-ode system (6), we have to insure that ρ_-^{n+1}, ρ_+^{n+1} and v_l^{n+1} fulfil the interface conditions (6b)–(6c). Figure 2 illustrates the discrete situation.

As in the previous section, we can determine the states ρ_-^{n+1}, ρ_+^{n+1} and v_l^{n+1} by a solution of the following half Riemann problem (see also Fig. 2):

Find

1. $\big(\rho_-^{n+1},v_-^{n+1}\big)^T$ on the (forward) 1-wave curve $\mathcal{L}_1^g(\sigma;u_-^n)$ through the given state $u_-^n:=\big(\rho_-^n,v_l^n\big)$ and
2. $\big(\rho_+^{n+1},v_+^{n+1}\big)^T$ on the (backward) 2-wave curve $\mathcal{L}_2^g(\sigma;u_+^n)$ through the given state $u_+^n:=\big(\rho_+^n,v_l^n\big)$ such that
3. $v_-^{n+1}=v_+^{n+1}=v_l^{n+1}$ is given by (11a), (11b) or (11c).

Case 1 (explicit): As v_l^{n+1} is given by the initial states ρ_-^n and ρ_+^n, it is completely determined by

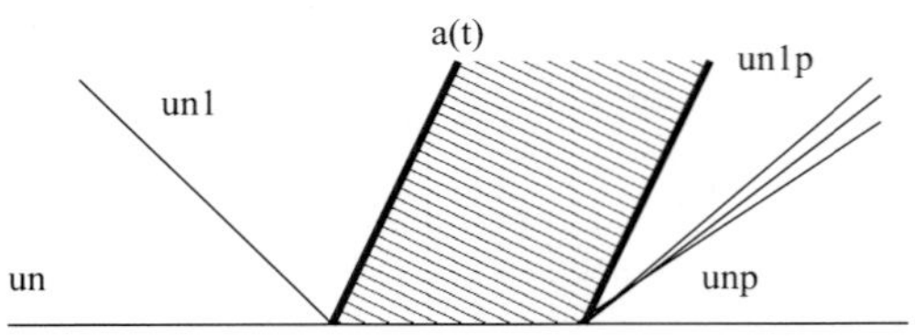

Figure 2. Illustration of the notation for the Riemann problem at the interface after discretization of the ordinary differential equation.

the states of the previous time step. We have thus to determine only σ_1^+ and σ_2^+ such that

$$v_1^g(\sigma_1^+) = v_l^{n+1} \tag{12a}$$

and

$$v_2^g(\sigma_2^+) = v_l^{n+1}. \tag{12b}$$

This yields then $\rho_-^{n+1} = \rho_1^g(\sigma_1^+)$ and $\rho_+^{n+1} = \rho_2^g(\sigma_2^+)$. Using the inverse function theorem, one can easily see that equations (12) can be solved uniquely at least locally.

Case 2 (implicit & mixed): Here $v_l^{n+1} := v_l^{n+1}(\sigma_1^+, \sigma_2^+)$, given by (11b) or (11c), has to be determined simultaneously with the states $\rho_-^{n+1} := \rho_1^g(\sigma_1^+)$ and $\rho_+^{n+1} := \rho_2^g(\sigma_2^+)$ such that $u_-^{n+1} := (\rho_-^{n+1}, v_l^{n+1})^T = \mathcal{L}_1^g(\sigma_1^+; u_-^n)$ and $u_+^{n+1} := (\rho_+^{n+1}, v_l^{n+1})^T = \mathcal{L}_2^g(\sigma_2^+; u_+^n)$. In other words, we have to find the states σ_1^+ and σ_2^+ such that

$$v_1^g(\sigma_1^+) = v_l^{n+1}(\sigma_1^+, \sigma_2^+) \tag{13a}$$

$$v_2^g(\sigma_2^+) = v_l^{n+1}(\sigma_1^+, \sigma_2^+). \tag{13b}$$

Using the implicit function theorem and the definition of the Lax curves (7), one can easily see that equations (13) have a unique solution locally around every given constant states $(\bar{\rho}_-, \bar{v})$ and $(\bar{\rho}_+, \bar{v})$ fulfilling the equation $p_g(\bar{\rho}_-) = p_g(\bar{\rho}_+)$.

Concerning the computational costs of the two different cases, observe that case 1 (explicit) consists in finding a solution to two decoupled nonlinear equations, whereas the nonlinear equations do not decouple in case 2 (implicit and mixed). We expect therefore higher computational costs for the solution of the half Riemann problem in case 2 than in case 1 (see also section 51).

Using the above described interface treatment of case 1 or 2, we can compute solutions to the one dimensional compressible-incompressible model by using standard algorithms for sharp interfaces such as the variant of the ghost fluid method described in (Jaegle and Schleper 2010) or other methods able to deal with moving boundaries.

3.2 *Higher space dimensions*

In higher space dimensions, the computational domain does in general not degenerate to two disjoint sets. Furthermore, the behaviour of the incompressible phase is no longer described by an ordinary differential equation but by a partial differential equation. More precisely, we have to deal with the following coupled system in the domain $\Omega = \Omega_g(t) \cup \Omega_l(t) \cup \Gamma(t)$.

$$\begin{cases} \partial_t \rho + \nabla_{\mathbf{x}} \cdot (\rho \mathbf{v}) = 0, & \text{gas phase} \\ \partial_t (\rho \mathbf{v}) + \nabla_{\mathbf{x}} \cdot (\mathbf{v} \otimes (\rho \mathbf{v})) + \nabla_{\mathbf{x}} p_i = 0, & \mathbf{x} \in \Omega_g(t) \end{cases} \tag{14a}$$

$$\begin{cases} \nabla_{\mathbf{x}} \cdot (\mathbf{v}) = 0, & \text{liquid phase} \\ \partial_t \mathbf{v} + \nabla_{\mathbf{x}} \cdot (\mathbf{v} \otimes \mathbf{v}) + \frac{1}{\rho} \nabla_{\mathbf{x}} p_i = 0, & \mathbf{x} \in \Omega_l(t) \end{cases} \tag{14b}$$

$$\begin{cases} (\mathbf{v}_g - \mathbf{v}_l) \cdot \mathbf{n} = 0, & \text{at the interface} \\ p_l(\rho_l) - p_g(\rho_g) = 0, & \mathbf{x} \in \Gamma(t) \end{cases} \tag{14c}$$

$$\gamma(t) = \mathbf{v}_g(t, \Gamma(t)-) \cdot \mathbf{n} \qquad \text{kinetic relation} \tag{14d}$$

where $\gamma(t)$ is the normal velocity of the interface $\Gamma(t)$ and $\mathbf{n}(t)$ is the normal vector of $\Gamma(t)$ *pointing into the incompressible region*. We illustrate the treatment of higher dimensional interfaces using the following simple two dimensional setting, see Figure 3. Hereby, we assume that the computational domain Ω is discretized into elements T_i (triangles or more general polygons or polyeder) and that each element T_i is either contained in (the discrete version of) Ω_g or Ω_l. The interface between liquid and gas phase is therefore given by a finite set of edges.

Solving the coupled equations (14) means that we have to determine the boundary states to the left and right of the (discretized) interface $\Gamma(t)$, such that the incompressibility condition $\nabla_{\mathbf{x}} \cdot \mathbf{v} = 0$ is fulfilled in Ω_l and such that the new boundary states in the gas phase are admissible for the compressible model.

We restrict the following discussion to discretization methods in the compressible phase $\Omega_g(t)$ that

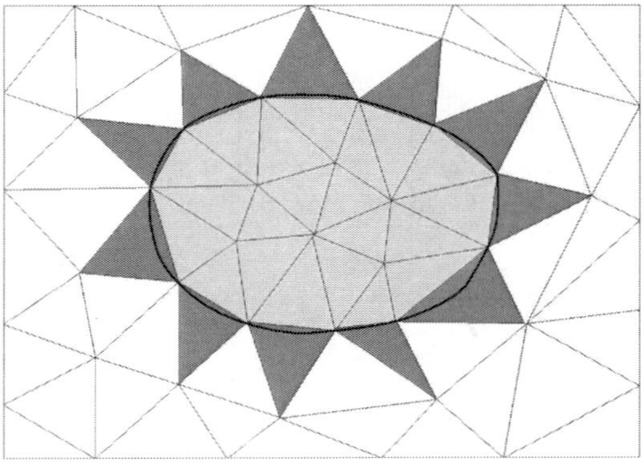

Figure 3. Illustration of a simple setting in two space dimensions. The light gray triangles show the elements with incompressible treatment (liquid phase), while the white and dark gray triangles show the elements with compressible treatment (gas). The dark gray triangles mark the elements of the compressible phase that are needed for the interface treatment.

are based on the solution of Riemann problems at the cell edges. In these methods, we can determine the boundary states of Ω_g by solving Riemann problems in direction normal to the boundary edges. Due to the orientation of the normal vector $\mathbf{n}(t)$, pointing into the incompressible region, we know that the interface separates a gas state u_g to the left and a liquid state u_l to the right in any Riemann problem in direction normal to the interface. The solution strategy is therefore essentially the same as in the one dimensional case.

In analogy to the one dimensional case, we present two different methods to compute solutions to the mixed model (14). Note first, that a solution in the incompressible region in uniquely determined by the boundary condition $\mathbf{v}(t,\Gamma(t))\cdot\mathbf{n}_{\Gamma(t)} = g(t)$ for a given function $g(t)$ together with suitable initial conditions p_l^0 and $\mathbf{v}_l^0$.

Case 1 (explicit): We use the normal velocity $\mathbf{v}_g^n\cdot\mathbf{n}$ of the compressible region at the interface as boundary condition of the incompressible region. Together with the previous state $(\mathbf{v}_l^n, p_l^n)$ this uniquely determines the solution $(\mathbf{v}_l^{n+1}, p_l^{n+1})$ in the incompressible region at ime t^{n+1}. In order to fulfil the interface conditions, we set now the boundary pressure in the compressible region to $p_g^{n+1} := p_l^{n+1}$. To obtain admissible boundary states for the compressible region, we have to find a value σ_i in each boundary cell T_i (marked dark gray in Fig. 3) connecting the new boundary state $u_{g,i}^{n+1} = (\rho_{g,i}^{n+1}, \mathbf{v}_{g,i}^{n+1}\cdot\mathbf{n}_i)^T := \mathbf{L}_1^g(\sigma_i; u_{g,i}^n)$ with the previous boundary state $u_{g,i}^n$ in the gas phase such that $p_g(\rho_{g,i}^{n+1}) = p_l^{n+1}$. This can be done separately in every element T_i and we have therefore to solve a certain number of decoupled half Riemann problems, whose solvability can be checked by the inverse function theorem as in the one dimensional case 1. The computational effort to solve these half Riemann problems is thus very low and corresponds to the effort of the solution of a standard Riemann problem by an exact Riemann solver.

Case 2 (implicit): The solution strategy in the above described Case 1 yields a violation of the interface condition $v_l^n = v_g^n$ at the interface for every n. Instead it fulfils $v_l^{n+1} = v_g^n$. While this can in general be neglected for small changes of v_g and ρ_g, it can lead to instabilities or wrong interface velocities in some cases. Therefore we present a second solution strategy, avoiding this defect at the expense of severe additional computational costs.

To fulfil all interface conditions simultaneously, we have to determine the boundary states ρ_g^{n+1}, $\mathbf{v}_g^{n+1}\cdot\mathbf{n} = \mathbf{v}_l^{n+1}\cdot\mathbf{n}$ and p_l^{n+1} such that $(\rho_g^{n+1}, v_g^{n+1}) = L_1^g(\sigma; u_g^n)$ for some σ and p_l^{n+1} is the solution of (14c) in $\Omega_l(t^{n+1})$ with initial data $u_l^n := (p_l^n, \mathbf{v}_l^n)$ and boundary condition $\mathbf{v}_l^{n+1}\cdot\mathbf{n} = \mathbf{v}_g^{n+1}\cdot\mathbf{n}$.

To show formally that this is possible, we denote by $T_1,\dots,T_m$ the interface grid cells of the mesh in the compressible region (dark gray cells in Fig. 3) and by E_i the corresponding interface edge with interface normal $\mathbf{n}_i$ pointing into the incompressible region. Let furthermore $F(u_l^n, \mathbf{v}_{g,1}^{n+1}\cdot\mathbf{n}_1,\dots,\mathbf{v}_{g,m}^{n+1}\cdot\mathbf{n}_m)$ be the solution operator in the incompressible region (light gray cells in Fig. 3) mapping the initial and boundary conditions to the solution $u_{l,j}^{n+1} := (p_{l,j}^{n+1}, v_{l,j}^{n+1})^T$ of the next time step. The operator F can be constructed using any valid discretization scheme of the incompressible Euler equations with given normal velocities at the boundary. Any solution u_l^{n+1} in the incompressible region fulfils therefore

$$\begin{pmatrix} p_l^{n+1} \\ \mathbf{v}_l^{n+1} \end{pmatrix} = F\left(p_l^n, \mathbf{v}_l^n, \mathbf{v}_{l,1}^{n+1}\cdot\mathbf{n}_1,\dots\mathbf{v}_{l,m}^{n+1}\cdot\mathbf{n}_m\right). \tag{15}$$

To fulfil the interface conditions (14c)

$$\mathbf{v}_{l,i}^{n+1}\cdot\mathbf{n}_i = \mathbf{v}_{l,i}^{n+1}\cdot\mathbf{n}_i$$
$$p_{l,i}^{n+1} = p_g(\rho_{l,i}^{n+1})$$

we substitute now the interface quantities $p_{l,i}^{n+1}$ and $\mathbf{v}_{l,i}^{n+1}\cdot\mathbf{n}_i$, $i = 1,\dots,m$ in (51) by the one parameter families $(p_g(\rho_{g,i}(\sigma_i)), (\mathbf{v}_{g,i}^{n+1}\cdot\mathbf{n}_i)(\sigma_i))$, where

$$\begin{pmatrix} \rho_{g,i}^{n+1}(\sigma_i) \\ \left(\mathbf{v}_{g,i}^{n+1}\cdot\mathbf{n}_i\right)(\sigma_i) \end{pmatrix} = L_1^g\left(\sigma_i; \begin{pmatrix} \rho_{g,i}^n \\ \mathbf{v}_{g,i}^n\cdot\mathbf{n}_i \end{pmatrix}\right).$$

Note that this substitution does not change the number of unknowns in (15). Computing now the derivative of (15) with respect to all unknowns ($\sigma_1,\dots\sigma_m$ and the remaining $u_{l,j}^{n+1}$ off from the boundary) and using the implicit function theorem, we can check if (15) is uniquely solvable locally. The result does strongly depend on the operator F and has to be checked for each discretization operator F of the incompressible Euler equations separately.

Once we know that the substituted version of (15) is solvable, we can determine the solution of this large coupled system by any solution routine for nonlinear systems (e.g. Newton's method to mention a simple and well-known example).

From a computational point of view, the implicit method is very expensive as we have to solve a large nonlinear system in every time step. This is due to the fact that all interface states $u_{g,i}^{n+1}$ in the gas phase are coupled to the solution of the incompressible system *in the whole incompressible domain* Ω_l. Especially in the case of several droplets in one computational domain Ω, the implicit method is unattractive and one should use the explicit version wherever possible.

4 EXAMPLES

To conclude this article we present a numerical comparison of the one dimensional compressible model (4) and of the implicit and explicit discretization of the mixed model (6).

We consider a droplet of size $L = 1$ and density $\rho_l = 500$, surrounded by gas. Initially, the droplet and the gas are at rest, i.e. $v_g(0,x) = v_l(0,x) = 0$. We assume that the gas can be described by the ideal gas law

$$p_g(\rho) = a^2\rho,$$

where we choose $a = 1$ in this example. Furthermore, the liquid is described by the Tait equation of state (2)

$$p_l(\rho) = k\left(\left(\frac{\rho}{500}\right)^7 - 1\right) + 1,$$

where we choose $k = \frac{7 \cdot 10}{500}$ such that the speed of sound in the liquid is approximately $\sqrt{10}$.

The initial conditions of the test example are chosen to

$$\begin{pmatrix} p_g(0,x) \\ v_g(0,x) \end{pmatrix} = \begin{cases} \begin{pmatrix} 1.5 \\ 0 \end{pmatrix}, & x \in [0,2[\\ \begin{pmatrix} 1.0 \\ 0 \end{pmatrix}, & x \in [3, x_{max}] \end{cases}$$

$$\begin{pmatrix} p_l(0,x) \\ v_l(0,x) \end{pmatrix} = \begin{pmatrix} 1 \\ 0 \end{pmatrix}, \quad x \in [2,3]$$

From these initial conditions, we obtain

$$\rho_g(0,x) = \begin{cases} 1.5, & \text{for } x \in [0,2[\\ 1, & \text{for } x \in]3, x_{max}] \end{cases}$$

and

$$\rho_l(0,x) = 500, \quad \text{for } x \in [2,3].$$

Note that the initial conditions are constructed such that the setting corresponds to a situation where a shock wave, carrying a pressure jump of 0.5 hits the droplet at time $t = 0$.

To see how the mixed model reacts on the interaction of one or more waves with the droplet, we use two test cases.

Test case 1: We assume that the gas can flow out of both boundaries unobstructed (i.e. we prescribe outflow boundary conditions at $x = 0$ and $x = 6$, (see e.g. (LeVeque 2002) for more details). Therefore, there is only the initial shock wave propagating inside the domain of calculation and we can observe how the droplet is accelerated due to the initial shock wave. We calculate the evolution of the droplet movement up to time $T = 200$ in the domain $\Omega = [0,20]$, discretized by 81 equally spaced grid points. The CFL constant is chosen to 0.9 and the corresponding time step size is determined adaptively in every time step.

For this test case, we compare the fully compressible model (4) and the mixed model (6) with a discretization of the ode by the implicit as well as the explicit Euler scheme in Table 1. The reference solution $(\bar{v}, \bar{p})$ is thereby given by the mixed model with implicit discretization. $\bar{r}$ is the location of the left interface in the reference solution.

Test case 2: We assume that the left domain boundary is a solid wall, reflecting every incoming wave without loss of momentum. The right domain boundary is kept as outflow boundary. This yields several shock and rarefaction waves hitting the droplet during the calculation time up to $T = 1000$ in the domain $\Omega = [0,6]$, discretized by 49 equally spaced grid points. The CFL constant is chosen to 0.9 and the corresponding time step size is determined adaptively in every time step.

Figure 4 shows a comparison of the resulting velocity field for the compressible model (4) and the mixed model (6) with ode discretization by implicit Euler. One can see that the difference of the two solutions is very small, even though the Tait equation of state with $k = 7 \cdot 10/500$ still allows a relevant compression of the droplet ($\approx 6\%$) for the pressure jump of the test case.

To see the influence of the compressible or incompressible treatment of the droplet on the solution, we also report the location of the interfaces in Figure 5. Here one can see that the solutions are indeed very close, but the differences increase with increasing computation time.

Table 1. Comparison of the mixed model with implicit ode discretization (implicit Euler) with the compressible model and with the mixed model with explicit ode discretization (explicit Euler). Relative error with respect to the l_2-Norm at time $T = 200$.

	$\frac{\|v-\bar{v}\|_2}{\|\bar{v}\|_2}$	$\frac{\|p-\bar{p}\|_2}{\|\bar{p}\|_2}$	$\frac{\|r-\bar{r}\|}{\bar{r}}$
Compressible	0.30%	1.43%	0.22%
Explicit	0.06%	0.92%	0.05%

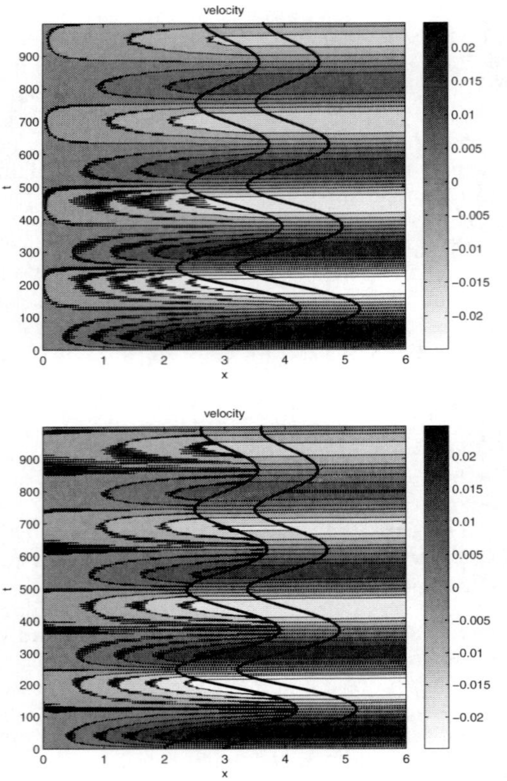

Figure 4. Comparison of the velocity field of the mixed model with implicit discretization of the ode (top) and the fully compressible model with sound speed $c_l \approx \sqrt{10}$ (bottom). The black solid lines indicate the position of the left and rightinterfaces.

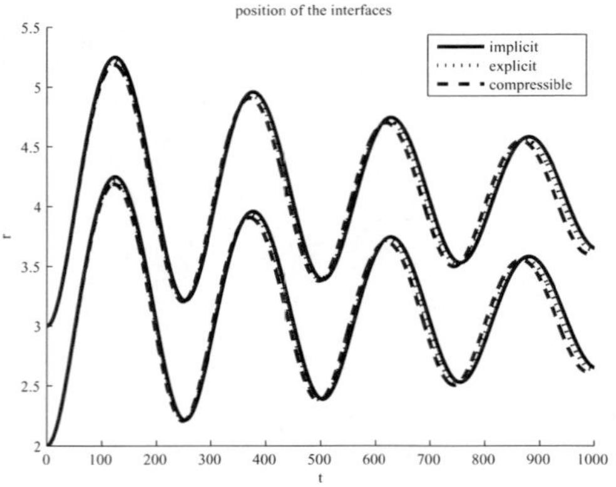

Figure 5. Comparison of the interface location for the fully compressible model (dashed line) and an implicit (solid line) and an explicit (dotted line) discretization of the ode in the mixed model.

Table 2. Comparison of the different models with respect to the computational effort: CPU time and number of time steps.

	Test case 1		Test case 2	
	Time	t-steps	Time	t-steps
Fully compressible	15 s	2878	330 s	28217
Incompressible (imp.)	5 s	955	295 s	9001
Incompressible (exp.)	5 s	955	85 s	9001

To conclude this section, we report the CPU time for the solution of the two different test cases in Table 2.

One can see that the number of time steps is drastically reduced by the use of the mixed model. However, the implicit interface discretization is computationally very expensive so that the advantage is only small. However, note that the speed of sound in the liquid phase of the fully compressible model was set to ≈3. For larger speed of sound, the number of time steps for the compressible model increases further, whereas the number of timesteps for the mixed model is constant, so that even the implicit mixed model is advantageous.

REFERENCES

Borsche, R., R.M. Colombo and M. Garavello 2012. Mixed systems: ODEs—balance laws, *J. Differential Equations 252*, 2311–2338.

Colombo, R.M. and V. Schleper 2012. Two-phase flows: Non-smooth well posedness and the compressible to incompressible limit, *Nonlinear Analysis: Real World Applications*, DOI 10.1016/j.nonrwa.2012.01.015.

Ferrari, A., C.-D. Munz and B. Weigand 2010. A High Order Sharp-Interface Method with Local Time Stepping for Compressible Multiphase Flows, *Commun. Comput. Phys. 9*, 205–230.

Jaegle, F. and V. Schleper 2010. Exact an approximate Riemann solvers at phase boundaries, *preprint*, available at http://www.ians.uni-stuttgart.de/am/Schleper/Preprints/Riemann.pdf.

LeVeque, R.J. 2002. *Finite volume methods for hyperbolic problems*, Cambridge Texts in Applied Mathematics, Cambridge: Cambridge University Press.

Liu, T.G., B.C. Khoo and C.W. Wang 2005. The ghost fluid method for compressible gas-water simulation, *J. Comput. Phys. 204*, 193–221.

IV Numerical methods in astrophysics

Numerical Methods for Hyperbolic Equations – Vázquez-Cendón et al. (eds)
© 2013 Taylor & Francis Group, London, ISBN 978-0-415-62150-2

Partially implicit high order Runge-Kutta methods for wave-like equations in spherical-type coordinates

Isabel Cordero-Carrión
Max-Planck-Institute for Astrophysics, Garching, Germany

ABSTRACT: In this paper I present new partially implicit Runge-Kutta high order methods in order to numerically evolve in time a set of partial differential equations. These methods are based on the ideas used in (Cordero-Carrión et al. 2011) to evolve numerically the hyperbolic metric equations in the so called Fully Constrained Formulation (Bonazzola et al. 2004) of Einstein equations, and are designed in order to overcome numerical instabilities, due, for example, to potential numerical unstable terms in the sources of the system or to effects of the chosen system of coordinates (as it was the case in (Cordero-Carrión et al. 2011)).

1 INTRODUCTION

The evolution in time of many complex systems, governed by Partial Differential Equations (PDE), leads—in most of the cases—to looking for the numerical solution of a system of Ordinary Differential Equations (ODEs). The most common methods to face on these systems of ODEs are the Runge-Kutta (RK) ones (see e.g. (Butcher 2003) for a general review of these methods and their properties). The family of RK methods is very large, depending on, e.g., their explicit/implicit structure or their order of convergence.

In this article, I am interested in Partially Implicit RK (PIRK) schemes. Implicit or partially implicit methods are used to deal with systems which require a special treatment in order to have a stable evolution. For example, the Implicit-Explicit (IMEX) RK methods (Asher et al. 1995; Asher et al. 1997; Pareschi 2001; Pareschi and Russo 2005) are used to evolve conservation laws with stiff terms or convection-diffusion-reaction equations. The method proposed here has been designed to face on a more general kind of systems. Moreover, due to the particular structure of the system I am interested in, the algorithm derived does not need of any matrix inversion, in contrast to a general implicit algorithm. Other implicit methods are the Explicit, Singly Diagonally Implicit RK (ESDIRK) methods (Kennedy and Carpenter 2003); they differ from the more traditional Singly Diagonally Implicit RK (SDIRK) methods (Alexander 1997; Hairer and Wanner 1996) by having an explicit first stage, something which goes in the opposite direction of one of the requirements of the schemes proposed in this paper (see next section).

Another important property of the algorithm derived is the following: it has to be a high order method with positive coefficients (see (Butcher 2003; Ruuth and Spiteri 2004) for more details about the implications of positive/negative coefficients on stability properties and storage requirements).

Initially, this work was motivated by the need of having a robust algorithm to evolve numerically Einstein equations in the Fully Constrained Formulation (FCF) (Cordero-Carrión et al. 2011; Bonazzola et al. 2004); in particular, one to be used to evolve numerically the hyperbolic sector of the metric equations. In this FCF formulation of Einstein equations, the resulting set of equations form a coupled hyperbolic-elliptic system of PDEs. The structure of the hyperbolic equations was analyzed in (Cordero-Carrión et al. 2008(a)); the gravitational radiation of the system is encoded in these equations. Transverse-traceless decomposition of metric tensors (Cordero-Carrión et al. 2008(b)) make possible a reformulation of the equations, providing local uniqueness properties to the elliptic ones. The hydrodynamics equations have to be solved, coupled with Einstein equations, in presence of matter.

The hyperbolic metric equations in FCF define a first-order system of 30 equations for the metric components and their first derivatives in space and time. It is a balance system of conservation laws, due to the presence of source terms. These sources are no stiff. In the simulations in (Cordero-Carrión et al. 2011), spherical orthonormal coordinates were used; in that system of coordinates, the components of the scalars, vectors and tensors involved in the equations are regular, both at the origin and at the pole. Using classical Explicit RK (ERK) methods from second to fourth-order of convergence, second to fourth-order finite differences for the spatial derivatives and several CFL (Courant-Friedrich-Lewis condition) values, it was not possible to obtain a stable evolution of the equations, even in a simple case as it is the evolution of Teukolsky waves with small amplitude in vacuum (Cordero-Carrión et al. 2009). The reason for the presence and growing of the instabilities is related with a $1/r$ factor in the source terms, when spherical orthonormal coordinates are used in sources involving spatial derivatives. Analytically, this factor is multiplied with a function, such that the whole term is regular at the origin $r = 0$; but numerical errors divided by the $1/r$ factor can be interpreted as stiff terms. This problem was cured with a partially implicit treatment of the sources (see (Cordero-Carrión et al. 2011) for more details), including the leading term responsible of these instabilities. As an illustrative example, I display in Figure 1 the profile of the time derivative of the component (θ, θ) of the metric tensor, in the evolution of Teukolsky waves in vacuum using both the ERK and PIRK methods described in (Cordero-Carrión et al. 2011). Although these methods were stable, due to the presence of implicit and explicit parts, the formal order of convergence of the methods reduces to around second-order. Therefore, a more rigorous analysis is necessary.

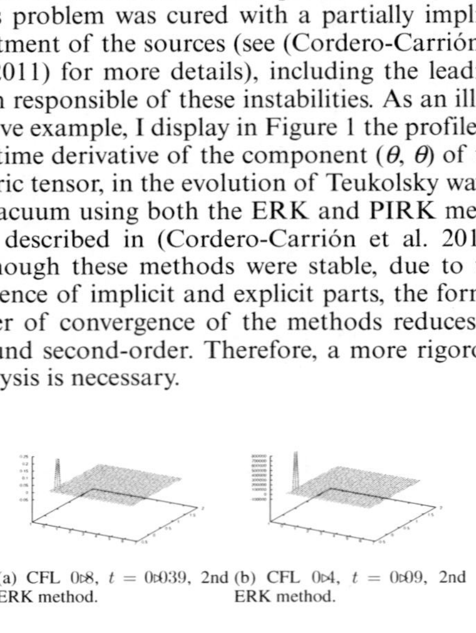

(a) CFL 0ɒ8, t = 0ɒ039, 2nd ERK method. (b) CFL 0ɒ4, t = 0ɒ09, 2nd ERK method.

(c) CFL 0ɒ8, t = 0ɒ7, 4th ERK method. (d) CFL 0ɒ4, t = 0ɒ7, 4th ERK method.

(e) CFL 0ɒ5, t = 2ɒ61, 2nd PIRK method. (f) CFL 0ɒ8, t = 2ɒ61, 4th PIRK method.

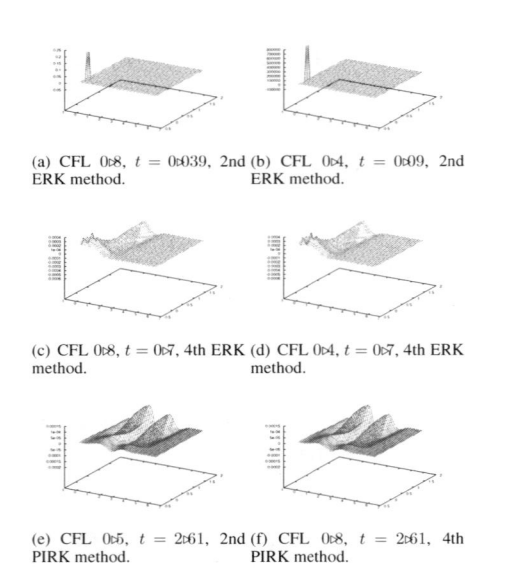

Figure 1. Radial profile of time derivative of the (θ, θ) component of the metric tensor for different CFL values using explicit and partially implicit 2nd and 4th RK methods. Red (green) lines correspond to numerical (analytical) values.

2 DERIVATION OF PIRK METHODS

In this section, I present the form of the system of equations, the requirements of the numerical schemes and finally the resulting PIRK methods from first to fourth-order of convergence.

2.1 *Structure of the system and requirements for the numerical algorithm*

Let us consider the following system of PDEs

$$\begin{cases} u_t = \mathcal{L}_1(u,v), \\ v_t = \mathcal{L}_2(u) + \mathcal{L}_3(u,v), \end{cases} \tag{1}$$

being $\mathcal{L}_1$, $\mathcal{L}_2$ and $\mathcal{L}_3$ general non-linear differential operators. Let us denote by L_1, L_2 and L_3 their discrete operators, respectively. There are no restrictions onto the discrete operators; they can correspond, for example, to a flux-conservative numerical scheme of a conservation law or a finite differences scheme of a general evolution system. On one hand, L_1 and L_3 will be treated into an explicit way, and on the other hand L_2 will be treated as much implicitly as possible but no analytical or numerical inversion will be necessary.

Numerical methods based on a nonlinear stability requirement are very desirable. Such methods were originally named as Total Variation Diminishing (TVD) and are also referred to as Strong Stability Preserving (SSP) methods. If $U = U(t)$ is a vector of discretized variables, i.e., $[U(t)]_j = U_j(t) = u(x_j,t)$, and u_j^n is the numerical approximation to $u(x_j,t_n)$, then TVD discretizations have the property that the total variation.

$$TV(U^n) = \sum \left| u_j^n - u_{j-1}^n \right| \tag{2}$$

of the numerical solution does not increase with time, i.e.,

$$TV(U^{n+1}) \leq TV(U^n). \tag{3}$$

A sequence U^n is said to be strongly stable in a given norm $||\cdot||$ provided that $||U^{n+1}|| \leq || U^n$ for all $n > 0$. The choice of the norm is arbitrary, being the TV-norm, and the infinity norm, two natural possibilities.

A s-stage ERK method for the equation $\partial_t U = L(U)$ can be written in the form:

$$\begin{aligned} U^{(0)} &= U^n, \\ U^{(i)} &= \sum_{k=0}^{i-1} \left(\alpha_{ik} U^{(k)} + \Delta t \beta_{ik} L(U^{(k)})\right), \\ & \quad i = 1,2,\ldots,s, \\ U^{n+1} &= U^{(s)}, \end{aligned} \tag{4}$$

where all the $\alpha_{ik} \geq 0$, and $\alpha_{ik} = 0$ only if $\beta_{ik} = 0$. TVD RK methods from first to fourth-order have been derived. The classical second-order one is:

$$
\begin{aligned}
U^{(0)} &= U^n, \\
U^{(1)} &= U^n + \Delta t\, L(U^n), \\
U^{n+1} &= \frac{1}{2}U^n + \frac{1}{2}U^{(1)} + \frac{1}{2}\Delta t\, L(U^{(1)}).
\end{aligned} \tag{5}
$$

The third-order one due to Shu and Osher (Shu and Osher 1988) is:

$$
\begin{aligned}
U^{(0)} &= U^n, \\
U^{(1)} &= U^n + \Delta t\, L(U^n), \\
U^{(2)} &= \frac{3}{4}U^n + \frac{1}{4}U^{(1)} + \frac{1}{4}\Delta t\, L(U^{(1)}), \\
U^{n+1} &= \frac{1}{3}U^n + \frac{2}{3}U^{(2)} + \frac{2}{3}\Delta t\, L(U^{(2)}),
\end{aligned} \tag{6}
$$

Gottlieb and Shu (Gottlieb and Shu 1998) proved that these second and third-order ERK schemes are the optimal two-stage second-order SSP RK (with CFL = 1) and three-stage third-order SSP RK (with CFL = 1) ones (the optimal adjective refers, for a given number of stages, to a maximization of the corresponding CFL value and the efficiency in the storage requirement). They also proved that a five-stage method is necessary in order to achieve fourth-order with positive coefficients. The numerically optimal five-stage fourth-order method was found in (Kraaijevanger 1991) and again independently in (Spiteri and Ruuth 2002), and guaranteed optimal in (Ruuth 2006):

$$
\begin{aligned}
U^{(0)} &= U^n, \\
U^{(1)} &= U^n + 0.391752226571890\,\Delta t\, L(U^n), \\
U^{(2)} &= 0.444370493651235\, U^n \\
&\quad + 0.555629506348765\, U^{(1)} \\
&\quad + 0.368410593050371\,\Delta t\, L(U^{(1)}), \\
U^{(3)} &= 0.620101851488403\, U^n \\
&\quad + 0.379898148511597\, U^{(2)} \\
&\quad + 0.251891774271694\,\Delta t\, L(U^{(2)}), \\
U^{(4)} &= 0.178079954393132\, U^n \\
&\quad + 0.821920045606868\, U^{(3)} \\
&\quad + 0.544974750228521\,\Delta t\, L(U^{(3)}), \\
U^{n+1} &= 0.517231671970585\, U^{(2)} \\
&\quad + 0.096059710526147\, U^{(3)} \\
&\quad + 0.386708617503269\, U^{(4)} \\
&\quad + 0.063692468666290\,\Delta t\, L(U^{(3)}) \\
&\quad + 0.226007483236906\,\Delta t\, L(U^{(4)}).
\end{aligned} \tag{7}
$$

Numerical values were taken from (Gottlieb et al. 2009).

The requirements for the numerical algorithm to solve the system (1) are the following ones:

1. PIRK method, with a diagonally implicit RK scheme for the implicit part.
2. Positive coefficients.
3. The explicit parts of the numerical method, L_1 and L_3, have to coincide with the TVD optimal ERK methods already mentioned.
4. Maximizing of the implicit parts, starting from first stage to next ones.

2.2 *First-order method*

In order to obtain the numerical methods for high order (third and fourth-order of convergence), let us start with the simplest case: the first-order one. Taking into account the first three requirements, the one-stage first-order method can be written as:

$$
\begin{cases}
u^{n+1} = u^n + \Delta t\, L_1(u^n, v^n), \\
v^{n+1} = v^n + \Delta t\big[(1-c_1)L_2(u^n) + c_1 L_2(u^{n+1}) \\
\qquad\qquad + L_3(u^n, v^n)\big].
\end{cases} \tag{8}
$$

A Taylor expansion of (u^{n+1}, v^{n+1}) and the operators evaluated there, in terms of (u^n, v^n) and their spatial derivatives, was carried out to guarantee the order of convergence of the method. Imposing the last requirement, i.e., maximizing the implicit part in the first (and unique) stage, $c_1 = 1$, and the resulting method is:

$$
\begin{cases}
u^{n+1} = u^n + \Delta t\, L_1(u^n, v^n), \\
v^{n+1} = v^n + \Delta t\big[L_2(u^{n+1}) + L_3(u^n, v^n)\big].
\end{cases} \tag{9}
$$

2.3 *Second-order method*

Let us consider now a two-stage second-order method. Taking into account the first three requirements, the method can be written as:

$$
\begin{cases}
u^{(1)} = u^n + \Delta t\, L_1(u^n, v^n), \\
v^{(1)} = v^n + \Delta t\big[(1-c_1)L_2(u^n) + c_1 L_2(u^{(1)}) \\
\qquad\qquad + L_3(u^n, v^n)\big],
\end{cases} \tag{10}
$$

$$
\begin{cases}
u^{n+1} = \frac{1}{2}\big[u^n + u^{(1)} + \Delta t\, L_1(u^{(1)}, v^{(1)})\big], \\
v^{n+1} = v^n + \frac{\Delta t}{2}\big[L_2(u^n) + 2c_2 L_2(u^{(1)}) \\
\qquad\qquad + (1-2c_2)L_2(u^{n+1}) \\
\qquad\qquad + L_3(u^n, v^n) + L_3(u^{(1)}, v^{(1)})\big]
\end{cases} \tag{11}
$$

A Taylor expansion of $(u^{(1)},v^{(1)})$, (u^{n+1},v^{n+1}) and the operator evaluated there, in terms of (u^n, v^n) and their spatial derivatives, was carried out to guarantee the order of convergence of the method. Maximizing the implicit part in the first stage, $c_1 = 1$. Maximizing the implicit part in the second and last stage, and imposing positiveness of the coefficients, $c_2 = 0$. The resulting method is:

$$\begin{cases} u^{(1)} = u^n + \Delta t\, L_1(u^n,v^n), \\ v^{(1)} = v^n + \Delta t\left[L_2(u^{(1)}) + L_3(u^n,v^n)\right], \end{cases} \tag{12}$$

$$\begin{cases} u^{n+1} = \frac{1}{2}\left[u^n + u^{(1)} + \Delta t\, L_1(u^{(1)},v^{(1)})\right], \\ v^{n+1} = v^n + \frac{\Delta t}{2}\left[L_2(u^n) + L_2(u^{n+1}) \right. \\ \qquad \left. + L_3(u^n,v^n) + L_3(u^{(1)},v^{(1)})\right]. \end{cases} \tag{13}$$

2.4 *Third-order method*

Let us consider now a higher order of convergence. Taking into account the first three requirements for a three-stage third-order method, it can be written as:

$$\begin{cases} u^{(1)} = u^n + \Delta t\, L_1(u^n,v^n), \\ v^{(1)} = v^n + \Delta t\left[(1-c_1)L_2(u^n) + c_1 L_2(u^{(1)}) \right. \\ \qquad \left. + L_3(u^n,v^n)\right], \end{cases} \tag{14}$$

$$\begin{cases} u^{(2)} = \frac{1}{4}\left[3u^n + u^{(1)} + \Delta t\, L_1(u^{(1)},v^{(1)})\right], \\ v^{(2)} = v^n + \frac{\Delta t}{4}\left[2(c_1 + 2c_2)L_2(u^n) + 4c_2 L_2(u^{(1)}) \right. \\ \qquad + 2(1 - c_1 - 4c_2)L_2(u^{(2)}) \\ \qquad \left. + sL_3(u^n,v^n) + L_3(u^{(1)},v^{(1)})\right], \end{cases} \tag{15}$$

$$\begin{cases} u^{n+1} = \frac{1}{3}\left[u^n + 2u^{(2)} + 2\Delta t\, L_1(u^{(2)},v^{(2)})\right], \\ v^{n+1} = v^n \\ \qquad + \frac{\Delta t}{6}\left[(1 + \frac{3c_3}{2}(1 + 3c_1 - 4c_2))L_2(u^n) \right. \\ \qquad + L_2(u^{(1)}) \\ \qquad + (4 + 3c_3(c_1 - 1 + 4c_2))L_2(u^{(2)}) \\ \qquad + \frac{3c_2}{2}(1 - c_1 - 4c_2)L_2(u^{n+1}) \\ \qquad + L_3(u^n,v^n) + L_3(u^{(1)},v^{(1)}) \\ \qquad \left. + 4L_3(u^{(2)},v^{(2)})\right]. \end{cases} \tag{16}$$

A Taylor expansion of $(u^{(1)},v^{(1)})$, $(u^{(2)},v^{(2)})$, (u^{n+1},v^{n+1}) and the operators evaluated there, in terms of (u^n, v^n) and their spatial derivatives, was carried out to guarantee the order of convergence of the method. Maximizing the implicit part in the first stage, $c_1 = 1$. Since positive coefficients are required, $c_2 = 0$. Coefficients in the second stage are then already fixed. Maximizing the implicit part in the third and last stage, $c_3 = 0$. The resulting method is:

$$\begin{cases} u^{(1)} = u^n + \Delta t\, L_1(u^n,v^n), \\ v^{(1)} = v^n + \Delta t\left[L_2(u^{(1)}) + L_3(u^n,v^n)\right], \end{cases} \tag{17}$$

$$\begin{cases} u^{(2)} = \frac{1}{4}\left[3u^n + u^{(1)} + \Delta t\, L_1(u^{(1)},v^{(1)})\right], \\ v^{(2)} = v^n + \frac{\Delta t}{4}\left[2L_2(u^n) \right. \\ \qquad \left. + L_3(u^n,v^n) + L_3(u^{(1)},v^{(1)})\right] \end{cases} \tag{18}$$

$$\begin{cases} u^{n+1} = \frac{1}{3}\left[u^n + 2u^{(2)} + 2\Delta t\, L_1(u^{(2)},v^{(2)})\right], \\ v^{n+1} = v^n + \frac{\Delta t}{6}\left[L_2(u^n) + L_2(u^{(1)}) + 4L_2(u^{(2)}) \right. \\ \qquad + L_3(u^n,v^n) + L_3(u^{(1)},v^{(1)}) \\ \qquad \left. + 4\,L_3(u^{(2)},v^{(2)})\right]. \end{cases} \tag{19}$$

Notice that, in this cases, second and third stage result to be explicit. Moreover, in the expression of u^{n+1} and v^{n+1} the coefficients of the L_2 and L_3 operators, evaluated in the corresponding stages, coincide.

2.5 *Fourth-order method*

Let us finish with a fourth-order method. First of all, let me remind the reader that a five-stage method is required according to previous considerations and the third requirement. Let us assume a similar contribution of the L_2 and L_3 operators in the first stage of the method for $v^{(1)}$, i.e.

$$\begin{cases} u^{(1)} = u^n + \Delta t\, L_1(u^n,v^n), \\ v^{(1)} = v^n + \Delta t\left[c_0 L_2(u^n) + c_1 L_2(u^{(1)}) \right. \\ \qquad \left. + \overline{c}_1 L_3(u^n,v^n)\right], \end{cases} \tag{20}$$

where $c_0 + c_1 = \overline{c}_1$ Then, the five-stage fourth-order method can be written as:

$$\begin{cases} u^{(1)} = u^n + \beta_{10}\,\Delta t\, L_1(u^n,v^n), \\ v^{(1)} = v^n + \Delta t\left[(\beta_{10} - c_1)L_2(u^n) + c_1\, L_2(u^{(1)}) \right. \\ \qquad \left. + \beta_{10}\, L_3(u^n,v^n)\right], \end{cases} \tag{21}$$

$$\begin{cases} u^{(2)} = u^n + \Delta t\left[B_{20}\, L_1(u^n,v^n) \right. \\ \qquad \left. + \beta_{21}\, L_1(u^{(1)},v^{(1)})\right], \\ v^{(2)} = v^n + \Delta t\left[(B_{20} + \beta_{21} - c_3 - c_2)L_2(u^n) \right. \\ \qquad + c_2\, L_2(u^{(1)}) + c_3\, L_2(u^{(2)}) \\ \qquad + B_{20}\, L_3(u^n,v^n) \\ \qquad \left. + \beta_{21}\, L_3(u^{(1)},v^{(1)})\right], \end{cases} \tag{22}$$

$$\begin{cases} u^{(3)} = u^n + \Delta t\Big[B_{30}\, L_1(u^n,v^n) + B_{31}\, L_1(u^{(1)},v^{(1)}) \\ \qquad + \beta_{32}\, L_1(u^{(2)},v^{(2)})\Big], \\ v^{(3)} = v^n + \Delta t\Big[(B_{30} + B_{31} + \beta_{32} - \sum_{i=4}^{6} c_i) L_2(u^n) \\ \qquad + \sum_{i=4}^{6} c_i\, L_2(u^{(i-3)}) \\ \qquad + B_{30}\, L_3(u^n,v^n) + B_{31}\, L_3(u^{(1)},v^{(1)}) \\ \qquad + \beta_{32}\, L_3(u^{(2)},v^{(2)})\Big], \end{cases} \tag{23}$$

$$\begin{cases} u^{(4)} = u^n + \Delta t\Big[B_{40}\, L_1(u^n,v^n) + B_{41}\, L_1(u^{(1)},v^{(1)}) \\ \qquad + B_{42}\, L_1(u^{(2)},v^{(2)}) \\ \qquad + \beta_{43}\, L_1(u^{(3)},v^{(3)})\Big], \\ v^{(4)} = v^n + \Delta t\Big[(\sum_{i=0}^{2} B_{4i} + \beta_{43} - \sum_{i=7}^{10} c_i) L_2(u^n) \\ \qquad + \sum_{i=7}^{10} c_i\, L_2(u^{(i-6)}) \\ \qquad + B_{40}\, L_3(u^n,v^n) + B_{41}\, L_3(u^{(1)},v^{(1)}) \\ \qquad + B_{42}\, L_3(u^{(2)},v^{(2)}) \\ \qquad + \beta_{43}\, L_3(u^{(3)},v^{(3)})\Big], \end{cases} \tag{24}$$

$$\begin{cases} u^{n+1} = u^n + \Delta t\Big[B_{50}\, L_1(u^n,v^n) + B_{51}\, L_1(u^{(1)},v^{(1)}) \\ \qquad + B_{52}\, L_1(u^{(2)},v^{(2)}) \\ \qquad + B_{53}\, L_1(u^{(3)},v^{(3)}) \\ \qquad + \beta_{54}\, L_1(u^{(4)},v^{(4)})\Big], \\ v^{n+1} = v^n + \Delta t\Big[B_{50}\, L_2(u^n) + B_{51}\, L_2(u^{(1)}) \\ \qquad + B_{52}\, L_2(u^{(2)}) \\ \qquad + B_{53}\, L_2(u^{(3)}) \\ \qquad + \beta_{54}\, L_2(u^{(4)}) \\ \qquad + B_{50}\, L_3(u^n,v^n) + B_{51}\, L_3(u^{(1)},v^{(1)}) \\ \qquad + B_{52}\, L_3(u^{(2)},v^{(2)}) \\ \qquad + B_{53}\, L_3(u^{(3)},v^{(3)}) \\ \qquad + \beta_{54}\, L_3(u^{(4)},v^{(4)})\Big], \end{cases} \tag{25}$$

where $c_i \geq 0$, $B_{20}: = \alpha_{21}\beta_{10}$, $B_{30}: = \alpha_{32}\alpha_{21}\beta_{10}$, $B_{31}: = \alpha_{32}\beta_{21}$, $B_{40} = \alpha_{43}\alpha_{32}\alpha_{21}\alpha_{10}$, $B_{41} = \alpha_{43}\alpha_{32}\alpha_{21}$, $B_{42} = \alpha_{43}\beta_{32}$, $B_{50} = [\alpha_{52} + (\alpha_{53} + \alpha_{54}\,\alpha_{43})\,\alpha_{32}]\,\alpha_{21}\beta_{10}$, $B_{51}: = [\alpha_{52} + (\alpha_{53} + \alpha_{54}\,\alpha_{43})\,\alpha_{32}]$, β_{21} $B_{52}: = (\alpha_{53} + \alpha_{54}\,\alpha_{43})\beta_{32}$, $B_{53}: = \alpha_{54}\beta_{43} + \beta_{53}$, and the α_{ij} and β_{ij} coefficients correspond to the ones of the optimal five-stage fourth-order ERK (written previously). A Taylor expansion of $(u^{(1)}, v^{(1)})$, $(u^{(2)}, v^{(2)})$, $(u^{(3)}, v^{(3)})$, $(u^{(4)}, v^{(4)})$, (u^{n+1}, v^{n+1}) and the operators evaluated there, in terms of (u^n, v^n) and their spatial derivatives, was carried out to guarantee the order of convergence of the method. Notice that last stage is completely explicit, and coefficients of all L_i operators evaluated in the corresponding stage values coincide.

Moreover, there are 5 extra equations involving the c_i, $i = 1, \&, 10$, coefficients. The set of coefficients $\{c_3, c_6, c_7, c_9, c_{10}\}$ can be written in terms of the other ones as:

$$c_3 \approx 0.246 + 0.239\,c_1 - 0.668\,c_2, \tag{26}$$

$$c_6 \approx 0.427 - 0.967\,c_1 - 0.826\,c_4 \\ -1.235\,c_5, \tag{27}$$

$$c_7 \approx 0.72 - 1.173\,c_1 - 0.256\,c_2 \\ -1.214\,c_4 - 0.806\,c_5 - 0.664\,c_8, \tag{28}$$

$$c_9 \approx 0.142 + 1.142\,c_1 + 0.089\,c_2 \\ +1.002\,c_4 + 1.148\,c_5 - 0.289\,c_8, \tag{29}$$

$$c_{10} \approx 0.08 - 0.022\,c_1 + 0.062\,c_2 \\ -0.245\,c_4 - 0.202\,c_5. \tag{30}$$

This choice is motivated by the following simplifications when the implicit parts are maximized; similar results are obtained if other set of five coefficients are written in terms of the other ones. Maximizing the implicit part in the first stage, $c_1 = \beta_{10}$. Imposing positive coefficients and maximizing the implicit parts in the second, third and last stages, $c_2 = 0$, $c_4 = c_5 = 0$ and $c_8 = 0$, respectively.

Taking into account all the previous equations, the resulting method is:

$$\begin{cases} u^{(1)} = u^n + \beta_{10}\,\Delta t\, L_1(u^n,v^n), \\ v^{(1)} = v^n + \beta_{10}\,\Delta t\Big[L_2(u^{(1)}) + L_3(u^n,v^n)\Big], \end{cases} \tag{31}$$

$$\begin{cases} u^{(2)} = u^n + \Delta t\Big[B_{20}\, L_1(u^n,v^n) \\ \qquad + \beta_{21}\, L_1(u^{(1)},v^{(1)})\Big], \\ v^{(2)} = v^n + \Delta t\Big[(B_{20} + \beta_{21} - c_3) L_2(u^n) \\ \qquad + c_3\, L_2(u^{(2)}) \\ \qquad + B_{20}\, L_3(u^n,v^n) \\ \qquad + \beta_{21}\, L_3(u^{(1)},v^{(1)})\Big], \end{cases} \tag{32}$$

$$\begin{cases} u^{(3)} = u^n + \Delta t\Big[B_{30}\, L_1(u^n,v^n) + B_{31}\, L_1(u^{(1)},v^{(1)}) \\ \qquad + \beta_{32}\, L_1(u^{(2)},v^{(2)})\Big], \\ v^{(3)} = v^n + \Delta t\Big[(B_{30} + B_{31} + \beta_{32} - c_6) L_2(u^n) \\ \qquad + c_6\, L_2(u^{(3)}) \\ \qquad + B_{30}\, L_3(u^n,v^n) + B_{31}\, L_3(u^{(1)},v^{(1)}) \\ \qquad + \beta_{32}\, L_3(u^{(2)},v^{(2)})\Big], \end{cases} \tag{33}$$

$$
\begin{cases}
u^{(4)} = u^n + \Delta t\Big[B_{40}\, L_1(u^n,v^n) + B_{41}\, L_1(u^{(1)},v^{(1)}) \\
\qquad + B_{42}\, L_1(u^{(2)},v^{(2)}) \\
\qquad + \beta_{43}\, L_1(u^{(3)},v^{(3)}) \Big], \\
v^{(4)} = v^n + \Delta t\Big[\Big(\sum_{i=0}^{2} B_{4i} + \beta_{43} - \sum_{i=7,9,10} c_i \Big) L_2(u^n) \\
\qquad + \sum_{i=7,9,10} c_i\, L_2(u^{(i-6)}) \\
\qquad + B_{40}\, L_3(u^n,v^n) + B_{41}\, L_3(u^{(1)},v^{(1)}) \\
\qquad + B_{42}\, L_3(u^{(2)},v^{(2)}) \\
\qquad + \beta_{43}\, L_3(u^{(3)},v^{(3)}) \Big],
\end{cases}
\tag{34}
$$

where the numerical values of the remaining coefficients are

$$c_3 = 0.33982363786964287, \tag{35}$$

$$c_6 = 0.04790456763582634, \tag{36}$$

$$c_7 = 0.26072335865585455, \tag{37}$$

$$c_9 = 0.5897708491724376, \tag{38}$$

$$c_{10} = 0.07188904053430445, \tag{39}$$

and last stage has the same expression as in Eq. (25).

3 CONCLUSIONS

In this work I have focused on a particular form of a system of ODEs (see Eq. (1) for details). I have derived PIRK methods having accuracies from first to fourth-order, and able to deal (and cure) with stability problems in the numerical evolution of a system of PDEs. The PIRK methods obtained have a part treated explicitly and another part, treated partially implicitly, for which no numerical or analytical inversion is needed. The algorithm for the explicit part keeps the good properties of the optimal ERK methods previously derived, and, on the other hand, I have tried to maximize implicit parts, from first to next stages, in the corresponding RK method according to the order of convergence and the number of stages. Moreover, I have kept the constraint of positiveness for the coefficients.

The choice of the coefficients, for the different methods, according to the strategy of maximizing the implicit parts, from first to next stages, have to be support by stability criteria. Probably, after an stability analysis of the methods, other values for the coefficients will be more convenient. The PIRK methods derived here have also to be tested numerically in order to check the behavior of these methods for different systems of ODEs, and also the formal order of convergence of the different methods. The analysis of the stability of the methods and the optimal CFL values in numerical evolutions, taking into account the implicit parts, is beyond the aim of this article. These aspects are of crucial importance and will be studied in forthcoming papers.

ACKNOWLEDGEMENTS

I. C.-C. acknowledges support from the Alexander von Humboldt Foundation. I also thank J.M. Ibáñez and A. Marquina for their useful comments.

REFERENCES

Alexander, R., 1997. Diagonally implicit RungeKutta methods for stiff O.D.E.s. *SIAM J. Numer. Anal.* **14**, 1006.

Asher, U.M., Ruuth, S.J. & Wetton, B.T.R., 1995. Implicit-explicit methods for time-dependent Pde's. *SIAM J. Numer. Anal.* **32**, 797.

Asher, U.M., Ruuth, S.J. & Spiteri, R.J., 1997. Implicit-explicit Runge-Kutta methods for time-dependent partial differential equations. *Appl. Numer. Math.* **25**, 151.

Bonazzola S., Gourgoulhon E., Grandclément, P. & Novak, J., 2004. Constrained scheme for the Einstein equations based on the Dirac gauge and spherical coordinates. *Phys. Rev. D* **70**, 104007.

Butcher, J.C., 2003. Numerical Methods for Ordinary Differential Equations, J. Wiley, Chichester.

Cordero-Carriòn, I., 2009. Evolution formalisms of Einstein equations: numerical and geometrical issues, PhD Thesis, University of Valencia, Valencia.

Cordero-Carrión, I., Cerdá-Durán, P., Dimmelmeier, H., Jaramillo, J.L., Novak, J. & Gourgoulhon, E., 2008. Improved constrained scheme for the Einstein equations: An approach to the uniqueness issue. *Phys. Rev. D* **77**, 024017.

Cordero-Carrión, I., Ibáñez, J.M., Gourgoulhon, E., Jaramillo, J.L. & Novak, J., 2008. Mathematical issues in a fully constrained formulation of the Einstein equations. *Phys. Rev. D* **77**, 084007.

Cordero-Carrión, I., Cerdá-Durán, P. & Ibááez, J.M., 2011. *Gravitational waves in dynamical spacetimes with matter content in the Fully Constrained Formulation.* arXiv:1108.0571.

Gottlieb, S. & S. Shu, C.W., 1998. Total variation diminishing Runge-Kutta schemes. *Math. Comput.* **67**, 73.

Gottlieb, S., Ketcheson, D.I. and Shu, C.W., 2009. High order strong stability preserving time discretizations. *J. Sci Comput.* **38**, 251.

Hairer, E. & Wanne,r G., 1996. Solving Ordinary Differential Equations II, Stiff and Differential Algebraic Problems, 2nd Edition, Springer-Verlag, Berlin.

Kennedy, C.A. & Carpenter, M.H., 2003. Additive Runge-Kuttaschemesforconvection-diffusion-reaction equations. *Appl. Numer. Math.* **44**, 139–181.

Kraaijevanger J.F.B.M., 1991. Contractivity of Runge-Kutta methods. *BIT* **31**, 482.

Pareschi, L., 2001. Central differencing based numerical schemes for hyperbolic conservation laws with relaxation terms. *SIAM J. Numer. Anal.* **39**, 1395.
Pareschi, L. & Russo, G., 2005. Implicit-explicit Runge-Kutta methods and applications to hyperbolic systems with relaxation. *J. Sci. Comput.* **25**, 129.
Ruuth, S.J. & Spiteri, R.J., 2004. High-order strong-stability-preserving Runge-Kutta methods with downwind-biased spatial discretizations. *SIAM J. Numer. Anal.* **42**, 974.
Ruuth, S.J., 2006. Global optimization of explicit strong-stability-preserving Runge-Kutta methods. *Math. Comput.* **75**, 183.
Shu, C.W. & Osher, S., 1988. Efficient implementation of essentially non-oscillatory shock-capturing schemes. *J. Comput. Phys.* **77**, 439.
Spiteri, R.J. and Ruuth, S.J., 2002. A new class of optimal high-order strong-stability-preserving time discretization methods. *SIAM J. Numer. Anal.* **40**, 469.

Numerical Methods for Hyperbolic Equations – Vázquez-Cendón et al. (eds)
© 2013 Taylor & Francis Group, London, ISBN 978-0-415-62150-2

Approximate Harten-Lax-van Leer Riemann solvers for relativistic magnetohydrodynamics

A. Mignone
Dipartimento di Fisica Generale, Universitá di Torino, Italy

G. Bodo
INAF, Osservatorio Astronomico di Torino, Italy

M. Ugliano
Max Planck Institut für Astrophysik, Garching, Deutschland

ABSTRACT: We review a particular class of approximate Riemann solvers in the context of the equations of ideal relativistic magnetohydrodynamics. Commonly prefixed as Harten-Lax-van Leer (HLL), this family of solvers approaches the solution of the Riemann problem by providing suitable guesses to the outermost characteristic speeds, without any prior knowledge of the solution. By requiring consistency with the integral form of the conservation law, a simplified set of jump conditions with a reduced number of characteristic waves may be obtained. The degree of approximation crucially depends on the wave pattern used in representing the Riemann fan arising from the initial discontinuity breakup. In the original HLL scheme, the solution is approximated by collapsing the full characteristic structure into a single average state enclosed by two outermost fast magnetosonic speeds. On the other hand, HLLC and HLLD improves the accuracy of the solution by restoring the tangential and the Alfvén modes therefore leading to a representation of the Riemann fan in terms of 3 and 5 waves, respectively.

1 INTRODUCTION

The numerical modeling of high-energy astrophysical plasma demands, in several circumstances, a relativistic formulation since the typical wave signals are comparable to the speed of light. For highly nonlinear flows, shock-capturing schemes have established a solid foundation for a correct description of complex flow structures with discontinuous waves. The key ingredient for such schemes is the solution of the Riemann problem which tackles the decay of a discontinuity initially separating two constant states. In the case of Relativistic Magnetohydrodynamics (RMHD) the solution consists of a seven waves self-similar pattern including pairs of fast, slow and Alfvén modes separated by a tangential discontinuity. The same patterns is also found in classical MHD. Nevertheless, due to the high degree of intrinsic nonlinearity, the solution of the Riemann problem in RMHD is considerably more elaborated than its classical counterpart. Thus, although an exact solution to the problem has been found by (Giacomazzo and Rezzolla 2006), approximate methods of solution are generally preferred because of their reduced computational cost.

In the present context we review a special class of approximate schemes, known as Harten-Lax-van Leer (Harten et al. 1983), which does not make use of the full characteristic information available from the decomposition of the flux Jacobian in eigenvectors and eigenvalues. HLL-type Riemann solvers have gained increasing popularity because of their ease of implementation and robustness. In its original form, the HLL scheme approximates the Riemann fan structure by a single averaged state, enclosed by two outermost fast magnetosonic speeds. Not surprisingly, this solver has considerably large numerical diffusion due to its inability to resolve the intermediate structure. As we shall see, the situation can be improved by restoring some of the missing waves in the representation of the Riemann fan. This comes at the cost of imposing additional jump conditions for the restored waves, therefore augmenting the number of equations and unknowns. Attempts to restore the central tangential discontinuity yields the so called HLLC scheme originally suggested by (Toro et al. 1994) in the context of the classical Euler equation and more recently extended to the RMHD equations by (Mignone et al. 2005), (Mignone and Bodo 2006) (paper I henceforth) and (Honkkila and Janhunen 2007). Additional inclusion of the rotational Alfvén waves leads to the HLLD scheme by (Mignone et al. 2009) (paper II henceforth) which can be considered as an extension of the classical scheme originally proposed by (Miyoshi and Kusano 2005).

In the present work we give a brief overview of the HLL, HLLC and HLLD formulation and provide some simple numerical benchmarks aiming at a quantitative comparison.

2 NUMERICAL FORMULATION

The equations of RMHD are derived under the physical assumptions of constant magnetic permeability and infinite conductivity, appropriate for a perfectly conducting fluid (Lichnerowicz 1967), (Anile 1989). We describe the fluid in terms of its rest-mass density ρ, three-velocity v^k ($k = x, y, z$, in units of the speed of light), gas pressure p_g and magnetic field B^k. By restricting our attention to one-dimensional flows, we write particle number and energy-momentum conservation as

$$\frac{\partial \mathbf{U}}{\partial t} + \frac{\partial \mathbf{F}}{\partial x} = 0, \tag{1}$$

Where **U** and **F** are the conserved variables and corresponding fluxes:

$$\mathbf{U} = \begin{pmatrix} D \\ m^k \\ E \\ B^k \end{pmatrix}, \quad \mathbf{F} = \begin{pmatrix} Dv^x \\ w\gamma^2 v^k v^x - b^k b^x + p\delta_{kx} \\ w\gamma^2 v^x - b^0 b^x \\ B^k v^x - B^x v^k \end{pmatrix} \tag{2}$$

In the previous equations, $D = \rho\gamma$ is the lab-frame density, γ the Lorentz factor, $m^k = w\gamma^2 v^k - b^0 b^k$ is the momentum density, $E = w\gamma^2 - p - b^0 b^0$ is the total energy density and b^μ is the covariant magnetic field four-vector with temporal and spatial components given respectively by $b^0 = \gamma \mathbf{v}\cdot\mathbf{B}$ and $b^k = B^k/\gamma + b^0 v^k$. Total pressure p and enthalpy w are written in terms of thermal and magnetic contributions: $p = p_g + b^2/2$, $w = w_g + b^2$, where $b^2 = b_\mu b^\mu$. An ideal Equation of State (EoS) is used to close the system,

$$w_g = \rho + \frac{\Gamma}{\Gamma - 1} p_g, \tag{3}$$

where Γ is the ratio of specific heats. Equation 1 is a system of hyperbolic partial differential equations allowing the propagation of 7 waves corresponding to three pairs of fast magnetosonic waves, slow magnetosonic waves and Alfvén waves separated by a tangential discontinuity.

A typical finite-volume conservative discretization of Equation 1 over a time step Δt yields

$$\mathbf{U}_i^{n+1} = \mathbf{U}_i^n - \frac{\Delta t}{\Delta x}\left(\mathbf{f}_{i+\frac{1}{2}} - \mathbf{f}_{i-\frac{1}{2}}\right), \tag{4}$$

where Δx is the mesh spacing and $\mathbf{f}_{i+1\backslash 2}$ is the upwind numerical flux computed at zone faces $x_{i+1\backslash 2}$ by solving, for $t^n < t < t^{n+1}$, the initial value problem defined by Equation 1 together with the initial condition

$$\mathbf{U}(x, t^n) = \begin{cases} \mathbf{U}_L & \text{for} \quad x < x_{i+\frac{1}{2}}, \\ \mathbf{U}_R & \text{for} \quad x > x_{i+\frac{1}{2}}, \end{cases} \tag{5}$$

where $\mathbf{U}_L$ and $\mathbf{U}_R$ are discontinuous left and right constant states on either side of the interface. This is also known as the Riemann problem. For a first order scheme, the input states are simply the cell-averages themselves, i.e. $\mathbf{U}_L = \mathbf{U}_i$ and $\mathbf{U}_R = \mathbf{U}_{i+1}$.

The decay of the initial discontinuity given by Equation 5 gives rise to a self-similar wave pattern in the $x - t$ plane where all seven wave modes can develop. Fast Waves (FW) are always located at the extremities of the Riemann fan and enclose two Alfvén Waves (AW). Inside, a pair of slow waves (SW) bound a Tangential Discontinuity (TD). A similar pattern is also found in classical MHD. Depending on the pressure jump and the norm of the magnetic field, FW and SW can be either shocks or rarefaction waves, where primary flow quantities (i.e. density, velocity, magnetic field and pressure) change smoothly or with a discontinuous transition, respectively. Across the AW, thermodynamic quantities remain continuous although, in contrast with the classical case, the tangential components of magnetic field trace ellipses instead of circles and the normal component of the velocity has a jump (Komissarov 1997). Finally, through the contact mode, only density exhibits a jump while thermal pressure, velocity and magnetic field remain continuous.

In a standard explicit numerical approach, all waves can be treated as discontinuities (Toro 1997) across which state and flux vectors must satisfy the jump conditions

$$[\mathbf{F} - \lambda\mathbf{U}]_{\lambda+} = [\mathbf{F} - \lambda\mathbf{U}]_{\lambda-} \tag{6}$$

where λ_+ and λ_- denote, respectively, the states immediately ahead and behind the (discontinuous) wave front with speed λ. In components, Equation 6 yields a set of 7 nonlinear equations and a consistent solution to the Riemann problem must *simultaneously* satisfy jump conditions for all waves. Physically relevant solutions should ensure continuity (i.e. zero jump) of velocity, magnetic field and total pressure across the TD as well as continuity of thermodynamics quantities across the AW. However, depending on the degree of approximation and on the number of waves included in the solution only some of these physical constraints can be actually imposed on the solution. This will be discussed in the next sections.

2.1 *The HLL Riemann solver*

If only the two outermost FW are used to approximate the Riemann fan (see Fig. 1, panel a), one has, from Equation 6, a total of 14 available equations and no particular constraint to satisfy. In this case, the single state can be described by 14 unknowns given by the components of **U** and **F** considered as independent quantities. In other words, $\mathbf{F} \neq \mathbf{F}(\mathbf{U})$. Thus the numerical flux can be computed as

$$\mathbf{f} = \begin{cases} \mathbf{F}_L & \text{if} \quad \lambda_L > 0 \\ \mathbf{F}^{\text{hll}} & \text{if} \quad \lambda_L \leq 0 \leq \lambda_R \\ \mathbf{F}_R & \text{if} \quad \lambda_R < 0 \end{cases} \tag{7}$$

where $\mathbf{F}_L \equiv \mathbf{F}(\mathbf{U}_L)$, $\mathbf{F}_R \equiv \mathbf{F}(\mathbf{U}_R)$ and

$$\mathbf{F}^{\text{hll}} = \frac{\lambda_R \mathbf{F}_L - \lambda_L \mathbf{F}_R + \lambda_R \lambda_L (\mathbf{U}_R - \mathbf{U}_L)}{\lambda_R - \lambda_L} \tag{8}$$

is obtained by solving the two jump conditions given by Equation 6 with $\lambda = \lambda_L$ and $\lambda = \lambda_R$. Notice that from the same solution we also have

$$\mathbf{U}^{\text{hll}} = \frac{\lambda_R \mathbf{U}_R - \lambda_L \mathbf{U}_L + \mathbf{F}_L - \mathbf{F}_R}{\lambda_R - \lambda_L} \tag{9}$$

Equations 9 and 8 can be regarded as the state and flux integral averages of the solution to the Riemann problem.

The outermost wave speeds should be estimated using the initial left and right input states. Here we follow the guess given by (Davis 1988), according to which

$$\begin{aligned} \lambda_L &= \min\left[\lambda_{cf-}(\mathbf{U}_L), \lambda_{cf-}(U_R)\right] \\ \lambda_R &= \min\left[\lambda_{cf+}(\mathbf{U}_L), \lambda_{cf+}(U_R)\right] \end{aligned} \tag{10}$$

where λ_{cf-} and λ_{cf+} are the minimum and maximum roots of the quartic equation

$$\begin{aligned} w_g\left(1-c_s^2\right)\gamma^4(\lambda - v_x)^4 &= \left(1-\lambda^2\right) \\ &\times\left[\left(b^2 - w_g c_s^2\right)\gamma^2(\lambda - v_x)^2 - c_s^2\left(b^x - \lambda b^0\right)^2\right] \end{aligned} \tag{11}$$

where $c_s = (\Gamma p_g / w_g)^{1\backslash 2}$ is the sound speed for an ideal EoS.

The HLL approach is simple to implement, cost-effective and requires only a guess to the outermost fast speed without any particular knowledge of the solution.

2.2 *The HLLC Riemann solver*

Following the approach outlined in Paper I, the initial input states $\mathbf{U}_L$ and $\mathbf{U}_R$ are now connected to each other by assuming a three-wave pattern representing, clockwise, a FW (λ_L), a TD (λ_c) and a FW (λ_R). The TD in the middle moves at the fluid speed and separates two unknown states $\mathbf{U}_{cL}$ and $\mathbf{U}_{cR}$, as schematically shown in Figure 1. Through the outermost waves, the jump conditions (Eq. 6) still give 14 equations to be satisfied. Across the TD, we impose 6 additional conditions requiring continuity of total pressure, velocity and tangential magnetic field. This leaves the freedom of choosing 20 independent unknowns (10 per state) which can be used to express the conservative variables and flux vectors in the cL and cR regions. We choose

$$\mathcal{P} = \left\{D, v^x, v^y, v^z, B^y, B^z, m^y, m^z, E, p\right\} \tag{12}$$

as our set of independent unknowns and use the fact that $m^x = (E + p)v^x - (\mathbf{v} \cdot \mathbf{B})B^x$ holds between conservative and primitive quantities. Therefore, **U** and **F** in the cL and cR regions can no longer be considered completely independent but should be expressed in terms of $\mathcal{P}$:

$$\mathbf{U}_c(\mathcal{P}) = \begin{pmatrix} D \\ (E+p)v^x - (\mathbf{v}\cdot\mathbf{B})B^x \\ m^t \\ B^t \\ E \end{pmatrix} \tag{13}$$

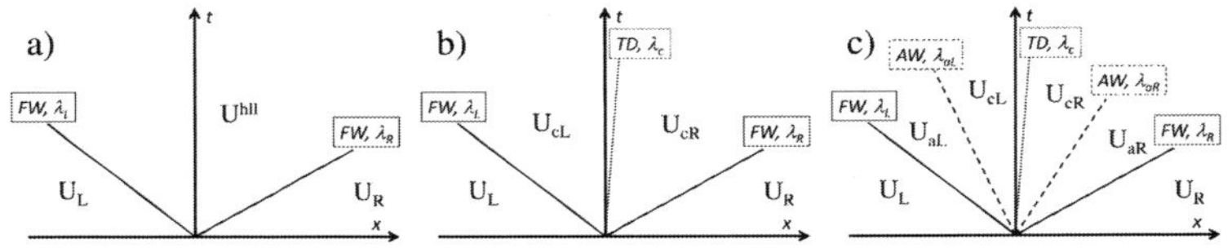

Figure 1. Approximate structures of the Riemann fan introduced by the different Riemann solvers, a) HLL, b) HLLC and c) HLLD. The different waves have been labeled with FW (fast waves with speeds λ_L and λ_R), AW (Alfvén waves with speeds λ_{aL} and λ_{aR}) and TD (tangential discontinuity, with speed λ_c).

$$\mathbf{F}_c(\mathcal{P}) = \begin{pmatrix} Dv^x \\ (E+p)\left(v^x\right)^2 - \left(\frac{B^x}{\gamma}\right)^2 + p \\ m^t v^x - B^x B^y - B^x v^t (\mathbf{v}\cdot\mathbf{B}) \\ B^t v^x - B^x v^t \\ (E+p)v^x - (\mathbf{v}\cdot\mathbf{B})B^x \end{pmatrix} \tag{14}$$

with $c = cL$ or $c = cR$ and where $t = y, z$ is used to label the transverse velocity and magnetic field components. Note also that the normal component of magnetic field should be continuous and treated as a parameter in the solution.

The numerical flux can then be computed as

$$f = \begin{cases} \mathbf{F}_L & \text{if } \lambda_L > 0 \\ \mathbf{F}_L + \lambda_L\left(\mathbf{U}_{cL} - \mathbf{U}_L\right) & \text{if } \lambda_L \le 0 \le \lambda_c \\ \mathbf{F}_R + \lambda_R\left(\mathbf{U}_{cR} - \mathbf{U}_R\right) & \text{if } \lambda_c \le 0 \le \lambda_R \\ \mathbf{F}_R & \text{if } \lambda_R < 0 \end{cases} \tag{15}$$

In order to express the states $\mathbf{U}_{cL}$ and $\mathbf{U}_{cR}$ in terms of $\mathcal{P}$ we conveniently combine the jump conditions (6) across the three waves to obtain the consistency conditions,

$$\frac{(\lambda_c - \lambda_L)\mathbf{U}_{cL} + (\lambda_R - \lambda_c)\mathbf{U}_{cR}}{\lambda_R - \lambda_L} = \mathbf{U}^{\text{hll}}, \tag{16}$$

and

$$\frac{\lambda_R(\lambda_c - \lambda_L)\mathbf{F}_{cL} + \lambda_L(\lambda_R - \lambda_c)\mathbf{F}_{cR}}{\lambda_R - \lambda_L} = \lambda_c \mathbf{F}^{\text{hll}}. \tag{17}$$

From the continuity of vector fields across the TD, the transverse components of magnetic field are simply found to be $B^t_{cL} = B^t_{cR} = B^{t,\text{hll}}$, while the normal $v^x_{cL} = v^x_{cR} = v^x_c$ velocity can be shown, from Equations 16 and 17, to satisfy the negative branch of the quadratic equation

$$a\left(v_c^x\right)^2 + bv_c^x + c = 0 \tag{18}$$

where $a = F_E^{\text{hll}} - \mathbf{B}^{t,\text{hll}} \cdot F_{\mathbf{B}^t}^{\text{hll}}, b = -F_{m^x}^{\text{hll}} - E^{\text{hll}} + \left|\mathbf{B}^{t,\text{hll}}\right|^2 + \left|\mathbf{F}_{\mathbf{B}^t}^{\text{hll}}\right|^2$ and $c = m^{x,\text{hll}} - \mathbf{B}^{t,\text{hll}} \cdot \mathbf{F}_{\mathbf{B}^t}^{\text{hll}}$ puted using the state and flux integral averages given by Equations 9 and 8. The transverse components of velocity, $v^t_{cL} = v^t_{cR} = v^t_c$ can be found from the magnetic field flux components in Equation 17, $B^x v_c^t = B_c^{t,\text{hll}} v_c^x - F_{B^t}^{\text{hll}}$. Similarly, the momentum and energy components of Equation 17 can be manipulated to give

$$E^{\text{hll}} v_c^x + p_c v_c^x - B^x\left(\mathbf{v}_c \cdot \mathbf{B}_c\right) = m^{x,\text{hll}} \tag{19}$$

which is finally used to recover the total pressure p_c.

The remaining quantities in the cL or cR regions are readily obtained from the jump conditions across the corresponding outermost FW:

$$D_c = \frac{\lambda - v^x}{\lambda - v_c^x} D \tag{20}$$

$$m_c^t = \frac{-B^x\left[B_c^t / \left(\gamma_c^2\right) + \left(\mathbf{v}_c \cdot \mathbf{B}_c\right)v_c^t\right] + R_{m^t}}{\lambda - v_c^x} \tag{21}$$

$$E_c = \frac{R_E + p_c v_c^x - \left(\mathbf{v}_c \cdot \mathbf{B}_c\right)B^x}{\lambda - v_c^x} \tag{22}$$

where we take $\lambda = \lambda_L$ for $c \equiv cL$ or $\lambda = \lambda_R$ for $c \equiv cR$. Similarly, R_{m^t} and R_E are known quantities corresponding to the transverse momentum and energy components of the array

$$\mathbf{R} = \lambda\mathbf{U} - \mathbf{F}(\mathbf{U}) \tag{23}$$

obtained from either the left ($\mathbf{R}_L = \lambda_L \mathbf{U}_L - \mathbf{F}_L$) or right ($\mathbf{U}_R = \lambda_R \mathbf{U}_R - \mathbf{F}_R$) input state.

In the limit of vanishing normal component of magnetic field ($B^x \to 0$), the previous derivation breaks down since flux components involving transverse velocities are unbounded for genuinely three-dimensional flows (see paper I). This is a direct consequence of the fact that the Riemann problem becomes degenerate, since AW and SW propagate at the velocity of the tangential mode for $B_x = 0$. Indeed, the assumption of continuity of tangential velocity and magnetic field components in our derivation is no longer valid and the number of unknowns can be reduced from 10 to 8 by replacing Equation 12 with $\mathcal{P} = \{D, v^x, m^t, B^t, E, p\}$. In this case one can show that Equation 18 still retains its form but with coefficients now given by $a = F_E^{\text{hll}}, b = -F_{m^x}^{\text{hll}} - E^{\text{hll}}$ and $c = m^{x,\text{hll}}$, while Equations 19 and 20–22 should be evaluated by taking $B^x = 0$. In this limiting case, HLLC gives a better representation of the exact solution which is now described in terms of 3 waves rather than 7. We also note that the alternative approach proposed by (Honkkila and Janhunen 2007) does not seem to suffer from this drawbacks.

2.3 *The HLLD Riemann solver*

Following paper II, we now divide the Riemann fan into 4 states separated by 5 waves, namely, two outermost FW (λ_R and λ_L) enclosing a pair of AW (λ_{aR} and λ_{aL}) separated by a TD λ_c, see panel c) in Figure 1. As before, the initial states $\mathbf{U}_L$ and $\mathbf{U}_R$ are inputs to the problem and a consistent solution must be sought in terms of a well-balanced set of equations and unknowns. Besides continuity of velocity, total pressure and magnetic field across the TD we further require that scalar quantities carry zero-jump through the AW. This leaves the freedom of expressing conservative quantities and fluxes in each of the 4 regions in terms of 8 unknowns given by the set $\mathcal{P} = \{D, v^x, v^y, v^z, B^y, B^z, w, p\}$. The solution can then be written as

$$\mathbf{f} = \begin{cases} \mathbf{F}_L & \text{if} \quad \lambda_L > 0 \\ \mathbf{F}_{aL} & \text{if} \quad \lambda_L \le 0 \le \lambda_{aL} \\ \mathbf{F}_{aL} + \lambda_{aL}\left(\mathbf{U}_{cL} - \mathbf{U}_{aL}\right) & \text{if} \quad \lambda_{aL} \le 0 \le \lambda_c \\ \mathbf{F}_{aR} + \lambda_{aR}\left(\mathbf{U}_{cR} - \mathbf{U}_{aR}\right) & \text{if} \quad \lambda_c \le 0 \le \lambda_{aR} \\ \mathbf{F}_{aR} & \text{if} \quad \lambda_{aR} \le 0 \le \lambda_R \\ \mathbf{F}_R & \text{if} \quad \lambda_R < 0 \end{cases} \tag{24}$$

where

$$\begin{aligned} \mathbf{F}_{aL} &= \mathbf{F}_L + \lambda_L\left(\mathbf{U}_{aL} - \mathbf{U}_L\right) \\ \mathbf{F}_{aR} &= \mathbf{F}_R + \lambda_R\left(\mathbf{U}_{aR} - \mathbf{U}_R\right) \end{aligned} \tag{25}$$

follows from the jump conditions across the FW. Note that Equation [16] in paper II contains a misprint since the wave speeds in the third and fourth expressions should be λ_{aL} and λ_{aR} and the fluxes in the aL and aR regions are correctly given by Equation 25.

Relativistic AW do not preserve continuity of the normal velocity as in classical MHD. As a consequence, only total pressure p can be assumed constant throughout the Riemann fan. From these considerations, a closed-form solution can be achieved by treating p as a free parameter and by reducing the jump conditions down to a scalar nonlinear equation which must solved iteratively. Hereafter we summarize the procedure and refer the reader to paper II for the detailed derivation.

We begin from the states immediately behind the outermost FW. For the sake of exposition, we drop the indices aL or aR in the unknowns (i.e. $\mathcal{P} = P_{aL,aR}$) and use λ to denote either λ_L or λ_R. With these definitions, the following expressions for the velocities can be derived:

$$v^x = \frac{B^x\left(AB^x + \lambda C\right) - \left(A + G\right)\left(p + R_{m^x}\right)}{X}, \tag{26}$$

$$v^t = \frac{QR_{m^t} + R_{B^t}\left[C + B^x\left(\lambda R_{m^x} - R_E\right)\right]}{X}, \tag{27}$$

where $t = y, z$, while

$$A = R_{m^x} - \lambda R_E + p\left(1 - \lambda^2\right), \tag{28}$$

$$G = R_{B^y}R_{B^y} + R_{B^z}R_{B^z}, \tag{29}$$

$$C = R_{m^y}R_{B^y} + R_{m^z}R_{B^z}, \tag{30}$$

$$Q = -A - G + \left(B^x\right)^2\left(1 - \lambda^2\right), \tag{31}$$

$$X = B^x\left(A\lambda B^x + C\right) - \left(A + G\right)\left(\lambda p + R_E\right). \tag{32}$$

Here the different R's are again computed from Equation 23.

Having defined the three components of velocity through the relations 26–27, one immediately obtains the transverse magnetic field, total enthalpy, density and energy from the jump conditions across the FW:

$$B^t = \frac{R_{B^t} - B^x v^t}{\lambda - v^x}, \quad w = p + \frac{R_E - \mathbf{v}\cdot\mathbf{R}_m}{\lambda - v^x} \tag{33}$$

$$D = \frac{R_D}{\lambda - v^x}, \quad E = \frac{R_E + pv^x - \left(\mathbf{v}\cdot\mathbf{B}\right)B^x}{\lambda - v^x} \tag{34}$$

while the momentum follows from the definition, $m^k = (E + p)v^k - (\mathbf{v}\cdot\mathbf{B})B^k$.

Next, we focus on the AW and take advantage of the fact that the expressions

$$K^k_{cL} = K^k_{aL} = \left(\frac{R_{m^k} + p\delta_{kx} - R_{B^k}S_x\sqrt{w}}{\lambda p + R_E - B^x S_x\sqrt{w}}\right)_L, \tag{35}$$

$$K^k_{cR} = K^k_{aR} = \left[\frac{R_{m^k} + p\delta_{kx} + R_{B^k}S_x\sqrt{w}}{\lambda p + R_E + B^x S_x\sqrt{w}}\right]_R, \tag{36}$$

are respectively invariant across λ_{aL} and λ_{aR} (Anile 1989) and that $K^x_{aL} = \lambda_{aL}$, $K^x_{aR} = \lambda_{aR}$ define the AW velocities. In the previous expressions $S_x = \text{sign}(B^x)$ and the R's are still the components of Equation 23 computed at the outermost FW using either $\lambda = \lambda_L$ or $\lambda = \lambda_R$, respectively, for the left or right state.

Imposing continuity of the normal velocity across the TD, $v^x_{cL} - v^x_{cR} = 0$, results in

$$\left(K^x_{aR} - K^x_{aL}\right) = B^x\Delta Y, \tag{37}$$

where

$$\Delta Y = \frac{1 - \mathbf{K}^2_R}{S_x\sqrt{w_R} - \mathbf{K}_R\cdot\mathbf{B}_c} + \frac{1 - \mathbf{K}^2_L}{S_x\sqrt{w_L} + \mathbf{K}_L\cdot\mathbf{B}_c}, \tag{38}$$

is a function of the total pressure p alone while $\mathbf{B}_c = \mathbf{B}_{cL} = \mathbf{B}_{cR}$ is the magnetic field in proximity of the contact wave, defined by

$$B^k_c = \frac{\left[B^k\left(\lambda - v^x\right) + B^x v^k\right]_{aR}}{\lambda_{aR} - \lambda_{aL}} - \frac{\left[B^k\left(\lambda - v^x\right) + B^x v^k\right]_{aL}}{\lambda_{aR} - \lambda_{aL}} \tag{39}$$

Equation 37 is a nonlinear equation in the total pressure variable and should be solved by means of a standard root-finder method. Once p has been found with sufficient accuracy, the velocities across the TD can be found by inverting the relation that holds between K^k and the velocity v^k. The final result is

$$v^k = K^k - \frac{B^k\left(1-\mathbf{K}^2\right)}{\pm S_x\sqrt{w} - \mathbf{K}\cdot\mathbf{B}}, \tag{40}$$

for $k = x, y, z$. Finally, density, energy and momentum are recovered from the jump conditions across λ_{aL} and λ_{aR} similarly to what done after Equation 34.

Although Equation 37 may have, in some circumstances, more than one root, the rationale for choosing the physically relevant solution is based on positivity of density and on preserving the correct eigenvalue order, i.e. $v^x_{aL} > \lambda_L, v^x_{cL} > \lambda_{aL}$ for the left state and $v^x_{aR} < \lambda_R, v^x_{cR} < \lambda_{aR}$ for the right state. In the sporadic occasions where one or more of these conditions cannot be met, we revert to the simpler HLL solver.

3 BENCHMARKS AND APPLICATIONS

A quantitative comparison between the HLL, HLLC and HLLD Riemann solvers has been presented by the authors in paper I and paper II. Similar investigations may also be found in the work of (Antón et al. 2010) and (Beckwith and Stone 2011).

Here we discuss some original and complementary aspects aiming at strenghtening important differences between the selected schemes. We use the PLUTO code for astropysical fluid dynamics with the 2nd order Corner-Transport-Upwind scheme (Mignone et al. 2007; Mignone et al. 2011) with the constrained transport method to preserve the divergence-free condition of magnetic field.

3.1 *One dimensional shock tube*

In the first test we consider the breakup of a discontinuity separating two constant states given by $(\rho, v^y, v^z, B^y, B^z, p)L = (1, 0.3, 0.4, 6, 2, 5)$ for $x < 0.5$ and $(\rho, v^y, v^z, B^y, B^z, p)R = (0.9, 0, 0, 5, 2, 5.3)$ otherwise. The normal component of velocity and magnetic field are initially constants everywhere and equal to $v_x = 0$ and $B_x = 1$, respectively. An ideal EoS is used with $\Gamma = 5/3$. This configuration, originally proposed by (Giacomazzo and Rezzolla 2006), has been also presented in paper II for a 1st-order scheme. Here, we show the results obtained with the 2nd-order scheme using 1600 zones and a Courant number of 0.8.

The exact solution (Giacomazzo and Rezzolla 2006) consists, from left to right, of a fast rarefaction wave, an Alfvén discontinuity, a slow shock, a contact discontinuity, a slow shock, an Alfvén discontinuity and a fast shock. In Figure 2 we show an enlargement of the regions around the left and right-going AW and plot the y components of magnetic field obtained with the three different solvers. The HLLD solver gives the sharpest resolution of the discontinuous fronts, whereas the HLL solver presents the largest smearing due to the inherent lack of information in the representation of the Riemann fan. The same trend, already estblished in paper II, is considerably larger for a 1st order scheme. Still, the example demonstrates the effectiveness and strength of adopting a more complete Riemann solver when describing the rich and complex features arising in relativistic magnetized flows.

3.2 *CP Alfvén waves*

We now consider the propagation of large amplitude, Circularly Polarized (CP) Alfvén waves on a two-dimensional unit square domain, as in (Del Zanna et al. 2007). The initial condition consists of a uniform medium with constant density and pressure, $\rho = p = 1$. The wave front is formed by rotating along the main diagonal the magnetic field and velocity vectors

$$\mathbf{B} = B_0\left(1, \eta\cos\phi, \eta\sin\phi\right),$$
$$\mathbf{v} = -v_A\left(0, \frac{B^y}{B_0}, \frac{B^z}{B_0}\right), \tag{41}$$

where $\phi = 2\pi(x + y)$, B_0 is the (constant) magnetic field component in the direction of propagation, η is the amplitude and the Alfvén velocity v_A is computed from

$$v_A^2 = \frac{2\alpha}{1+\sqrt{1-4\eta^2\alpha^2}}, \quad \alpha = \frac{B_0^2}{w_g + B_0^2\left(1+\eta^2\right)} \tag{42}$$

thus corresponding to $v_A \approx 0.382$ for our parameter choice. Equations 41–42 set the initial conditions

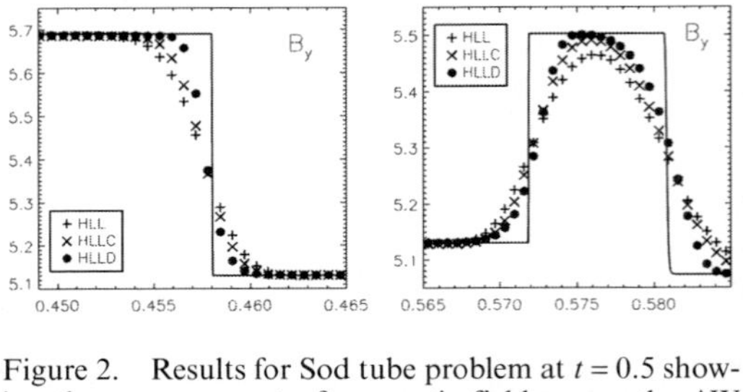

Figure 2. Results for Sod tube problem at $t = 0.5$ showing the y component of magnetic field across the AW waves. The exact solution is shown as a solid line while the numerical solutions computed on 1600 grid zones with the HLL, HLLC and HLLD solvers are overplotted using, respectively, plus signs, crosses and filled circles.

for an exact wave solution of the RMHD equations provided $\phi \rightarrow \phi - wt$, where $w = \sqrt{8\pi} v_A$ is the angular frequency and are thus valid for arbitrary amplitude η. Following (Del Zanna et al. 2007) we choose $B_0 = \eta = 1$ and carry out the integration with Courant number $C_a = 0.4$ using $N_x \times N_x$ grid zones. The ideal EoS (Eq. 3) is employed with $\Gamma = 4/3$.

In the left panel of Figure 3 we measure, as a function of the resolution N_x, the accuracy of the selected Riemann solvers by computing, after one period $T = 1\left(\sqrt{2} v_A\right)$, the L_1 norm errors of the vertical component of velocity v^z. All solvers yield 2nd-order accurate solutions although HLLC and HLLD exhibit somewhat smaller errors when compared to HLL.

In the right panel of Figure 3 we check the dissipative properties of the three solvers by measuring the decay of the wave amplitude, defined as δv^z = max(vz) − min(vz) (normalized to its initial value) up to ten revolutions, using 8 and 16 zones per wavelength. At the end of the computation, the results obtained with HLL, HLLC and HLLD at the lowest resolution show the wave amplitude has been reduced, respectively, to $\sim 10^{-5}$, $\sim 2.5 \times 10^{-3}$ and $\sim 2 \times 10^{-2}$ of the nominal value. By increasing the resolution to 16 zones, we notice a drastic improvement although the HLLD scheme still gives the lowest dissipation (~0.55 of the nominal value) followed by HLLC (~0.46) and HLL (~0.3).

These results are relevant, for example, in the context turbulence modeling where the amount of numerical dissipation is crucial in the formation of small scale structure and vortex generation. We also note that further studies should address the directionally-biased dissipation at smaller angles since the error observed in an oblique propagation is usually minimized at 45° since contributions coming from different directions have comparable magnitude.

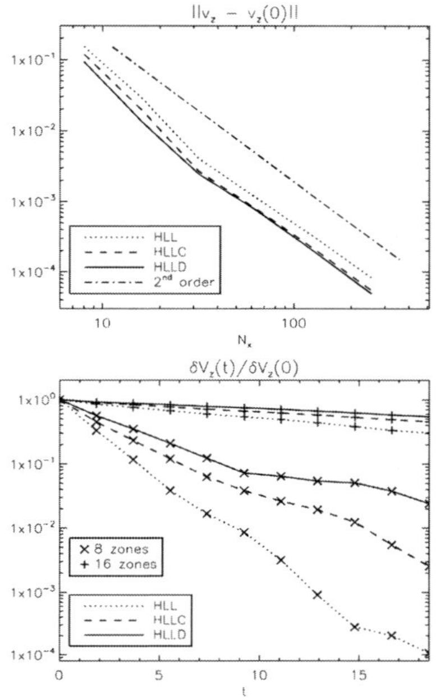

Figure 3. Results for the circularly polarized AW test. Top: L_1 error norm of v_z after one period for HLL, HLLC and HLLD Riemann solvers as a function of the number of zones. Bottom: normalized amplitude of v_z as a function of time for the three solvers (dotted, dashed and solid lines) using 8 (plus signs) and 16 (x signs) cells per wavelength.

3.3 *Spine sheath jets*

As a final application, we discuss the propagation of a three-dimensional relativistic magnetized spine-sheath jet (Mizuno et al. 2007), initially confined in the unit cylinder $r = \left(x^2 + y^2\right)^{1/2} < 1$ and flowing in the positive z direction with velocity $v_j = \left(1 - 1/\gamma_j^2\right)^{1/2}$ where $\gamma_j = 2.5$ is the Lorentz factor. Inside the jet, we set $\rho_j = 1$ while pressure and vertical magnetic field are given by

$$p_j = \frac{c_{sj}^2 \rho_j}{\Gamma - c_{sj}^2 \dfrac{\Gamma}{\Gamma - 1}}, \quad B_{zj} = \frac{V_{Aj}}{c_{sj}} \sqrt{\frac{\Gamma p_j}{1 - V_{Aj}^2}} \qquad (43)$$

where $\Gamma = 13/9$ while $c_{sj} = 0.511$ and $V_{Aj} = 0.064$ are the sound and Alfvén speeds inside the jet. Outside the beam, where $\rho_e = 0.5$, the environment is at rest with sound speed $c_{se} = 0.574$ and the vertical magnetic field B_{ze} is given from the condition of total pressure balance, $p_j + B_{zj}^2/2 = p_e + B_{ze}^2/2$. At the lower z boundary flow variables are held constant to their inflow values except for $r < 1$ where we impose a jet precession by adding a transverse radial velocity component with constant amplitude 0.01 and angular frequency of rotation $\omega = 0.4 v_j$. Free outflow is assumed on the remaining sides. This corresponds to the weakly magnetized RHDbn case of (Mizuno et al. 2007).

We carry out simulations with the HLL and HLLD Riemann solvers on the computational Cartesian domain [−6, 6] × [−6, 6] × [0, 60] using 10 points per jet radius, for a total of 128×128×640 grid zones. The Courant number if set to $C_a = 0.4$. Results are shown in the two panels of Figure 4, showing a volume rendering of density and magnetic field structure at $t = 60$ for the two schemes. The chosen configuration is unstable to helical Kelvin-Helmholtz modes which grow during the propagation. At $t = 60$, the instability has become disruptive showing large flow and magnetic field distortions. Computation performed with HLLD shows, however, that the instability

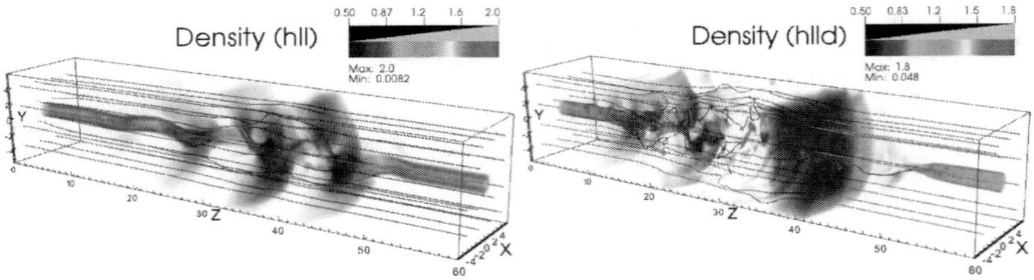

Figure 4. Volume rendering of density for the spine-sheath jet at $t = 60$. The left and right panels refer to computations obtained using the HLL and HLLD solver, respectively. Magnetic field lines are overplotted.

has grown at a faster rate as the beam rupture is in a more advanced stage and closer to the injection region. On the contrary, computations obtained with HLL present a larger amount of numerical dissipation which tend to delay or suppress the growth of small scale modes (i.e. large wavenumbers).

4 SUMMARY

A review of Harten-Lax-van Leer (HLL) Riemann solver types has been presented. Three different schemes based on increasingly more sophisticated approximations of the Riemann fan have been summarized, namely HLL, HLLC and HLLD. The three solvers describe the structure of the Riemann problem in terms of 2, 3 and 5 waves representing, progressively, two outermost fast magnetosonic waves, tangential discontinuity and Alfvén waves.

Selected numerical tests have shown that HLLD and HLLC provide better accuracy and disclose higher resolution of small-scale structures together with reduced amount of numerical dissipation.

REFERENCES

Anile, A.M. (1989). *Relativistic fluids and magneto-fluids: with applications in astrophysics and plasma physics.*

Antón, L., J.A. Miralles, J.M. Martí, J.M. Ibáñez, M.A. Aloy, and P. Mimica (2010, May). Relativistic Magnetohydrodynamics: Renormalized Eigenvectors and Full Wave Decomposition Riemann Solver. *ApJS 188*, 1–31.

Beckwith, K. and J.M. Stone (2011, March). A Second-order Godunov Method for Multi-dimensional Relativistic Magnetohydrodynamics. *ApJS 193*, 6.

Davis, S.F. (1988). Simplified second-order godunov-type methods. *SIAM Journal on Scientific and Statistical Computing 9*(3), 445–473.

Del Zanna, L., O. Zanotti, N. Bucciantini, and P. Londrillo (2007, October). ECHO: a Eulerian conservative high-order scheme for general relativistic magnetohydrodynamics and magnetodynamics. A&A *473*, 11–30.

Giacomazzo, B. and L. Rezzolla (2006, September). The exact solution of the Riemann problem in relativistic magnetohydrodynamics. *Journal of Fluid Mechanics 562*, 223–259.

Harten, A., P.D. Lax, and B. van Leer (1983). On upstream differencing and godunov-type schemes for hyperbolic conservation laws. *SIAM Review 25*(1), 35–61.

Honkkila, V. and P. Janhunen (2007, May). HLLC solver for ideal relativistic MHD. *Journal of Computational Physics 223*, 643–656.

Komissarov, S.S. (1997, February). On the properties of Alfvén waves in relativistic magnetohydrodynamics. *Physics Letters A 232*, 435–442.

Lichnerowicz, A. (1967). *Relativistic Hydrodynamics and Magnetohydrodynamics.*

Mignone, A. and G. Bodo (2006, May). An HLLC Riemann solver for relativistic flows—II. Magnetohydrodynamics. MNRAS *368*, 1040–1054.

Mignone, A., G. Bodo, S. Massaglia, T. Matsakos, O. Tesileanu, C. Zanni, and A. Ferrari (2007, May). PLUTO: A Numerical Code for Computational Astrophysics. *ApJS 170*, 228–242.

Mignone, A., S. Massaglia, and G. Bodo (2005, Novem ber). Relativistic MHD Simulations of Jets with Toroidal Magnetic Fields. *Space Sci. Rev. 121*, 21–31.

Mignone, A., M. Ugliano, and G. Bodo (2009, March). A five-wave Harten-Lax-van Leer Riemann solver for relativistic magnetohydrodynamics. *MNRAS 393*, 1141–1156.

Mignone, A., C. Zanni, P. Tzeferacos, B. van Straalen, P. Colella, and G. Bodo (2011, October). The PLUTO Code for Adaptive Mesh Computations in Astrophysical Fluid Dynamics. *ArXiv e-prints.*

Miyoshi, T. and K. Kusano (2005, September). A multistate HLL approximate Riemann solver for ideal magnetohydrodynamics. *Journal of Computational Physics 208*, 315–344.

Mizuno, Y., P. Hardee, and K.-I. Nishikawa (2007, June). Three-dimensional Relativistic Magnetohydrodynamic Simulations of Magnetized Spine-Sheath Relativistic Jets. *ApJ 662*, 835–850.

Toro, E. (1997). *Riemann solvers and numerical methods for fluid dynamics: a practical introduction.* Springer.

Toro, E.F., M. Spruce, and W. Speares (1994, July). Restoration of the contact surface in the HLL-Riemann solver. *Shock Waves 4*, 25–34.

Numerical Methods for Hyperbolic Equations – Vázquez-Cendón et al. (eds)
© 2013 Taylor & Francis Group, London, ISBN 978-0-415-62150-2

MUSTA schemes in magnetohydrodynamics and neutrino transfer: Application to core-collapse supernovae and gamma-ray bursts

Martin Obergaulinger & Miguel Ángel Aloy
Departament d'Astronomia i Astrofísica, Universitat de València, Spain

ABSTRACT: The influence of magnetic fields on the evolution of core-collapse supernovae and gamma-ray bursts is not fully understood. In many cases, the initial fields are dynamically insignificant and have to be amplified, e.g., by unstable, turbulent flows. In order to follow these flows, the numerical modelling of these processes requires highly accurate methods. A further complexity arises from the large range of additional physical effects that may have an influence on the explosion, chief among them the transport of neutrinos and their interaction with the stellar matter. Neutrino transport is numerically a very challenging problem because of the high dimensionality of the distribution function of the neutrinos. It is possible to reduce the computational requirements by solving the equations governing the evolution of the first few moments of the neutrino distribution function. The equations for the zeroth and first moments can be written as a hyperbolic system, and therefore can be treated numerically with standard methods for hyperbolic systems.

We describe a code employing high-order reconstruction methods and multistage (MUSTA) approximate Riemann solvers to solve the MHD equations coupled to the system of neutrino moments and show results of simulations addressing the issue of field amplification in core-collapse supernovae and gamma-ray bursts.

1 INTRODUCTION

The evolution of compact astrophysical objects such as neutron stars and black holes, in particular their explosive formation events, viz. core-collapse supernovae (SNe) and Gamma-Ray Bursts (GRBs, i.e., explosions characterised by strong gamma-ray emission and highly relativistic ejecta), is affected by a broad range of physical effects. Neutron stars concentrate a mass around $M_{NS} \sim 1.5\ M_{\odot}$[1] into a radius of about $R_{NS} \gtrsim 12$ km, and, therefore, their structure is the result of the interplay between general relativistic gravity and the Equation of State (EOS) of matter above nuclear density, $\rho_{nuc} \sim 2 \times 10^{14}$ g cm^{-3}; for a review of the physics of the EOS of neutron stars and the current uncertainties, see Lattimer 2010. Relativistic gravity is even more important for black holes.

Neutron stars and stellar-mass black holes are formed in some of the most energetic explosions in the universe. When a star of more than ~8 $M_{\odot}$ has exhausted its nuclear fuel, its inner core of a radius of $R_{ic} \sim 1500$ km, consisting of $M_{ic} \sim 1.5\ M_{\odot}$ of heavy elements synthesised by nuclear reactions in previous evolutionary stages, becomes unstable against gravity and collapses. It leaves behind a hot, dense Proto-Neutron Star (PNS) ($R_{PNS} \sim 40$ km) that will subsequently radiate away its excess thermal energy and contract to a cold neutron star or a black hole. The gravitational energy liberated in the collapse is partially used to eject the outer layers of the star in a supernova or a long GRB (i.e., one of a duration of more than 2 seconds) and to power the electromagnetic emission associated with these explosions. The details of this energy conversion, involving hydrodynamics and neutrino transport are not well understood, and current models have difficulties in explaining the observed properties of most SN explosions.

An alternative channel for the formation of stellar-mass black holes is the merger of two neutron stars. The merger remnant will most likely have a mass exceeding the upper mass limit for neutron stars. Therefore it is prone to collapse to a black hole. This event as well as mergers in which one or both of the neutron stars are replaced by a black hole are accompanied by a short GRB. Again the details of the acceleration of the gas to high kinetic energies and the emission of non-thermal radiation are still the topic of intense investigation.

The gas flows in these explosions are described very well by one-fluid hydrodynamics. Under some conditions, in particular if the objects rotate rapidly, magnetic fields may have a strong influence on

[1] 1 $M_{\odot} = 1.989 \times 10^{3}$ g is the mass of the sun.

the dynamics, and we have to use the equations of Magnetohydrodynamics (MHD) for modelling. In all these objects, the dimensionless numbers measuring the importance of non-ideal effects, viz. the hydrodynamic and magnetic Reynolds numbers, are very large, and therefore turbulence is very common, often triggered by one of various instabilities. Even if the Reynolds numbers have nominally very small values, turbulent flows may be affected by formally very small non-ideal effects such as viscosity and resistivity. Thus, numerical simulations need to have a high accuracy, and should in many situations include physical viscosity and resistivity. If limited computational resources do not allow for numerical grids sufficiently fine to resolve the smallest scales of turbulent flows (a condition that can almost never be met in astrophysics), sub-grid models could be used to model the flows on these small spatial scales. However, for the purpose of MHD flows in compact objects few suitable sub-grid models exist, and, thus, auxiliary large-scale simulations are needed to develop and calibrate such models.

The computationally most expensive ingredient in SN simulations is the treatment of neutrino radiation. Neutrinos are produced in large amounts in the hot dense matter of SNe and GRBs. Due to the tiny cross sections for reactions between neutrinos and matter, they do in fact carry away by far most of the gravitational binding energy released in SN core collapse, which would by far be sufficient to power a SN, and only a small fraction (on the percent level) is transferred to the matter. Because the success of a SN simulation hinges on energy conversions with such a low efficiency, high accuracy in the neutrino transport can be quite crucial.

SN cores contain matter both at supra-nuclear densities that is opaque to neutrinos and at much lower densities transparent to neutrinos. Neutrinos diffuse out of the opaque core until they reach the neutrinosphere where the neutrino-optical depth is unity. The semi-transparent region around the neutri-nosphere is the location of the most important heating reactions transferring energy from the neutrinos to the matter (the *gain region*). Consequently, numerical schemes for neutrino transport in SNe should treat the neutrino field equally well at all optical depths, the opaque and transparent regions where the diffusion and free-streaming approximation are applicable, respectively, as well as the intermediate region where both approximations break down.

In addition to this, the interaction rates between matter and neutrinos are a function of neutrino energy and neutrino flavour. The resulting need for a multi-energy multi-flavour transport further increases the computational demands.

These requirements can be met by different numerical techniques. The costs for the most elaborate transport, viz. an S_N method, are still prohibitively high; nevertheless, several groups (Mezzacappa and Bruenn 1993; Liebendörfer, Messer, Mezzacappa, Bruenn, Cardall, and Thielemann 2004) have implemented this method. Other codes are based on the formal solution of the Boltzmann equation underlying radiative transfer (e.g., Rampp and Janka 2002), on the flux-limited diffusion approximation (e.g., (Livne, Burrows, Walder, Lichtenstadt, and Thompson 2004)), or on the decomposition of the neutrino field into a free-streaming and a trapped population (Liebendörfer, Whitehouse, and Fischer 2009). In an alternative approach to radiative transfer, (Pons, Ibnez, and Miralles 2000) have suggested to solve the hyperbolic two-moment system of transport using standard methods for hyperbolic systems such as *High-Resolution Shock Capturing* (HRSC) schemes.

We note, however, that there is a wide variety of much simpler approximations that can be used in simulations addressing only specific questions. The overall dynamics of a SN core as seen in detailed simulations can be reproduced at least qualitatively when one approximates the cooling and heating of matter due to the emission and absorption of neutrinos by simple source terms depending on the position and on thermodynamic variables such as the gas density, electron fraction and temperature.

2 OUR NUMERICAL METHODS

Our goal is to develop a numerical code for simulations of a wide range of MHD flows in highly dynamic phases of the evolution of compact objects. Consequently, we have to include the physics described in the previous section with sufficient accuracy. The code should be able to evolve the coupled system of (viscous, resistive) MHD and (energy-dependent) neutrino transport, but also include simpler approximations for neutrino cooling and heating.

We solve the MHD equations in a Eulerian finite-volume discretisation using HRSC methods for hyperbolic system. One of the most important properties of the MHD system is the divergence constraint: the magnetic field, $\vec{b}$, has to satisfy the condition $\vec{\nabla} \cdot \vec{b} = 0$. Unless this constraint is built into the numerical methods explicitely, small numerical errors will inevitably lead to the growth of $\vec{\nabla} \cdot \vec{b}$, which in turn will cause spurious forces acting on the fluid and therefore degrade the quality of the numerical solution (Brackbill and Barnes 1980). To avoid these errors, we apply the *Constrained Transport* (CT) scheme (Evans and Hawley 1988) limiting the divergence error to the level of round-off errors by a discretisation that is consistent with the integral version of the induction equation, i.e., the evolution equation for the magnetic field,

$$\partial_t \vec{b} = \vec{\nabla} \times \vec{E}, \tag{1}$$

where $\vec{E}$ is the electric field; for ideal MHD, $\vec{E} = \vec{v} \times \vec{b}$. Applying Stokes' theorem to derive the integral version of the induction equation, CT schemes evolve surface averages of the magnetic field over the interfaces between grid cells, i.e., the magnetic flux, rather than volumetric averages. Since the conserved variables of hydrodynamics (the densities of mass, momentum, and total energy), on the other hand, are discretised as volumetric averages over grid cells, CT codes have to deal with a dual set of variables, viz. volume averages and surface averages. To compute the electric field, hydrodynamic variables have to be transferred to cell interfaces, and to compute the magnetic forces acting on the fluid, volume averages of the magnetic field have to be reconstructed from the surface averages.

Certainly the most critical part in terms of the computational costs, the neutrino transport demands for a compromise between accuracy and numerical effort. The radiation (neutrino) field is described by the radiative intensity $I(t, \vec{r}, \vec{n}, \in)$ measuring the energy flux transmitted through a unit surface into the direction of the unit vector $\vec{n}$ at time t and position $\vec{r}$ by neutrinos of a particle energy $\in$ (see Mihalas and Weibel Mihalas 1984). Since the intensity is a function of seven independent variables, a direct discretisation of its evolution equation, the Boltzmann equation of radiative transfer, is not feasible except for few simple problems. The dimensionality can be reduced by an expansion of I into moments $I^k(t, \vec{r}, \in) = \int d\Omega(\vec{n})^k I$, $k = 0, 1, 2, \ldots$. The kth moment is a tensor of rank k based on the kth tensorial product of the unit vector.

Though in principle I is the sum of an infinite series of moments, truncating the series after the first few moments yields an accuracy sufficient for most applications. The evolution equation for the kth moment involves the moment of rank $k + 1$ in the flux terms. Therefore, a numerical scheme based on the first k moments must close the system of equations with an assumption for the moment of rank $k + 1$. The simplest moment system evolves only a scalar equation for the 0th moment, i.e., the radiation energy density; together with a (flux-limited) diffusion closure, this leads to a parabolic equation. We include the 1st moment, i.e., the neutrino energy flux (which is equal to the neutrino momentum density) $\vec{F}$, and apply an analytic closure to the (tensorial) 2nd moment, $P^{ij}(\varepsilon, \vec{F})$, the pressure tensor. In the multi-dimensional case, the tensor has the form (Audit, Charrier, Chièze, and Dubroca 2002).

$$\mathcal{P}^{ij} = \frac{1-X}{2}\delta^{ij} + \frac{3X-1}{2}n^i n^j \tag{2}$$

The closure depends on a single scalar function X, the *variable Eddington factor*; $\vec{n} = \vec{\mathcal{F}} / |\vec{\mathcal{F}}|$ is unit vector pointing into the direction of the neutrino flux. The Eddington factor has to go to 1/3 and 1 for small and large values of the normalised flux factor, $|\vec{\mathcal{F}}|/(c\varepsilon)$. The result is a system of one scalar and one vector conservation law, which has hyperbolic character for suitable choices of the Eddington factor (Pons, Ibãñez, and Miralles 2000).

The interchange of energy and momentum between neutrinos and matter (emission, absorption, and scattering) introduces source terms into the respective equations. For dense matter opaque to neutrinos, these sources terms are stiff. In the opaque limit, the energy flux has to be equal to the diffusive flux, $\mathcal{F}_{\text{diff}} = c/3\kappa \vec{\nabla}\varepsilon$, where k is the opacity of the gas. Numerical schemes have to obey this constraint.

The two-moment system with analytic closure yields a good approximation of the neutrino field from the opaque (free-streaming) to the transparent (diffusion) limit. In terms of accuracy, it is inferior to a two-moment system using an Eddington factor determined from an additional evolutionary equation such as the model Boltzmann equation used by Rampp and Janka 2002. On the other hand, the analytic closure has a much lower complexity and computational costs.

We are using standard techniques for hyperbolic systems to deal with both the MHD and the neutrino transport parts of our code. Following HRSC schemes (see, e.g., LeVeque 2002), we apply a two-step procedure to compute the fluxes required to update the conserved variables of these systems.

First, we reconstruct the volume averages of the hydrodynamic conserved variables to zone interfaces. We have implemented reconstruction methods of different formal orders of accuracy, viz. piecewise constant (1st order), total variation diminishing piecewise linear (2nd order), and monotonicity-preserving methods of up to 9th order (Suresh and Huynh 1997).

In the second step, we obtain the fluxes by applying a Riemann solver to the reconstructed variables. We are applying HRSC schemes to systems for which the exact Riemann solutions are unknown or rather involved. Therefore, we are using approximate Riemann solvers (Lax-Friedrichs, HLL, GFORCE (Toro and Titarev 2006), and HLLD (Miyoshi and Kusano 2005)) that do not rely only on the (minimum and maximum) characteristic velocities of the system rather than on a decomposition of the Riemann problem into characteristic variables. Approximate solvers tend to have considerably larger numerical dissipation or diffusion than exact ones. With only little computational effort, this shortcoming can be amended by the *multi-stage* (MUSTA) scheme due to Toro and Titarev 2006. A MUSTA scheme opens the Riemann fan at each interface by a predictor-corrector approach:

1. We apply the approximate Riemann solver to the reconstructed (left and right) interface states to get a flux through that interface.
2. Using this flux, we evolve the interface states for one virtual time step, yielding intermediate predictor states.
3. In the last step, we determine the final interface fluxes by applying the Riemann solver again to the predictor states.

Steps 1 and 2 can be repeated arbitrary times, each iteration reducing the error of the MUSTA solver w.r.t. an exact solver. Usually one iteration is sufficient to obtain a quality of the solution comparable to that of an exact solver.

In a HRSC scheme, the hyperbolic system is transformed into a time-dependent semi-discrete system of ordinary partial differential equations, which we solve using explicit Runge-Kutta integrators of 2nd, 3rd, or 4th order and a method-of-lines approach for multidimensional simulations.

Besides the hyperbolic terms, the two systems of MHD and neutrino transport contain additional terms:

1. We use a finite-difference discretisation of the derivatives of the velocity and magnetic fields appearing in the viscous and resistive terms, respectively, and include these effects into the explicit time integration.
2. Local, potentially stiff, source terms such as the ones describing the coupling between matter and neutrinos are solved in an operator-split way using implicit time integrators.

Since the neutrino distribution function and the neutrino-matter reaction rates depend on particle energy and on neutrino flavour, we evolve the spectral (multi-energy), multi-flavour transport system. In practice, this means that we have to solve one set of moment equations for each flavour and each energy bin. Different energy bins and different flavours are coupled by terms depending on the fluid velocities (blue- and redshift) as well as by the reactions with matter.

3 TESTS AND APPLICATIONS

The new code has been subject to standard tests for MHD and radiative transfer. We will discuss selected tests in this section, starting with the MHD part and present a few applications. For more tests, see Obergaulinger 2008.

A first set of tests compares the accuracy of different reconstruction techniques and Riemann solvers.

This comparison is most conveniently done in one-dimensional simple problems. These tests confirm the gain in accuracy that can be achieved by using MUSTA solvers. For all Riemann solvers we used, the MUSTA version exhibits a considerably lower error than the corresponding non-MUSTA solver. The largest gain in accuracy can be found for the most diffusive of the solver, viz. Lax-Friedrichs, but also the more accurate HLL and GFORCE profit from a combination with the MUSTA scheme. Similar comparisons using different reconstruction methods show that for higher-order methods the potential gain from a MUSTA scheme is less pronounced, but can still be found.

We show results for one transport test, proposed by Pons, Ibáñez, and Miralles 2000, that can serve as a toy model for a star. The test involves a spherically symmetric opacity profile that is constant for radii $r < 1$ (the core) and goes to zero at radii $1 < r < 10$ (the envelope of the star). The core is surrounded by a cloud, i.e., a region of high constant opacity at radii $15 < r < 17$. Initially the core contains a high radiative energy starting to be emitted from the core. See Figure 1 for an overview of the

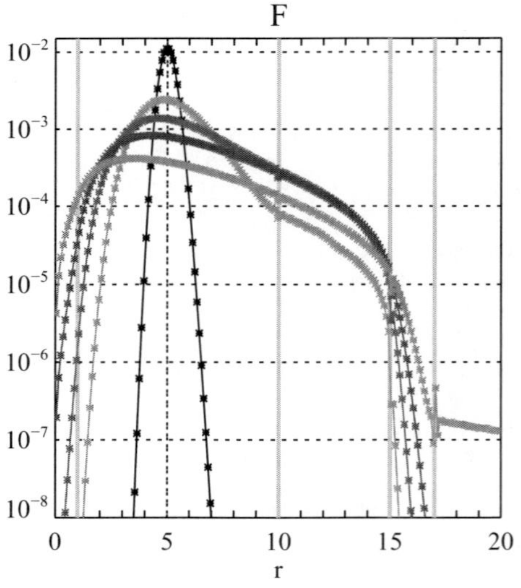

Figure 1. Profiles of the radiation momentum density in a stellar toy model. The lines show the results of a simulation at times $t \approx 0$, 5, 100, 250, 500, and 1000 with black, blue, light green, dark green, red, and orange lines, respectively. Yellow vertical lines show the borders of the different opacity regimes.

time evolution of the radiative energy flux. In the centre, the core is opaque, and radiation diffuses, while the decreasing opacity in the envelope allows for the transition from diffusion to free streaming. Outside the star, the radiation streams freely until it encounters the opaque cloud at $r = 15$. Since the diffusion time through the cloud is of the order our simulation time, the radiation just starts to leak out of the outer edge of the cloud when we finish the test at $t = 1000$. This test demonstrates that the hyperbolic two-moment transport performs well in the transparent and opaque as well as in the intermediate semi-transparent regime.

Our code has been employed for simulations of highly dynamic phases of the evolution of stars and compact objects. We will present selected applications and begin the discussion with simulations of MHD instabilities in simplified settings that do not include the complex microphysics of the nuclear EOS or neutrino transport. Instabilities that eventually lead to turbulence may affect the evolution of both core-collapse SNe and neutron-star mergers. Global simulations of these objects may lack the resolution required to follow the evolution of small-scale features in the flow and the magnetic field, which makes it hard to determine, e.g., the saturation level of the instabilities from these simulations.

Magnetic fields affect the evolution of core-collapse SNe if they possess an energy density comparable to the kinetic energy of the flow. Since the pre-collapse field strength is expected to be rather weak (Meier, Epstein, Arnett, and Schramm 1976), efficient amplification of the field has to take place during and after the collapse. Several mechanisms for field amplification involve a rapid rotation of the core. Among these, the Magneto-Rotational Instability (MRI; Balbus and Hawley 1991; Balbus and Hawley 1998) leading to an exponential growth of a weak seed field may be most promising. Under conditions typical for a SN core, the MRI grows fastest on very short scale; furthermore, its saturation, and hence the magnetic energy that can be reached, depends crucially on physics at small spatial scales. Thus, the MRI is studied extensively by means of local (*shearing-box*) simulations (Hawley and Balbus 1992).

In order to study the physics of the MRI, we started a series of local simulations (Obergaulinger, Cerdá-Durán, Müller, and Aloy 2009). These differ from similar computations performed before by the fact that they focus on conditions typical for SN cores: they study, e.g., the interplay between the MRI and convection. The goal of these models is to find the dependence of the saturated state on, e.g., the initial field strength. Although simulations in small local boxes can use comparably fine grid resolutions, highly accurate numerical methods are nevertheless important since they have a reduced numerical viscosity and resistivity, and therefore allow to explore a larger range of (magnetic) Reynolds numbers. We show the structure of the magnetic field of a three-dimensional box simulation in Figure 2. In the non-linear saturated phase of the instability, the flow and magnetic field exhibit features of a wide range of spatial scales. The field is organised in magnetic flux tubes elongated in the direction of the fluid rotation (the green structures in the figure). The mean magnetic energy of the turbulent field is low, but in many simulations coherent large-scale structures develop recurrently, the so called *channel modes*, that correspond to a strongly increased magnetic field strength.

We investigated a related issue in the context of neutron-star mergers (Obergaulinger, Aloy, and Müller 2010). The two merging neutron stars touch each other in a contact surface where strong shear flows lead to the growth of Kelvin-Helmholtz (KH) instabilities. Large-scale global simulations of mergers (Price and Rosswog 2006; Rezzolla, Giacomazzo, Baiotti, Granot, Kouveliotou, and Aloy 2011) found that the magnetic field of the neutron stars is amplified very strongly there. Due to the limited resolution of the simulations, no converged saturation level of the field strength could be determined. Therefore, we performed local box simulations of idealised models of magnetised KH instabilities to investigate the dependence of the saturated magnetic field on the seed field and the parameters of the shear flow. We were able to determine laws for the scaling of the final field strength and energy with the most important parameters of the model. Our results

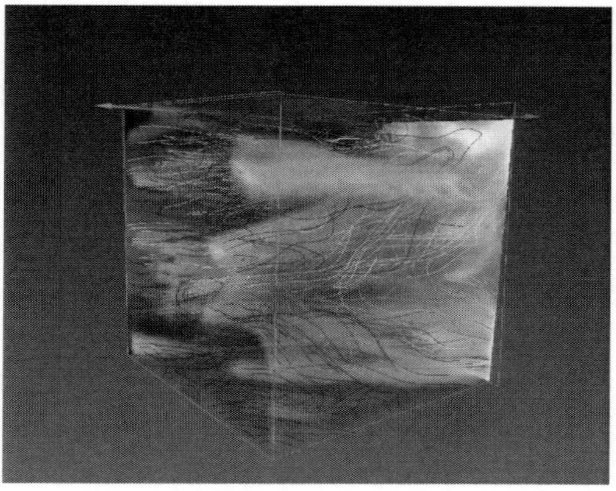

Figure 2. The structure of 3d MRI model in the nonlinear saturated state: a volume rendering of the magnetic field strength and magnetic field lines. The red and green arrows are pointing in the radial and Φ (rotational) direction, respectively.

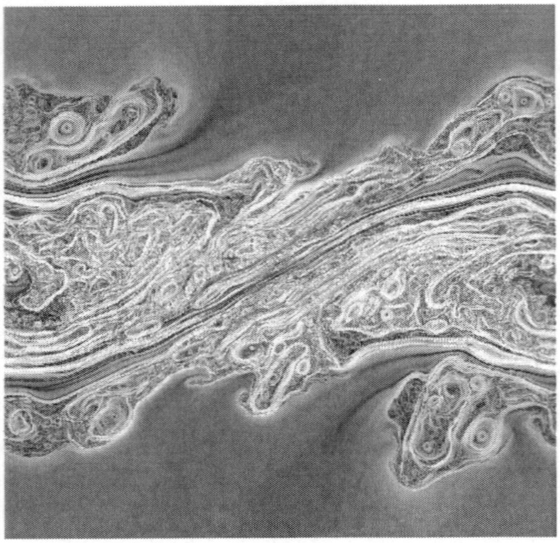

Figure 3. The structure of the magnetic field in a two-dimensional KH simulation.

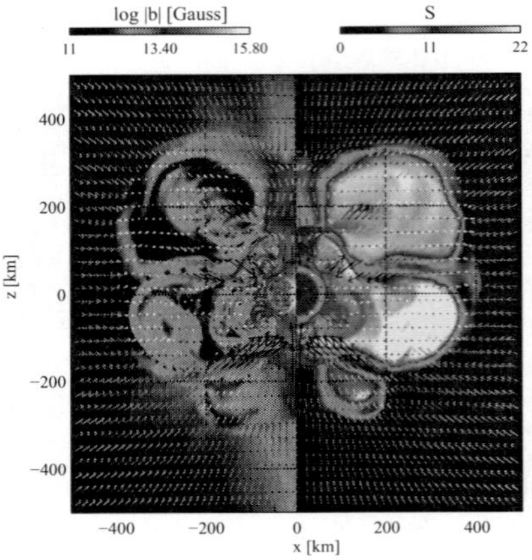

Figure 4. A magnetised SN core in the post-collapse phase. The left and right halves of the figure show colour-coded the magnetic field strength and the entropy per baryon, respectively, and the velocity field (arrows). The model is symmetric about the *z*-axis.

confirm local equipartition of the magnetic field with the kinetic energy of the flow as an upper limit to the amplification.

While rotation is the main ingredient of most field amplification mechanisms in SN cores, some mechanisms do not depend on rotation. Various hydrodynamic instabilities operate in the post-collapse phase (convection, the standing accretion shocks instability; see, e.g., Scheck, Janka, Foglizzo, and Kifonidis 2008). These instabilities and the associated turbulence can amplify the field exponentially, though the possible amount of amplification is unclear. Guilet, Foglizzo, and Fromang 2011 have suggested an additional effect leading to field growth: MHD Alfvén waves may be amplified if they propagate in the accretion flow surrounding the PNS. Since at least an approximate treatment of neutrino transport is required for modelling these effects, we have used the full coupled neutrino-MHD code to perform simulations of the collapse and the post-collapse evolution of non-rotating cores endowed with magnetic fields of different initial strength. In order to reduce the computational costs, we use a limited set of the dynamically most important neutrino-matter reactions only. We tested our code against reference simulations of spherically symmetric SN simulations with a state-of-the-art neutrino-hydrodynamics code (Rampp and Janka 2002) and found a good agreement.

We present a snapshot from one of the models (the most strongly magnetised of our cores) in Figure 4. At this instant (about 350 ms after collapse), the initial spherical symmetry of the core is broken by the convection and the standing accretion shock instability. The shock wave, visible best in the entropy (r.h.s. of the figure), is deformed strongly. Matter continues to fall through the shock and flows towards the PNS (the innermost ≈ 40 km) in a few narrow downflows. It accumulates in the layers surrounding the PNS, where it is heated by neutrinos. Neutrino heating drives convection, and heated bubbles (yellow regions) rise between the downflows.

The structure of the magnetic field (l.h.s. of the figure) traces these flow patterns. Together with the gas, magnetic flux is accreted towards the PNS and accumulates there, leading to an increased magnetic energy at the base of the hot bubble region. The magnetic field tends to be strong in the accretion down-flows. This effect is strongest along the polar axis where the constraint of axisymmetry of this model enforces converging flows that concentrate the magnetic flux in a small region and therefore increase the field strength. We can find indications of the amplification of Alfvén waves in these regions. In total, we find final magnetic fields that are in the range of the ones observed in typical neutron stars if we start our simulations with moderate initial fields, while strong initial fields translate into fields of the order of the strongest magnetised neutron stars (*magnetars*, Duncan and Thompson 1992).

4 CONCLUSIONS

In order to simulate the formation and evolution of compact astrophysical objects, we developed a numerical code solving the coupled equations of magnetohydrodynamics and neutrino transport. Since many evolutionary phases of compact objects are dominated by instabilities and turbulent flows, one of our main objectives was a high accuracy of the methods. On the other hand, the wide range of physics involved in these systems and the wide range of relevant spatial scales demanded high flexibility and low computational costs.

The last requirement has consequences in particular for the most expensive part, viz. the neutrino transport. As a consequence between accuracy and numerical costs, we used an expansion of the neutrino transfer equation in angular moments and truncated the resulting infinite series of equations after the first moments, evolving only the equations for the energy and momentum densities of the neutrinos. This leads to a hyperbolic system with stiff source terms, which we solve analogously to the MHD equations using HRSC methods. In both subsystems, we employ high-order monotonicity-preserving reconstruction methods and approximate Riemann solvers in the multistage framework.

We used our code for different kinds of simulations modelling compact objects, in particular the evolution of MHD instabilities and field amplification.

1. Local box simulations of magnetohydrodynamic instabilities. These simulations do not involve any neutrino physics but rely heavily on a high accuracy of the MHD solver. We investigated the dependence of the saturation level of the Kelvin-Helmholtz instability in remnants of neutron star mergers and the magneto-rotational instability in differentially rotating SN cores. Though even on the finest grids, our models are far from reproducing the physical (magnetic) Reynolds numbers of these objects, we could use them to study the saturation mechanisms by secondary instabilities.
2. Global simulations of MHD instabilities in jets. Again, no neutrino transport was used in these models, but a high accuracy of was required by the large dynamic range of the flow.
3. Global simulations of the amplification of magnetic fields in SN cores. Here, the inclusion of neutrino transport was essential to reproduce the dynamics of a post-bounce core, viz. the stall of the shock wave and the heating of the gas around the PNS by interactions with neutrinos. Studying non-rotating cores, we found that field strengths typical for neutron stars can be amplified by the flow from pre-collapse fields on the upper end of the range predicted by current stellar evolution models.

ACKNOWLEDGEMENTS

We thank E. Müller, Th. Janka, A. Marek, R. Moll, K. Kifonidis and O. Just for their collaboration. We also thank the organisers of the conference for their support and acknowledge support from the European Research Council (grant CAMAP-259276).

REFERENCES

Audit, E., P. Charrier, J. Chièze, and B. Dubroca (2002, June). A radiation-hydrodynamics scheme valid from the transport to the diffusion limit. *ArXiv Astrophysics e-prints*.

Balbus, S.A. and J.F. Hawley (1991, July). A powerful local shear instability in weakly magnetized disks. I—Linear analysis. II—Nonlinear evolution. *Astrophys. J. 376*, 214–233.

Balbus, S.A. and J.F. Hawley (1998, January). Instability, turbulence, and enhanced transport in accretion disks. *Reviews of Modern Physics 70,* 1–53.

Brackbill, J. and D.C. Barnes (1980). The Effect of Nonzero $\nabla \cdot B$ on the Numerical Solution of the Magnetohydrodynamic Equations. *J. Computat. Phys. 35*, 426.

Duncan, R.C. and C. Thompson (1992, June). Formation of very strongly magnetized neutron stars—Implications for gamma-ray bursts. *Astrophys. J. Letters 392*, L9–L13.

Evans, C.R. and J.F. Hawley (1988, September). Simulation of magnetohydrodynamic flows—A constrained transport method. *Astrophys. J. 332*, 659–677.

Guilet, J., T. Foglizzo, and S. Fromang (2011, March). Dynamics of an Alfvén Surface in Core Collapse Supernovae. *Astrophys. J. 729*, 71–+.

Hawley, J.F. and S.A. Balbus (1992, December). A powerful local shear instability in weakly magnetized disks. III—Long-term evolution in a shearing sheet. IV—Nonaxisymmetric perturbations. *Astrophys. J. 400*, 595–621.

Lattimer, J.M. (2010, March). Neutron star equation of state. *New Astronomy Reviews 54,* 101–109.

LeVeque, R.J. (2002). *Finite Volume Methods for Hyperbolic Problems*. Cambridge University Press.

Liebendörfer, M., O.E. B. Messer, A. Mezzacappa, S.W. Bruenn, C.Y. Cardall, and F. Thielemann (2004, January). A Finite Difference Representation of Neutrino Radiation Hydrodynamics in Spherically Symmetric General Relativistic Spacetime. *Astrophys. J. Suppl. 150*, 263–316.

Liebendörfer, M., S.C. Whitehouse, and T. Fischer (2009, June). The Isotropic Diffusion Source Approximation for Supernova Neutrino Transport. *Astrophys. J. 698*, 1174–1190.

Livne, E., A. Burrows, R. Walder, I. Lichtenstadt, and T.A. Thompson (2004, July). Two-dimensional, Time-dependent, Multigroup, Multiangle Radiation Hydrodynamics Test Simulation in the Core-Collapse Supernova Context. *Astrophys. J. 609*, 277–287.

Meier, D.L., R.I. Epstein, W.D. Arnett, and D.N. Schramm (1976, March). Magnetohydrodynamic phenomena in collapsing stellar cores. *Astrophys. J. 204*, 869–878.

Mezzacappa, A. and S.W. Bruenn (1993, March). A numerical method for solving the neutrino Boltzmann equation coupled to spherically symmetric stellar core collapse. *Astrophys. J. 405,* 669–684.

Mihalas, D. and B. Weibel Mihalas (1984). *Foundations of radiation hydrodynamics*. New York: Oxford University Press, 1984.

Miyoshi, T. and K. Kusano (2005, September). A multistate HLL approximate Riemann solver for ideal magnetohydrodynamics. *Journal of Computational Physics 208*, 315–344.

Obergaulinger, M. (2008, January). *Astrophysical magnetohydrodynamics and radiative transfer: numerical methods and applications.* Ph. D. thesis, Technische Universität München.

Obergaulinger, M., M.A. Aloy, and E. Müller (2010, June). Local simulations of the magnetized Kelvin-Helmholtz instability in neutron-star mergers. *Astron. & Astrop. 515*, A30+.

Obergaulinger, M., P. Cerdá-Durán, E. Müller, and M.A. Aloy (2009, April). Semi-global simulations of the magneto-rotational instability in core collapse supernovae. *Astron. & Astrop. 498*, 241–271.

Pons, J.A., J.M. Ibáñez, and J.A. Miralles (2000, September).Hyperbolic character of the angular moment equations of radiative transfer and numerical methods. *Mon. Not. Roy. Astron. Soc. 317,* 550–562.

Price, D.J. and S. Rosswog (2006, May). Producing Ultrastrong Magnetic Fields in Neutron Star Mergers. *Science 312*, 719–722.

Rampp, M. and H.-T. Janka (2002, December). Radiation hydrodynamics with neutrinos. Variable Eddington factor method for core-collapse supernova simulations. *Astron. & Astrop. 396*, 361–392.

Rezzolla, L., B. Giacomazzo, L. Baiotti, J. Granot, C. Kouve-liotou, and M.A. Aloy (2011, May). The Missing Link: Merging Neutron Stars Naturally Produce Jet-like Structures and Can Power Short Gamma-ray Bursts. *Astrophys. J. Letters 732*, L6.

Scheck, L., H. Janka, T. Foglizzo, and K. Kifonidis (2008, January). Multidimensional supernova simulations with approximative neutrino transport. II. Convection and the advective-acoustic cycle in the supernova core. *Astron. & Astrop. 477*, 931–952.

Suresh, A. and H. Huynh (1997). Accurate monotonicity-preserving schemes with runge-kutta time stepping. *J. Computat. Phys. 136,* 83–99.

Toro, E.F. and V.A. Titarev (2006, August). MUSTA fluxes for systems of conservation laws. *J. Computat. Phys. 216*, 403–429.

V Seismology and geophysics modelling

Numerical Methods for Hyperbolic Equations – Vázquez-Cendón et al. (eds)
© 2013 Taylor & Francis Group, London, ISBN 978-0-415-62150-2

IFCP Riemann solver: Application to tsunami modelling using GPUs

M.J. Castro, M. de la Asunción, J. Macías & C. Parés
Departamento Análisis Matemático. Universidad de Málaga, Spain

E.D. Fernández-Nieto
E.T.S. Arquitectura. Universidad de Sevilla, Spain

J.M. González-Vida
E.T.S. Telecomunicación. Universidad de Málaga, Spain

T. Morales
Departamento de Matemáticas. Universidad de Córdoba, Spain

ABSTRACT: In this work, we present a simplified two-layer model of Savage-Hutter type to simulate tsunamis generated by landslides (see (Fernández et al. 2008)). A layer composed of fluidized granular material is assumed to flow within an upperlayer composed of an inviscid fluid (e.g. water). The sediment layer is modelled by a Savage-Hutter type model where buoyancy effects have been considered. The system is discretized using IFCP finite volume scheme. The first order IFCP scheme was introduced in (Fernández et al. 2011) and it is constructed by using a suitable decomposition of a Roe matrix by means of a parabolic viscosity matrix, that captures information of the intermediate fields (Intermediate Field Capturing Parabola). Its extension to highorder and two-dimensional domains is straightforward. To conclude, some numerical examples are presented.

1 INTRODUCTION

Let us consider a simplified two-layer Savage-Hutter type system that can be use to model tsunamis generated by landslides. This model is a simplified version of the one introduced in (Fernández et al. 2008): a layer composed of fluidized granular material is assumed to flow within an upper layer composed of an inviscid fluid (e.g. water). The sediment layer is modelled by a Savage-Hutter type model where buoyancy effects have been considered:

$$\begin{cases} \dfrac{\partial h_1}{\partial t} + \dfrac{\partial q_1}{\partial x} = 0, \\ \dfrac{\partial q_1}{\partial t} + \dfrac{\partial}{\partial x}\left(\dfrac{q_1^2}{h_1} + \dfrac{g}{2}h_1^2\right) = -gh_1\dfrac{\partial h_2}{\partial x} - gh_1\dfrac{db}{dx} \\ \qquad\qquad +\tau_i, \\ \dfrac{\partial h_2}{\partial t} + \dfrac{\partial q_2}{\partial x} = 0, \\ \dfrac{\partial q_2}{\partial t} + \dfrac{\partial}{\partial x}\left(\dfrac{q_2^2}{h_2} + \dfrac{g}{2}h_2^2\right) = -rgh_2\dfrac{\partial h_1}{\partial x} - gh_2\dfrac{db}{dx} \\ \qquad\qquad -r\tau_i - \tau_b \end{cases} \tag{1}$$

In these equations, index 1 makes reference to the upper layer (water) and index 2 to the lower one (granular material). The coordinate x refers to the axis of the channel, t is time, and g is the acceleration due to gravity. $b(x)$ represents the bathymetry. Each layer is assumed to have a constant density, ρ_i, $i = 1,2$ $(\rho_1 < \rho_2)$, and $r = \rho_1/\rho_2$. The unknowns $q_i(x,t)$ and $h_i(x,t)$ represent respectively the mass-flow and the thickness of the i-th layer. τ_i is the friction term between the two layers and here it is written as

$$\tau_i = c_i \frac{h_1 h_2}{rh_1 + h_2} | u_1 - u_2 | (u_2 - u_1),$$

and τ_b denotes the Coulomb friction term. This term must be understood as:

$$\text{If } |\tau_b| \geq \sigma_c \quad \Rightarrow \quad \tau_b = g(1-r)h_2 \frac{u_2}{|u_2|}\tan(\delta_0), \tag{2}$$

$$\text{If } |\tau_b| < \sigma_c \quad \Rightarrow \quad u_2 = 0. \tag{3}$$

where $\sigma_c = g(1-r)h_2\tan(\delta_0)$, being δ_0 the Coulomb friction angle.

Observe that the presence of the term $(1-r)$ in the definition of the Coulomb friction term is due to the buoyancy effects, that must be taken into account only in the case that the sediment layer is submerged in the fluid, otherwise this term must be replaced by 1.

Notice that system (1) can be written in the following form:

$$w_t + F(w)_x + B(w)\cdot w_x = S(w)b_x + \tau, \tag{4}$$

where

$$w(x,t) = \begin{bmatrix} h_1(x,t) \\ q_1(x,t) \\ h_2(x,t) \\ q_2(x,t) \end{bmatrix}, \quad F(w) = \begin{bmatrix} q_1 \\ \dfrac{q_1^2}{h_1} + \dfrac{g}{2}h_1^2 \\ q_2 \\ \dfrac{q_2^2}{h_2} + \dfrac{g}{2}h_2^2 \end{bmatrix},$$

$$S(w) = \begin{bmatrix} 0 \\ -gh_1 \\ 0 \\ -gh_2 \end{bmatrix}, \quad B(w) = \begin{bmatrix} 0 & 0 & 0 & 0 \\ 0 & 0 & gh_1 & 0 \\ 0 & 0 & 0 & 0 \\ grh_2 & 0 & 0 & 0 \end{bmatrix},$$

$$\tau = \left[0, \tau_i, 0, -r\tau_i - \tau_b\right]^T.$$

The vector w takes values in the set:

$$O = \left\{[h_1, q_1, h_2, q_2]^T \in \mathbb{R}^4, \quad h_1 \ge 0, h_2 \ge 0\right\},$$

as the thickness of the layers may vanish in practical applications when one or the two layers disappear in part of the domain. Let us also define the matrix $A(w)$ given by

$$A(w) = J(w) + B(w), \quad \text{being } J(w) = \frac{\partial F}{\partial w}(w).$$

The characteristic equation of $A(w)$ is:

$$(\lambda^2 - 2u_1\lambda + u_1^2 - gh_1)(\lambda^2 - 2u_2\lambda + u_2^2 - gh_2) = rg^2 h_1 h_2. \tag{5}$$

It is easy to check that the condition under which one of the eigenvalues vanishes is:

$$G^2 = F_1^2 + F_2^2 - (1-r)F_1^2 F_2^2 = 1, \tag{6}$$

where G is the so-called *composite Froude number*, and F_i for $i = 1, 2$ are the internal Froude numbers ($F_i^2 = u_i^2/g'h_i$, where g' is the *reduced gravity*, $g' = (1-r)g$). When this condition is achieved at a section of coordinate x, the flow is said to be *critical* at this point and the section x is called a *control*. When $G^2 < 1$, the flow is *subcritical*. Finally, when $G^2 < 1$, the flow is *supercritical*.

Observe that, when $r = 0$, the eigenvalues are those corresponding to each layer separately. Therefore, when $r \cong 0$, the coupling terms do not affect the nature of the system in an essential manner. The eigenvalues of A can be classified in two external and two internal eigenvalues. The external eigenvalues, $\lambda_{ext}^{\pm}$, are related to the propagation speed of barotropic perturbations and the internal ones $\lambda_{int}^{\pm}$, to the propagation of baroclinic perturbations. First order approximation of the eigenvalues can be found in (Schijf & Schonfeld 1953).

In most applications to geophysical flows, one has $\lambda_{ext}^{-} < 0$ and $\lambda_{ext}^{+} > 0$. Moreover, the internal eigenvalues depend on the reduced gravity g'. As a consequence the absolute value of internal eigenvalues are usually smaller than those of external ones, that is

$$|\lambda_{int}^{\pm}| < |\lambda_{ext}^{-}|, \quad |\lambda_{int}^{\pm}| < |\lambda_{ext}^{+}|.$$

This fact implies that first order numerical schemes that only use information concerning the external eigenvalues are in general too diffusive when applied to the simulation of internal waves. On the other hand, methods that use explicitly the eigenstructure of A, as it is the case of Roe method, are computationally expensive, as it does not exist any easy explicit expression of the eigenvalues and eigenvectors of this system. IFCP scheme is a computationally fast and precise method that uses information concerning the internal eigenvalues. The definition of the method is based on a suitable decomposition of a Roe matrix (see (Toumi 1992)) by means of a parabolic viscosity matrix (see (Degond et al. 1999)) that captures intermediate field information.

2 NUMERICAL SCHEME

Here, only the description of the 1D first and high order IFCP scheme is considered. Its extension to two-dimensional problems is straightforward following the procedure described in (Castro et al. 2009) and (Gallardo et al. 2011).

Friction terms $\mathcal{T}$ will be discretized semi-implicitly as described in (Fernández et al. 2008), so they are neglected at this point.

Solutions of (4) may develop discontinuities and, due to the non-divergence form of the equations, the notion of weak solution in the sense of distributions cannot be used. The theory introduced by Dal Maso, LeFloch, and Murat (Dal Maso et al. 1995) is followed here to define weak solutions of (4). This theory allows to define the nonconservative products as a bounded measure provided a family of Lipschitz continuous paths

$\Phi:[0,1]\times\Omega\times\Omega\to\Omega$ is prescribed, which must satisfy certain natural regularity conditions. Here, the family of straight segments is considered:

$$\Phi(s;w_L,w_R)=w_L+s(w_R-w_L).$$

In (Fernández et al. 2011) authors introduce a first order numerical scheme, named IFCP. IFCP numerical scheme is constructed by using a suitable decomposition of a Roe matrix of system (4) by means of a parabolic viscosity matrix (see (Degond et al. 1999)), that captures information of the intermediate fields. IFCP is a path-conservative scheme in the sense defined in (Parés 2006).

IFCP numerical scheme can be written as follows

$$w_i^{n+1}=w_i^n-\frac{\Delta t}{\Delta x}(D_{i-1/2}^{+}+D_{i+1/2}^{-}), \tag{7}$$

being $D_{i+1/2}^{\pm}=D_{i+1/2}^{\pm}(w_i,w_{i+1},b_i,b_{i+1})$ defined by

$$D_{i+1/2}^{\pm}=\frac{1}{2}(F(w_{i+1}^n)-F(w_i^n)+\mathcal{B}_{i+1/2}-\mathcal{S}_{i+1/2} \pm Q_{i+1/2}(w_{i+1}^n-w_i^n-A_{i+1/2}^{-1}\mathcal{S}_{i+1/2}) \tag{8}$$

where $\mathcal{B}_{i+1/2}=B_{i+1/2}(w_{i+1}-w_i)$ being

$$B_{i+1/2}=\int_0^1 B(\Phi(s;W_i,W_{i+1}))ds; \tag{9}$$

$\mathcal{S}_{i+1/2}=S_{i+1/2}(b_{i+1}-b_i)$, where

$$S_{i+1/2}=\int_0^1 S(\Phi(s;W_i,W_{i+1}))ds. \tag{10}$$

$A_{i+1/2}=J_{i+1/2}+B_{i+1/2}$, being $J_{i+1/2}$ a Roe linearization of the Jacobian of the flux F in the usual sense:

$$J_{i+1/2}\cdot(w_{i+1}-w_i)=F(w_{i+1})-F(w_i); \tag{11}$$

and $Q_{i+1/2}$ is a viscosity matrix.

Remark 2.1 Note that the numerical scheme depends on the choice of the family of path Φ. This scheme is a path-conservative numerical scheme in the sense introduced by Parés in (Parés 2006). In (Castro et al. 2008) and (Parés & Muñoz 2009) it has beenproved that, in general, the numerical solutions provided by a path-conservative numerical scheme converge to functions which solve a perturbed system in which an error source-term appears on the right-hand side. The appearance of this source term, which is a measure supported on the discontinuities, has been first observed in (Hou & LeFloch 1994) when a scalar conservation law is discretized by means of a nonconservative numerical method. Nevertheless, in certain special situations the convergence error vanishes for finite difference methods: this is the case for systems of balance laws (see (Muñoz & Parés 2011)). Moreover for more general problems, even when the convergence error is present, it may be only noticeable for very fine meshes, for discontinuities of large amplitude, and/or for large-time simulations: see (Castro et al. 2008), (Parés & Muñoz 2009) for details.

The key point is the definition of the matrix $Q_{i+1/2}$, that in the case of the IFCP is defined by:

$$Q_{i+1/2}=\alpha_0 Id+\alpha_1 A_{i+1/2}+\alpha_2 A_{i+1/2}^2, \tag{12}$$

where $\alpha_j, j=0,1,2$ are defined by:

$$\begin{aligned}\alpha_0 &= \delta_1\lambda_{4,i+1/2}\chi_{int}+\delta_{4,i+1/2}\lambda_{1,i+1/2}\chi_{int}\\&\quad+\delta_{int}\lambda_{1,i+1/2}\lambda_{4,i+1/2},\\ \alpha_1 &= -\lambda_1(\delta_4+\delta_{int})-\lambda_{4,i+1/2}(\delta_1+\delta_{int})\\&\quad-\chi_{int}(\delta_{1,i+1/2}+\delta_{4,i+1/2}),\\ \alpha_2 &= \delta_1+\delta_4+\delta_{int},\end{aligned} \tag{13}$$

being

$$\begin{aligned}\delta_1 &= \frac{|\lambda_{1,i+1/2}|}{(\lambda_{1,i+1/2}-\lambda_{4,i+1/2})(\lambda_{1,i+1/2}-\chi_{int})},\\ \delta_4 &= \frac{|\lambda_{4,i+1/2}|}{(\lambda_{4,i+1/2}-\lambda_{1,i+1/2})(\lambda_{4,i+1/2}-\chi_{int})},\\ \delta_{int} &= \frac{|\chi_{int}|}{(\chi_{int}-\lambda_{1,i+1/2})(\chi_{int}-\lambda_{4,i+1/2})},\end{aligned}$$

where $\lambda_{1,i+1/2}<\lambda_{2,i+1/2}<\lambda_{3,i+1/2}<\lambda_{4,i+1/2}$ are the eigenvalues of matrix $A_{i+1/2}$ and

$$\chi_{int}=\mathcal{S}_{ext}\max(|\lambda_{2,i+1/2}|,|\lambda_{3,i+1/2}|), \tag{14}$$

with

$$\mathcal{S}_{ext}=\begin{cases}\operatorname{sgn}(\chi_{ext}), & \text{if } (\chi_{ext})\neq 0,\\ 1, & \text{otherwise,}\end{cases} \tag{15}$$

where $\chi_{ext}=\lambda_{1,i+1/2}+\lambda_{4+i+1/2}$.

It can be proved that IFCP scheme is linearly L^∞ stable under the usual CFL condition

$$\max\left\{\left|\lambda_{l,i+1/2}\right|,1\le l\le 4,i\in\mathbb{Z}\right\}\frac{\Delta t}{\Delta x}=CFL\le 1. \tag{16}$$

Remark 2.2 Note that the coefficients α_i are defined in terms of the eigenvalues of the matrix $A_{i+1/2}$. Here, we use the first order approximations defined in (Schijf & Schonfeld 1953) to estimate the wave speeds $\lambda_{l,i+1/2}$, $l=1,\ldots,4$.

Remark 2.3 Note that if $\alpha_0=(1-\omega)\,\Delta x/\Delta t$, $\alpha_1=0$ and $\alpha_2=\omega\Delta x/\Delta t$, then the numerical scheme (7)–(8) coincides with the family introduced in (Castro et al. 2010). This family contains, as particular cases, a well-balanced extension of the Lax-Friedrichs

($\omega = 0$), Lax-Wendroff ($\omega = 1$), FORCE ($\omega = 1/2$), and GFORCE (($\omega = 1/1 + CFL$) methods.

Remark 2.4 Notice that in the definition of (8) the term

$$C = Q_{i+1/2} A_{i+1/2}^{-1} S_{i+1/2},$$

that can be interpreted as the upwinding part of the source term discretization, makes no sense if one of the eigenvalues of $A_{i+1/2}$ vanishes. In this case the problem is said to be resonant. Resonant problems exhibit an additional difficulty, as weak solutions may not be uniquely determined by their initial data, and the limits of the numerical solutions may depend both on the family of paths and the numerical scheme itself. The analysis of this difficulty is beyond the scope of this work. Here, we follow the strategy described in (Fernández et al. 2011) to get rid of this difficulty and to obtain well-balanced numerical schemes for a given set of stationary solutions.

2.1 *Extension to high order*

In order to define a high order numerical scheme for system (4), we follow the procedure described in (Castro et al. 2006). First, a high order reconstruction operator of the form $P_W^t(x) = (P_w^t(x), P_b(x))^T$ is considered, that is, an operator that associates, to a given sequence $\{W_i(t) = (w_i(t), b_i)^T\}$, two new sequences $\{W_{i+1/2}^-(t) = (w_{i+1/2}^-(t), b_{i+1/2}^-)^T\}$, $\{W_{i+1/2}^+(t) = (w_{i+1/2}^+(t), b_{i+1/2}^+)^T\}$ in such a way that, whenever

$$w_i(t) = \frac{1}{\Delta x}\int_{I_i} w(x,t)\,dx$$

$$b_i = \frac{1}{\Delta x}\int_{I_i} b(x)\,dx$$

for some regular function $W = (w, b)^T$, then

$$(w_{i+1/2}^{\pm}, b_{i+1/2}^{\pm}) = (w(x_{i+1/2}, t), b(x_{i+1/2}))^T + O(\Delta x^p),\quad \forall i \in \mathbb{Z}.$$

Here, we propose the following semi-discrete high order numerical scheme for (4) (see (Castro et al. 2006)):

$$\begin{aligned} w_i'(t) = &-\frac{1}{\Delta x}(D_{i-1/2}^+ + D_{i+1/2}^-) \\ &-\frac{1}{\Delta x}\Big(F(w_{i+1/2}^-(t)) - F(w_{i-1/2}^+(t))\Big) \\ &-\frac{1}{\Delta x}\int_{I_i} B(P_{w_i}^t(x))(P_{w_i}^t(x))_x\,dx \\ &+\frac{1}{\Delta x}\int_{I_i} S(P_{w_i}^t(x))(P_{b_i}(x))_x\,dx, \end{aligned} \tag{17}$$

with $D_{i+1/2}^{\pm} = D(w_{i+1/2}^-(t), w_{i+1/2}^+(t), b_{i+1/2}^-, b_{i+1/2}^+)$ and where $w_{i+1/2}^{\pm}(t)$ and $b_{i+1/2}^{\pm}$ are the reconstructed values at $x_{i+1/2}$ of $w(x,t)$ and $b(x)$, respectively. $P_{w_i}^t(x)$ and $P_{b_i}(x)$ are functions defined in I_i such that

$$\begin{aligned} &\lim_{x \to x_{i-1/2}^+} (P_{w_i}^t(x), P_{b_i}(x)) = (w_{i-1/2}^+(t), b_{i-1/2}^+), \\ &\lim_{x \to x_{i+1/2}^-} (P_{w_i}^t(x), P_{b_i}(x)) = (w_{i+1/2}^-(t), b_{i+1/2}^-). \end{aligned} \tag{18}$$

Fourth order Romberg quadrature formula is used to compute the integrals

$$\int_{I_i} B(P_{w_i}^t(x))(P_{w_i}^t(x))_x \text{ and } \int_{I_i} S(P_{w_i}^t(x))(P_{b_i}(x))_x.$$

Remark 2.5 Note that high order schemes for conservative systems only depend on $w_{i+1/2}^{\pm}$, where they depend on $P_{w_i}^t$ and P_{b_i} for nonconservative systems (see (Castro et al. 2006)).

Finally, a high order TVD-Runge-Kutta discretization can be used for the time-stepping (see (Gottlieb & Shu 1998)). Concerning the high order reconstruction operator, we usually use the PHM (piecewise hyperbolic method) introduced in (Marquina 1994). The extension to 2D systems is straightforward following (Castro et al. 2009) and (Gallardo et al. 2011).

3 NUMERICAL TESTS

In this section we present some numerical tests. In the first one, a battery of numerical tests is presented to study the dependency of the sediment layer profile and the generated tsunami with respect to the friction angle δ_0 and the ratio of densities, r. In the second one, the generation and propagation of tsunami on a real bathymetry is considered. In both cases, a GPU implementation of the previous scheme in two-dimensional domains has been used. Modern Graphics Processing Units (GPUs) offer hundreds of processing units optimized for massively performing floating point operations in parallel and have shown to be a cost-effective way to obtain a substantially higher performance in the applications related to shallow water flows due to the high exploitable parallelism which exhibits the finite volume schemes (see for example: (Castro et al. 2011),(Brodtkorb et al. 2011) or (Gallardo et al. 2011)). Here, a MPI-CUDA implementation like the one presented in [?] has been performed to increase the speed-up of the computations up to two-orders of magnitude using a cluster of NVIDIA GTX-490 graphics cards with respect to a mono-core implementation in a modern CPU (Intel Xeon E5430 (2.66 GHz 12MB L2 Cache)).

3.1 *Test 1*

A battery of numerical tests is presented here to study numerically the dependency of the sediment layer profile and the generated tsunami with respect to the friction angle δ_0 and the ratio of densities, r. The effective angle of repose of the sediment layer after an avalanche is also measured at the stationary state. Let us consider a square domain of 10 m side, centered at the origin, with a flat bottom topography, that is, $b = -2$. As initial condition, we set $u_1 = u_2 = 0$ and

$$h_2(x,y,0) = \begin{cases} 1, & \text{if } \|(x,y)\| < 1, \\ 0, & \text{otherwise,} \end{cases}$$

$h_1(x,y,0) = 2 - h_2(x,y,0)$. Free boundary conditions are imposed at both channels ends. The CFL parameter is set to 0.8.

In Figure 1 we compare the final stationary interface that we obtain for three different meshes with $\Delta x = \Delta y \in \{0.1, 0.05, 0.02\}$ for $r = 0.4$ and $\delta_0 = 20°$. Only some small differences near the "wet/dry" fronts can be observed. Table 1 shows the maximum and the mean effective angle of repose of the sediment layer after the landslide at the stationary state. As expected, the maximum value is under $\delta_0 = 20°$, while the mean value is close to 8.5°.

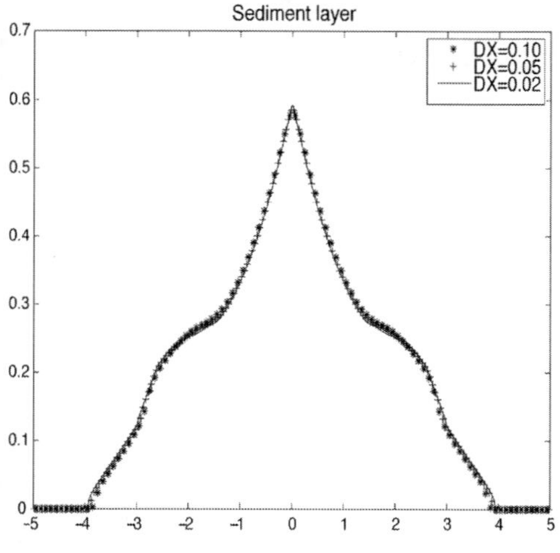

Figure 1. Sediment layer at stationary state $\Delta x = \Delta y \in \{0.1, 0.05, 0.02\}$, ($\delta_0 = 20°$, $r = 0.4$).

Table 1. Effective angle of repose ($r = 0.4$, $\delta_0 = 20°$).

Δx	Max	Mean	Δx	Max	Mean
0.1	18.03°	8.43°	0.05	19.16°	8.46°
0.02	19.69°	8.46°			

Figure 2 shows the profiles of the sediment layer at the stationary state for $r = 0.4$, $\Delta x = \Delta y = 0.05$ and $\delta_0 \in \{10°, 15°, 20°, 25°, 30°\}$and Table 2 shows the maximum and mean effective angle of repose of the sediment layer after the landslide. As expected, the maximum value is always under δ_0. Figure 3 shows the maximum of the free surface, $\eta = h_1 + h_2 - 2.0$, vs. δ_0. Figure 3 gives an idea of the amplitude of the generated tsunami. Note that the amplitude decreases for bigger values of the parameter δ_0.

Now, the parameter δ_0 is set to 20° and $r \in \{0.0, 0.1, 0.2, 0.3, 0.4\}$ Figure 4 shows the profiles of the sediment layer at the stationary state for $\Delta x = \Delta y = 0.05$ and Table 3 shows the maximum and mean effective angle of repose of the sediment layer after the landslide. Again, the maximum value is always under δ_0. Note that the maximum value decreases with r while the mean increases with respect to r. Nevertheless, the variations are not relevant. More differences can be observed in the stationary profile of the second layer (see Fig. 4), in particular the position of the front decreases with r, as well as the maximum height of the sediment layer. Figure 5 shows the maximum of the free surface, $\eta = h_1 + h_2 - 2.0$, vs. r. As expected, the

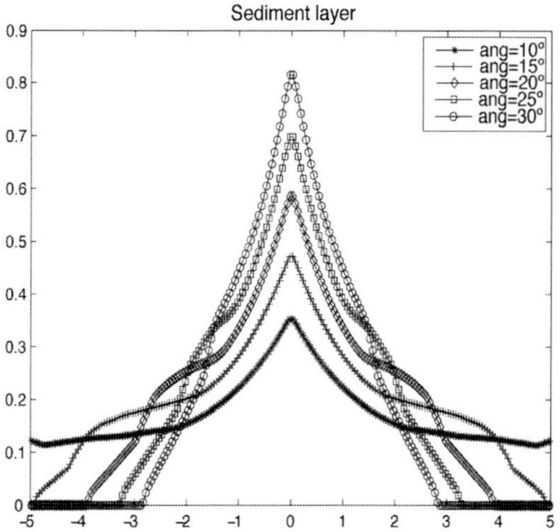

Figure 2. Sediment layer depth at the stationary state for $r = 0.4$ and $\delta_0 \in \{10°, 15°, 20°, 25°, 30°\}$.

Table 2. Effective angle of repose ($r = 0.4$)

δ_0	Max	Mean	δ_0	Max	Mean
10°	9.82°	2.84°	15°	14.51°	5.35°
20°	19.16°	8.46°	25°	23.90°	12.05°
30°	29.42°	16.13°			

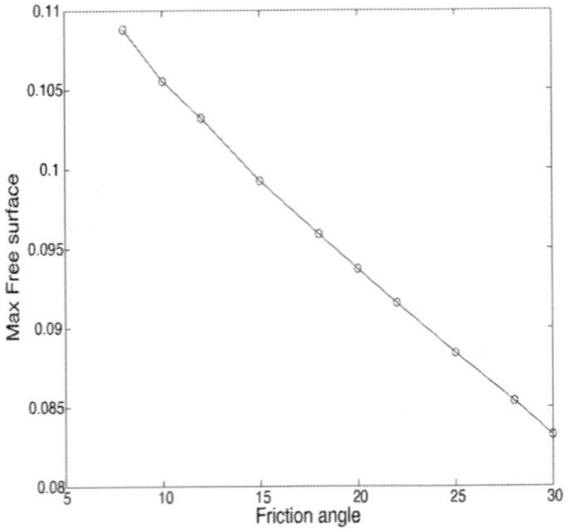

Figure 3. Maximal height of the free surface for $r = 0.4$ and $\delta_0 \in \{10°, 15°, 20°, 25°, 30°\}$.

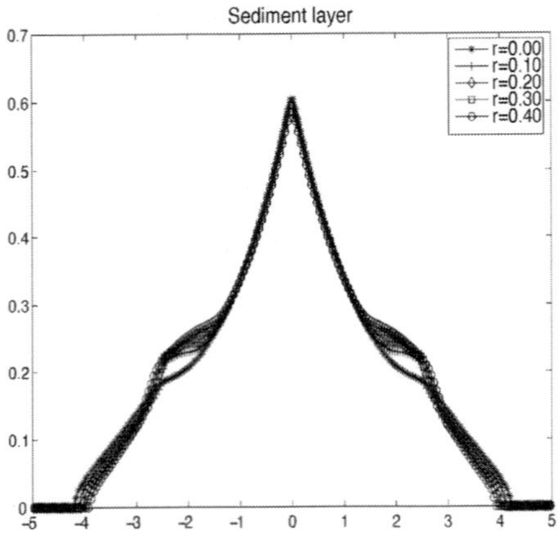

Figure 4. Sediment layer depth at the stationary state for $\delta_0 = 20°$ and $r \in \{0, 0.1, 0.2, 0.3, 0.4\}$.

Table 3. Effective angle of repose $\delta_0 = 20°$.

r	Max	Mean	r	Max	Mean
0.0	19.64°	8.07°	0.1	19.56°	8.22°
0.2	19.37°	8.30°	0.3	19.23°	8.41°
0.4	19.16°	8.46°			

amplitudes of the generated tsunami are bigger for smaller values of r.

3.2 *A tsunami generated by submarine landslide over real bathymetry*

There are geological evidences about a tsunami generated by a submarine landslide located in

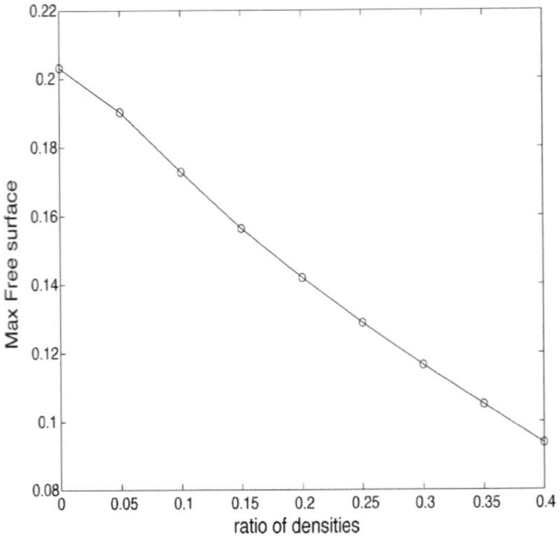

Figure 5. Maximal height of the free surface for $\delta_0 = 20°$ and $r \in \{0, 0.1, 0.2, 0.3, 0.4\}$.

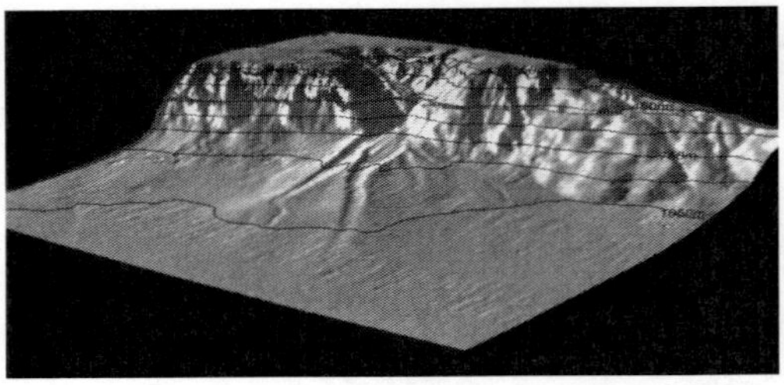

Figure 6. Current bathymetry.

the continental margin of Alboran island (western Mediterranean). The initial landslide area is located at south-west of Alboran island and covers an area about 9.5 km^2 where water depth range varies from 100 to 1000 m. Deposit covers an area of about 30 km^2. In this section we show an advanced numerical experiment that consist on starting from the current bathymetry, reconstruct the pre-tsunami paleo-bathymetry and then, simulate the landslide and the generated tsunami.

A rectangular 180 km × 190 km with $\Delta x = \Delta y = 25$ m grid has been considered (54720000 cells). The simulated time covers 3600s after the tsunami is triggered. We set, CFL = 0.9, $r = 0.55$, $c_i = 10^{-5}$, and $\delta_0 = 10°$.

Figures 6 and 7 show the current bathymetry and the reconstructed pre-tsunami original bathymetry.

Figures 8 and 9 show two different stages of the propagated tsunami. It can be observed how

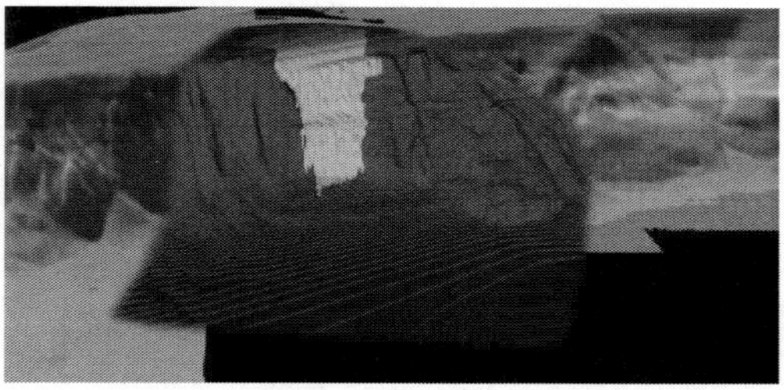

Figure 7. Supposed paleo-bathymetry before the landslide.

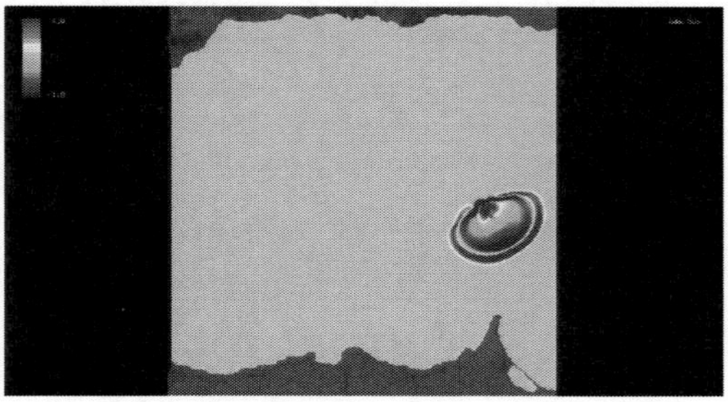

Figure 8. Free surface height at t = 4.00 min.

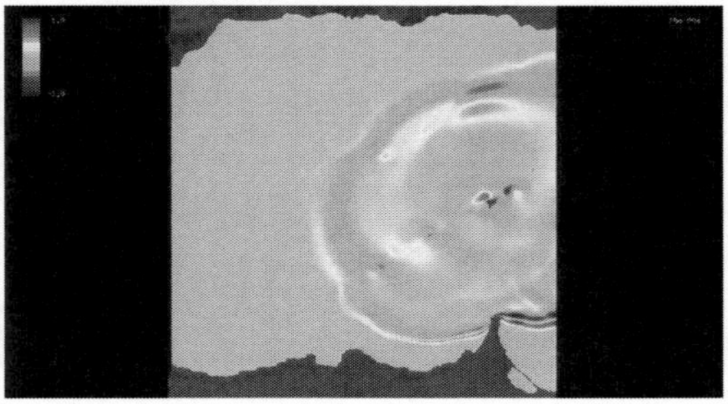

Figure 9. Free surface height at t = 15.00 min.

the shape of the tsunami varies according to the bathymetry; while the bathymetry is smoother southwards from the Alborani canyon, the shape of the tsunami wave is almost symmetric, while northwards, the sharper bathymetry effects are quite more visible.

Finally in Figure 10 it is shown a wave height time series extracted from a point near to Málaga (located in the north-west of the domain). We can observe that the tsunami wave arrives to this point about 32 mins after the tsunami is triggered.The first wave height is very small, just about 7–8 cm, then, two minutes later, a larger negative wave (about –40 cm height) arrives followed, 4 minutes

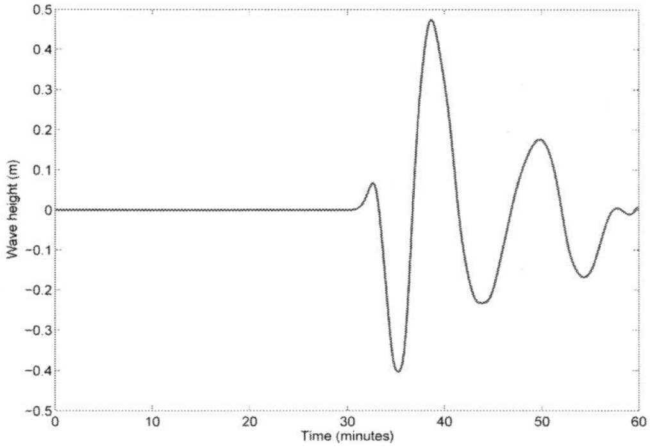

Figure 10. Wave height in a point near to Málaga (Iberian coast).

later, by a larger wave of about 50 cm height. After this wave, other waves appear with lower height. We can remark than due to the geometry of the Málaga bay, some of these waves are reflected producing resonant effects.

4 CONCLUSIONS

In this work, a high order extension of the IFCP scheme has been introduced to solve a two-layer Savage-Hutter type model to simulate tsunamis generated by landslides. IFCP is a method constructed by a suitable decomposition of a Roe matrix $A_{i+1/2}$, whose viscosity matrix is computed by a linear combination of the identity matrix, $A_{i+1/2}$ and $A^2_{i+1/2}$ and whose coefficients are given in terms of the eigenvalues of $A_{i+1/2}$. The resulting numerical scheme is linearly L_∞-stable and well-balanced for the water at rest solution. A GPU implementation has been performed to speed up the computations. Two numerical tests have been presented: in the first one, a study the dependency of the sediment layer profile and the generated tsunami with respect to the friction angle δ_0 and the ratio of densities, r has been performed. In the second one, the generation and propagation of tsunami in the Alboran See using a real bathymetry have been performed.

ACKNOWLEDGEMENTS

This research has been partially supported by the Spanish Government Research projects MTM2009-11923. The numerical computations have been performed at the Laboratory of Numerical Methods of the University of Málaga.

REFERENCES

Asunción, M. de la & Mantas, J.M. & Castro, M.J. & Fernández, E.D. 2011. An MPI-CUDA implementation of an improved Roe method for two-layer shallow water systems. *J. Parallel and Distributed Computing* (In Press, DOI:10.1016/j.jpdc.2011.07.012).

Brodtkorb, A. & Hagen, T. & Lie, K.A. & Natvig, J. 2011. Simulation and visualization of the saint-venant system using GPUs. *Computing and Visualization in Science*, 1–13.

Castro, M.J. & Gallardo, J.M. & Parés, C. 2006. High order finite volume schemes based on reconstruction of states for solving hyperbolic systems with nonconservative products. Applications to shallow-water systems. *Math. Comp*. 75:1103–1134.

Castro, M.J. & LeFloch, P.G. & Muñoz, M.L. & Parés, C. 2008. Why many theories of shock waves are necessary: Convergence error in formally path-consistent schemes. *J. Comp. Phys.* 3227:8107–8129.

Castro, M.J. & Fernández, E.D. & Ferreiro, A.M. & García, J.A. & Parés, C. 2009. High order extensions of Roe schemes for two dimensional nonconservative hyperbolic systems. *J. Sci. Comp.* 39:67–114.

Castro, M.J. & Pardo, A. & Parés, C. & Toro, E.F. 2010. On some fast well-balanced first order solvers for nonconservative systems. *Math. Comp.* 79:1427–1472.

Castro, M.J. & Ortega, S. & Asunción, M. de la & Mantas, J.M. & Gallardo, J.M. 2011. GPU computing for shallow water flow simulation based on finite volume schemes. *Comptes Rendus Mécanique* 339(2–3): 165–184, High Performance Computing.

Dal Maso, G. & LeFloch, P.G. & Murat, F. 1995. Definition and weak stability of nonconservative products. *J. Math. Pures Appl.* 74:483–548.

Degond, P. & Peyrard, P.F. & Russo, G. & Villedieu, Ph. 1999. Polynomial upwind schemes for hyperbolic systems. *C. R. Acad. Sci. Paris* 328:479–483.

Fernández, E.D. & Bouchut, F. & Bresh, D. & Castro, M.J. & Mangeney, A. 2008. A new Savage-Hutter type model for submarine avalanches and generated tsunami. *J. Comp. Phys.* 227:7720–7754.

Fernández, E.D. & Castro, M.J. & Parés, C. 2011. On an Intermediate Field Capturing Riemann Solver Based on a Parabolic Viscosity Matrix for the Two-Layer Shallow Water System. *J. Sci. Comput.* 48(1–3):117–140.

Gallardo, J.M. & Ortega, S. & Asunción, M. de la & Mantas, J.M. 2011. Two-dimensional compact third-order polynomial reconstructions. Solving nonconservative hyperbolic systems using GPUs. *J. Sci. Comput.* 48(1–3):141–163.

Gottlieb, S. & Shu, C.W. 1998. Total variation diminishing Runge-Kutta schemes. *Math. Comp.* 67:73–85.

Hou, T.Y. & LeFloch, P.G. 1994. Why nonconservative schemes converge to wrong solutions: error analysis. *Math. of Comput.* 62:497–530.

Marquina, A. 1994. Local piecewise hyperbolic reconstructions for nonlinear scalar conservation laws. *SIAM J. Sci. Comp* 15:892–915.

Muñoz, M.L. & Parés, C. 2011. On the convergence and well-balanced property of path-conservative numerical schemes for systems of balance laws. *J. Sci. Comp.* 48(1–3):274–295.

Parés, C. & Castro, M.J. 2004. On the well-balance property of Roe's method for nonconservative hyperbolic systems. Applications to Shallow-Water Systems. *M2AN* 38(5):821–852.

Parés, C. 2006. Numerical methods for nonconservative hyperbolic systems: a theoretical framework. *SIAM J. Num. Anal.* 44(1):300–321.

Parés, C. & Muñoz Ruíz, M.L. 2009. On some difficulties of the numerical approximation of nonconservative hyperbolic systems. *Boletín SEMA* 47:23–52.

Schijf, J.B. & Schonfeld, J.C. 1953. Theoretical considerations on the motion of salt and fresh water. *In Proc. of the Minn. Int. Hydraulics Conv.* 321–333. Joint meeting IAHR and Hyd. Div. ASCE.

Toumi, I. 1992. A weak formulation of Roe approximate Riemann solver. *J. Comp. Phys.* 102(2):360–373.

Numerical Methods for Hyperbolic Equations – Vázquez-Cendón et al. (eds)
© 2013 Taylor & Francis Group, London, ISBN 978-0-415-62150-2

Dispersive waves generated by an underwater landslide

Denys Dutykh
LAMA, UMR 5127 CNRS, Université de Savoie, Campus Scientifique, France

Dimitrios Mitsotakis
IMA, University of Minnesota, USA

Sonya Beisel & Nina Shokina
Institute of Computational Technologies, Siberian Branch of the Russian Academy of Sciences, Russia

ABSTRACT: In this work we study the generation of water waves by an underwater sliding mass. The wave dynamics are assumed to fell into the shallow water regime. However, the characteristic wavelength of the free surface motion is generally smaller than in geophysically generated tsunamis. Thus, dispersive effects need to be taken into account. In the present study the fluid layer is modeled by the Peregrine system modified appropriately and written in conservative variables. The landslide is assumed to be a quasi-deformable body of mass whose trajectory is completely determined by its barycenter motion. A differential equation modeling the landslide motion along a curvilinear bottom is obtained by projecting all the forces acting on the submerged body onto a local moving coordinate system. One of the main novelties of our approach consists in taking into account curvature effects of the sea bed.

1 INTRODUCTION

Extreme water waves can become an important hazard in coastal areas. Many geophysical mechanisms are related with underwater earthquakes and landslides. The former genesis mechanism has been intensively investigated since the Tsunami Boxing Day but also before (Okal 1988; Okal and Synolakis 2003; Okal and Synolakis 2004; Synolakis and Bernard 2006; Dutykh and Dias 2007; Dutykh and Dias 2009; Beisel, Chubarov, Didenkulova, Kit, Levin, Pelinovsky, Shokin, and Sladkevich 2009). The list of references is far from being exhaustive. In this study we focus on the latter mechanism—the underwater landslides which can cause some damage in the generation region. In general, the wavelength of landslide generated waves is much smaller than the length of transoceanic tsunamis. Consequently, the dispersive effects might be important. This consideration explains why we opt for a dispersive model (Dutykh, Katsaounis, and Mitsotakis 2011), which is able to simulate the propagation and run-up of weakly nonlinear weakly dispersive water waves on nonuniform beaches.

Most of the landslide models which are currently used in the literature can be conventionally divided into three big categories. The first category contains the simplest models where the landslide shape and its trajectory are known *a priori* (Tinti, Bortolucci, and Chiavettieri 2001; Todorovska, Hayir, and Trifunac 2002; Liu, Lynett, and Synolakis 2003; Didenkulova, Nikolkina, Pelinovsky, and Zahibo 2010). Another approach consists in assuming that the landslide motion is translational and the sliding mass follows the trajectory of its barycenter. The governing equation of the center of mass is obtained by projecting all the forces, acting on the slide, onto the horizontal direction of motion (Grilli and Watts 1999; Watts, Imamura, and Grilli 2000; Di Risio, Bellotti, Panizzo, and De Girolamo 2009). Finally, the third category of models describe the slide-water evolution as a two-layer system, the sliding mass being generally described by a Savage-Hutter type model (Fernández-Nieto, Bouchut, Bresch, Castro-Diaz, and Mangeney 2008). Taking into account all the uncertainties which exist in the modeling of the real-world events, we choose in this paper to study the intermediate level (i.e. the second category) which corresponds better to the precision of the available data.

The present study is organized as follows. In Section 3 we briefly describe the water wave model we use, while Section 2 contains a more detailed presentation of the landslide model. The test-case considered in our study along with numerical results are presented in Section 5. Finally, the main conclusions of this study are outlined in Section 5.

2 WATER WAVE MODEL

The water wave model we use in this study is based on the classical system derived by D.H. Peregrine (Peregrine 1967). However, the original derivation assumes that the bottom is stationary in time, i.e. $z = -d(x)$. Later, the bottom dynamics has been included into this system derivation by T. Wu (Wu 1981; Wu 1987). In order to simulate the wave run-up, a conservative form of this system has to be derived. In the static bottom case it was done recently (Dutykh, Katsaounis, and Mitsotakis 2011). The conservative system we use in the present study can be obtained in a similar way and can be written in the form:

$$H_t + Q_x = 0, \tag{1}$$

$$\left(1+\frac{1}{3}H_x^2-\frac{1}{6}HH_{xx}\right)Q_t-\frac{1}{3}H^2Q_{xxt}-\frac{1}{3}HH_xQ_{xt} +\left(\frac{Q^2}{H}+\frac{g}{2}H^2\right)_x = gHd_x+\frac{1}{2}Hd_{xtt}, \tag{2}$$

where $H(x,t) := d(x,t) + \eta(x,t)$ is the total water depth and $\eta(x,t)$ is the free surface elevation below the still water level. The horizontal mass flux is denoted by $Q(x,t) := H(x,t)u(x,t)$ where $u(x,t)$ is the depth-averaged horizontal velocity variable. The bottom motion enters into the momentum balance equation (2) through the source term $\frac{1}{2}Hd_{xtt}$. The mass conservation equation (1) keeps naturally its initial form. We underline that the linear dispersion relation of the modified Peregrine system (1), (2) is identical with that the original Peregrine model (Peregrine 1967) since these models differ only in nonlinear terms.

This modified Peregrine (m-Peregrine) system (1), (2) has several advantages. First of all, we note that the full water wave problem is invariant under the vertical translations (Benjamin and Olver 1982). The asymptotic expansion method around the mean water level breaks this symmetry. The introduction of conservative variables (H, Q) allows to recover this property. Another advantage is that the dispersive terms in the m-Peregrine system naturally vanish as the total water depth H along with its first derivative H_x tend to zero. This property is in complete agreement with the physical behavior of water waves which become more and more nonlinear to the detriment of the dispersion while approaching the shore.

In order to solve numerically the m-Peregrine system (1), (2) we choose the use of the finite volume method. Moreover, the run-up technique is well understood in the framework of Nonlinear Shallow Water equations (Dutykh, Poncet, and Dias 2011), which allows us to reuse this technology in the dispersive setting. The advective terms are discretized using the FVCF approach (Ghidaglia, Kumbaro, and Le Coq 2001) with UNO2 space reconstruction (Harten and Osher 1987). The dispersive terms are treated with the finite differences. For the time discretization we use the Bogacki-Shampine 3rd order Runge-Kutta scheme with adaptive time step control. Note that on each time step we have to solve a tridiagonal system of linear equations in order to determine the time derivative Q_t. We refer to (Dutykh, Katsaounis, and Mitsotakis 2011) for more details on the numerical method.

3 LANDSLIDE MODEL

In this section we briefly present a model of an underwater landslide motion. This process has to be addressed carefully since it determines the subsequent formation of water waves. In this study we will assume the moving mass to be a solid quasi-deformable body with a prescribed shape and known physical properties that preserves its mass and volume. Under these assumptions it is sufficient to compute the trajectory of the barycenter $x = x_c(t)$ to determine the motion of the whole body. In general, only uniform slopes are considered in the literature in conjunction with this type of landslide models (Pelinovsky and Poplavsky 1996; Grilli and Watts 1999; Watts, Imamura, and Grilli 2000; Di Risio, Bellotti, Panizzo, and De Girolamo 2009). However, a novel model, taking into account the bottom geometry and curvature effects, has been recently proposed (Khakimzyanov and Shokina 2010). Hereafter we will follow in great lines this study.

The static bathymetry is prescribed by a sufficiently smooth (at least of the class C^2) and single-valued function $z = -d_0(x)$. The landslide shape is initially prescribed by a localized in space function $z = \zeta_0(x)$. For example, in this study we choose the following shape function:

$$\zeta_0(x) = A\begin{cases}\frac{1}{2}\left(1+\cos\left(\frac{2\pi(x-x_0)}{\ell}\right)\right), & |x-x_0| \le \frac{\ell}{2} \\ 0, & |x-x_0| > \frac{\ell}{2},\end{cases} \tag{3}$$

where parameters A is the maximum height, ℓ is the length of the slide and x_0 is the initial position of its barycenter. Obviously, the model description given below is valid for any other reasonable shape.

Since the landslide motion is translational, its shape at time t is given by the function $z = \zeta(x, t) = \zeta_0(x - x_c(t))$. Recall that the landslide center is located at the point with abscissa $x = x_c(t)$. Then, the impermeable bottom for the water wave problem can be easily determined at any time by simply superposing the static and dynamic components:

$$z = -d(x, t) = -d_0(x) + \zeta(x, t).$$

To simplify the subsequent presentation, we introduce the classical arc-length parametrization, where the parameter $s = s(x)$ is given by the following formula:

$$s = L(x) = \int_{x_0}^{x} \sqrt{1 + \left(d_0'(\xi)\right)^2}\, d\xi. \tag{4}$$

The function $L(x)$ is monotonic and can be efficiently inverted to turn back to the original Cartesian abscissa $x = L^{-1}(s)$. Within this parametrization, the landslide is initially located at point with the curvilinear coordinate $s = 0$. The local tangential direction is denoted by τ and the normal by n.

The landslide motion is governed by the following differential equation obtained by a straightforward application of Newton's second law:

$$m\frac{d^2 s}{dt^2} = F_\tau(t),$$

where m is the mass and $F_\tau(t)$ is the tangential component of the forces acting on the moving submerged body. In order to project the forces onto the axes of local coordinate system, the angle $\theta(x)$ between τ and Ox can be easily determined:

$$\theta(x) = \arctan\left(d_0'(x)\right).$$

Let us denote by ρ_w and ρ_ℓ the densities of the water and sliding material correspondingly. If V is the volume of the slide, then the total mass m is given by this expression:

$$m := (\rho_\ell + c_w \rho_w)V,$$

where c_w is the added mass coefficient (Batchelor 2000). A portion of the water mass has to be added since it is entrained by the underwater body motion. For a cylinder, for example, the coefficient c_w is equal exactly to one. The volume V canbe computed as

$$V = W \cdot S = W\int_{\mathbb{R}} \zeta_0(x)dx,$$

where W is the landslide width in the transverse direction. The last integral can be computed exactly for the particular choice (3) of the landslide shape to give

$$V = \frac{1}{2}\ell A W.$$

The total projected force F_τ acting on the landslide can be conventionally represented as a sum of two different kind of forces denoted by F_g and F_d:

$$F_\tau = F_g + F_d,$$

where F_g is the joint action of the gravity and buoyancy, while F_d is the total contribution of various dissipative forces (to be specified below).

The gravity and buoyancy forces act in opposite directions and their horizontal projection F_g can be easily computed:

$$F_g(t) = (\rho_\ell - \rho_w)W_g\int_{\mathbb{R}} \zeta(x,t)\sin\left(\theta(x)\right)dx.$$

Now, let us specify the dissipative forces. The water resistance to the motion force F_r is proportional to the maximal transversal section of the moving body and to the square of its velocity:

$$F_r = -\frac{1}{2}c_d \rho_w A W \sigma(t)\left(\frac{ds}{dt}\right)^2,$$

here c_d is the resistance coefficient of the water and $\sigma(t) := \operatorname{sign}(ds/dt)$. The coefficient $\sigma(t)$ is needed to dissipate the landslide kinetic energy independently of its direction of motion. The friction force F_f is proportional to the normal force exerted on the body due to the weight:

$$F_f = -c_f\,\sigma(t)N(x,t).$$

The normal force $N(x,t)$ is composed of the normal components of gravity and buoyancy forces but also of the centrifugal force due to the variation of the bottom slope:

$$N(x,t) = (\rho_\ell - \rho_w)gW\int_{\mathbb{R}} \zeta(x,t)\cos\left(\theta(x)\right)dx + \rho_l W\int_{\mathbb{R}} \zeta(x,t)\,\kappa\,(x)\left(\frac{ds}{dt}\right)^2 dx,$$

where $\kappa(x)$ is the signed curvature of the bottom which can be computed by the following formula:

$$\kappa(x) = \frac{d_0''(x)}{\left(1 + \left(d_0'(x)\right)^2\right)^{\frac{3}{2}}}.$$

We note that the last term vanishes for a plane bottom since $\kappa(x) \equiv 0$ in this particular case.

Finally, if we sum up all the contributions of described above forces, we obtain the following second order differential equation:

$$(\gamma + c_w)S\frac{d^2s}{dt^2} = (\gamma - 1)g\big(I_1(t) - c_f\sigma(t)I_2(t)\big) - \sigma(t)\left(c_f\gamma I_3(t) + \frac{1}{2}c_d A\right)\left(\frac{ds}{dt}\right)^2, \quad (5)$$

where $\gamma := \rho_\ell/\rho_w > 1$ is the ratio of densities and integrals $I_{1,2,3}(t)$ are defined as:

$$I_1(t) = \int_{\mathbb{R}} \zeta(x,t)\sin(\theta(x))dx,$$
$$I_2(t) = \int_{\mathbb{R}} \zeta(x,t)\cos(\theta(x))dx,$$
$$I_3(t) = \int_{\mathbb{R}} \zeta(x,t)\kappa(x)dx,$$

Note also that equation (5) was simplified by dividing both sides by the width value W. In order to obtain a well-posed initial value problem, equation (5) has to be completed by two initial conditions:

$$s(0) = 0, \quad s'(0) = 0.$$

In order to solve numerically equation (5) we employ the same Bogacki-Shampine 3rd order Runge-Kutta scheme as we use to solve the Boussinesq equations (1), (2). The integrals $I_{1,2,3}(t)$ are computed using the trapezoidal rule. Once the landslide trajectory $s = s(t)$ is found, we use equation (4) to find its motion $x = x(t)$ in the initial Cartesian coordinate system.

4 NUMERICAL RESULTS

Let us consider the one-dimensional computational domain $\mathcal{I} = [a, b] = [-3, 10]$ composed from three regions: the left and right sloping beaches surrounding a complex generation region. Specifically, the static bathymetry function $d_0(x)$ is given by the following expression:

$$d_0(x) = \begin{cases} d_0 + 4\tan\delta \cdot x, & x \le 0, \\ d_0 + \tan\delta \cdot x + p(x), & 0 < x \le m, \\ d_0 + 4m\tan\delta - 3\tan\delta \cdot x, & x > m, \end{cases}$$

where the function $p(x)$ is defined as

$$p(x) = A_1 e^{-k_1(x-x_1)^2} + A_2 e^{-k_2(x-x_2)^2}.$$

Basically, this function represents a perturbation of the sloping bottom by two underwater bumps. We made this nontrivial choice in order to illustrate better the advantages of our landslide model, which was designed to handle general non-flat bathymetries. The values of all physical and numerical parameters are given in Table 1. The bottom profile for these parameters is depicted on Figure 1.

The landslide motion starts from the rest position under the action of the gravity force. We study its motion along with the waves of the free surface up to $T = 21$ s. The landslide barycenter

Table 1. Values of various parameters used in the numerical computations.

Parameter	Value
Gravity acceleration, g	1.0
Water depth at $x = 0$, d_0	1.0
Bottom slope, $\tan(\delta)$	0.15
Underwater bump amplitude, A_1	1.0
Underwater bump amplitude, A_2	1.6
Bump steepness, k_1	1.9
Bump steepness, k_2	3.9
Bump center position, x_1	1.2
Bump center position, x_2	2.6
Boundary between bottom regions, m	4.0
Number of control volumes, N	1000
Slide amplitude, A	0.3
Length of the slide, ℓ	2.0
Initial slide position, x_0	0.0
Added mass coefficient, c_w	1.0
Water drag coefficient, c_d	1.0
Friction coefficient, c_f	tan 1°
Ratio between water and slide densities, γ	1.5

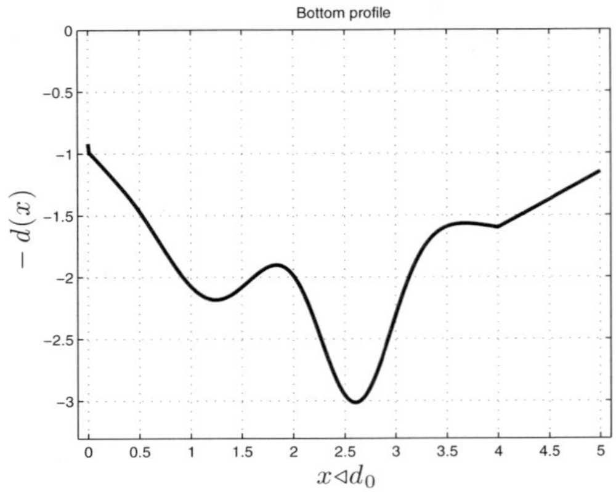

Figure 1. Bathymetry profile for parameters given in Table 1.

trajectory along with its speed and acceleration are shown in Figure 2. As it is expected, the landslide remains trapped in the second underwater bump, where it oscillates before stopping completely its motion.

One of the important parameters in shallow water flows is the Froude number defined as the ratio between the characteristic fluid velocity to the gravity wave speed. We computed also this parameter along the landslide trajectory:

$$\mathrm{Fr}(t) := \frac{|x_c'(t)|}{\sqrt{gd\left(x_c(t),t\right)}}.$$

The result is presented in Figure 3. We can see that in our case the motion remains subcritical during the experiment.

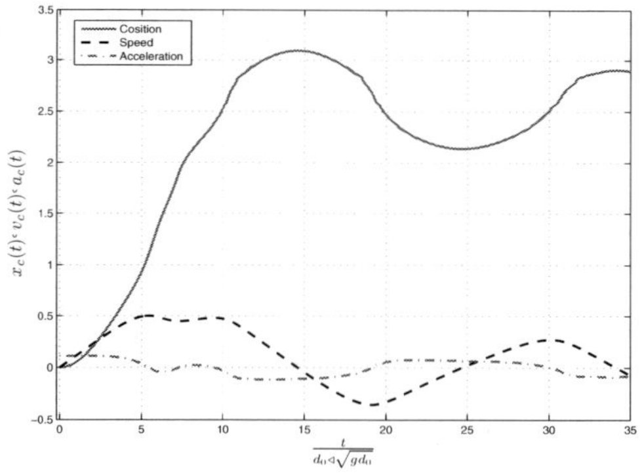

Figure 2. Barycenter position (blue solid line), velocity (black dashed line) and acceleration (blue dash-dotted line) during the landslide motion.

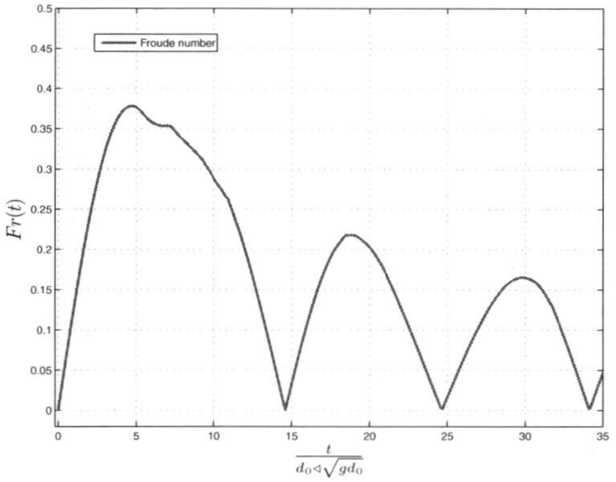

Figure 3. Local Froude number computed along the slide motion.

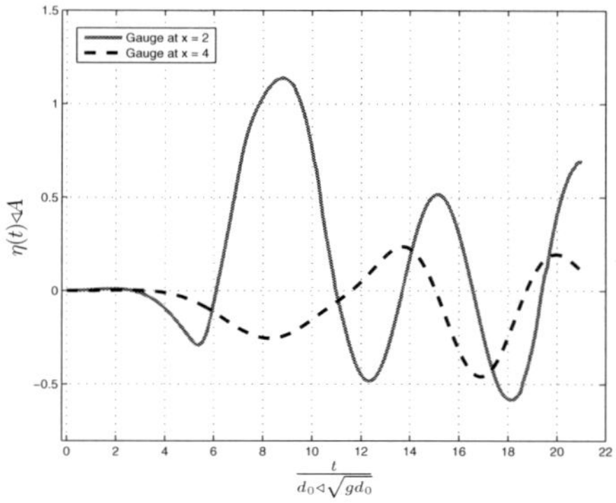

Figure 4. Synthetic wave gauge records located at $x = 2$ (blue solid line) and $x = 4$ (black dashed line). The vertical axis is relative to the landslide amplitude A.

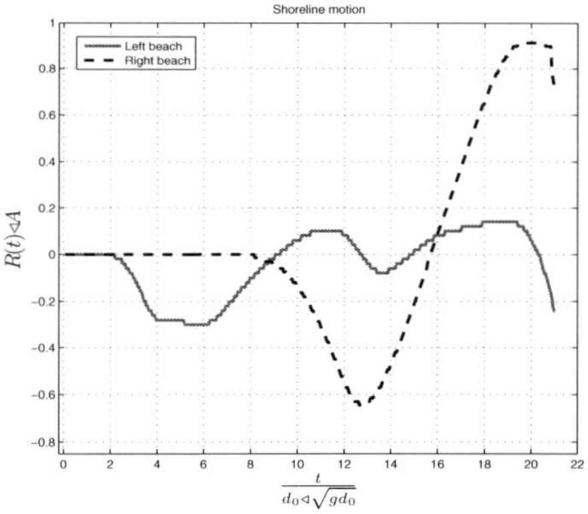

Figure 5. Shoreline motion on the left (blue solid line) and right (black dashed line) beaches.

We installed two synthetic wave gauges located at $x = 2$ (point located between two underwater bumps) and $x = m$ (endpoint of the generation region and the beginning of the right sloping beach, $m = 4$ in our simulations) in order to measure the magnitude of generated free surface motions. These synthetic records are presented in Figure 4. Finally, the wave run-up is simulated numerically on the left and right beaches. The shoreline motion is represented in Figure 5. We can see, that the proposed scenario provides higher run-up values on the opposite beach to the slope where the landslide takes place.

5 CONCLUSIONS

In this study we presented a novel model of a landslide motion over general curvilinear bottoms. This model takes into account the effects of bottom curvature, generally neglected in the literature (Pelinovsky and Poplavsky 1996; Grilli and Watts 1999; Watts, Imamura, and Grilli 2000; Di Risio, Bellotti, Panizzo, and De Girolamo 2009). Despite the inclusion of some new physical effects, the considered model is computationally inexpensive and can be potentially used in more operational context. The computed bottom motion is strongly coupled with a conservative Peregrine system (Dutykh, Katsaounis, and Mitsotakis 2011) which describes the propagation of nonlinear weakly dispersive water waves. The run-up of landslide generated waves on both beaches is simulated using the m-Peregrine system. The extension of this approach to three dimensions represents a natural perspective for future research along with investigation of other possible scenarios.

ACKNOWLEDGEMENTS

S. Beisel, N. Shokina and D. Dutykh acknowledge the support from the projects PICS CNRS No 5607 and RFBR No 10-05-91052. D. Dutykh also acknowledges the French Agence Nationale de la Recherche, project MathOcéan (ANR-08-BLAN-0301-01). The authors would like to thank Professors Leonid Chubarov and Gayaz Khakimzyanov for very helpful and stimulating discussions on the landslide modeling.

REFERENCES

Batchelor, G.K. (2000). *An introduction to fluid dynamics*, Volume 61 of *Cambridge mathematical library*. Cambridge University Press.

Beisel, S., L. Chubarov, I. Didenkulova, E. Kit, A. Levin, E. Pelinovsky, Y. Shokin, and M. Sladkevich (2009). The 1956 Greek tsunami recorded at Yafo, Israel, and its numerical modeling. *Journal of Geophysical Research—Oceans 114*, C09002.

Benjamin, T.B. and P. Olver (1982). Hamiltonian structure, symmetries and conservation laws for water waves. *J. Fluid Mech 125*, 137–185.

Di Risio, M., G. Bellotti, A. Panizzo, and P. De Girolamo (2009). Three-dimensional experiments on landslide generated waves at a sloping coast. *Coastal Engineering 56*(5–6), 659–671.

Didenkulova, I., I. Nikolkina, E. Pelinovsky, and N. Zahibo (2010). Tsunami waves generated by submarine landslides of variable volume: analytical solutions for a basin of variable depth. *Nat. Hazards Earth Syst. Sci. 10*, 2407–2419.

Dutykh, D. and F. Dias (2007). Water waves generated by a moving bottom. In A. Kundu (Ed.), *Tsunami and Nonlinear waves*, pp. 65–96. Springer Verlag (Geo Sc.).

Dutykh, D. and F. Dias (2009). Tsunami generation by dynamic displacement of sea bed due to dip-slip faulting. *Mathematics and Computers in Simulation 80(4)*, 837–848.

Dutykh, D., T. Katsaounis, and D. Mitsotakis (2011). Finite volume schemes for dispersive wave propagation and runup. *J. Comput. Phys 230*, 3035–3061.

Dutykh, D., R. Poncet, and F. Dias (2011). The VOLNA code for the numerical modeling of tsunami waves: Generation, propagation and inundation. *Eur. J. Mech. B/Fluids 30*(6), 598–615.

Fernández-Nieto, E.D., F. Bouchut, D. Bresch, M.J. Castro-Diaz, and A. Mangeney (2008). A new Savage-Hutter type models for submarine avalanches and generated tsunami. *J. Comput. Phys. 227*(16), 7720–7754.

Ghidaglia, J.-M., A. Kumbaro, and G. Le Coq (2001). On the numerical solution to two fluid models via cell centered finite volume method. *Eur. J. Mech. B/Fluids 20*, 841–867.

Grilli, S.T. and P. Watts (1999). Modeling of waves generated by a moving submerged body. Applications to underwater landslides. *Engineering Analysis with boundary elements 23*, 645–656.

Harten, A. and S. Osher (1987). Uniformly high-order accurate nonscillatory schemes, I. *SIAM J. Numer. Anal. 24*, 279–309.

Khakimzyanov, G.S. and N.Y. Shokina (2010). Numerical modelling of surface water waves arising due to a movement of the underwater landslide on an irregular bottom. *Computational technologies 15*(1), 105–119.

Liu, P.L.-F., P. Lynett, and C. Synolakis (2003). Analytical solutions for forced long waves on a sloping beach. *J. Fluid Mech. 478*, 101–109.

Okal, E.A. (1988). Seismic Parameters Controlling Far-field Tsunami Amplitudes: A Review. *Natural Hazards 1*, 67–96.

Okal, E.A. and C.E. Synolakis (2003). A theoretical comparison of tsunamis from dislocations and landslides. *Pure and Applied Geophysics 160*, 2177–2188.

Okal, E.A. and C.E. Synolakis (2004). Source discriminants for near-field tsunamis. *Geophys. J. Int. 158*, 899–912.

Pelinovsky, E. and A. Poplavsky (1996). Simplified model of tsunami generation by submarine landslides. *Physics and Chemistry of the Earth 21*(12), 13–17.

Peregrine, D.H. (1967). Long waves on a beach. *J. Fluid Mech. 27*, 815–827.

Synolakis, C.E. and E.N. Bernard (2006). Tsunami science before and beyond Boxing Day 2004. *Phil. Trans. R. Soc. A 364*, 2231–2265.

Tinti, S., E. Bortolucci, and C. Chiavettieri (2001). Tsunami Excitation by Submarine Slides in Shallow-water Approximation. *Pure appl. geophys. 158*, 759–797.

Todorovska, M.I., A. Hayir, and M.D. Trifunac (2002). A note on tsunami amplitudes above submarine slides and slumps. *Soil Dynamics and Earthquake Engineering 22*, 129–141.

Watts, P., F. Imamura, and S.T. Grilli (2000). Comparing model simulations of three benchmark tsunami generation cases. *Science of Tsunami Hazards 18*(2), 107–123.

Wu, T.Y. (1981). Long Waves in Ocean and Coastal Waters. *Journal of Engineering Mechanics 107*, 501–522.

Wu, T.Y.T. (1987). Generation of upstream advancing solitons by moving disturbances. *Journal of Fluid Mechanics 184*, 75–99.

Numerical Methods for Hyperbolic Equations – Vázquez-Cendón et al. (eds)
© 2013 Taylor & Francis Group, London, ISBN 978-0-415-62150-2

Finite volume schemes for balance laws on time-dependent surfaces

Jan Giesselmann & Maria Wiebe
IANS, University of Stuttgart, Germany

ABSTRACT: In the paper at hand we consider scalar hyperbolic balance laws on manifolds with time-dependent Riemannian metric. We propose a family of finite volume schemes to solve such problems and give an error estimate of order $h^{1/4}$ for such schemes, provided the entropy solution has bounded total variation for finite times. In addition we present numerical experiments showing the performance of the proposed schemes for balance laws posed on moving surfaces in $\mathbb{R}^3$. The experimental orders of convergence obtained in these experiments confirm the rate of convergence of the error estimate, in fact they usually converge with an order of at least $h^{1/2}$.

1 INTRODUCTION

In many applications conservation or balance laws are not posed in open subsets of Euclidean space but on surfaces or more generally Riemannian or Lorentzian manifolds. We will consider the case of a balance law on a manifold, whose geometry changes in time. In particular such a situation occurs when we study balance laws which are posed on moving surfaces in Euclidean space. Important examples are processes taking place on the surface of (biological) cells, e.g. receptor clustering, or on the interfacial manifold between different phases in three dimensional multiphase or multi component flow, (Dreyer et al.), (Alt 2009).

In most engineering examples the problems are posed on moving hyper surfaces of $\mathbb{R}^3$ Let us shortly explain how they are linked to problems on manifolds with time dependent Riemannian metric. Given a balance law on some moving, closed surface $M = M(t) \subset \mathbb{R}^3$, which does not change its diffeomorphism type we can map the surface at each time to a reference configuration M_0 using a smooth family of diffeomorphisms $\Phi(t): M(t) \rightarrow M_0$. We define a Riemannian metric $g(t)$ on the reference manifold M_0 at each time t by imposing the map $\Phi(t)$ to be an isometry. Thereby we obtain a balance law on the fixed, closed manifold M_0 equipped with a family of Riemannian metrics $\{g(t)\}_{t\in\mathbb{R}_+}$ changing in time.

The well-posedness theory for conservation laws on stationary Riemannian manifolds, i.e. manifolds with fixed Riemannian metric, goes back to the seminal paper (Ben-Artzi and LeFloch 2007). The theoretical investigation of their solution by finite volume schemes was started in (Amorim et al., 2008) and convergence rates were obtained in (Giesselmann 2009; LeFloach et al., 2009). Different generalizations of the above results were considered: most notably the notion of finite volume schemes based on differential forms, see (LeFloch and Okutmustur 2008), which even includes the theory of conservation laws on Lorentzian manifolds. In this context also a convergence rate has been proven (Amorim et al.), but source terms and not divergence free fluxes were not considered. To cope with the source terms and not geometry compatible fluxes, i.e. fluxes whose divergence does not vanish for constant solution u, we pursue the more tangible setting of conservation laws based on vector-fields here. The main new difficulty apart from the source terms and the terms arising from the non-zero divergence of the flux, lies in the fact that the cells and faces change their volumes/areas during timesteps. Furthermore the cut-off functions which have to be used in the proof of the error estimate are time dependent.

The aim of this paper is twofold: On the one hand we will prove a convergence rate for finite volume schemes, detailed below, for balance laws on manifolds with Riemannian metric changing in time, see Theorem 4.1; on the other hand we will complement these theoretical results with numerical simulations. These display a convergence rate of approximately $h^{1/2}$ for linear problems and convergence rates between $h^{1/2}$ and h for nonlinear problems. Thus they confirm the theoretical order of convergence. The findings here are in accordance with the situation in the Euclidean setting where the same orders of convergence as here can be proven and are obtained in numerical simulations.

2 STATEMENT OF THE PROBLEM

To state the problem under consideration we need to introduce some notation. Let M be some closed, oriented smooth manifold of dimension d. We will write T_xM for the tangent space of M at some point $x \in M$. Let M be equipped with some a priori prescribed smooth family $g = g(t)$. of Riemannian metrics. Furthermore $u(x,t) \in \mathbb{R}$ is the unknown. The flux f satisfies $f(x,\cdot,\cdot):\mathbb{R}_+ \times \mathbb{R} \to T_xM$ for every $x \in M$, and we have a source term $q: M \times \mathbb{R}_+ \times \mathbb{R} \to$ v. With these data a scalar balance law on a manifold with changing geometry takes the following integral form:

$$\frac{d}{dt}\int_{\Omega} u(x,t)dv_{g(t)}(x) = \int_{\Omega} q(x,t,u(x,t))dv_{g(t)}(x) + \int_{\partial\Omega} f(x,t,u(x,t))n_{g(t)}(x,t)de_{g(t)}, \tag{1}$$

where Ω is an arbitrary open subset of $M, dv_{g(t)}$, $de_{g(t)}$ are volume and surface forms induced by $g(t)$ respectively and $n_{g(t)}(x,t)$ is the unit outer normal to $\partial\Omega$ at x with respect to $g(t)$.

If one wants to write (1) as a differential equation one has to observe that now the integral on the left hand side of (1) changes due to the changes of not only u but also $dv_{g(t)}$ in time. The latter is given by

$$\frac{d}{dt}dv_{g(t)} = \frac{1}{2}\mathrm{tr}_{g(t)}(g_t(t))dv_{g(t)}. \tag{2}$$

where $\mathrm{tr}_{g(t)}(g_t(t))$ denotes the trace of the time derivative $g_t(t)$ with respect to $g(t)$, see e.g. (Kühnel 2002) for details. Thus the differential formulation of (1) reads

$$u_t + \nabla_{g(t)} \cdot f(x,t,u) = -\frac{1}{2}\mathrm{tr}_{g(t)}(g_t(t))u + q(x,t,u) \text{ in } M \times \mathbb{R}_+, u(\cdot,0) = u_0 \text{ on } M. \tag{3}$$

Where $u_0 \in L^\infty(M)$ is the initial data and $\nabla_{g(t)}$ is the divergence operator induced by $g(t)$.

Let us comment briefly on how equation (3) looks like on some surface $\Gamma = \Gamma(t) \subset \mathrm{R}^3$ moving with speed $\mathbf{w} \in \mathrm{R}^3$ There are two terms which would look different. The partial time derivative in (3) is computed at some fixed point in the reference configuration, i.e. $u_t = u_t(x \in M,t)$. When we consider the situation on some moving surface in R^3 this fixed point on the reference manifold corresponds to a curve in R^3. Hence,

$$u_t(x \in M,t) = \frac{d}{dt}u(x(t),t) = u_t + w\nabla u =: D_t^\Gamma u \tag{4}$$

where ∇ is the Euclidean gradient in R^3 The quantity D_t^Γ defined in (4) is usually called surface time derivative. Furthermore, as shown in e.g. (Muller 1985) the change of surface area $1/2\,\mathrm{tr}_{g(t)}(g_t(t))$ is explicitly given by

$$\mathrm{div}_{\Gamma(t)}(\mathbf{W}_t) - 2\,\kappa w_v, \tag{5}$$

where W_t, Wv are the tangential and normal part of $\mathbf{w}$ respectively, $\mathrm{div}_{\Gamma(t)}$ is the surface divergence on $\Gamma(t)$ and κ is the pointwise mean curvature of $\Gamma(t)$. The other terms would essentially remain unchanged and (3) takes the equivalent form

$$u_t + \mathbf{w}\nabla u + (div_{\Gamma(t)}(\mathbf{w}_t) - \kappa\omega_v)u + \mathrm{div}_{\Gamma(t)}(f(x,t,u)) - q(x,t,u) = 0 \times \text{on } \{(x,t) \in \mathbb{R}^3 \times \mathbb{R}_+ : x \in \Gamma(t)\}, \tag{6}$$

Let us now return to the general setting. As in the Euclidean case one cannot in general expect to find smooth solutions for long times and weak solution will not be unique. However the definition of Kruzkov entropy solutions for balance laws in Euclidean space, cf. e.g. (Chainais-Hillairet and Champier 2001) can be generalized straightforwardly:

Definition 2.1 (Entropy solution). *A function $u \in L^\infty(M \times [0,T])$ for all $T > 0$ is called an entropy solution of (3), if it satisfies*

$$\begin{aligned}0 \le &\int_{M\times R+} |u(x,t)-\kappa|\,\varphi_t(x,t)dv_{g(t)}dt \\ &+\int_{M\times R+}(f(x,t,u(x,t)\mathrm{T}\kappa) - f(x,t,u(x,t)\perp\kappa)) \nabla_{g(t)}\varphi(x,t)dv_{g(t)}dt \\ &-\int_{M\times R+}[\mathrm{sgn}(u(x,t)-\kappa)\varphi(x,t)(\nabla_{g(t),x}\cdot f(x,t,\kappa) \\ &-q(x,t,u(x,t)) + \frac{u(x,t)}{2}\mathrm{tr}_{g(t)}(g_t(t)))]dv_{g(t)}dt \\ &+\int_M |u_0(x)-\kappa|\,\varphi(x,0)dv_{g(0)}\end{aligned} \tag{7}$$

for all $\kappa \in R$ and $\varphi \in C_0^\infty(M \times R_+, R_+)$, where $\nabla_{g(t),x} \cdot f$ denotes the surface divergence of f with respect to the space variable, i.e. for u fixed, and for $a, b \in \mathrm{R}$ *we denote the maximum and minimum of a* and b by $a\mathrm{T}b$ and $a \perp b$ respectively.

In what follows we impose the subsequent conditions on the problem (3), which relate to those in the Euclidean setting:

$1. u_0 \in L^\infty(M) \cap \mathrm{BV}(M, g(0)),$

$2. f \in C^1(M \times \mathbb{R}_+ \times \mathbb{R}, TM), \text{s.t} f(x,\cdot,\cdot) \in T_x M,$

$3. \frac{\partial f}{\partial s}(x,t,s)$ is locally Lipschitz continuous,

$4. \nabla_{g(t),x} \cdot f(x,t,s)$ is locally Lipschitz continous

and $\left| \frac{\partial}{\partial s} \nabla_{g(t),x} \cdot f(x,t,s) \right|$ is locally bounded

$5. q \in C^1(M \times \mathbb{R}_+ \times \mathbb{R}, \mathbb{R}),$

and $\mathrm{tr}_{g(t)}(gt(t) \in C^1(M \times \mathbb{R}_+, \mathbb{R}).$

$$6. \exists Q > 0 : \left| \frac{d}{dt} g(t)(\upsilon, w) \right| \le Q \|\upsilon\|_{g(t)} \|w\|_{g(t)} \tag{8}$$

$\forall t \in \mathbb{R}_+, \upsilon, w \in T_x M, x \in M,$

where BV is the space of function of bounded total variation and we denote the tangent bundle of M by TM: $= \{(x,v): x \in M, \upsilon \in T_x M\}$. Furthermore we will assume the existence of a unique entropy solution to (3) having bounded total variation on $M \times [0,T]$ for finite times T>0. This assumption seems reasonable, although a rigorous well-posedness analysis only exists for the case of fluxes with vanishing divergence and without source terms (LeFloch and Okutmustur 2008). The well-posedness theory in the setting pursued here, is a question of its own right.

3 THE NUMERICAL SCHEME

To define a finite volume scheme we first need to decompose the manifold into cells. We will call such a decomposition a triangulation although the cells can have more than three faces. For completeness we give the notion of triangulation on a curved manifold, although it basically coincides with the definition in (Giesselmann 2009).

Definition 3.1 *A* **curved polyhedron** *K is an open subset of M such that the boundary of K is the union of finitely many hypersurfaces with boundary e of M, which are called the faces of K. We impose $e1 \cap e2$ is empty or a submanifold of M with dimension at most d–2. Furthermore we demand $(\overline{K})^\circ = K$ to rule out any degeneracy of K. A* **triangulation** *on M is a set T of curved polyhedra K on M such that $M = \cup_T \overline{K}$. We impose $K1 \cap K2$ to be a common face of K1, K2 or a submanifold of dimension ≤d–2.*

We denote the set of faces e of a polyhedron K by ∂K and K_e is the unique polyhedron sharing the face e with K. By $n_{K,e}(x,t) \in T_x M$ we denote the unit outer conormal to a polyhedron K in a point $x \in e$ with respect to $g(t)$ These definitions are illustrated in Figure 1. Finally $m_t(K), m_t(e)$ denote the d- and $(d–1)$-dimensional Hausdorff measures of K and e induced by $g(t)$ respectively.

We need several technical assumptions on the triangulation, which basically coincide with those in the case of a stationary Riemannian manifold, see (Giesselmann 2009) for details. The new feature here is that we must guarantee that these conditions are satisfied uniformly in the time interval under consideration. This reflects the fact that if the mesh becomes to distorted over time one has to consider remeshing and calculating on the new mesh. However as all quantities depend on the Riemannian metric in a continuous way, we can, provided we have some bound for the change of the Riemannian metric, guarantee certain time intervals in which a mesh which is admissible at time t=0 stays admissible.

Note: In particular we need—as in the stationary case—that the curvatures of the faces are bounded, which is needed for error estimates of the fluxes. It is essentially not satisfied under grid refinement for grids using parts of latitudes as faces near the poles. This happens in the classical latitude longitude mesh (which also has several other numerical drawbacks) and in one of the grids proposed in (Calhoun et al., 2008). However this condition is satisfied for example by geodesic grids like (Ronchi et al., 1996; Giraldo 2006) and all but one of the grids proposed in (Calhoun et al., 2008; Berger et al., 2009).

Now we will describe the finite volume schemes. We start by fixing a triangulation T of M and a time sequence $0=t^0<t^1<\ldots$ which we assume is equidistant, i.e. $t^n = nk$ for some fixed $k>0$. We define the grid size h by

$$h := \max_t \max_{K \in T} \mathrm{diam}_{g(t)}.$$

and assume that there is some constant $\alpha>0$ such that

$$m_t(K) > \alpha h^d, \sum_{e \in \partial K} m_t(e) \le \frac{h}{\alpha}$$

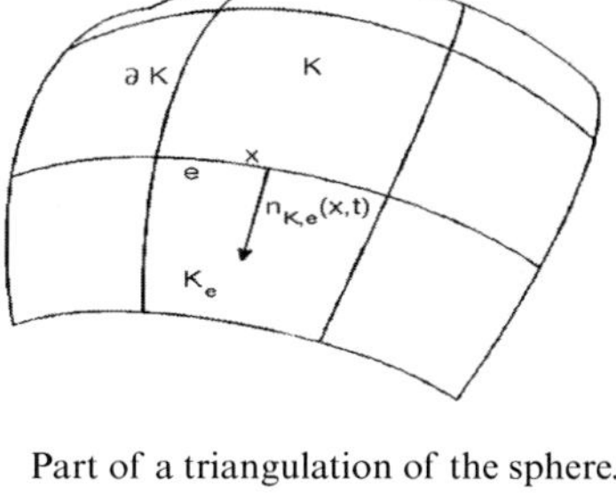

Figure 1. Part of a triangulation of the sphere.

for all considered times. Then we consider a family of numerical fluxes $\{f_K^n,e\}K \in T, e \in \partial K, n \in \mathbb{N}$ such that $f_K^n,e : \mathbb{R}^2 \to \mathbb{R}$ satisfies the classical monotonicity, consistency, conservation and Lipschitz continuity conditions. Additionally we will rely on the following CFL condition, which will guarantee the stability of the scheme

$$k \le \frac{\alpha^2 h}{8L}, \tag{9}$$

where L is the uniform Lipschitz constant of the numerical fluxes.

Now we are in position to state our finite volume scheme, which is based on the integral formulation (1).

$$\begin{aligned}
u_K^0 &= \frac{1}{m_0(K)} \int_K u_0(x) dv_{g(0)}(x), \\
u_K^{n+1} &= u_K^n - \sum_{e \in \partial K} k m_t n(e) \frac{f_K^n,e(u_K^n, u_{Ke}^n)}{m_t n+1(K)} \\
&\quad + q_K^n - g_K^n, \\
q_K^n &= \frac{1}{m_t n+1(K)} \int_{t^n}^{t^{n+1}} \int_K q(x,t,u_K^n) dv_{g(t)} dt, \\
g_K^n &= u_K^n \frac{m_t n+1(K) - m_t n(K)}{m_t n+1(K)}, \\
u^h(x,t) &= u_K^{n+1} \text{for}\, t \in [t^n, t^{n+1}], x \in K.
\end{aligned} \tag{10}$$

Note that we included the effects of the change of the Riemannian metric into the source term.

4 AN ERROR ESTIMATE

In this section we will present an error estimate for the approximate solution given by (2). The proven estimate is in agreement with the results obtained for balance laws in Euclidean space (Chainais-Hillairet and Champier 2001) on the one hand and for conservation laws on general spacetimes in (Amorim et al.) on the other hand.

Theorem 1. *Let u^h be the approximate solution defined by the finite volume scheme* (10) *and u an entropy solution of* (3) *having bounded total variation in space and time on $M\times[0,T]$ then there is a constant $C>0$ depending on T, M, u_0 and $\{g(t)\}_{t\in[0,T]}$ such that the following error estimate holds*

$$\int_{M\times[0,T]} | u^h(x,t) - u(x,t) | dv_{g(t)} dt \le C h^{\frac{1}{4}}. \tag{11}$$

We will only give a sketch of the proof. Similar to the Euclidean case one first has to prove a stability estimate introducing suitable L^∞-bounds for the approximate solution for finite times. Based on the L^∞-stability a weak BV estimate, essentially showing that the total variation (in space and time) of the approximate solution is bounded by $h^{-\frac{1}{2}}$ can be established. As the last prerequisite for the proof of the error estimate one shows a weak entropy inequality for the approximate solution. It is similar to the entropy inequality (7) satisfied by the entropy solution, but the right hand side is nonzero, but consists of terms, which have to be shown to be small.

The proof of the actual error estimate is based on the classical doubling of variables technique used in the Euclidean setting in (Eymard et al., 1998) and (Chainais-Hillairet and Champier 2001) for conservation and balance laws respectively. A generalization of this technique to general Riemannian manifolds was presented in (Giesselmann 2009). The proof of the theorem at hand relies heavily on this generalization. The main new difficulty in the analysis lies in the fact that time dependent cut-off functions have to be used and the properties of their time-derivatives have to be controlled. Apart from that the estimates are rather similar to the ones employed in (Giesselmann 2009) and the main generalization lies in the fact that all the geometric estimates must be ensured uniformly in time.

5 NUMERICAL EXPERIMENTS

In this section we will present the results of numerical experiments obtained using scheme (10) for different moving surfaces and fluxes. All experiments

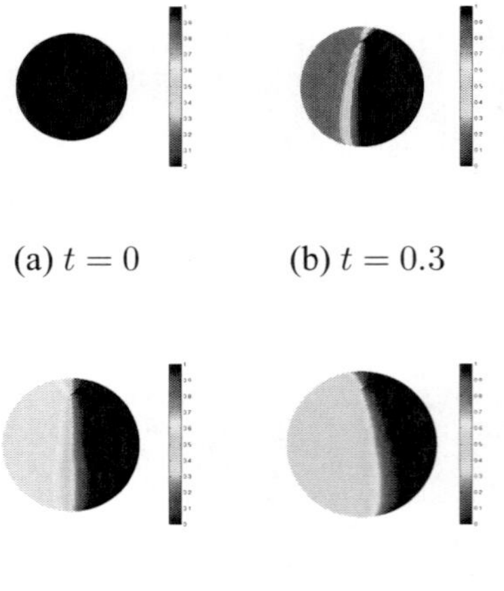

(a) $t = 0$ (b) $t = 0.3$

(c) $t = 0.6$ (d) $t = 0.9$

Figure 2. Numerical solution of (3), (12) using a cubed sphere grid with 9600 cells and the upwind flux.

were performed for two-dimensional surfaces in $\mathbb{R}^3$. It is a well-known feature of monotone finite volume schemes in Euclidean space (for d > 1) that while one can only prove a convergence rate of $h^{\frac{1}{4}}$ numerical experiments yield a convergence rate of $h^{\frac{1}{2}}$ for linear problems and convergence rates between $h^{\frac{1}{2}}$ and h for nonlinear problems.

For our numerical experiments we consider surfaces, which are diffeormorphic to spheres such that we can map grids on spheres on these surfaces. We use grids based on the grids from (Ronchi et al., 1996), which we will call "cubed sphere" grids, even if we consider them on non-spherical surfaces, and grids based on the ones in (Berger et al., 2009), which we will call BCHL grids.

5.1 *Advection on an expanding sphere*

In this case we consider an expanding sphere with radius $r = r(t) = 1+0.4 \cdot t$ and the following initial data, linear flow in latitudinal direction and no source term

$$u_0(x) = \begin{cases} 1 : x_2 > 0 \\ 0 : x_2 \le 0 \end{cases},$$
$$f(x,u) = \begin{pmatrix} -x_2 \\ x_1 \\ 0 \end{pmatrix} u, q \equiv 0. \qquad 12$$

In these and all subsequent images the colours correspond to different values of u. Here blue means diate values. Problem (3), (12) allows an entropy solution. In spherical coordinates i.e.

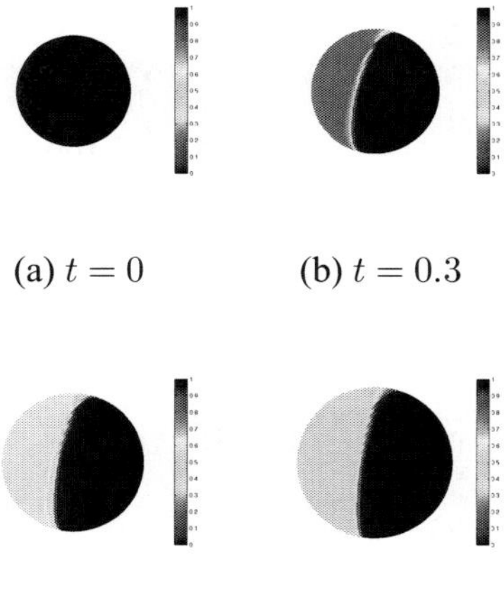

(a) $t = 0$ (b) $t = 0.3$

(c) $t = 0.6$ (d) $t = 0.9$

Figure 3. Numerical solution of (3), (13) on the cubed sphere grid having 9600 cells using the upwind flux.

$$x = r(t)\cdot(\cos\varphi\cos\theta, \sin\varphi\cos\theta, \sin\theta)^T$$

it is given by

$$u(\varphi,\theta,t) = \frac{1}{(r(t))^2} u_0(\varphi - t, \theta).$$

In the following table we compare the approximate solutions obtained using the finite volume scheme (10) for different mesh sizes on BCHL grids to the exact solution. We used the upwind flux.

# cells	h	Rel. L^1-error	Eoc
200	0.5517	0.0533	
800	0.2904	0.0391	0.48
1800	0.1973	0.01319	0.52
3200	0.1497	0.0273	0.56
5000	0.1205	0.0240	0.58
7200	0.1009	0.0215	0.61

5.2 *Burgers like flux on an expanding sphere*

In our second example we consider a nonlinear Burgers-like flux with the same initial data and surface as in (12) above, but

$$f(x,t,u) = \begin{pmatrix} -x_2 \\ x_1 \\ 0 \end{pmatrix} \frac{u^2}{2}, q(x,t,u) = 0. \qquad (13)$$

The table below displays the experimental errors in the L^1-norm comparing the numerical results using (10) with the upwind flux on the BCHL grid to the exact solution u which is given as follows. Let v be the solution of

$$v_t + v v_\varphi = 0 \,\text{in}\, [-\pi,\pi] \times \mathbb{R}_+$$
$$v(\varphi,0) = u_0((\cos\varphi, \sin\varphi, o)^T) \,\text{on}\, [-\pi,\pi]$$

with periodic boundary conditions, which consists of a shock wave with speed 1/2 starting in 0 and a rarefaction wave emanating from $-\pi$, then

$$u(\varphi,\theta,t) = \frac{1}{r^2(t)} v\left(\varphi, \int_0^t \frac{1}{r^2(s)} ds\right),$$

in spherical coordinates. We want to point out that the BCHL grid and the cubed sphere grid give very similar results.

These first two numerical test cases display very similar results to what we observe for Riemann problems for advection and Burgers flux in Euclidean space. They were mainly chosen because they enable a comparison of the numerical results

# cells	h	Rel.L^1-error	Eoc
200	0.5517	0.0239	
800	0.2904	0.0175	0.49
1800	0.1973	0.0139	0.59
3200	0.1497	0.0116	0.66
5000	0.1205	0.0099	0.69
7200	0.1009	0.0087	0.72

to exact solutions. In the next examples we will consider nonlatitudinal flow and see how the changes in geometry can lead to interesting flux patterns.

5.3 *Non-latitudinal flow*

Here we consider a flux whose direction is u dependent, in spherical coordinates it reads

$$f(\varphi,\theta,t,u)=\frac{u^2}{2}r(t)\begin{pmatrix}-\sin\varphi\cos\theta\\ \cos\varphi\cos\theta\\ 0\end{pmatrix} + ur(t)\cos\theta\begin{pmatrix}-\cos\varphi\sin\theta\\ -\sin\varphi\sin\theta\\ \cos\theta\end{pmatrix} \tag{14}$$

on an expanding sphere with radius

$r(t)=1+.2t$

with no source term and initial data

$$u_0(x)=\begin{cases}1: x_1>0.8\\ 0: x1\leq 0.8\end{cases}. \tag{15}$$

We compute the EOC using BCHL grids, the Lax-Friedrichs flux and a refernce solution computed on 131072 cells. The stronger numerical dissipation (compared to the upwind flux) can be seen in the simulation results.

# cells	h	Rel. L^1-error	Eoc
32	1.1000	1.0713	
128	0.6182	0.7009	0.74
512	0.3277	0.4052	0.86
2048	0.1701	0.2365	0.82
8192	0.0868	0.1257	0.94

5.4 *Ellipsoid with shrinking half-axis*

The example at hand will be the first one to display novel flow patterns induced by an interaction of the changing geometry and the flux. We consider

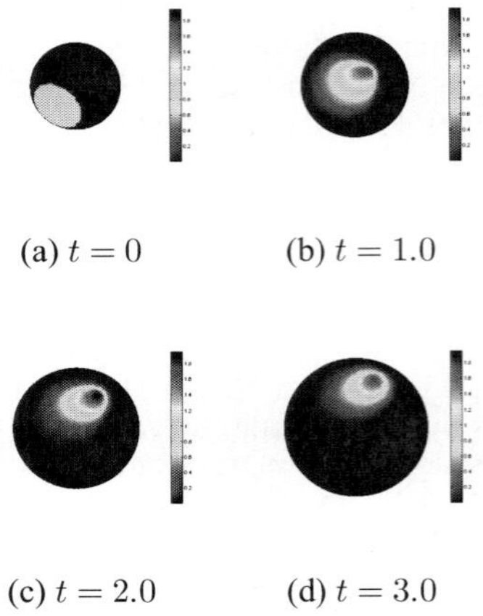

(a) $t=0$ (b) $t=1.0$ (c) $t=2.0$ (d) $t=3.0$

Figure 4. Numerical solution of (3), (14) on the cubed sphere grid having 9600 cells and time step .01 using the Lax-Friedrichs flux.

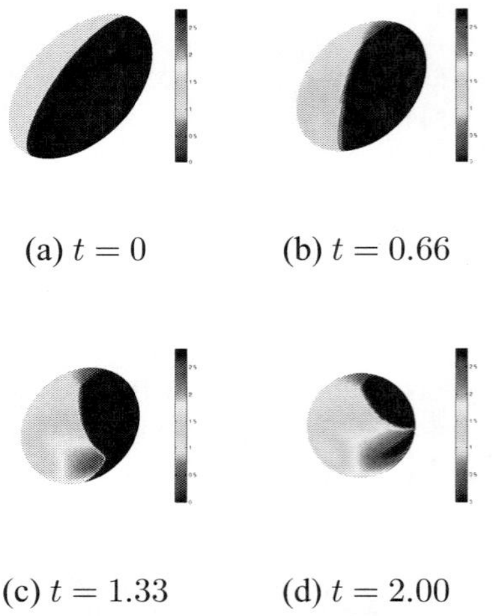

(a) $t=0$ (b) $t=0.66$ (c) $t=1.33$ (d) $t=2.00$

Figure 5. Numerical solution of (3), (16) on the cubed sphere grid having 9600 cells with time step size .01 using the upwind flux.

an ellipsoid with time dependent half-axis given as the zero set of

$$\left(\frac{x_1}{2-0.5\cdot t}\right)^2+x_2^2+x_3^2-1.$$

On this surface the flow is given by

$$f(x,t,u)=\frac{u^2}{2}\begin{pmatrix}-(2-0.5\cdot t)x_2\\ \dfrac{x_1}{2-0.5\cdot t}\\ 0\end{pmatrix} \tag{16}$$

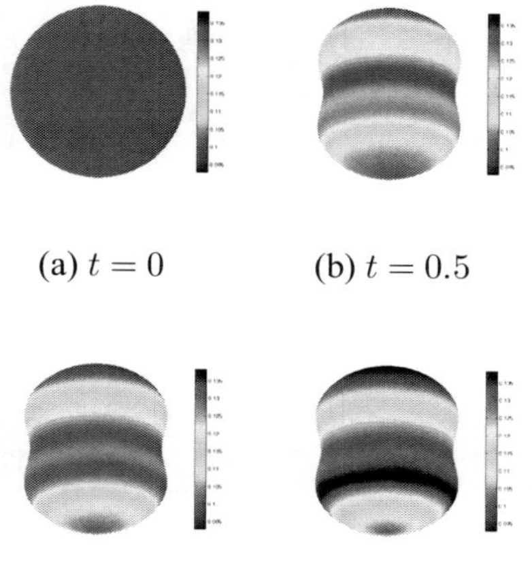

(a) $t = 0$ (b) $t = 0.5$

(c) $t = 1.0$ (d) $t = 1.5$

Figure 6. Numerical solution of (3), (17) on the cubed sphere grid having 9600 cells with time step size .01 using the upwind flux.

# cells	h	Rel. L^1-error	Eoc
32	1.7355	0.3248	
128	1.0088	0.1799	1.09
512	0.5311	0.0938	1.01
2048	0.2747	0.0474	1.04
8192	0.1386	0.0223	1.10

and we consider the situation with no source term and initial data as in subsection 5.1.

The errors displayed below where computed using the upwind flux on a BCHL grid. We used a numerical reference solution computed on 131072 cells.

5.5 *Deformation inducing a shock wave*

In this example the moving surface in space and time $M = M(t)$ under consideration is determined by a deformation of the unit sphere S^2 given by

$$\Psi(\cdot,t): S^2 \to M(t),$$
$$\Psi(x,t) = \begin{cases} \begin{pmatrix} h(x,t)x_1 \\ h(x,t)x_2 \\ x_3 \end{pmatrix} : t \le 0.5 \\ \Psi(x,0.5): t > 0.5 \end{cases}, \tag{17}$$

where $h(x,t) = 1 - \frac{t}{2} e^{-5x_3^2}$.

We consider constant initial data and a Burgers-like flux towards the south pole of the surface

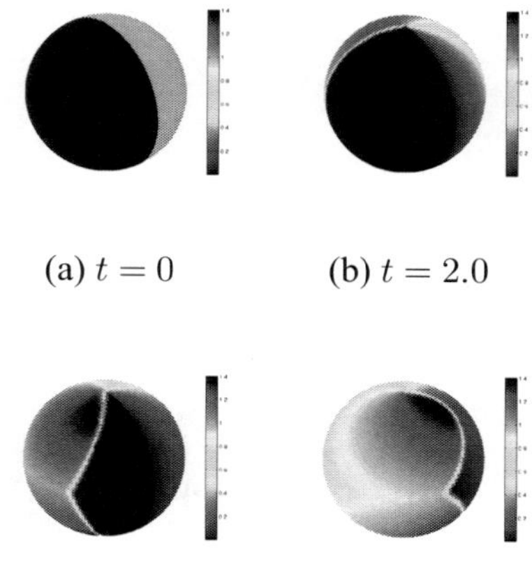

(a) $t = 0$ (b) $t = 2.0$

(c) $t = 4.0$ (d) $t = 6.0$

Figure 7. Numerical solution of (3), (18) on the cubed sphere grid having 9600 cells with time step size .01 using the upwind flux.

$$u_0(x) = 0.1, f(x,t,u) = \frac{u^2}{2}(x_3^2 - 1)\frac{\Psi_\theta}{\|\Psi_\theta\|}, q \equiv 0.$$

We performed numerical experiments on a latitude longitude grid. Comparing the approximate solutions to a reference solution on a grid having 65536 cells.

# cells	h	Rel. L^1-error	Eoc
16	1.4142	0.0188	
64	0.8327	0.0144	0.49
256	0.4333	0.0060	1.34
1024	0.2188	0.0029	1.08
4096	0.1097	0.0014	1.03

5.6 *Solution dependent source term*

Finally we consider a test-case in which we have a space and solution dependent source term on the stationary unit sphere with the same initial data as in (12). Flux and source are given as follows

$$f(x,t,u) = \frac{u^2}{2}\begin{pmatrix} -x_2 \\ x_1 \\ 0 \end{pmatrix}, \; q(x,t,u) = \frac{u}{10}|x_3|. \tag{18}$$

We determined the EOC on a BCHL grid using the upwind flux. We used a numerical reference solution computed on 131072 cells.

# cells	h	rel. L^1-error	eoc
32	1.0000	0.2845	
128	0.5620	0.1455	1.16
512	0.2979	0.0751	1.04
2048	0.1547	0.0381	1.04
8192	0.0790	0.0180	1.12

6 CONCLUSIONS

We have introduced a family of finite volume schemes for balance laws on manifolds with Riemannian metric depending on time. We have generalized the results from the Euclidean setting to this setting, obtaining the same convergence rates.

Furthermore we have presented several numerical test-cases showing that the schemes are able to deal with different physical fluxes, geometries and source terms. In some cases we were able to compare the numerical solutions to exact solutions. In the other cases we compared them to numerical reference solutions on very fine grids. In all cases the numerical experiments exhibit better convergence rates, than those predicted by the error estimate. The convergence rates for linear problems were not as good as those for nonlinear problems. These findings are in agreement with the situation in the Euclidean case.

ACKNOWLEDGEMENT

Both authors would like to thank the German Research Foundation (DFG) for financial support of the project within the Cluster of Excellence in Simulation Technology (EXC 310/1) at the University of Stuttgart.

REFERNCES

Alt, H.W. (2009). The entropy principle for interfaces. Fluids and solids. *Adv. Math. Sci. Appl. 19*(2), 585–663.

Amorim, P., LeFloch, P.G. and Neves, W. A geometric approach to error estimates for conservation laws posed on a spacetime. http://arxiv.org/abs/1002.3137.

Amorim, P., LeFloch, P.G., and Okutmustur, B. (2008). Finite volume schemes on Lorentzian manifolds. *Commun. Math. Sci. 6*(4), 1059–1086.

Ben-Artzi, M. and LeFloch, P.G. (2007). Well-posedness theory for geometry-compatible hyperbolic conservation laws on manifolds. *Ann. Inst. H. Poincaré Anal. Non Linéaire 24*(6), 989–1008.

Berger, M.J., Calhoun, D.A., Helzel, C. and LeVeque, R.J. (2009). Logically rectangular finite volume methods with adaptive refinement on the sphere. *Philos. Trans. R. Soc. Lond. Ser. A Math. Phys. Eng. Sci. 367*(1907), 4483–4496.

Calhoun, D.A., Helzel, C. and LeVequ, R.J. (2008). Logically rectangular grids and finite volume methods for PDEs in circular and spherical domains. *SIAM Rev. 50*(4), 723–752.

Chainais-Hillairet, C. and Champier, S. (2001). Finite volume schemes for nonhomogeneous scalar conservation laws: error estimate. *Numer. Math. 88*(4), 607–639.

Dreyer, W., Giesselmann, J., Kraus, C. and Ro-hde, C. Asymptotic analysis for korteweg models. SimTech Preprint 2010-67, http://www.simtech.uni-stuttgart.de/publikationen/preprints10.php, submitted to Interfaces and Free Boundaries.

Eymard, R., Gallouet, T., Ghilani, M. and Herbin, R. (1998). Error estimates for the approximate solutions of a nonlinear hyperbolic equation given by finite volume schemes. *IMA J. Numer. Anal. 18*(4), 563–594.

Giesselmann, J. (2009). A convergence result for finite volume schemes on Riemannian manifolds. *M2AN Math. Model. Numer. Anal. 43*(5), 929–955.

Giraldo, F.X. (2006). High-order triangle-based discontinuous galerkin methods for hyperbolic equations on a rotating sphere. *J. Comput. Phys. 214*(2), 447–465.

Kühnel, W. (2002). *Differential geometry*, Volume 16 of *Student Mathematical Library*. Providence, RI: American Mathematical Society. Curves—surfaces—manifolds, Translated from the 1999 German original by Bruce Hunt.

LeFloch, P.G. and Okutmustur, B. (2008). Hyperbolic conservation laws on spacetimes. A finite volume scheme based on differential forms. *Far East J. Math. Sci. (FJMS) 31*(1), 49–83.

LeFloch, P.G., Okutmustur, B. and Neves, W. (2009). Hyperbolic conservation laws on manifolds. An error estimate for finite volume schemes. *Acta Math. Sin. (Engl. Ser.) 25*(7), 1041–1066.

Muller, I. (1985). *Thermodynamics*. Pitman.

Ronchi, C., Iacono, R. and Paolucci, P.S. (1996). The cubed sphere: A new method for the solution of partial differential equations in spherical geometry. *J. Comput. Phys. 124*(1), 93–114.

Numerical Methods for Hyperbolic Equations – Vázquez-Cendón et al. (eds)
© 2013 Taylor & Francis Group, London, ISBN 978-0-415-62150-2

On multilayer shallow water systems

T. Chacón Rebollo
Departamento Ecuaciones Diferenciales y Análisis Numérico, Facultad de Matemática, Universidad de Sevilla, Spain

E.D. Fernández-Nieto & El Hadji Koné
Departamento Matemática Aplicada I, E.T.S. Arquitectura, Universidad de Sevilla, Spain

ABSTRACT: The work is devoted to the derivation of a multilayer shallow water model. Subdividing the 2D fluid domain into layers and averaging the horizontal velocity in each layer, we derive a 1D horizontal model using a variational formulation. That model keeps the mass exchanges property between the neighboring layers by considering weak solutions of Navier-Stokes equations. Next, some meaningful numerical simulations are presented.

1 INTRODUCTION

Considering flows with large friction coefficients, with significants water depth or with important wind effects, the horizontal velocity can hardly be approximated by a vertically constant velocity as in the classical shallow water system. A multilayer shallow water model, where each layer is described by its own height and its own velocity, is then often used to drop that limitation. In (Audusse et al. 2011) a multilayer shallow water model is proposed, by considering a constant profile of the horizontal velocity at each layer and by including mass and momentum exchange terms between the layers. See also (Sainte Marie 2011) for the case of Euler system with non-hydrostatic pressure.

In this work, we start from the 2D unsteady incompressible Navier-Stokes equations and in a similar manner as in (Amara et al. 2008), we derive a multilayer model from a variational formulation of the problem. Actually, we combine the weak formulation of the Navier-Stokes equations with the standard shallow water technique of the horizontal velocity averaging in each layer. This procedure yields, in a natural way, the transmission conditions for the mass and momentum balances as the equations of the interfaces between the layers. The model that we introduce here, is derived in a hydrostatic pressure framework. Then the unknowns are the height of the fluid and the horizontal velocities in each layer, similarly to the model proposed in (Audusse et al. 2011). However, the expressions of the momentum transference terms are different for these two models.

Finally, some numerical simulations are done to observe the behavior of the multilayer shallow water model in a dam break situation. The dam break is an interesting case due to the strong variations in the vertical direction, that is recovered by the simulations realized.

2 THE MULTILAYER MODEL

We start by recalling the 2D non-stationary incompressible Navier-Stokes equations. At each time $t > 0$, let us denote the fluid domain by $\Omega_F(t)$ and by $I_F(t)$, its projection on the horizontal line. For given density $\rho \in \mathbb{R}$, dynamic viscosity $\mu \in \mathbb{R}$ and gravity acceleration $g \in \mathbb{R}$, the unknown velocity $\vec{u} := (u,w)^t \in \mathbb{R}^2$ and pressure $p \in \mathbb{R}$ functions satisfy the equations:

$$\begin{cases} \operatorname{div} \vec{u} = 0 \\ \rho\partial_t \vec{u} + \rho\vec{u} \cdot \nabla\vec{u} - \operatorname{div} \sum_T = \rho\vec{g}, \end{cases} \tag{1}$$

where $\vec{g} = (0,-g)^t \in \mathbb{R}^2$, $\nabla\vec{u} = \begin{pmatrix} \partial_x u & \partial_z u \\ \partial_x w & \partial_z w \end{pmatrix}$ and $\sum_T = -pI + \mu(\nabla\vec{u} + (\nabla\vec{u})^t)$ is the total stresses tensor.

In order to introduce the multilayer modeling, the fluid domain is decomposed in the vertical direction into $N \in \mathbb{N}^*$ layers of thickness $h_\alpha(t,x)$ with $N+1$ interfaces defined by $z_{\alpha+\frac{1}{2}}(t,x)$ for $\alpha = 0,1,\ldots,N$. That is $\Omega_F(t) = \bigcup_{\alpha=1}^{N} \Omega_\alpha(t)$, where we denote

$$\Omega_\alpha(t) = \left\{(x,z); x \in I_F(t) \text{ and } z_{\alpha-\frac{1}{2}} < z < z_{\alpha+\frac{1}{2}}\right\}$$

$$\partial\Omega_\alpha(t) = \Gamma_{\alpha-\frac{1}{2}} \sqcup \Gamma_{\alpha+\frac{1}{2}} \sqcup \Theta_\alpha, \text{ with}$$

$$\Gamma_{\alpha+\frac{1}{2}}(t) = \{(x,z); x \in I_F(t) \text{ and } z = z_{\alpha+\frac{1}{2}}\},$$

$$\Theta_\alpha(t) = \left\{(x,z); x \in \partial I_F(t) \text{ and } z_{\alpha-\frac{1}{2}} < z < z_{\alpha+\frac{1}{2}}\right\}.$$

Setting $z_B = z_{1/2}$ and $z_S = z_{N+1/2}$, we have $z_{\alpha+1/2} = z_B + \sum_{\beta=1}^{\alpha} h_\beta$ for $\alpha = 1, \ldots, N$ and then $h = z_S - z_B = \sum_{\alpha=1}^{N} h_\alpha$. We introduce the velocities $\mathbf{\vec{u}}_\alpha := (u_\alpha, w_\alpha)^t := \mathbf{\vec{u}}_{|\Omega_\alpha(t)}$, for $\alpha = 1, \ldots, N$, leading to the following governing equations in each layer $\Omega_\alpha(t)$:

$$\begin{cases} \operatorname{div} \vec{u}_\alpha = 0 \\ \rho_\alpha \partial_t \vec{u}_\alpha + \rho_\alpha \vec{u}_\alpha \cdot \nabla \vec{u}_\alpha - div \sum_{T,\alpha} = \rho_\alpha \vec{g}, \end{cases} \tag{2}$$

where $\sum_{T,\alpha} = -p_\alpha I + \mu_\alpha(\nabla \vec{u}_\alpha + (\nabla \vec{u}_\alpha)^t)$ and $p_\alpha = p_{|\Omega\alpha(t)}$.

Definition 1. (Averaged horizontal velocity)
For all $\alpha = 1, \ldots, N$, the horizontal velocity u_α is defined by

$$u_\alpha(t,x) := \frac{1}{h_\alpha} \int_{z_{\alpha-\frac{1}{2}}}^{z_{\alpha+\frac{1}{2}}} u(t,x,z)dz.$$

2.1 *The mass balance equation*

The vertically averaged velocity assumption for the horizontal velocity leads to the mass equation

$$\partial_t h_\alpha + \partial_x(h_\alpha u_\alpha) = G^-_{\alpha+\frac{1}{2}} - G^+_{\alpha-\frac{1}{2}} \tag{3}$$

for $\alpha = 1, \ldots, N$, where we denote $G^\pm_{\alpha+\frac{1}{2}} = \partial_t z_{\alpha+\frac{1}{2}} + u^\pm_{\alpha+\frac{1}{2}} \partial_x z_{\alpha+\frac{1}{2}} - w^\pm_{\alpha+\frac{1}{2}}$ with $u^-_{\frac{1}{2}} = w^-_{\frac{1}{2}} = u^+_{N+\frac{1}{2}} = w^+_{N+\frac{1}{2}} = 0$. Indeed the expression of the horizontal velocity u_α yields

$$\begin{aligned} \partial_x(h_\alpha u_\alpha) &= \partial_x \int_{z_{\alpha-\frac{1}{2}}}^{z_{\alpha+\frac{1}{2}}} u(t,x,z)dz \\ &= u^-_{\alpha+\frac{1}{2}} \partial_x z_{\alpha+\frac{1}{2}} - u^+_{\alpha-\frac{1}{2}} \partial_x z_{\alpha-\frac{1}{2}} \\ &\quad + \int_{z_{\alpha-\frac{1}{2}}}^{z_{\alpha+\frac{1}{2}}} \partial_x u(t,x,z)dz. \end{aligned}$$

Since $\partial_x u + \partial_z w = 0$, we have

$$\begin{aligned} \partial_x(h_\alpha u_\alpha) &= u^-_{\alpha+\frac{1}{2}} \partial_x z_{\alpha+\frac{1}{2}} - u^+_{\alpha-\frac{1}{2}} \partial_x z_{\alpha-\frac{1}{2}} \\ &\quad - \int_{z_{\alpha-\frac{1}{2}}}^{z_{\alpha+\frac{1}{2}}} \partial_z w(t,x,z)dz \\ &= u^-_{\alpha+\frac{1}{2}} \partial_x z_{\alpha+\frac{1}{2}} - u^+_{\alpha-\frac{1}{2}} \partial_x z_{\alpha-\frac{1}{2}} \\ &\quad - (w^-_{\alpha+\frac{1}{2}} - w^+_{\alpha-\frac{1}{2}}). \end{aligned}$$

Hence adding the equality $\partial_t h_\alpha = \partial_t z_{\alpha+\frac{1}{2}} - \partial_t z_{\alpha-\frac{1}{2}}$ to the previous one, we get the equation (3). In addition the mass flux transmission at the interfaces $\Gamma_{\alpha+\frac{1}{2}}(t)$, for $\alpha = 0, 1, \ldots, N$, is given by

$$[\rho]_{|\Gamma_{\alpha+\frac{1}{2}}(t)} \partial_t z_{\alpha+\frac{1}{2}} + [\rho u]_{|\Gamma_{\alpha+\frac{1}{2}}(t)} \partial_x z_{\alpha+\frac{1}{2}} - [\rho w]_{|\Gamma_{\alpha+\frac{1}{2}}(t)} = 0,$$

where the bracket $[\cdot]_{|\Gamma_{\alpha+\frac{1}{2}}(t)}$ denotes the jump at the interface $\Gamma_{\alpha+\frac{1}{2}}(t)$. That is $\rho_\alpha G^-_{\alpha+\frac{1}{2}} = \rho_{\alpha+1} G^+_{\alpha+\frac{1}{2}}$.

Hypothesis 1. For all $\alpha = 1, \ldots, N, [\rho]_{|\Gamma_{\alpha+\frac{1}{2}}(t)} = 0$. That means there is no jump to the density.[2]

With that assumption, the mass flux transmission condition at the interfaces leads to the equality $G^-_{\alpha+\frac{1}{2}} = G^+_{\alpha+\frac{1}{2}}$. We denote then $G_{\alpha+\frac{1}{2}} := G^\pm_{\alpha+\frac{1}{2}}$. Therefore, the equation (3) is written, for $\alpha = 1, \ldots, N$, as

$$\partial_t h_\alpha + \partial_x(h_\alpha u_\alpha) = G_{\alpha+\frac{1}{2}} - G_{\alpha-\frac{1}{2}}. \tag{4}$$

2.2 *The variational formulation*

The aim is to determine Hilbert spaces $\mathbf{V}$, $\mathbf{V}^0 \subseteq \mathbf{V}$, M and $M^0 \subseteq M$ and to look for a solution $(\mathbf{\vec{u}}, p) \in \mathbf{V} \times M$ such that for all $(\mathbf{\vec{v}}, q) \in \mathbf{V}^0 \times M^0$ we have

$$\int_{\Omega_F} q \operatorname{div} \mathbf{\vec{u}} d\Omega = \sum_{\alpha=1}^{N} \int_{\Omega_\alpha} q_\alpha \operatorname{div} \mathbf{\vec{u}}_\alpha d\Omega = 0, \tag{5}$$

for the incompressibility property and for the momentum law we write

$$\begin{aligned} &\sum_{\alpha=1}^{N} \Big\{ \int_{\Omega\alpha} \rho \partial_t \vec{u}_\alpha \cdot \vec{v}_\alpha d\Omega + \int_{\Omega\alpha} \rho (\vec{u}_\alpha \cdot \nabla \vec{u}_\alpha) \vec{v}_\alpha d\Omega \\ &\quad + \int_{\Omega\alpha} \mu_\alpha \left(\nabla \vec{u}_\alpha + (\nabla \vec{u}_\alpha)^t \right) \cdot\cdot \nabla \vec{v}_\alpha d\Omega \\ &\quad - \int_{\Omega\alpha} p_\alpha div \vec{v}_\alpha d\Omega \Big\} - \sum_{\alpha=1}^{N} \int_{\partial\Omega\alpha} \left(\sum_{T,\alpha} \vec{n}\alpha \right) \cdot \vec{v}_\alpha d\Gamma \\ &\quad = \sum_{\alpha=1}^{N} \int_{\Omega\alpha} \rho \vec{g} \cdot \vec{v}_\alpha d\Omega. \end{aligned} \tag{6}$$

$\vec{n}_\alpha$ denotes the outward unit normal vector at $\partial\Omega_\alpha(t)$.

2.3 *Interfaces conditions*

For a given layer $\Omega_\alpha(t)$, $\forall \alpha = 1, \ldots, N$ the outward unit normal vector at $\partial\Omega_\alpha(t)$ is given by

$$\vec{n}\alpha = \begin{cases} \dfrac{1}{\sqrt{1 + (\partial_x z_{\alpha+\frac{1}{2}})^2}} \left(-\partial_x z_{\alpha+\frac{1}{2}}, 1 \right)^t \\ \text{at } \Gamma_{\alpha+\frac{1}{2}}(t) \\ \dfrac{1}{\sqrt{1 + (\partial_x z_{\alpha-\frac{1}{2}})^2}} \left(\partial_x z_{\alpha-\frac{1}{2}}, -1 \right)^t \\ \text{at } \Gamma_{\alpha-\frac{1}{2}}(t) \\ (\mp 1, 0)^t \quad \text{respectively on the left an} \\ \text{the right at } \Theta_\alpha(t). \end{cases}$$

Denoting $\vec{\mathbf{u}}^{\pm}_{\alpha+1/2} := (u^{\pm}_{\alpha+1/2}, w^{\pm}_{\alpha+1/2})'$, for $\alpha = 0, 1, ..., N$, the kinematic equations at $\Gamma_{\alpha+1/2}(t)$ are then $G_{\alpha+1/2} = \partial_t z_{\alpha+1/2} - \sqrt{1+(\partial_x z_{\alpha+1/2})^2}\,\vec{\mathbf{n}}_\alpha \cdot \vec{\mathbf{u}}^{-}_{\alpha+1/2}$. Let us introduce the notations $\widetilde{G_{\alpha+\frac{1}{2}}} := \partial_t z_{\alpha+\frac{1}{2}} - G_{\alpha+}$ and $\vec{\tilde{\mathbf{u}}}^{\pm}_{\alpha+1/2} := \vec{\mathbf{u}}^{\pm}_{\alpha+1/2} - (0, G_{\alpha+1/2})'$, we can rewrite the kinematic equations as $\vec{\mathbf{n}}_\alpha \cdot \vec{\tilde{\mathbf{u}}}^{\pm}_{\alpha+1/2} = 0$ at $\Gamma_{\alpha+1/2}(t)$ noticing that $\vec{\mathbf{n}}_{\alpha+1} = -\vec{\mathbf{n}}_\alpha$ at that interface. We can see that the interface conditions yield $\vec{\mathbf{n}}_\alpha \cdot (\vec{\mathbf{u}}^{+}_{\alpha+1/2} - \vec{\mathbf{u}}^{-}_{\alpha+1/2}) = 0$ at $\Gamma_{\alpha+1/2}(t)$. Concerning the stresses, with the tensors $\Sigma_{T,\alpha}$ for $\alpha = 0, 1, ..., N$, we consider the transmission conditions $\Sigma_{T,\alpha+1}\vec{\mathbf{n}}_\alpha = \Sigma_{T,\alpha}\vec{\mathbf{n}}_\alpha$ at $\Gamma_{\alpha+1/2}(t)$. Since $\vec{\mathbf{n}}_{\alpha+1} = -\vec{\mathbf{n}}_\alpha$ and $\vec{\mathbf{n}}_\alpha \cdot (\vec{\mathbf{u}}^{+}_{\alpha+1/2} - \vec{\mathbf{u}}^{-}_{\alpha+1/2}) = 0$ at $\Gamma_{\alpha+1/2}(t)$ $\forall \alpha = 0, 1, ..., N$, we set for $\alpha = 1, ..., N$:

$$\Sigma_{T,\alpha}\vec{\mathbf{n}}_\alpha = \begin{cases} -\widetilde{p_{\alpha+\frac{1}{2}}}\vec{\mathbf{n}}_\alpha + c_{\alpha+\frac{1}{2}}\left(\vec{\mathbf{u}}^{+}_{\alpha+\frac{1}{2}} - \vec{\mathbf{u}}^{-}_{\alpha+\frac{1}{2}}\right) \\ \quad \text{at } \Gamma_{\alpha+\frac{1}{2}}(t) \\ -\widetilde{p_{\alpha-\frac{1}{2}}}\vec{\mathbf{n}}_\alpha - c_{\alpha-\frac{1}{2}}\left(\vec{\mathbf{u}}^{+}_{\alpha-\frac{1}{2}} - \vec{\mathbf{u}}^{-}_{\alpha-\frac{1}{2}}\right) \\ \quad \text{at } \Gamma_{\alpha-\frac{1}{2}}(t), \end{cases}$$

where $\widetilde{p_{\alpha+1/2}} = \widetilde{p_{\alpha+1/2}}(t,x)$ and $c_{\alpha+1/2}$ are respectively the dynamic pressure and the frictions coefficient at the interfaces $\Gamma_{\alpha+1/2}(t)$. Hence we summarize the interfaces conditions for the multilayer system as:

- at $\Gamma_S(t) = \Gamma_{N+1/2}(t)$, $\vec{\mathbf{n}}_N \cdot \vec{\tilde{\mathbf{u}}}_{N+1/2} = 0$, $p_N = p_S$ and $\Sigma_{T,N}\vec{\mathbf{n}}_N = -p_S\vec{\mathbf{n}}_N + c_S\vec{\tilde{\mathbf{u}}}_{N+1/2}$
- at $\Gamma_B(t) = \Gamma_{1/2}(t)$, $\vec{\mathbf{n}}_1 \cdot \vec{\tilde{\mathbf{u}}}_{1/2} = 0$ and $\Sigma_{T,1}\vec{\mathbf{n}}_1 = -p_B\vec{\mathbf{n}}_1 + c_B\vec{\tilde{\mathbf{u}}}_{1/2}$
- For $\alpha = 1, ..., N-1$, *at* $\Gamma_{\alpha+1/2}(t)$, $\vec{\mathbf{n}}_\alpha \cdot \vec{\tilde{\mathbf{u}}}^{\pm}_{\alpha+1/2} = 0$ and $\Sigma_{T,\alpha}\vec{\mathbf{n}}_\alpha = -\widetilde{p_{\alpha+1/2}}\vec{\mathbf{n}}_\alpha + c_{\alpha+1/2}(\vec{\mathbf{u}}^{+}_{\alpha+1/2} - \vec{\mathbf{u}}^{-}_{\alpha+1/2})$
- For $\alpha = 1, ..., N$, *at* $\Theta_\alpha(t)$ $\vec{\mathbf{n}}_\alpha \cdot \vec{\mathbf{u}}_\alpha = k_\alpha$ and $\Sigma_{T,\alpha}\vec{\mathbf{n}}_\alpha = \vec{f}_\alpha$.

k_α and $\vec{f}_\alpha$ are respectively given scalar and vector functions, where $\vec{f}_\alpha$ represents the wind's effect in each layer $\Omega_\alpha(t)$.

2.4 *The multilayer shallow water model*

From now on, we place ourselves in the framework of the classical shallow water system in each layer. This translates into hydrostatic pressure and neglected vertical velocity. Using the hydrostatic assumption, we have for $\alpha = 1, ..., N$ and $z_{\alpha-\frac{1}{2}} \leq zz \leq_{\alpha+\frac{1}{2}}$,

$$p_\alpha(x,z) = p_{\alpha+\frac{1}{2}}(x) + \rho g(z_{\alpha+\frac{1}{2}} - z), \tag{7}$$

with

$$p_{\alpha+\frac{1}{2}}(x) = p_S(x) + \rho g \sum_{\beta=\alpha+1}^{N} h_\beta(x). \tag{8}$$

Moreover, at the interfaces, the horizontal components of $\vec{\mathbf{u}}^{-}_{\alpha+\frac{1}{2}}$ and $\vec{\mathbf{u}}^{+}_{\alpha-\frac{1}{2}}$ are set equal to u_α, for $\alpha = 1, ..., N$. First we do straightforward calculations using an integration, with respect to the variable z, of the horizontal component of the equation (6). Next we identify the horizontal component of the test vector functions, which we denote v_α. That leads to the following variational equation for the horizontal velocity in each layer Ω_α, $\alpha = 1, ..., N$:

$$\begin{aligned} &\int_{I_F} \rho h_\alpha(\partial_t u_\alpha + u_\alpha \partial_x u_\alpha) v_\alpha\, dx \\ &+ \int_{I_F} h_\alpha\left(2\mu_\alpha \partial_x u_\alpha - p_{\alpha+\frac{1}{2}} - \rho g \frac{h_\alpha}{2}\right)\partial_x v_\alpha\, dx \\ &+ \int_{I_F} c_{\alpha-\frac{1}{2}}\sqrt{1+(\partial_x z_{\alpha-\frac{1}{2}})^2}(u_\alpha - u_{\alpha-1}) v_\alpha\, dx \\ &- \int_{I_F} c_{\alpha+\frac{1}{2}}\sqrt{1+(\partial_x z_{\alpha+\frac{1}{2}})^2}(u_{\alpha+1} - u_\alpha) v_\alpha\, dx \\ &= \int_{I_F} \left(\widetilde{p_{\alpha+\frac{1}{2}}}\partial_x z_{\alpha+\frac{1}{2}} - \widetilde{p_{\alpha-\frac{1}{2}}}\partial_x z_{\alpha-\frac{1}{2}}\right) v_\alpha\, dx, \end{aligned} \tag{9}$$

where we have denoted $u_{N+1} = u_0 = 0$, $c_{N+\frac{1}{2}} = -c_S$ and $c_{\frac{1}{2}} = -c_B$.

The 1D boundary value problem corresponding to the previous 1D variational formulation is derived here, in order to compare it with the standard shallow water equations and the multilayer system derived in (Audusse et al. 2011). Therefore we have

$$\begin{aligned} &\rho h_\alpha(\partial_t u_\alpha + u_\alpha \partial_x u_\alpha) \\ &- \partial_x\left(h_\alpha\left(2\mu_\alpha \partial_x u_\alpha - p_{\alpha+\frac{1}{2}} - \rho g \frac{h_\alpha}{2}\right)\right) \\ &+ c_{\alpha-\frac{1}{2}}\sqrt{1+(\partial_x z_{\alpha-\frac{1}{2}})^2}(u_\alpha - u_{\alpha-1}) \\ &- c_{\alpha+\frac{1}{2}}\sqrt{1+(\partial_x z_{\alpha+\frac{1}{2}})^2}(u_{\alpha+1} - u_\alpha) \\ &= \widetilde{p_{\alpha+\frac{1}{2}}}\partial_x z_{\alpha+\frac{1}{2}} - \widetilde{p_{\alpha-\frac{1}{2}}}\partial_x z_{\alpha-\frac{1}{2}}. \end{aligned}$$

In addition using the expression (7) of the kinematic pressure and successively the equalities $h_\alpha \partial_t u_\alpha = \partial_t(h_\alpha u_\alpha) - u_\alpha \partial_t h_\alpha$, next

$\partial_t h_\alpha = -\partial_x(h_\alpha u_\alpha) + G_{12} - G_{\alpha-1/2}$ from (4) and later $u_\alpha \partial_x(h_\alpha u_\alpha) + h_\alpha u_\alpha \partial_x u_\alpha = \partial_x(h_\alpha u_\alpha^2)$, we obtain the final equation

$$\begin{aligned}
&\rho\partial_t(h_\alpha u_\alpha) + \rho\partial_x(h_\alpha u_\alpha^2) - 2\mu_\alpha \partial_x(h_\alpha \partial_x u_\alpha) \\
&\quad - \rho u_\alpha\left(G_{\alpha+\frac{1}{2}} - G_{\alpha-\frac{1}{2}}\right) \\
&\quad + c_{\alpha-\frac{1}{2}}\sqrt{1+(\partial_x z_{\alpha-\frac{1}{2}})^2}\,(u_\alpha - u_{\alpha-1}) \\
&\quad - c_{\alpha+\frac{1}{2}}\sqrt{1+(\partial_x z_{\alpha+\frac{1}{2}})^2}\,(u_{\alpha+1} - u_\alpha) \\
&= -\rho g h_\alpha \partial_x(z_B + h) + W_\alpha,
\end{aligned} \tag{10}$$

where

$$\begin{aligned}
W_\alpha = &\left(\widetilde{p_{\alpha+\frac{1}{2}}} - \rho g \sum_{\beta=\alpha+1}^{N} h_\beta\right)\partial_x h_\alpha \\
&+ \left(\widetilde{p_{\alpha+\frac{1}{2}}} + \rho g h_\alpha - \widetilde{p_{\alpha-\frac{1}{2}}}\right)\partial_x z_{\alpha-\frac{1}{2}} \\
&- \partial_x(h_\alpha p_S).
\end{aligned}$$

3 NUMERICAL APPROACH

A finite volume method is applied to solve the multilayer model derived previously. The aim is to express the system in the form

$$\partial_t w + \partial_x F(w) + B(w)\partial_x w = S(w)\partial_x H, \tag{11}$$

where $w \in \mathbb{R}^n$ $(n \in \mathbb{N}^*)$ is the unknown vector, F is a regular vector function from $\mathbb{R}^n$ to itself, B is a matrix function from $\mathbb{R}^n$ to $\mathcal{M}_n(\mathbb{R})$, this one denotes the space of real square matrices of order n, $\mathbb{S}$ is a vector function from $\mathbb{R}^n$ to itself and H is a scalar function from $\mathbb{R}^n$ to itself. The form (11) constitutes a classical simplified model type for multiphase or multilayer flows in the literature. In the following we point out the expressions of the different terms in (11). For the sake of simplicity, we neglect the viscosity of the fluid and the frictions between the layers. Due to the hydrostatic assumption, we neglect too the difference between the dynamic and the kinematic pressures. Then the system made up of (4) and (10) becomes, for $\alpha = 1, \ldots, N$,

$$\begin{cases}
\partial_t h_\alpha + \partial_x(h_\alpha u_\alpha) = G_{\alpha+\frac{1}{2}} - G_{\alpha-\frac{1}{2}} \\
\partial_t(h_\alpha u_\alpha) + \partial_x(h_\alpha u_\alpha^2) - u_\alpha\left(G_{\alpha+\frac{1}{2}} - G_{\alpha-\frac{1}{2}}\right) \\
\quad = -g h_\alpha \partial_x(z_B + h).
\end{cases} \tag{12}$$

Summing up the equations (4) for $1 \le \beta \le \alpha \le N$, we have $G_{\alpha+1/2} - G_{1/2} = \partial_t \sum_{\beta=1}^{\alpha} h_\beta + \partial_x \sum_{\beta=1}^{\alpha} h_\beta u_\beta$. At the bottom, the impermeability kinematic condition gives $G_{\frac{1}{2}} = 0$. Then we get for $\alpha = 1, \ldots, N$,

$$G_{\alpha+\frac{1}{2}} = \partial_t \sum_{\beta=1}^{\alpha} h_\beta + \partial_x \sum_{\beta=1}^{\alpha} h_\beta u_\beta.$$

Especially since $G_{N+1/2} = 0$, due to the kinematic condition at the free surface, we have

$$\partial_t \sum_{\alpha=1}^{N} h_\alpha + \partial_x \sum_{\alpha=1}^{N} h_\alpha u_\alpha = 0.$$

From now on, we consider layers with heights proportional to the total fluid height h. That is $h_\alpha = l_\alpha h$, with l_α a positive constant coefficient for $\alpha = 1, \ldots, N$ (i.e $\sum_{\alpha=1}^{N} l_\alpha = 1$). Then we get the global continuity equation

$$\partial_t h + \partial_x\left(h \sum_{\alpha=1}^{N} l_\alpha u_\alpha\right) = 0. \tag{13}$$

Furthermore we can write $\partial_t h_\beta = l_\beta \partial_t h = -\sum_{\gamma=1}^{N} l_\beta l_\gamma \partial_x (hu_\gamma)$ and that leads to the following expression of the mass fluxes:

$$G_{\alpha+\frac{1}{2}} = \sum_{\gamma=1}^{N} \xi_{\alpha,\gamma}\partial_x(hu_\gamma), \tag{14}$$

where for $\alpha, \gamma = 1, \ldots, N$, we have denoted

$$\xi_{\alpha,\gamma} = \sum_{\beta=1}^{\alpha} (\delta_{\beta\gamma} - l_\beta) l_\gamma.$$

$\delta_{\beta\gamma}$ is the symbol of Kronecker. Since $\sum_{\beta=1}^{N} l_\beta = 1$, we have $\xi_{N,\gamma} = 0, \forall\, \gamma = 1, \ldots, N$. Hence we find again $G_{N+1/2} = 0$. Moreover, we can see that $\xi_{\alpha,\gamma} = \xi_{\alpha-1,\gamma} + (\delta_{\alpha\gamma} - l_\alpha)l_\gamma, \forall\, \alpha, \gamma = 1, \ldots, N$, with the setting $\xi_{0,\gamma} = 0, \forall\, \gamma = 1, \ldots, N$. Therefore we get

$$G_{\alpha+\frac{1}{2}} - G_{\alpha-\frac{1}{2}} = \sum_{\gamma=1}^{N} (\delta_{\alpha\gamma} - l_\alpha) l_\gamma \partial_x(hu_\gamma). \tag{15}$$

Let us denote $q_\alpha = hu_\alpha, \forall\, \alpha = 1, \ldots, N$. Then using the expression (15), we define the following system of $(N+1)$ equations from the equations (13) and (12):

$$\begin{cases}
\partial_t h + \partial_x\left(\sum_{\beta=1}^{N} l_\beta q_\beta\right) = 0, \\
\partial_t q_\alpha + \partial_x\left(\dfrac{q_\alpha^2}{h} + \dfrac{gh^2}{2}\right) + \sum_{\beta=1}^{N} \dfrac{(l_\alpha - \delta_{\alpha\beta}) l_\beta q_\alpha}{l_\alpha h}\partial_x q_\beta \\
= -gh\partial_x z_B, \quad \forall\, \alpha = 1, \ldots, N.
\end{cases} \tag{16}$$

Actually the system (16) has the form (11), where the unknown vector is $w = (h, q_1, q_2, \ldots, q_N)^T$ and the functions $H \in \mathbb{R}, F(w) = (F_\alpha(w))_{\alpha=0,1,\ldots,N} \in \mathbb{R}^{N+1}, S(w) = (S_\alpha(w))_{\alpha=0,1,\ldots,N} \in \mathbb{R}^{N+1}$ and $B(w) = (B_{\alpha,\beta}(w))_{\alpha,\beta=0,1,\ldots,N} \in \mathcal{M}_{(N+1)}(\mathbb{R})$ are given by: $H = z_B$,

$$F_\alpha(\boldsymbol{w}) = \begin{cases} \sum_{\beta=1}^{N} l_\beta q_\beta & \text{if } \alpha = 0 \\ \dfrac{q_\alpha^2}{h} + \dfrac{gh^2}{2} & \text{if } \alpha = 1,\ldots,N, \end{cases}$$

$$S_\alpha(w) = \begin{cases} 0 & \text{if } \alpha = 0 \\ -gh & \text{if } \alpha = 1,\ldots,N \end{cases}$$

and

$$B_{\alpha,\beta}(w) = \begin{cases} 0 & \text{if } (\alpha,\beta) \in \{0\} \times \{0,1,\ldots,N\} \\ & \quad \cup \{1,\ldots,N\} \times \{0\} \\ \dfrac{(l_\alpha - \delta_{\alpha\beta}) l_\beta q_\alpha}{l_\alpha h} & \text{if } \alpha,\beta = 1,\ldots,N. \end{cases}$$

It is noteworthy that the equation (11) can be rewritten in the form $\partial_t w + A(w)\partial_x w = S(w)\partial_x H$, where $A(w) = B(w) + J(w)$ with $J(w) = \partial F(w)/\partial w$ the Jacobian matrx of F. Eventually, the matrix $J(w) = (J_{\alpha,\beta}(w))_{\alpha,\beta=0,1,\ldots,N} \in \mathcal{M}_{(N+1)}(\mathbb{R})$ is given by

$$J_{\alpha,\beta}(w) = \begin{cases} 0 & \text{if } (\alpha,\beta) = (0,0) \\ l_\beta & \text{if } (\alpha,\beta) \in \{0\} \times \{1,\ldots,N\} \\ gh - \dfrac{q_\alpha^2}{h^2} & \text{if } (\alpha,\beta) \in \{1,\ldots,N\} \times \{0\} \\ \dfrac{2q_\alpha}{h} \delta_{\alpha\beta} & \text{if } \alpha,\beta = 1,\ldots,N. \end{cases}$$

We simulate a dam break in a 2D channel of horizontal width $L = 10$ which we discretize with 20 nodes x_i. The vertical direction is discretized using $N = 10$ layers with a CFL number = 0.8 and we consider the following initial data: $q_\alpha(t=0) = 0.1 \times (\alpha - 1)^2$,

$$h(t=0) = \begin{cases} 5 - z_B & \text{if } x \in [4,6] \\ 3 - z_B & \text{else.} \end{cases}$$

The bottom is given by $z_B = \begin{cases} 1 & \text{if } x \in [2,3] \\ 0 & \text{else.} \end{cases}$

In Figures 1–4, we show some instants of the evolution of the height of the channel. As a matter of fact, we can observe that the shock caused by the bottom elevation is well captured by the height pattern. The figures ??–7 introduce at the same instants, the patterns of the discharges $q_\alpha = hu_\alpha$, in all the layers $\alpha = 1, \ldots, N = 10$. In

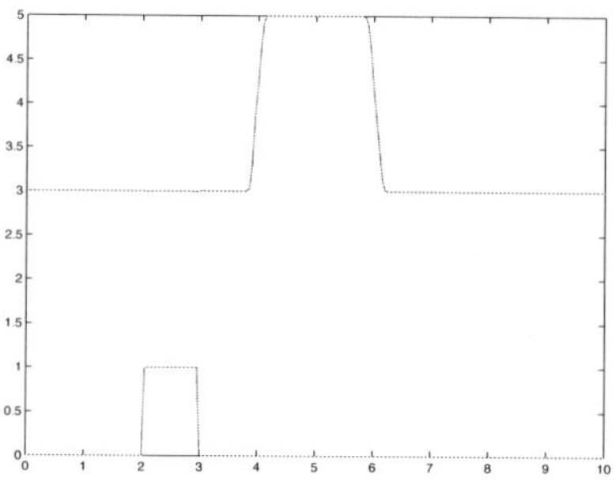

Figure 1. Water free surface at the time ($t = 0.01\ s$).

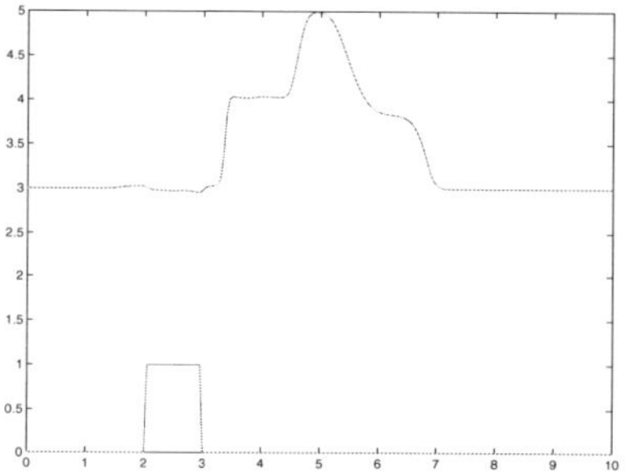

Figure 2. Water free surface at the time $t = 0.1\ s$.

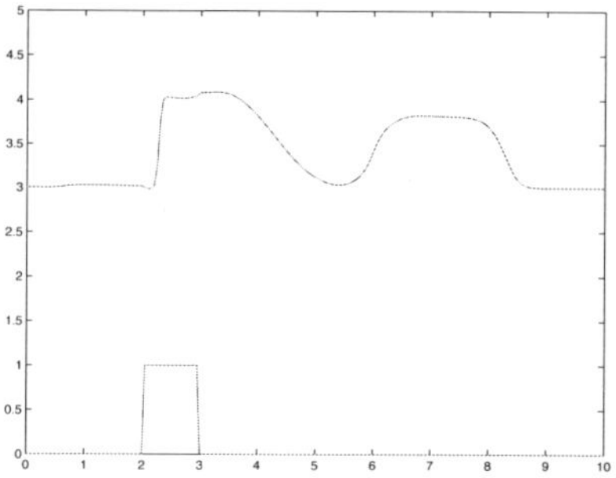

Figure 3. Water free surface at the time $t = 0.3\ s$.

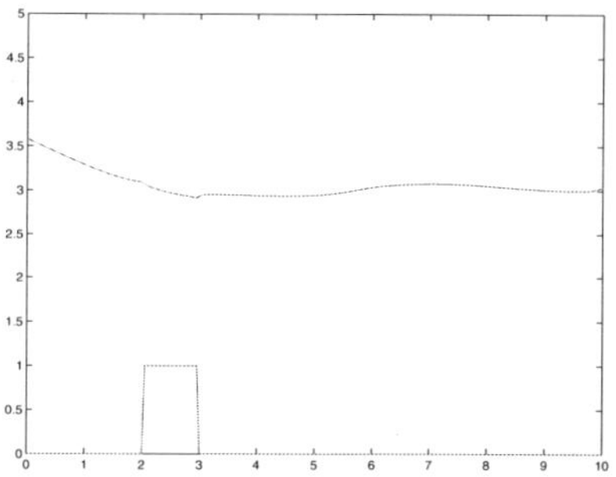

Figure 4. Water free surface at the time ($t = 1\ s$).

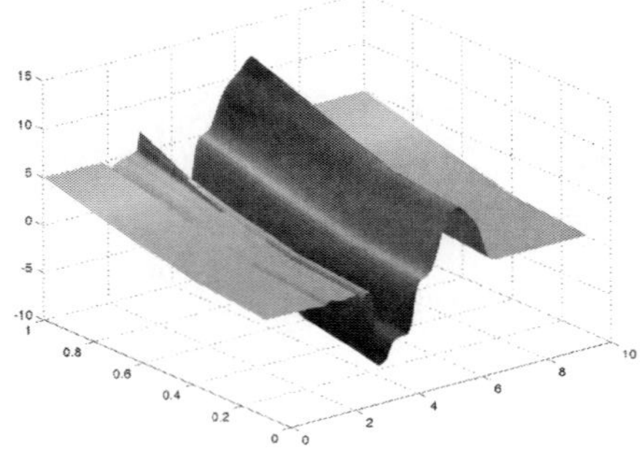

Figure 5. Discharges pattern at the time $t = 0.1\ s$.

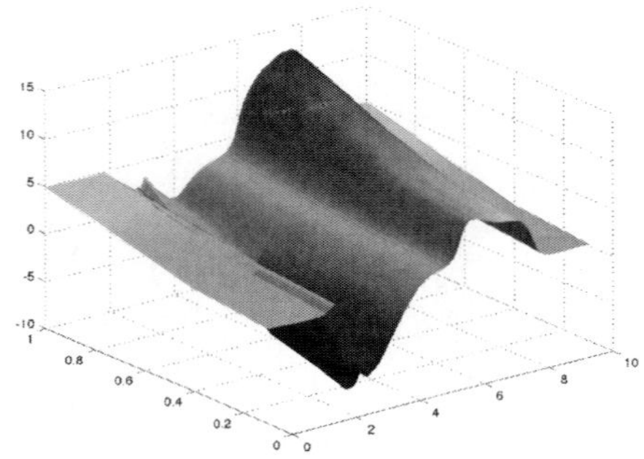

Figure 6. Discharges pattern at the time $t = 0.3\ s$.

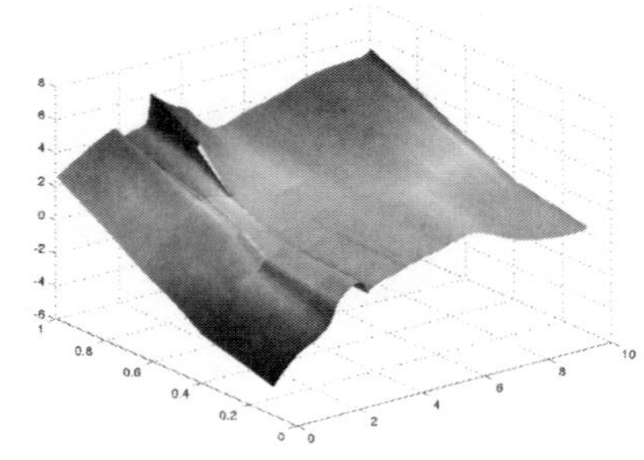

Figure 7. Discharges pattern at the time $(t = 1\ s)$.

those figures, the horizontal plane corresponds to the fluid domain where the water height is scaled down to 1.

4 CONCLUSIONS

In this preliminary work, we derive a multilayer shallow water system considering hydrostatic pressure. The system is quite similar to the one exhibited and analyzed in (Audusse et al. 2011), assuming hydrostatic pressure too and using an asymptotic development technique. Instead, we use here the weak formulation of the Navier-Stokes equations to establish the transmission conditions at the interfaces. In particular, the friction coefficients of the actual model can be defined so that the corresponding friction terms of both systems coincide. However, apart from the modeling technique leading to the two systems, their fundamental difference lies in the treatment of the mass fluxes exchanges at the interfaces of the layers. Especially, the leading velocities at those interfaces are taken into account in different ways. The current modeling strategy further will be used to derive a multilayer system in a general framework that consider non-hydrostatic pressure. That more general model still need to be analyzed and validated with numerical tests.

ACKNOWLEDGEMENTS

This work as been partially suported by Spanish Ministerio de Educación y Ciencia Research Project MTM2009-07719.

REFERENCES

Amara, M., D. Capatina, and D. Trujillo (2008). Variational approach for the multiscale modeling of an estuarian river. part1: Derivation and numerical approximation of a 2d horizontal model. *Report, INRIA RR-6742*.

Audusse, E., M.-O. Bristeau, B. Perthame, and J. Sainte-Marie (2011). A multilayer Saint-Venant system with mass exchanges for shallow water flows. Derivation and numerical validation. *ESAIM Math. Model. Numer. Anal.* 45 (1), 169–200.

Sainte-Marie, J. (2011). Vertically averaged models for the free surface non-hydrostatic Euler system: derivation and kinetic interpretation. *Math. Models Methods Appl. Sci.* 21(3), 459–490.

VI High order methods for hyperbolic conservation laws

Numerical Methods for Hyperbolic Equations – Vázquez-Cendón et al. (eds)
ISBN 978-0-415-62150-2

Approximate solutions of generalized Riemann problems: The Toro-Titarev solver and the LeFloch-Raviart expansion

Claus R. Goetz & Armin Iske
Department of Mathematics, University of Hamburg, Germany

ABSTRACT: This work concerns the solution of generalized Riemann problems. To this end, we consider the ADER scheme of Titarev & Toro (2002), which relies on a generalization of the classical Godunov scheme. Another solution method is the power series expansion of LeFloch & Raviart (1988). We analyze the two resulting approximation schemes, where we show that for scalar 1d problems the Toro-Titarev solver and the LeFloch-Raviart expansion yield the same Taylor series expansions in time. The full analysis for the Burgers equation is finally provided.

1 INTRODUCTION

We consider a system of hyperbolic conservation laws

$$\partial_t q(x,t) + \partial_x f(q(x,t)) = 0 \quad \text{for} \quad x \in \mathbb{R},\ t > 0, \tag{1}$$

in one spatial dimension, where $q : \mathbb{R} \times [0,\infty) \to \mathbb{R}^n$ is a vector of conserved quantities and where the flux $f : \mathbb{R}^n \to \mathbb{R}^n$ is a smooth vector-valued function.

We are interested in a Cauchy problem for (1), with discontinuous initial data of the form

$$q(x,0) = \begin{cases} \hat{q}_L(x) & \text{for } x < 0, \\ \hat{q}_R(x) & \text{for } x > 0, \end{cases} \tag{2}$$

where the two functions $\hat{q}_L$ and $\hat{q}_R$ are smooth. The Cauchy problem (1)–(2) is called *generalized Riemann problem*.

Due to the pioneering work of (van Leer 1979) and (Ben-Artzi and Falcovitz 1984), the generalized Riemann problem has been successfully used in the construction of high order extensions to the classical Godunov scheme. A state of the art variant of this approach is the ADER scheme of (Titarev and Toro 2002; Toro and Titarev 2006). The basic idea of the ADER scheme is to use a high order spatial reconstruction of the solution from cell averages and to use the generalized Riemann problem to design a time discretisation of matching order. The key strategy in the Toro-Titarev solver is the reduction of the generalized Riemann problem to a series of classical Riemann problems.

On the other hand, (LeFloch and Raviart 1988) have shown that, for $t > 0$ sufficiently small, the solution q of the generalized Riemann problem can be expanded into a power-series of self-similar functions,

$$q(x,t) = \sum_{k \geq 0} t^k u^k\left(\frac{x}{t}\right) \tag{3}$$

with polynomial functions $\xi \mapsto u^k(\xi)$. LeFloch & Raviart have given an explicit method to construct the expansion (3). We re-interpret the method of Toro & Titarev in the context of that LeFloch-Raviart series expansion for the generalized Riemann problem.

The outline of this paper is as follows. In Section 2 we briefly review generalized Godunov schemes and the Toro-Titarev solver for the ADER scheme. Then, we recall well-known results on the solution of classical and generalized Riemann problems in Section 3. In Section 4, we discuss the key steps of the LeFloch-Raviart expansion, where we show that for scalar 1d problems the Toro-Titarev solver and the LeFloch-Raviart expansion yield the same the Taylor series expansions in time. The full analysis for a scalar 1d example concerning Burgers equation is finally provided in Section 6.

2 THE ADER SCHEME

2.1 *Generalized godunov schemes*

To numerically solve the Cauchy problem

$$\left.\begin{array}{lll} \partial_t q + \partial_x f(q) = 0 & \text{for} & x \in \mathbb{R}, t > 0, \\ q(x,0) = \hat{q}(x) & \text{for} & x \in \mathbb{R}, \end{array}\right\}$$

we use a Godunov-type finite volume scheme. To this end, we work with control volumes (cells) of the form

$$[x_{i-1/2}, x_{i+1/2}] \times [t^n, t^{n+1}] \quad \text{for} \quad i \in \mathbb{Z}, n \in \mathbb{Z}.$$

For the sake of simplicity, we assume uniform grids, so that $x_{i+1/2} = (i + 1/2)\Delta x, i \in \mathbb{Z}$, and $t^n = n\Delta t$ with Δx, $\Delta t > 0$. The cell average in the i-th cell at initial time $t^0 = 0$ is then given by

$$q_i^0 = \frac{1}{\Delta x}\int_{x_{i-1/2}}^{x_{i+1/2}} \hat{q}(x)\,\mathrm{d}x.$$

Now, the generalized Godunov scheme works as follows. At any time step $t^n \to t^{n+1}$, for $n \geq 0$,

- reconstruct a piecewise smooth function

  $$v^n(x) = \mathcal{R}\left(\{q_i^n\}_{i\in\mathbb{Z}}\right)(x)$$

  from the cell averages $\{q_i^n\}_i$, where $\mathcal{R}$ is a suitable conservative nonlinear reconstruction operator, e.g., WENO reconstruction. Denote the restriction of v^n to the cell $[x_{i-1/2}, x_{i+1/2}]$ by v_i^n;
- use v^n as initial data and evolve for one time step

  $$\tilde{v}^{n+1}(x) = \varepsilon(\Delta t)v^n(x),$$

  where ε is the exact entropy evolution operator associated with (1);
- update the cell averages by averaging $\tilde{v}^{n+1}$,

  $$q_i^{n+1} = \mathcal{A}_i \tilde{v}^{n+1},$$

where $\mathcal{A}_i$ is the cell averaging operator, given as

$$\mathcal{A}_i v = \frac{1}{\Delta x}\int_{x_{i-1/2}}^{x_{i+1/2}} v(x)\,\mathrm{d}x.$$

In a finite volume framework, evolution and averaging can be done in one step by the update formula

$$q_i^{n+1} = q_i^n - \frac{\Delta t}{\Delta x}\left(\bar{f}_{i+1/2}^n - \bar{f}_{i-1/2}^n\right),$$

if we can compute the flux f through the cell boundaries exactly, i.e., we need to compute the integral

$$\bar{f}_{i+1/2}^n = \frac{1}{\Delta t}\int_{t^n}^{t^{n+1}} f\left(\varepsilon(\tau)v^n(x_{i+1/2})\right)\mathrm{d}\tau \qquad (4)$$

exactly. However, this may be exceedingly complicated, if not impossible. Therefore, we are looking for an approximation to (4), being based on the approximate solution of a generalized Riemann problem.

To obtain a numerical flux, we use ADER *state expansion*, i.e., a Taylor expansion of the solution q around time $t = t^n$ at the cell interface $x_{i+1/2}$,

$$q(x_{i+1/2}, \tau) \approx q(x_{i+1/2}, 0_+) + \sum_{k=1}^{r-1} \partial_t^k q(x_{i+1/2}, 0_+)\frac{\tau^k}{k!}, \qquad (5)$$

where $r > 1$ is a fixed integer, $\tau = t - t^n$ is the local time, $0_+ = \lim_{\tau\searrow 0} \tau$, and $q(x_{i+1/2}, 0_+)$ evaluates the solution of the generalized Riemann problem

$$\left.\begin{array}{ll} \partial_t q + \partial_x f(q) = 0 & \text{for } x \in \mathbb{R}, \tau > 0, \\ q(x,0) = \begin{cases} v_i^n(x) & \text{for } x < x_{i+1/2}, \\ v_{i+1}^n(x) & \text{for } x > x_{i+1/2}, \end{cases} & \end{array}\right\}$$

right at the cell interface for time $\tau = 0_+$.

Recall that the solution q may contain discontinuities. But for fixed $x_{i+1/2}$, the function $q(x_{i+1/2}, \cdot)$ (of the time variable) is smooth for small $\tau > 0$. To solve the generalized Riemann problem, we work with a numerical flux approximating the time integral in (4) by a Gaussian quadrature of the form

$$f_{i+1/2}^n = \sum_{\gamma=1}^{N} \omega_\gamma f(q(x_{i+1/2}, \tau_\gamma)),$$

where $\omega_\gamma, \tau_\gamma$ are the Gaussian weights and nodes, and N is the number of nodes. The values $q(x_{i+1/2}, \tau_\gamma)$ are determined through (5).

2.2 *The Toro-Titarev solver*

We now describe how to compute the coefficients in (5), according to (Toro and Titarev 2006). The key idea is to reduce the solution of the generalized Riemann problem to a series of classical Riemann problems. To find the sought value $q(x_{i+1/2}, 0_+)$, we solve a classical Riemann problem

$$\left.\begin{array}{ll} \partial_t q + \partial_x f(q) = 0 & \text{for } x \in \mathbb{R}, \tau > 0, \\ q(x,0) = \begin{cases} \hat{q}_L^0 & \text{for } x < x_{i+1/2}, \\ \hat{q}_R^0 & \text{for } x > x_{i+1/2}, \end{cases} & \end{array}\right\} \qquad (6)$$

with the extrapolated values

$$\hat{q}_L^0 = \lim_{x \nearrow x_{i+1/2}} \hat{q}_L(x) \quad \text{and} \quad \hat{q}_R^0 = \lim_{x \searrow x_{i+1/2}} \hat{q}_R(x).$$

This problem has a similarity solution that we denote by $q^0((x-x_{i+1/2})/\tau)$. The leading term of the expansion (5) is then given by $q(x_{i+1/2},0_+)=q^0(0)$, called the *Godunov state* of (6). For nonlinear systems of conservation laws, computing the complete solution of the Riemann problem can be a quite difficult task, and so we may need to employ a numerical (approximative) Riemann solver to compute the leading term. However, as we are mainly interested in the analytical aspects of the scheme, we assume that the Godunov state of (6) can be computed exactly.

For the higher order terms we perform a standard Cauchy-Kowalewskaya-type procedure to express all time derivatives as functions of lower order spatial derivatives, relying on a recursive mapping

$$\partial_t^k q = \Phi^k\left(q, \partial_x q, \dots, \partial_x^k q\right) \quad \text{for} \quad k = 0, \dots, r-1.$$

Recall that for piecewise smooth initial data, the classical Cauchy-Kowalewskaya theorem does not apply. But to illustrate the basic ideas, we assume that q is smooth. In this case, the following equations can be obtained by simple manipulations of derivatives.

Using Φ^k we can can compute the expansion (5), provided that we can find the spatial derivatives

$$q^{(k)}(x,t) = \partial_x^k q(x,t).$$

To do so, we first compute the one-sided derivatives

$$\hat{q}_L^k = \lim_{x \nearrow x_{i+1/2}} \partial_x^k \hat{q}_L(x) \text{ and } \hat{q}_R^k = \lim_{x \searrow x_{i+1/2}} \partial_x^k \hat{q}_R(x).$$

Then, we use these values as initial conditions for classical Riemann problems. For the evolution equations of the spatial derivatives we take inhomogeneous equations of the form

$$\partial_t q^{(k)} + A(q)\partial_x q^{(k)} = H^k(q^{(0)}, \dots, q^{(k)}), \tag{7}$$

where $A(q) = Df(q)$ is the Jacobian of the flux. Again, if the solution q was smooth, equation (7) could be derived by straight forward computation. Note, however, that we do not have yet a rigorous analysis whether these equations also can be used for discontinuous solutions.

Now we simplify the given problem as follows. Firstly, we neglect the source terms and secondly, we linearise the equations, so that we work with

$$\left.\begin{array}{ll} \partial_t q^{(k)} + A_{LR}\partial_x q^{(k)} = 0 & \text{for } x \in \mathbb{R}, \tau > 0, \\ q(x,0) = \begin{cases} \hat{q}_L^k & \text{for } x < x_{i+1/2}, \\ \hat{q}_R^k & \text{for } x > x_{i+1/2}, \end{cases} & \end{array}\right\}$$

where $A_{LR} = A(q(x_{i+1/2},0_+))$. Then the self-similar solutions $q^k((x-x_{i+1/2})/\tau)$ of these *linear* problems can be easily computed. Note that for all k we have the same A_{LR}.

These simplifications appear to be reasonable and, in fact, they have already been used in many practical applications. However, to the best of our knowledge, no theoretical justification concerning these simplifications has been given so far. Therefore, we show in the following analysis, that for nonlinear, scalar, 1d problems the proposed simplifications lead to a method, whose solution agrees with a series expansion of the exact solution. Thereby, we show that the resulting method does not reduce the accuracy order.

3 GENERALIZED RIEMANN PROBLEMS

To review some well-known results for the solution to the classical and the generalized Riemann problem, let us consider the system

$$\partial_t q + \partial_x f(q) = 0 \quad \text{for} \quad x \in \mathbb{R}, t > 0, \tag{8}$$

which we assume to be strictly hyperbolic, i.e., the Jacobian $A(q) = Df(q)$ has n distinct real eigenvalues

$$\lambda_1(q) < \lambda_2(q) < \dots < \lambda_n(q) \quad \text{for all} \quad q \in \mathbb{R}^n.$$

We further assume that all eigenvalues $\lambda_i(q(x,t))$ are uniformly bounded in a neighbourhood of the origin.

We then choose bases of left and right eigenvectors, $\{\ell_1(q), \dots, \ell_n(q)\}$ and $\{r_1(q), \dots r_n(q)\}$, i.e.,

$$\ell_i(q)^T A(q) = \lambda_i(q)\ell_i(q)^T, \quad A(q)r_i(q) = \lambda_i(q)r_i(q),$$

for $i = 1, \dots, n$ and all $q \in \mathbb{R}^n$, here normalized as

$$\ell_j(q) \cdot r_i(q) = \begin{cases} 1 \text{ for } i = j, \\ 0 \text{ for } i \neq j, \end{cases} \quad \text{for all} \quad q \in \mathbb{R}^n.$$

We assume f to be a smooth function and thereby, all λ_i, ℓ_i, r_i have the same regularity.

We restrict our analysis to systems, where we assume that their characteristic fields are, for any $1 \le i \le n$, either genuinely nonlinear in the sense of (Lax 1957),

$$\nabla\lambda_i(q)\cdot r_i(q) \neq 0 \quad \text{for all} \quad q \in \mathbb{R}^n,$$

or linearly degenerate,

$$\nabla\lambda_i(q)\cdot r_i(q) \equiv 0 \quad \text{for all} \quad q \in \mathbb{R}^n.$$

Under these assumptions, we have the following well-known result: Given two states $\hat{q}_L^0, \hat{q}_R^0 \in \mathbb{R}^n$ with $|\hat{q}_R^0 - \hat{q}_L^0|$ sufficiently small, the classical Riemann problem

$$\left.\begin{aligned} &\partial_t q^0 + \partial_x f(q^0) = 0 \qquad \text{for} \quad x \in \mathbb{R}, t > 0, \\ &q^0(x,0) = \begin{cases} \hat{q}_L^0 & \text{for} \quad x < 0, \\ \hat{q}_R^0 & \text{for} \quad x > 0, \end{cases} \end{aligned}\right\}$$

permits a unique entropy admissible weak solution that is self-similar,

$$q^0(x,t) = u^0\left(\frac{x}{t}\right).$$

The solution consists of at most $(n+1)$ constant states, separated by rarefaction waves, shock waves or contact discontinuities. For a comprehensive analysis on the classical Riemann problem and the properties of its solution, see (Bressan 2000).

Assume that the initial data

$$q(x,0) = \begin{cases} \hat{q}_L(x) & \text{for} \quad x < 0, \\ \hat{q}_R(x) & \text{for} \quad x > 0, \end{cases} \tag{9}$$

is piecewise smooth but discontinuous at $x = 0$. For the generalized Riemann problem (8)–(9) it is well-known that taking $\hat{q}_L^0 = \hat{q}_L(0), \hat{q}_R^0 = \hat{q}_R(0)$, for sufficiently small $|\hat{q}_R^0 - \hat{q}_L^0|$, there exists a neighbourhood around the origin in which (8)–(9) has a unique entropy admissible weak solution, see (Li and Yu 1985). Moreover, the solution q consists of at most $(n+1)$ open domains of smoothness $D_i, 0 \leq i \leq n$, separated either by smooth curves $x = \gamma_j(t)$ passing through the origin, or by rarefaction zones of the form

$$R = \{(x,t) \in \mathbb{R} \times [0,\infty) \,|\, \underline{\gamma}_j(t) < x < \overline{\gamma}_j(t)\},$$

where $x = \underline{\gamma}_j(t), x = \overline{\gamma}_j(t), 1 \leq j \leq n$, are smooth characteristic curves passing through the origin. In either case, we assume that these curves are defined for $t > 0$ sufficiently small. Then q has a shock or contact discontinuity across each curve $x = \gamma_j(t)$ and is continuous across the characteristic curves $x = \underline{\gamma}_j(t)$, $x = \overline{\gamma}_j(t)$. The solution of the generalized Riemann problem and the solution of

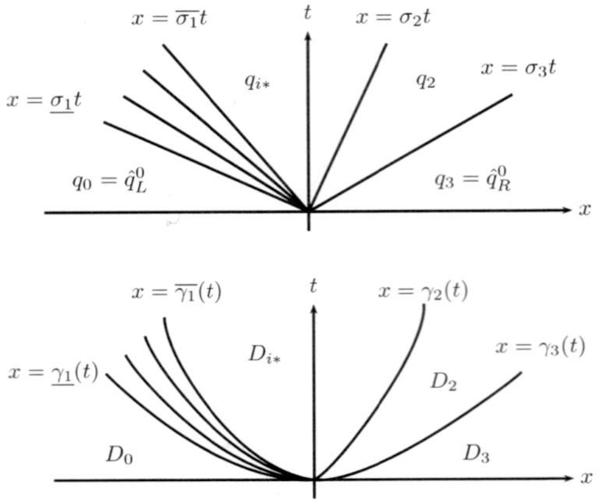

Figure 1. Corresponding wave patterns.

the corresponding classical Riemann problem for q^0 with the initial states $\hat{q}_L^0 = \hat{q}_L(0)$ and $\hat{q}_R^0 = \hat{q}_R(0)$ have the same wave structure, at least for small time $t > 0$. That is, if the solution of the classical Riemann problem contains an i-shock moving to the right, the same is true for the solution of the generalized Riemann problem and so on. For illustration, a typical configuration of corresponding wave patterns is shown in Figure 1.

We finally remark that the generalized Riemann problem has been the subject of ongoing research. Special emphasis has been put on the global existence and the structural stability of solutions. We refer to (Chen, Huang, and Han 2009; Chen, Han, and Zhang 2009; Kong 2003; Kong 2005) and references therein for an up-to-date account on the generalized Riemann problem. However, for the analysis of the numerical schemes under consideration in this paper, we can rely the results on local existence and local structural stability.

4 ASYMPTOTIC EXPANSION

4.1 *Preliminary discussion*

To further study the solution of the generalized Riemann problem

$$\left.\begin{aligned} &\partial_t q + \partial_x f(q) = 0 \qquad \text{for} \quad x \in \mathbb{R}, t > 0, \\ &q(x,0) = \begin{cases} \hat{q}_L(x) & \text{for} \quad x < 0, \\ \hat{q}_R(x) & \text{for} \quad x > 0, \end{cases} \end{aligned}\right\}$$

we follow (LeFloch and Raviart 1988), where we want to find an asymptotic expansion of the form

$$q(x,t) = \sum_{k \geq 0} t^k u^k(\xi) \tag{10}$$

with $\xi = x/t$. This is possible in any domain of smoothness D_i, by simply taking a Taylor expansion, so that every u^k is a polynomial of degree k. We will further discuss this in Section 22.

It can be shown that such a series expansion can also be constructed inside a rarefaction zone R. However, for our numerical scheme we only need detailed information about the solution along the line segment $\{x = 0\} \times [0, \Delta t]$ (in local coordinates). We assume that the solution does *not* contain a transonic rarefaction wave. In that case, the solution along that line segment is given by some function q_{i*} inside a domain of smoothness $D_{i*}, 0 \le i* \le n$, and we do not need the explicit construction of the expansion inside a rarefaction zone.

The above construction may be summarized as follows: Take a Taylor expansion in regions, where the solution q is smooth. Then, investigate the jump conditions at the boundaries of the smoothness domains. As we are looking for an expansion in terms of self-similar functions, it is useful to change the variables to $\xi = x/t$. We let $\tilde{q}(\xi,t) = q(\xi t,t)$ and see that

$$\partial_x = \frac{1}{t}\partial_\xi \quad \text{and} \quad \partial_t q = \partial_t \tilde{q} - \frac{\xi}{t}\partial_\xi \tilde{q}. \tag{11}$$

4.2 *Step I: Derivation of the differential equations*

To derive an explicit construction of the functions u^k in (10), we first analyze the equations satisfied by these functions inside the domains of smoothness. By using (11), the conservation law (1) can be written as

$$t\partial_t \tilde{q} - \xi \partial_\xi \tilde{q} + \partial_\xi f(\tilde{q}(\xi,t)) = 0.$$

Note that the expansion

$$\tilde{q}(\xi,t) = \sum_{k \ge 0} t^k u^k(\xi)$$

gives

$$t\partial_t \tilde{q} - \xi \partial_\xi \tilde{q} = -\xi \frac{du^0}{d\xi} + \sum_{k \ge 1} t^k \left(k u^k - \xi \frac{du^k}{d\xi} \right). \tag{12}$$

Inserting this expansion into the physical flux yields

$$f(\tilde{q}(\xi,t)) = f(u^0) + \sum_{k \ge 1} t^k (A(u^0)u^k + f^k(U^{k-1})). \tag{13}$$

Here, the function f^k depends only on the previous terms $U^{k-1} = (u^0, \ldots, u^{k-1})$. We get f^k by a Taylor expansion of the physical flux, so that f^k contains all higher order terms in that expansion, i.e., all but $A(u^0)u^k$. By using that Taylor expansion of f in powers of t, it is easy to see that f^k is a polynomial of degree at most k, if every u^ℓ is a polynomial (in ξ) of degree at most ℓ, for $0 \le \ell \le k-1$.

Combining (12) and (13), we get

$$-\xi \frac{du^0}{d\xi} + \frac{d}{d\xi} f(u^0) + \sum_{k \ge 1} t^k \left(k u^k - \xi \frac{du^k}{d\xi} + \frac{d}{d\xi}(A(u^0)u^k + f^k) \right) = 0.$$

For $k = 0$, we have

$$-\xi \frac{du^0}{d\xi} + \frac{d}{d\xi} f(u^0) = 0 \tag{14}$$

and for $k \ge 1$, we have

$$k u^k - \xi \frac{du^k}{d\xi} + \frac{d}{d\xi}(A(u^0)u^k + f^k) = 0.$$

Letting

$$h^k(\xi) = -\frac{d}{d\xi} f^k(u^0(\xi), \ldots, u^{k-1}(\xi)),$$

this becomes

$$k u^k - \xi \frac{du^k}{d\xi} + \frac{d}{d\xi}(A(u^0)u^k) = h^k. \tag{15}$$

Since f^k is a polynomial in ξ of degree at most k, the function h^k is a polynomial of degree at most $k-1$.

4.3 *Step II: Jump conditions*

The above construction is valid wherever q is smooth. So next we need to investigate the jump conditions satisfied by u^k at the points of discontinuity of q. Take a curve $x = \gamma(t)$ that separates two domains of smoothness of q, thus either a shock curve, a contact discontinuity or the boundary of a rarefaction zone. Since these curves are all smooth, we can use a Taylor expansion to write

$$\gamma(t) = \sigma^0 t + \sigma^1 t^2 + \ldots + \sigma^{k-1} t^k + \ldots.$$

It follows from (10) that

$$q(\gamma(t),t) = \sum_{k \ge 0} t^k u^k \left(\frac{\gamma(t)}{t} \right) = \sum_{k \ge 0} t^k u^k \left(\sum_{\ell \ge 0} t^\ell \sigma^\ell \right).$$

In fact, the solution q is smooth, not only in D_i, but also in the closure $\overline{D_i}$, see (Li and Yu 1985). So again we can use a Taylor expansion in powers of t around the origin to obtain

$$q(\gamma(t),t) = u^0(\sigma^0) + \sum_{k\geq 1} t^k \left(u^k(\sigma^0) + \sigma^k \frac{du^0}{d\xi}(\sigma^0) \right) + \sum_{k\geq 0} t^k z^k(\Sigma^{k-1}, U^{k-1}). \quad (16)$$

Similar to the f^k in (13), the functions z^k depend only on $\Sigma^{k-1} = (\sigma^0, \ldots, \sigma^{k-1})$ and U^{k-1}. Again, we insert all higher order terms in an Taylor expansion into z^k.

We denote the jump of a function u at a point x_0 by

$$[u](x_0) = u(x_{0,+}) - u(x_{0,-}),$$

so that in the case where q is continuous across the curve $x = \gamma(t)$ we simply get

$$[u^0](\sigma^0) = 0 \quad \text{for} \quad k = 0 \quad (17)$$

from (16), whereas for $k \geq 1$ we get

$$\left[u^k + \sigma^k \frac{du^0}{d\xi} + z^k(\Sigma^{k-1}, U^{k-1}) \right](\sigma^0) = 0.$$

Note that u^0 is continuous at the point σ^0, whereas, for $k \geq 1$, u^k is in general discontinuous at σ^0.

Now let q have a jump across the curve $x = \gamma(t)$. Then, by the Rankine-Hugoniot conditions, we have

$$\dot{\gamma}(t)[q](x) = [f(q)](x) \quad \text{for} \quad x = \gamma(t).$$

Then we take expansions of both $\dot{\gamma}(t)q(\gamma(t),t)$ and $f(q(\gamma(t),t))$. By a similar technique as above, we find

$$\sigma^0[u^0] = [f(u^0)] \quad \text{at } \sigma^0, \quad (18)$$

for $k = 0$, whereas for $k \geq 1$ we get

$$\left[(A(u^0) - \sigma^0)u^k \right] + \sigma^k \left[(A(u^0) - \sigma^0)\frac{du^0}{d\xi} \right] - \sigma^k \left[(k+1)u^0 \right] + \left[w^k \right] = 0$$

with a function w^k depending only on Σ^{k-1}, U^{k-1}. For further details on the (rather technical) proof we refer to (LeFloch and Raviart 1988).

We remark that (by finite speed of propagation) for $|\xi|$ large enough, say $|\xi| \geq \xi_0$, we have

$$u^0(\xi) = \begin{cases} \hat{q}_R^0 & \text{for} \quad \xi > \xi_0, \\ \hat{q}_L^0 & \text{for} \quad \xi < -\xi_0. \end{cases} \quad (19)$$

We summarize our above construction as follows.

Proposition. The function u^0 satisfies the relations (14), (17), (18) and (19), which characterize the piecewise continuous self-similar entropy solution $q^0(x,t) = u^0(\xi)$ of the classical Riemann problem

$$\left. \begin{aligned} \partial_t q^0 + \partial_x f(q^0) = 0 \quad & \text{for} \quad x \in \mathbb{R}, t > 0, \\ q^0(x,0) = \begin{cases} \hat{q}_L^0 & \text{for} \quad x < 0, \\ \hat{q}_R^0 & \text{for} \quad x > 0, \end{cases} & \end{aligned} \right\} \quad (20)$$

with initial states $\hat{q}_L^0 = \hat{q}_L(0)$ and $\hat{q}_R^0 = \hat{q}_R(0)$.

This shows that the solution strategy of the ADER method sets up "the right problem" for computing the leading term of the expansion.

4.4 *Step III: Higher order terms*

Assume that the solution of (20) contains no transonic rarefaction wave. Then line segment $\{x = 0\} \times [0, \Delta t]$ is contained in a domain of smoothness, say in D_{i*}, and u_{i*}^0 is the Godunov state of the Riemann problem (20). Since we do not explicitly need the expansion inside the rarefaction zones, we only consider the simplified case, where the solution u^0 contains only shock waves or contact discontinuities. The full problem requires similar techniques, although some of the details are more involved, see (LeFloch and Raviart 1988) for the full construction.

In situations, where we only have shocks and contact discontinuities, the solution u^0 of (20) has the form

$$u^0(\xi) = \begin{cases} u_0^0 = \hat{q}_L^0 & \text{for} \quad \xi \in (-\infty, \sigma_1^0), \\ u_i^0 & \text{for} \quad \xi \in (\sigma_i^0, \sigma_{i+1}^0), 1 \leq i < n, \\ u_n^0 = \hat{q}_R^0 & \text{for} \quad \xi \in (\sigma_n^0, \infty). \end{cases}$$

In case of an i-shock, we have

$$\lambda_i(u_i^0) \geq \sigma_i^0 \geq \lambda_i(u_{i+1}^0),$$

and for an i-contact discontinuity, we have

$$\sigma_i^0 = \lambda_i(u_i^0) = \lambda_i(u_{i+1}^0).$$

Now consider the domains

$$D_i^0 = \{(x,t) \mid \sigma_i^0 < \xi < \sigma_{i+1}^0\} \quad \text{for} \quad i = 0, \ldots, n,$$

in which u^0 takes the constant value u_i^0. As a convention, we let $\sigma_0^0 = -\infty$ and $\sigma_{n+1}^0 = +\infty$. Then Equation (15) in D_i^0 becomes

$$ku^k + \left(A(u_i^0) - \xi\right)\frac{d}{d\xi}u^k = h^k. \tag{21}$$

Recall that h^k is a polynomial of degree at most k–1. It is straightforward to show that the general solution of (21) is given by

$$u^k(\xi) = \left(\xi - A(u_i^0)\right)^k u_i^k + p_i^k(\xi), \tag{22}$$

see Lemma 2 in (LeFloch and Raviart 1988), where $u_i^k \in \mathbb{R}^n$ is an arbitrary vector and $p_i^k : \mathbb{R} \to \mathbb{R}^n$ is a polynomial of degree at most k–1 with coefficients that depend only on $U^{k-1} = (u^0, \ldots, u^{k-1})$.

5 CONNECTING SOLVER AND EXPANSION

Now let us take a look at the Taylor expansion that we used to define the functions u^k. We consider the domains

$$D_i = \{\xi \in \mathbb{R} \mid \gamma_{i-1}(t)/t < \xi < \gamma_i(t)/t\}.$$

Since we have $\gamma_i(0) = 0, \gamma_i'(0) = \sigma_i^0$, the domains remain close to the domains D_i^0 in which u^0 is constant, for small $t > 0$. In every domain of smoothness D_i we can take some (x_0, t_0) close to the origin and write

$$q(x,t) = q(x_0,t_0) + \sum_{k=1}^{\infty}\sum_{\ell=0}^{k} \frac{\partial^\ell}{\partial x^\ell}\frac{\partial^{k-\ell}}{\partial t^{k-\ell}} q(x_0,t_0)\frac{(x-x_0)^\ell (t-t_0)^{k-\ell}}{\ell!(k-\ell)!}.$$

Let $i_* \in \{1, \ldots, n\}$ be the index for which the line segment $\{x=0\}\times[0,\Delta t]$ is contained in D_{i_*}. Inside D_{i_*} we may take the limit $(x_0,t_0) \to (0,0_+)$ and thus the Taylor expansion around the origin gives

$$q(x,t) = \sum_{k=0}^{\infty}\sum_{\ell=0}^{k} \frac{\partial^\ell}{\partial x^\ell}\frac{\partial^{k-\ell}}{\partial t^{k-\ell}}\frac{q(0,0_+)}{\ell!(k-\ell)!}x^\ell t^{k-\ell}$$

$$= u_{i_*}^0 + \sum_{k=1}^{\infty} t^k \underbrace{\sum_{\ell=0}^{k} \frac{\partial^\ell}{\partial x^\ell}\frac{\partial^{k-\ell}}{\partial t^{k-\ell}}\frac{q(0,0_+)}{\ell!(k-\ell)!}\cdot\left(\frac{x}{t}\right)^\ell}_{=u^k(x/t)}$$

Thus, the vector $u_{i_*}^k$ in (22), which gives the leading coefficient of this polynomial, defines the value $\partial_x^k q(0,0_+)$.

To determine the vectors u_i^k, we first describe u_0^k and u_n^k. Using the notation from Section 2, we can write for the initial data

$$\hat{q}_L(x) = \hat{q}_L^0 + \sum_{k=1}^{r-1}\frac{\hat{q}_L^k}{k!}x^k \text{ and } \hat{q}_R(x) = \hat{q}_R^0 + \sum_{k=1}^{r-1}\frac{\hat{q}_R^k}{k!}x^k.$$

In D_0, the solution is given by the functions

$$u^k(\xi) = (\xi - A(u_0^0))^k u_0^k + p_0^k(\xi).$$

Since p_0^k is a polynomial of degree at most k–1,

$$\lim_{\substack{t\to 0,\\ x<\gamma_1(t)}} t^k u^k\left(\frac{x}{t}\right) = x^k u_0^k.$$

Hence, it follows

$$q(x,0) = \lim_{\substack{t\to 0,\\ x<\gamma_1(t)}} q(x,t) = u_0^0 + \sum_{k=1}^{r-1} u_0^k x^k.$$

Therefore, $u_0^k = \hat{q}_L^k/k!$, and likewise, $u_n^k = \hat{q}_R^k/k!$, for $k = 0, \ldots, r-1$.

Now consider the scalar case. For a strictly convex flux, $f'' > 0$, we only have two domains of smoothness. In that case, all coefficients $u_i^k, i = 0,1$, and $k = 1, \ldots, r$–1 are uniquely determined by the initial data and its derivatives. Assuming that there is no transonic wave, solving linear Riemann problems merely means picking the left or the right side, depending on the sign of the coefficient in the evolution equation. Thus, to build the expansion, we first have to solve one nonlinear Riemannproblem to determine which domain of smoothness contains the line segment $\{x=0\}\times[0,\Delta t]$. Then we use the data from that side, which is equivalent to solving linear Riemann problems. Therefore, the Toro-Titarev solver reproduces the first r–1 terms of the LeFloch-Raviart expansion exactly, so that we can finally conclude:

Theorem: Consider the generalized Riemann problem for a scalar, nonlinear hyperbolic conservation law with strictly convex flux in one spatial dimension. Let the initial data consist of piecewise polynomials of degree r–1. Assume that the solution does not contain a transonic wave. Then the numerical flux constructed with the solver of Toro and Titarev is accurate of order $O(\Delta t^r)$ as $\Delta t \to 0_+$, in the sense that

$$\left| f(q(0,\tau)) - f\left(\sum_{k=0}^{r-1} u^k(0)\tau^k \right) \right| = O(\Delta t^r).$$

6 BURGERS EQUATION

Consider Burgers equation,

$$\partial_t q + \partial_x (q^2/2) = 0$$

with initial data

$$q(x,0) = \begin{cases} \hat{q}_L(x) = x^2 + 2x + 1 & \text{for } x < 0, \\ \hat{q}_R(x) = 2x^2 - 4x + 2 & \text{for } x > 0. \end{cases}$$

In this case, we have $\hat{q}_L^0 = 1 < 2 = \hat{q}_R^0$. Therefore, the classical Riemann problem for the leading term contains a rarefaction wave. Therefore, the solution of the generalized Riemann problem is given as

$$q(x,t) = \begin{cases} q_0(x,t) & \text{for } x \le \underline{\gamma}(t), \\ x/t & \text{for } \underline{\gamma}(t) < x < \overline{\gamma}(t), \\ q_1(x,t) & \text{for } \overline{\gamma}(t) \le x. \end{cases}$$

Using the method of characteristics, we obtain

$$q_0(x,t) = \frac{2t(x+1) + 1 - \sqrt{4t(x+1)+1}}{2t^2},$$
$$q_1(x,t) = \frac{4t(x-1) + 1 - \sqrt{8t(x-1)+1}}{4t^2}.$$

The boundaries of the rarefaction zone are given by the head-characteristic $\overline{\gamma}(t) = 2t$ and the tail-characteristic $\underline{\gamma}(t) = t$.

For sufficiently small time $t > 0$, we approximate the solution along the t-axis by

$$q(0,t) \approx q(0,0_+) + \partial_t q(0,0_+)t + \partial_t^2 q(0,0_+)\frac{t^2}{2}.$$

Note that the t-axis is contained in the domain D_0, so in the subsequent analysis we only need to consider the function q_0. We have

$$\partial_t q_0(0,0_+) = -2 \quad \text{and} \quad \partial_t^2 q_0(0,0_+) = 10.$$

We now compute the terms of the LeFloch-Raviart expansion up to the function $u^2(\xi)$. At first, consider the expansion of the flux around $t > 0$,

$$f(\tilde{q}(\xi,t)) \approx f(u^0) + tf'(u^0)u^1 + t^2\left(f'(u^0)u^2 + \frac{1}{2}f''(u^0)(u^1)^2 \right).$$

Then, u^0 is the solution of the Riemann problem

$$\left. \begin{aligned} \partial_t u^0 + \partial_x((u^0)^2/2) = 0 \quad \text{for } x \in \mathbb{R}, t > 0, \\ u^0(x,0) = \begin{cases} 1 & \text{for } x < 0, \\ 2 & \text{for } x > 0, \end{cases} \end{aligned} \right\}$$

in which case $u^0(\xi) = 1$ for all $\xi \in D_0$. Then, the equation for u^1 is

$$u_0^1 + (1-\xi)\frac{d}{d\xi}u_0^1 = 0,$$

and the solution consistent with the initial data is

$$u_0^1(\xi) = 2(\xi - 1).$$

Therefore, we have

$$h^2(\xi) = -\frac{f''(u_0^0)}{2}(u^1(\xi))^2 = -4(\xi - 1).$$

Note that the inhomogeneous equation for u^2,

$$2u_0^2 + (1-\xi)\frac{d}{d\xi}u_0^2 = -4(\xi - 1),$$

has the solution $u_0^2(\xi) = (\xi-1)^2 - 4(\xi-1)$. Thus,

$$\tilde{q}(\xi,t) \approx 1 + 2(\xi-1)t + ((\xi-1)^2 - 4(\xi-1))t^2,$$
$$q(x,t) \approx x^2 + 2x - 6xt + 5t^2 - 2t + 1,$$

and in particular $q(0,t) \approx 1 - 2t + 5t^2$.

Now we use the solver of Toro and Titarev, where we find the leading term $q(0,0_+)$ by solving

$$\left. \begin{aligned} \partial_t q + \partial_x (q^2/2) = 0 \quad \text{for } x \in \square, t > 0, \\ q(x,0) = \begin{cases} \hat{q}_L^0 = 1 & \text{for } x < 0, \\ \hat{q}_R^0 = 2 & \text{for } x > 0. \end{cases} \end{aligned} \right\}$$

This gives $q(0,0_+) = 1$.

Next, the Cauchy-Kowaleskaya procedure leads to

$$\partial_t q = -q\partial_x q, \quad \partial_t^2 q = 2(\partial_x q)^2 + q\partial_x^2 q,$$
$$\partial_t(\partial_x q) = -(\partial_x q)^2 - q\partial_x^2 q, \qquad (23)$$
$$\partial_t(\partial_x^2 q) = -3\partial_x q\partial_x^2 q - q\partial_x^3 q.$$

For $q^{(1)} = \partial_x q$ and $q^{(2)} = \partial_x^2 q$, we have the evolution equations

$$\partial_t q^{(1)} + q\partial_x q^{(1)} = -(q^{(1)})^2$$

and

$$\partial_t q^{(2)} + q\partial_x q^{(2)} = -3q^{(1)}q^{(2)}$$

with the initial conditions

$$q^{(1)}(x,0) = \begin{cases} \hat{q}_L^{(1)} = 2 & \text{for } x < 0, \\ \hat{q}_R^{(1)} = -4 & \text{for } x > 0, \end{cases} \qquad (24)$$

and

$$q^{(2)}(x,0) = \begin{cases} \hat{q}_L^{(2)} = 2 & \text{for } x < 0, \\ \hat{q}_R^{(2)} = 4 & \text{for } x > 0. \end{cases} \qquad (25)$$

We drop the source terms and linearise around $q(0,0_+) = 1$, so that we have

$$\partial_t q^{(k)} + \partial_x q^{(k)} = 0 \quad \text{for } k = 1,2,$$

together with the initial conditions (24) and (25), respectively. These *linear* problems are readily solved, where we find the Godunov states

$$q^{(1)}(0,0_+) = 2 \quad \text{and} \quad q^{(2)}(0,0_+) = 2.$$

By the Cauchy-Kowaleskaya procedure (23) we get

$$\partial_t q(0,0_+) = -q(0,0_+)q^{(1)}(0,0_+) = -2,$$
$$\partial_t^2 q(0,0_+) = 2(q^{(1)}(0,0_+))^2 + q(0,0_+)q^{(2)}(0,0_+) = 10,$$

and thus again we find

$$q(0,t) \approx 1 - 2t + 5t^2.$$

REFERENCES

Ben-Artzi, M. and Falcovitz, J. (1984). A second-oder Go-dunov type scheme for compressible fluid dynamics. *J.Comput. Phys. 55*, 1–32.

Bressan, A. (2000). *Hyperbolic systems of conservation laws. The one-dimensional Cauchy problem.* Oxford University Press, Oxford.

Chen, S., Han, X. and Zhang, H. (2009). The generalized Rie-mann problem for first order quasilinear hyperbolic systems of conservation laws II. *Acta Appl. Math. 108*, 235–277.

Chen, S., Huang, D. and Han, X. (2009). The generalized Riemann problem for first order quasilinear hyperbolic systems of conservation laws I. *Bull. Korean Math. Soc. 46*(3), 409–434.

Kong, D.-X. (2003). Global structure stability of Riemann solutions of quasilinear hyperbolic systems of conservation laws: Shocks and contact discontinuities. *J. Differential Equations 188*, 242–271.

Kong, D.-X. (2005). Global structure instability of Riemann solutions of quailinear hyperbolic systems of conservation laws: Rarefaction waves. *J. Differential Equations 219*, 421–450.

Lax, P. (1957). Hyperbolic systems of conservation laws, II. *Comm. Pure Appl. Math. 10*, 537–556.

LeFloch, P. and Raviart, P.A. (1988). An asymptotic expansion for the solution of the generalized Riemann problem. Part I: General theory. *Ann. Inst. H. Poincare Anal. Non Lineaire 5*(2), 179–207.

Li, T. and Yu, W. (1985). *Boundary value problems for quasilinear hyperbolic systems.* Duke University Press, Durham, NC.

Titarev, V.A. and Toro, E.F. (2002). ADER: Arbitrary high order Godunov approach. *J. Sci. Comput. 17*, 609–618.

Toro, E.F. and Titarev, V.A. (2006). Derivative Riemann solvers for systems of conservation laws an ADER methods. *J. Comput. Phys.* 212, 150–165.

van Leer, B. (1979). Towards the ultimate conservative difference scheme. IV: A second order sequel to Godunov's method. *J. Comput. Phys. 32*, 101–136.

Numerical Methods for Hyperbolic Equations – Vázquez-Cendón et al. (eds)
 ISBN 978-0-415-62150-2

CFL-Number-dependent TVD-Limiters

Friedemann Kemm
Brandenburg University of Technology, Cottbus, Germany

ABSTRACT: In this paper, we show how algebraic limiting and limiters which take into account the local CFL-number lead to a significant increase in accuracy. Furthermore, we show how these limiters can be easily adapted to the nonlinear TVD-condition by Jeng and Payne.

1 INTRODUCTION

TVD schemes are widely used, although ENO and WENO schemes comprise alternatives of higher order. This may be explained by the simple implementation, and—compared to ENO and WENO—the compact stencil and the robustness of the resulting schemes. Unfortunately, only few limiters are discussed in the literature in greater detail. Most of them are (1) limiters which do not take into account the local CFL number, or (2) symmetric limiter functions, where the information of the upwind direction is discarded. For implicit schemes or for the computation of steady state solutions this is not critical, as the choice of the limiter function is restricted to the above mentioned cases without compromising (too much) on the quality of the results. For explicit codes (in algebraic limiting), such as the wave-propagation-method[1] (Leveque 2002), there is much more freedom in the choice of limiters. Here it is possible to use CFL-dependent limiters, which allow for a better reproduction of amplitudes and even for third-order schemes.

In this paper, we present several limiters which take into account the local CFL-number. Beside the old limiters—most of them by Roe—we show some new variants. One of them is a slightly less compressive modification of Roe's original Ultrabee, which allows to reduce the squaring effect. Another limiter is based on a generalization of the Power-limiter by Serna and Marquina (Serna and Marquina 2004) which, in turn, is based on a generalization of the harmonic mean. It turns out to be a good choice for compressive nonlinear waves. We also discuss some variations of Roe's third-order TVD-limiter. It has to be stressed that the above mentioned limiters are originally constructed for 1D linear advection (Kemm 2011). However, in (Jeng and Payne 1995) Jeng and Payne deduce a TVD-condition for nonlinear hyperbolic conservation laws. For the linear advection, this condition reduces to the sufficient condition given by Harten (Harten 1983). We show how this condition can be incorporated in the construction of limiters, and compare the results obtained with these limiters to standard limiters. The test cases include linear and nonlinear 1D-problems.

[1]The numerical results in this paper are obtained with clawpack (R. J. LeVeque and M. J. Berger 2004), a code which implements the wave-propagation-method.

2 LIMITERS FOR LINEAR WAVES

For linear waves, the theory on how to construct TVD-limiters is complete and can be found in textbooks like (Toro 1999). As a starting point, linear schemes of second-order for the (linear) advection equation

$$qt + aq_x = 0, \tag{1}$$

with advection speed a, are formulated as correction of the standard upwind scheme

$$q_i^{n+1} = q_i^n - \left[|\nu| + \frac{1}{2}|\nu|(1-|\nu|)\right) \cdot \left(\frac{\varphi(r_i^n)}{r_i^n} - \varphi(r_{i-1}^n)\right]\Bigg)\Delta q^n_{\text{up}\,i} \tag{2}$$

with CFL-number $\nu = \frac{a\Delta x}{\Delta x}$ and the ratio of upwind—and downwind-jump

$$r = \frac{\Delta q_{up}}{\Delta q_{down}}. \tag{3}$$

For $\varphi(r) \equiv 1$ this resembles the Lax-Wendroff scheme, a 3-point central scheme, whereas for $\varphi(r) = r$ the Beam-Warming scheme, which is based on a full upwind 3-point stencil. According to the results by Roe (Roe 1981), these two schemes span the whole space of linear, second-order schemes for the (linear) advection Equation (1). Thus, we can obtain any second order scheme when we replace φ by some mean of 1 and r. For

$$\begin{aligned}\varphi_3(r) &= 1-\frac{1+|\nu|}{3}+\frac{1+|\nu|}{3}r \\ &= 1+\frac{1+|\nu|}{3}(r-1),\end{aligned} \quad (4)$$

we even get a third-order scheme (Roe 1981). Unfortunately, all schemes with constant weights lead to an increase of the total variation of the solution, which can be seen in Figure 1. Thus, we have to resort to non-constant weights, i.e., to nonlinear schemes. For these schemes, Harten (Harten 1983) gives the following sufficient condition

Theorem 1. *Let a numerical scheme for a scalar hyperbolic conservation law be written in the form*

$$q_i^{n+1} = q_i^n - C_{i-1/2}^n \Delta q_{i-1/2}^n + D_{i+1/2}^n \Delta q_{i+1/2}^n. \quad (5)$$

If for all $i \in \mathbb{Z}$ and all non-negative integers n the coefficients satisfy the following two conditions

$$C_{i+1/2}^n, D_{i+1/2}^n \geq 0, \quad (6)$$

$$C_{i+1/2}^n + D_{i+1/2}^n \leq 1, \quad (7)$$

then the scheme is TVD.

When applied to scheme (2), this leads to the *General TVD-condition*:

$$-\frac{2}{1-|\nu|} \leq \frac{\varphi(r)}{r} - \varphi(R) \leq \frac{2}{|\nu|}, \quad \forall r, R \in \mathbb{R}. \quad (8)$$

If we set $\varphi(r) = 0$ or $\varphi(R) = 0$, the general TVD-condition (8) yields

$$\varphi(R) \leq \frac{2}{1-|\nu|}, \quad (9)$$

$$\varphi(r) \leq \frac{2}{|\nu|} r \quad (10)$$

Together with

$$\varphi(r) = 0 \quad \text{for} \quad r \leq 0, \quad (11)$$

this is sufficient for a scheme to be TVD. We call the limiter function φ_{ub} with

$$\varphi_{ub}(r) = \max\left\{0, \min\left\{\frac{2}{|\nu|} r, \frac{2}{1-|\nu|}\right\}\right\} \quad (12)$$

the upper bound for the limiters. This will play an important role in the construction of high quality limiters. The region enclosed by $\varphi_{ub}(r)$ and the real axis is called the *TVD-region*. Another mean function which we will frequently use in the construction of limiters is φ_3, which defines the above mentioned third-order scheme.

Since $\varphi_{ub}(r)$ is no mean of 1 and r, it leads to a first-order scheme. As Figure 2 shows, this does not imply high diffusion. The solution consists only of square waves. This squaring is unphysical for linear advection, it is only expected for nonlinear compressive waves. Therefore, if one limiter tends more to smearing, while the other one tends more to squaring, we call the second one more compressive than the first one. The tuning of this artificial compression is essential for the quality of the numerical results.

First we consider piecewise linear limiter functions. They switch between different second-order schemes, or even the third-order scheme, depending on the local ratio r defined in Equation (3).

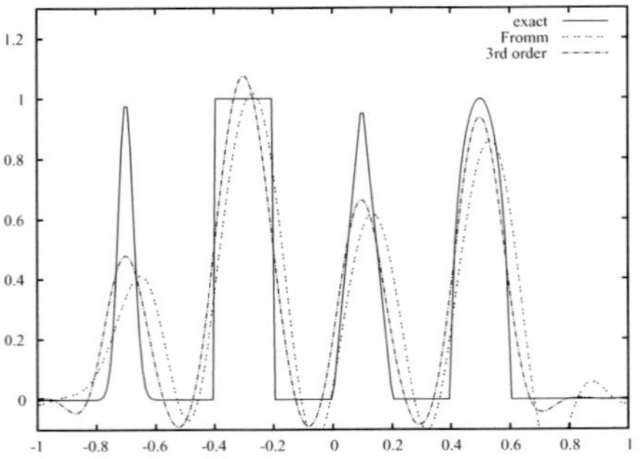

Figure 1. Application of linear schemes to Zalesak test with CFL-number 0.1 ($t = 20$).

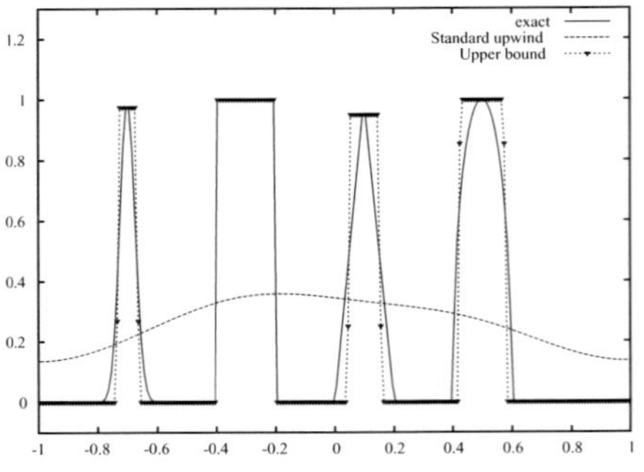

Figure 2. Zalesak test with first-order schemes at time $t = 20$, the standard upwind and the compressive upper bound scheme.

2.1 *Limiters based on the third-order scheme*

Roe (Roe and Baines 1982; Roe 1985) points out that it is advantageous to limit the third-order scheme to the TVD-region. Since there are many limiters connected to Roe's name, and the resulting limiter is the core of the paper of Arora and Roe (Arora and Roe 1997), we refer to it as the *Arora-Roe-Limiter*:

$$\varphi_{AR}(r) = \max\{0, \min\{\varphi_{ub}(r), \varphi_3(r)\}\}. \quad (13)$$

A variant of this, which goes back to an idea by van Leer, is to multiply the term $\varphi_{ub}(r)$ with some $\theta \in [1/2,]$. Especially in the case of strongly compressive waves, it is not advisable to fully exploit the freedom given by the original limiter. Thus, a bound $\theta \geq 1/2$ ensures that the limiter stays above the minmod limiter

$$\varphi_{\text{minmod}}(r) = \text{minmod}(r,1) = \max\{0, \min\{1,r\}\}, \quad (14)$$

which is the lower bound for second-order limiters that vanish for negative r, and are non-decreasing.

By introducing a parameter $1/2 \leq \theta \leq 1$, the freedom gained in the TVD-condition might be used to extend the area in which third-order is obtained to a part of the negative r-axis. For positive r the limiter satisfies

$$\varphi(r) \leq \theta \frac{2}{1-|\nu|}, \qquad \frac{\varphi(r)}{r} \leq \theta \frac{2}{|\nu|}. \quad (15)$$

If for some R equality holds in the first relation of (15), then the general TVD-condition turns into

$$-\frac{2}{1-|\nu|} \leq \frac{\varphi(r)}{r} - \theta \frac{2}{1-|\nu|} \leq \frac{2}{|\nu|}, \quad (16)$$

from which we conclude

$$-(1-\theta)\frac{2}{1-|\nu|} \leq \frac{\varphi(r)}{r} \leq \frac{2}{|\nu|} + \theta \frac{2}{1-|\nu|}. \quad (17)$$

Thus, for negative r the limiter has to satisfy

$$\varphi(r) \leq -(1-\theta)\frac{2}{1-|\nu|} r. \quad (18)$$

If for some r equality holds in the second relation of (15), then the general TVD-condition turns into

$$-\frac{2}{1-|\nu|} \leq \theta \frac{2}{|\nu|} - \varphi(R) \leq \frac{2}{|\nu|}, \quad (19)$$

which leads to the condition

$$\varphi(R) \geq -(1-\theta)\frac{2}{\nu}. \quad (20)$$

The optimized limiter, therefore reads as

$$\begin{aligned}\varphi_\theta(r) = \min\{&\theta \frac{2}{1-\nu}, \\ &\max\{-(1-\theta)\frac{2}{1-\nu} r, \theta \frac{2}{\nu} r\}, \\ &\max\{-(1-\theta)\frac{2}{\nu}, \varphi_3(r)\}\}.\end{aligned} \quad (21)$$

The limiters are presented in Figure 3; numerical results for the application of the limiters of this family are displayed in Figure 4.

It appears that a third-order scheme not necessarily guarantees a good representation of the amplitude. The gain in quality by allowing for reconstruction at extrema, i.e. with $\theta < 1$ instead of the original Arora-Roe limiter, is small compared to the loss due to restrictions in monotone regions. This is mainly because in a well resolved solution only few extrema occur compared to the full number of grid cells.

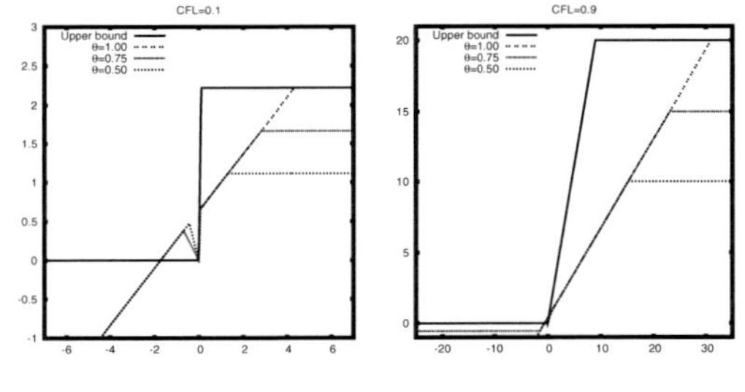

Figure 3. Normal forms of the Arora-Roe limiter ($\theta = 1$), and the $\theta = 0.75$ – and $\theta = 0.5$-versions for CFL-numbers 0.1 and 0.9.

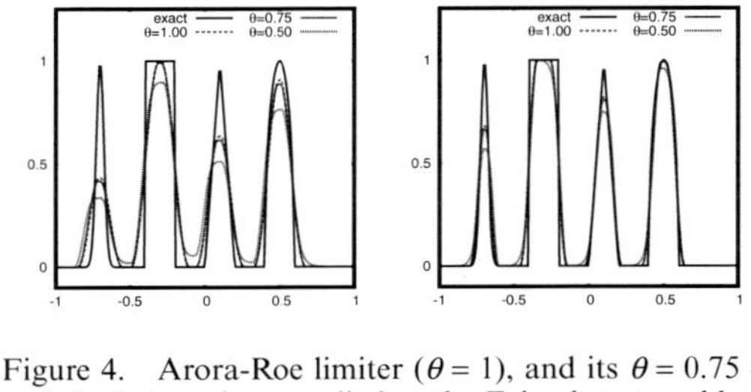

Figure 4. Arora-Roe limiter ($\theta = 1$), and its $\theta = 0.75$ – and $\theta = 0.5$-versions applied to the Zalesak test problem at time $t = 20$ for CFL-numbers 0.1 and 0.9.

2.2 *Limiters with higher compression*

Figure 2 shows that the upper bound limiter φ_{ub} is highly compressive. To obtain a compressive second-order limiter, it is of advantage to stick to the upper bound as long as it can be written as a mean of 1 and r with positive weights. If this is not the case, as in Figure 2, we should resort to the maximum of 1 and r. This results in a limiter, which was first proposed by Roe (Roe and Baines 1984):

$$\varphi_{\mathrm{cs}}(r) = \min\{\varphi_{\mathrm{ub}}, \max\{1, r\}\}. \tag{22}$$

Due to its close relation to the Superbee limiter, we refer to it as *CFL-Superbee*.

Bokanowski and Zidani (Bokanowski and Zidani 2007) proved that for piecewise linear data, this limiter satisfies some optimality condition. In (Després and Lagoutiére 2001), Després and Lagoutière showed that the upper bound limiter is optimal for piecewise constant data. While the upper bound leads to a first-order, and the CFL-superbee to a second-order scheme, we would expect for piecewise smooth data a third-order scheme to be optimal. But there is no third-order scheme within the considered class, and the Arora-Roe-type schemes reproduce the amplitudes of the waves rather poorly. Therefore, we have to look somewhere between Arora-Roe and CFL-superbee.

We introduce a parameter β and define the new limiter as

$$\begin{aligned}\varphi(r) = \max\{0, \min\{&\varphi_{ub}(r),\\ &\max\{1 + (\varphi'_3 - \beta/2)(r-1),\\ &1 + (\varphi'_3 + \beta/2)(r-1)\}\}\}.\end{aligned} \tag{23}$$

For $\beta = 0$ this resembles the Arora-Roe limiter, and for the CFL-number $\nu = 1/2$ and $\beta = 1$ it would coincide with the original CFL-Superbee. The question remains, how to choose β. If we set this parameter as constant, the maximal value that allows us to stay below CFL-superbee is $\beta = 2/3$. But we could also employ some shock switch to increase compression near discontinuities. The switch introduced by Harten in (Harten 1983) is on the same stencil as in our schemes and can be expressed in terms of the ratio r. A possibility would be to use a constant value plus the switch times a constant for β. In our tests we therefore employ

$$\beta = \frac{1}{3} + \frac{2}{3} \cdot \frac{|1-r|}{|1+r|}. \tag{24}$$

The normal forms of the limiters in this section are displayed in Figure 5. Results in Figure 6, however, allow the conclusion that the constant setting is a better compromise between order and compression than the variable choice with the shock switch.

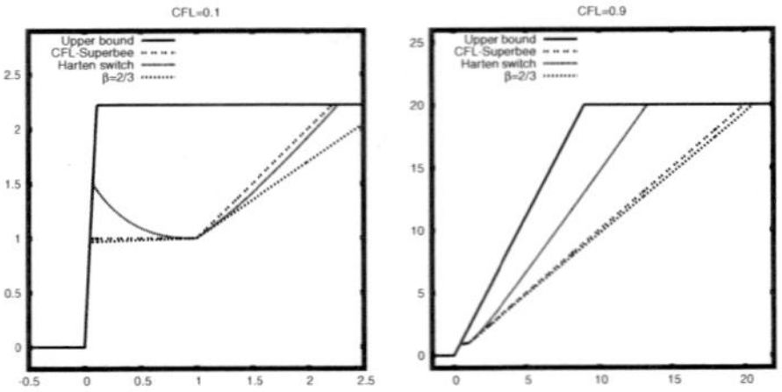

Figure 5. Normal forms of compressive limiters for CFL-numbers 0.1 and 0.9.

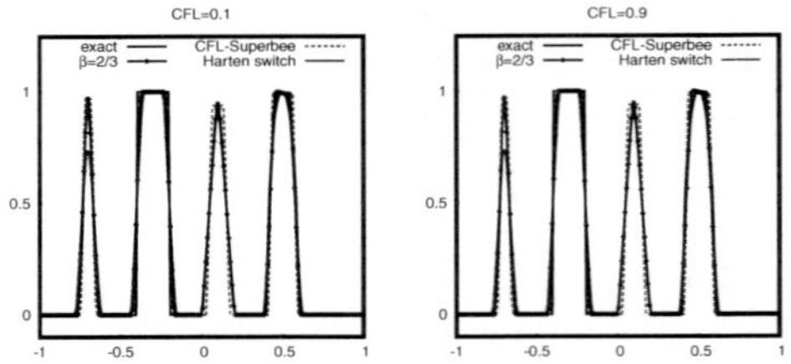

Figure 6. Compressive limiters applied to the Zalesak test problem at time $t = 20$ for CFL-numbers 0.1 and 0.9.

2.3 *Smooth limiters*

In this section we consider limiter functions which are smooth at least on the positive r-axis. Since they are always far enough away from the upper bound, smooth limiter functions are a good choice for compressive nonlinear waves. Here we show two examples which take into account the local CFL-number and are tangential to the third-order scheme at $r = 1$.

The *hyperbee limiter* was introduced by Roe (Roe 1985) mainly for theoretical purposes. By using hyperbee, it is possible to advect an exponential initial distribution exactly. The name is due to the fact that it can be written by use of hypergeometric functions. The normal form reads as

$$\varphi_{hyp}(r) = \begin{cases} 0, & r \le 0; \\ 1, & r = 1; \\ \frac{2r}{\nu(1-\nu)} \cdot \frac{\nu(r-1)+(1-r^{\nu})}{(r-1)^2}, & \text{otherwise}. \end{cases} \tag{25}$$

The function is non-decreasing and we find the following limits:

$$\lim_{r\to\infty} \varphi_{hyp}(r) = \tfrac{2}{1-|\nu|}, \qquad \lim_{r\to 0} \varphi'_{\mathrm{hyp}}(r) = \tfrac{2}{|\nu|}. \tag{26}$$

The TVD-condition (8) is sharply satisfied (Roe 1985). Furthermore, the slope in $r = 1$ coincides with the third-order slope φ_3. To avoid division by zero, in equation (25), we replace ν

by $\max\{\varepsilon, \min\{1 - \varepsilon, \nu\}\}$ for some small positive number ε. To circumvent the removable singularity of φ_{hyp} in $r = 1$, in a computer code, it should be replaced by φ_3 in a small neighbourhood of $r = 1$.

In (Kemm 2011), we introduced as an alternative to hyperbee the *superpower limiter*, which is based on the power limiters by Serna and Marquina (Serna and Marquina 2004)

$$\varphi_{sp}(r) = \max\{0, \varphi_3(r)(1 - \left|\frac{1-|r|}{1+|r|}\right|^{p(r)})\} \tag{27}$$

with

$$p(r) = \begin{cases} \frac{2}{|\nu|} \cdot 2(1 - \varphi_3'), & r \le 1; \\ \frac{2}{|1-\nu|} \cdot 2\varphi_3', & r \ge 1. \end{cases} \tag{28}$$

Again, the TVD—condition (8) is sharply satisfied, and the slope in $r = 1$ coincides with the third-order slope φ_3.

The normal forms of Hyperbee and Superpower are presented in Figure 7. Since these limiters are mainly intended for use on nonlinear waves, we do not show numerical results. As can be seen from their definition, for a wide range around $r = 1$ on the r-axis, they are even less compressive than the θ-limiters of Section 2.

3 MODIFICATION FOR NONLINEAR WAVES

The theory for nonlinear scalar conservation laws was accomplished by Jeng and Payne (Jeng and Payne 1995). They found that if we do not intend reconstruction at extremal points, i.e., we set $\varphi \equiv 0$ on the negative real axis, the nonlinear TVD-region is given by

$$0 \le \varphi(r) \le \min\left\{\frac{2}{|\nu|} \cdot \frac{1-\nu_{\text{up}}}{1-|\nu|} r, \frac{2}{1-|\nu|}\right\}. \tag{29}$$

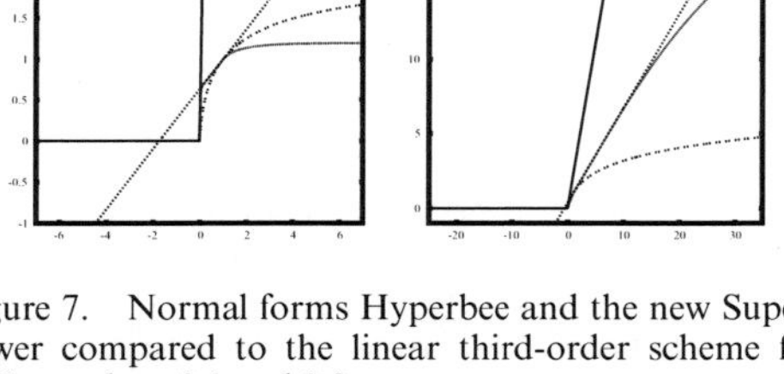

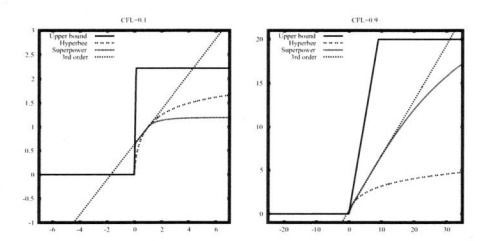

Figure 7. Normal forms Hyperbee and the new Superpower compared to the linear third-order scheme for CFL-numbers 0.1 and 0.9.

The only difference to the linear case is that, in the construction of limiters, the term $2/\nu\ r$ has to be replaced by $2/\nu . 1-\nu_{\text{up}}/1-\nu\ r$. For the upper bound limiter, this would lead to the nonlinear version

$$\varphi_{\text{ub nl}} = \max\left\{0, \min\left\{\frac{2}{|\nu|} \cdot \frac{1-|\nu_{\text{up}}|}{1-|\nu|} r, \frac{2}{1-|\nu|}\right\}\right\}. \tag{30}$$

Unfortunately, the derivation of the formulas for limiters which allow the reconstruction at extrema, as given in Section 2, does not easily translate to the nonlinear case (due to the nonlocal CFL-numbers), and thus will not be the scope of this study.

If we apply the nonlinear theory to the *Arora-Roe-Limiter*, things are simple. We only have to replace the upper bound φ_{ub} in (13) by its nonlinear counterpart $\varphi_{ub\ nl}$. For the *CFL-Superbee*, we have to do the same in (22) and for the β-limiters in (23).

For *Superpower*, things are still simple. We just have to replace $2/\nu\ r$ by $2/4 . 1-\nu_{\text{up}}/1-\nu\ r$ in (28). Then, the nonlinear condition (29) is still sharply satisfied and the slope in $r = 1$, by construction, coincides with the slope of the third-order slope φ_3. Unfortunately, this is not possible for *Hyperbee*. To satisfy condition (29), we have to replace for $0 \le r \le 1$, the local CFL-number ν by a modified CFL-number

$$\tilde{\nu} = \nu \cdot \frac{1-|\nu|}{1-|\nu_{\text{up}}|} \tag{31}$$

As a consequence, the left hand slope at $r = 1$ coincides with the third-order slope for $\tilde{\nu}$; the right hand slope with the third-order slope for ν. Thus, the limiter is in general no longer smooth at $r = 1$. For our numerical tests, we, therefore, only use Superpower as an example of a smooth limiter.

4 NUMERICAL RESULTS

In this section, we present numerical results for nonlinear equations obtained with given limiters. For numerical results for the linear advection equation the reader is referred to Section ?. Since, for linear degenerate waves in nonlinear systems, the CFL-number also varies in space and time, we always use the nonlinear and, thus, nonlocal version of the limiters. When we apply a piecewise linear limiter to a compressive nonlinear wave in a system like the Euler equations, we use $(1 - \varepsilon)\ \varphi_{ub\ nl}$ with small positive ε instead of $\varphi_{ub\ nl}$ in the limiter to ensure that we stay within the nonlinear TVD-region. For our tests, we used $\varepsilon = 0.05$. In the case of a scalar conservation law like the Burgers equation, this modification is not necessary. First, we show the effect of the nonlinear correction for the Burgers equation. Then, we discuss results for the 1d-Euler equations.

4.1 *Results for the Burgers equation*

Results in this section are for the inviscid Burgers equation

$$u_t + (\frac{u^2}{2}) = 0 . \qquad (32)$$

We start with a sinusoidal distribution. Due to compression and decompression, this leads over time to a jigsaw type solution, consisting of linear parts and jumps.

Figure 8 shows that without the nonlinear correction both the Superpower and the CFL-Superbee limiter overshoot at the peak. But with the correction, no overshooting is observed, even for the highly compressive CFL-Superbee.

4.2 *Results for 1D Euler equations*

Here, we show some results for the one-dimensional Euler equations. As test cases, we employ the blast wave problem by Woodward and Colella (Woodward and Colella 1984) and the shock-entropy-interaction problem by Shu and Osher (Shu and Osher 1989).

The *blast wave problem of Woodward and Colella* (Woodward and Colella 1984) is chosen because it includes highly compressive nonlinear waves, and, thus is very sensitive to compressive limiters. The problem consists of two colliding blast waves of different strength. The interval is set to [0,1] with the blast waves starting at $x = 0.1$ and $x = 0.9$. Initially, the density is constant and set equal to one, the velocity is zero, and the pressure in the three areas is

$$p_l = 10^3 , \qquad p_m = 10^{-2} , \qquad p_r = 10^2 . \qquad (33)$$

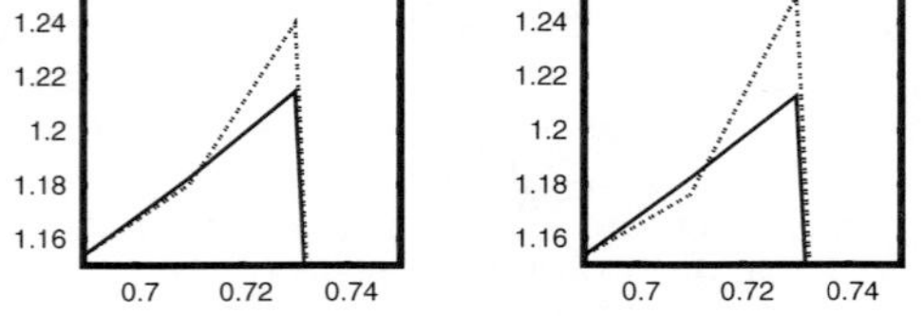

Figure 8. Detail of Burgers with sinusoidal initial data.

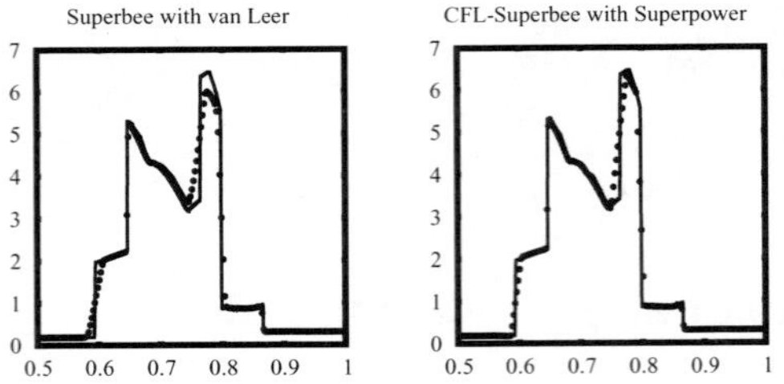

Figure 9. Zoom of Woodward-Colella blast wave problem at $t = 0.038$ with 400 points; mixed use of limiters.

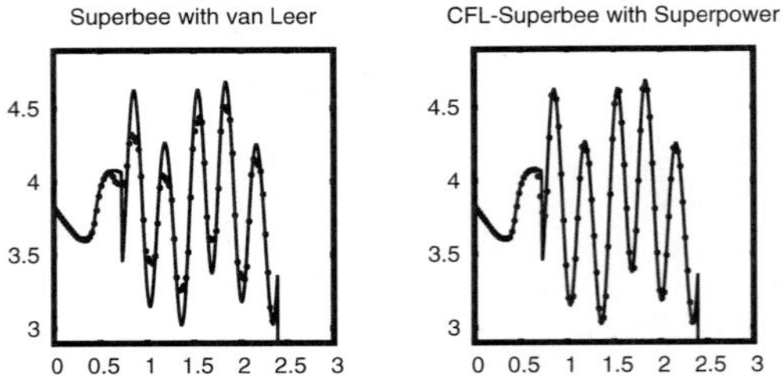

Figure 10. Zoom of Shu-Osher blast shock-entropy interaction at $t = 1.8$ with 400 points; mixed use of limiters.

The reference solution is done with the linear version of the $\theta = 0.75$ limiter on a grid with 25 600 cells. In Figure 9, we present a comparison of the mixed use of Superpower (for nonlinear waves) and CFL-Superbee (for linear waves) with their CFL-number independent counterpart, a mixed use of van-Leer and Superbee (left panel in Figure 9). It turns out that the representation of the solution is much better with the CFL-dependent limiters, while the high compression of CFL-superbee has no adverse effect.

The *shock entropy wave interaction problem of Shu and Osher* (Shu and Osher 1989) involves a moving shock interacting with density fluctuations of small amplitude. The initial data are like in (Shu and Osher 1989), results are presented for $t = 1.8$.

In Figure 10 we show a comparison of the same limiters as for the blast-wave problem. Here, an even greater accuracy is observed than for the blast wave problem. While, for the CFL-independent limiters, there is both, squaring and loss of amplitude, the solution for the CFL-dependent limiters doesn't show either of them. So, is it necessary to use different limiters for different waves? Figure 11 shows results for the Arora-Roe and the β= 2/3-limiter. It turns out that, for moderate grid resolutions, it is advantageous to resort to the more compressive β= 2/3-limiter, as it almost reaches the quality

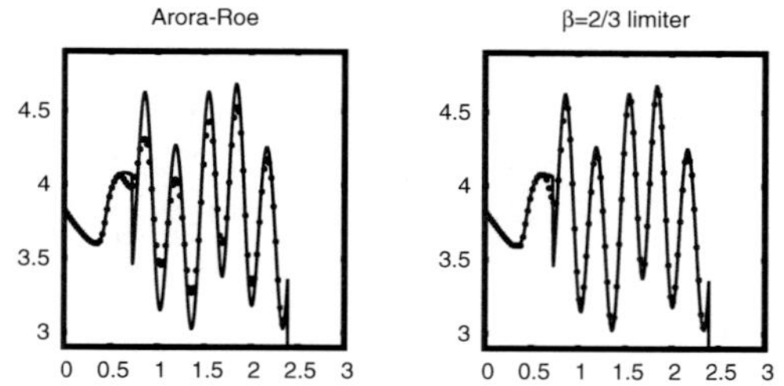

Figure 11. Zoom of Shu-Osher blast shock-entropy interaction at $t = 1.8$ with 400 points; same limiter for all waves.

obtained with the mixed use of Superpower and CFL-Superbee.

5 CONCLUSIONS

In this study we have shown that, in the context of algebraic limiting, TVD-schemes of very high quality can be obtained if, in the construction of limiters, we take into account the local CFL-number. Previous limiters, which are constructed for linear advection, can be further enhanced when using the nonlinear TVD-condition by Jeng and Payne (Jeng and Payne 1995). This allows us to avoid unphysical overshoots at shocks, especially for scalar conservation laws like the Burgers equation. Even for the Shu-Osher problem, which usually is considered to be intractable by TVD-schemes, the exact solution is almost perfectly reproduced. On the other hand, the high compression of the limiters does not lead to any difficulties with highly compressive shocks as they occur in the Woodward-Colella blast wave problem.

REFERENCES

Arora, M. and Roe, P.L. (1997). A well-behaved TVD limiter for high-resolution calculations of unsteady flow. *J. Comput. Phys.* (1), 3–11.

Bokanowski, O. and Zidani, H. (2007). Anti-dissipative schemes for advection and application to Hamilton-Jacobi-Bellmann equations. *J. Sci. Comput.* (1), 1–33.

Després, B. and Lagoutière, F. (2001). Contact discontinuity capturing schemes for linear advection and compressible gas dynamics. *J. Sci. Comput.* (4), 479–524.

Harten, A. (1983). High resolution schemes for hyperbolic conservation laws. *J. Comput. Phys.*, 357–393.

Jeng, Y.N. and Payne, U.J. (1995). An adaptive TVD limiter. *J. Comput. Phys.* (2), 229–241.

Kemm, F. (2011). A comparative study of TVD limiters—well known limiters and an introduction of new ones. *Internat. J. Numer. Methods Fluids* 67 (4) 404–440.

Leveque, R.J. (2002). *Finite volume methods for hyperbolic problems.* Cambridge Texts in Applied Mathematics. Cambridge: Cambridge University Press.

LeVeque, R.J. and Berger, M.J. (2004, June). Clawpack software 4.3. http://www.clawpack.org.

Roe, P. and Baines, M. (1982). Algorithms for advection and shock problems. Numerical methods in fluid mechanics, Proc. 4th GAMM-Conf., Paris 1981, Notes Numer. Fluid Mech. 5, 281–290 (1982).

Roe, P. and Baines, M. (1984). Asymptotic behaviour of some non-linear schemes for linear advection. In Pandolfi, M. and Piva, R. (Eds.), *Proceedings of the Fifth GAMM-Conference on Numerical Methods in Fluid Mechanics*, pp. 283–290. Vieweg.

Roe, P.L. (1981). Numerical algorithms for the linear wave equation. Technical Report 81047, Royal Aircraft Establishement.

Roe, P.L. (1985). Some contributions to the modelling of discontinuous flows. Large-scale computations in fluid mechanics, Proc. 15th AMS-SIAM Summer Semin. Appl. Math., La Jolla/Calif. 1983, Lect. Appl. Math. 22, Pt. 2, 163–193 (1985).

Serna, S. and Marquina, A. (2004). Power ENO methods: A fifth-order accurate weighted power ENO method. *J. Comput. Phys.* (2), 632–658.

Shu, C.-W. and Osher, S. (1989). Efficient implementation of essentially non-oscillatory shock-capturing schemes. *J. Comput. Phys.* (2), 439–471.

Toro, E.F. (1999). Riemann solvers and numerical methods for fluid dynamics. A practical introduction. 2nd ed. Berlin: Springer.

Woodward, P. and Colella, P. (1984). The numerical simulation of two-dimensional fluid flow with strong shocks. *J. Comput. Phys.*, 115–173.

Numerical Methods for Hyperbolic Equations – Vázquez-Cendón et al. (eds)
© 2013 Taylor & Francis Group, London, ISBN 978-0-415-62150-2

Arbitrary high order schemes for transport problems

B. Latorre
Soil and Water Department, Estación Experimental de Aula Dei, CSIC, Spain

P. García-Navarro
Fluid Mechanics, University of Zaragoza, Spain

ABSTRACT: In this work, a method based on Legendre polynomials is presented for the simulation of the passive transport of a solute. The formulation is conservative, explicit and made in a single step. The spatial accuracy is achieved by means of cell polynomial approximations using Legendre series. This kind of spatial representation is also found in finite element discretization and allows for information on the variation of the fields at the sub-grid scale. The time resolution of the transport is based on both a numerical estimation of the displacement at the advection speed and a grid deformation, according to the semi-Lagrangian rules, followed by a projection of the solution on the fixed initial grid.

First, the resolution of the scalar transport of a concentration field is presented. The main interest is focused on the analysis of the accuracy and the efficiency of the method when moving from order 1 to order 20 as compared to standard methods of virtual reconstruction. The interest in this work is the study of the computational saving that can be achieved if the required accuracy is medium or low. This is possible thanks to the sub-grid information that offers the possibility to solve problems with enough accuracy using only a few grid cells and high order polynomials. Furthermore, this enables the use of large time steps hence leading to low computational times.

In a second part, the method is applied to the resolution of the passive transport of a solute in shallow water flows. A technique has been developed to couple Legendre schemes to any conservative method used for the resolution of the shallow water equations. The coupling offers the possibility to combine solvers of different order of accuracy, always enforcing conservative and monotone behavior in the numerical solution of the solute concentration. This strategy is interesting since it is possible to require high order of accuracy only in the solute transport simulation, hence concentrating the computational effort in the component with more numerical error. The coupled method is applied to solve transport problems including bed level variations (source terms in the flow equations), water depth discontinuities and different regimes in order to analyze the performance of the proposed coupling technique.

1 INTRODUCTION

This work aims to develop and study efficient tools for the simulation of the transport and mixing of a passive solute in free surface flows.

The numerical resolution of passive transport is more demanding than the calculation of water flow. This phenomenon is justified from different perspectives. The characteristic space (Rusanov 1963) associated to pressure waves that dominate the water flowis compressive presenting certain numerical advantages such as the possibility to obtain good numerical results with a first order approximation. However, passive transport takes place in degenerate characteristic fields. Furthermore, water flow is pseudo-steady in most of the problems whereas transport processes are usually transient. Hence, the requirement on the order of approximation increases at least to second order (Bermudez and Vázquez-Cendón 1994; Vázquez—Cendón 1999).

For these reasons, first-order methods that have been used successfully for the simulation of the shallow water equations lead to high errors and low convergence when solving passive transport. It is therefore necessary to develop specific strategies thatprovide an accurate and efficient resolution of this process.

It is widely accepted that high order methods are an efficient choice when low errors are required in the solutions (Dumbser and Kaeser 2007). Among the different methodologies available for the development of high order methods, we focus on the finite volume formulation. A new family of methods, based on polynomial sub-cell representation, is presented and its properties are analyzed and compared with the ADER (Toro,Millington, and Nejad 2001) methodology. For this purpose, different problems are numerically solved studying the dependence of the accuracy and efficiency of the methods on the discrete parameters.

Instead of integrating the resolution of water flow and passive transport in a single method, separate strategies have been developed for the calculation of these two processes applying more precision in the passive transport.

Finally, the coupled resolution of the solute transport and the water flow must comply with certain physical principles: preservation of a constant concentration state and a monotone behavior of the concentration field (Burguete, Murillo, and García-Navarro 2008; Murillo, García-Navarro, and Burguete 2009). Anew mechanism for coupling the two processes has been developed that respects these conditions. The properties of the technique are verified solving the transport of a passive solute in a complex flow.

2 LEGENDRE SERIES

The technique to approximate a function, $\tilde{\phi}(x)$, as linear combination of Legendre polynomials, $P_n(x)$, is presented:

$$\phi(x)=\sum_{n=1}^{N} {}^{n}\phi\, P_n(x)\approx\tilde{\phi}(x) \tag{1}$$

where the coefficients, ${}^{n}\phi$, containing the information of the approximation, are calculated solving the projection of the function over different normalized polynomials (Hilbert 1912):

$${}^{n}\phi=\frac{2n-1}{2}\int_{-1}^{1}\tilde{\phi}(x)P_n(x)dx \tag{2}$$

where $[-1,1]$ represents the orthogonality domain of the polynomials.

This technique will be used in the proposed numerical method to represent the solute distribution in each cell and time step, adopting the following notation:

$$\phi_i(x,t)=\sum_{n=1}^{N} {}^{n}\phi_i^t\, P_n(x) \tag{3}$$

where index i refers to the cell and t represents time. The space coordinates are assumed to be adapted, matching the orthogonality interval of the Legendre polynomials and the domain of each cell.

2.1 *Conservative approximation*

Among the different sets of orthogonal polynomials (Chebyshev, Hermite, Jacobi, Laguerre and Legendre) only Legendre polynomials present a constant weight function:

$$w(x) = 1 \tag{4}$$

providing thus a relationship between the first coefficient of the series (1) and the average value of the function:

$${}^{1}\phi=\frac{1}{2}\int_{-1}^{1}\tilde{\phi}(x)dx \tag{5}$$

The remaining elements of the Legendre basis, due to the principle of orthogonality, present no net area in the domain and do not contribute to the integral value of the approximate function:

$$\int_{-1}^{1}P_n(x)dx=0 \quad \text{if} \quad n>1 \tag{6}$$

Inserting (5) and (6) in (1) it can be demonstrated that the integral of the approximation matches the integral of the initial function:

$$\int_{-1}^{1}\phi(x)dx=\int_{-1}^{1}\tilde{\phi}(x)dx \tag{7}$$

Equation (7) represents the conservation property of this discretization technique and allows to apply Legendre series in the design of a numerical method without affecting the conservation of the variables.

Finite volume methods describe the problems by means of the average value of the variables in each cell. Therefore, this representation technique can be framed as a first order Legendre series. Higher order techniques in the finite volume formulation apply virtual reconstruction of the variables involving information from other cells. On the other hand, Legendre series develop a local calculation of the variable distribution avoiding the reconstruction step.

Legendre methods were introduced by Latorre et al. (Latorre and García-Navarro 2011) for the resolution of the scalar transport equation. These methods, in which the order of approximation is a parameter, produce spatial variations whose scale is smaller than the cell size. This implies an increase in the computational efficiency compared to high order finite volume methods. Transport resolution is performed computing first, by numerical integration techniques, the movement of the mesh nodes, then stretching and cutting the polynomials contained in each cell. The resulting schemes are explicit and maintain a constant stability criterion not conditioned by the order of approximation. Legendre methods present common characteristics with the discretization in finite elements and the temporal resolution applied in conservative semi-Lagrangian methods.

The Legendre method formulation is presented considering, for simplicity, the transport of a concentration field in one dimension with constant and positive velocity ($v > 0$):

$$\vec{\phi}_i^{\,t+\Delta t} = \mathbf{A}(c)\,\vec{\phi}_{i-1}^{\,t} + \mathbf{B}(c)\,\vec{\phi}_i^{\,t} \tag{8}$$

where:

$$\vec{\phi}_i = \begin{pmatrix} {}^1\phi_i \\ \vdots \\ {}^N\phi_i \end{pmatrix}, A = \begin{pmatrix} A_{11} & \cdots & A_{1N} \\ \vdots & \ddots & \vdots \\ A_{N1} & \cdots & A_{NN} \end{pmatrix} \tag{9}$$

Scheme (8) represents a set of N equations and contains the influence of the previous state on the solution at each cell through the action of matrices **A** and **B** depending on the CFL dimensionless number:

$$c = \frac{|v|\Delta t}{\Delta x} \tag{10}$$

The expression of the numerical method (8) in the particular case of the second order approximation is next presented:

$$\begin{pmatrix} {}^1\phi_i^{\,t+\Delta t} \\ {}^2\phi_i^{\,t+\Delta t} \end{pmatrix} = c\begin{pmatrix} 1 & 1-c \\ 3(c-1) & c^2 - 3(1-c)^2 \end{pmatrix}\begin{pmatrix} {}^1\phi_{i-1}^{\,t} \\ {}^2\phi_{i-1}^{\,t} \end{pmatrix} + (1-c)\begin{pmatrix} 1 & -c \\ 3c & (1-c)^2 - 3c^2 \end{pmatrix}\begin{pmatrix} {}^1\phi_i^{\,t} \\ {}^2\phi_i^{\,t} \end{pmatrix} \tag{11}$$

For any order of approximation, the first row of the Legendre schemes matches the ADER method assuming an arbitrary reconstruction of the variable. The contribution of Legendre methods is the representation and resolution of the remaining degrees of freedom, avoiding the reconstruction step using information from other cells.

2.2 *Case 1: Accuracy and costs*

The accuracy and efficiency of Legendre methods is analyzed and compared with those of the ADER methodology. The transport with constant velocity $v = 1$ of a Gaussian profile:

$$\phi(x) = exp\left(-(x-0.5)^2/0.02\right) \tag{12}$$

is considered in a periodic domain of size $x \in [0,1]$.

This problem is solved using Legendre and ADER methods of different order of approximation and different domain partitions. The error of the numerical results is characterized using the following expression:

$$L_1 = \frac{0.286}{N\Delta x}\sum_{i=1}^{N}\int\left|\phi_i(x) - \tilde{\phi}(x)\right|dx \tag{13}$$

where N denotes the number of cells, the integral runs along the domain associated with the cell, $\phi_i(x)$ represents the numerical results and $\tilde{\phi}(x)$ the exact solution. The error has been normalized so that the unit value corresponds to the worst result provided by the first order method, which transforms the numerical diffusion profile in a constant state.

The transport of the Gaussian profile is solved during $t = 10$ units and a constant time step under $CFL = 0.5$ condition. ADER schemes of different order of approximation are first considered using different domain partitions. The error of the results and the number of operations required in each case is represented in logarithmic scale in Figure 1-a. Results from the same order of approximation

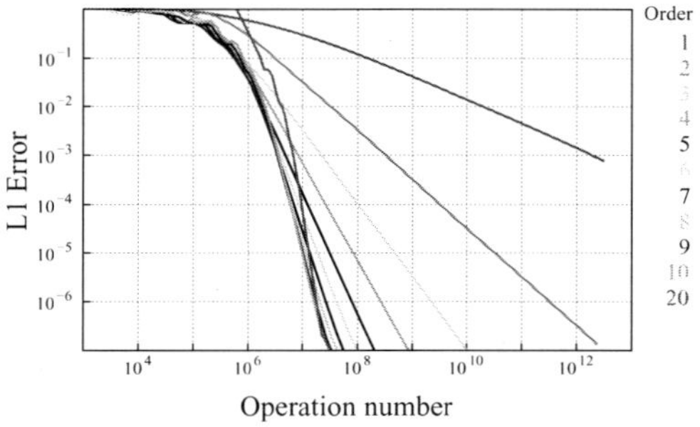

a)

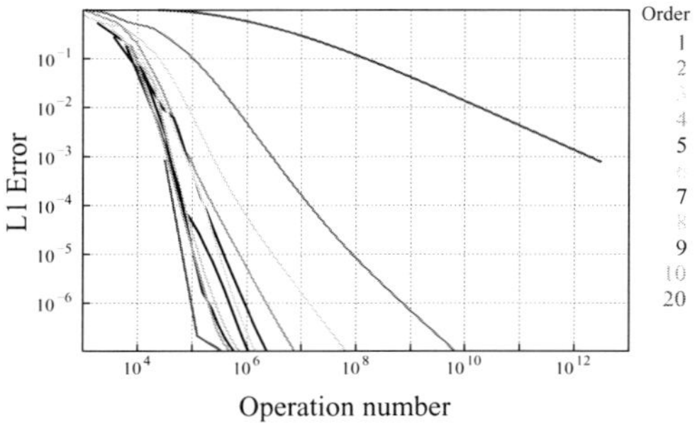

b)

Figure 1. Accuracy and computational costs of ADER a) and Legendre b) methods solving the transport of the Gaussian profile. (t = 10, CFL = 0.5).

are drawn under the same line, whose value can be found in the legend on the right of the figure. Results of Legendre methods using the same discrete parameters are presented in Figure 1-b.

Figures 1-a and 1-b show that Legendre schemes are more efficient than ADER methods in the resolution of this problem. Comparing the two families for a certain maximum error in the solution, the Legendre methods result 100 times faster than the ADER family for a wide range of errors. Legendre schemes are able to solve transport problems accurately and efficiently using fewer cells and higher orders, situation in which it is not possible to build reconstructions.

2.3 *Case 2: Smolarkiewicz flow*

The propagation of a Gaussian function on a square domain, $x \in [0,6]$, $y \in [0,6]$, under the following velocity field, defined by Smolarkiewicz in 1982 (Smolarkiewicz 1982), is studied:

$$\left(v_x, v_y\right) = \frac{\pi}{5}\left(sin(\pi x)sin(\pi y), cos(\pi x)cos(\pi y)\right) \quad (14)$$

This periodic distribution consists of square domains with closed trajectories and alternate senses of rotation. The following initial condition of the concentration field is considered:

$$\phi(r) = exp\left(-r^2/0.1\right) \quad (15)$$

where r represents the distance from the center of the domain:

$$r = \sqrt{(x-3)^2 + (y-3)^2}$$

The transport of the initial distribution is solved during two simulation times (t = 5, 10) using a grid of 120 × 120 cells and a constant time step under $CFL = 0.5$ condition. Legendre schemes of different order of approximation are considered. The representation of the velocity field and the resolution of the movement of the mesh nodes is made in each method using numerical techniques of the same order of accuracy.

Results obtained using Legendre schemes of order 1, 2 and 3 are shown in Figure 2 using a color map. The Gaussian distribution is transported and deformed in two rotational flows with opposite sense of rotation. The first order scheme presents high numerical diffusion and a divergent calculation of the trajectories, shifting the concentration outside the squares. Theses errors are reduced when considering higher order methods.

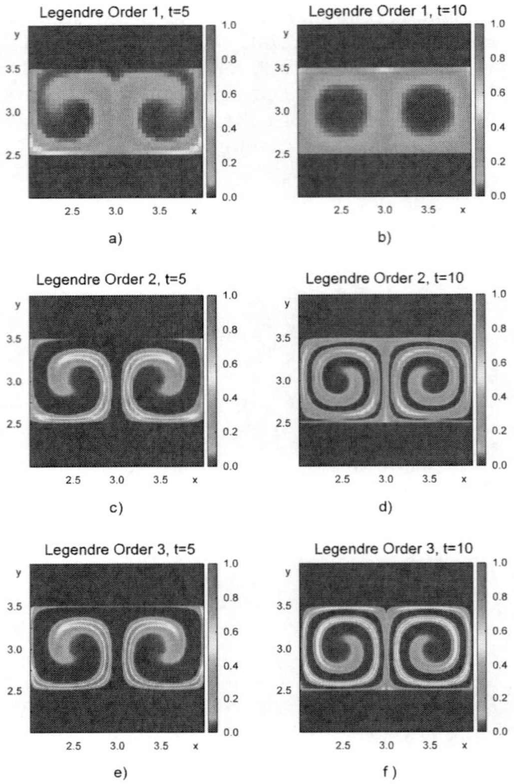

Figure 2. Numerical results corresponding to the transport of the Gaussian profile in the Smolarkiewicz flow using Legendre methods of order 1, 2 and 3. (t = (5, 10), CFL = 0.5, 120 × 120 cells (40 × 40 represented)).

3 DEPTH-AVERAGED TRANSPORT

The shallow water model in one dimension, assuming for simplicity a frictionless rectangular channel of unit width, is presented:

$$\frac{\partial \vec{u}(x,t)}{\partial t} + \frac{\partial \vec{f}(x,t)}{\partial x} = \vec{s}(x) \quad (16)$$

This system of equations describes the evolution of the variables that characterize the flow in terms of the spatial variations of the physical flux vector $\vec{f}$ and the action of the source terms contained in the vector $\vec{s}$. These terms depend on the water depth h, the unit discharge q and the channel bed elevation z as follows:

$$\vec{u} = \begin{pmatrix} h \\ q \end{pmatrix}, \vec{f} = \begin{pmatrix} q \\ q^2 h + gh^2 2 \end{pmatrix}, \vec{s} = \begin{pmatrix} 0 \\ -gh\frac{\partial z}{\partial x} \end{pmatrix} \quad (17)$$

This work is focused on explicit numerical methods. In this case, the conservative resolution of the shallow water equations can be formulated using the following general expression:

$$\vec{u}_i^{t+\Delta t} = \vec{u}_i^t - \frac{\Delta t}{\Delta x}\left(\vec{f}_{i+\frac{1}{2}} - \vec{f}_{i-\frac{1}{2}} - \vec{s}_i\right) \tag{18}$$

where Δt represents the time step, Δx denotes the cell size and $\vec{f}_{i+\frac{1}{2}}$ is the numerical flux in the interface between cells i and $i+1$. The latter represents a numerical approximation of the physical mass and momentum exchange across the interface.

Upwind discretization of the bed slope term $\vec{s}$ (Vázquez-Cendón 1999) introduces a non-conservative contribution in the momentum equation, represented as numerical source term $\vec{s}_i$, and a conservative contribution to the mass equation, associated to the interface and grouped in the numerical flux $\vec{f}_{i+\frac{1}{2}}$. The components of these two vectors are denoted as follows:

$$\vec{f}_{i+\frac{1}{2}} = \begin{pmatrix} f^h \\ f^q \end{pmatrix}_{i+\frac{1}{2}}, \quad \vec{s}_i = \begin{pmatrix} 0 \\ s^q \end{pmatrix}_i \tag{19}$$

Any conservative numerical method can be formulated following (20), grouping the conservative contribution from the upwind discretization of the bed slope term in the first component of the numerical flux. This becomes important when solving the transport of a solute so that a physical coupling of the two methods is required in those cases.

This procedure is illustrated presenting the contribution to the mass equation from the upwind discretization of the bed slope term in the numerical scheme of Roe (Roe 1986). In the case of supercritical flow this contribution is nil but, when considering subcritical flow, the following can be written:

$$f^h_{i+\frac{1}{2}} = {}^{Roe}f^h_{i+\frac{1}{2}} - \frac{1}{2}\sqrt{g\frac{(h_i + h_{i+1})}{2}}(z_{i+1} - z_i) \tag{20}$$

The additional resolution of the passive transport of a solute is considered, described by the following conservation law:

$$\partial_t m + \partial_x\left(\frac{qm}{h}\right) = 0 \tag{21}$$

where $m = m(x,t)$ represents the solute mass and $\phi = m/h$ the associated concentration.

The technique to couple first order methods for the resolution of passive transport and water flow is presented. The proposed method maintains the form of the shallow water scheme (19) extending the vectors of conserved variables $\vec{u}_i$, numerical flux $\vec{f}_{i+\frac{1}{2}}$ and source term $\vec{s}_i$ as follows:

$$\vec{u}_i = \begin{pmatrix} h \\ q \\ m \end{pmatrix}_i, \vec{f}_{i+\frac{1}{2}} = \begin{pmatrix} f^h \\ f^q \\ f^m \end{pmatrix}_{i+\frac{1}{2}}, \vec{s}_i = \begin{pmatrix} 0 \\ s^q \\ 0 \end{pmatrix}_i \tag{22}$$

All the discrete approximations in (23) are already defined except the numerical flux of solute mass $f^m_{i+\frac{1}{2}}$. This term is estimated using the method proposed by Latorre et al. (Latorre and García-Navarro 2009), based on the physical properties of the mixing of two fluids with different concentrations. The numerical flux $f^h_{i+\frac{1}{2}}$ introduces variations of the water mass in the cells i and $i+1$ that are interpreted as physical exchanges of water volumes through the interface. Assuming that the transported water keeps the concentration of the cell of origin, the following expression for the numerical flux of solute mass is obtained:

$$f^m_{i+\frac{1}{2}} = f^h_{i+\frac{1}{2}}\phi_{i+\frac{1}{2}} \tag{23}$$

where the concentration $\phi_{i+\frac{1}{2}}$, dependig on the sign of the water mass numerical flux, is calculated as follows:

$$\phi_{i+\frac{1}{2}} = \begin{cases} \left(\frac{m}{h}\right)_i & \text{if } f^h_{i+\frac{1}{2}} > 0 \\ \left(\frac{m}{h}\right)_{i+1} & \text{if } f^h_{i+\frac{1}{2}} < 0 \end{cases} \tag{24}$$

Once the numerical solute mass flux is defined, the expression of the solute numerical scheme is:

$$m_i^{t+\Delta t} = m_i^t - \frac{\Delta t}{\Delta x}\left(f^m_{i+\frac{1}{2}} - f^m_{i-\frac{1}{2}}\right) \tag{25}$$

The presented method solves the solute transport in first order of approximation coupled to any numerical scheme for the resolution of the water flow, ensuring a physical behavior (denoted as *P* property) of the concentration field associated with the solute: preservation of a constant concentration state and monotone behavior. The formulation (19), for the water mass numerical flux, warrants this behavior also in presence of source terms including the effect of the upwind discretization in the passive transport.

The coupling technique is extended to high order Legendre methods computing first, from the water mass numerical flux, $f^h_{i+\frac{1}{2}}$, the volume that is transferred through every interface in one time step. This amount is represented as a proportion of the total volume of water contained in each cell at the initial time in the region near the wall. The water volumes are assumed to move through the interface, complying with the net transfer described by the numerical flux. Two different volumes, corresponding to two depths, are obtained in each cell, whose sum corresponds to the volume calculated by the water flow scheme. The volumes are then assumed to evolve without mixing, adapting their limits and depth to reach the final average state in each cell. This process, whose details can be found in (Latorre 2011), determines the conservative transformation to be applied to the solute mass in the Legendre schemes.

To illustrate the procedure, the formulation considering solute mass transport in a one-dimensional domain and positive water mass numerical fluxes is presented:

$$\vec{m}_i^{t+\Delta t} = \mathbf{A}\,\vec{m}_{i-1}^{t} + \mathbf{B}\,\vec{m}_i^{t} \tag{26}$$

where the matrices **A** and **B** depend on the following dimensionless numbers:

$$c_1 = \frac{\Delta t\, f^h_{i-\frac{1}{2}}}{\Delta x\, h^t_{i-1}},\; c_2 = \frac{\Delta t\, f^h_{i-\frac{1}{2}}}{\Delta x\, h_i^{t+\Delta t}},\; c_3 = \frac{\Delta t\, f^h_{i+\frac{1}{2}}}{\Delta x\, h^t_i} \tag{27}$$

The proposed scheme (27) solves the passive transport of a solute in the water flow combining the Legendre scheme of arbitrary order of approximation, for the calculation of the transport, and any method for the resolution of the shallow water equations. This coupled method ensures the physical behavior of the concentration field (*P* property). This condition also holds in presence of source terms thanks to the formulation (19), that includes the effect ofupwind discretization in the transport.

3.1 *Case 3: Transport in a channel*

The transport of a set of solute pulses in a 20 m long rectangular channel is considered (Fernández-Nieto and Narbona-Reina 2008). The bottom of the channel contains a parabolic profile given by the expression:

$$z(x) = max\left(0.2 - 0.05(x-10)^2, 0\right) \tag{28}$$

The water flow is characterized by a constant discharge, $q = 0.18\ m^2/s$, and a constant water depth at the outlet, $h = 0.33\ m$. Under these boundary conditions, the flow is subcritical at the entrance, is accelerated to change regime in the slope of the obstacle and then becomes subcritical again, showing a hydraulic jump in the transition. The instantaneous injection of solute is performed every five

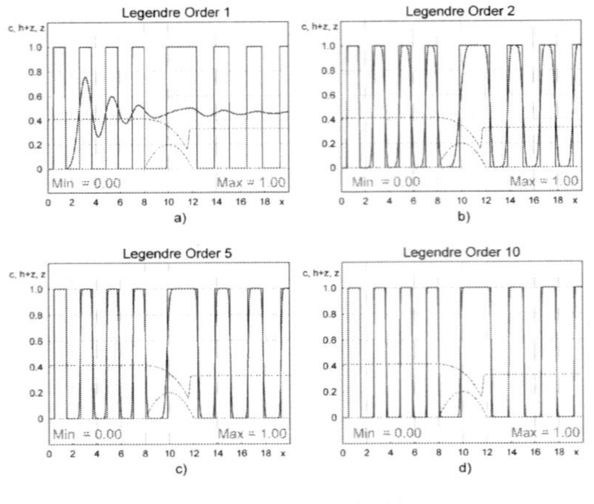

Figure 3. Numerical results and exact solution of solute transport in test case 3. Water flow is solved using a TVD second order method and solute transport using Legendre schemes of order of approximation 1, 2, 5 and 10. (t = 40, CFL = 0.9, 200 cells).

seconds setting the concentration of cells that meet $0.5 < x < 1.5$ to unit value.

The steady flow in the channel is computed with a second order TVD numerical scheme (Burguete and García-Navarro 2001) using a regular division of the domain in 200 cells and a time step according to the condition $CFL = 0.9$. The transport of the discontinuous profiles of solute mass is coupled by means of (27) using Legendre schemes of order of approximation 1, 2, 5 and 10.

Results for the simulation time $t = 40$ s are shown in Figure 3. In each graph the exact solution (continuous black line) and the numerical results (continuous blue line) of the concentration field are represented. The numerical concentration has been obtained as the ratio of the sub-cell solute mass distribution between the constant water depth value supplied by the second order flow scheme. The presented coupled method solves the transport of the solute preserving the *P* property in presence of bed slope terms and through the hydraulic jump.

4 CONCLUSIONS

A method for the simulation of the passive transport of a solute based on Legendre polynomials has been presented. The properties of this technique have been studied and compared with the ADER methodology solving the transport of a Gaussian profile using orders of approximation from 1 to 20. Legendre schemes are more efficient than ADER methods in the resolution of this problem due to the polynomial sub-cell resolution.

In a second part, a modular technique has been introduced to couple Legendre schemes to any conservative method used for the resolution of the shallow water equations. The coupling technique has been applied to solve the transport of a set of pulses in a steady channel flow including bed level variations, water depth discontinuities and different regimes. Results show the ability of this technique to ensure a physical behavior of the concentration field.

ACKNOWLEDGEMENTS

This research work has been partially funded by the Spanish Ministry of Science and Education under research project CGL2005-07059-C02-02.

REFERENCES

Bermudez, A. and Vázquez-Cendón, M. (1994). Upwind methods for hyperbolic conservation laws with source terms. *Computers and Fluids* 23(8), 1049–1071.

Burguete, J. and García-Navarro, P. (2001). Efficient construction of high-resolution TVD conservative schemes for equations with source terms: application to shallow water flows. *International Journal for Numerical Methods in Fluids* (2), 209–248.

Burguete, J., Murillo, J. and García-Navarro, P. (2008). Preserving bounded and conservative solutions of transport in one-dimensional shallow-water flow with upwind numerical schemes: Application to fertigation and solute transport in rivers. *International Journal for Numerical Methods in Fluids* (9), 1731–1764.

Dumbser, M. and Kaeser, M. (2007). Arbitrary high order non-oscillatory finite volume schemes on unstructured meshes for linear hyperbolic systems. *Journal of Computational Physics* (2).

Fernández-Nieto, E. and Narbona-Reina, G. (2008). Extension of WAF Type Methods to Non-Homogeneous ShallowWater Equations with Pollutant. *Journal of Scientific Computing*, 193–217.

Hilbert, D. (1912). Grundzüge einer allgemeinen theorie der linearen integralgleichungen. *Teubner, B.G., Leipzig*.

Latorre, B. (2011). Simulación eficiente del transporte pasivo en flujos de superficie libre. Ph. D. thesis, University of Zaragoza, Spain.

Latorre, B. and García-Navarro (2009). An efficient and conservative model for solute transport in unsteady shallow water flow. In P. López-Jiménez (Ed.), *Proceedings of the Int. Workshop on Environmental Hydraulics*, pp. 114–119.

Latorre, B. and García-Navarro, P. (2011). Accurate and efficient simulation of transport in multidimensional flow. *International Journal for Numerical Methods in Fluids* (4), 405–431.

Murillo, J., García-Navarro, P. and Burguete, J. (2009). Conservative numerical simulation of multi-component transport in two-dimensional unsteady shallow water flow. *Journal of Computational Physics* (15), 5539–5573.

Roe, P.L. (1986). Upwind differencing schemes for hyperbolic conservation laws with source terms. *Proceedings of the 1st international Congress on Hyperbolic Problems*, 41–51.

Rusanov, V. (1963). Characteristics of the general equations of gas dynamics. *Zhurnal Vychislitelnoi Mathematiki Mathematicheskoi Fiziki*, 508–527.

Smolarkiewicz, P. (1982). The multi-dimensional Crowley advection scheme. *Mon. Wea. Rev.*, 1050–1065.

Toro, E., Millington, R. and Nejad, L. (2001). *Towards very high order Godunov schemes*. Kluwer/Plenum Academic Publishers.

Vázquez-Cendón, M. (1999). Improved treatment of source terms in upwind schemes for the shallow water equations in channels with irregular geometry. *Journal of Computational Physics* (2), 497–526.

Numerical Methods for Hyperbolic Equations – Vázquez-Cendón et al. (eds)
© 2013 Taylor & Francis Group, London, ISBN 978-0-415-62150-2

Efficient numerical solution of the model kinetic equations

Vladimir A. Titarev
Dorodnicyn Computing Centre of Russian Academy of Sciences, Moscow, Russia

ABSTRACT: The paper is devoted to the evaluation of a recent deterministic framework for modelling of three-dimensional rarefied gas flows on the basis of the numerical solution of the Boltzmann kinetic equation with the model collision integrals. Performance of the framework is demonstrated on a gas flow into vacuum as well as an external supersonic flow over a rocket. Comparisons with the DSMC calculations are provided.

1 INTRODUCTION

Past few years have seen rapid development of numerical methods and associated computer codes for solving the Boltzmann kinetic equation with the exact or model collision integrals in three space dimensions. In the existing approaches for solving kinetic problems in complex geometries (Li and Zhang 2003; Kolobov et al. 2007; Kloss et al. 2008) explicit time marching methods are typically used for both steady-state and unsteady calculations, with the only exception reported in (Arkhipov and Bishaev 2007) for plasma thruster modelling. However, despite these significant advances in computational rarefied gas dynamics, there is still much room for improvement in terms of convergence in steady-state calculations, accuracy andversatility of spatial discretisation as well as parallel scalability.

A recent numerical framework (Titarev 2010; Titarev 2012a) for obtaining three-dimensional solutions of the Boltzmann kinetic equation aims at circumventing the deficiencies of the existing methods. The framework consists of a high-order accurate implicit advection scheme on hybrid unstructured meshes, conservative procedure for the calculation of the model collision integral and a simple and efficient implementation on modern high-performance clusters. Accuracy and efficiency of the proposed algorithms across a wide range of Knudsen numbers as well as good scalability for up to 512 cores on a modern HPC machine were demonstrated by computing slow internal rarefied gas flows.

The aim of the present work is to summarize the developments and to apply the method to more demanding computational test cases, including both slow subsonic and fast supersonic flows. Two example problems are considered. The first test problem concernsrarefied gas flow through a circular pipe into vacuum (Shakhov 1996; Varoutis et al. 2008). The computational difficulties in this problem include the presence of very low density regions as well as very different solution gradients inside the pipe and near its exit. The accurate resolution of the combination of large and small flow gradients is required in order to achieve acceptable accuracy in computing not only flow rate, but also flow pattern in the low density region. The second test case is an external supersonic flow over a rocket-like geometry with two wings and four stabilizers. Here the challenge is to handle complex geometry as well as flow gradients resulting in the high-speed flow. Of particular interest is also the low-density region in the wake.

The rest of the paper is organized as follows. In Section 2 the governing kinetic equation is presented, along with the boundary conditions. In Section 3 the general method of solution is outlined, including conservative approximation of the collision integral, accurate spatial reconstruction on an unstructured mesh as well as an efficient one-step implicit time evolution. Parallel implementation of the method is also discussed. Computational results are presented in Section 4. Conclusions are drawn in Section 5.

2 GOVERNING EQUATIONS

A steady three-dimensional state of the rarefied gas is determined by the velocity distribution function $f(x,\xi)$, where $x = (x_1, x_2, x_3) = (x,y,z)$ is the spatial coordinate vector, $\xi = (\xi_x, \xi_y, \xi_z)$ is the molecular velocity vector. Let l_*, p_*, T_*, μ_* are characteristic scales length, pressure, temperature and viscosity, respectively; $\sqrt{2RT_*}$ is used as the characteristic scale of velocity. In the non-dimensional variables the Boltzmann equation with the S-model collision integral (Shakhov 1968a; Shakhov 1968b) for the distribution function f has the following form

$$\xi_x \frac{\partial f}{\partial x} + \xi_y \frac{\partial f}{\partial y} + \xi_z \frac{\partial f}{\partial z} = \nu (f^{(S)} - f),$$
$$\nu = \frac{nT}{\mu}\delta, \quad f_M = \frac{n}{(\pi T)^{3/2}}\exp\left(-c^2\right),$$
$$f^{(S)} = f_M\left[1 + \frac{4}{5}(1-\mathrm{Pr})Sc\left(c^2 - \frac{5}{2}\right)\right], \tag{1}$$
$$\upsilon_i = \xi_i - u_i, \quad c_i = \frac{\upsilon_i}{\sqrt{T}}, \quad S_i = \frac{2q_i}{nT^{3/2}}.$$

Here rarefaction parameter

$$\delta = \frac{l_* p_*}{\mu_* \sqrt{2kT_*/m}}$$

defines the degree of gas rarefaction and is inversely proportional to the Knudsen number Kn; m molecular mass, k the Boltzmann constant. For a monatomic gas the Prandtl number Pr = 2/3. The hard-sphere intermolecular interaction $\mu = \sqrt{T}$ is used in all calculations.

The non-dimensional macroscopic quantities are defined as the integrals of the velocity distribution function with respect to the molecular velocity:

$$\begin{pmatrix} n \\ nu \\ n\left(\frac{3}{2}T + u^2\right) \\ q \end{pmatrix} = \int \begin{pmatrix} 1 \\ \xi \\ \xi^2 \\ \frac{1}{2}\upsilon\upsilon^2 \end{pmatrix} f d\xi. \tag{2}$$

The non-dimensional pressure is given by $p = nT$.

The kinetic equation (1) has to be augmented with the boundary conditions. Let $n = (n_x, n_y, n_z)$ be the unit normal vector to a boundary surface, pointing in the outward direction to the surface, $\xi_n = (\xi\mathbf{n})$ projection of the molecular velocity on the normal. The diffuse molecular scattering boundary condition with complete thermal accommodation to the surface temperature T_w is given by:

$$f(x,\xi) = \frac{n_w}{(\pi T_w)^{3/2}}\exp\left(-\frac{\xi^2}{T_w}\right), \quad \xi_n > 0. \tag{3}$$

The density of reflected molecules n_w is found from impermeability condition stating that the mass flux through the walls is equal to zero:

$$n_w = N_i/N_r, \quad N_i = -\int_{\xi_n<0} \xi_n f d\xi,$$
$$N_r = +\int_{\xi_n>0} \xi_n \frac{1}{(\pi T_w)^{3/2}}\exp\left(-\frac{\xi^2}{T_w}\right)d\xi. \tag{4}$$

On the inflow and outflow boundaries the distribution function of the incoming molecules is prescribed as the locally-Maxwellian one with the corresponding values of the macroscopic variables.

3 NUMERICAL ALGORITHM

3.1 *Framework*

The steady-state solution of the six-dimensional boundary-value problem (1) is found by means of the implicit time-marching algorithm, proposed by the author in a sequence of publications (Titarev 2007; Titarev 2010; Titarev 2012b; Titarev 2012a). The first step in the numerical solution procedure is to replace the infinite domain of integration in the molecular velocity space ξ by a finite computational domain $|\xi_x|, |\xi_y| |\xi_y|, |\xi_z| \le \xi_0$ which is then discretized using the non-uniform mesh with $N_{\xi x} \cdot N_{\xi y} N_{\xi z} \equiv N_\xi$ cells. The velocity distribution function is then defined in centres ξ_α of the resulting velocity mesh. The kinetic equation (1) is replaced by a system of N_ξ time-dependent advection equations for each of f_α:

$$\frac{\partial}{\partial t} f_\alpha = -\xi_\alpha \nabla f_\alpha + J(f_\alpha),$$
$$J(f_\alpha) = \nu (f^{(S)} - f)_\alpha, \tag{5}$$

which are connected by the macroscopic parameters in the function $f^{(s)}$ from the model collision integral J. Here ∇ is the gradient operator in the physical space (x,y,z).

3.2 *Approximation of the model collision integral*

The direct approximation of expressions (2) for macroscopic quantities yields a non-conservative numerical method that violates the discrete mass, momentum, and energy conservation laws (Titarev 2007). In order to circumvent this difficulty, the present work uses a recent approach for calculating macroscopic quantities in model kinetic equations (Titarev 2003; Titarev 2007). Let ω_α be the weights of the second order composite quadrature rule used for integration in ξ space. In order to compute the vector of primitive variables

$$U = (n, u_1, u_2, u_3, T, q_1, q_2, q_3)^T$$

for each spatial cell the conservative procedure gives the following system of equations

$$\sum_\alpha \begin{pmatrix} 1 \\ \xi \\ \xi^2 \\ vv^2 \end{pmatrix}_\alpha (f_\alpha^{(S)} - f_\alpha)\omega_\alpha + \begin{pmatrix} 0 \\ 0 \\ 0 \\ 2\mathrm{Pr}q \end{pmatrix} = 0. \tag{6}$$

Here subscript i of the spatial mesh is omitted for simplicity. The eight equations (6) are solved using the Newton iterations the initial guess for which is provided by the direct (non-conservative) approximation for (2). Usually, one or two Newton iterations are sufficient for convergence. In the special case Pr = 1 (BGK model (Bhatnagar et al. 1954)) the function $f^{(s)}$ no longer contains the heat flux vector making last three equations in (6) not necessary. If these three equations are omitted, the procedure (6) for macroscopic parameters coincide with the ones proposed in (Mieussens 2000) from different considerations for the BGK model.

3.3 *Advection scheme*

The next step is to describe a numerical method to solve each of the kinetic equations (5) assuming the model collision integral is known. Introduce in the physical variables $x = (x_1, x_2, x_3) = (x, y, z)$ a computational mesh consisting of elements (spatial cells) V_i. The total number of spatial cells is N_{space}. Let $|V_i|$ be the cell volume, $|A|_{il}$ area of face l, $\Delta t = t^{n+1} - t^n$ time step, $f_{\alpha i}^n$ the spatial average of the velocity distribution function in the cell V_i at time t^n for the molecular velocity ξ_α, $\delta_{\alpha i}^n = f_{\alpha i}^{n+1} - f_{\alpha i}^n$ time increment of the distribution function. Also denote by $\sigma_l(i)$ the cell index of the cell adjacent to the face l of cell V_i.

For the given index α of the velocity mesh the implicit finite-volume method takes the form of a system of N_{space} linear equations for $f_{\alpha i}^n$:

$$\left(1 + \nu_i^n \Delta t + \frac{\Delta t}{|V_i|} \sum_{l, \xi_{\alpha n l} > 0} \xi_{\alpha n l} \, | A_{il} | \right) \delta_{\alpha i}^n + \frac{\Delta t}{|V_i|} \sum_{l, \xi_{\alpha n l} < 0} \xi_{\alpha n l} \, | A_{il} | \, \delta_{\alpha, \sigma_l(i)}^n = \Delta t R_{\alpha i}^n, \tag{7}$$

where the discrete right-hand side is defined as

$$R_{\alpha i}^n = -(\xi_\alpha \nabla f_\alpha^n)_i + J_{\alpha i}^n,$$

$\xi_{\alpha n l}$ is the projection of the vector ξ_α onto the unit normal to the face l of the cell V_i.

In the right-hand side of the scheme the advection operator acting on the values of the velocity distribution function on the lower time level, is approximated as a sum of fluxes through cell faces:

$$\begin{aligned} \left(\xi_\alpha \nabla f_\alpha^n\right)_i &= \frac{1}{|V_i|} \sum_l \Phi_{\alpha i l}^n, \\ \Phi_{\alpha i l}^n &= \frac{1}{2} \xi_{\alpha n l} \, | A_{il} | \, f^*, \\ f^* &= f^- + f^+ - \operatorname{sign}(\xi_{\alpha n l})(f^+ - f^-), \\ f^- &= f_{\alpha i l}^n, \quad f^+ = f_{\alpha, \sigma_l(i), l_1}^n. \end{aligned} \tag{8}$$

Here l_1 is the number of the face of the cell $\sigma_l(i)$, adjacent to the face l of the cell i, the face averages $f_{\alpha i l}^n$ of the function g for each cell i are computed to high order of accuracy by means of the reconstruction procedure, described below. The general formula (8) is modified if the face l is adjacent to a boundary surface and $\xi_{\alpha n l} < 0$ by applying the corresponding boundary condition.

The calculation of the numerical fluxes $\Phi_{\alpha i l}^n$ with high-order of accuracy requires the knowledge of the face averages of the distribution function $f_{\alpha i l}^n$. For the first-order accurate method it is sufficient to set these face values equal to the cell value $f_{\alpha i l}^n = f_{\alpha i}^n$. It is well known, however, that the first-order method is quite inaccurate. In the present work a fully multi-dimensional piece-wise linear reconstruction in local coordinates based on citeTitarev:2010e,Titarev:2012c is used for computing the face averages with the high order of spatial accuracy. In each spatial cell V_i the distribution function f_α is approximated locally by the reconstruction polynomial. The reconstruction is carried out in a local coordinate system, which avoids scaling effects for skewed cells (Dumbser and Käser 2007). The piece-wise linear reconstruction polynomial $p_{\alpha i}(\hat{x})$ is given by the expansion over the basis functions $e_{ik}(\hat{x})$:

$$p_{\alpha i}(\hat{x}) = f_{\alpha i}^n + \sum_{k=1}^{3} a_{\alpha i k}^n e_{ik}(\hat{x}), \tag{9}$$

where the basis functions $e_{ik}(\hat{x})$ are chosen in such a way that the reconstruction is conservative. The coefficients of the polynomial are computed using the values of f in the cell V_i and a sufficient number of its neighbours, which form the so-called spatial reconstruction stencil. These coefficients can be expressed directly as the linear combination of the values of the distribution function in the stencil. For the linear second-order method it is sufficient to set face values of the distribution function $f_{\alpha i l}$ equal to the face averages $p_{\alpha i l}$ of the reconstruction polynomial, which are expressed in terms of the polynomial coefficients and face averages of the basis functions e_{ikl} as

$$p_{\alpha i l} = f_{\alpha i}^n + \sum_{k=1}^{3} a_{\alpha i k}^n e_{ikl}.$$

The resulting linear numerical method is prone to non-physical (spurious) oscillations at discontinuities (Godunov 1959). In order to avoid oscillations (Kolgan 1972; Kolgan 2011) the reconstruction procedure is made non-linear (solution adaptive) by introducing the so-called slope limiter $\psi_{\alpha i}$. This is equivalent to replacing the coefficients $a_{\alpha i k}^n$ by the modified coefficients $\tilde{a}^n_{\alpha ik} = \psi_{\alpha i} a^n_{\alpha ik}$. A good choice of the slope limiter for steady-state calculations is found in (Venkatakrishnan 1993). The final face averages f_{il} of the distribution function, used in the actual calculations, are then given by

$$f^n_{\alpha il} = f^n_{\alpha i} + \psi_{\alpha i} \cdot (p_{\alpha il} - f^n_{\alpha i}).$$

The first-order scheme is recovered by setting $\psi_{\alpha i} \equiv 0$ whereas $\psi_{\alpha i} \equiv 1$ leads to a linear (oscillatory) spatially second-order method.

The described reconstruction procedure can be extended to polynomial reconstructions of any order of spatial accuracy. However, the test calculations have shown that the piece-wise linear (second-order) representation is the best compromise between the accuracy and computational cost. Further details can be found in (Titarev 2012a) and are omitted here.

3.4 *Solution of the equations on the upper time level*

The direct numerical solution of the linear system (7) is a very slow operation with the computational cost proportional to N^3_{space}. Therefore, an approximate factorization of the system is carried out using the approach suggested in ortciteMenshov:2000a. Regrouping of (7) yields

$$\delta^n_{\alpha i} + \sum_l \Delta t z_{i,\sigma_l(i)} \delta^n_{\alpha\sigma_l(i)} = \frac{\Delta t}{\lambda_i} R^n_{\alpha i}, \tag{10}$$

where the coefficients $z_{i,\sigma_l(i)}$ and λ_i are given by

$$b_{\alpha i} = \sum_l \xi_{\alpha nl}(1 + \text{sign}\,\xi_{\alpha nl}) \frac{|A_{il}|}{2|V_i|},$$

$$c_{\alpha,i,\sigma_l(i)} = \xi_{\alpha nl}(1 - \text{sign}\,\xi_{\alpha nl}) \frac{|A_{il}|}{2|V_i|},$$

$$\lambda_{\alpha i} = 1 + \Delta t \, \nu^n_i + \Delta t b_{\alpha i}, \quad z_{\alpha i\sigma_l(i)} = \frac{1}{\lambda_{\alpha i}} c_{\alpha i\sigma_l(i)},$$

or in the matrix form

$$(I + \Delta t Z_\alpha) \cdot \delta^n_\alpha = \Delta t \Lambda^{-1}_\alpha \cdot R^n_\alpha. \tag{11}$$

Next, the matrix Z_α is approximately factorized into the product of a low-triangular L_α and upper-triangular U_α matrices according to the expression

$$l_{ij} = \begin{cases} \Delta t z_{ij}, & j < i \\ 0, & j > i \end{cases}, \quad u_{ij} = \begin{cases} 0, & j < i \\ \Delta t z_{ij}, & j > i \end{cases},$$

and

$$l_{ii} = u_{ii} = 1$$

so that

$$I + \Delta t Z_\alpha = L_\alpha \cdot U_\alpha + O(\Delta t^2)$$

The implicit method (7) takes its final form:

$$L_\alpha \cdot U_\alpha \cdot \delta^n_\alpha = \Delta t \Lambda^{-1}_\alpha \cdot R^n_\alpha, \quad f^{n+1}_{\alpha i} = f^n_{\alpha i} + \delta^n_{\alpha i}. \tag{12}$$

The computational cost of solving (12) is linearly proportional to N_{space}. As a result, the cost of one time step of the implicit method is only 25% larger than the computational cost of an explicit method with the same spatial reconstruction procedure and the conservative calculation of macroscopic parameters. If the available computer memory allows to store the matrices L_α, U_α and Λ_α from (12) rather than recalculate them at each time step, the implicit method becomes practically as fast as the explicit one.

In calculations, the value of the time step Δt is evaluated according to the expression

$$\Delta t = C \min_i d_i / \xi_0,$$

where C is the prescribed CFL number, d_i the characteristic linear size of the cell V_i.

The convergence of the solution to the steady-state is verified by calculating the global residual in the macroscopic conservation laws. The numerical solution is deemed as converged to the steady state if this residual drops by four orders of magnitude.

3.5 *Parallel implementation*

For large-scale problems such as the ones reported here the calculations are carried out on modern high-performance clusters using Message Passing Interface (MPI). Unlike conventional approaches, in which the spatial mesh (in physical coordinates) is split into blocks (see e.g. (Dumbser, Käser, Titarev, and Toro 2007)), in the present method the decomposition of the ξ_z and ξ_x directions of the rectangular molecular velocity mesh is implemented. The present approach to parallel implementation is similar to the one used earlier for the Boltzmann equation with the exact collision integral (Aristov and Zabelok 2002). For each block the kinetic equation is solved using the implicit method (7). The sequential algorithm is then modified to include inter-processor communications in the calculation of integral sums with respect to the molecular velocity mesh as well as data exchange for the boundary conditions.

The advantages of the present approach to parallelization over existing methods (Kolobov,

Arslanbekov, Aristov, Frolova, and Zabelok 2007; Kloss, Cheremisin, Khokhlov, and Shurygin 2008) are its simplicity and fast converge to steady state due to the use of the implicit time marching method (7). The parallel version of the algorithm was shown to scale well up to 512 cores (Titarev 2012a). In the present work the HPC "Chebyshev" of Lomonosov Moscow State University and MIPT-60 machine of Moscow Institute of Physics and Technology, Russia, were utilised. Up to 256 cores were used for a single run in the present work.

4 NUMERICAL EXAMPLES

4.1 *Pipe flow into vacuum*

The formulation of the problem follows (Shakhov 1996; Varoutis et al. 2008). Consider a rarefied gas flow through a circular pipe of length L and radius R, connecting two infinitely large reservoirs (volumes) filled with the same monatomic gas. Gas in reservoir 1 is kept under pressure p_1 and temperature T_1, whereas in reservoir 2 is kept under vacuum conditions $p_2 = 0$. The complete accommodation of momentum and energy of molecules occurs at the pipe surface, which is kept under the same constant temperature $T_w = T_1$. In calculations, the values of pressure, temperature and viscosity in reservoir 1 are used as p_*, T_*, μ_*, whereas the tube radius R is taken as the characteristic linear dimension l_*.

The solution to the problem is determined by two parameters: the length to radius ration L/R and the rarefaction parameter. The main computed characteristic of the flow is the reduced mass flow rate Q, which is defined as the ratio of the mass flow rate $\dot{M}$ at given δ and L/R to its values $\dot{M}_0$ at the free-molecular regime $ta = 0$ in case of the orifice flow $L/R = 0$. In the non-dimensional variables Q is computed at any position z inside the pipe according to the formula

$$Q = \frac{\dot{M}}{\dot{M}_0}, \quad \dot{M}_0 = \frac{\sqrt{\pi}}{2}, \qquad (13)$$
$$\dot{M} = \int_A n(x,y,z)w(x,y,z)dxdy.$$

In the present work the calculations are carried out for $L/R = 10$, using purely hexahedral unstructured spatial meshes. Figure 1 illustrates one typical spatial mesh consisting of $N_{space} = 1.3 \times 10^5$ cells. In the velocity space thethree-dimensional mesh is constructed using the polar arrangement of ξ_x, ξ_y components and typically consisted of 20^3 cells. No other special measures or changes to the three-dimensional computational method have been carried out to guaranteethe radial symmetry of the solution.

Table 1 contains the comparison of the computed values of the reduced flow rate Q with the well-resolved DSMC values from (Varoutis, Sazhin, Valougeorgis, and Sharipov 2008). There is a generally good agreement between the kinetic data of the present work and that from (Varoutis, Sazhin, Valougeorgis, and Sharipov 2008). The discrepancy is of the order of 1–2%, which is well within the computational error of the present calculations. Figure 2 shows isolines of density and

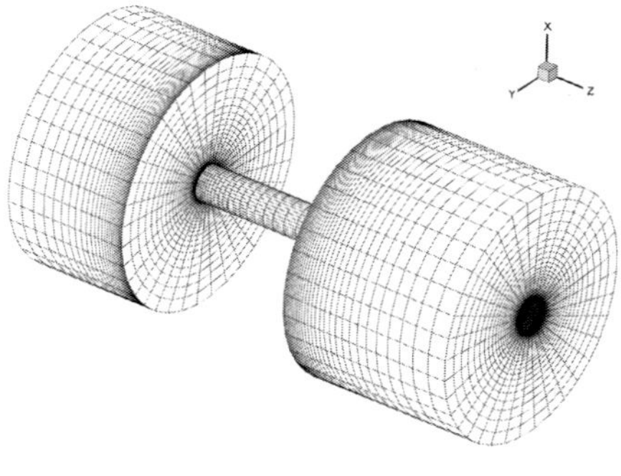

Figure 1. Spatial mesh for micro pipe flow.

Table 1. Reduced flow rate S06-CH33_titarev-E152.eps defined in (13).

δ	0	0.1	1	5	10
Varoutis et. al	0.192	0.190	0.198	0.258	0.335
present	0.188	0.190	0.199	0.258	0.330

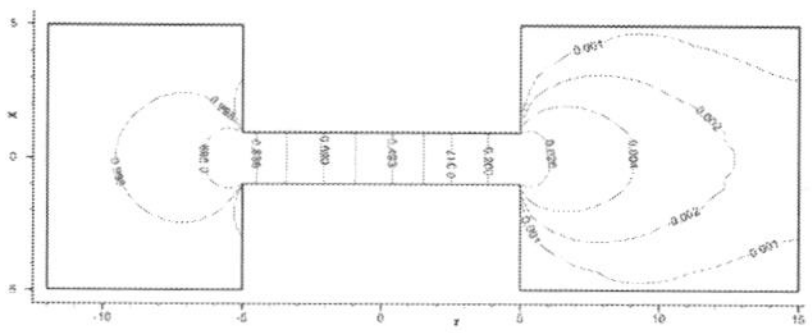

(a) Density

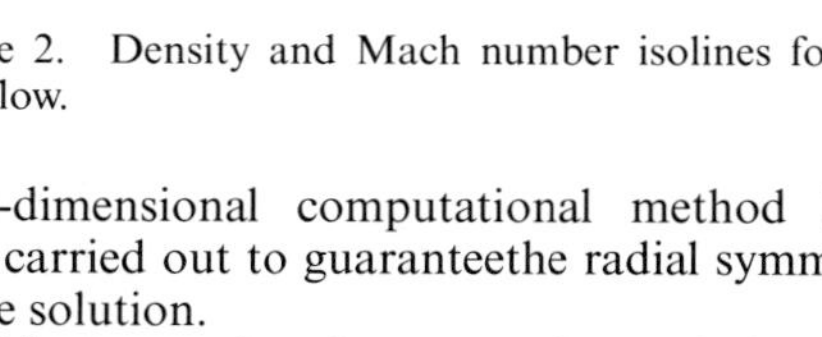

(b) Mach number

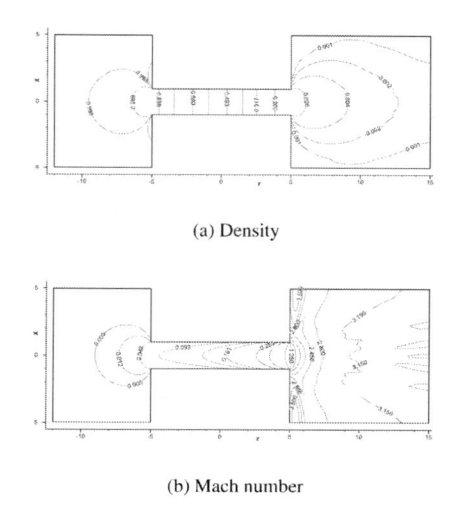

Figure 2. Density and Mach number isolines for the pipe flow.

Mach number in the whole computational domain for $\delta_1 = 1$. It can be seen that near the entrance and especially near the exit the flow is essentially non-linear and gets supersonic in the vacuum region whereas in the central part of the tube density is constant at each cross section.

4.2 *External supersonic flow*

Consider a supersonic flow over a rocket-like geometry, shown on Figure 3. Calculations were carried out for the free-stream Mach number $M_\infty = 3$ and rarefaction parameter $\delta = 1$ under zero angle of attack. The rocket diameter is chosen as the spatial scale l_*, whereas the free-stream values of pressure and temperature are set as p_*, T_*. The surface temperature was fixed and equal to the free-stream temperature.

The kinetic equation is solved using a hybrid tetra-prismatic mesh, which allows to capture all flow features without spending considerable effort on generation of a multi-block structured mesh. The details of the mesh are shown on Figures 4–5, including the prismatic layer near the surface. The total number of spatial cells is $N_{space} = 3.3 \times 10^5$. The velocity mesh consisted of $25 \times 16 \times 32$ nodes. The total number of cells in the 6-dimensional mesh is thus approximately 4.2×10^9.

On 256 Xeon cores one time step of the second-order implicit TVD scheme requires around 180 seconds and convergence to steady state was achieved in 1500 time steps. Figures 6, 7 provide a general representation of the flow field. Shown are the isolines of density and Mach numbers in x-z and y-z planes. Overall, a typical flow pattern of a supersonic rarefied gas flow over a cold body is observed. The flow is characterized by smoothly changing profiles of density and temperature on the symmetry line as well as sharp drop in pressure and

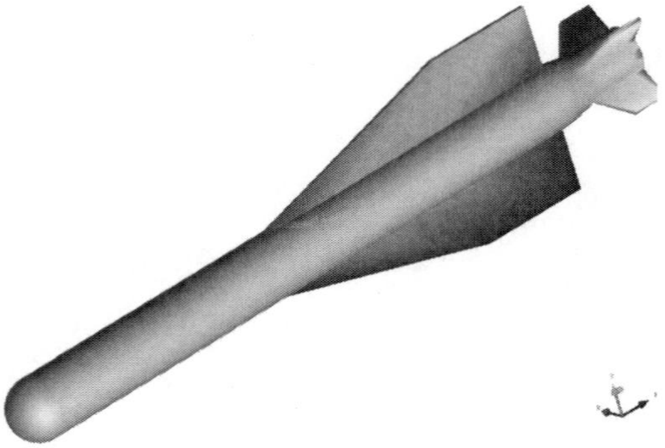

Figure 3. Rocket geometry.

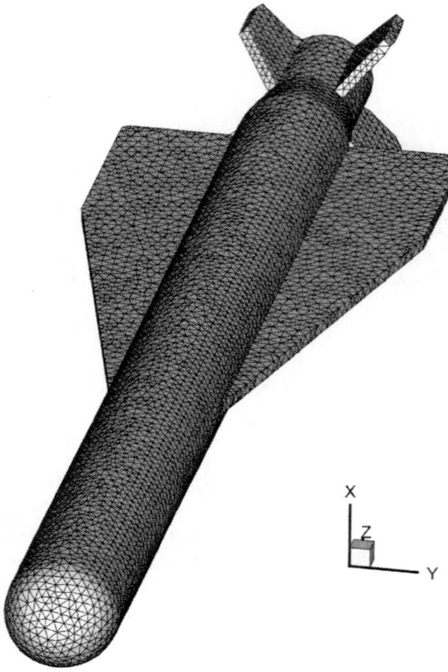

Figure 4. Surface spatial mesh for the rocket geometry.

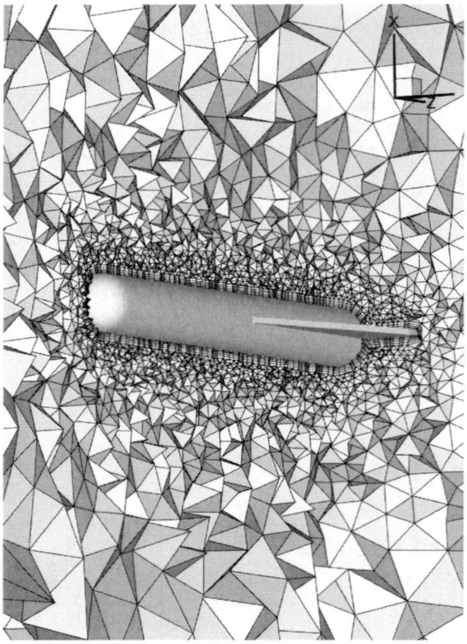

Figure 5. Volume spatial mesh for the rocket geometry.

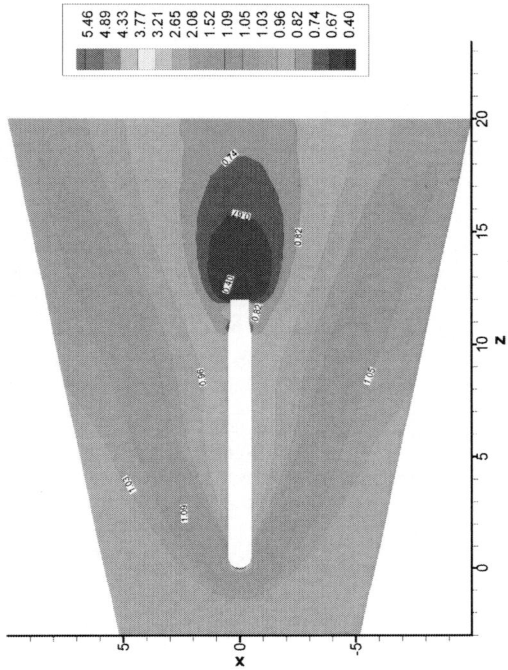

(a) x-z plane

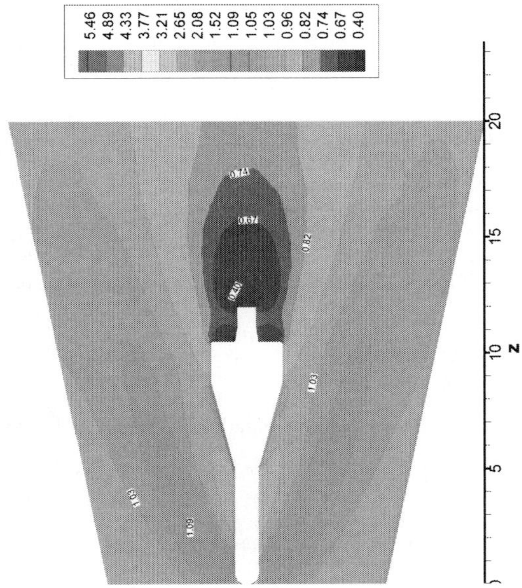

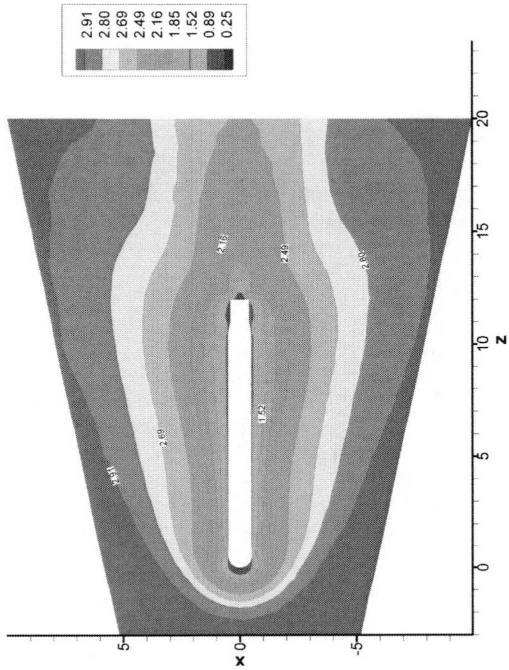

(a) x-z plane

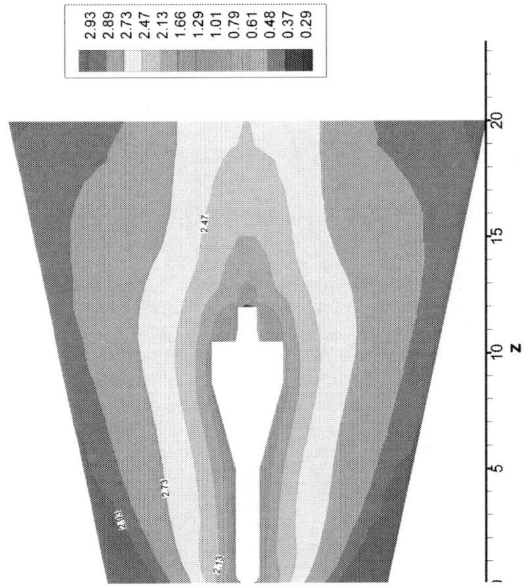

Figure 6. Density isolines.

Figure 7. Mach number isolines.

density in the weak close to the tail of the rocket. Along the surface pressure and density change quite slowly except at the wings and stabilizers.

5 CONCLUSIONS

A numerical framework for modelling the three-dimensional steady rarefied gas flows on the basis of the Boltzmann kinetic equation with the model collision integrals has been reviewed and applied to two difficult test problems. The problems correspond to main applications of the computational rarefied gas dynamics: slow internal micro-scale flows and external high-speed flows over complex shapes. The presented results demonstrate accuracy and robustness of the proposed numerical approach for kinetic equations. Good agreement with the results of DSMC in the case of flow into vacuum is also shown. Future work include the extension of the method to diatomic gases and gas mixtures.

ACKNOWLEDGEMENTS

The author would like to thank Dr V. Garanzha and Mrs. L. Kudryavtseva for providing the model geometry. This work was supported by the Russian Foundation for Basic Research, project no. 10–01–00721-a. The supercomputer support of Lomonosov Moscow State University and Moscow Institute of Physics and Technology, Russia, is gratefully acknowledged.

REFERENCES

Aristov, V. and Zabelok, S. (2002). A deterministic method for solving the Boltzmann equation with parallel computations. *Comp. Math. Math. Phys.* (3), 406–418.

Arkhipov, A. and Bishaev, A. (2007). Three-dimensional numerical simulation of the plasma plume from a stationary plasma thruster. *Computational Mathematics and Mathematical Physics* (3), 472–486.

Bhatnagar, P., Gross, E. and Krook, M. (1954). A model for collision processes in gases. I. Small amplitude processes in charged and neutral one-component systems. *Phys. Rev.* (511), 1144–1161.

Dumbser, M. and Käser, M. (2007). Arbitrary high order non-oscillatory finite volume schemes on unstructured meshes for linear hyperbolic systems. *J. Comput. Phys.* (2), 693–723.

Dumbser, M., Käser, M., Titarev, V. and Toro, E. (2007). Quadrature-free non-oscillatory finite volume schemes on unstructured meshes for nonlinear hyperbolic systems. *J. Comput. Phys.*, 204–243.

Godunov, S. (1959). A finite difference method for the computation of discontinuous solutions of the equations of fluid dynamics. *Mat. Sbornik*, 357–393.

Kloss, Y., Cheremisin, F., Khokhlov, N. and Shurygin, B. (2008). Programming and modelling environment for studies of gas flows in micro—and nanostructures based on solving the Boltzmann equation. *Atomic Physics* (4), 270–279.

Kolgan, V. (1972). Application of the minimum-derivative principle in the construction of finite-difference schemes for numerical analysis of discontinuous solutions in gas dynamics. *Transactions of the Central Aerohydrodynamics Institute* (6), 68–77. in Russian.

Kolgan, V. (2011). Application of the principle of minimizing the derivative to the construction of finite-difference schemes for computing discontinuous solutions of gas dynamics. *J. Comput. Phys.* (7), 2384–2390.

Kolobov, V., Arslanbekov, R., Aristov, V., Frolova, A. and Zabelok, S. (2007). Unified solver for rarefied and continuum flows with adaptive mesh and algorithm refinement. *J. Comp. Phys.*, 589–608.

Li, Z.-H. and Zhang, H.-X. (2003). Numerical investigation from rarefied flow to continuum by solving the Boltzmann model equation. *International Journal for Numerical Methods in Fluids* (4), 361–382.

Men'shov, I. and Nakamura, Y. (2000). On implicit Godunov's method with exactly linearized numerical flux. *Computers and Fluids* (6), 595–616.

Mieussens, L. (2000). Discrete-velocity models and numerical schemes for the Boltzmann-BGK equation in plane and axisymmetric geometries. *J. Comput. Phys.* (2), 429–466.

Shakhov, E. (1968a). Approximate kinetic equations in rarefied gas theory. *Fluid Dynamics* (1), 156–161.

Shakhov, E. (1968b). Generalization of the Krook kinetic relaxation equation. *Fluid Dynamics* (5), 142–145.

Shakhov, E. (1996). The axisymmetric non-linear steady flow of a rarefied gas in a pipe of circular cross section. *Comp. Maths. Math Phys.* (8), 1123–1131.

Titarev, V. (2003). Towards fully conservative numerical methods for the nonlinear model Boltzmann equation. In *Preprint NI03031-NPA*, pp. 13. Isaac Newton Institute for Mathematical Sciences, University of Cambridge, Cambridge, UK.

Titarev, V. (2007). Conservative numerical methods for model kinetic equations. *Computers and Fluids* (9), 1446–1459.

Titarev, V. (2010). Implicit numerical method for computing three-dimensional rarefied gas flows using unstructured meshes. *Computational Mathematics and Mathematical Physics* (10), 1719–1733.

Titarev, V. (2012a). Efficient deterministic modelling of three-dimensional rarefied gas flows. *Communications in Computational Physics*. in press.

Titarev, V. (2012b). Rarefied flow in a long planar microchannel of finite length. *J. Comput. Phys.* (1), 109–134.

Varoutis, S., Sazhin, O., Valougeorgis, D. and Sharipov, F. (2008). Rarefied gas flow through short tubes into vacuum. *J. Vac. Sci. Technol.* (1), 228–238.

Venkatakrishnan, V. (1993). On the accuracy of limiters and convergence to steady-state solutions. In *AIAA paper 93–0880, 31st Aerospace Science Meeting & Exhibit, January 11–14, 1993, Reno, NV*.

Numerical Methods for Hyperbolic Equations – Vázquez-Cendón et al. (eds)
© 2013 Taylor & Francis Group, London, ISBN 978-0-415-62150-2

High order approximations for hyperbolic conservation laws with random initial data

Svetlana Tokareva, Siddhartha Mishra & Christoph Schwab
Seminar for Applied Mathematics, ETH Zurich, Switzerland

ABSTRACT: We construct the high order Stochastic Finite Volume (SFV) method applicable to quantify the uncertainty in hyperbolic conservation laws with random initial data. We assess the efficiency of the constructed method for the numerical solution of the stochastic conservation laws. We apply the SFV method to solve numerically the Riemann problem for the one-dimensional Euler equations with random initial discontinuity location. We develop a theory of the probabilistic shock profiles and prove that the statistical quantities such as the mean, variance and moments are more regular than the deterministic path-wise solution. We show that the same holds true for the coefficients of the Generalized Polynomial Chaos (gPC) expansion. We apply the Stochastic Galerkin method for the numerical solution of the linearized Euler equations equipped with Riemann initial data and demonstrate that higher convergence rates can be obtained for the gPC moments as compared to path-wise simulations.

1 INTRODUCTION

In this paper, we concentrate on the analysis of the stochastic hyperbolic conservation laws of the form

$$\frac{\partial \mathbf{U}}{\partial t}+\sum_{k=1}^{d}\frac{\partial \mathbf{F}_k(\mathbf{U})}{\partial x_k}=0, \mathbf{x}=(x_1,\ldots,x_d), t>0; \qquad (1)$$

with random initial data depending on $\omega \in \Omega$:

$$\mathbf{U}(\mathbf{x},0,\omega)=\mathbf{U}_0(\mathbf{x},\omega), \mathbf{x}\in\mathbb{R}^d, \omega\in\Omega. \qquad (2)$$

Here Ω denotes the set of all elementary events in the probability space $(\Omega,\mathcal{F},\mathbb{P})$ on which the randomness of the initial data in (1)–(2) is modeled, and the random solution $\omega \mapsto \mathbf{U}=[u_1,\ldots,u_p]^T$ is a measurable mapping from F into a suitable space of vector-valued functions, defined on $\mathbb{R}^d\times[0,\infty]$ and taking values in a set of states $\mathbb{S}\subset\mathbb{R}^p$.

The flux functions $\mathrm{F}_k, 1\le k\le d$ in (1) are assumed to be known smooth functions from $\mathbb{S}$ into $\mathbb{R}^p$.

Many efficient numerical methods have been developed to approximate the entropy solutions of systems of conservation laws (Godlewski and Raviart 1995; LeVeque 1992), e.g. finite volume or discontinuous Galerkin methods. The classical assumption in designing efficient numerical methods is that the initial data U_0 is known *exactly*. However, in many practical applications it is not always possible to obtain exact initial data due to, for example, measurement or modeling errors.

In the present paper, we follow (Mishra and Schwab 2012) and describe incomplete information in the initial data (2) mathematically as random fields. Such initial data are described in terms of statistical quantities of interest like the mean, variance, higher statistical moments; in some cases the distribution law of the stochastic initial data is also assumed to be known. In any of these situations one needs a mathematical formulation of (1)–(2) allowing *random initial data*.

A mathematical framework of *random entropy solutions* for scalar conservation laws has been developed in (Mishra and Schwab 2012), where the random entropy solution has been defined and the existence of a unique random entropy solution to (1)–(2) has been proven for scalar hyperbolic conservation laws, also in multiple dimensions. Furthermore, the existence of the statistical quantities of the random entropy solution such as the statistical mean and k-point spatial and temporal correlation functions under suitable assumptions on the random initial data have been proven.

There exist several techniques to quantify the uncertainty (i.e. determine the mean flow and its statistical moments). One of the most well-known approaches in the Monte-Carlo (MC) method, which is based on sampling: a number of realizations of the randomvariable are generated and special estimators are applied to define the mean flow. The main advantage of the MC method is that it is non-intrusive, i.e., existing solvers of the corresponding deterministic problems can be reused directly. Hence, the MC method is very simple to implement. However, a crucial drawback of this method is the order of convergence of 1/2 with respect to the number of samples. A recent improvement of the MC method is the Multi-Level Monte Carlo (MLMC) method (Mishra and Schwab 2012; Mishra et al. 2012), which is build on the idea of application of an adaptive sampling on nested meshes. The MLMC method converges much faster as compared to MC method and is also non-intrusive. Nevertheless, the MLMC estimators quite often lead to the generation of negative variances, which can be critical in real industrial simulations.

A different class of approaches is represented by the Stochastic Galerkin method. It is designed on the representation of the solution in terms of the truncated Generalized Polynomial Chaos (gPC) expansion (Xiu and Karniadakis 2002; Xiu and Karniadakis 2003) over the basis of stochastic polynomials and the application of the Galerkin scheme to the governing equations using the chosen stochastic basis, see also (Troyen et al. 2010a; Troyen et al. 2010b; Ernst et al. 2011).

In this paper, we present a new approach to the uncertainty quantification in the conservation laws, which we refer to as the Stochastic Finite Volume method (SFVM) and which is based on the Finite Volume framework. The SFVM is formulated to solve numerically the system of conservation laws with sources of randomness in both flux coefficients and initial data.

2 STOCHASTIC FINITE VOLUME METHOD

2.1 *SFVM for a general stochastic hyperbolic conservation law*

Consider the hyperbolic system of conservation laws with random flux coefficients

$$\partial_t \mathbf{U} + \nabla_x \cdot \mathbf{F}(\mathbf{U}, \omega) = 0, t > 0; \tag{3}$$

$\mathbf{x} = (x_1, x_2, x_3) \in D_x \subset \mathbb{R}^3$, $\mathrm{U} = [u_1, \ldots, u_p]^T$, $\mathbf{F} = [\mathbf{F}_1, \mathbf{F}_2, \mathbf{F}_3]$ $\mathbf{F}_k = [f_1, \ldots, f_p]^T$, $k = 1,2,3$ and random initial data

$$\mathbf{U}(\mathbf{x}, 0, \omega) = \mathbf{U}_0(\mathbf{x}, \omega), \omega \in \Omega. \tag{4}$$

We parametrize the equations (3)–(4) using the random variable $\mathrm{y} = \mathrm{Y}(\omega)$ which takes values in $D_x \subset \mathbb{R}^q$ and therefore consider the parametric conservation law

$$\partial_t \mathbf{U} + \nabla_x \cdot \mathbf{F}(\mathbf{U}, \mathbf{y}) = 0, \mathbf{x} \in D_x \subset \mathbb{R}^3, t > 0; \tag{5}$$

$$\mathbf{U}(\mathbf{x}, 0, \mathbf{y}) = \mathbf{U}_0(\mathbf{x}, \mathbf{y}), \mathbf{y} \in D_y \subset \mathbb{R}^q. \tag{6}$$

Let $T_X = \cup_{i=1}^{N_x} K_x^i$ be the triangulation of the computational domain D_x in the physical space and $C_y = \cup_{j=1}^{N_y} K_y^j$ be the cartesian grid in the domain D_y of the parametrized probability space.

We further assume the existence of the probability density function $\mu(\mathbf{y})$ and compute the expectation of the n-th solution component of the conservation law (5)–(6) as follows:

$$E[u_n] = \int_{D_y} u_n \mu(\mathbf{y}) d\mathbf{y}, n = 1, \ldots, p$$

The scheme of the Stochastic Finite Volume method (SFVM) can be obtained from the integral form of the equations (5)–(6):

$$\int_{K_y^j} \int_{K_x^i} \partial_t \mathbf{U}\, \mu(\mathbf{y}) d\mathbf{x} d\mathbf{y} + \int_{K_y^j} \int_{K_x^i} \nabla_x \cdot \mathbf{F}(\mathbf{U}, \mathbf{y}) \mu(\mathbf{y}) d\mathbf{x} d\mathbf{y} = 0.$$

Introducing the cell average

$$\bar{\mathrm{U}}_{ij}(t) = \frac{1}{|K_x^i||K_y^j|} \int_{K_y^j} \int_{K_x^i} \mathbf{U}(\mathbf{x}, t, \mathbf{y}) \mu(\mathbf{y}) d\mathbf{x} d\mathbf{y}$$

with the cell volumes

$$|K_x^i| = \int_{K_x^i} d\mathbf{x}, \quad |K_y^j| = \int_{K_y^j} \mu(\mathbf{y}) d\mathbf{y}$$

and performing the partial integration over K_x^i we get

$$\frac{d\bar{\mathrm{U}}_{ij}}{dt} + \frac{1}{|K_x^i||K_y^j|} \int_{K_y^j} \left[\int_{K_X^i} \mathbf{F}(\mathbf{U}, \mathbf{y}) \cdot \mathbf{n}\, dS \right] \mu(\mathbf{y}) d\mathbf{y} = 0$$

Next, we use any standard numerical flux approximation $\hat{\mathbf{F}}(\tilde{\mathbf{U}}_L(\mathbf{x}, t, \mathbf{y}), \tilde{\mathbf{U}}_R(\mathbf{x}, t, \mathbf{y}), \mathbf{y})$ to replace the discontinuous flux through the element interface $\mathbf{F}(\mathbf{U}, \mathbf{y}) \cdot \mathbf{n}$. Here $\tilde{\mathbf{U}}_{L,R}$ denote the boundary extrapolated solution values at the edge of the cell K_x^i, obtained by the high order reconstruction from the cell averages. The complete numerical flux is then approximated by a suitable quadrature rule as

$$\begin{aligned} \bar{\mathbf{F}}_{ij}(t) &= \frac{1}{|K_y^j|} \int_{K_y^j} \left[\int_{K_x^i} \hat{\mathbf{F}}(\tilde{\mathbf{U}}_L, \tilde{\mathbf{U}}_R, \mathbf{y}) \right] \mu(\mathbf{y}) d\mathbf{y} \\ &\approx \frac{1}{|K_y^j|} \sum_{\mathbf{m}} \hat{F}(t, \mathbf{y}_{\mathbf{m}}) \mu(\mathbf{y}_{\mathbf{m}}) w_{\mathbf{m}}, \end{aligned} \tag{7}$$

where we have denoted the flux integral over the physical cell as $\hat{\mathcal{F}}$, $\mathbf{m} = (m_1,\ldots, m_q)$is the multi-index, $\mathbf{y_m}$ and $w_\mathbf{m}$ are quadrature nodes and weights, respectively.

The SFV method then results in the solution of the following ODE system:

$$\frac{d\bar{\mathbf{U}}_{ij}}{dt} + \frac{1}{|K_x^i|}\bar{\mathbf{F}}_{ij}(t) = 0, \tag{8}$$

for all $i = 1,\ldots,N_x, j = 1,\ldots, N_y$. Therefore, to obtain the high-order scheme we first need to provide the high-order flux approximation based, for example, on the ENO/WENO reconstruction in the physical space. Second, we have to guarantee the high-order integration in (12) also by applying the ENO/WENO reconstruction in the stochastic space and choosing the suitable quadrature rule. Finally, we need the high-order time-stepping algorithm to solve the ODE system (13), such as Runge-Kutta method.

2.2 *SFVM for 1D conservation law with random initial data*

Consider the one-dimensional system of conservation laws with random initial data:

$$\frac{\partial \mathbf{U}}{\partial t} + \frac{\partial \mathbf{F}(\mathbf{U})}{\partial x} = 0, x \in (x_L, x_R), t > 0; \tag{9}$$

$$\mathbf{U}(x,0,\omega) = \mathbf{U}_0(x,\omega), \omega \in \Omega. \tag{10}$$

Assume that the random initial condition can be parametrized using one stochastic variable $y = Y(\omega)$ and therefore takes the form

$$\mathbf{U}(x,0,y) = \mathbf{U}_0(x,y), y \in (y_L, y_R). \tag{11}$$

Introduce the uniform grid in both physical and stochastic variables, $x_i = x_L + (i-1/2)\Delta x$, $y_i = y_L + (j-1/2)\Delta y$, and denote the cell averages as

$$\mathbf{U}_{ij}(t) = \frac{1}{\Delta x\,|\Delta y|}\int_{x_{i-1/2}}^{x_{i+1/2}}\int_{y_{j-1/2}}^{y_{j+1/2}} \mathbf{U}(x,t,y)\mu(y)\,dxdy,$$

where the volume of the cell in y-direction is defined as

$$|\Delta y| = \int_{y_{j-1/2}}^{y_{j+1/2}} \mu(y)\,dy.$$

Applying the SFVM scheme (8) to the equations (9), (11) we obtain the following system of ODE with respect to the cell averages

$$\frac{d\mathbf{U}_{ij}(t)}{dt} + \frac{1}{\Delta x}[\mathbf{F}_{i+1/2}(t) - \mathbf{F}_{i-1/2}(t)] = 0, \tag{12}$$

$$\mathbf{U}_{ij}(0) = \frac{1}{\Delta x\,|\Delta y|}\int_{x_{i-1/2}}^{x_{i+1/2}}\int_{y_{j-1/2}}^{y_{j+1/2}} \mathbf{U}_0(x,y)\mu(y)\,dxdy.$$

We use the Gauss quadrature of appropriate order to compute the numerical fluxes in (2):

$$\mathbf{F}_{i+1/2}(t) = \frac{1}{|\Delta y|}\int_{y_{j-1/2}}^{y_{j+1/2}} \mathbf{F}(\mathbf{U}(x_{i+1/2},t,y))\mu(y)\,dy \approx \sum_{m=0}^{M-1} \hat{\mathbf{F}}(\tilde{\mathbf{U}}_{i+1/2}^L(t,y_m), \tilde{\mathbf{U}}_{i+1/2}^R(t,y_m))w_m, \tag{13}$$

where $\hat{\mathbf{F}}(a,b)$ is any standart approximation of the flux (Godunov, Lax-Friedrichs, Rusanov, HLLC flux, etc.) and $\tilde{\mathbf{U}}_{i+1/2}^L(t,y_m)$ and $\tilde{\mathbf{U}}_{i+1/2}^R(t,y_m)$ are the high order reconstructed solution values on the left and right side of $x = x_{i+1/2}$, taken at the quadrature points y_m.

2.3 *Convergence analysis*

We perform the convergence analysis of the SFVM for a simple linear advection equation with uncertain phase initial condition

$$u_t + au_x = 0, \quad x \in (0,1),$$
$$u(x,0) = \sin\big(2\pi(x + 0.1Y(\omega))\big).$$

The random variable $y = Y(\omega)$ is assumed to be distributed uniformly on [0,1].

In Figures 1–4, we plot the $L^1(0,1)$ error for the expectation and the variance of u with respect

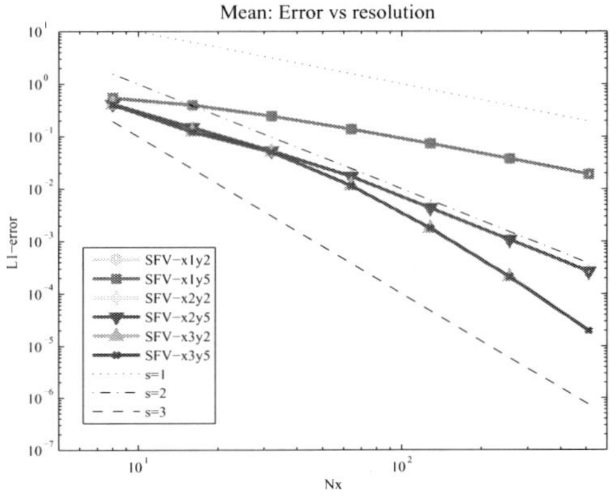

Figure 1. Mean: dependence of the error on the mesh resolution.

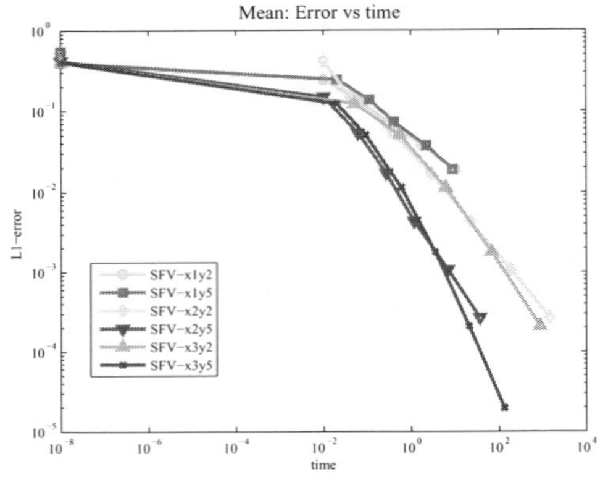

Figure 2. Mean: dependence of the error on the computational time.

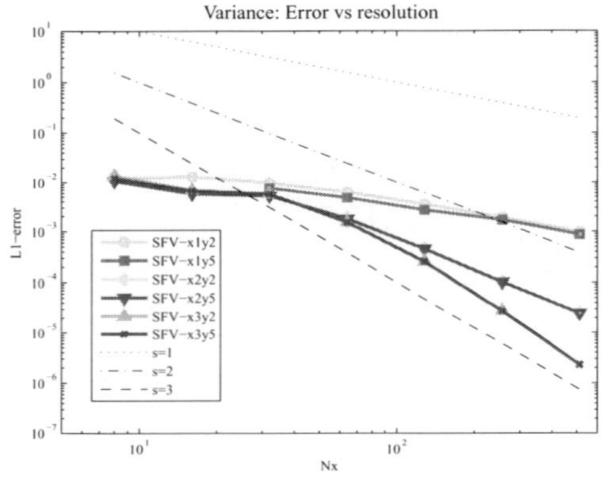

Figure 3. Variance: dependence of the error on the mesh resolution.

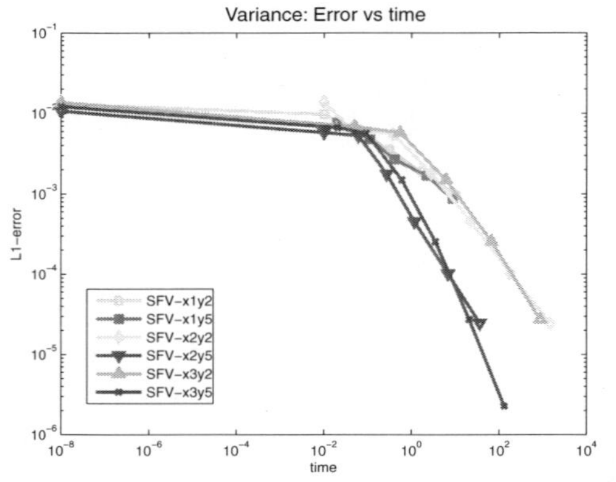

Figure 4. Variance: dependence of the error on the computational time.

to mesh size and the computational time. We investigate the influence of different reconstruction orders in spacial and stochastic variables on the convergence rates and therefore present the convergence plots for the SFVM based on different combinations of ENO/WENO reconstruction in x and y. We compare the SFVM with 1st, 2nd and 3rd order of accuracy in physical space combined with3rd and 5th order reconstruction in stochastic variable. The results show that, while the convergence rate is dominated by the order of accuracy in x , the algorithms with higher order reconstruction in y are more efficient computationally since the same error can be reached with less overall computational time as compared to the lower order reconstruction in y.

2.4 *Stochastic Sod's shock tube problem*

Consider the Riemann problem for the Euler equations

$$\frac{\partial \mathbf{U}}{\partial t}+\frac{\partial \mathbf{F}(\mathbf{U})}{\partial x}=0, \quad x \in (0,2), \tag{14}$$

$$\mathrm{U}(x,0,y)=\mathrm{U}_0(x,y)=\begin{cases} \mathrm{U}_L, \ x<Y(\omega); \\ U_R, \ x>Y(\omega); \end{cases} \tag{15}$$

with $y = Y(\omega)$, $\omega \in \Omega$ and

$$\mathbf{U}=[\rho,\rho u,E]^T, \mathbf{F}=[\rho u,\rho u^2+p,\rho u(E+p)]^T.$$

The initial data is set in primitive variables as

$$\begin{aligned}\mathbf{W}_0(x,\omega) &= [\rho_0(x,\omega),u_0(x,\omega),p_0(x,\omega)]^T \\ &= \begin{cases}[1.0,0.0,1.0] & \text{if} \quad x<Y(\omega), \\ [0.125,0.0,0.1] & \text{if} \quad x>Y(\omega).\end{cases}\end{aligned}$$

We apply the SFVM to solve the system (4)–(6) with $Y(\omega)$ uniformly distributed on [0.95,1.05]. The results are presented in Figures 5–7, in which the solution mean (solid line) as well as mean plus/minus standard deviation (dashed lines) are presented.

The typical deterministic solution of the Sod's shock tube problem with the given initial conditions consists of the left-traveling rarefaction wave and the right-traveling shock wave separated by the contact discontinuity. However, a continuous transition between the intermediate states instead of the discontinuities is observed in the mean flow. This effect is unrelated to the diffusion of the numerical scheme and is due to the smoothing properties of the probabilistic shock profile.

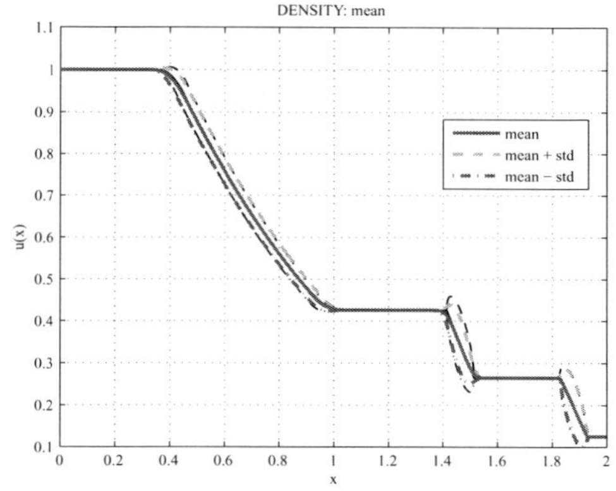

Figure 5. Sod's shock tube problem with random shock location: density.

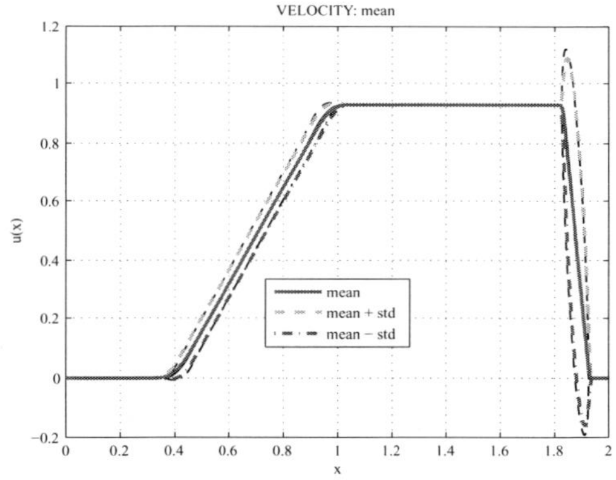

Figure 6. Sod's shock tube problem with random shock location: velocity.

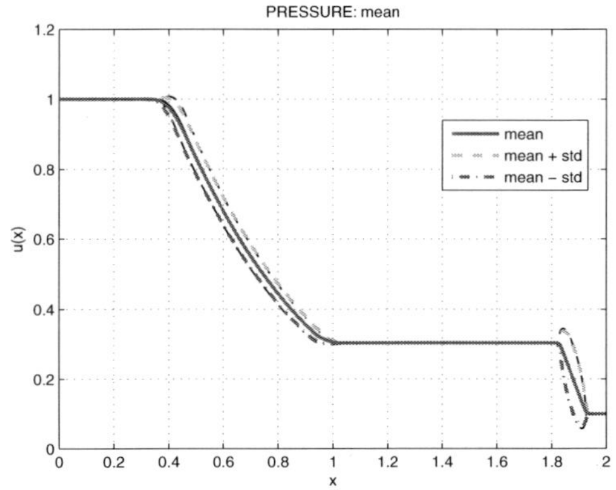

Figure 7. Sod's shock tube problem with random shock location: pressure.

3 REGULARITY OF THE PROBABILISTIC SHOCK PROFILES

3.1 *Regularity analysis*

Consider an arbitrary 1D scalar conservation law with random location initial data:

$$u_t + f_x(u) = 0, (x,t) \in \mathbb{R} \times \mathbb{R}_+,$$

$$u_0(x,\omega) = \begin{cases} u_L, \text{if} & x < Y(\omega); \\ u_R, \text{if} & x > Y(\omega). \end{cases} \tag{16}$$

The exact path-wise solution to (6) can be written as

$$u(x,t,\omega) = u_L + (u_R - u_L)H(x - St - Y(\omega)),$$

where H is the Heaviside step function and

$$S = \frac{f(u_R) - f(u_L)}{u_R - u_L}$$

is the deterministic shock speed. Let $\mu(y)$ be the distribution density function of the random variable $y = Y(\omega)$. The expectation of the solution can be expressed through the cumulative distribution function $P(x)$ as follows:

$$E[u] = u_L + (u_R - u_L)P(x - St).$$

The following theorem holds for the expectation (Schwab and Tokareva 2011).

Theorem. Let $\mu \in (L^1 \cap L^\infty)$. Then the following holds.

1. If $\mu \in C^k_{pw}(\mathbb{R})$, then $\mathbb{E}[u] \in (C^{k+1}_{pw} \cap C^0)(\mathbb{R})$ for any $t > 0$.
2. If $\mu \in C^\infty_{pw}(\mathbb{R})$, then $\mathbb{E}[u] \in (C^\infty_{pw} \cap C^0)(\mathbb{R})$ for any $t > 0$.
3. If $\mu \in C^{k,\alpha}_{pw}(\mathbb{R})$, then $\mathbb{E}[u] \in (C^{k+1,\alpha}_{pw} \cap C^0)(\mathbb{R})$ for any $t > 0$.
4. $\mathbb{E}[u] \in W^{1,\infty}(\mathbb{R})$ for any $t > 0$.

Consider next the random amplitude initial data:

$$u_0(x,\omega) = \begin{cases} u_L + Y(\omega), & \text{if} \quad x < 0; \\ u_R, & \text{if} \quad x > 0. \end{cases}$$

If $u_L + Y(\omega) > u_R$ for all $\omega \in \Omega$, then the exact solution of (6) is a shock wave

$$u(x,t,Y(\omega)) = u_L + Y(\omega) + \\ +(u_R - u_L - Y(\omega))H(x - S(\omega)t),$$

where the random shock wave speed is defined by

$$S(\omega)=\frac{f(u_R)-f(u_L+Y(\omega))}{u_R-u_L-Y(\omega)}.$$

Let the function $\mu(y)$ be the the distribution density function of the random variable $y = Y(\omega)$ and $P(x)$ be its cumulative distribution function, and assume the flux $f(u)$ to be strictly convex. Then the solution mean is

$$\mathbb{E}[u]=u_L+E_1+\int_{-\infty}^{S^{-1}(\frac{x}{t})}P(y)dy+ \\ +\left(u_R-u_L-S^{-1}(\frac{x}{t})\right)P\left(S^{-1}(\frac{x}{t})\right),$$

where

$$E_1\equiv\int_{-\infty}^{\infty}y\mu(y)dy=\text{const}.$$

Theorem. Let the flux $f(u)$ be strictly convex, i.e. $f''(u) > 0$ and the probability density function μ be non-atomic. Then the following holds.

1. If $\mu\in C^k_{pw}(\mathbb{R})$, then $\mathbb{E}[u]\in(C^{k+1}_{pw}\cap C^0)(\mathbb{R})$ for any $t>0$.
2. If $\mu\in C^\infty_{pw}(\mathbb{R})$, then $\mathbb{E}[u]\in(C^\infty_{pw}\cap C^0)(\mathbb{R})$ for any $t>0$.
3. If $\mu\in C^{k,\alpha}_{pw}(\mathbb{R})$, then $\mathbb{E}[u]\in(C^{k+1,\alpha}_{pw}\cap C^0)(\mathbb{R})$ for any $t>0$.
4. If $\mu\in L^\infty(\mathbb{R})$, then $\mathbb{E}[u]\in W^{1,\infty}(\mathbb{R})$ for any $t>0$.

Similar theorems hold for the statistical moments and coefficients of the gPC expansion. As seen in the previous section, the smoothness results also hold for nonlinear systems.

3.2 *Stochastic Galerkin method*

In this section, we describe the construction of the Stochastic Galerkin Finite Volume method (sGFVM) for one-dimensional system of hyperbolic conservation laws.

$$\frac{\partial\mathbf{U}}{\partial t}+\frac{\partial\mathbf{F}(\mathbf{U})}{\partial x}=0, x\in(a,b), t>0; \tag{17}$$

$$\mathbf{U}(x,0,y)=\mathbf{U}_0(x,y), y=Y(\omega)\in D_y\subset\mathbb{R}. \tag{18}$$

The sGFVM aims to approximate the coefficients of the truncated gPC expansion

$$u(x,t,y)=\sum_{k=0}^{N}\mathbf{U}_k(x,t)\varphi_k(y),$$

where $\{\varphi_k\}$, $k = 0,\ldots,N$ is a system of basis polymonials on D_y. Note that these polynomials are not necessarily orthogonal with respect to the probability density function $\mu(y)$.

Multiplying the equations (6)–(7) by the basis function $\varphi_i(y)$ and integrating the result over D_y we obtain

$$\sum_{k=0}^{N}\frac{\partial\mathbf{U}_k}{\partial t}M_{ki}+\frac{\partial}{\partial x}\int_{D_y}\mathbf{F}(\mathbf{U})\varphi_i(y)\mu(y)dy=0, \tag{19}$$

$$\sum_{k=0}^{N}\mathbf{U}_k^{(0)}M_{ki}=\int_{D_y}\mathbf{U}_0(x,y)\varphi_i(y)\mu(y)dy. \tag{20}$$

where $\mathbf{U}_k^{(0)}=\mathbf{U}_k(x,0)$ and M_{ki} are the mass matrix components,

$$M_{ki}=\int_{D_y}\varphi_k(y)\varphi_i(y)\mu(y)dy.$$

The resulting form of the Galerkin system (8)–(9) depends on the concrete form of the flux function $\mathbf{F}(\mathbf{U})$.

3.3 *Convergence analysis for linearized Euler equations*

We apply the sGFVM for the numerical solution of the linearized Euler equations to demonstrate the obtained regularity results (Schwab and Tokareva 2011). In particular, we show the improvement of the convergence rates for the gPC coefficients produced by sGFVM due to the increased regularity of the solution statistics.

Consider the linearized Euler equations with respect to density ρ, velocity u and pressure p of an inviscid gas:

$$\frac{\partial\mathbf{U}}{\partial t}+\mathbf{A}\frac{\partial\mathbf{U}}{\partial x}=0,$$

where

$$\mathbf{U}=\begin{bmatrix}\rho\\ u\\ p\end{bmatrix},\qquad \mathbf{A}=\begin{bmatrix}\bar{u}&\bar{\rho}&0\\ 0&\bar{u}&\frac{1}{\bar{\rho}}\\ 0&\gamma\bar{p}&\bar{u}\end{bmatrix}.$$

We set $\gamma=7/5$, $\bar{\rho}=1$, $\bar{u}=1$ and $\bar{p}=1/\gamma$. Taking in (19) $\mathbf{F}(\mathbf{U}) = \mathbf{A}\mathbf{U}$ we get

$$\sum_{k=0}^{N}\frac{\partial\mathbf{U}_k}{\partial t}M_{ki}+\sum_{k=0}^{N}\mathbf{A}\frac{\partial\mathbf{U}_k}{\partial x}M_{ki}=0.$$

Hence, the sGFVM for linearized Euler equations consists of solving $(N + 1)$ advection equations to determine the coefficients $U_k(x,t)$, $k = 0,\ldots,N$:

$$\frac{\partial \mathbf{U}_k}{\partial t} + \mathbf{A}\frac{\partial \mathbf{U}_k}{\partial x} = 0,$$

$$\sum_{k=0}^{N} \mathbf{U}_k^{(0)} M_{ki} = \int_{D_y} \mathbf{U}_0(x,y)\varphi_i(y)\mu(y)dy.$$

According to the theorem mentioned above, the gPC coefficients are smoother than the path-wise solution of the conservation law with Riemann initial data with uncertain shock location; in particular, the regularity of the coefficients is determined by theregularity of the given probability density function for the initial shock position.

To demonstrate the implications of the obtained regularity results on high-order approximations of gPC coefficients we solve (10)–(11) with the following Riemann initial data:

$$\mathbf{U}_0(x,y) = \begin{cases} \mathbf{U}_L, x < y; \\ \mathbf{U}_R, x > y, \end{cases}$$

where $U_L = [1.0, 0.0, 1.5]^T$, $\mathbf{U}_R = [0.1, 0.0, 0.2]^T$.

Assume $y = Y(\omega)$ is distributed with probability density

$$\mu(\xi) = \frac{1}{A} e^{-\frac{1}{1-\xi^2}},$$

where A is the normalization constant. Note that $\mu \in C^\infty(-1,1)$, therefore, under the theory developed in (Schwab and Tokareva 2011) we expect the improvement of the convergence rates in the computation of the statistical quantities based on the gPC coefficients compared to the approximation of these quantities based on the simulation of a number of path-wise solutions (e.g. in Monte Carlo type methods).

Consider the approximation of the gPC coefficients obtained using the sGFVM with six gPC terms. The convergence plots are shown in Figs. 8–10. In this computation, the 5th order WENO scheme was used to approximate the solution of (10)–(11). We note that the average convergence rate reaches the value of 5 asymptotically, while the path-wise simulations of shock solutions would result in the first order of accuracy as a maximum.

Hence, we have demonstrated how the smoothness of the gPC coefficients for problems with Riemann initial data and random discontinuity position affects the convergence rates of the

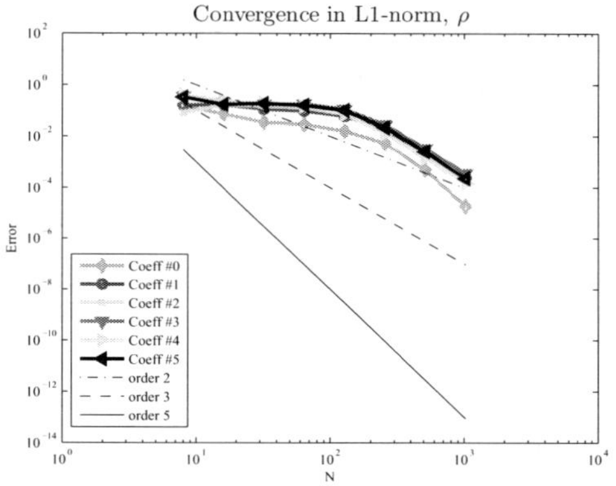

Figure 8. Stochastic Galerkin method for linearized Euler equations: density.

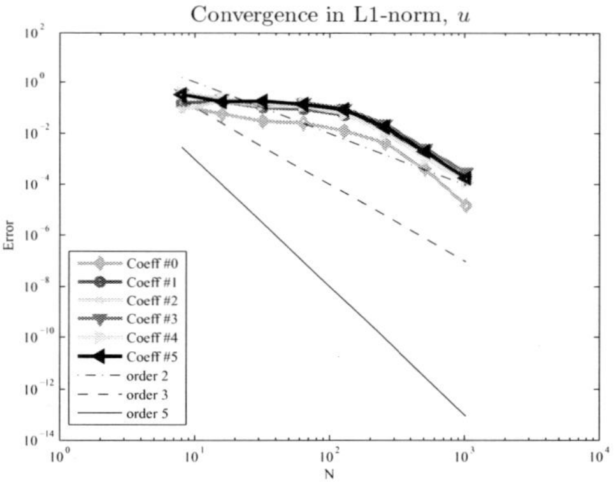

Figure 9. Stochastic Galerkin method for linearized Euler equations: velocity.

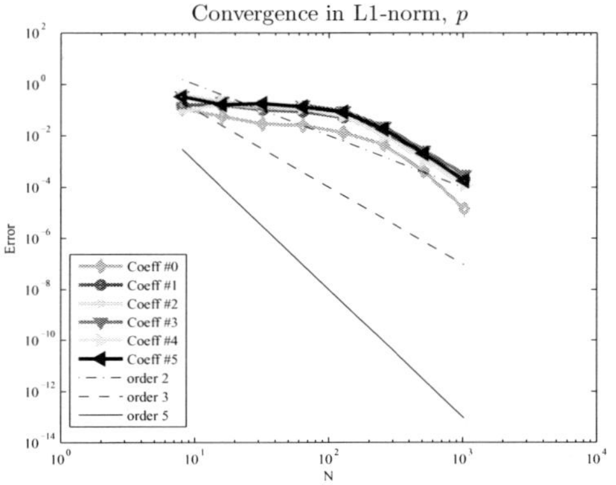

Figure 10. Stochastic Galerkin method for linearized Euler equations: pressure.

sGFVM method, thus making the gPC based computations more efficient as compared to the methods based on the simulation of the path-wise shock solutions.

4 CONCLUSIONS

We have constructed the SFV method for the numerical solution of the conservation laws with random flux coefficients and random initial data. It has been shown that the algorithm based on the high order reconstruction in the stochastic space is needed forbetter computational efficiency of the method.

We have investigated the regularity properties of random entropy solutions to hyperbolic conservation laws with Riemann initial data with either uncertain amplitude or uncertain shock location and proven that the resulting probabilistic shock profiles exhibit additional smoothness related to the probability density function of the distribution law of the random variable in the initial data, except for the special case of an atomic density function. Moreover, regularized solutions are also observed in the case of random flux coefficients. This smoothing effect is unrelated to viscosity and is found even in cases where the path-wise entropy solutions develop shocks. This holds true also for statistical moments like the mean, variance, higher order spatio-temporal correlation functions as well as the coefficients in the gPC expansion. The improved regularity was shown numerically to imply absence of discontinuities in hyperbolic systems obtained by stochastic Galerkin projections onto truncated generalized polynomial chaos expansions of random entropy solutions. This opens the possibility for developing high-order convergent numerical schemes by combining intrusive gPC-type Galerkin discretizations in stochastic space with standard high-order finite volume discretizations in physical space.

REFERENCES

[1] Ernst, O.G., Mugler, A., Starkloff, H.J. and Ullmann, E. (2011). On the convergence of generalized polynomial chaos expansions. *To appear in M2 AN*.

[2] Godlewski, E. and Raviart, P. (1995). *Hyperbolic systems of conservation laws*. Paris: Ellipses Publ.

[3] LeVeque, R. (1992). *Numerical methods for conservation laws*. Birkhaeuser Verlag.

[4] Mishra, S. and Schwab, C. (2012). Sparse tensor multi-level monte carlo finite volume methods for hyperbolic conservation laws with random intitial data. *To appear in Math. Comp.*.

[5] Mishra, S., Schwab, C. and Šukys, J. (2012). Multi-level monte carlo finite volume methods for nonlinear systems of conservation laws in multidimensions. *To appear in J. Comp. Phys.*.

[6] Schwab, C. and Tokareva, S. (2011). High order approximation of probabilistic shock profiles in hyperbolic conservation laws with uncertain initial data. *SAM Research report No. 2011–53*. http://www.sam.math.ethz.ch/reports/2011/53.

[7] Troyen, J., tre, O.L.M., Ndjinga, M. and Ern, A. (2010a). Intrusive galerkin methods with upwinding for uncertain nonlinear hyperbolic systems. *J. Comp. Phys.*, 6485–6511.

[8] Troyen, J., tre, O.L.M., Ndjinga, M. and Ern, A. (2010b). Roe solver with entropy corrector for uncertain hyperbolic systems. *J. Comp. Phys.*, 491–506.

[9] Xiu, D. and Karniadakis, G.E. (2002). Modeling uncertainty in steady state diffusion problems via generalized polynomial chaos. *Comput. Methods Appl. Mech. Engrg*, 4927–4948.

[10] Xiu, D. and Karniadakis, G.E. (2003). Modeling uncertainty in flow simulations via generalized polynomial chaos. *J. Comp. Phys.*, 137–167.

VII Finite volume and discontinous Galerkin schemes for stiff source term problems

Numerical Methods for Hyperbolic Equations – Vázquez-Cendón et al. (eds)
© 2013 Taylor & Francis Group, London, ISBN 978-0-415-62150-2

Some comments on the numerical approximation of hyperbolic nonconservative systems

M.L. Muñoz-Ruiz
Departamento Matemática Aplicada, Universidad de Málaga, Spain

C. Parés
Departamento Análisis Matemático, Universidad de Málaga, Spain

ABSTRACT: This work deals with the numerical approximation of nonconservative hyperbolic systems, which has become a very active front of research in recent years, as this class of PDE systems arise in many flow models. We are particularly interested in the design of high-order well-balanced shock-capturing numerical methods for nonconservative hyperbolic problems. We expect the approximated solutions to be consistent with the physics of the real flows to be simulated, in the following sense: on the one hand, they should satisfy the conservation properties prescribed by the physics of the problem and, on the other hand, their discontinuities should satisfy some jump conditions consistent with the real phenomena to be simulated. For the case of conservative systems, it is possible toconstruct numerical schemes satisfying these two properties. The question addressed here is whether or not it is possible to construct numerical schemes satisfying these two requirements for nonconservative problems, as some important difficulties arise in certain nonconservative cases. In this paper we will discuss these difficulties and state some conclusions.

1 INTRODUCTION

In this work we deal with the numerical approximation of a hyperbolic system of PDE of the form:

$$w_t + f(w)_x + \mathcal{B}(w)w_x = S(w)\sigma_x, \quad x \in \mathbb{R}, t > 0, \quad (1)$$

where the unknown $w(x,t)$ takes values in an open convex set $\mathcal{O}$ of $\mathbb{R}^N$; f is a regular function from $\mathcal{O}$ to $\mathbb{R}^N$; $\mathcal{B}$ is a regular matrix function from $\mathcal{O}$ to $\mathcal{M}_{N\times N}(\mathbb{R})$; S, a function from $\mathcal{O}$ to $\mathbb{R}^N$; and $\sigma(x)$, a known function from $\mathbb{R}$ to $\mathbb{R}$.

System (1) includes as particular cases systems of conservation laws, when $\mathcal{B} = 0$ and $S = 0$, as well as systems of balance laws, when $\mathcal{B} = 0$. A number of models having this form have been introduced in fluid dynamics to serve as simplified flow models. The system of partial differential equations governing the one-dimensional flow of two superposed immiscible layers of shallow water fluids through a straight channel with constant rectangular cross-section (see (Castro et al., 2001)) is a particular case of (1). These systems also appear in models of turbulent shallow water, two-phase flows, sediment transport, turbidity currents, avalanches, submarine avalanches, etc.

We consider an initial condition

$$w(x,0) = w_0(x), \quad x \in \mathbb{R}, \quad (2)$$

and we study the Cauchy problems (1)–(2).

System (1) can be considered as a particular case of the family of PDE systems

$$W_t + \mathcal{A}(W)W_x = 0, \quad x \in \mathbb{R}, t > 0, \quad (3)$$

in which the unknown $W(x,t)$ takes values in an open convex set Ω of $\mathbb{R}^M$, and $\mathcal{A}$ is a smooth locally bouded map from Ω to $\mathcal{M}_{M\times M}(\mathbb{R})$. In effect, following (LeFloch 1989), the artificial unknown σ and the equation

$$\sigma_t = 0 \quad (4)$$

are added to (1), so the augmented system can be written in the form (3) with $M = N + 1$,

$$W = \begin{bmatrix} w \\ \sigma \end{bmatrix} \in \Omega = \mathcal{O} \times \mathbb{R},$$

and $\mathcal{A}(W)$ the matrix-valued function whose block structure is given by

$$\mathcal{A}(W) = \left[\begin{array}{c|c} Df(w) + \mathcal{B}(w) & -S(w) \\ \hline 0 & 0 \end{array}\right], \quad (5)$$

where $Df(w)$ is the Jacobian matrix of f. It can be easily verified that (1)–(2) is equivalent to (3) with initial condition

$$W(x,0) = W_0(x) = \begin{bmatrix} w_0(x) \\ \sigma(x) \end{bmatrix}, \quad x \in \mathbb{R}. \tag{6}$$

The appearance of discontinuities in the solution even for smooth initial conditions is a common feature for systems of the form (3). These discontinuities are related to real phenomena in which some of the variables present a sharp variation in a very thin region, as it is the case of shocks of transonic airplanes, or hydraulic jumps for free surface water flows.

The design of numerical methods with good properties for problems of the form (1) or the particular cases corresponding to $\mathcal{B} = 0$ and/or $S = 0$ constitutes a very active front of research. The following properties are usually required to the numerical schemes in order to obtain reasonable numerical solutions:

- The solutions provided by the numerical schemes have to be accurate approximations of the solutions in the smoothness regions.
- The numerical solutions have to sharply capture the discontinuities of the solution, avoiding unphysical oscillations near them.
- The smooth nontrivial stationary solutions (or at least a family of them) have to be exactly or accurately enough approximated in order to avoid unphysical oscillations near equilibria.

Numerical schemes satisfying these properties are called high-order well-balanced shock-capturing methods. In recent years, several methods with good properties for hyperbolic systems with source terms and/or nonconservative products have been developed.

The approximated solutions must also satisfy the following properties, in order to be consistent with the physics of the real flows to be simulated:

(P1) The numerical solutions have to satisfy all the conservation properties prescribed by the physics of the problem.

(P2) The appearance and the propagation of discontinuities have to be consistent with the real phenomena that is simulated.

Property (P1) may seem paradoxical for nonconservative systems, but notice that system (1) may contain a conservative subsystem. This is the case, for instance, for the two mass conservation equations in the two-layer shallow-water system. The numerical solutions are thus expected to satisfy the mass conservation principle.

For the particular case of conservative systems, conservative methods meet the two requirements (P1) and (P2). Some well-known examples of conservative methods are centered scheme, Lax-Friedrichs method, Godunov method, and Roe method (for a review of these schemes and references, see (Toro 1999). Unfortunately, for general nonconservative systems (3), to determine whether or not it is possible to design numerical schemes having both properties is a difficult task. In the following we will discuss which are the main difficulties involved.

2 WEAK SOLUTIONS OF NONCONSERVATIVE HYPERBOLIC SYSTEMS

We consider first order quasi-linear PDE systems

$$w_t + \mathcal{A}(w) w_x = 0, \quad x \in \mathbb{R}, t > 0, \tag{7}$$

in which the unknown $w(x,t)$ takes values in an open convex set $\mathcal{O}$ of $\mathbb{R}^N$, and $\mathcal{A}$ is a smooth locally bounded map from $\mathcal{O}$ to $\mathcal{M}_{N\times N}(\mathbb{R})$. The system is supposed to be strictly hyperbolic and the characteristic fields $R_i(w)$, $i = 1,\ldots,N$, are supposed to be either genuinely nonlinear:

$$\nabla\lambda_i(w)\cdot R_i(w) \neq 0, \quad \forall\, w \in \mathcal{O},$$

or linearly degenerate:

$$\nabla\lambda_i(w)\cdot R_i(w) = 0, \quad \forall\, w \in \mathcal{O},$$

being $\lambda_1(w),\ldots,\lambda_N(w)$ the eigenvalues of $\mathcal{A}(w)$ (in increasing order) and $R_1(w),\ldots,R_N(w)$ a set of associated eigenvectors.

The formulation (3) of a system of the form (1) is a particular case of (7).

The first difficulty to define weak solutions for system (7) comes from the fact that the usual procedure to obtain a variational formulation (multiply by a regular enough test function and then integrate by parts) does not allow one to pass all the derivatives to the test function.

Hence, we try the alternative way that consists in obtaining the integral equation satisfied by a smooth solution in an arbitrary rectangle $[a,b] \times [t_0,t_1] \subset \mathbb{R} \times [0,\infty]$:

$$\int_a^b w(x,t_1)\,dx = \int_a^b w(x,t_0)\,dx - \int_{t_0}^{t_1}\int_a^b \mathcal{A}(w(x,t))\,w_x(x,t)\,dx\,dt. \tag{8}$$

Nevertheless, a new difficulty arises: the integrand of the last term is not defined for discontinuous functions w. At a discontinuity, the product $\mathcal{A}(w)w_x$ is expected to produce a Dirac measure whose mass must be related with the jumps of both w and $\mathcal{A}(w)$. Unfortunately, the expression of such a measure is not determined in all the cases.

In order to give a sense to these integrals for discontinuous functions we use the theory introduced by Dal Maso, LeFloch, and Murat (Dal Maso et al., 1995). Using this theory, the nonconservative product $\mathcal{A}(w)w_x$ can be defined as a bounded measure for functions w with bounded variation, under some hypotheses of regularity for $\mathcal{A}$, and provided a family of Lipschitz continuous paths $\Phi:[0,1]\times\mathcal{O}\times\mathcal{O}\rightarrow\mathcal{O}$ is prescribed. This family of paths must satisfy certain regularity and compatibility conditions. In particular,

$$\Phi(0;w_l,w_r)=w_l, \qquad \Phi(1;w_l,w_r)=w_r, \tag{9}$$

and

$$\Phi(s;w,w)=w. \tag{10}$$

For a rigorous and complete presentation of this theory the interested reader is addressed to (Dal Maso et al., 1995). Here, the family of paths will be just understood as a tool to give a sense to integrals of the form

$$\int_a^b \mathcal{A}(v(x))v_x(x)dx,$$

for functions v with jump discontinuities. More precisely, given a bounded variation function $v:[a,b]\rightarrow\mathbb{R}$, we define

$$\begin{aligned}⨍_a^b \mathcal{A}(v(x))v_x(x)dx &= \int_a^b \mathcal{A}(v(x))v_x(x)dx\\ &+\sum_m \int_0^1 \mathcal{A}(\Phi(s;v_m^-,v_m^+))\frac{\partial\Phi}{\partial s}(s;v_m^-,v_m^+)ds,\end{aligned} \tag{11}$$

where v_m^- and v_m^+ represent, respectively, the limits of v to the left and right of its mth discontinuity (remember that the set of discontinuities of a bounded variation function is countable). Observe that the family of paths has been used in (11) to determine the Dirac measures placed at the discontinuities of v.

According to this meaning for the integral, a weak solution of the system is defined as a bounded variation function satisfying

$$\begin{aligned}\int_a^b w(x,t_1)dx &= \int_a^b w(x,t_0)dx\\ &-\int_{t_0}^{t_1} ⨍_a^b \mathcal{A}(w(x,t))w_x(x,t)dx\,dt,\end{aligned} \tag{12}$$

for every $[a,b]\times[t_0,t_1]\subset\mathbb{R}\times[0,\infty]$. Across a discontinuity, weak solutions are shown to satisfy the generalized Rankine-Hugoniot condition

$$\xi[w]=\int_0^1 \mathcal{A}(\Phi(s;w^-,w^+))\frac{\partial\Phi}{\partial s}(s;w^-,w^+)ds, \tag{13}$$

where ξ is the speed of propagation of the discontinuity, w^- and w^+ are the left and right limits of the solution at the discontinuity, and $[w]w^+ - w^-$. In the particular case of a system of conservation laws, that is, when $\mathcal{A}(w)$ is the Jacobian matrix for some function $f(w)$, (13) reduces to the standard Rankine-Hugoniot condition

$$\xi[w]=[f(w)]. \tag{14}$$

As it occurs in the conservative case, a weak solution for system (7) with initial condition (2) is not necessarily unique, so a concept of entropy is required. Given an entropy pair (η,g), i.e. a pair of regular functions from $\mathcal{O}$ to $\mathbb{R}$, η being convex, such that

$$\nabla g(w)=\nabla\eta(w)\cdot A(w), \quad \forall w\in\mathcal{O},$$

a weak solution is said to be an entropy solution if it satisfies the inequality

$$\eta(w)_t+g(w)_x\leq 0 \tag{15}$$

in the distributions sense. Any smooth solution satisfies the conservation law

$$\eta(w)_t+g(w)_x=0, \quad x\in\mathbb{R}, t>0, \tag{16}$$

exactly, and it can be verified again that a piecewise continuous solution is an entropy solution if the jump condition

$$\xi[\eta(w)]+[g(w)]\leq 0 \tag{17}$$

is satisfied across a discontinuity.

A conservation property satisfied by weak solutions of (7) with initial condition (2) is obtained by taking $a=-\infty$ and $b=\infty$ in (12):

$$\begin{aligned}\int_R w(x,t_1)dx &= \int_R w(x,t_0)dx\\ &-\int_{t_0}^{t_1} ⨍_{\mathbb{R}} \mathcal{A}(w(x,t))w_x(x,t)dx\,dt,\end{aligned} \tag{18}$$

for every $[t_0,t_1]\subset[0,\infty]$

It can also be shown that the Riemann problem composed by (7) and the initial condition

$$w_0(x)=\begin{cases} w_l & \text{if} \quad x<0,\\ w_r & \text{if} \quad x>0,\end{cases} \tag{19}$$

has a unique self-similar weak solution, when $|w_r - w_l|$ is sufficiently small, composed by at most N simple waves (rarefaction waves, contact discontinuities or shock waves).

Unfortunately, the concept of weak solution depends on the family of paths, which is a priori arbitrary. The crucial question is thus how to choose the 'good' family of paths. In fact, when the hyperbolic system is the vanishing-viscosity limit of the parabolic problems

$$w_t^\varepsilon + A(w^\varepsilon)w_x^\varepsilon = \varepsilon(R(w^\varepsilon)w_x^\varepsilon)_x, \tag{20}$$

where the second order term is elliptic (for instance R may be a constant symmetric positive defined matrix), the adequate family of paths should be related to the viscous profiles. Let us suppose that the existence of a viscous profile is adopted as a criterion of admissibility for a discontinuity linking the states w^-, w^+ at speed ξ. In this case, a viscous profile is a travelling wave

$$w^\varepsilon(x,t) = v\left(\frac{x-\xi t}{\varepsilon}\right), \tag{21}$$

which is a solution of (20) satisfying

$$\lim_{\chi\to\pm\infty} v(\chi) = w^\pm, \quad \lim_{\chi\to\pm\infty} v'(\chi) = 0. \tag{22}$$

It can be easily verified that v has to solve now the equation

$$-\xi v' + \mathcal{A}(v)v' = (\mathcal{R}(v)v')'. \tag{23}$$

By integrating (23) from $-\infty$ to ∞ and taking into account (22), we obtain the jump condition

$$\xi[w] = \int_{-\infty}^{\infty} \mathcal{A}(v(\chi))v'(\chi)d\chi. \tag{24}$$

Comparing this jump condition with (13), it seems clear that, in this case, the good choice for the path connecting the states w^- and w^+ would be, after a reparameterization, the viscous profile v. For instance, the path

$$\Phi(s;w^-,w^+) = v\big(\tan(\pi(s-1/2))\big), \quad s\in[0,1],$$

(or any other parameterization, as the definition of weak solutions is invariant for reparametrizations of the paths) would be a natural choice. Every choice of viscous term $\mathcal{R}$ leads now to different jump conditions, while for conservative systems the Rankine-Hugoniot conditions (14) are always recovered independently of the choice of the viscous term. Unfortunately, the computation of viscous profiles for complex hyperbolic systems is far from being an easy task and thus the computation of a family of paths based on them can be very difficult in practice.

3 PATH-CONSERVATIVE METHODS

Let us denote now by w_i^n the approximation at the cell I_i at time t_n of the cell average of the exact solution. The initial cell values are given by

$$w_i^0 = \frac{1}{\Delta x}\int_{x_{i-1/2}}^{x_{i+1/2}} w_0(x)dx. \tag{25}$$

The following equality can be deduced from (12):

$$\frac{1}{\Delta x}\int_{x_{i-1/2}}^{x_{i+1/2}} w(x,t_{n+1})dx = \frac{1}{\Delta x}\int_{x_{i-1/2}}^{x_{i+1/2}} w(x,t_n)dx$$
$$-\frac{\Delta t}{\Delta x}\frac{1}{\Delta t}\int_{t_n}^{t_{n+1}} ⨍_{x_{i-1/2}}^{x_{i+1/2}} \mathcal{A}(w(x,t))w_x(x,t)dx\,dt.$$

In order to define a numerical scheme we think in a formula such that

$$w_i^{n+1} = w_i^n - \frac{\Delta t}{\Delta x}⨍_{x_{i-1/2}}^{x_{i+1/2}} \mathcal{A}(w^n(x))w_x^n(x)dx, \tag{26}$$

where w^n represents the piecewise constant function whose value at the cell I_i is the approximation w_i^n. But now the meaning of the weak integral in (26) is ambiguous: as w^n is piecewise constant, its weak integral only consists of the Dirac measures placed at the boundaries of the cells:

$$⨍_{-\infty}^{\infty} \mathcal{A}(w^n(x))w_x^n(x)dx$$
$$= \sum_i \int_0^1 A(\Phi(s;w_i^n,w_{i+1}^n))\frac{\partial\Phi}{\partial s}(s;w_i^n,w_{i+1}^n)ds. \tag{27}$$

What is thus the meaning of the restriction of this integral to the cell I_i? Should the Dirac measure placed in $x_{i+1/2}$ contribute to the weak integral in the cell I_i or to that in its neighbor I_{i+1}? The general idea is to decompose the total mass of this Dirac measure into two summands $D_{i+1/2}^\pm$, one contributing to the cell I_i and the other to the cell I_{i+1}. This idea leads to the following definition:

Definition 3.1 Given a family of paths Φ, a numerical scheme is said to be Φ-conservative if it can be written in the form:

$$w_i^{n+1} = w_i^n - \frac{\Delta t}{\Delta x}(D_{i-1/2}^{n,+} + D_{i+1/2}^{n,-}), \tag{28}$$

where

$$D^{n,\pm}_{i+1/2} = D^{\pm}(w^n_{i-q},\dots,w^n_{i+p}),$$

D^- and D^+ being two Lipschitz continuous functions from $\mathcal{O}^{p+q+1}$ to $\mathcal{O}$ satisfying

$$D^{\pm}(w,\dots,w) = 0, \quad \forall w \in \mathcal{O}, \tag{29}$$

and

$$D^-(w_{-q},\&,w_p) + D^+(w_{-q},\dots,w_p) = \int_0^1 \mathcal{A}(\Phi(s;w_0,w_1))\frac{\partial\Phi}{\partial s}(s;w_0,w_1)\,ds, \tag{30}$$

for every set $\{w_{-q},\dots, w_{-p}\}\subset\mathcal{O}$.

The concept of path-conservative numerical scheme generalizes that of conservative scheme. In effect, if the system is conservative, then every path-conservative numerical scheme is equivalent to a conservative scheme with numerical flux

$$F(w_{-q},\&,w_p) := D^-(w_{-q},\&,w_p) + f(w_0) := -D^+(w_{-q},\&,w_p) + f(w_1),$$

where, due to (30), the two definitions of the numerical flux coincide. Conversely, a conservative numerical scheme is path-conservative for any family of paths: notice that by adding and subtracting $f(w_i)$ in

$$w_i^{n+1} = w_i^n - \frac{\Delta t}{\Delta x}(F^n_{i+1/2} - F^n_{i-1/2}), \tag{31}$$

where $F^n_{i+1/2}$ is some approximation of the averaged flux through $x_{i+1/2}$,

$$F^n_{i+1/2} \cong \frac{1}{\Delta t}\int_{t_n}^{t_{n+1}} f(w(x_{i+1/2},t))\,dt,$$

a conservative scheme can be written in the form (28) by defining:

$$D^-(w_{-q},\dots,w_p) := F(w_{-q},\dots, w_p) - f(w_0),$$
$$D^+(w_{-q},\dots,w_p) := f(w_1) - F(w_{-q},\dots,w_p).$$

It can be trivially verified using these definitions that (29) and (30) are satisfied for any family of paths.

Path-conservative numerical schemes satisfy a (P1) property: if a Φ-conservative scheme is applied to the initial cell values (25), the equality

$$\int_{\mathbb{R}} w^{n+1}(x)dx = \int_{\mathbb{R}} w^n(x)dx - \Delta t \int_{\mathbb{R}} \mathcal{A}(w^n(x))w^n_x(x)dx, \tag{32}$$

(which is the discrete analogous to (18)), can be obtained by summing up in (28) and taking (30) into account. In particular, it can be easily verified that, if the nonconservative system (7) has a conservative subsystem, a path-conservative numerical scheme is conservative in the usual sense for that subsystem.

Some examples of path-conservative numerical schemes that extend to nonconservative systems very well known examples of conservative methods (in the sense that, if the system is conservative, they are equivalent to their conservative counterpart independently of the choice of the family of paths Φ) are: centered scheme, Lax-Friedrichs method (as well as the Rusanov or local Lax-Friedrichs method), Godunov method, and Roe method (see (Parés & Muñoz-Ruiz 2009) for details) for details).

If the nonconservative system (7) comes from a PDE system (1), it is always possible to rewrite any of these methods in a closer form to the original formulation of the problem, in which some discretizations of the flux, the source terms, and the nonconservative products appear explicitly. The advantage of this global formulation is that it ensures the stability under usual CFL conditions and that it makes easier the analysis of the well-balanced properties (see (Parés & Castro 2004), (Parés 2006)).

As it occurs in the conservative case, first order path-conservative numerical schemes can be extended to high-order either by using reconstruction operators (see (Castro et al., 2006), (Parés 2006)), discontinuous Galerkin methods (see (Shu 2006)) or ADER techniques (see (Dumbser et al., 2008)).

3.1 *Convergence property*

Concerning property (P2), the following result can be obtained (see (Castro et al., 2008)): consider a nonconservative hyperbolic system (7) together with a given family of paths Φ. Consider also a Φ-conservative method. Let $\{w^{\Delta x}\}$ be a sequence of piecewise constant approximated solutions generated by the scheme that satisfies the inequality

$$\sup_{\mathbb{R}} |w^{\Delta x}(\cdot,t)| + TV_{\mathbb{R}}(w^{\Delta x}(\cdot,t)) \le const. \tag{33}$$

uniformly in time. Let us also suppose that $w^{\Delta x}$ converges to a function *w uniformly in the sense of graphs* (see (Dal Maso et al., 1995) for a review on this concept of convergence). Then it can be shown that w is a weak solution of the system according to the family of paths Φ and thus its discontinuities satisfy the jump conditions (13). To ensure that w is also an entropy solution, an extra requirement has to be imposed to the numerical scheme: the discrete entropy inequality

$$\eta(w_i^{n+1}) \le \eta(w_i^n) - \frac{\Delta t}{\Delta x}(G_{i+1/2}^n - G_{i-1/2}^n) \qquad (34)$$

is enough, as it occurs in the conservative case. In particular, Godunov and Lax-Friedrichs methods satisfy such an inequality under the condition

$$\Delta t \max_{i,n}\{|\lambda_j(w_i^n)|, j=1,\ldots,N\} = cfl \cdot \Delta x, \qquad (35)$$

with $cfl \in (0,1/2)$ (see (Muñoz-Ruiz & Parés 2007) and (Castro et al., 2010) respectively) but Roe methods also require the use on an entropy-fix technique (see (Harten & Hyman 1983), (Castro et al., 2010)).

This theoretical result seems to give a positive answer to the question about whether or not path-conservative schemes satisfy a (P2) property, but in practice this property may fail.

The reason is that, although this notion of convergence is weaker than the usual uniform convergence, it is still too strong to expect the numerical solutions provided by a finite difference-type scheme to converge in this sense. If the sequence $w^{\Delta x}$ is assumed to converge in a weaker but more realistic sense (as it is the case in the Lax-Wendroff convergence theorem forconservative methods and conservative systems), a convergence result cannot be established for general problems. For instance, if $w^{\Delta x}$ converges almost everywhere to a function w, by extending the arguments in Hou and LeFloch (Hou & LeFloch 1994) and using the stability results established in (Dal Maso et al., 1995), it has been proved in (Castro et al., 2008) that the limit function w solves the following hyperbolic system with source term

$$w_t + [A(w)w_x]_\Phi = \nu_w, \qquad (36)$$

where ν_w is a bounded measure supported on the discontinuities of w, called the *convergence error measure*. Due to the appearance of this measure, the limit function will be a classical solution of (7) in $C(w)$ but itsdiscontinuities could not satisfy the generalized Rankine-Hugoniot conditions (13) corresponding to the family of paths Φ.

What are then the Rankine-Hugoniot conditions satisfied at the discontinuities of the limit function? The modified equations are useful to answer this question. For simplicity, let us consider the path-conservative Lax-Friedrichs scheme defined by

$$\begin{aligned} D_{LF}^{\pm}(w_l, w_r) &= \frac{1}{2}\int_0^1 A(\Phi(s;w_l,w_r))\frac{\partial \Phi}{\partial s}(s;w_l,w_r)\,ds \\ &\qquad \pm \frac{\Delta x}{2\Delta t}(w_r - w_l), \end{aligned} \qquad (37)$$

which can also be rewritten as follows:

$$\begin{aligned} &\frac{1}{\Delta t}\left(w_i^{n+1} - \frac{1}{2}\left(w_{i-1}^n + w_{i+1}^n\right)\right) \\ &+\frac{1}{2\Delta x}\Big(\int_0^1 A\left(\Phi(s;w_{i-1}^n,w_i^n)\right)\frac{\partial \Phi}{\partial s}(s;w_{i-1}^n,w_i^n)\,ds \\ &+\int_0^1 A\left(\Phi(s;w_i^n,w_{i+1}^n)\right)\frac{\partial \Phi}{\partial s}(s;w_i^n,w_{i+1}^n)\,ds\Big) = 0. \end{aligned}$$

A formal Taylor expansion shows that the numerical solutions solve up to second order the modified equation:

$$\begin{aligned} w_t + A(w)w_x &= \frac{\Delta x^2}{2\Delta t}\Bigg(w_{xx} - \frac{\Delta t^2}{\Delta x^2}\left(A^2(w)w_x\right)_x \\ &\quad - \frac{\Delta t^2}{\Delta x^2}\Big(DA(w)\left(A(w)w_x, w_x\right) \\ &\quad - DA(w)\left(w_x, A(w)w_x\right)\Big)\Bigg) \\ &\quad - \frac{\Delta x}{2}I_1(w), \end{aligned} \qquad (38)$$

where

$$\begin{aligned} I_1(w) &= \int_0^1 DA(w)\left(D_{w_l}\Phi \cdot w_x, D_{w_l}\Phi_s \cdot w_x\right)ds \\ &\quad + \int_0^1 DA(w)\left(D_{w_r}\Phi \cdot w_x, D_{w_r}\Phi_s \cdot w_x\right)ds. \end{aligned}$$

The discontinuities of the limit function will then satisfy a generalized Rankine-Hugoniot condition (13) corresponding to a family of paths Ψ related to the viscous profiles of this modified equation. Even if the hyperbolic model is the vanishing viscosity limit of a family of parabolic problems (20) and the family of paths has been constructed on the basis of the viscous profiles of these regularized problems, in general it is not expectable that both the viscous profiles of the modified equation and the regularized problems coincide.

Moreover, as the second order terms of the modified equations depend both on the chosen family of paths Φ (notice that in the case of the Lax-Friedrichs scheme Φ only appears in the term

$I_1(w)$) and on the specific form of the viscous terms of the numerical scheme, different numerical schemes based on a same family of paths may produce limit functions satisfying different jump conditions.

In fact, this difficulty potentially affects to any method having some numerical viscosity, which are almost all. Random choice methods, as Glimm or the front tracking schemes are viscosity free and, in effect, it can be proved that the numerical solutions provided by these methods converge uniformly in the sense of graphs (see (LeFloch & Liu 1993)). An example of such a method is the following: once the approximations w_i^n at time t_n have been obtained, the new approximations are given by

$$w_i^{n+1} = \tilde{u}(x_{i-1/2} + \theta\Delta x, t_{n+1}),$$

where $\tilde{u}$ is the entropy solution of the problem

$$\begin{cases} \tilde{u}_t + \mathcal{A}(\tilde{u})\tilde{u}_x = 0, & x \in \mathbb{R}, t > t_n, \\ \tilde{u}(x,t_n) = w^n(x), & x \in \mathbb{R}, \end{cases} \tag{39}$$

and θ is a random number in [0,1] If again the CFL condition (35) is imposed with $cfl \in (0,1/2)$, $\tilde{u}$ can be obtained by solving the Riemann problems at the intercells. Notice that the average stage of Godunov-type methods, which is the source of the numerical viscosity, is avoided. Nevertheless, due to the random stage, this method will not satisfy a global property (P1). Moreover, its implementation may be very difficult or very costly in practice, as the exact solutions of Rieman problems have to be explicitly known.

In certain special situations, the convergence error measure is found to vanish identically. This is the case for systems of balance laws if the family of paths satisfies some compatibility condition with the integral curves of the linearly degenerate characteristic fields. In particular, if the path connecting two states that belong to the same integral curve of a linearly degenerate field is a parametrization of the arc of that integral curve connecting both states, then all of the discontinuities are correctly approximated and the scheme does converge to exact solutions (see (Muñoz-Ruiz & Parés 2011) for details). To characterize the nonconservative systems for which this convergence error vanishes would be an important issue.

Nevertheless, the construction of numerical schemes having both properties (P1) and (P2) seems to be a difficult task. (P1), indeed, usually requires an averaging stage. This implies the appearance of certain numerical diffusion, that affects to the viscous profiles captured by the numerical scheme and thus to the jump conditions satisfied by the limits of numerical solutions. In spite of it, for some particular problems the numerical results given by a path-conservative scheme may be acceptable even when the convergence error measure is present.

4 CONCLUDING REMARKS

When a hyperbolic system with nonconservative products and genuinely nonlinear fields is discretized, we should take the following steps in order to be sure that the numerical approximations converge to a function which is a classical solution where it is smooth and whose discontinuities are in good agreement with the physics of the problem: choose a regularization of the system which is consistent with the physics of the problem, determine the family of paths consistent with this regularization, and, finally, design a numerical scheme whose solutions converge to weak solutions associated with this family of paths.

Nevertheless, this strategy may be difficult to follow in practice, since the actual calculation of a family of paths requires calculating viscous profiles. On the other hand, the convergence of the numerical solutions to the correct weak solutions is known for random choices method only, but these methods do not satisfy good conservation properties and their implementation can be difficult and time consuming since they require the explicit knowledge of the corresponding Riemann solver. In fact, when the nonconservative model under consideration is a simplified version of a more complex (but conservative) model, the above strategy may end up being more costly than solving directly the more complex problem. In these cases, the use of a numerical strategy based on a direct discretization of the nonconservative system by means of a finite difference scheme which is path-conservative is advisable and may have the following advantages:

- The numerical solutions satisfy global properties similar to those satisfied by the exact weak solutions.
- The approximations of the shocks provided by the schemes are consistent with a regularization of the system with higher order terms that vanish as Δx tends to 0.
- In addition, this strategy is extendable to high-order methods or to multidimensional problems.

Obviously, the main drawback is that the actually used regularization depends both on the chosen family of paths and on the numerical scheme itself. Nevertheless, the following facts should also be considered:

- The convergence error is not present in every nonconservative system.

- Even when the convergence error is present, it may only be noticeable for very fine meshes, for discontinuities of great amplitude, and/or for large-time simulations. Otherwise, it may be masked by the discretization errors.
- The convergence error should also be compared with the experimental error: in practice it is very difficult to accurately measure the speed of propagation and the amplitude of shocks in real complex flows.

In the case of the two-layer shallow water system, the shocks captured by Roe scheme and the family of straight segments have been found (Castro et al., 2004) to be in good agreement with the experimental measurements of internal bores in the Strait of Gibraltar, despite of the simplicity of the chosen family of paths (the family of segments).

Some animations corresponding to simulations of geophysical flows obtained by our group of research on the basis of this type of models can be found in the web page http://anamat.cie.uma.es/animaciones

ACKNOWLEDGMENT

This research has been partially supported by the Spanish Government Research projects MTM2009–11923.

REFERENCES

Castro, M.J., Gallardo, J.M. & Parés, C. 2006. High order finite volume schemes based on reconstruction of states for solving hyperbolic systems with nonconservative products. Applications to shallow-water systems. *Math. Comp.* 75: 1103–1134.

Castro, M.J., García, J.A., González, J.M., Macías, J., Parés, C. & Vázquez, M.E. 2004. Numerical simulation of two layer shallow water flows through channels with irregular geometry. *J. Comput. Phys.* 195: 202–235.

Castro, M.J., LeFloch, P.G., Muñoz-Ruiz, M.L. & Parés, C. 2008. Why many theories of shock waves are necessary: Convergence error in formally path-consistent schemes. *J. Comput. Phys.* 3227: 8107–8129.

Castro, M.J., Macías, J. & Parés, C. 2001. AQ-Scheme for a class of systems of coupled conservation laws with source term. Application to a two-layer 1-D shallow water system. *Math. Model. Numer. Anal.* 35: 107–127.

Castro, M.J., Pardo, A., Parés, C. & Toro, E.F. 2010. On some fast well-balanced first order solvers for nonconservative systems. *Math. Comp.* 79: 1427–1472.

Dal Maso, G., LeFloch, P.G., Murat, F. 1995. Definition and weak stability of nonconservative products. *J. Math. Pures Appl.* 74: 483–548.

Dumbser, M., Enaux, C. & Toro, E.F. 2008. Finite volume schemes of very high order of accuracy for stiff hyperbolic balance laws. *J. Comput. Phys.* 227: 3971–4001.

Harten, A., Hyman, J.M. 1983. Self-adjusting grid methods for one-dimensional hyperbolic conservation laws. *J. Comput. Phys.* 50: 235–269.

Hou, T.Y. & LeFloch, P.G. 1994. Why nonconservative schemes converge to wrong solutions: error analysis. *Math. Comp.* 62: 497–530.

LeFloch, P.G. 1989. Shock waves for nonlinear hyperbolic systems in nonconservative form. Institute for Math. and its Appl., Minneapolis, Preprint # 593.

LeFloch, P.G. & Liu, T.P. 1993. Existence theory for nonlinear hyperbolic systems in nonconservative form. *Forum Math.* 5: 261–280.

Muñoz-Ruiz, M.L. & Parés, C. 2007. Godunov method for nonconservative hyperbolic systems. *Math. Model. Numer. Anal.* 41: 169–185.

Muñoz-Ruiz, M.L. & Parés, C. 2011.On the convergence and well-balanced property of path-conservative numerical schemes for systems of balance laws. *J. Sci. Comp.* 48: 274–295.

Parés, C. 2006. Numerical methods for nonconservative hyperbolic systems: a theoretical framework. *SIAM J. Numer. Anal.* 44: 300–321.

Parés, C. & Castro, M.J. 2004. On the well-balance property of Roe's method for nonconservative hyperbolic systems. Applications to Shallow-Water Systems. M2 AN, 38: 821–852.

Parés, C. & Muñoz-Ruiz, M.L. 2009. On some difficulties of the numerical approximation of nonconservative hyperbolic systems. *Boletín SEMA*, 47: 23–52.

Toro, E.F. 1999. Riemann Solvers and Numerical Methods for Fluid Dynamics.Springer.

Xing, Y. & Shu, C.W. 2006. High order well-balanced finite volume WENO schemes and discontinuous Galerkin methods for a class of hyperbolic systems with source terms. *J. Comput. Phys.* 214: 567–598.

Numerical Methods for Hyperbolic Equations – Vázquez-Cendón et al. (eds)
© 2013 Taylor & Francis Group, London, ISBN 978-0-415-62150-2

An explicit discontinuous Galerkin scheme based on a Runge-Kutta predictor with application to magnetohydrodynamics

Christoph Altmann, Gregor Gassner & Claus-Dieter Munz
Institut für Aerodynamik und Gasdynamik, University of Stuttgart, Germany

ABSTRACT: In recent years the discontinuous Galerkin schemes became more and more popular, since they combine the flexibility in handling complex geometries, h/p-adaptivity and efficient parallel computing. These advantages put them into position to be anideal candidate for numerical calculations in various fields of interest, especially in computational fluid mechanics. The explicit discontinuous Galerkin scheme proposed in this paper may be considered as a predictor corrector approach. The predictor isbased on a solution in the small, which takes into account the time evolution within the local grid cell only. The locality of this approach allows the introduction of a time-consistent local time-stepping in a natural way. We show the use of this feature for a divergence cleaning based on a hyperbolic correction equation which enables an adaptive control of divergence errors. In regions with high divergence errors, the correction is enhanced by a high propagation rate which may be interpreted as a localsub-cycling.

1 INTRODUCTION

Discontinuous Galerkin (DG) methods were first proposed in the early 1970's by Reed and Hill in (Reed and Hill 1973). Later on, Cockburn and Shu showed that these methods may be a powerful computational tool for the solution of systems of conservation laws. A comprehensive description of the development of discontinuous Galerkin schemes and their applications can be found in (Cockburn, Karniadakis, and Shu 2000) and in the recent text book by Hesthaven and Warburton (Hesthaven andWarburton 2008). The extension to problems of viscous gas dynamics was initiated by Bassi and Rebay, (Bassi and Rebay 1997; Bassi and Rebay 2002).

The scheme, proposed in this paper, is an explicit discontinuous Galerkin scheme which starts from a space-time variational formulation, but is kept explicit by a predictor. For the predictor, it is sufficient to take into account the local time evolution within the grid cell only. Starting with a one-dimensional setting, the basic concept is described in (Löorcher, Gassner, and Munz 2007) and for the multi-dimensional case in (Gassner, Lörcher, and Munz 2008). We show in the following that the predictor can be chosen to be a Runge Kutta method which is much more general than that used in (Gassner, Lörcher, and Munz 2008). The whole procedure can be easily applied to general systems. Here, we apply this scheme to the MHD equations. The main property of the scheme, namely to allow local time steps in a time consistent way, is used here to define a local sub-cycling of the divergence cleaning.

The structure of our paper is the following: We first describe the principles and layout of our discontinuous Galerkin approach. This method is modified with respect to efficiency taking advantage of the local time-stepping feature of the scheme. These modifications and their corresponding settings are described. Afterward give a brief introduction into the governing MHD equations together with a short description of their structure and peculiarities. For divergence correction, the hyperbolic GLM divergence correction from Dedner et al. (Dedner, Kemm, Kröner, Munz, Schnitzer, and Wesenberg 2002) is described. Numerical results along with convergence studies are given and discussed. In our conclusions a short summary of the work done is given as well as an outlook on further activities.

2 PRINCIPLES OF THE SPACE-TIME EXPANSION DISCONTINUOUS GALERKIN SCHEME

We will now summarize the basics of the Space-Time Expansion Discontinuous Galerkin (STE-DG) scheme and its local time stepping functionality, including the handling of the diffusive terms as needed for an artificial diffusion shock capturing. For a detailed view on these topics the reader is referred to the corresponding articles (Gassner, Lörcher, and Munz 2008) and (Lörcher, Gassner, and Munz 2007).

2.1 *Weak formulation including artificial viscosity*

Without the loss of generality, let us consider a system of nonlinear advection-diffusion system of the form

$$U_t + \nabla \cdot \left(\vec{F}(U) - \nu(\vec{x})(\vec{\nabla} U)\right) = 0, \tag{1}$$

where U denotes the vector of conservative variables and $\vec{F}(U)$ the advection flux. The gradient of U in our case represents an artificial viscosity term that is added for shock capturing purpose, while $\nu = \nu(\vec{x})$ is the amount of added element-wise viscosity.

The above PDE system could be the representation of any nonlinear equation system like the Euler equations or MHD equations, both extended with an artificial viscosity term.

The detailed procedures for deriving the weak formulation are not shown here. There are only minor changes to the handling of viscous terms for an artificial diffusion limiting process as they do not influence any critical parts of the system's diffusion matrix. The reader will find a complete step-by-step explanation for equation systems and handling of viscosity in (Gassner, Lörcher, and Munz 2008) and (Lörcher, Gassner, and Munz 2008) and we therefore skip the detailed derivation. Additionally, this also gives a possiblity to an extension to viscous MHD which is out of the scope of this paper.

We will next introduce the DG framework by defining a numerical approximate solution $U_h = U_h(\vec{x},t)$. In an arbitrary space-time cell $\Omega_i^n := Q_i \times [t_n, t_{n+1}]$ the approximation is given by

$$U|_{\Omega_i^n} \approx U_h(\vec{x},t) := \sum_{j=1}^{N} \hat{U}_j(t)\varphi_j(\vec{x}) \quad \text{for all} \quad x \in Q_i. \tag{2}$$

The $\hat{U}$ are the time dependent degrees of freedom (DOF) and the basis functions $\{\varphi_j(x)\}_{j=1,\ldots,N}$ span the space of polynomials with degree $\leq P$ over the spatial grid cell Q_i. This ansatz will then be inserted into the weak formulation and tested with each of the test functions φ_j that are chosen to be the same as the basis functions.

To keep things simple, we will in the following omit the h from the numerical solution U_h within Q_i as long as no misunderstanding can occur. We will also suppress the index i in the notation of U^i and Φ^i.

To derive the weak formulation, we first multiply the equation system (1) by a test function $\Phi = \Phi(\vec{x})$, integrate over an arbitrary space-time cell and perform a spatial integration by parts:

$$\int_{\Omega_i^n} U_t \Phi \, d\vec{x}dt + \int_{\partial\Omega_i^n} \left(\vec{F}(U) - \nu\vec{\nabla} U\right) \cdot \vec{n}\Phi \, dsdt - \int_{\Omega_i^n} \left(\vec{F}(U) - \nu\vec{\nabla} U\right) \cdot \vec{\nabla}\Phi \, d\vec{x}dt = 0. \tag{3}$$

For the artificial viscosity part of our equation which contains second order derivatives, we proceed with a second integration by parts of the viscous volume integral part in (3):

$$\begin{aligned} \int_{\Omega_i^n} \nu\vec{\nabla} U \cdot \vec{\nabla}\Phi \, d\vec{x}dt &= \int_{\Omega_i^n} \vec{\nabla} U \cdot \nu\vec{\nabla}\Phi \, d\vec{x}dt \\ &= \int_{\partial\Omega_i^n} U\nu\vec{\nabla}\Phi \cdot \vec{n} dsdt - \int_{\Omega_i^n} U\nabla \cdot \left(\nu\vec{\nabla}\Phi\right) d\vec{x}dt. \end{aligned} \tag{4}$$

The integration by parts is used forth and back, under the assumption that the volume integral is calculated from data inside Ω_i^n and only the surface integral covers the interaction between the grid cells. The objective is to lift a jump at the boundary between the functional values from the interior and the interface state into the discrete variational formulation, see (Gassner, Lörcher, and Munz 2008) in addition.

We then end up with the weak formulation

$$\begin{aligned} &\int_{\Omega_i^n} U_t \Phi \, d\vec{x}dt + \int_{\partial\Omega_i^n} \left(\vec{F}(U) - \nu\vec{\nabla} U\right) \cdot \vec{n}\Phi \, dsdt \\ &+ \int_{\partial\Omega_i^n} \left(\nu U\vec{\nabla}\Phi - \left[\nu U\vec{\nabla}\Phi\right]_{INT}\right) \cdot \vec{n} \, dsdt \\ &- \int_{\Omega_i^n} \left(\vec{F}(U) - \nu\vec{\nabla} U\right) \cdot \vec{\nabla}\Phi \, d\vec{x}dt = 0. \end{aligned} \tag{5}$$

The identity approach allows us to combine the volume integral portion of both the advective volume integral and the volume integral part obtained by the treatment of the diffusive terms for the numerical implementation.

With the use of orthonormal basis functions which are constructed via the Gram-Schmidt orthogonalization algorithm, this discretization yields a diagonal mass matrix, even for elements with curved boundaries.

Because our approximation U_h is discontinuous across element interfaces, we have to introduce numerical flux functions to guarantee both the stability and consistency of the discretization. For high order MHD calculations we are mainly using an HLLC flux described in (Li 2005) or a simple HLL or Rusanov flux.

2.2 *The space-time expansion approach*

Equation (5) contains space-time integrals which have to be approximated by an appropriate quadrature rule, e.g. Gaussian quadrature. For its evaluation, solution values at intermediate time levels are needed. We will use the strong form of Equation (1) and build a local predictor for the intermediate solution that does not depend on neighboring information. The correction is then performed by taking into account the predictions from neighboring cells.

To obtain the strong form of Equation (1), we start from Equation (3) and again perform a backward integration by parts. As long as only data of the element U_i is involved, this is an equivalence transformation. Data from the elementsinterior is again marked with a subscript "INT", which stands for "internal". We now get

$$\begin{aligned}&\int_{\Omega_i^n} \left(U_t + \nabla \cdot \vec{F}(U) - \nu(\vec{\nabla} U)\right)\Phi \, d\vec{x}dt \\ &\quad + \int_{\partial\Omega_i^n} \left(\vec{F}(U) - \nu\vec{\nabla} U\right)\cdot \vec{n}\Phi \, dsdt \\ &\quad - \int_{\partial\Omega_i^n} \left(\vec{F}(U_{INT}) - \nu\vec{\nabla} U_{INT}\right)\cdot \vec{n}\Phi \, dsdt \\ &\quad + \int_{\partial\Omega_i^n} \left(\nu U\vec{\nabla}\Phi - \left[\nu U\vec{\nabla}\Phi\right]_{INT}\right)\cdot \vec{n}\, dsdt = 0.\end{aligned} \tag{6}$$

In a local time stepping setting, a predictor will not be able to take neighboring data into account. We therefore solve the so-called local Cauchy problem within our update-cell and neglect all neighboring data. Instead, we extent our DG polynomial that holds as initial condition for the Cauchy problem onto $\mathbb{R}^n$. We therefore explicitly do not take neighboring information into account. The local Cauchy problem then reads as

$$\int_{U_i^n} W_t\Phi \, d\vec{x}dt = \int_{U_i^n} \nabla \cdot \vec{F}(W) - \nu(\vec{\nabla} W)\Phi \, d\vec{x}dt, \tag{7}$$

with $W_{U_i^n}(t=0) = U_{U_i^n}(t=0)$, the DG polynomial as initial condition. After using appropriate integration rules, we can now solve this ODE in time with standard schemes, providing the scheme is able to offer a time series of the solution of the Cauchy problem. We need that series to evaluate the integrals using Gaussian quadrature. Once the solution is obtained, we move back to the strong form, now inserting the predictors of our cell and its neighbors into the surface integral. We end up with our predictor-corrector formulation being derived as

$$\begin{aligned}&\int_{\Omega_i^n} \left(U_t + \nabla \cdot \vec{F}(W) - \nu(\vec{\nabla} W)\right)\Phi \, d\vec{x}dt \\ &\quad + \int_{\partial\Omega_i^n} \left(\vec{F}(W) - \nu\vec{\nabla} W\right)\cdot \vec{n}\Phi \, dsdt \\ &\quad - \int_{\partial\Omega_i^n} \left(\vec{F}(W_{INT}) - \nu\vec{\nabla} W_{INT}\right)\cdot \vec{n}\Phi \, dsdt \\ &\quad + \int_{\partial\Omega_i^n} \left(\nu W\vec{\nabla}\Phi - \left[\nu W\vec{\nabla}\Phi\right]_{INT}\right)\cdot \vec{n}\, dsdt = 0.\end{aligned} \tag{8}$$

We can see that the correction is done via the surface integral, using local predictions of neighboring cells. By using a continuous extension Runge-Kutta (CERK) scheme, introduced by (Owren and Zennaro 1992), we can efficiently advance the Cauchy problem in time. This special Runge-Kutta scheme provides a time series solution and reads as

$$\hat{W}^{n+1} = \hat{U}^n + \Delta t \sum_{i=1}^{n_{stages}} b_i \hat{K}_i,$$

where

$$\hat{K}_i = \mathbb{R}(\hat{W}_i^n), \tag{9}$$

with

$$\hat{W}_i^n = \hat{U}^n + \sum_{j=1}^{n_{stages}} a_{ij} \hat{K}_j.$$

We then get a time series for our predictive solution

$$\hat{W}(t) = \sum_{i=1}^{p_t+1} c_i^t t_j, \tag{10}$$

where

$$c_i^t = \sum_{j=1}^{n_{stages}} b_{ij}^t \hat{K}_j.$$

Here, n_{stages} and the CERK coefficients a_{ij}, b_i and b_{ij}^t depend on the desired time order.

Other methods are possible as well, especially Continuous and Discontinuous Galerkin methods, as described in (Gassner, Dumbser, Hindenlang, and Munz 2011), giving the framework excellent adaptation abilities for various numerical problems. This space-time setting now allows us to perform a high order time accurate and fully conservative explicit local time stepping, where each cell within the computational domain runs with its own optimal time step. Via the outlined predicitve time polynomial, necessary values at the cell boundaries can be reconstructed in time. It is therefore ensured that no information that has to be exchanged with the neighbors will be lost during the time updates. In the following, the methodology of the local time stepping is briefly outlined.

A sequence of four time steps with three adjacent grid cells in Figure 1starting from a common time level $t^0 = 0$ holds as a starting point.

After the determination of the local time steps, which are assumed to be different in our example due to the local stability restriction, the space-time Taylor expansions are calculated in each grid cell. This results in a predictive approximate solution in the space-time cells $Q_i \times [t_i^0, t_i^1]$—in our example for $i = 1,2,3$. These space-time polynomials are stored. We note that after this step the DOF $\vec{\hat{U}}_i^0$ at the time level t_i^0 are not needed any longer and may be overwritten. First, the volume integrals are calculated for each element Q_i. They rely only on the local space-time polynomials. The contribution of these terms are added to the DOF of the old time level. We call these values $\vec{\hat{U}}_i^*$. In the end, wehave for each grid element a space-time polynomial already containing the volume integral contributions. Next, the surface flux contributions involving neighboring grid cells have to be considered.

The local time stepping algorithm relies on the following evolve condition. The update of the DOF can only be completed, if

$$t_i^{n+1} \leq \min\left\{t_j^{n+1}\right\}, \forall j : Q_j \cap Q_i \neq \emptyset \tag{11}$$

is satisfied. This condition guarantees that all the data for the interface fluxes are available. In our example, the first grid cell satisfying this condition is Q_2. So Q_2 can now be evolved to t_2^1. To do so, the flux contributions at the right and left cell interface have to be computed and its contribution is then added to the local $\hat{\vec{U}}_2^*$. The flux integrals are calculated from $t = t_2^0$ to $t = t_2^1$ at the right interface $\partial Q_{2+\frac{1}{2}}$ and the left interface $\partial Q_{2+\frac{1}{2}}$. The arguments for the numerical flux functions at the time Gaussian points are obtained from the left and right space-time polynomials.

In order to keep this calculation exactly conservative as well as efficient, the flux contributions computed for the evolution of Q_2 are added simultaneously to the corresponding neighbors $\vec{\hat{U}}_1^*$ and $\vec{\hat{U}}_3^*$ with the minus sign. Then the update for Q_2 is completed and the DOF at the new time level t_2^1 are known. We can now start the procedure again: A new space-time polynomial is constructed in $Q_2 \times [t_2^1, t_2^2]$ and the volume integral contribution is added tothe local DOF $\vec{\hat{U}}_2^1$, now named by $\vec{\hat{U}}_2^*$.

If boundary values at the left are given, now Q_1 satisfies the evolve condition and can be advanced to t_1^1. As before, the volume integral contribution is already added. But in this case, also a part of the flux contributions has already beenadded to the $\vec{\hat{U}}_1^*$ during the previous evolution of Q_2. Thus, only the missing flux contributions, which are sketched in the lower left corner of Figure 1, have to be added to the $\vec{\hat{U}}_1^*$ in order to get the $vec\hat{U}_1^1$. Namely, on the interface $\partial Q_{1+\frac{1}{2}}$, the flux integral has to be computed with a quadrature formula from t_2^1 to t_1^1. As before, the flux integral computed on this shared interface is not only added to $\vec{\hat{U}}_1^*$,

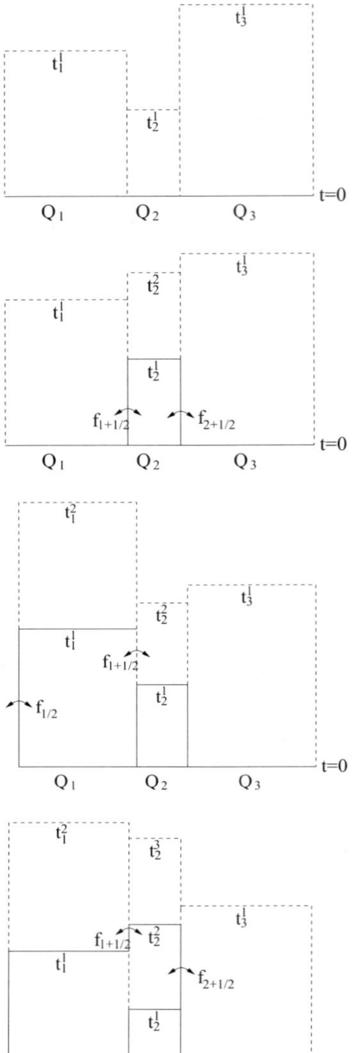

Figure 1. Sequence of steps 1–4 of a computation with 3 different elements and local time-stepping.

but also to $\vec{\hat{U}}_2^*$. The time interval, for which the flux contribution at the interface shared by an element Q_i and an adjacent element Q_j has to be computed when evolving $\vec{\hat{U}}_i^*$ to $\vec{\hat{U}}_i^{n+1}$, is generally

$$[t_{ij}^\star, t_i^{n+1}] = [max(t_i^n, t_j^n), t_i^{n+1}]. \tag{12}$$

In this manner, the algorithm continues by searching for elements satisfying the evolve condition (11). At each time, the interface fluxes are defined uniquely for both adjacent elements, making the scheme exactly conservative.

The presented local time stepping algorithm minimizes the total number of time steps for a computation with fixed end time. However, in some cases where the difference of time levels of adjacent grid cells are very small compared to the local time steps, the efficiency of the presented algorithm may decrease. To overcome this deficiency we locally synchronize the time levels of those cells. It is also not difficult to introduce some common global time levels as needed for example at the end of the computation. Please note that this procedure has absolutely no influence on the accuracy of the underlying numerical scheme, as convergence tests e.g. in (Lörcher, Gassner, and Munz 2007) and (Gassner, Lörcher, and Munz 2008) verify.

3 EQUATIONS

The equations of magnetohydrodynamics (MHD) describe the interaction of a gas with magnetic fields. They are based on hydrodynamic conservation laws and Maxwell equations with several simplifications such as electrical neutrality. These equations play an important role in various fields of applications: From computation of planetary system genesis or solar wind up to the design of fusion reactors.

3.1 *The ideal MHD equations*

The ideal MHD equations can be expressed in conservative form in compact notation as

$$\begin{aligned}
\frac{\partial}{\partial t}\rho &= -\nabla\cdot(\rho\mathbf{v}) \\
\frac{\partial}{\partial t}(\rho\mathbf{v}) &= -\nabla\cdot\left(\rho\mathbf{v}\mathbf{v}^t - \mathbf{B}\mathbf{B}^t + \left(p+\frac{1}{2}|\mathbf{B}|^2\right)I\right) \\
\frac{\partial}{\partial t}E &= -\nabla\cdot\left((E+p)\mathbf{v} + \left(\frac{1}{2}|\mathbf{B}|^2\, I - \mathbf{B}\mathbf{B}^t\right)\cdot\mathbf{v}\right) \\
\frac{\partial}{\partial t}\mathbf{B} &= -\nabla\cdot\left(\mathbf{B}\mathbf{v}^t - \mathbf{v}\mathbf{B}^t\right), \qquad (13)
\end{aligned}$$

together with the divergence constraint $\nabla \cdot \mathrm{B} = 0$. The superscript $\mathbf{u}^t$ nominates a transposed vector or matrix $\mathbf{u}$. We use the usual nomination: ρ, $\mathbf{v}$, B, E, and p to denote density, velocity, magnetic induction, total energy per unit volume, and pressure,respectively. Please note that for simplicity reasons, the magnetic permeability constant μ_0 is set to 1 throughout the paper.

The equation of state of a perfect gas is assumed with adiabatic exponent γ. The pressure p can therefore be calculated from the conserved variables as

$$p = (\gamma-1)\left(E - \frac{1}{2}\rho|\mathbf{v}^2| - \frac{\mathbf{B}^2}{2}\right). \qquad (14)$$

It can be easily seen that the MHD system of equations reduces to the Euler equations in the case of a vanishing magnetic field $\mathrm{B} = 0$.

3.2 *The augmented MHD equations for divergence cleaning*

It is well known that $\nabla \cdot \mathrm{B} = 0$ condition is satisfied for all times, if the initial conditions and consistent boundary conditions are given, due to the fact that $\nabla \cdot$ *curl* is zero. This is valid with respect to the exact solution, but the numerical methods do in general not satisfy this relation exactly but with some orders of accuracy. These numerically induced divergence errors may have severe impact on the accuracy of the solution. The exact relation of such vector quantities can be established by using discrete operators with special grid arrangements. An other approach is to use an explicit divergence cleaning. For Maxwell's equations Munz et al. (Munz, Omnes, Schneider, Sonnendrücker, and Voß 2000) proposed to replace the hyperbolic evolution equations in the Maxwell system by an augmented system that allows charge conservation errors, but has the property to diminish them when time evolves. Existence, uniqueness and relation to the Maxwell equations were discussed in this paper. The augmented system is more robust for numerical simulations and includes several charge correction approaches proposed earlier. The first application of this method to the divergence-free constraint of the magnetic field in MHD flow was done by Dedner et al. (Dedner, Kemm, Kröner, Munz, Schnitzer, and Wesenberg 2002). They presented a hyperbolic divergence cleaning strategy that is rather simple as well as easy to implement. It adds a hyperbolic correction system to the MHD equations, carried by an additional variable that controls the divergence of the magnetic field B. The correction system can be solved in a very fast and straight-forward way splitted from the rest. Since the effect of that additional equation on the whole system is similar to an Lagrangian multiplier, they called this method the Generalized Lagrange Multiplier (GLM) divergence correction method. Shown for just the one-dimensional part of the ideal MHD equations, this method yields the following extended equation system:

$$\frac{\partial}{\partial t}\begin{pmatrix}\rho\\ \rho u\\ \rho v\\ \rho w\\ \varepsilon\\ B_x\\ B_y\\ B_z\\ \psi\end{pmatrix}+\frac{\partial}{\partial x}\begin{pmatrix}\rho u\\ \rho u^2+p+\mathbf{B}^2/2-B_x^2\\ \rho uv-B_xB_y\\ \rho uw-B_xB_z\\ (\varepsilon+p+\mathbf{B}^2/2)u-B_x(v\cdot\mathbf{B})\\ \psi\\ (uB_y-vB_x)\\ -(wB_x-uB_z)\\ c_h^2B_x\end{pmatrix}=\begin{pmatrix}0\\0\\0\\0\\0\\0\\0\\0\\0\\-\frac{c_h^2}{c_p^2}\end{pmatrix}. \tag{15}$$

The additional variable ψ is introduced and propagates the divergence error out of the computational domain. With the addition of $-c_h^2/c_p^2$ for the ψ equation on the right hand side of (15), we will not only transport the errors out of the computational domain but also damp them. The damping effect can be scaled by setting the value of c_p. We are generally using the value of 0.18 for c_p, as proposed in (Dedner, Kemm, Kröner, Munz, Schnitzer, and Wesenberg 2002), which has been proven as a good compromise between damping and hyperbolic transport. With that modification, the divergence cleaning method is then called "mixed GLM method". For calculating MHD equations, it was proposed in (Dedner, Kemm, Kröner, Munz, Schnitzer, and Wesenberg 2002 to set the hyperbolic transport speed to the maximum of the fastest system wave, to guarantee that errors are at least spread with the same velocity as they may be generated. This way it is ensured that the correction subsystem is not affecting the time step restriction of the overlying MHD calculation. In some cases, where large errors locally occur or when the problem is sensitive with respect to the fastest waves, the amplitudes of the moving waves correcting the errors may be too large. They are not damped fast enough resulting in an interaction with the physical waves and generating spurious effects.

It is important to mention that when using the shock capturing method with MHD equations, the artificial viscosity from the limiting routine does not necessarily need to be added to the additional ninth variable of the divergence correction system. Nevertheless, doing so would result in an additional dissipation of the errors, leading to a more robust algorithm.

4 HIGH PERFORMANCE COMPUTING

Besides the promising fundamental properties of the DG framework, its main advantages are based on its "high performance computing" capability. The DG algorithm is inherently parallel, since all elements do communicate only with their direct neighbors via solution and flux exchange. Independent of the local polynomial degree, only exchange of surface data between direct von Neumann neighbors is necessary. Note that the DG operator can be split into the two building blocks, namely the volume integral—solely depending on element local DOF—and the surface integral, where neighbor information is needed. This fact can help to hide communication latency by exploiting local element operations and will further reduce the negative influence of data transfer on efficiency. It is therefore possible to send surface data while simultaneously performing volume data operations.

To examine the parallel efficiency, weak scaling test with different setups were used to determine the code behavior. Scaling results are shown in Figure 2. In a first run, the number of elements per processor were kept constant while the polynomial degree was increased. Here, scaling improved for higher polynomial degree. This is due to the fact that high order elements are computationally more expensive, therefore a better computation-communication ratio was achieved. For the second run, the polynomial degree was kept constant to $N = 6$ while the number of elements per processor varied between 5^3 and 9^3. Here, we can see the same effect, only topped by the 8^3 run. This test is very important to find out the optimal load for a computation at a fixed order of accuracy. In the third run, the per-processor number of DG degrees of freedom (DoF) was kept constant. This test aims at determining the best setup for a required accuracy: The primary goal was to determine whether one should perform higher order calculations on large grid cells, or reduce the polynomial degree in favor of more but smaller grid cells. We can see that again the trend goes towards high order calculations.

5 NUMERICAL RESULTS

In this chapter we focus on numerical results of the proposed DG scheme for the ideal MHD equations.

5.1 *Convergence test*

We will start with a two-dimensional wave propagation example stated in (Balsara 2004). Although it was initially not intended to, we use this test problem to perform convergence tests of the STE-DG scheme since it has the nice property of an analytical solution easily available at any time. It consists of fluctuations in both the magnetic field

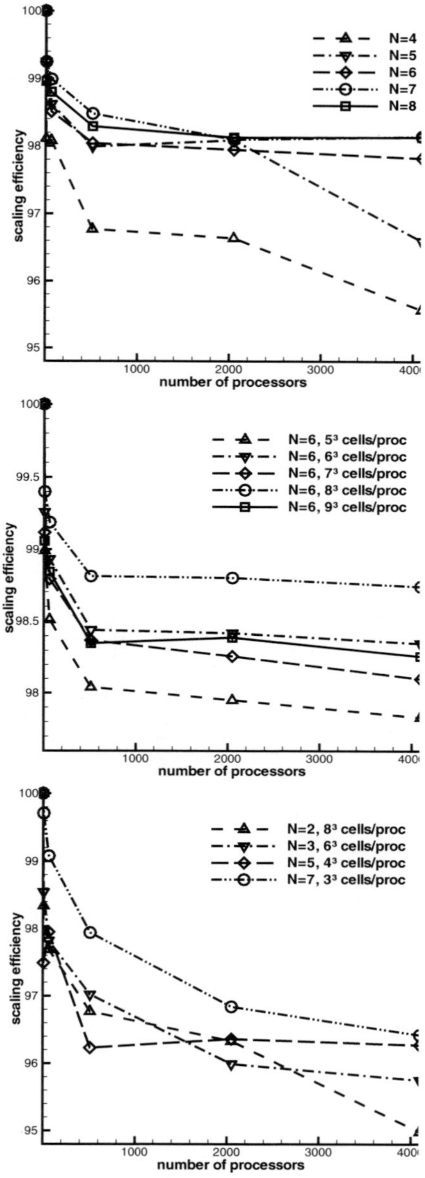

Figure 2. Weak scaling (constant load per processor) for the DG code with different polynomial orders (up), element numbers (middle) and constant DoF load (down).

and the velocity and is called Alfven wave decay. Its original primary is to examine the dissipation of torsional Alfven waves that are made to propagate at a small angle to the mesh. The computational domain spans $[-r/2,r/2] \times [-r/2,r/2]$ in the xy-plane with $r = 6$. A uniform density, $\rho_0 = 1$, and pressure, $p_0 = 1$, are initialized on the mesh. The unperturbed velocity is set to $v_0 = 0$, and the unperturbed magnetic field to $B_0 = 1$. The amplitude of the Alfven wave fluctuation is parametrized in terms of the velocity fluctuation which has a value of $\varepsilon = 0.2$ in our calculations. The Alfven wave is made to propagate at an angle of $\tan^{-1}(1/r) = \tan^{-1}(1/6) = 9.462$ with respect to the y-axis. The direction of wave propagation is along the unit vector

$$\hat{n} = n_x \hat{i} + n_y \hat{j} = \frac{1}{\sqrt{r^2+1}} \hat{i} + \frac{r}{\sqrt{r^2+1}} \hat{j}. \tag{16}$$

The phase of the wave is taken to be

$$\phi = \frac{2\pi}{n_y}\left(n_x x + n_y y - V_a t\right), \tag{17}$$

where $V_a = \frac{B_0}{\sqrt{4\pi\rho_0}}$. The velocity is given by

$$\begin{aligned} \mathbf{v} = & \left(v_0 n_x - \varepsilon n_y \cos\phi\right)\hat{i} \\ & + \left(v_0 n_y + \varepsilon n_x \cos\phi\right)\hat{j} \\ & + (\varepsilon \sin\phi)\hat{k}. \end{aligned} \tag{18}$$

The magnetic field is given by

$$\begin{aligned} \mathbf{B} = & \left(B_0 n_x + \varepsilon n_y \sqrt{2\pi\rho_0} \cos\phi\right)\hat{i} \\ & + \left(B_0 n_y - \varepsilon n_x \sqrt{4\pi\rho_0} \cos\phi\right)\hat{j} \\ & - \left(\varepsilon \sqrt{4\pi\rho_0} \sin\phi\right)\hat{k}. \end{aligned} \tag{19}$$

Meshes up to 32×32 zones were used. The following tables show convergence rates for 5th and 6th order schemes. As variable, the x-component of the magnetic field was used.

This problem also holds as a three-dimensional convergence test case. No modifications to the initial conditions had to be performed, the computational domain was simply extended to the third dimension. First, convergence rates for the two-dimensional ideal MHD case are shown. These are compared to a calculation without the divergence cleaning mechanism. As clearly visible, the L_2 error results show a degradation of up to one order of magnitude, in some cases even resulting in a degraded order of convergence.

For the three-dimensional extension of this problem, convergence rates of polynomial degree $\mathcal{P}4$ and $\mathcal{P}5$ with divergence cleaning are illustrated to show the desired behavior.

5.2 *Magnetic rotor*

This second example, also presented in (Balsara 2004), consists of a spinning dense circle inside an initially stationary fluid. Inside the circle, we have a

Table 1. Experimental order of convergence for the two-dimensional Alfven wave test problem for ideal MHD on an equidistant grid with divergence cleaning.

Nb cells^2	**Nb DOF**	$\|B_x\|L_2$ **w divB corr.**	$\mathcal{O}_{L_2}$
		$\mathcal{P}4$ **STE-DG**	
4	**240**	2.93E-04	
8	**960**	9.21E-06	**5.0**
16	**3840**	3.00E-07	**4.9**
32	**15360**	9.55E-09	**5.0**
		$\mathcal{P}5$ **STE-DG**	
2	**84**	2.86E-03	
4	**336**	4.09E-05	**6.1**
8	**1344**	7.02E-07	**5.9**
16	**5376**	1.09E-08	**6.0**

Table 2. Experimental order of convergence for the two-dimensional Alfven wave test problem for ideal MHD on an equidistant grid without divergence cleaning.

Nb cells^2	**Nb DOF**	$\|B_x\|L_2$ **w divB corr.**	$\mathcal{O}_{L_2}$
		$\mathcal{P}4$ **STE-DG**	
4	**240**	3.54E-04	
8	**960**	1.33E-05	**4.7**
16	**3840**	4.61E-07	**4.8**
32	**15360**	1.50E-08	**4.9**
		$\mathcal{P}5$ **STE-DG**	
2	**84**	3.38E-03	
4	**336**	5.40E-05	**6.0**
8	**1344**	1.20E-06	**5.5**
16	**5376**	3.86E-08	**5.0**

density of $\rho_0 = 10$ while set to $\rho_0 = 1$ in the ambient fluid. Pressure and magnetic field are uniquely defined as $p = 1$ and $\mathbf{B} = (2.5/\sqrt{4\pi}, 0, 0)$. Within the circle, we set a uniform angular velocity up to a radius of $r = 0.1$ and a value of $v_a = 1$ at that position. The ratio of specific heats is $\gamma = \frac{5}{3}$. Periodic boundaries were used and this example was calculated up to time $t = 0.29$, which corresponds to calculations found in the literature. Figure 3 shows the density distribution and Figure 4 the magnetic

Table 3. Experimental order of convergence for the three-dimensional Alfven wave test problem for ideal MHD on an equidistant grid.

Nb cells^2	**Nb DOF**	$\|B_x\|L_2$ **w divB corr.**	$\mathcal{O}_{L_2}$
		$\mathcal{P}4$ **STE-DG**	
2	**280**	1.84E-02	
4	**2240**	4.12E-04	**5.5**
8	**17920**	1.06E-05	**5.3**
16	**143360**	3.32E-07	**5.0**
		$\mathcal{P}5$ **STE-DG**	
2	**448**	7.78E-03	
4	**3584**	5.01E-05	**7.3**
8	**28672**	1.11E-06	**5.5**
16	**229367**	1.75E-08	**6.0**

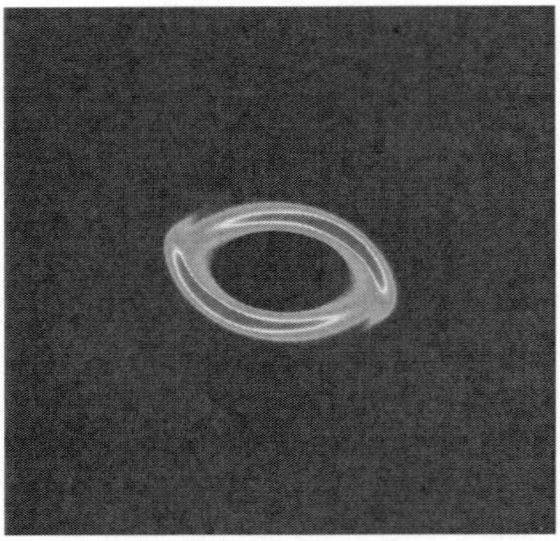

Figure 3. Plot of the density distribution for the magnetic rotor problem at time $t = 0.29$.

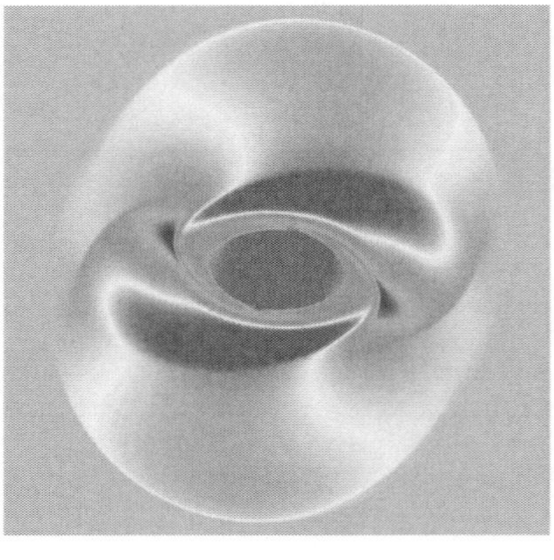

Figure 4. Plot of the magnetic field magnitude for the magnetic rotor problem at time $t = 0.29$.

field magnitude of a 5th order calculation on a 200×200 grid. Divergence cleaning was performed as described previously.

5.3 *Orszag-Tang Vortex*

This vortex system of Orszag and Tang (Orszag and Tang 1979) which was studied extensively in (Picone and Dahlburg 1991) and (Dahlburg and Picone 1989) is an ambitious test problem for any numerical scheme. The computational domain is $[0;1]\times[0;1]$ with periodic boundaries. The initial condition of the problem is given by

$$
\begin{aligned}
\rho &= \gamma e \\
u &= -sin(2\pi y) \\
v &= sin(2\pi x) \\
e &= \frac{10}{24}\pi \\
B_x &= -\frac{1}{\sqrt{(4\pi)}} sin(2\pi y) \\
B_y &= \frac{1}{\sqrt{(4\pi)}} sin(4\pi x).
\end{aligned}
\tag{20}
$$

with $\gamma = 5/3$. The calculation is usually run up to t = 0.5. By then, several shocks have crossed the computational domain and a vortex system has built up near the center. The pictures from Figure 5 at $t = 0.1$ and Figure 6 at $t = 0.5$ on a 200×200 grid show a good agreement with the reference results in the literature.

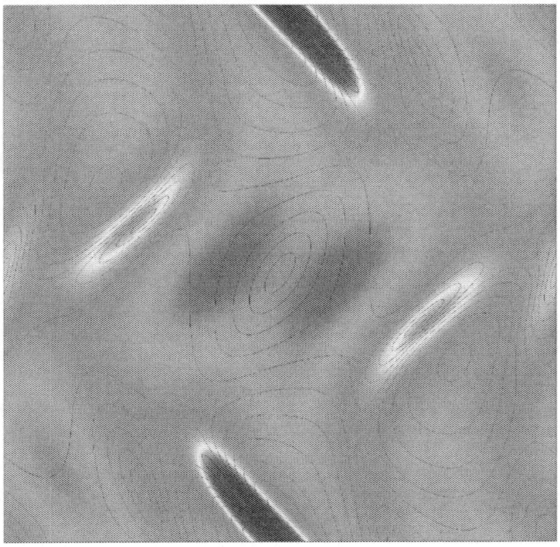

Figure 5. Contour plot of the density for the Orszag-Tang vortex at time level $t = 0.1$. Contour levels of magnetic field magnitude are also shown.

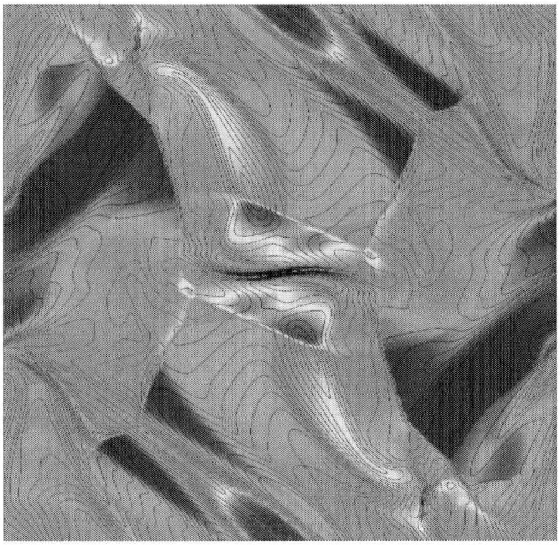

Figure 6. Contour plot of the density for the Orszag-Tang vortex at time level $t = 0.5$. Contour levels of magnetic field magnitude are also shown.

6 CONCLUSION

In this paper we have shown the extension of the recently developed STE-DG scheme to handle ideal MHD equations. Utilizing the scheme's local time stepping feature we have introduced an advancement of the mixed GLM divergence correction method which uses variational propagation speeds for performing the correction that may be higher than the system speeds. Convergence tests had proven the expected order of convergence for the scheme including divergence correction. The calculations performed showed that this implementation can significantly decrease the divergence error while only minimally increasing the computational costs. The adjustment of the correction speed with an appropriate divergence sensor was also shown. Nevertheless, the sensor settings are slightly problem dependent. This issue is still under investigation. The numerical results also showed that even very strong shocks as they often appear in MHD calculations are capturing nicely using artificial viscosity and that we are able to resolve the shock profile even within one single grid cell when calculating with sufficiently high polynomial order.

ACKNOWLEDGEMENTS

This project is kindly supported by the Deutsche Forschungsgemeinschaft (DFG) within SPP 1276: MetStroem, Cluster of Excellence SimTech, and by the Bundesministerium für Bildung und Forschung

(BMBF, Federal Ministry for Education and Research) in the HPC Software Initiative Projekt "STEDG: Hocheffiziente und skalierbare Software für die Simulation turbulenter Strömungen in komplexen Geometrien".

REFERENCES

[1] Balsara, D.S. (2004). Second-Order Accurate Schemes for Magnetohydrodynamics with Divergence-Free Reconstruction. *The Astrophysical Journal Supplement Series*, 149–184.

[2] Bassi, F. and Rebay, S. (1997). A high-order accurate discontinuous finite element method for the numerical solution of the compressible Navier-Stokes equations. *J. Comput. Phys.*, 267–279.

[3] Bassi, F. and Rebay, S. (2002). Numerical evaluation of two discontinuous Galerkin methods for the compressible Navier-Stokes equations. *International Journal for Numerical Methods in Fluids*, 197–207.

[4] Cockburn, B., Karniadakis, G.E. and Shu, C.-W. (2000). *Discontinuous Galerkin Methods*. Lecture Notes in Computational Science and Engineering. Springer.

[5] Dahlburg, R.B. and Picone, J.M. (1989). Evolution of the Orszag-Tang vortex system in a compressible medium. I. Initial average subsonic flow. *Phys. Fluids B*, 2153–2171.

[6] Dedner, A., Kemm, F., Kröner, D., Munz, C.-D., Schnitzer, T. and Wesenberg, M. (2002). Hyperbolic divergence cleaning for the MHD equations. *J. Comput. Phys.175*(2), 645–673.

[7] Gassner, G., Dumbser, M., Hindenlang, F. and Munz, C. (2011). Explicit one-step time discretizations for discontinuous galerkin and finite volume schemes based on local predictors. *J. Comput. Phys.*

[8] Gassner, G., Lörcher, F. and Munz, C.-D. (2008). A discontinuous Galerkin scheme based on a space-time expansion. II. Viscous flow equations in multi dimensions. *J. Sci. Comp.* (3), 260–286.

[9] Hesthaven, J. and Warburton, T. (2008). Nodal Discontinuous Galerkin Methods: Algorithms, Analysis, and Applications. Springer Verlag, New York.

[10] Li, S. (2005). An HLLC Riemann solver for magneto-hydrodynamics. *J. Comput. Phys. 203*(1), 344–357.

[11] Lörcher, F., Gassner, G. and Munz, C.-D. (2007). A discontinuous Galerkin scheme based on a space-time expansion. I. Inviscid compressible flow in one space dimension. *Journal of Scientific Computing* (2), 175–199.

[12] Lörcher, F., Gassner, G. and Munz, C.-D. (2008). An explicit discontinuous Galerkin scheme with local time-stepping for general unsteady diffusion equations. *J. Comput. Phys.* (11), 5649–5670.

[13] Munz, C.-D., Omnes, P., Schneider, R., Sonnendrücker, E. and Voß, U. (2000, July). Divergence correction techniques for maxwell solvers based on a hyperbolic model. *J. Comput. Phys.*, 484–511.

[14] Orszag, S.A. and Tang, C.M. (1979). Small-scale structure of two-dimensional magnetohydrodynamic turbulence. *Journal of Fluid Mechanics*, 90–129.

[15] Owren, B. and Zennaro, M. (1992). Derivation of efficient continuous explicit Runge-Kutta methods. *SIAM J. Sci. Stat. Comput.*, 1488–1501.

[16] Picone, J.M. and Dahlburg, R.B. (1991). Evolution of the Orszag-Tang vortex system in a compressible medium. II. Supersonic flow. *Phys. Fluids B*, 29–44.

[17] Reed, W. and Hill, T. (1973). Triangular mesh methods for the neutron transport equation. Technical Report LA-UR-73–479, Los Alamos Scientific Laboratory.

VIII Methods and models for biomedical problems

Numerical Methods for Hyperbolic Equations – Vázquez-Cendón et al. (eds)
© 2013 Taylor & Francis Group, London, ISBN 978-0-415-62150-2

Simulation of an epidemic model with nonlinear cross-diffusion

Stefan Berres
Departamento de Ciencias Matemáticas y Físicas, Facultad de Ingeniería, Universidad Católica de Temuco, Chile

Ricardo Ruiz-Baier
Modeling and Scientific Computing CMCS-MATHICSE-SB, École Polytechnique Fédérale de Lausanne, Switzerland

ABSTRACT: A spatially two-dimensional epidemic model is formulated by a reaction-diffusion system. The spatial pattern formation is driven by a cross-diffusion corresponding to a non-diagonal, upper-triangular diffusion matrix. Whereas the reaction terms describe the local dynamics of susceptible and infected species, the diffusion terms account for the spatial distribution dynamics. For both self-diffusion and cross-diffusion nonlinear constitutive assumptions are suggested. To simulate the pattern formation two finite volume formulations are proposed, which employ a conservative and a non-conservative discretization, respectively. Numerical examples illustrate the impact of the cross-diffusion on the pattern formation.

1 INTRODUCTION

The knowledge of spreading dynamics of infectious diseases helps to design prevention measures. A generic model category for the quantitative description of the epidemic evolution dynamics by an ordinary differential equation are the so-called SIR models, which classify a population into 'susceptible' (S), 'infected' (I) and 'recovered' (R) subgroups and balance the changes between these. One very early and simple prototype of a SIR-model is due to Kermack and McKendrick 1927. It describes the population evolution by the system of ordinary differential equations

$$\frac{dS}{dt} = -\alpha SI, \quad \frac{dI}{dt} = \alpha SI - \beta I, \quad \frac{dR}{dt} = \beta I,$$

where $\alpha > 0$ is the infection rate and $\beta > 0$ the recovery rate. There are several suggestions for improving the specification of these ODE-dynamics (Kim et al. 2010, Li et al. 2010), and structural modifications like SIR-models in networks (Liu & Zhang 2010). A key issue in epidemic modeling is the formation of spatial patterns. Based on a general setting in the two-dimensional reaction-diffusion framework for epidemic processes (Webb 1981), there are several suggestions for the combination of the system of ordinary differential equations of the SIR-model with a spatially two-dimensional diffusion equation of the involved variables (He & Stone 2003, Milner & Zhao 2008, Li & Zou 2009, Sun et al. 2009). Moreover, several contributions have been proposed to study pattern formation induced by cross-diffusion (Ni 2004, Bendahmane et al. 2009b, Tian et al. 2010). In addition to a fundamental existence proof for general reaction-diffusion systems (Crandall et al. 1987), there are several approaches to analyze reaction-diffusion equations with one single "cross-diffusion" that lead to a system with upper triangular diffusion matrix (Badraoui 2006, Daddiouaissa 2008). The structure of an upper triangular diffusion matrix has also been utilized in the existence analysis for systems of convection-diffusion equations with both Dirich-let and Neumann boundary conditions (see e.g. Frid & Shelukhin 2004, Frid & Shelukhin 2005, Berres et al. 2006). Besides numerous contributions to the development of numerical methods to solve reaction-diffusion equations in related contexts (Wong 2008, Phongthanapanich & Dechaumphai 2009), convergence proofs of associated finite volume schemes (Bendahmane & Sepúlveda 2009, Andreianov et al. 2011) and finite element formulations (Galiano et al. 2003, Barrett & Blowey 2004) have been provided.

This contribution is a condensed version of Berres & Ruiz-Baier 2011. The goal is, on the one hand, to generate pattern formation in an epidemic model by a cross-diffusion term, and, on the other hand, to prevent blow-up by a non-linear limitation of the cross-diffusion. These assumptions are designed to qualitatively reflect psychological behavior. The cross-diffusion term has the interpretation that the susceptible population moves away from increasing gradients of the infected population. In addition, it is assumed that the cross-diffusion effect depends on the local population density. For the nonlinear cross-diffusion it is assumed that there exists *carelessness* at a small and *fatalism* at a high total population number. At carelessness and fatalism the susceptible population decreases its tendency to avoid agents of the infected population. Such an avoidance is most effective for intermediate (neither too small nor too large) population numbers.

2 CONCEPTUAL MODEL

The two-dimensional reaction-diffusion system describing spatial epidemic dynamics with cross diffusion is written as

$$u_t = f(u,v) + \nabla\cdot(a(u)\nabla u) + \nabla\cdot(c(u,v)\nabla v),$$
$$v_t = g(u,v) + \nabla\cdot(b(v)\nabla v), \quad (1)$$

in $\Omega_T = \Omega\times(0,T)$, where u and v denote the populations of susceptible and infected persons, respectively. No external input is imposed, therefore on the physical domain boundary $\delta\Omega$ the Neumann boundary condition is assumed to hold:

$$(a(u)\nabla u + c(u,v)\nabla v)\cdot n = 0, \quad (b(v)\nabla u)\cdot n = 0,$$

where n is the outer normal vector to the physical domain boundary. In the system (1), an additional equation for the recuperated population is omitted because the model does not consider their feedback on the susceptible or infected population. With the notation

$$u = \begin{pmatrix} u \\ v \end{pmatrix}, \quad f(u) = \begin{pmatrix} f(u,v) \\ g(u,v) \end{pmatrix},$$
$$a(u) = \begin{pmatrix} a(u) & c(u,v) \\ 0 & b(v) \end{pmatrix},$$

the system (1) can be written in compact form as

$$\frac{d}{dt}u = f(u) + \nabla\cdot(a(u)\nabla u).$$

The model enforeces phase separation since the susceptible species avoid the infected population by a cross-diffusion term $\nabla\cdot(c(u,v)\nabla v)$. The cross-diffusion term directs the flow in the opposite direction of the gradient ∇v. Whenever there is an increase of the amount of the infected population then the susceptible aents move away from the direction of the increasing gradient.

The reaction terms are considered to be given by the following specifications (see e.g. Su et al. 2009)

$$f(u,v) = ru(1-u/K) - \beta\frac{uv}{u+v},$$
$$g(u,v) = \beta\frac{uv}{u+v} - kv, \quad (2)$$

where the model parameters are the carrying capacity of the susceptible species K, r is the intrinsic birth rate, β is the rate of disease transmission, and k represents the recovery rate of the infected species.

The equilibrium points are pairs (u,v) such that $f(u,v) = 0$ and $g(u,v) = 0$. For (2), the equilibrium points are (0,0) (trivial equilibrium), $(K,0)$, which corresponds to the disease-free point, and (u^*,v^*), which corresponds to an endemic stationary state that is explicitly given by

$$(u^*,v^*) = \left(\frac{K(r-\beta+k)}{r}, \frac{K(r-\beta+k)(\beta-k)}{rk}\right).$$

In Figure 1, the phase portrait of the ODE system associated to (1), (2). From now on, as model variants we will consider diffusion terms which are linear (Model 1) and nonlinear (Model 2). For better comparison, in Model 2 the same reaction kinetics (2) as for Model 1 are used.

For Model 1 the diffusion terms are given by the constants

$$a(u) = a_0, \quad b(v) = b_0, \quad c(u,v) = c_0, \quad (3)$$

which adopts the setting of Sun et al. 2009. In Model 2 we propose nonlinear model variants of the parametric functions. The self-diffusion terms are chosen as

$$a(u) = a_0 u^m, \quad b(v) = b_0 v^m. \quad (4)$$

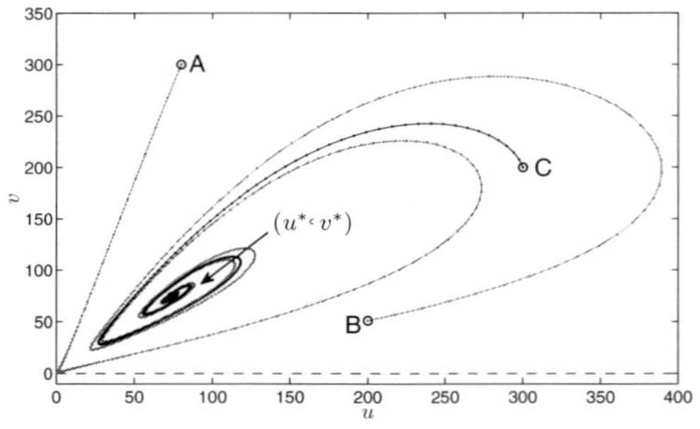

Figure 1. Phase portrait for the ODE system associated to (1), (2). Three trajectories are displayed starting from the states $A = (80,300)$, $B = (200,50)$ and $C = 300,200)$ and reaching the equilibrium point (u^*, v^*). The parameters correspond to those used in Example 1.

By the assumption that $m \in (-1,0)$, a degressive growth is described since

$$\partial_u a(u) < 0, \quad \partial_v b(v) < 0 \quad \text{for all} \quad u,v.$$

The biological interpretation is that the tendency to avoid crowds reduces with higher numbers as the population "gets used" to them. Using the notation of the Laplace operator,

$$\Delta A(u) = \nabla \cdot (a(u)\nabla u), \quad \Delta B(v) = \nabla \cdot (b(v)\nabla v), \qquad (5)$$

we have

$$A(u) = \frac{a_0}{m+1} u^{m+1}, \quad B(v) = \frac{b_0}{m+1} v^{m+1}, \qquad (6)$$

where $A(u)$ and $B(v)$ are sub-linear functions. For $m = 0$ the functions (4) of Model 2 boil down to those of the linear Model 1 (3) as

$$A(u) = a_0 u, \quad B(v) = b_0 v.$$

The construction of a conservative or nonconservative discretization depends roughly speaking on whether the finite differences are based on the formulation of the left-hand or right-hand side of (5), respectively.

A nonlinear cross-diffusion function implements a situation-dependent tendency of the susceptible population to avoid the infected population. This situation-dependent behavior reflects an average psychological disposition. An approach to model such a disposition is as follows. From the perspective of the susceptible population, avoidance is pursued whenever there is an awareness, i.e., when there is a detectable fraction of the infected population. For a small number of populations the necessary awareness has not been matured or is temporarily not active, since then there is no vital urgency for self-protection. In the other extreme, at large population numbers, such a selective detection is neither possible nor makes sense, since there is less, or even no chance to avoid infection in the crowd. The population number affects the conscious disposition of avoidance. Therefore, the cross-diffusion coefficient is designed to be negligible for imposed by the following constraints both small and large number of populations, which is imposed by the following constraints

$$c(u,0) = 0, \quad c(0,v) = 0, \quad \text{for all} \quad u,v \in \mathbb{R},$$
$$c(u,v) = 0 \quad \text{for} \quad v \geq V(u),$$

where V is a Lipschitz continuous monotonically decreasing function with a zero for $u > 0$. For example, one might choose $V(u) = c_1 - u$ with $c_1 > 0$. By these constraints, the reaction-diffusion equation with cross-diffusion (1) degenerates into an equation without cross-diffusion outside the domain

$$\tilde{\Omega} := \{(u,v) : u,v \geq 0, v < V(u)\}.$$

The constraints (7) are satisfied, for example, by a function which is quadratic in the domain $u, v \geq 0$, $u + v = c_1$

$$c(u,v) = c_0 uv(c_1 - u - v), \quad c_0, c_1 > 0, \qquad (8)$$

and vanishes ($c(u,v) = 0$) otherwise. This quadratic function is convex and takes its global maximum in

$$(\hat{u},\hat{v}) = \left(\frac{c_1}{3}, \frac{c_1}{3}\right). \qquad (9)$$

In the sequel some supporting arguments for the constraints (7) and in particular for the nonlinear model (8) are summarized. First of all, the constraint $c(u,0) = 0$ for all v corresponds to carelessness; during the absence, and also in the case of a small number of infected persons, the consciousness of the danger of the disease is not sufficiently present, even though there might be are some single dangerous intercourses. The constraint $c(0,v) = 0$ for all u is not only set for symmetry reasons, since, at a small number of susceptible agents, they have little chance to form a group consciousness on the importance of a separation from the infected population; instead, the susceptible population is absorbed by the infected population. The maximum $(\hat{u},\hat{v})$ has the interpretation that there is most avoidance when there is a fairly equal mixing of susceptible and infected population, whereas at small population numbers there is less need and at large total population numbers no possibility for avoidance.

In the situation of a large concentration of persons there is little possibility of a selective avoidance. Since the infected species is present anywhere and thus cannot be sustainably avoided in the crowd, there is small to no possibility to keep distance from the infected species. Thus, fatalism rules above a certain threshold 'upper' population number. This fatalism is modelled by the assumption that the cross-diffusion coefficient vanishes above this threshold population number. This upper population bound is set by the function V such that c_1 corresponds to a maximum population, where, in the case that $c_1 = u + v$, total fatalism rules.

A formal property of the constraints (7) is the that the cross-diffusion is switched off at a certain finite total population. By this setting, it is prevented that the population attains (unrealistic) local population peaks. In fact, by this limitation of the cross-diffusion, a maximum principle for the system (1) is imposed. The switch-off of the diffusion term can be seen in the context of equations with strongly degenerate diffusion, where the diffusion function is set to zero on an interval as proposed in (Bürger & Ruiz-Baier 2009, Bürger et al. 2010).

3 NUMERICAL ANALYSIS

In order to provide a space-adaptive numerical scheme, we apply the technique of fully adaptive multiresolution (Bendahmane et al. 2009a, Bürger et al. 2010) constructed on the basis of a reference finite volume approximation (Eymard et al. 2000, see also e.g. Müller 2003 for a survey on multiresolution methods for PDEs). The success of this approach mainly relies on the strategy used for storing only the relevant information. The numerical approximation obtained in each time step is represented (and also computed) using a dynamically evolving adaptive mesh which is generated from a sequence of nested grids. An appropriate smoothness analysis of the solution is performed using wavelet decomposition and such information on the local smoothness is used to locally adapt the mesh and the numerical scheme. Essentially, positions related to small wavelet coefficients may be discarded, allowing for substantial data and CPU-time compression.

Admissible rectangular meshes

These ideas are made precise, first by introducing a nested hierarchy of grids $T^0 \subset \cdots \subset T^H$, where each grid T^ℓ, $\ell = 0, \ldots, H$ is assumed to be an *admissible rectangular mesh*. The index $\ell = 0$ corresponds to the coarsest and $\ell = H$ to the finest resolution level, which is fixed and chosen large enough at the beginning of the algorithm. That is, a partition of Ω formed by control volumes K^ℓ (open rectangles of maximum diameter h_K^{ℓ}), constrained by the condition that the segment joining the centers of two neighboring control volumes x_K^{ℓ} and x_L^{ℓ} must be orthogonal to the corresponding interface $\sigma = \sigma(K^\ell, L^\ell)$. The interface length is denoted by $|\sigma| = |\sigma(K^\ell, L)|$. By $\varepsilon(K^\ell)$ we denote the set of edges of K^ℓ, $\varepsilon_{\text{int}}(K^\ell)$ corresponds to those is in the interior of T^ℓ and $\varepsilon_{\text{ext}}(K^\ell)$ is the set of edges of K^ℓ lying on the boundary $\partial\Omega$, i.e.,

$$\varepsilon(K^\ell) = \varepsilon_{\text{int}}(K^\ell) \cup \varepsilon_{\text{ext}}(K^\ell),$$
$$\varepsilon_{\text{int}}(K^\ell) \cap \varepsilon_{\text{ext}}(K^\ell) = \varnothing \quad \text{for all } K^\ell \in T^\ell.$$

By $\varepsilon_{\text{int}}^{\ell}$ and $\varepsilon_{\text{ext}}^{\ell}$ we will denote the sets of all edges in the interior of T^ℓ and lying on the boundary $\partial\Omega$, respectively. For a given finite volume K^ℓ, we denote by $N(K^\ell)$ the set of neighbors of K^ℓ which share a common edge with K^ℓ. For all $L^\ell \in N(K^\ell)$, $d(K^\ell, L^\ell)$ denotes the distance between x_K^{ℓ} and x_L^{ℓ}.

Two one-level finite volume methods

In order to define the discrete marching formula for (1), we choose an admissible discretization of Ω_T consisting of an admissible mesh T^ℓ of Ω and a time step size $\Delta t > 0$. We may choose $N > 0$ as the smallest integer such that $N\,\Delta\, t \geq T$, and set $t^n := n\,\Delta\, t$ for $n \in \{0, \ldots, N\}$.

We denote the cell averages of u and v on $K^\ell \in T^\ell$ at time $t = t^n$ by the respective expressions

$$u_{K^\ell}^{n} := \frac{1}{|K^\ell|} \int_{K^\ell} u(x, t^n)\, dx,$$
$$v_{K^\ell}^{n} := \frac{1}{|K^\ell|} \int_{K^\ell} v(x, t^n)\, dx.$$

Furthermore, we define the coefficients

$$f_{K^\ell}^{n} := f\left(u_{K^\ell}^{n}, v_{K^\ell}^{n}\right), g_{K^\ell}^{n} := g\left(u_{K^\ell}^{n}, v_{K^\ell}^{n}\right),$$
$$a_{K^\ell}^{n} := a\left(u_{K^\ell}^{n}\right), b_{K^\ell}^{n} := b\left(v_{K^\ell}^{n}\right),$$
$$c_{K^\ell}^{n} := c\left(u_{K^\ell}^{n}, v_{K^\ell}^{n}\right).$$

For constant coefficient functions (3) one has an $a_{K^\ell}^{n} = a_0, b_{K^\ell}^{n} = b_0, c_{K^\ell}^{n} = c_0$ on all cells K^ℓ and time steps n. The computation starts from the initial cell averages

$$u_{K^\ell}^{0} = \frac{1}{|K^\ell|} \int_{K^\ell} u_0(x)\, dx, \quad v_{K^\ell}^{0} = \frac{1}{|K^\ell|} \int_{K^\ell} v_0(x)\, dx.$$

The resulting finite volume scheme for the approximation of (1), defined on the multiresolution level l assumes values $u_{K^\ell}^{n}$ and $v_{K^\ell}^{n}$ for all $K^\ell \in T^\ell$ at time $t = t^n$ and determines $u_{K^\ell}^{n+1}$ and $v_{K^\ell}^{n+1}$ for all $K^\ell \in T^\ell$ at time $t = t^{n+1} = t^n + \Delta t$ by a marching formula. For linear coefficients (3) the system (1) is discretized as

$$|K^\ell| \frac{u_{K^\ell}^{n+1} - u_{K^\ell}^{n}}{\Delta t} =$$
$$|K^\ell| f_{K^\ell}^{n} + \sum_{\sigma \in \varepsilon_{\text{int}}(K^\ell)} \frac{\left|\sigma(K^\ell, L^\ell)\right|}{d(K^\ell, L^\ell)} \Big\{ a_0 \left(u_{L^\ell}^{n} - u_{K^\ell}^{n}\right)$$
$$+ c_0 \left(v_{L^\ell}^{n} - v_{K^\ell}^{n}\right) \Big\},$$

$$|K^\ell|\frac{v^{n+1}_{K^\ell}-v^n_{K^\ell}}{\Delta t}=$$
$$|K^\ell|g^n_{K^\ell}+\sum_{\sigma\in\mathcal{E}_{\text{int}}(K^\ell)}\frac{|\sigma(K^\ell,L^\ell)|}{d(K^\ell,L^\ell)}b_0\left(u^n_{L^\ell}-u^n_{K^\ell}\right). \quad (10)$$

This marching formula is valid for all cells and in particular for the boundary cells. The no-slip boundary condition is considered automatically by not considering boundary fluxes, such that they are automatically set to zero.

For nonlinear coefficient functions the generalization of (10) is not uniquely determined. Therefore, two versions are suggested, which are denoted by scheme A and scheme B, respectively. Scheme A has the form

$$|K^\ell|\frac{u^{n+1}_{K^\ell}-u^n_{K^\ell}}{\Delta t}=|K^\ell|f^n_{K^\ell}+$$
$$\sum_{\sigma\in\mathcal{E}_{\text{int}}(K^\ell)}\frac{|\sigma(K^\ell,L^\ell)|}{d(K^\ell,L^\ell)}\left\{\frac{\left(a^n_{L^\ell}+a^n_{K^\ell}\right)}{2}\left(u^n_{L^\ell}-u^n_{K^\ell}\right)+\right.$$
$$\left.\frac{2c^n_{K^\ell}C^n_{L^\ell}}{C^n_{K^\ell}+C^n_{L^\ell}}\left(v^n_{L^\ell}-v^n_{K^\ell}\right)\right\},$$
$$|K^\ell|\frac{v^{n+1}_{K^\ell}-v^n_{K^\ell}}{\Delta t}=|K^\ell|g^n_{K^\ell}+$$
$$\sum_{\sigma\in\mathcal{E}_{\text{int}}(K^\ell)}\frac{|\sigma(K^\ell,L^\ell)|}{d(K^\ell,L^\ell)}\frac{\left(b^n_{L^\ell}+b^n_{K^\ell}\right)}{2}\left(v^n_{L^\ell}-v^n_{K^\ell}\right). \quad (11)$$

Whereas the coefficient functions $a(u,v)$ and $b(u,v)$ are averaged, for the coefficient function $c(u,v)$ the exchange coefficient in the cross-diffusion term is computed by the following harmonic mean formula (see e.g. Eymard et al. 2000).

$$c\left(\left(u^n_{K^\ell},v^n_{K^\ell}\right);\left(u^n_{L^\ell},v^n_{L^\ell}\right)\right):=\frac{2c^n_{K^\ell}c^n_{L^\ell}}{c^n_{K^\ell}+c^n_{L^\ell}}. \quad (12)$$

We note that (12) is consistent in the sense that $c((u,v);(u,v))=c(u,v)$.

Even though scheme A looks reasonable, it is not conservative. For the discretization of conservation laws, it is well known that a non-conservative discretization might converge to a wrong solution (see e.g. Hayes & LeFloch 1998). For non-conservative equations a possible remedy is the formulation of path-conservative schemes (Castro et al. 2006, Parés 2006). For parabolic equations (as treated here) there is a similar situation, which demands a careful consideration. In Bürger et al. 2000, Figure 2, it is demonstrated that a non-conservative discretization of the parabolic term

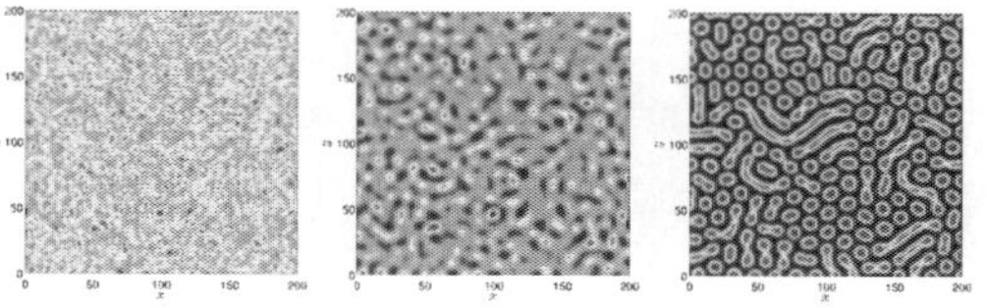

Figure 2. Numerical solution for the susceptible species u (top) and leaves of the corresponding tree data structure (bottom) at time instants $t = 10$, $t = 100$ and $t = 1500$ (Example 1).

can produce spurious solutions; our scheme A is a two-dimensional version of the nonconservative discretization specified in their formula (12). Therefore, we try to mimic in scheme B the conservative discretization, see e.g. their formula (13). In other words, scheme B avoids averaged transmission coefficients, instead finite differences are calculated in terms of the antiderivatives A, B, C,

$$\frac{\partial A(u)}{\partial u}=a(u),\quad \frac{\partial B(v)}{\partial v}=b(v),$$
$$\frac{\partial C(u,v)}{\partial v}=c(u,v),$$

that retain the local nonlinear properties. Differencing with respect to these antiderivatives gives to the resulting scheme a conservative form. Scheme B has the form

$$|K^\ell|\frac{u^{n+1}_{K^\ell}-u^n_{K^\ell}}{\Delta t}=|K^\ell|f^n_{K^\ell}+$$
$$\sum_{\sigma\in\mathcal{E}_{\text{int}}(K^\ell)}\frac{|\sigma(K^\ell,L^\ell)|}{d(K^\ell,L^\ell)}\left\{\left(A^n_{L^\ell}-A^n_{K^\ell}\right)+\left(C^n_{L^\ell}-C^n_{K^\ell}\right)\right\},$$
$$|K^\ell|\frac{v^{n+1}_{K^\ell}-v^n_{K^\ell}}{\Delta t}=|K^\ell|g^n_{K^\ell}+ \quad (13)$$
$$\sum_{\sigma\in\mathcal{E}_{\text{int}}(K^\ell)}\frac{|\sigma(K^\ell,L^\ell)|}{d(K^\ell,L^\ell)}\left(B^n_{L^\ell}-B^n_{K^\ell}\right),$$

with $A^n_{K^\ell}:=A\left(u^n_{K^\ell}\right)$, $A^n_{L^\ell}:=A\left(u^n_{L^\ell}\right)$, $B^n_{K^\ell}:=B\left(u^n_{K^\ell}\right)$, $B^n_{L^\ell}:=B\left(u^n_{L^\ell}\right)$ which is justified due to the equalities

$$\Delta A(u)=a(u)\Delta u,\quad \Delta B(v)=b(v)\Delta v.$$

With respect to the definition of the coefficient C, there is the difficulty that

$$\nabla C=\frac{\partial C}{\partial u}\Delta u+\frac{\partial C}{\partial v}\Delta v,$$

i.e. there remains the unresolvable term $\frac{\partial C}{\partial u}\nabla u$. Therefore a semi-averaged form of $C^n_{K^\ell}$ is built as

$$C^n_{K^\ell} := C\left(\bar{u}, v^n_{K^\ell}\right), \quad C^n_{L^\ell} := C\left(\bar{u}, v^n_{L^\ell}\right),$$
$$\bar{u} = \bar{u}_{L^\ell K^\ell} = \frac{u^n_{L^\ell} + u^n_{K^\ell}}{2}.$$

The antiderivative of $c(u, v)$ as defined in (8) with respect to the variable v is calculated as

$$C(u,v) = c_0 u \tilde{v}^2 \left[(c_1 - u)/2 - \tilde{v}/3\right],$$

with $\tilde{v} = \min(v, c_1 - u)$. Scheme B should be more accurate than scheme A since the nonlinear functions are better approximated; more information is retained when differencing instead of simply calculating averages. For linear coefficient functions, when

$$A^n_{K^\ell} := a_0 u^n_{K^\ell}, \quad B^n_{K^\ell} := b_0 u^n_{K^\ell}, \quad C^n_{K^\ell} := c_0 u^n_{K^\ell},$$

both schemes (11) and (13) are the same and reduce to (10).

4 EXAMPLES

In Example 1, Model 1 is simulated, where the parameters are chosen according to Sun et al. 2009. The simulation is performed using a Cartesian mesh of $N = 262{,}144$ control volumes in the highest resolution level $H = 9$ and the time stepping is explicit with fixed time step $\Delta t = 0.01$. The model parameters are set to $K = 1000$, $\beta = 0.5$, and the constant self—and cross-diffusion coefficients are chosen to be $a_0 = 0.1$, $b_0 = 2$, $c_0 = 0.02$. The reference tolerance for the multiresolution algorithm is $\varepsilon_R = 0.001$. As initial data we assume that the density of both species is a random perturbation around the endemic stationary state (u^*, v^*). That is,

$$u(x,0) = u^* + u(x)\delta, v(x,0) = v^* + v(x)\delta, \quad x \in \Omega,$$

where $w(x)_\delta \in [0, 1]$ is a normally distributed variable, $w \in \{u, v\}$. In this contribution, two examples for the linear diffusion model are shown, in Berres & Ruiz-Baier 2011 there are two more examples. For Example 1 we set $d = 0.25$, $r = 0.27$, which gives $(u^*, v^*) = (74.0741, 74.0741)$. The computational domain for Example 1 is the square $\Omega = (0, 200)^2$.

In Figure 2 In Example 1, "islands" of high concentration of susceptible individuals are formed. This reflects the phase separation triggered by the susceptible species avoiding the infected species.

In Example 2, Model 2 is simulated, where the parameters for the reaction equation are the same as in Example 1. The parameters for Model 2 are calibrated such that they quantitatively recover the orders of magnitudes of Model 1. More specifically we choose $a_0 = 0.5$, $b_0 = 3$, $c_1 = 3u^*$,

$$c_0 = 0.02\left(u^* v^* \left(c_1 - u^* - v^*\right)\right)^{-1},$$

and the remaining parameters as in Example 1. The initial condition is now

$$u(x,0) = u^* + \rho u_\delta, v(x,0) = v^* + \rho v_\delta \quad x \in \Omega,$$

where $\rho = 1e^{-4}$.

From Figure 3 the result of a qualitative comparison between Schemes A and B for Model 2 is given; we also notice that the solution recovers the same scaling as in Example 1.We have computed a numerical solution of a one-dimensional problem using both schemes with a maximal resolution of 512 control volumes. Even though scheme A is based on a nonconservative discretization, one can see that the two solutions are almost indistinguishable. On the other hand, in terms of computational effort, scheme B has found to be more efficient than scheme A. Therefore, the numerical solution for the two-dimensional case has been computed using scheme B enhanced with the multiresolution strategy. In this setting, we also observe the formation of spatial patterns (see Figure 4). Notice however, that in contrast with the results related to Model 1, here the "islands" of high concentration values of the susceptible species are surrounded by a layer of low concentration values (also noticeable from Figure 3). This behavior is in well accordance with previous contributions in the field of

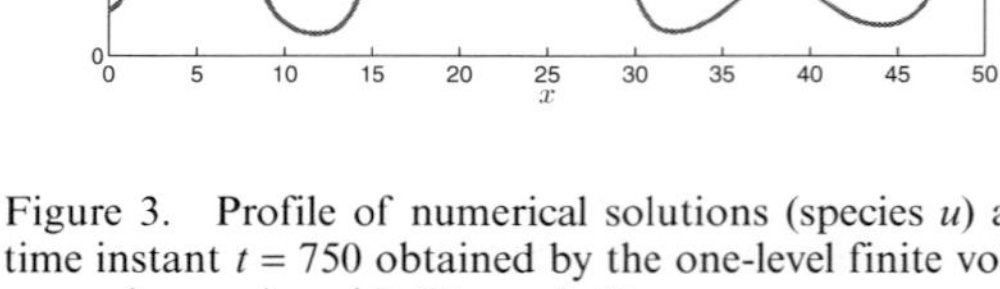

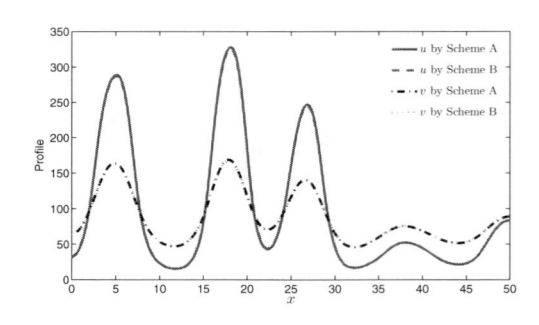

Figure 3. Profile of numerical solutions (species u) at time instant $t = 750$ obtained by the one-level finite volume schemes A and B (Example 2).

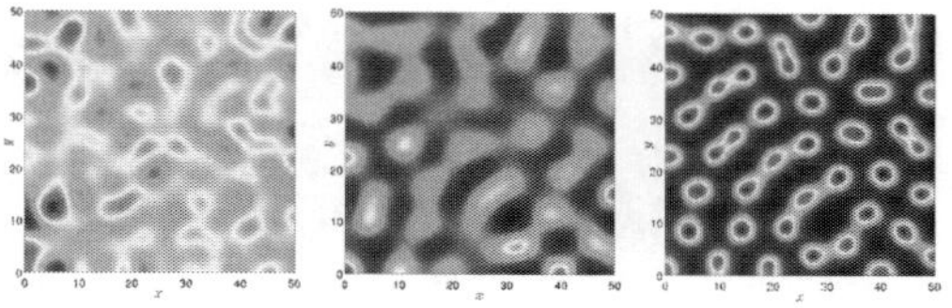

Figure 4. Numerical solution for u (top) and leaves of the corresponding tree data structure (bottom) at time instants $t = 50$, $t = 250$ and $t = 750$ (Example 2).

numerical simulation of cross-diffusion systems (see e.g. Andreianov et al. 2011, Galiano et al. 2003, Gambino et al. 2009).

5 CONCLUSIONS

This contribution is a condensed version of Berres & Ruiz-Baier 2011, where an efficient multiresolution method for the simulation of a nonlinear crossdiffusion model for epidemic dynamics is proposed. The ordinary differential equations involving the proposed reaction terms of our epidemic model are asymptotically stable in the sense that if the initial data are chosen close to the equilibrium then the solution converges to the equilibrium. The numerical examples of the two-dimensional reaction-diffusion equation show that there is a spatial phase separation in spite of the convergence behavior of the pure reaction terms. This means that the cross-diffusion in the parabolic terms of the reaction-diffusion equation is "stronger" than the attraction of the reaction terms. This is a remarkable property in comparison to other cross-diffusion models where the ordinary differential equations represented by the reaction part only show Lyapunov stability in the sense that initial data chosen in a close neighborhood of the equilibrium then the solution remains in this neighborhood.

For the simulation both a conservative and a non-conservative discretization have been proposed; both produce the same limit solution. The proposed schemes work stable both for the linear and nonlinear equations. The fully adaptive numerical method is particularly efficient to resolve phase interfaces due to the adaptive strategy.

From the numerical viewpoint, possible straightforward improvements include the use of time adaptive strategies such as local time stepping or Runge-Kutta-Fehlberg methods (Bendahmane et al. 2009a), or the use of a multiresolution analysis defined on general unstructured meshes.

From the model point of view one further issue is to choose different reaction kinetics in order study quantitatively how the pattern formation produced by the cross-diffusion term can be compensated by a stronger asymptotical stability of the reaction ODEs.

ACKNOWLEDGEMENTS

The first author is supported by Conicyt (Chile) through Fondecyt project #1120587. The work of the second author is supported by the European Research Council Advanced Grant "Mathcard, Mathematical Modelling and Simulation of the Cardiovascular System", Project ERC-2008-AdG 227058.

REFERENCES

Andreianov, B. Bendahmane, M. Ruiz-Baier, R. 2011. Analysis of a finite volume method for a crossdiffusion model in population dynamics. *Math. Meth. Models Appl. Sci.* 21: 307–344.

Badraoui, S. 2002. Existence of global solutions for systems of reaction-diffusion equations on unbounded domains. *Electron. J. Diff. Eqn.* 2002: 1–10.

Badraoui, S. 2006. Asymptotic behavior of solutions to a 2×2 reaction-diffusion system with a cross diffusion matrix on unbounded domains. *Electron. J. Diff. Eqn.* 2006: 1–13.

Barrett, J.W. & Blowey J.F. 2004. Finite element approximation of a nonlinear cross-diffusion population model. *Numer. Math.* 98: 195–221.

Bendahmane, M. B¨urger, R. Ruiz-Baier, R. Schneider, K. 2009a. Adaptive multiresolution schemes with local time stepping for two-dimensional degenerate reaction-diffusion systems. *Appl. Numer. Math.* 59: 1668–1692.

Bendahmane, M. Lepoutre, T. Marrocco, A. Perthame, B. 2009b. Conservative cross diffusions and pattern formation through relaxation. *J. Math. Pures Appl.* 92: 651–667.

Bendahmane, M. & Sepúlveda, M. 2009. Convergence of a finite volume scheme for nonlocal reaction-diffusion systems modelling an epidemic disease. *Disc. Cont. Dynam. Syst. - Series B 11*: 823–853.

Berres, S. Bürger, R. Frid, H. 2006. Neumann problems for quasi-linear parabolic systems modeling polydisperse suspensions. *SIAM J. Math. Anal.* 38: 557–573.

Berres, S. & Ruiz-Baier, R. 2011. A fully adaptive numerical approximation for a two-dimensional epidemic model with nonlinear cross-diffusion. *Nonlinear Anal., Real World Appl.* 12: 2888–2903.

Bürger, R. Evje, S. Karlsen, K.H. Lie, K.-A 2000. Numerical methods for the simulation of the settling of flocculated suspensions. *Chem. Eng. J.* 80: 91–104.

Bürger & R. Ruiz-Baier, R. 2009. Multiresolution simulation of reaction-diffusion systems with strong degeneracy. *Bol. Soc. Esp. Mat. Apl. S⃗eMA* 47: 73–80.

Bürger, R. Ruiz-Baier, R. Schneider, K. 2010. Adaptive multiresolution methods for the simulation of waves in excitable media. *J. Sci. Comput.* 43: 261–290.

Bürger, R. Ruiz-Baier, R. Schneider, K. Sepúlveda, M. 2008. Fully adaptive multiresolution schemes for strongly degenerate parabolic equations in one space dimension. *M2 AN Math. Model. Numer. Anal.* 42: 535–563.

Castro, M.J. Gallardo, J.M. Parés, C. 2006. High-order finite volume schemes based on reconstruction of states for solving hyperbolic systems with nonconservative products. Applications to shallow-water systems. *Math. Comp.* 75: 1103–1134.

Crandall, M. Pazy, A. Tartar, L. 1987. Global existence and boundedness in reaction-diffusion systems. *SIAM J. Math. Anal.* 18: 744–761.

Daddiouaissa, E.H. 2008. Existence of global solutions for a system of reaction-diffusion equations having a triangular matrix. *Electron. J. Diff. Eqn.* 2008: 141.

Eymard, R. Gallouët, T. Herbin, R. 2000. *Finite Volume Methods*. In: P.G. Ciarlet, J.L. Lions (eds.), Handbook of Numerical Analysis, vol. VII, North-Holland, Amsterdam: 713–1020.

Frid H. & Shelukhin, V. 2004. A quasi-linear parabolic system for three-phase capillary flow in porous media. *SIAM J. Math. Anal.* 35: 1029–1041.

Frid, H. & Shelukhin, V. 2005. Initial boundary value problems for a quasi-linear parabolic system in three-phase capillary flow in porous media. *SIAM J. Math. Anal.* 36: 1407–1425.

Galiano, G. Garzón, M.L. Jüngel, A. 2003. Semi-discretization and numerical convergence of a nonlinear cross-diffusion population model. *Numer. Math.* 93: 655–673.

Gambino, G. Lombardo, M.C. Sammartino, M. 2009. A velocity-diffusion method for a Lotka-Volterra system with nonlinear cross and self-diffusion. *Appl. Num. Math.* 59: 1059–1074.

Hayes, B.T. and LeFloch, P.G. 1998. Non classical shocks and kinetic relations: Finite difference schemes. *SIAM J. Numer. Anal.* 35: 2169–2194.

He, D. and Stone, L. 2003. Spatio-temporal synchronization of recurrent epidemics. *Proc. Roy. Soc. Lond. B* 270: 1519–1526.

Kermack, W.O. & McKendrick A.G. 1927. A contribution to the mathematical theory of epidemics. *Proc. Roy. Soc. Lond. A* 115: 700–721.

Kim, K.I. Lin, Z. Zhang, L. 2010. Avian-human influenza epidemic model with diffusion. *Nonl. Anal.: Real World Appl.* 11: 313–322.

Li, J. & Zou, X. 2009. Modeling spatial spread of infectious diseases with a fixed latent period in a spatially continuous domain. *Bull. Math. Biol.* 71: 2048–2079.

Li, K. Small, M. Zhang, H. Fu, X. 2010. Epidemic outbreaks on networks with effective contacts. *Nonl. Anal.: Real World Appl.* 11: 1017–1025.

Liu, J. & Zhang, T. 2010. Analysis of a nonautonomous epidemic model with density dependent birth rate. *Appl. Math. Model.* 34: 866–877.

Milner, F.A. & Zhao, R. 2008. S-I-R model with directed spatial diffusion. *Math. Popul. Stud.* 15: 160–181.

Müller, S. 2003. Adaptive Multiscale Schemes for Conservation Laws. Berlin: Springer-Verlag.

Ni, W.-M. 2004. Diffusion and cross-diffusion in pattern formation. *Rend. Mat. Acc. Lincei* 15: 197–214.

Parés, C. 2006. Numerical methods for nonconservative hyperbolic systems: A theoretical framework. *SIAM J. Numer. Anal.* 44: 300–321.

Phongthanapanich, S. & Dechaumphai, P. 2009. Finite volume element method for analysis of unsteady reaction-diffusion problems. *Acta Mech. Sin.* 25: 481–489.

Sun, G.-Q. Jin, Z. Liu, Q.-X. Li, L. 2009. Spatial pattern in an epidemic system with cross-diffusion of the susceptible. *J. Biol. Syst.* 17: 141–152.

Tian, C. Lin, Zh. Pedersen, M. 2010. Instability induced by cross-diffusion in reaction-diffusion systems. *Nonl. Anal.: Real World Appl.* 11: 1036–1045.

Webb, G. 1981. A reaction-diffusion model for a deterministic diffusive epidemic. *J. Math. Anal. Appl.* 84: 150–161.

Wong, J.C. 2008. The Galerkin finite element method for the solution of some spatio-temporally dependent reaction-diffusion systems. *Appl. Numer. Math.* 58: 352–375.

Numerical Methods for Hyperbolic Equations – Vázquez-Cendón et al. (eds)
© 2013 Taylor & Francis Group, London, ISBN 978-0-415-62150-2

A simple model of filtration and macromolecule transport through microvascular walls

L. Facchini, A. Bellin & E.F. Toro
University of Trento, Italy

ABSTRACT: Multiple Sclerosis (MS) is a disorder that usually appears in adults in their thirties. It has a prevalence that ranges between 2 and 150 per 100 000. Epidemiological studies of MS have provided hints on possible causes for the diseaseranging from genetic, environmental and infectious factors to other factors of vascular origin. Despite the tremendous effort spent in the last few years, none of the hypotheses formulated so far has gained wide acceptance and the causes of the disease remain unknown. From a clinical point of view, a high correlation has been recently observed between MS and Chronic Cerebro-Spinal Venous Insufficiency (CCSVI) in a statistically significant number of patients. In this pathological situation CCSVI may induce alterations of blood pressure in brain microvessels, thereby perturbing the exchange of small hydrophilic molecules between the blood and the external cells. In the presence of large pressure alterations it cannot be excluded also the leakage of macromolecules that otherwise would not cross the vessel wall. All these disorders may trigger immune defenses with the destruction of myelin as a side effect. In the present work we investigate the role of perturbed blood pressure in brain microvessels as driving force for an altered exchange of small hydrophilic solutes and leakage of macromolecules into the interstitial fluid. With a simplified, yet realistic, model we obtain closed-form steady-state solutions for fluid flow and solute transport across the microvessel wall. Finally, we use these results (i) to interpret experimental data available in the literature and (ii) to carry out a preliminary analysis of the disorder in the exchange processes triggered by an increase of blood pressure, thereby relating our preliminary results to the hypothesised vascular connection to MS.

1 INTRODUCTION

Multiple Sclerosis is an autoimmune neurodegenerative disorder of unknown origin that damages the myelin, a fatty layer that envelops and protects the axons. The damage of the myelin causes neuron electrical impulses to travel slowly along their axons, leading to a variety of symptoms with debilitating consequences. The repeated damage of the myelin causes the loss of the remyelination capacity of oligodendrocytes and produces scar-like lesions around damaged axons. From a clinical point of view, these lesions are demonstrated to be localised in the white matter and to be venocentric (i.e., these plaques are always found around venules).

Recent clinical evidence suggests an association of MS with CCSVI (Zamboni 2006; Singh & Zamboni 2009; Zamboni et al., 2009). However, the evolution of the process from venous stenosis to local hypertension and leakage of hematic substance, which may trigger the immune response as ultimate cause of demyelination and neurodegeneration typical of MS has been the subject of an intense debate with often opposed views. This is a controversial hypothesis, yet it is consistent with the predominantly venocentric orientation of the MS inflammatory lesions and with the otherwise unexplained perivenular iron deposition observed in many clinical cases (Adams 1988). A possible path connecting cerebrospinal venous stenosis to chronic fatigue and MS was proposed by Tucker (2011) on the basis of qualitative considerations of elementary fluid mechanics.

From a physiological point of view, the microvessel wall plays an important role in maintaining the equilibrium between intravascular and extravascular fluid compartments. Under normal conditions the vessel walls are nearly impermeable to macromolecules, while lipophilic species and small hydrophilic substances are allowed to cross the wall and reach the surrounding tissue. Fluid flow and transport of dissolved molecules across the walls depend on the permeability and diffusivity of the membrane composingthe wall. Therefore, alterations of the blood pressure may lead to impaired exchange processes and, in extreme cases, to leakage of hematic fluid. Several alterations of these exchange processes have been observed, mainly in compartments other than the brain, resulting in leakage of macromolecules, which is typically attributed to reduction of osmotic pressure, or inflammatory processes that alter the endothelial structure.

In the present work we investigate the role of an altered blood pressure as the driving force for alteration of the exchange processes and the leakage of macromolecules. In particular, we analyze through a simplified, yet realistic, flow and transport model, the impact of alterations in the hydrostatic blood pressure on transport of molecules across the microvessel wall. The microvessel wall is assumed to be composed of two layers with different permeability and porosity, as assumed in previous studies on fluid flow and macromolecules transport in heteroporous membranes. The inner layer represents the glycocalyx, a membrane composed of extracellular polymeric material which is believed to exert an important sieving effect on macromolecules, while the external layer represents the combined effect of the endothelial cells, the basal membrane and the external astrocyte feet.

With this model we obtain closed-form steady-state solutions for the fluid flow and solute transport through the microvessel walls, which can be used for a preliminary analysis of the leakage of macromolecules due to an increase of blood pressure in CCSVI/MS patients.

2 MICROVESSEL ANATOMY

In this section we summarise anatomic features of mammalian blood vessels useful to describe the geometry of the computational domain used in the present work.

The circulatory system is composed by vessels of size ranging from centimetres in the main ones to a few microns in the capillary bed. The structure of the vessel wall differs between arteries and veins and also between large vessels and capillaries. Eachsegment of circulation shows an optimal combination of size, wall composition, thickness and cross-sectional area that best fulfils its function. For example, arteries are more muscular than veins because they have to bear the pumping force of the heart.

Large vessels are formed of three layers: the endothelium, the middle layer composed by smooth muscle cells and the connective layer. On the contrary, small vessels such as capillaries, venules and arterioles are only one-cell thick, in order to optimize the exchange of small hydrophilic molecules from the blood stream to the interstitial volume before crossing the cell membrane.

Molecules dissolved in water are driven through the vessel wall by the gradient of the net pressure p, which is given by the difference between the hydrostatic P and osmotic Π pressure: $p = P - \sigma\Pi$, where σ is the reflection coefficient. σ depends on the ratio between the Stokes radius of the molecule and the pore radius, or the size of the cleft between adjacent endothelial cells. When the size of the molecule is comparable with the pore size (or the aperture), the vessel wallbehaves as a perfect membrane and $\sigma \to 1$. On the other hand, when the molecules are much smaller than the pore size, the membrane effect vanishes and $\sigma \to 0$. In the latter case, transport across the vessel wall is controlled by the gradient of the hydrostatic pressure. For a given pore (or cleft) size, the role of the osmotic pressure increases with the size of molecule. In the present work we consider two layers, one represented by the glycocalyx and the other by the cleft. The sieving effect of glycocalyx on macromolecules is represented by a σ value that approaches 1, while in the stratum representing the endothelial cells σ is typically smaller, to reflect the larger aperture of the tight junctions connecting the two sides of the cleft at the border between adjacent cells (Levick 2010).

3 CONCEPTUAL MODEL

Let us approximate the microvessel geometry as a rigid circular cylinder, infinitely long in the z-direction, i.e. in the direction of the blood stream. We assume the vessel wall composed by one or more permeable layers of a given thickness. Physical properties, such as permeability and molecular diffusion are assumed constant within a layer, but may vary across the layers. The porosity is assumed the same in all layers. Molecules of a given Stokes radius are dissolved into the blood plasma at a concentration that does not modify its density and viscosity. Furthermore, to simplify the analysis we assume that the pressure gradient is small in the longitudinal direction, such that blood flow through the vessel lumen can be decoupled from the filtration through its wall. In general, the osmotic pressure changes with the solute concentration c. For small concentrations, the following linear relationship is often considered: $\Pi = \phi RTc$, where ϕ is a parameter that depends on the Stokes radius of the molecule, R is the gas constant and T is the absolute temperature. Consequently, the flow and transport equations are coupled through the concentration c that feeds back through Π to the flow. This leads Levick and Michel (2010) to conclude that microvessels cannot absorb fluid from the interstitial space, as is often argued. However, this feedback is important mainly when hydrostatic pressure is abruptly reduced, as in the Landis experiment (Landis 1932), whereas here we are interested in the increase of hydrostatic pressure. We therefore neglect this feedback andsolve the flow and transport equations separately.

Under the above assumptions, mass balance of the solvent and the solute leads to the following governing equations for the pressure $p = p(x,y,z,t)$

$$\frac{\partial p}{\partial t} = \frac{k\rho g}{\mu S_s}\nabla^2 p, \tag{1}$$

and for the concentration $c = c(x,y,z,t)$

$$\frac{\partial c}{\partial t} + \frac{\mathbf{q}}{n}\cdot\nabla c = \nabla\cdot(\mathbf{D}\cdot\nabla c). \tag{2}$$

Where k is the wall permeability, ρ is the blood density, g is the acceleration due to gravity, μ is the blood dynamic viscosity, S_s is the specific storage of the porous material, n is the porosity of the material and $\mathbf{D}$ is the diffusion tensor.

The specific water (solvent) discharge $\mathbf{q} = \mathbf{q}(x, y, z, t)$ is proportional to the net pressure gradient through the Starling equation (Levick 2010)

$$\mathbf{q} = -\frac{K}{\rho g}\nabla p, \tag{3}$$

where

$$K = \frac{k\rho g}{\mu}. \tag{4}$$

Finally, the mass flux of the solute $\mathbf{f}_m = \mathbf{f}_m(x,y,z,t)$ is given by

$$\mathbf{f}_m = (1-\sigma)\mathbf{q}c - n\mathbf{D}\cdot\nabla c. \tag{5}$$

The above equations written in cylindrical coordinates (r, θ, z) and assuming radial symmetry take the following form

$$\frac{\partial p}{\partial t} = \frac{k\rho g}{\mu S_s}\frac{1}{r}\frac{\partial}{\partial r}\left(r\frac{\partial p}{\partial r}\right), \tag{6}$$

$$q = -\frac{K}{\rho g}\frac{\partial p}{\partial r}, \tag{7}$$

$$\frac{\partial c}{\partial t} + \frac{q}{n}\frac{\partial c}{\partial r} = \left(\frac{d}{r} + \frac{\partial d}{\partial r}\right)\frac{\partial c}{\partial r} + d\frac{\partial^2 c}{\partial r^2}, \tag{8}$$

$$f_m = (1-\sigma)qc - nd\frac{\partial c}{\partial r}. \tag{9}$$

Since the initial and boundary conditions are independent from the coordinates z and θ, we only consider the radial component d of the diffusion tensor $\mathbf{D}$.

Furthermore, we assume that d is given by the sum of the molecular diffusion d_m and the hydrodynamic dispersion $d_h = Aq$ where A is the dispersivity and q is the radial component of the specific discharge $\mathbf{q}$.

In the next section we consider the steady-state solution of the above flow and transport equations.

4 ANALYSIS

The steady-state equations for the solvent and for the solute in cylindrical coordinates assume the following form

$$0 = \frac{\partial}{\partial r}\left(r\frac{\partial p}{\partial r}\right), \tag{10}$$

$$\frac{q}{n}\frac{\partial c}{\partial r} = \left(\frac{d}{r} + \frac{\partial d}{\partial r}\right)\frac{\partial c}{\partial r} + d\frac{\partial^2 c}{\partial r^2}. \tag{11}$$

4.1 *Steady-state solutions for a single-layer vessel wall*

We now consider a geometrical situation as depicted in Figure 1(a), which shows a cylinder whose inner surface of the endothelial cells is represented by radius r_1 and whose outer wall is determined by radius r_2.

In this case, we obtain two generic solutions

$$p(r) = \alpha + \beta \ln r, \qquad r \in [r_1, r_2] \tag{12}$$

$$c(r) = \delta + \gamma h(r), \qquad r \in [r_1, r_2] \tag{13}$$

each of them depending on two parameters which can be computed by imposing the boundary

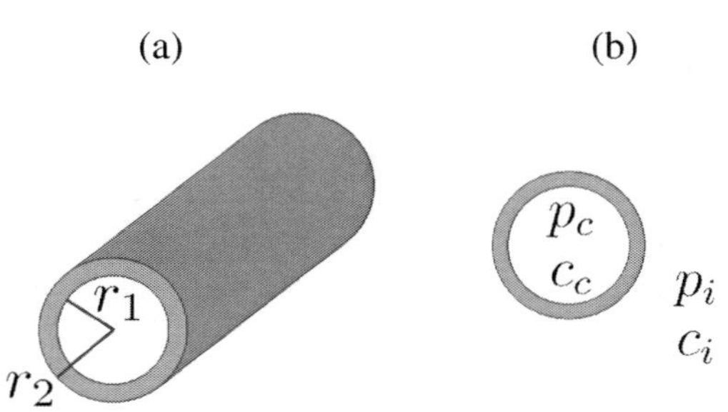

Figure 1. (a) The domain is an infinitely long hollow cylinder composed by one layer only, whose inner and outer radii are r_1 and r_2. (b) Cross section depicting boundary conditions, where $p_c = P_c - \sigma\Pi_c$ and c_c refer to the net blood pressure and solute blood concentration, respectively and $p_i = P_i - \sigma\Pi_i$ and c_i indicate the blood pressure and solute concentration in the interstitial fluid.

conditions, where $h(r)$ is an auxiliary function depending on the permeability and on the boundary conditions

$$h(r)=\begin{cases}-\dfrac{\mu n}{k\beta}\left[\mu d_m r-k\beta A\right]^{-\dfrac{k\beta}{n\mu d_m}}, & \text{if } k(p_c-p_i)\neq 0\\ \dfrac{\ln(\mu d_m r)}{d_m}, & \text{if } k(p_c-p_i)=0\end{cases} \tag{14}$$

for $r\in[r_1,r_2]$, with

$$\beta=\frac{p_c-p_i}{\ln r_1-\ln r_2}. \tag{15}$$

We suppose that the boundary conditions are independent from the coordinates z and θ. So we set constant pressures and concentrations at the boundary, as depicted in Figure 1(b),

$$p(r_1)=p_c, \tag{16}$$

$$p(r_2)=p_i, \tag{17}$$

$$c(r_1)=c_c, \tag{18}$$

$$c(r_2)=c_i, \tag{19}$$

where $p_c=P_c-\sigma\Pi_c$ refers to the net blood pressure, c_c to the solute blood concentration, $p_i=P_i-\sigma\Pi_i$ indicates the blood pressure in the interstitial fluid and c_i the interstitial solute concentration. So we obtain closed-form steady-state solutions for $r\in[r_1,r_2]$, given by

$$p(r)=\frac{p_c\ln(r_2/r)+p_i\ln(r/r_1)}{\ln(r_2/r_1)}, \tag{20}$$

$$q(r)=\frac{K}{\ln r_1-\ln r_2}\frac{p_c-p_i}{\rho g r}, \tag{21}$$

$$c(r)=\frac{c_c[h(r)-h(r_2)]+c_i[h(r_1)-h(r)]}{h(r_1)-h(r_2)}. \tag{22}$$

The mass flux depends on the values of the permeability and on the boundary conditions

$$f_m(r)=\begin{cases}-\dfrac{K\beta}{\rho g r}\dfrac{c_i h(r_1)-c_c h(r_2)}{h(r_1)-h(r_2)}, & \text{if } k(c_c-c_i)\neq 0\\ -\dfrac{n}{r}\dfrac{c_c-c_i}{h(r_1)-h(r_2)}, & \text{if } k(c_c-c_i)=0\end{cases} \tag{23}$$

for $r\in[r_1,r_2]$, with

$$\beta=\frac{p_c-p_i}{\ln r_1-\ln r_2}. \tag{24}$$

4.2 *Steady-state solutions for a vessel wall composed by two layers*

We now consider a more complex case in which the vessel wall is composed by two layers (r_1,r_2) and (r_2,r_3), as depicted in Figure 2(a), with different values of permeability, diffusivity and reflection coefficient.

The general case of m layers can be treated similarly.

The generic solution for the net pressure assumes the following form

$$\begin{cases}p_1(r)=\alpha_1+\beta_1\ln r, & r\in[r_1,r_2]\\ p_2(r)=\alpha_2+\beta_2\ln r, & r\in[r_2,r_3]\end{cases} \tag{25}$$

while the solute concentration is given by

$$\begin{cases}c_1(r)=\delta_1+\gamma_1 h_1(r), & r\in[r_1,r_2]\\ c_2(r)=\delta_2+\gamma_2 h_2(r), & r\in[r_2,r_3]\end{cases} \tag{26}$$

where, similarly to the previous case, $h_j(r)$ is a function that depends on the geometry and the permeability of the layer

$$h_j(r)=\begin{cases}-\dfrac{\mu n}{k_{eq}B}\left[\mu d_{m_j} r-k_{eq}BA_j\right]^{-\dfrac{k_{eq}B}{n\mu d_{m_j}}}, & \text{if } k_1k_2(p_c-p_i)\neq 0\\ \dfrac{\ln\left(\mu d_{m_j} r\right)}{d_{m_j}}, & \text{if } k_1k_2(p_c-p_i)=0\end{cases} \tag{27}$$

for $r\in[r_j,r_{j+1}]$, recalling that d_{m_j} is the molecular diffusion and A_j is the dispersivity of the j-th layer. B and k_{eq} are now defined as

$$B=p_c-p_i, \tag{28}$$

$$k_{eq}=\frac{k_1k_2}{k_1(\ln r_2-\ln r_3)+k_2(\ln r_1-\ln r_2)}. \tag{29}$$

The constants appearing in the above solutions are obtained by imposing suitable boundary conditions for both the net pressure and solute concentration, as depicted in Figure 2(b)

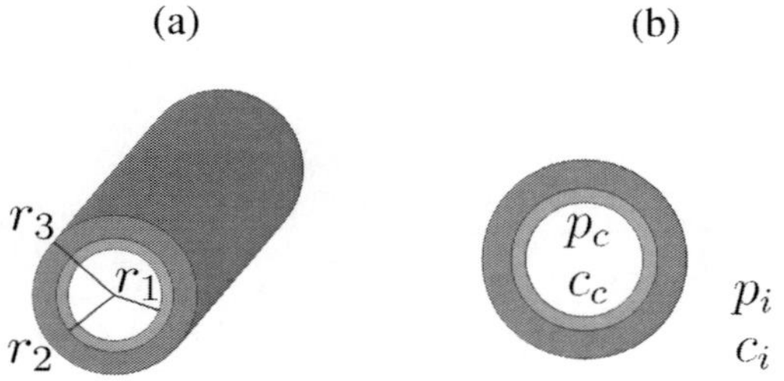

Figure 2. (a) The domain is an infinitely long hollow cylinder composed by two layers (r_1, r_2) and (r_2, r_3). (b) Cross section depicting boundary conditions, where $p_c = P_c - \sigma_1 \Pi_c$ and c_c refer to the net blood pressure and solute blood concentration, respectively and $p_i = P_i - \sigma_2 \Pi_i$ and c_i indicate the blood pressure and solute concentration in the interstitial fluid.

$$p_1(r_1) = p_c, \tag{30}$$

$$p_2(r_3) = p_i, \tag{31}$$

$$c_1(r_1) = c_c, \tag{32}$$

$$c_2(r_3) = c_i, \tag{33}$$

where $p_c = P_c - \sigma_1 \Pi_c$ refers to the net blood pressure, c_c to the solute blood concentration, $p_i = P_i - \sigma_2 \Pi_i$ indicates the blood pressure in the interstitial fluid and c_i the interstitial solute concentration. These boundary conditions should be supplemented by the conditions resulting from imposing the continuity of the specific discharge and the solute flux at the interface between the two layers at $r = r_2$

$$q_1(r_2) = q_2(r_2), \tag{34}$$

$$f_{m,1}(r_2) = f_{m,2}(r_2), \tag{35}$$

and that both pressure and solute concentration are continuous at $r = r_2$

$$p_1(r_2) = p_2(r_2), \tag{36}$$

$$c_1(r_2) = c_2(r_2). \tag{37}$$

With all these conditions the pressure within the first and second layer are given by

$$p_1(r) = \frac{k_2\left[p_i \ln(r_1/r) - p_c \ln(r_2/r)\right] + p_c k_1 \ln(r_2/r_3)}{k_1 \ln(r_2/r_3) + k_2 \ln(r_1/r_2)} \tag{38}$$

and

$$p_2(r) = \frac{k_1\left[p_i \ln(r_2/r) - p_c \ln(r_3/r)\right] - p_i k_2 \ln(r_2/r_1)}{k_1 \ln(r_2/r_3) + k_2 \ln(r_1/r_2)} \tag{39}$$

respectively.

The resulting expression for the specific discharge is the same in the two regions, indeed

$$q(r) = -\frac{k_j}{\mu} \cdot \frac{\partial p_j}{\partial r}(r) = = -\frac{k_1 k_2 (p_c - p_i)}{\mu[k_1 \ln(r_2/r_3) + k_2 \ln(r_1/r_2)]} \frac{1}{r}, \tag{40}$$

for $r \in [r_1, r_3]$, where $j \in \{1,2\}$ indicates the layer we are considering.

Similarly, under steady-state conditions, the solute concentration assumes the following expression

$$c(r) = \begin{cases} c_1(r) = \dfrac{S_1 + T_1 h_1(r)}{V}, & r \in [r_1, r_2] \\ c_2(r) = \dfrac{S_2 + T_2 h_2(r)}{V}, & r \in [r_2, r_3] \end{cases} \tag{41}$$

where the parameters S_1, T_1, S_2, T_2, V depend on the value of $k_1 k_2 (p_c - p_i)$. Indeed, these parameters are defined as

$$S_1 = \begin{cases} c_i h_1(r_1) - c_c[h_1(r_2) - h_2(r_2) + h_2(r_3)], \\ \qquad \text{if} \quad k_1 k_2 (p_c - p_i) = 0 \\ c_c(1 + \sigma_1 - \sigma_2) h_1(r_2) h_2(r_3) - c_i h_1(r_1) h_2(r_2) \\ + c_c(\sigma_2 - \sigma_1) h_1(r_2) h_2(r_2), \\ \qquad \text{if} \quad k_1 k_2 (p_c - p_i) \neq 0 \end{cases} \tag{42}$$

$$T_1 = \begin{cases} c_c - c_i, & \text{if} \quad k_1 k_2 (p_c - p_i) = 0 \\ \left[c_i - c_c(1 - \sigma_1 + \sigma_2)\right] h_2(r_2) + c_c(\sigma_2 - \sigma_1) h_2(r_3), \\ \qquad \text{if} \quad k_1 k_2 (p_c - p_i) \neq 0 \end{cases} \tag{43}$$

$$S_2 = \begin{cases} c_i[h_1(r_1) - h_1(r_2) + h_2(r_2)] - c_c h_2(r_3), \\ \qquad \text{if} \quad k_1 k_2 (p_c - p_i) = 0 \\ c_c h_1(r_2) h_2(r_3) + c_i(\sigma_2 - \sigma_1) h_1(r_2) h_2(r_2) + \\ - c_i(1 - \sigma_1 + \sigma_2) h_1(r_1) h_2(r_2), \\ \qquad \text{if} \quad k_1 k_2 (p_c - p_i) \neq 0 \end{cases} \tag{44}$$

$$T_2 = \begin{cases} c_c - c_i, & \text{if} \quad k_1 k_2 (p_c - p_i) = 0 \\ \left[c_i(1 + \sigma_1 - \sigma_2) - c_c\right] h_1(r_2) + c_i(\sigma_2 - \sigma_1) h_1(r_1), \\ \qquad \text{if} \quad k_1 k_2 (p_c - p_i) \neq 0 \end{cases} \tag{45}$$

$$V = \begin{cases} h_1(r_1) - h_1(r_2) + h_2(r_2) - h_2(r_3), \\ \qquad \text{if} \quad k_1 k_2 (p_c - p_i) = 0 \\ (\sigma_2 - \sigma_1)[h_1(r_2)h_2(r_2) - h_1(r_1)h_2(r_3)] + \\ -(1 - \sigma_1 + \sigma_2)h_1(r_1)h_2(r_2) \\ +(1 + \sigma_1 - \sigma_2)h_1(r_2)h_2(r_3), \\ \qquad \text{if} \quad k_1 k_2 (p_c - p_i) \neq 0. \end{cases} \tag{46}$$

The resulting solute flux is the following

$$f_{m,j}(r) = \begin{cases} -\frac{nT_j}{V}\frac{1}{r}, \quad \text{if} \quad k_1 k_2 (p_c - p_i) = 0 \\ \frac{k_{eq}B}{\mu r}\left[\sigma_j \frac{T_j}{V} h_j(r) - (1 - \sigma_j)\frac{S_j}{V}\right], \\ \qquad \text{if} \quad k_1 k_2 (p_c - p_i) \neq 0 \end{cases} \tag{47}$$

for $r \in [r_j, r_{j+1}]$, where $j \in \{1,2\}$ indicates the layer we are considering and σ_j represents the reflection coefficient that may be different in the two layers.

Similar expressions may be obtained for three and more layers.

4.3 *The travel time through the vessel wall*

An important quantity in the exchange process is the time a single solute molecule takes to cross the vessel wall. We call this time the travel time τ, in analogy with transport in porous media.

For the single layer case, τ may be approximated by neglecting the diffusive component of the mass flux

$$\tau = \int_{r_1}^{r_2} \frac{n}{(1-\sigma)q(r)} \mathrm{d}r = \\ = -\frac{n\rho g}{p_c - p_i} \frac{\ln r_1 - \ln r_2}{K(1-\sigma)} \frac{r_2^2 - r_1^2}{2}. \tag{48}$$

5 PRELIMINARY RESULTS

The structure of the vessels is very specialized in relation to their functionality and this specialization results in different permeability and reflection coefficients of the vessel wall. Table 1 shows typical values of the geometrical properties of microvessels together with the hydraulic conductivity to serum albumin and the reflection coefficient. Although the permeability of the venules is expected to be larger than the permeability of the arterioles, in the absence of specific data, and for illustration purposes in the subsequent exercise we assumed the same permeability for both microvessels.

Table 1. Typical values of the parameters used in the computation. K is the hydraulic conductivity for serum albumin, σ is the reflection coefficient for serum albumin, n is the porosity, r is the mean radius of the vessel and Δx is the vessel thickness. A refers to the arteriolar end of the capillary bed, while V to the venous end.

Parameter [unit]	Value	Reference
K [kg sec^{-3} (cm H_2O)$^{-1}$]	$2.49 \cdot 10^{-12}$	Michel & Curry 1999
σ	0.85	Michel 1980
n	0.5	Robinson 1988
r_A [μ m]	15	Silverthorn 2010
r_V [μ m]	10	Silverthorn 2010
Δx_A [μ m]	6	Silverthorn 2010
Δx_V [μ m]	1	Silverthorn 2010

In addition, venules and arterioles are subjected to different internal hydrostatic pressures and external osmotic pressures. Table 2 shows the typical mean pressures in different microvessels.

The difference of the net pressure P between the internal (subscript c) and the external side (subscript i) of the microvessels, i.e.

$$\Delta p = p_c - p_i = (P_c - \sigma\Pi_c) - (P_i - \sigma\Pi_i), \tag{49}$$

provides a first rough quantification of the expected flux through the vessel wall per unit area, i.e. the specific discharge. In Table 2, we observe that Δp is positive for arterioles (12.38 cm H_2O) and negative for venules (-10.88 cmH_2O). This leads to a tendency for absorpion at the venular end of the capillary bed, which may be contrasted by the parallel increase of the osmotic pressure within the clefts just downstream the glycocalyx, the membrane coating the internal surface of the endothelial cells (Levick 2010). As mentioned before, in the present work we neglect this feedback mechanism.

We start by considering the microvessel wall composed by a single layer. Figure 3 shows the specific discharge q crossing the vessel wall as a function of the hydrostatic pressure P_c for both arterioles and venules.

In the case in which the vessel wall is composed by only one layer, we can study the behavior of the discharge per unit length and of the travel time of a molecule, assuming that the external pressures P_c and Π_i and the internal osmotic pressure Π_c are constant. The internal hydrostatic pressure P_c is the residual pressure, controlled by the cardiac pressure, so we can represent our quantities with respect to it.

For typical values of venular pressure (see the black bullet on the thick straight line in Figure 3), the specific discharge is negative, meaning that

Table 2. Mean pressures in human body, taken from Boron & Boulpaep 2005. P represents the hydrostatic pressure, while Π is the osmotic pressure and σ is the reflection coefficient. The subscript c refers to the pressure measured inside the vessel, while the subscript i is measured just outside the vessel. Δp is defined as the difference of the net pressure p between the internal and the external side of the microvessels, i.e. $\Delta p = p_c - p_i = (P_c - \sigma\Pi_c) - (P_i - \sigma\Pi_i)$.

	P_c	P_i	$\sigma\Pi_c$	$\sigma\Pi_i$	Δp
Location	cm H_2O	cm H_2O	cm H_2O	cm H_2O	cm H_2O
arteriolar end of capillary	47.62	−2.72	38.10	0.14	12.38
venular end of capillary	20.41	−2.72	38.10	4.08	−10.88

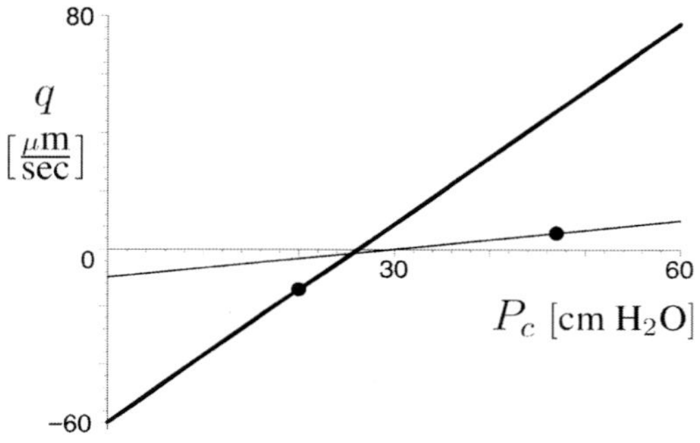

Figure 3. Discharge per unit length depending on the internal hydrostatic pressure, in the arteriolar case (thin straight line) and in the venular case (thick straight line). The dots represent the typical values of internal blood pressure in both cases.

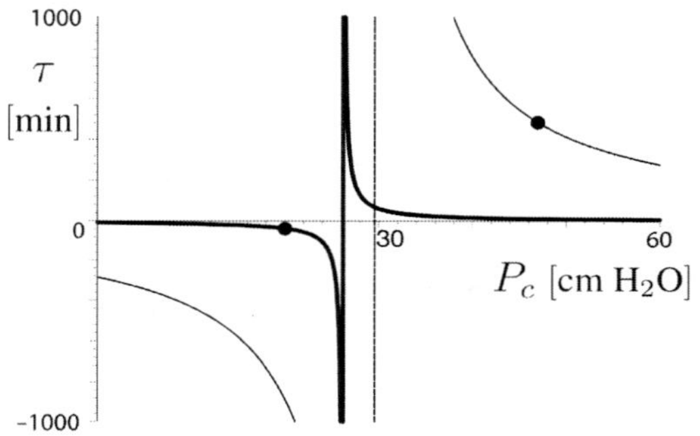

Figure 4. Travel time of a molecule of serum albumin, in arterioles (thin curve) and in venules (thick curve). The black bullets represent the typical values of internal blood pressure in both cases.

venules absorb fluid and the dissolved molecules from the interstitial volume. On the other hand, arteriolar pressure is positive letting the oxygen and the nutrients nourish the surrounding tissues.

In Figure 4, the travel time τ of a target molecule (in this case, serum albumin) is depicted with respect to the internal hydrostatic pressure P_c for arteriolar (thin curve) and venular end (thick curve) of the capillary bed.

For typical values of internal pressure (see the black bullets in Figure 4), τ is positive for arterioles and negative for venules, reflecting the opposite direction of the flow in the two cases. An increase of the hydrostatic pressure leadsto a reduction of τ for the arterioles. In the case of venules, the same increase leads to a larger travel time $|\tau|$. Both occurrences may induce a significant alteration of the exchange mechanisms between the interstitial fluid and the cells. If the hydrostatic pressure increases above a given threshold (about 26 cm H_2O, in the present case) the flux is inverted across the venule wall and the travel time becomes positive, thereby leading to leakage of hematic fluid from the venules into the interstitial volume. Close to this threshold τ is large, but it reduces rapidly as the hydrostatic pressure increases further. This may provide a plausible explanation for streaks of blood observed in the histology of MS brain plaques (Singh & Zamboni 2009).

Finally, we observe that our simple model is in agreement with the early experiments conducted by Landis (Landis 1932) in frog mesenteric capillaries.

6 CONCLUSIONS

We have presented a simplified analytical model of steady-state flow and transport of a target molecule through the wall of microvessels. The advantage of this model is that it allows us to easily explore the explicit influence of the many parameters controlling the process and thereby avoiding, for the time being, the use of numerical methods. With this model we have performed a preliminary analysis of the flux across arterioles and venules by using parameters taken from existing studies on mesenteric capillaries. In both cases we computed the time a target molecule (with a given reflection coefficient) spends crossing the wall, which may provide an indication of the alteration of exchange mechanisms due to modification of the hydrostatic pressure at the arteriolar and venular ends.

An increase of the hydrostatic pressure above the value observed in normal conditions leads to an increase of the flux crossing the wall of arterioles and a corresponding reduction of the travel time. In this condition more hydrophilic molecules are released in the interstitial fluid surrounding the vessel, thereby potentially reducing downstream the availability of such substances needed for the cell metabolism. On the venular side a threshold hydrostatic pressure separates two different ways of functioning. For hydrostatic pressures below such a threshold the flux is negative and the venule absorbs fluid from the interstitial space, while above this threshold the venule leaks hematic fluid to the interstitial space. An increase of the hydrostatic pressure has thena different impact according to the reference hydrostatic pressure. For low reference pressure (i.e., below the threshold) an increase of the hydrostatic pressure leads to a reduction of the absorption and a parallel increase of the travel time. However, if the reference pressure is larger than this threshold the venule behaves similarly to the arteriole and leaks hematic fluid to the interstitial space with a travel time that reduces rapidly with the increase of the hydrostatic pressure. The impact of these alterations on the cell metabolism may be significant and potentially may be responsible of the suffering status of oligodendrocytes of patients affected by MS and CCSVI.

REFERENCES

Adams, C.W. 1988. Perivascular iron deposition and other vascular damage in multiple sclerosis, *J Neurol Neurosurg Psychiatry* 51(2):260–265.

Barnett, M.H. & Sutton, I. 2006. The pathology of multiple sclerosis: a paradigm shift, *Curr Opin Neurol* 19(3):242–7.

Boron, W.F. & Boulpaep, E.L. 2005. *Medical Physiology* (2nd Edition) Philadelphia: Saunders Elsevier.

Landis, E.M. 1932. Factors controlling the movement of fluid through the human capillary wall, *Yale Journal of Biology and Medicine* 5(3):201–225.

Levick, J.R. 2010. *Introduction to Cardiovascular Physiology* (5th Edition) London: Hodder Arnold.

Levick, J.R. & Michel, C.C. 2010. Microvascular fluid exchange and the revised Starling principle, *Cardiovascular Research* 87:198–210.

Michel, C.C. 1980. Filtration coefficients and osmotic reflexion coefficients of the walls of single frog mesenteric capillaries, *Journal of Physiology* 309:341–355.

Michel, C.C. & Curry, F.E. 1999. Microvascular Permeability, *Physiol Rev* 79(3):703–761.

Robinson, T.C. 1988. Arterial graft prosthesis, *United States Patent* 4 731 073.

Silverthorn, D.U. 2010. *Human Physiology: An Integrated Approach* (5th Edition) San Francisco: Benjamin Cummings.

Singh, A.V. & Zamboni, P. 2009. Anomalous venous blood flow and iron deposition in multiple sclerosis, *J Cereb Blood Flow Metab* 29(12):1867–78.

Tucker, T.W. 2011. A physics link between venous stenosis and multiple sclerosis, *Med Hypotheses*, 77(6): 1074–8.

Zamboni, P. 2006. The big idea: iron-dependent inflammation in venous disease and proposed parallels in multiple sclerosis, *J R Soc Med* 99(11):589–93.

Zamboni, P., Galeotti, R., Menegatti, E., Malagoni, A.M., Tacconi, G., Dall'Ara, S., Bartolomei, I., Salvi, F. 2009. Chronic cerebrospinal venous insufficiency in patients with multiple sclerosis, *J Neurol NeurosurgPsychiatry* 80(4):392–399.

Numerical Methods for Hyperbolic Equations – Vázquez-Cendón et al. (eds)
 ISBN 978-0-415-62150-2

Some issues in modelling venous haemodynamics

Lucas O. Müller, Gino I. Montecinos & Eleuterio F. Toro
Laboratory of Applied Mathematics, University of Trento, Italy

ABSTRACT: We adopt a classical one-dimensional mathematical model for blood flow in large to medium-sized arteries and veins and study two possible formulations of the equations, a conservative and a non-conservative one. We solve exactly the Riemann problem for both formulations and assess their suitability for various scenarios. In addition we discuss the source terms present in both formulations and investigate their potential stiffness, with the associated numerical complications. Finally, we deploy the high-order ADER approach to solve the equations and point out the efficiency benefits of using modern non-linear methods of very high order of accuracy in both space and time.

1 INTRODUCTION

One-dimensional mathematical models are widely applied in cardiovascular modelling, especially for arteries (Blanco, Leiva, Feijóo, and Buscaglia 2011; Formaggia and Lamponi, Daniele, Quarteroni 2003; Alastruey, Khir, Matthys, Segers, Sherwin, Verdonck, Parker, and Peiró 2011; Sherwin, Franke, Peiró, and Parker 2003), to study pressure wave propagation and overall flow characteristics. One-dimensional models offer a valid alternative to both the oversimplified 0D models and the often unaffordable 3D models. Moreover, 1D models play a crucial role in more sophisticated multi-scale models comprising coupled 0D, 1D and 3D models. In this paper we identify some aspects of 1D models that deserve consideration. This is particularly the case for veins, which compared to arteries, have received less attention both from the mathematical modelling and numerical analysis points of view. This paper is mainly concerned with numerical aspects of one-dimensional models for veins. Background for the steady case is found in the classical works of Shapiro (Shapiro 1977) and Pedley *et al.* (Pedley, Brook, and Seymour 1996), for example. For the time-dependent case see, for instance, Brook *et al.* (Brook, Falle, and Pedley 1999), Brook and Pedley (Brook and Pedley 2002) and more recently, Fullana and Zaleski (Fullana and Zaleski 2009) and Marchandise and Flaud (Marchandise and Flaud 2010).

In this paper we adopt a well-known, widely accepted basic mathematical model for one-dimensional blood flow in collapsible tubes and study two main issues: (a) the formulation of the equations and (b) their numerical approximation. On the first issue we re-examine two possible formulations of the equations, namely a conservative and a non-conservative formulation. The latter is generally valid for smooth solutions (e.g., no elastic jumps, no contact discontinuities) and is widely used. Curiously, these equations can also be expressed in conservation-law form, in which one of the *conserved variables* is the particle velocity. From the physical point of view such equations are not conservation equations, though they have a mathematical conservative form. We study the formulations in detail, at the analytical level, and solve exactly the Riemann problem for both conservative and non-conservative formulations. Then, regarding point (b) on numerical methods to solve the equations, we identify two main issues that require attention. First we study the nature of source terms in the equations and how these may become stiff, posing a serious challenge to numerical solvers. In addition we consider the question of efficiency of numerical algorithms. This is a crucial aspect if large-scale realistic ambitious simulations are to be carried out. Large computational facilities are not enough, the design of efficient numerical methods is also important.

The rest of the paper is structured as follows. In section 2 we state the 1D model. In section 3 we solve exactly the Riemann problem for both formulations. In section 4 we address issues regarding stiffness of source terms. In section 5 we deal with numerical methods, in section 6 we show numerical results and conclusions are drawn in section 7.

2 MATHEMATICAL MODEL

Here we adopt the following mathematical model for 1D blood flow

$$\begin{cases} \partial_t A + \partial_x (uA) = 0, \\ \partial_t (uA) + \partial_x \left(\hat{\alpha} A u^2\right) + \dfrac{A}{\rho} \partial_x p = Ag - f. \end{cases} \tag{1}$$

$A(x,t)$ is the cross-sectional area of the vessel, $u(x,t)$ is the cross-sectional averaged axial velocity, $p(x,t)$ is the average internal pressure over the cross-section, g is the acceleration due to gravity and $f(x,t)$ is the friction force per unit length of the tube. The coefficient $\hat{\alpha}$ is determined from the velocity profile. Here we assume $\hat{\alpha} = 1$, which corresponds to a blunt velocity profile. For a full description of the derivation of the model see Formaggia *et al.* (Formaggia, Quarteroni, and Veneziani 2009).

The internal pressure $p(x,t)$ can be expressed as

$$p(x,t) = p_e + \Psi(x,t), \tag{2}$$

where p_e is the external pressure, assumed constant here, and $\Psi(x,t)$ is the transmural pressure assumed of the form

$$\Psi(x,t) = K(x)\left[\left(\frac{A(x,t)}{A_0(x)}\right)^m - \left(\frac{A(x,t)}{A_0(x)}\right)^n\right]. \tag{3}$$

Here m and n are real numbers to be specified. $A_0(x)$ is the equilibrium cross-sectional area ($\Psi = 0$) and $K(x)$ is the bending stiffness of the vessel wall

$$K(x) = \frac{E(x)}{12(1-\nu^2)}\left[\frac{h_0(x)}{R_0(x)}\right]^3. \tag{4}$$

$E(x)$ is the Young modulus whereas $h_0(x)$ and $R_0(x)$ are the equilibrium values for wall thickness and vessel radius, respectively. In this paper we shall assume constant material properties, that is K and A_0 are constant along the length of the vessel.

As written, equations (1) may be expressed in conservative form, when material properties are constant. The resulting system will be called the conservative formulation. The momentum equation in (1) can also be written in terms of velocity u, for smooth solutions. This will give rise to a non-conservative formulation, even though, this can also mathematically be written in conservation-law form. Note that a conservation principle for velocity does not make sense, physically. It is well known that the non-conservative formulation will fail to describe weak solutions containing elastic jumps (shocks). The formation of elastic jumps in model (1) has been reported by Shapiro (Shapiro 1977) for steady state solutions and by Brook *et al.* (Brook and Pedley 2002) for unsteady problems. In these references elastic jumps are found in the case of forced expiration manoeuvres (in airways) and for the case of a giraffe jugular vein.

3 EXACT SOLUTION OF THE RIEMANN PROBLEM

As the non-conservative formulation is often used to solve problems that include elastic jumps, we would like to assess the errors that may be introduced by this formulation. Accordingly, in this section we solve the Riemann problem exactly for both formulations and identify critical features that may arise when modelling collapsible vessels, such as veins.

3.1 *Solution for the conservative formulation*

This problem was previously solved by Brook *et al.* (Brook, Falle, and Pedley 1999) and for an augmented system to account for variable material (even discontinuous) properties by Toro and Siviglia (Toro and Siviglia 2011). Here we collect their results very succinctly. The homogeneous version of system (1) can be written in conservation-law form as

$$\partial_t \mathbf{Q} + \partial_x \mathbf{F}(\mathbf{Q}) = 0, \tag{5}$$

where the conserved variables and flux are respectively

$$\mathbf{Q} = \begin{bmatrix} q_1 \\ q_2 \end{bmatrix} = \begin{bmatrix} A \\ Au \end{bmatrix} \tag{6}$$

and

$$\mathbf{F}(\mathbf{Q}) = \begin{bmatrix} Au \\ Au^2 + C \end{bmatrix}. \tag{7}$$

Here $C = \int_{A_0}^{A} c(\tau)^2 d\tau$, is a primitive of the *wave speed c*, yet to be defined. We are taking as the reference state in the integral to be the equlibrium area A_0. The Jacobian of the system is

$$\mathbf{A}(\mathbf{Q}) = \begin{bmatrix} 0 & 1 \\ \dfrac{A}{\rho}\Psi_A - u^2 & 2u \end{bmatrix}. \tag{8}$$

The eigenvalues of (8) are $\lambda_1 = u - c$ and $\lambda_2 = u + c$, with the *wave speed c* given as

$$c = \sqrt{\frac{A}{\rho}\frac{\partial \Psi}{\partial A}}. \tag{9}$$

The Riemann invariants, for later use, are

$$\begin{cases} \Gamma_1 = u - \int_{A_0}^{A} \frac{c(\tau)}{\tau} d\tau, \\ \Gamma_2 = u + \int_{A_0}^{A} \frac{c(\tau)}{\tau} d\tau. \end{cases} \quad (10)$$

We find the exact solution $\mathbf{Q}_{LR}(x/t)$ of the Riemann problem for system (5) with initial conditions

$$\mathbf{Q}(x,0) = \begin{cases} \mathbf{Q}_L & \text{if } x < 0, \\ \mathbf{Q}_R & \text{if } x > 0. \end{cases} \quad (11)$$

We first find the constant state $\mathbf{Q}^*$ between the non-linear waves in the x–t plane. See (Toro 2001) for details on the solution strategy. Appropriate functions connecting $\mathbf{Q}^*$ to the data states $\mathbf{Q}_L$ (left) and $\boldsymbol{Q}_R$ (right) give rise to the non-linear equation

$$f(A) = f_R(A, A_R) + f_L(A, A_L) + u_R - u_L = 0, \quad (12)$$

where f_L and f_R are

$$f_K = \begin{cases} \int_{A_K}^{A_*} \frac{c(\tau)}{\tau} d\tau & \text{if} \quad A_* \le A_K, \\ \sqrt{B_K \frac{A_* - A_K}{A_* A_K}} & \text{if} \quad A_* > A_K. \end{cases} \quad (13)$$

$$B_K = \frac{K}{\rho}\left(\frac{m}{m+1} \frac{A_*^{m+1} - A_K^{m+1}}{A_0^m}\right) - \frac{K}{\rho}\left(\frac{n}{n+1} \frac{A_*^{n+1} - A_K^{n+1}}{A_0^n}\right). \quad (14)$$

The root of the nonlinear equation (12) yields A_*. Finally, the speed u_* is computed as (see (Toro 2001) for details)

$$u_* = \frac{1}{2}(u_L + u_R) + \frac{1}{2}\left[f_R(A_*, A_R) - f_L(A_*, A_L)\right]. \quad (15)$$

The velocity of propagation of the elastic jump is

$$S_L = u_L - \frac{M_L}{A_L}, \quad S_R = u_R + \frac{M_R}{A_R}, \quad (16)$$

where the mass flux is given by

$$M_K = \sqrt{B_K \frac{A_* A_K}{A_* - A_K}}. \quad (17)$$

We omit the details for the solution inside rarefaction fans.

3.2 *Solution for the non-conservative formulation*

For smooth solutions system (1) can be written in terms of A and u, still expressed in (mathematical) conservation-law form (5). Now the vector of unknowns is

$$\mathbf{Q} = \begin{bmatrix} q_1 \\ q_2 \end{bmatrix} = \begin{bmatrix} A \\ u \end{bmatrix} \quad (18)$$

and the flux vector is

$$\mathbf{F(Q)} = \begin{bmatrix} Au \\ \frac{1}{2}u^2 + \frac{p}{\rho} \end{bmatrix}, \quad (19)$$

where p is the internal pressure (2). The Jacobian of the system is

$$\mathbf{A(Q)} = \begin{bmatrix} u & A \\ \frac{A}{\rho}\Psi_A - u^2 & u \end{bmatrix}. \quad (20)$$

The eigenvalues and wave speed for (20) are identical to those of Jacobian (8), but the eigenvectors of (20) are different. However the resulting Riemann invariants coincide.

Now the functions f_L and f_R in (12) are

$$f_K = \begin{cases} \int_{A_K}^{A_*} \frac{c(\tau)}{\tau} d\tau & \text{if} \quad A_* \le A_K, \\ \sqrt{\frac{2}{\rho} \frac{p_* - p_K}{A_*^2 - A_K^2}}(A_* - A_K) & \text{if} \quad A_* > A_K. \end{cases} \quad (21)$$

The speed u_* is computed from (15), with f_L and f_R from (21). The expression defining the elastic jump speed has the form of (16), but in this case mass fluxes are

$$M_K = \sqrt{\frac{2}{\rho} \frac{p_* - p_L}{A_*^2 - A_K^2}} A_* A_K. \quad (22)$$

3.3 *Shock speed prediction*

The elastic jump speeds for both formulations differ, as can be seen by comparing (17) and (22). Moreover, functions defining the solution of the Riemann problem also differ, since relations (13) and (21) are not equivalent. Hence, in general solutions are not equivalent, at the analytical level.

In particular, there is a parameter that strongly influences how solutions differ, namely the wave speed (9). The wave speed varies considerably with the ratio $\alpha = A/A_0$, as shown in Figure 1. For elastic jumps the shock speeds in both formulations are different and they are very different as the shock strength increases and this is particularly the case for nearly collapsed states (Figure 2). This consideration is supported by the fact that for low α we may still have large A_*/A ratios, as in the case of a nearly collapsed vessels. In other words, the magnitude of errors computed by using the non-conservative formulation is large for the case of nearly collapsed states, which can generate large elastic-jump strengths. This limitation of the non-conservative formulation will become evident when showing results on a numerical test for the collapse of a giraffe jugular vein, see section 6.3.

We have pointed out possible errors that may be introduced in the numerical solution of the problem if an inappropriate formulation of the mathematical model is used. Moreover, errors may be larger for collapsible tubes. Another important conclusion is that in collapsed states, the wave speed (9) will decrease considerably, making the transition from subcritical to supercritical flow regimes feasible.

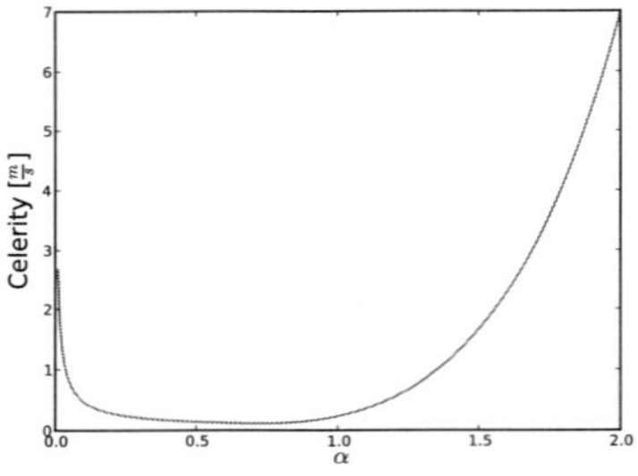

Figure 1. Wave speed (9). Parameters: $A_0 = 0.001\ m^2$, $m = 10$, $n = -1.5$, $\rho = 1050\ kg/m^3$, $K = 5\ Pa$.

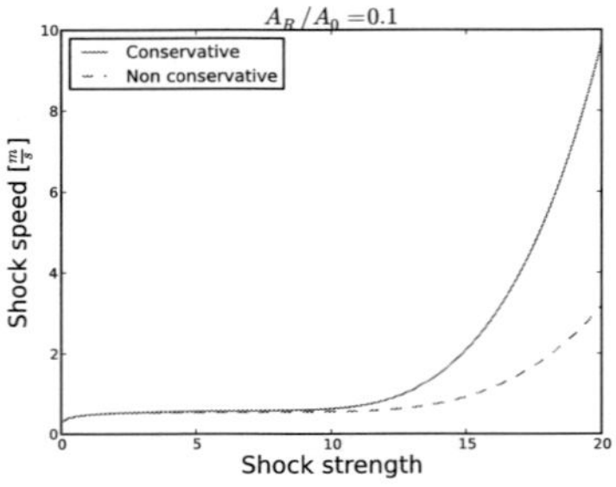

Figure 2. Elastic jump speeds obtained using the conservative and non-conservative formulations. Parameters: $A_0 = 0.001\ m^2$, $m = 10$, $n = -1.5$, $\rho = 1050\ kg/m^3$, $K = 5\ Pa$.

These considerations indicate that the numerical scheme should be able to capture elastic jumps, or sharp gradients in general, in a correct manner and to deal with supercritical as well as subcritical regimes. This last property is not common in one-dimensional numerical models for blood flow in arteries, but seems to be mandatory for models applied to collapsible vessels, such as veins.

In the next section we analyse a further aspect of the mathematical model which may have profound consequences in the design of a robust and accurate numerical scheme for solving system (1), namely the source term.

4 SOURCE TERM STIFFNESS

If we consider source terms, such as gravitational forces and dissipation due to viscous forces, system (1) can be rewritten as

$$\partial_t \mathbf{Q} + \partial_x \mathbf{F}(\mathbf{Q}) = \mathbf{S}(\mathbf{Q}), \tag{23}$$

where $\mathbf{Q}$ is given in (6) and the flux vector $\mathbf{F}(\mathbf{Q})$ is (7). The source term is given by

$$\mathbf{S}(\mathbf{Q}) = \begin{bmatrix} 0 \\ Ag - Ru \end{bmatrix}. \tag{24}$$

The function R is obtained from an assumed velocity profile; here we adopt the one proposed in (Pedley, Brook, and Seymour 1996)

$$R = \frac{8\pi\nu A_{ref}^{\frac{1}{2}}}{A^{\frac{1}{2}}}, \tag{25}$$

where ν is the kinematic viscosity and A_{ref} is a reference cross-sectional area. If $A_{ref} = A$ one recovers the friction term for tubes which when collapsing maintain a circular shape. When setting $A_{ref} = A_0$, R represents the friction term for a tube that assumes an elliptical shape during collapse (as for highly compliant vessels, such as veins).

In this section we concentrate our attention on the study of the stiffness of the source term. A source term can be considered stiff if

$$\Delta t \max_i \{|\beta_i|\} > 1, \quad i = 1, \ldots, N, \tag{26}$$

where β_i is the i-th eigenvalue of the Jacobian of (24), $\partial \mathrm{S}(\mathbf{Q})/\partial \mathbf{Q}$, and N is the number of unknowns of the system.

Assuming a CFL-number $CFL = 1$ we can express relation (26) in terms of ratios between the eigenvalues of the dissipative/productive process and the characteristic speeds for the advective process, namely

$$\Delta x \frac{max_i\{|\beta_i|\}}{max_i\{|\lambda_i|\}} > 1. \tag{27}$$

See (Dumbser, Enaux, and Toro 2008). Replacing (25) in source term (24) gives

$$\mathbf{S(Q)} = \begin{bmatrix} 0 \\ gq_1 - 8\pi\nu A_0^{\frac{1}{2}} \frac{u}{A^{\frac{1}{2}}} \end{bmatrix}. \tag{28}$$

Its Jacobian is

$$\frac{\partial \mathbf{S(Q)}}{\partial \mathbf{Q}} = \begin{bmatrix} 0 & 0 \\ g + 12\pi\nu A_0^{\frac{1}{2}} \frac{u}{A^{\frac{3}{2}}} & -8\pi\nu A_0^{\frac{1}{2}} \frac{1}{A^{\frac{3}{2}}} \end{bmatrix}. \tag{29}$$

The eigenvalues of (29) are

$$\beta_1 = 0, \quad \beta_2 = -8\pi\nu A_0^{\frac{1}{2}} \frac{1}{A^{\frac{3}{2}}} = -8\pi\nu \frac{1}{\alpha^{\frac{1}{2}} A}. \tag{30}$$

We carry out an approximate analysis of the order of magnitude of ratio (27) for physiological values of the parameters involved with $A : O(10^{-4} \div 10^{-7})$, $u : O(10^0)$, $c : O(10^0)$, $\nu : O(10^{-6})$. We obtain

$$\Delta x \frac{8\pi\nu \frac{1}{\alpha^{\frac{1}{2}} A}}{u + c} \approx \Delta x \frac{\frac{O(10^1)O(10^{-6})}{\alpha^{\frac{1}{2}} O(10^{-4} \div 10^{-7})}}{O(10^0)} \approx O(10^{-1} \div 10^2) \frac{\Delta x}{\alpha^{\frac{1}{2}}}. \tag{31}$$

The resulting ratio corresponds to a source term that may become stiff. The parameter α in (31) makes it evident that the collapse of a vessel may easily lead to an increase of (31) by one order of magnitude.

Considerations made in this section show us that the model applied to collapsible tubes may have stiff source terms, which means that one must choose appropriate numerical schemes for the modelling of collapsible tubes to treat stiff source terms correctly. This task is not trivial and only few schemes available in the literature are potential candidates.

5 HIGH ORDER NUMERICAL SCHEMES

The main conclusions from the previous sections can be summarised as follows. The numerical scheme should be implemented in conservative formulation; it should allow for trans-critical flow regimes to occur and treat stiff source terms appropriately. Moreover, high order of accuracy is desirable to increase the efficiency of the algorithms. This is bound to be a mandatory requirement for very complex models in ambitious applications. We note however that high accuracy and stiffness are contradictory requirements and necessitate of the appropriate choice of numerical methods.

Here we adopt the ADER framework, see Toro *et al.* (Toro, Millington, and Nejad 2001), for constructing non-linear numerical schemes of high order of accuracy in space and time. This is a unified approach that allows for arbitrary order of accuracy in both space and time on unstructured meshes and including source terms. The order of accuracy is a parameter open to choice. The ADER finite volume methods are fully discrete, perform a non-linear spatial reconstruction and then solve a generalised (or high order) Riemann problem (GRP) at each cell interface. The solution of the GRP relies on a time series expansion right at the interface followed by solutions of derivative Riemann problems to complete the solution. The numerical flux is obtained from numerical integration of the physical flux function evaluated at the solution of the GPR at the interface, see Toro & Titarev (Toro and Titarev 2002). More details on the original ADER schemes are found in Toro *et al.* (Toro, Millington, and Nejad 2001), Toro & Titarev (Toro and Titarev 2002), Titarev & Toro (Titarev and Toro 2005) and Castro & Toro (Castro and Toro 2008), for instance.

Here we use the ADER-type scheme proposed by Dumbser *et al.* (Dumbser, Enaux, and Toro 2008). In this version the solution of the GRP is carried out by numerically evolving the data either side of the interface using an implicit space-time Discontinuous Galerkin finite element method and then interacting these evolved data at specified *integration points* via a classical Riemann solver. The numerical flux for the finite volume scheme is again obtained by numerical integration of the GRP solution at the interface. In Castro & Toro (Castro and Toro 2008) it is shown that for linear systems with constant coefficients the two approaches are identical when the data evolution is performed through a time series expansion on each side of the interface. The version adopted here, Dumbser *et al.* (Dumbser, Enaux, and Toro 2008), has the advantage of reconciling high accuracy with stiffness of the source terms. See also Hidalgo and Dumbser (Hidalgo and Dumbser 2011). The final scheme is of the finite volume type and locally implicit.

6 NUMERICAL RESULTS

Here we show sample numerical results. We first carry out a convergence rate study of the schemes with implementations for first to fifth order of accuracy in space and time. Then we apply the schemes to solve a problem with stiff source terms.

6.1 *Empirical convergence rates*

Here we carry out a convergence rate study of the ADER numerical schemes. We consider schemes from first to fifth order of accuracy in space and time. The study is empirical, for which a reference, ideally exact, solution is needed. We *manufacture* an exact reference solution for the homogeneous system using the following procedure. We prescribe functions

$$A(x,t) = A_0 + a_0 \sin\left(2\frac{\pi}{L}x\right)\cos\left(2\frac{\pi}{T_0}t\right), \quad (32)$$

and

$$(Au)(x,t) = (Au)_0 - \frac{a_0 L}{T_0}\cos\left(2\frac{\pi}{L}x\right)\sin\left(2\frac{\pi}{T_0}t\right), \quad (33)$$

for the cross-sectional area $A(x,t)$ and for the mass flux $A(x,t)\ u(x,t)$ respectively. Obviously these are not solutions of the original equations. Substitution of these functions into the original equations produces the modified system

$$\partial_t \mathbf{Q} + \partial_x \mathbf{F}(\mathbf{Q}) = \tilde{\mathbf{S}}(x,t). \quad (34)$$

The resulting system is not the original homogenous system, but a modified system with a source term. The good news is that the functions (32) and (33) are exact solutions of the modified system (34). This is the system used to carry out the convergence rate study.

Table 1 shows empirical convergence rates for implementations ranging from second to fifth order. The error is measured in three norms, namely L_∞, L_1 and L_2. Empirical convergence rates correspond to the expected theoretical accuracy of the scheme in all cases. We note however that for certain types of physiological data we observe difficulties in achieving the high convergence rates for schemes above third order for the finest meshes used. This is due to the fact that the error in those cases can be too small to be represented by the machine (double precision used here).

Efficiency of numerical schemes is of fundamental importance if large-scale simulations are to be performed. By efficiency we mean this: given an error ε deemed acceptable, which scheme will attain that error at the smallest computational time. A low order scheme, first or second order, will attain that error on a mesh M_L, which will be very fine if ε is sufficiently small. The associated CPU time will be T_L. A good high order method will attain the error on a mesh M_H, much coarser than M_L, with a CPU time T_H, much lower than T_L. This is what we mean by efficiency.

Table 1. Empirical convergence rates for the ADER scheme applied to blood flow in veins with parameters: $CFL = 0.9$, $A_0 = 10^{-4}\ m^2$, $a_0 = 10^{-5}\ m^2$, $L = 1\ m$, $T_0 = 3\ s$, $Au_0 = 10^{-7}\ m^3/s$.

Scheme	N	$\mathcal{O}(L_\infty)$	$\mathcal{O}(L_1)$	$\mathcal{O}(L_2)$
ADER 2	8	–	–	–
	16	1.33	1.83	1.72
	32	2.34	1.96	2.11
	64	1.73	1.96	1.96
	128	1.97	1.99	2.01
ADER 3	8	–	–	–
	16	2.53	2.67	2.64
	32	3.07	3.26	3.24
	64	2.96	2.99	2.98
	128	3.01	3.03	3.03
ADER 4	8	–	–	–
	16	3.33	3.92	3.77
	32	4.09	3.94	4.13
	64	3.53	3.91	3.8
	128	3.61	3.61	3.64
ADER 5	8	–	–	–
	16	4.56	4.75	4.71
	32	5.13	5.26	5.25
	64	4.84	4.92	4.92
	128	0.57	0.96	0.92

Figure 3 shows CPU time versus error for a first and a fifth order the ADER schemes. As an example take $\varepsilon = 10^{-06}$, then the CPU time T_L necessary to attain that error will be more than two orders of magnitude larger than T_H.

6.2 *Stiff source term*

The following test is based on a problem formulation presented by Dumbser *et al.* (Dumbser, Enaux, and Toro 2008). The PDEs to be solved are (1) without gravity and using coefficient (25) for the frictional source term. The spatial domain is divided into two regions with different dynamic viscosities. Initial conditions are

$$A(x,0) = \begin{cases} A_0 & \text{if} \quad x \le 0.15m, \\ 0.05A_0 & \text{if} \quad x > 0.15m, \end{cases} \quad (35)$$

with $u(x,0) = 0m/s$ and viscosity

$$\mu(x) = \begin{cases} 0 & \text{if} \quad x \le 0.15, \\ 0.004\,Pas & \text{if} \quad x > 0.15. \end{cases} \quad (36)$$

Vessel dimensions and mechanical characteristics are those for small human veins (radius in

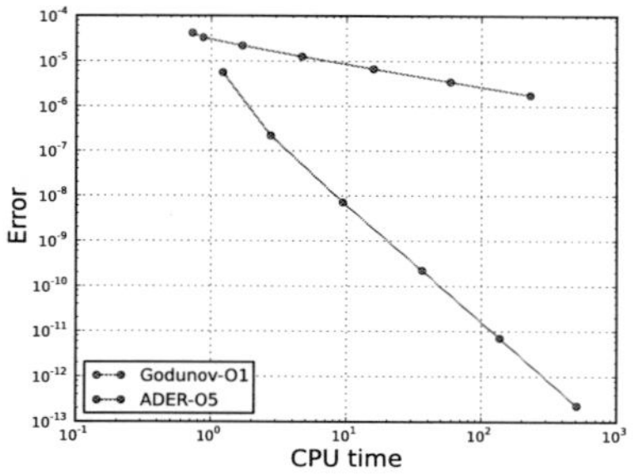

Figure 3. Error versus CPU time for a first order Godunov scheme and for a fifth order ADER scheme.

unloaded configuration $R_0 = 1.0\ mm$). A priori analysis of ratio (27) results in

$$\frac{8\pi\nu\frac{1}{\alpha^{\frac{1}{2}}A}}{u+c}\Delta x = \frac{8\pi\nu\frac{1}{0.05^{\frac{1}{2}}1.57\cdot 10^{-7}}}{0+2.59}\Delta x \approx O(10^3)\Delta x \qquad (37)$$

We use a domain $\Omega = [0,1]\ m$ and 200 cells, so that $\Delta x = 0.005\ m$. Ratio (27) will be ≈ 2, ensuring that the source term will be stiff in the portion of the domain where there is viscosity.

This problem set up resembles the interface of an inviscid liquid and a porous medium. The diffusive process should be dominant in comparison to advective terms. The Courant number was set to $CFL = 0.9$ and the final simulation time is $t_{end} = 6.0\ s$.

Figure 4 shows results for the ADER method, in its 5th order version. The ADER scheme is able to capture the dominant diffusive nature of the solution, as shown by the agreement with a reference solution obtained on a very fine grid (2000 cells). We emphasize that the current version of the ADER scheme automatically takes into account the source term stiffness; this is without changing the CFL-stability condition and remaining efficient in regions where stiffness does not arise.

6.3 *Collapse of a giraffe jugular vein*

This test was proposed by Pedley *et al.* (Pedley, Brook, and Seymour 1996) for the stationary case and solved later for the unsteady case by Brook *et al.* (Brook and Pedley 2002).

The test is very useful because it addresses several issues that we have identified as crucial for the development of numerical schemes for flow in collapsible tubes. The tube collapses and the transient phase involves the transition through a critical point from supercritical to subcritical regime via an elastic jump. The scheme will thus have to deal with different regimes and elastic jumps. Moreover, the solution of elastic jumps accentuates errors

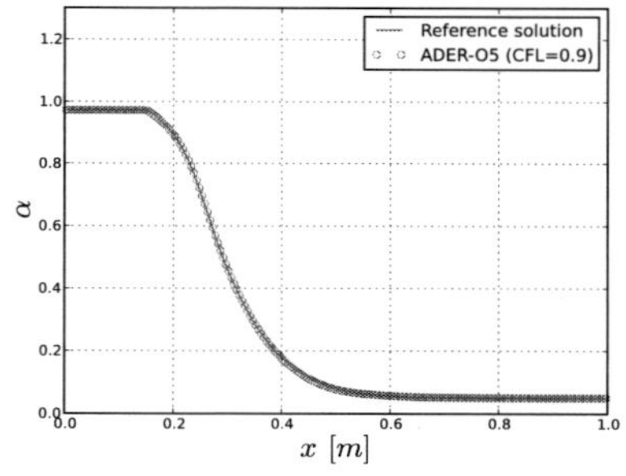

Figure 4. Test with stiff source term. Solution at $t_{end} = 6.0s$. Reference solution obtained using a fine grid of 2000 cells. The fifth order ADER scheme reproduces the reference solution without source term related restrictions on the CFL number.

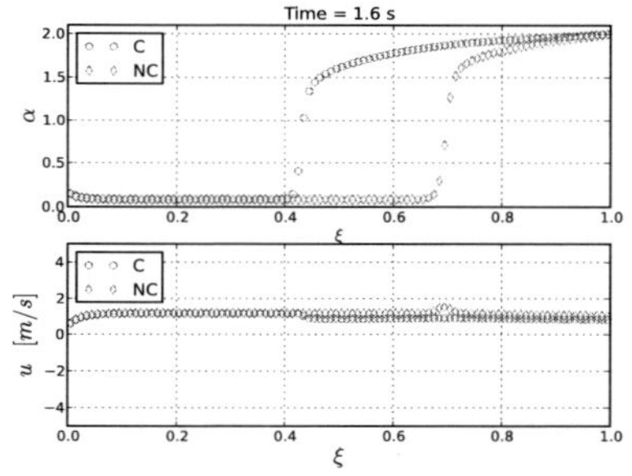

Figure 5. Giraffe jugular vein collapse test. Solution at output time $t = 1.6\ s$, during the transient phase. C: conservative formulation; NC: non-conservative formulation.

produced by the non-conservative formulation, since we have a strongly collapsed vessel. This condition was identified as critical in section 3.3.

Test parameters are: domain lenght: $L = 2\ m$; area at rest $A_0 = 0.0005\ m^2$; wall stiffness $K = 5Pa$; tube law coefficients $m = 10$ and $n = -1.5$ initial conditions: $A(x,0) = (0.2 + 1.8\xi)A_0$, $(Au)(x,0) = (Au)_0 = 40\ ml/s$ and boundary conditions: $A(2,t) = 2A_0$, $(Au)(0,t) = (Au)_0$.

The development of the transient phase can be described as follows: two forces act in contrasting ways, gravity tends to empty the vessel, while downstream boundary conditions tend to inflate it. The emptying due to gravity generates a supercritical flow in the upstream portion of the vessel, while downstream boundary conditions impose subcritical flow. Connection of both conditions is achieved via an elastic jump. The problem was solved using a second order ADER scheme with a CFL number $CFL = 0.9$ for both formulations, the conservative and the non-conservative ones. Figure 5 shows discrepancies between numerical solutions during the transient phase, deriving from observations made in section 3.3. The same discrepancies

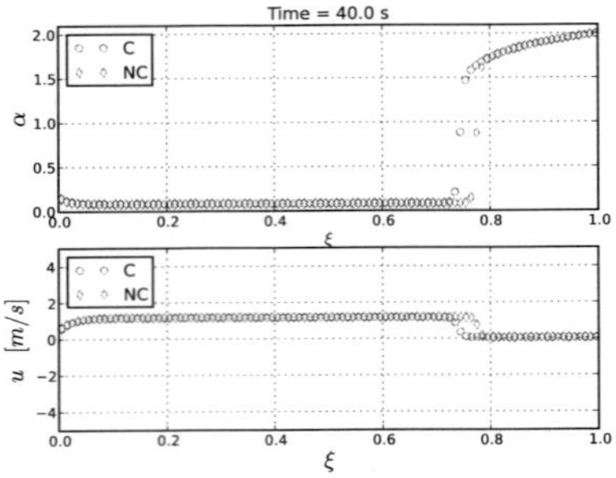

Figure 6. Giraffe jugular vein collapse test. Solution at output time $t = 40.0s$, corresponding to the steady solution. C: conservative formulation; NC: non-conservative formulation.

are found when the stationary solution is reached (Figure 6).

7 SUMMARY AND CONCLUSIONS

We have adopted a classical one-dimensional mathematical model for blood flow in large to medium-sized arteries and veins. Assuming constant material properties we have studied two possible formulations of the equations, a conservative and a non-conservative one. Then we have solved exactly the Riemann problem for both formulations and have assessed their suitability for various scenarios. We have in addition discussed the source terms present in both formulations and their potential stiffness, with the associated numerical complications. Regarding the choice of numerical methods to solve the equations we have implemented a high-order ADER-type finite volume scheme suitable for treating stiff source terms. We have carried carried out a convergence rate study of the methods and have verified their high accuracy form schemes from second to fifth order. We have also discussed the efficiency benefits of using modern non-linear methods of very high order of accuracy in both space and time, illustrating this point via an example. The methods have been assessed on test problems with subcritical and supercritical regions, with vessel collapse, with elastic jumps and with stiff source terms. In all cases we have obtained satisfactory results.

REFERENCES

Alastruey, J., Khir, A.W., Matthys, K.S., Segers, P., Sherwin, S.J., Verdonck, P.R., Parker, K.H. and Peiró, J. (2011, August). Pulse wave propagation in a model human arterial network: Assessment of 1-D visco-elastic simulations against in vitro measurements. *Journal of Biomechanics 44*(12), 2250–8.

Blanco, P., Leiva, J., Feijóo, R. and Buscaglia, G. (2011, March). Black-box decomposition approach for computational hemodynamics: One-dimensional models. *Computer Methods in Applied Mechanics and Engineering 200*(13–16), 1389–1405.

Brook, B.S., Falle, S. and Pedley, T.J. (1999). Numerical solutions for unsteady gravity-driven flows in collapsible tubes: evolution and roll-wave instability of a steady state. *Journal of Fluid Mechanics 396*(1), 223–256.

Brook, B.S. and Pedley, T.J. (2002). A model for time-dependent flow in (giraffe jugular) veins: uniform tube properties. *Journal of Biomechanics 35*, 95–107.

Castro, C.E. and Toro, E.F. (2008, February). Solvers for the high-order Riemann problem for hyperbolic balance laws. *Journal of Computational Physics 227*(4), 2481–2513.

Dumbser, M., Enaux, C. and Toro, E.F. (2008, April). Finite volume schemes of very high order of accuracy for stiff hyperbolic balance laws. *Journal of Computational Physics 227*(8), 3971–4001.

Formaggia, L. and Lamponi, Daniele, A. Quarteroni (2003). One-dimensional models for blood flow in arteries. *Journal of Engineering Mathematics 47*(3/4), 251–276.

Formaggia, L., Quarteroni, A. and Veneziani, A. (2009). *Cardiovascular Mathematics: Modeling and simulation of the circulatory system*, Volume 1. Springer Verlag.

Fullana, J. and Zaleski, S. (2009). A branched one-dimensional model of vessel networks. *Journal of Fluid Mechanics 621*, 183–204.

Hidalgo, A. and Dumbser, M. (2011, October). ADER Schemes for Nonlinear Systems of Stiff AdvectionDiffusionReaction Equations. *Journal of Scientific Computing 48*(1), 173–189.

Marchandise, E. and Flaud, P. (2010, January). Accurate modelling of unsteady flows in collapsible tubes. *Computer Methods in Biomechanics and Biomedical Engineering 13*(2), 279–90.

Pedley, T.J., Brook, B.S. and Seymour, R.S. (1996). Blood Pressure and Flow Rate in the Giraffe Jugular Vein. *Philosophical Transactions: Biological Sciences 351*, 855–866.

Shapiro, A.H. (1977). Steady Flow in Collapsible Tubes. *Journal of Biomechanical Engineering 99*, 126–147.

Sherwin, S., Franke, V., Peiró, J. and Parker, K. (2003, December). One-dimensional modelling of a vascular network in space-time variables. *Journal of Engineering Mathematics 47*(3/4), 217–250.

Titarev, V.A. and Toro, E.F. (2005, April). ADER schemes for three-dimensional non-linear hyperbolic systems. *Journal of Computational Physics 204*(2), 715–736.

Toro, E.F. (2001). Shock-capturing methods for free-surface shallow flows. Wiley and Sons Ltd.

Toro, E.F., Millington, R. and Nejad, L. (2001). Towards very high order Godunov schemes. In E. Toro (Ed.), *Godunov Methods. Theory and Applications*, Number July, New York, Boston and London, pp. 905–938. Kluwer/Plenum Academic Publishers.

Toro, E.F. and Siviglia, A. (2011). Simplified blood flow model with discontinous vessel properties. In G. Ambrosi, Davide; Quarteroni, Alfio; Rozza (Ed.), *Modeling of Physiological Flows*. Springer.

Toro, E.F. and Titarev, V.A. (2002, February). Solution of the generalized Riemann problem for advection-reaction equations. *Proceedings of the Royal Society A: Mathematical, Physical and Engineering Sciences 458*(2018), 271–281.

Numerical Methods for Hyperbolic Equations – Vázquez-Cendón et al. (eds)
© 2013 Taylor & Francis Group, London, ISBN 978-0-415-62150-2

Efficient algorithms for the solution of fluid-structure interaction problems in haemodynamic applications

Matteo Pozzoli
MOX, Dipartimento di Matematica, Politecnico di Milano, Milano, Italy

Christian Vergara
Dipartimento di Ingegneria dell'Informazione e Metodi Matematici, Univerità di Bergamo, Dalmine (BG), Italy

Fabio Nobile
MATHICSE, EPFL, Lausanne, Switzerland

ABSTRACT: In this work we deal with the numerical solution of the fluid-structure interaction problem arising in the haemodynamic environment. In particular, we consider Newmark time discretization schemes, and we study different methods for the treatment of the fluid-structure interface position, focusing on partitioned algorithms. We consider explicit and implicit schemes, and new hybrid methods. We study numerically the performances and the accuracy of these schemes, highlighting the best solutions for haemodynamic applications.

1 INTRODUCTION

Building efficient strategies for the solution of the fluid-structure interaction (FSI) problem is a major issue in *computational haemodynamics*. In particular here we are interested in the FSI problem arising by the interaction between the blood flow and the vessel wall deformation (see, e.g., (Perktold et al., 1994), (Bazilevs et al., 2006), (Figueroa et al., 2006), (Tezduyar et al., 2007), (Badia et al., 2008), (Formaggia et al., 2009)). The main difficulties related to the numerical solution of the FSI problem are: (i) the treatment of the *interface position*, since the fluid domain is an unknown of the problem (*geometrical non-linearity*); (ii) the treatment of the *interface continuity conditions*, which enforce continuity of velocities and normal stresses between fluid and structure; (iii) the fact that the subproblems could be non-linear (*physical non-linearities*). These features make the FSI problem a strongly non-linear coupled problem, as there is a substantial amount of energy exchanged between fluid and structure in each cardiac beat. This non-linear behaviour is essentially related to points (i) and (ii) above. Therefore, in this work we focus mainly on these two points. Regarding the third point, we consider just the fluid non-linearity due to the convective term in the Navier-Stokes equations, and we consider a linear structure.

Concerning the first point, we can mainly detect two strategies: an *implicit treatment* of the interface position or an *explicit treatment*, thanks to extrapolations of the solution at previous time steps.

After a suitable linearization of the physical non-linearities, whichever of the two strategies is adopted for the treatment of the interface position (implicit or explicit), one has to deal with a *linearized* FSI problem (in the sense that we have eliminated the geometrical and physical non-linearities). However, this problem is still coupled through the interface continuity conditions. For the solution of this linearized FSI problem we consider partitioned schemes, where one solves the fluid and structure subproblems in an iterative framework, until fulfillment of the interface continuity conditions (see, e.g., (Piperno and Farhat 2000), (Causin et al., 2005), (Deparis et al., 2006), (Forster et al., 2007), (Badia et al., 2008)).

The goal of this work is to compare the accuracy and performances of different treatments of the FS interface position, when partitioned procedures are considered for the enforcement of the continuity conditions. To this aim, we consider an application of such schemes to a patient-specific case.

2 THE FSI PROBLEM

2.1 *The continuos problem*

Let us consider an open domain $\Omega_f^t \subset \mathbb{R}^3$ like the one represented in Figure 1 (on the left). This represents the lumen of a vessel and it is function of time t. Inflow and outflow sections are denoted by $\Sigma_{f,i}^t$ (three in Figure 1). Blood velocity is denoted by $\boldsymbol{u}_f(x,t)$, the pressure by $p_f(x,t)$. The incompressible Navier-Stokes equations for a Newtonian fluid are assumed to hold in Ω_f^t. Let T_f be the related Cauchy stress tensor defined by

$$\boldsymbol{T}_f(\boldsymbol{u}_f, p_f) := -p_f \boldsymbol{I} + \mu(\nabla \boldsymbol{u}_f + (\nabla \boldsymbol{u}_f)^T).$$

Since we work in a moving domain, the fluid problem is stated in an *Arbitrary Lagrangian-Eulerian* (ALE) framework (see e.g., (Hughes et al., 1981), (Donea 1982)). The ALE map $\mathcal{A}$ is defined by an appropriate lifting of the structure displacement at the FS interface Σ^t, and defines the displacement of the points of the fluid domain η_m and their velocity $\boldsymbol{u}_m$. For any function v living in the current fluid configuration, we denote by $\tilde{v} := v \circ \mathcal{A}$ its counterpart in the reference configuration. A classical choice in haemodynamic applications to define the ALE map is to consider a harmonic extension operator in the reference domain (see, e.g., (Nobile 2001)).

The vessel wall is denoted by Ω_s^t, which is an open subset of $\mathbb{R}^3$ (see Figure 1, right). The intersection of Ω_s^t and Ω_f^t is empty, and $\Sigma^t := \bar{\Omega}_s^t \cup \bar{\Omega}_f^t$ is the FS interface. On Σ^t we define a normal unit vector $\boldsymbol{n}$ poiting outward of the solid domain and inward to the fluid domain. The inflow/outflow sections (three in Figure 1) are denoted by $\Sigma_{s,i}^t$.

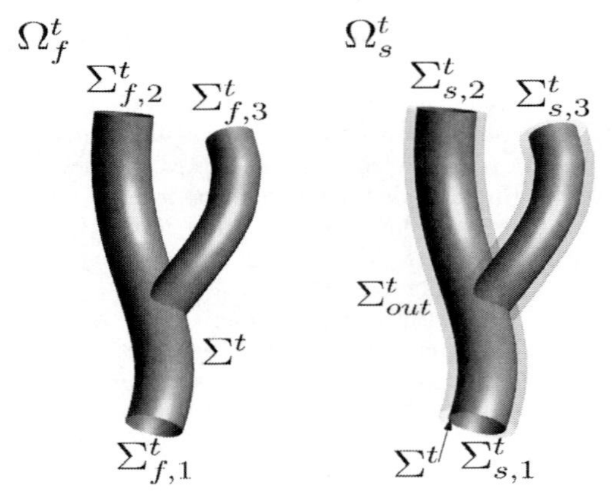

Figure 1. Representation of the domain of the FSI problem: fluid domain on the left, structure domain on the right.

With Σ_{out}^t we denote the external surface of the structure domain. We denote by $\eta_s(\boldsymbol{x},t)$the wall displacement. We assume that the solid is a linear elastic material, characterized by the following Piola-Kirchhoff stress tensor

$$\tilde{\boldsymbol{T}}_s = \frac{E}{2(1+\nu)}\varepsilon(\tilde{\eta}_s) + \frac{E\nu}{(1+\nu)(1-2\nu)}\mathrm{tr}(\varepsilon(\tilde{\eta}_s))\boldsymbol{I},$$

where $\varepsilon(\eta) := (\nabla\eta + (\nabla\eta)^T)/2, E$ is the Young modulus, and ν is the Poisson ratio. To describe the structure kinematics we adopt a purely Lagrangian approach, where $\mathcal{L}$ is the Lagrangian map. For any function g defined in the current solid configuration Ω_s^t, we denote by $\tilde{g} := g \circ \mathcal{L}$ its counterpart in the reference domain.

The strong formulation of the FSI problem, including the computation of the ALE map, reads therefore as follows

1. *Fluid-Structure problem.* Given the (unknown) fluid domain velocity $\boldsymbol{u}_m$ and fluid domain Ω_f^t, find, at each time $t \in (0,T]$, fluid velocity $\boldsymbol{u}_f$, pressure p_f and structure displacement η_s such that

$$\begin{cases} \rho_f \dfrac{D^A \boldsymbol{u}_f}{Dt} + \rho_f((\boldsymbol{u}_f - \boldsymbol{u}_m)\cdot\nabla)\boldsymbol{u}_f & \\ -\nabla\cdot \boldsymbol{T}_f(\boldsymbol{u}_f, p_f) = \boldsymbol{f}_f & \text{in } \Omega_f^t, \\ \nabla\cdot\boldsymbol{u}_f = 0 & \text{in } \Omega_f^t, \\ \rho_s \dfrac{\partial^2 \tilde{\eta}_s}{\partial t^2} - \nabla\cdot\tilde{\boldsymbol{T}}_s(\tilde{\eta}_s) = \tilde{\boldsymbol{f}}_s & \text{in } \Omega_s^0, \\ \boldsymbol{u}_f = \dfrac{\partial \eta_s}{\partial t} & \text{on } \Sigma^t, \\ \boldsymbol{T}_s(\eta_s)\boldsymbol{n} - \boldsymbol{T}_f(\boldsymbol{u}_f, p_f)\boldsymbol{n} = 0 & \text{on } \Sigma^t, \\ \alpha_e \tilde{\eta}_s + \tilde{\boldsymbol{T}}_s(\tilde{\eta}_s)\tilde{\boldsymbol{n}} = P_{ext}\tilde{\boldsymbol{n}}, & \text{on } \Sigma_{out}^0, \end{cases} \quad (1)$$

where D^A/Dt is the ALE time derivative, ρ_f and ρ_s are the fluid and structure densities, μ is the constant blood viscosity, $\boldsymbol{f}_f$ and $\boldsymbol{f}_s$ the forcing terms;

2. *Geometry problem.* Given the (unknown) interface structure displacement $\tilde{\eta}_s|_{\Sigma^0}$, find the displacement of the points of the fluid domain η_m such that

$$\begin{cases} -\Delta\tilde{\eta}_m = 0 & \text{on } \Omega_f^0, \\ \tilde{\eta}_m = \tilde{\eta}_s & \text{on } \Sigma^0, \end{cases} \quad (2)$$

and then find accordingly the fluid domain velocity $\tilde{\boldsymbol{u}}_m := \frac{\partial\tilde{\eta}_m}{\partial t}$, and the new points $\boldsymbol{x}_f^t$ of the fluid domain by moving the points $\tilde{\boldsymbol{x}}_f$ of the reference domain Ω_f^0:

$$x_f^t = \tilde{x}_f + \tilde{\eta}_m.$$

The two matching conditions enforced at the interface are the *continuity of velocities* $(1)_4$ and the *continuity of normal stresses* $(1)_5$. The fluid and structure are also coupled by the geometry problem, leading to a highly nonlinear system of partial differential equations. Equations (1) and (2) have to be endowed with suitable boundary conditions on $\Omega_f^t \setminus \Sigma^t$ and $\Omega_s^0 \setminus \Sigma^0 \cup \Sigma_{out}^0$, and with suitable initial conditions. We prescribe the Robin boundary condition $(1)_6$ on Σ_{out}^0, with the aim of modeling the presence of a surrounding tissue around the vessel. This choice corresponds to consider an elastic behaviour of this tissue, where α_e is the corresponding elastic coefficient (see (Moireau et al., 2011), (Liu et al., 2007)).

2.2 *Time discretization of FSI problem*

Let Δt be the time discretization parameter and $t^n := n\Delta t, n = 0,1,\ldots$. For a generic function z, with z^n we denote the approximation of $z(t^n)$. In this work we consider the *Newmark scheme* for the structure in combination with the *theta-method* for the fluid. We propose in what follows the discrete-in-time formulation of the time discrete problem (1)–(2).

1. *Fluid-Structure problem.* Given the (unknown) fluid domain velocity u_m^{n+1} and the fluid domain Ω_f^{n+1}, the parameters $\beta_{f,i}$ $(i = 0,1)$, $\chi_f, \beta_{s,i}$ $(i = 0,1), \sigma_s, \zeta_s, \xi_{s,i}$ $(i = 0,1), \chi_s, \kappa_s$, the solution at previous time steps, and func tions f_f^{n+1}, f_s^{n+1} and P_{ext}, find fluid velocity u_f^{n+1}, pressure p_f^{n+1} and structure displacement η_s^{n+1} such that (Nobile et al., 2011)

$$\begin{cases} \rho_f \dfrac{\beta_{f,0}}{\Delta t} u_f^{n+1} \\ \quad +\rho_f((u_f^{n+1} - u_m^{n+1})\cdot\nabla)u_f^{n+1} \\ \quad -\nabla\cdot T_f^{n+1} = f_f^{n+1} + \rho_f f_{f,W}^{n+1} & \text{in } \Omega_f^{n+1}, \\ \nabla\cdot u_f^{n+1} = 0 & \text{in } \Omega_f^{n+1}, \\ \rho_s \dfrac{\xi_{s,0}}{\Delta t^2}\tilde{\eta}_s^{n+1} - \nabla\cdot\tilde{T}_s^{n+1}(\tilde{\eta}_s^{n+1}) \\ \quad = \tilde{f}_s^{n+1} + \rho_s \tilde{f}_{s,W}^{n+1} & \text{in } \Omega_s^0, \\ u_f^{n+1} = u_s^{n+1} & \text{on } \Sigma^{n+1}, \\ T_s^{n+1}(\eta_s^{n+1})n \\ \quad - T_f^{n+1}(u_f^{n+1}, p_f^{n+1})\, n = 0 & \text{on } \Sigma^{n+1}, \\ \alpha_e \tilde{\eta}_s^{n+1} + \tilde{T}_s^{n+1}(\tilde{\eta}_s^{n+1})\tilde{n} = P_{ext}\tilde{n} & \text{on } \Sigma_{out}^0, \end{cases} \quad (3)$$

where

$$f_{s,U}^{n+1} := \frac{\beta_{s,1}}{\Delta t}\eta_s^n + \chi_s u_s^n + \Delta t \kappa_s w_s^n,$$

$$f_{s,W}^{n+1} := \frac{\xi_{s,1}}{\Delta t^2}\eta_s^n + \frac{\sigma_s}{\Delta t} u_s^n + \zeta_s w_s^n,$$

$$f_{f,W}^{n+1} := \frac{\beta_{f,1}}{\Delta t} u_f^n + \chi_f w_f^n,$$

are the forcing terms coming from the time discretization. In problem 1 we have also introduced the structure velocity $u_s^n := \beta_{s,0}/\Delta t\,\eta_s^n - f_{s,U}^n$, the structure acceleration $w_s^n := \xi_{s,0}/\Delta t^2\,\eta_s^n - f_{s,W}^n$, and the fluid acceleration $w_f^n := \beta_{f,0}/\Delta t\, u_f^n - f_{f,W}^n$.

2. *Geometry problem.* Given the (unknown) interface structure displacement $\tilde{\eta}_s^{n+1}|_{\Sigma^0}$, solve a harmonic extension problem

$$\begin{cases} -\Delta\tilde{\eta}_m^{n+1} = 0 & \text{in } \Omega_f^0, \\ \tilde{\eta}_m^{n+1} = \tilde{\eta}_s^{n+1} & \text{on } \Sigma^0, \end{cases} \quad (4)$$

and then find accordingly the discrete fluid domain velocity

$$\tilde{u}_m^{n+1} := \frac{\beta_{s,0}}{\Delta t}\tilde{\eta}_m^{n+1} - \tilde{f}_{m,U}^{n+1}, \quad (5)$$

and the points x_f^{n+1} of the new fluid domain by $x_f^{n+1} = \tilde{x}_f + \tilde{\eta}_m^{n+1}$. Here $\tilde{f}_{m,U}^{n+1}, \tilde{w}_m^{n+1}$ and $\tilde{f}_{m,W}^{n+1}$ (the last two quantities are needed for the computation of $\tilde{f}_{m,U}^{n+1}$) are obtained using the same formulae as for $f_{s,U}, w_s$ and $f_{s,W}$. Observe that $(4)_2$ guarantees that the displacement of the fluid interface coincides with that of the structure (geometrical conformity), whereas (5) guarantees that also the mesh and structure velocities coincide at the FS interface.

2.3 *A Lagrange multipliers-based formulation*

In order to introduce suitable algorithms for the numerical solution of (3) and (4), we consider here an equivalent formulation based on the introduction of three Lagrange multipliers living at the FS interface, representing the fluid and structure normal stresses λ_f and λ_s, and the normal derivative of the fluid mesh displacement λ_m (see (Nobile et al., 2011). These new unknowns are introduced just to simplify the expression of the three interface continuity conditions $(3)_{4\text{–}5}$ and $(4)_2$, and the derivation of the partitioned algorithms. However, we have not introduced them in our practical implementation of the algorithms to avoid extra costs.

We start by introducing some new notations. For the sake of notation we remove the temporal index[n+1]. Given a space W, we denote with W^* its dual, with Σ_f^D and Σ_m^D we denote the parts of the boundary $\partial\Omega_f \backslash \Sigma$ where Dirichlet boundary conditions are prescribed for the fluid subproblem and for the harmonic extension problem, respectively, and with $\Sigma_s^{D,0}$ the part of $\partial\Omega_s^0 \backslash \Sigma^0$ where Dirichlet conditions are prescribed for the structure subproblem. Then, we define the following spaces

$$
\begin{aligned}
&V_f := \{v \in H^1(\Omega_f) : v|_{\Sigma_f^D} = 0\}, \qquad Q := L^2(\Omega_f),\\
&V_s := \{v \in H^1(\Omega_s^0) : v|_{\Sigma_s^{D,0}} = 0\},\\
&V_m := \{v \in H^1(\Omega_f^0) : v|_{\Sigma_m^{D,0}} = 0\}.
\end{aligned}
$$

Let $F : [V_f]^3 \times Q \times [V_m]^3 \rightarrow ([V_f]^3 \times Q)^*$ be the fluid operator and G_f be the operator related to the right hand side of the fluid equations. Analogously, for the structure subproblem we introduce the operator $S : [V_s]^3 \rightarrow ([V_s]^3)^*$ and G_s. Finally, for the harmonic extension, we introduce the operator $H : [V_m]^3 \rightarrow ([V_m]^3)^*$. For the definitions of the above operators, we refer the reader to (Nobile et al., 2011). We also define the following trace operators

$$
\begin{aligned}
&\tilde{\gamma}_f : [V_f]^3 \rightarrow [H^{1/2}(\Sigma^0)]^3, \tilde{\gamma}_f \boldsymbol{v} := \tilde{\boldsymbol{v}}\big|_{\Sigma^0},\\
&\tilde{\gamma}_s : [V_s]^3 \rightarrow [H^{1/2}(\Sigma^0)]^3, \tilde{\gamma}_s \boldsymbol{\mu} := \tilde{\boldsymbol{\mu}}\big|_{\Sigma^0},\\
&\tilde{\gamma}_m : [V_m]^3 \rightarrow [H^{1/2}(\Sigma^0)]^3, \tilde{\gamma}_m z := \tilde{z}\big|_{\Sigma^0},
\end{aligned}
\tag{6}
$$

and the related adjoint operators.

We are now ready to rewrite problem 1–2 as follows

$$
\begin{cases}
H(\tilde{\eta}_m) + \tilde{\gamma}_m^* \tilde{\lambda}_m = 0 & \text{in } ([V_m]^3)^*,\\
\tilde{\gamma}_m \tilde{\eta}_m = \tilde{\gamma}_s \tilde{\eta}_s & \text{on } \Sigma^0,\\
\mathcal{F}(\boldsymbol{u}_f, p_f, \boldsymbol{u}_m) + \tilde{\gamma}_f^* \tilde{\lambda}_f = \mathcal{G}_f & \text{in } ([V_f]^3)^*,\\
\alpha_f \tilde{\gamma}_f \boldsymbol{u}_f + \tilde{\lambda}_f & \\
\quad = \alpha_f \tilde{\gamma}_s \left(\dfrac{\beta_{s,0}}{\Delta t} \tilde{\eta}_s - \tilde{\boldsymbol{f}}_{s,U} \right) - \tilde{\lambda}_s & \text{on } \Sigma^0,\\
\alpha_s \tilde{\gamma}_s \dfrac{\beta_{s,0}}{\Delta t} \tilde{\eta}_s + \tilde{\lambda}_s & \\
\quad = \alpha_s \tilde{\gamma}_f \boldsymbol{u}_f - \tilde{\lambda}_f + \alpha_s \tilde{\gamma}_s \tilde{\boldsymbol{f}}_{s,U} & \text{on } \Sigma^0,\\
S(\tilde{\eta}_s) + \tilde{\gamma}_s^* \tilde{\lambda}_s = \mathcal{G}_s & \text{in } ([V_s]^3)^*,
\end{cases}
\tag{7}
$$

where the interface continuity conditions $(7)_{4-5}$ are linear combinations of conditions $(3)_{4-5}$, through the introduction of two functions in $L^\infty(\Sigma^0)$, $\alpha_f \neq \alpha_s$. This will be useful to derive partitioned procedures based on Robin interface conditions (Robin-Robin (RR) schemes, see (Badia et al., 2008), (Badia et al., 2009), (Astorino et al., 2009), (Gerardo Giorda et al., 2010)). This approach has good convergence properties, independent of the added-mass effect (which is very high in haemodynamic contexts, see (Causin et al., 2005)) when the parameters α_f and α_s are suitably chosen, as shown in (Badia et al., 2008), (Gerardo Giorda et al., 2010).

In (Nobile et al., 2011) it has been shown that the Lagrange multiplier λ_f and λ_s have the physical meaning of the fluid and structure normal stress at the FS interface *in the reference configuration.*

3 NUMERICAL ALGORITHMS

For the solution of the FSI problem 5, we propose to use a general preconditioned Richardson method

$$
\widehat{F}(y^k)\delta y^{k+1} = -F(y^k), \tag{13}
$$

where y^k denotes the FSI solution $[\tilde{\eta}_m^k, \tilde{\lambda}_m^k, \boldsymbol{v}_f^k, \tilde{\lambda}_f^k, \tilde{\lambda}_s^k, \tilde{\eta}_s^k]$ at the generic subiteration k, with $\boldsymbol{v}_f := (\boldsymbol{u}_f, p_f), \delta y^{k+1}$ is the increment of the FSI solution at the new iteration $k+1$ with respect to y^k, $F(y) = 0$ corresponds to problem 5, and $\widehat{F}$ is a suitable preconditioner.

In this work, we consider quasi-Newton methods. In particular, we consider the following approximation of the exact jacobian (Nobile et al., 2011).

$$
\widehat{F} = \left[\begin{array}{cc|cc|cc}
\mathcal{H} & \tilde{\gamma}_m^* & & & & \\
\tilde{\gamma}_m & & & & & -\tilde{\gamma}_s \\
\hline
 & & \widehat{\nabla}_{v_f} F & \tilde{\gamma}_f^* & & \\
 & & \alpha_f \tilde{\gamma}_f & I & I & -\alpha_f \dfrac{\beta_{s,0}}{\Delta t} \tilde{\gamma}_s \\
\hline
 & & -\alpha_s \tilde{\gamma}_f & I & I & \alpha_s \dfrac{\beta_{s,0}}{\Delta t} \tilde{\gamma}_s \\
 & & & & \tilde{\gamma}_s^* & S
\end{array}\right],
$$

where $\widehat{\nabla}_{v_f} F$ is obtained from $\nabla_{v_f} F$ by skipping the term $(\boldsymbol{\delta u}_f \cdot \nabla)\boldsymbol{u}_f$ (Nobile et al., 2011). Moreover, we do not consider the shape derivatives $\nabla_{u_m} F$. This leads to the Oseen approximation of the Navier-Stokes problem obtained by using as convective term previous solutions $\boldsymbol{u}_f$ and $\boldsymbol{u}_m$.

We are ready now to derive from $\widehat{F}$ another preconditioner, leading to suitable algorithms for the numerical solution of (7). Since the structure is linear, we report these algorithms in non-incremental form.

3.1 *Double-Loop Algorithm*

We consider the following *two block Gauss-Seidel* preconditioner

$$
\hat{J}_{DL}=\left[\begin{array}{cc|cc|cc}
\mathcal{H} & \tilde{\gamma}_m^* & & & & \\
\tilde{\gamma}_m & & & & & \\
\hline
 & & \hat{\nabla}_{v_f}F & \tilde{\gamma}_f^* & & \\
 & & \alpha_f\tilde{\gamma}_f & I & I & -\alpha_f\dfrac{\beta_{s,0}}{\Delta t}\tilde{\gamma}_s \\
\hline
 & & -\alpha_s\tilde{\gamma}_f & I & I & \alpha_s\dfrac{\beta_{s,0}}{\Delta t}\tilde{\gamma}_s \\
 & & & & \tilde{\gamma}_s^* & S
\end{array}\right],
$$

which corresponds to the sequential solution of the harmonic extension and of a linearized FSI problem. For the solution of the latter, since we are interested in partitioned algorithms, we use the following preconditioner (see (Badia et al., 2008)).

$$
\hat{P}_{RR}=\left[\begin{array}{cc|cc}
\hat{\nabla}_{v_f}\mathcal{F} & \tilde{\gamma}_f^* & & \\
\alpha_f\tilde{\gamma}_f & I & & \\
\hline
-\alpha_s\tilde{\gamma}_f & I & I & \alpha_s\dfrac{\beta_{s,0}}{\Delta t}\tilde{\gamma}_s \\
 & & \tilde{\gamma}_s^* & S
\end{array}\right].
$$

This corresponds to consider two nested loops, an external one for the treatment of the interface position through a fixed-point (quasi-Newton) scheme, and an internal one for the treatment of the interface continuity conditions through the RR scheme. In particular, we have the following algorithm:

Given the solution at iteration k, solve at the current iteration $k+1$ until convergence (we omit the superscript $^{k+1}$)

1. The harmonic extension

$$
\begin{cases}
-\Delta\tilde{\eta}_m=0 & \text{in } \Omega_f^0,\\
\tilde{\gamma}_m\tilde{\eta}_m=\tilde{\gamma}_s\tilde{\eta}_s^k & \text{on } \Sigma^0,
\end{cases}
$$

obtaining the new fluid domain Ω_f and the fluid domain velocity $\boldsymbol{u}_m$.

2. The linearized FSI problem. In particular, given the solution at subiteration $l-1$, solve at the current subiteration l until convergence
   a. The fluid subproblem with a Robin condition at the FS interface

$$
\begin{cases}
\rho_f\dfrac{\beta_{f,0}}{\Delta t}\boldsymbol{u}_{f,l} & \\
\quad+\rho_f((\boldsymbol{u}_f^k-\boldsymbol{u}_m)\cdot\nabla)\boldsymbol{u}_{f,l} & \\
\quad-\nabla\cdot\boldsymbol{T}_{f,l}=\boldsymbol{f}_f+\rho_f\boldsymbol{f}_{f,W} & \text{in } \Omega_f,\\
\nabla\cdot\boldsymbol{u}_{f,l}=0 & \text{in } \Omega_f,\\
\alpha_f\gamma_f\boldsymbol{u}_{f,l}+\boldsymbol{T}_{f,l}\boldsymbol{n}=\boldsymbol{T}_{s,l-1}\boldsymbol{n} & \\
\quad+\alpha_f\gamma_s\left(\dfrac{\beta_{s,0}}{\Delta t}\eta_{s,l-1}-\boldsymbol{f}_{s,U}\right) & \text{on } \Sigma;
\end{cases}
$$

   b. The structure subproblem with a Robin condition at the FS interface

$$
\begin{cases}
\rho_s\dfrac{\xi_{s,0}}{\Delta t^2}\tilde{\eta}_{s,l}-\nabla\cdot\tilde{\boldsymbol{T}}_{s,l} & \\
\quad=\tilde{\boldsymbol{f}}_s+\rho_s\tilde{\boldsymbol{f}}_{s,W} & \text{in } \Omega_s^0,\\
\alpha_s\dfrac{\beta_{s,0}}{\Delta t}\tilde{\gamma}_s\tilde{\eta}_{s,l}-\tilde{\boldsymbol{T}}_{s,l}\tilde{\boldsymbol{n}}=-\tilde{\boldsymbol{T}}_{f,l}\tilde{\boldsymbol{n}} & \\
\quad+\alpha_s\left(\tilde{\gamma}_f\tilde{\boldsymbol{u}}_{f,l}+\tilde{\gamma}_s\tilde{\boldsymbol{f}}_{s,U}\right) & \text{on } \Sigma^0.
\end{cases}
$$

The use of two different loops for the geometrical/physical non-linearities and for the imposition of the interface continuity conditions makes this scheme more robust with respect to the use of just a single loop, as shown in (Nobile et al., 2011).

3.2 *Inexact solutions*

In order to improve the performances of the Double-loop scheme in terms of CPU time, we report here a family of algorithms drawn from the *Double-loop* scheme and introduced in (Nobile et al., 2011) In particular, we consider the *geometrical and convective inexact schemes-m* (GCIS-m), obtained from Double-loop by performing at most *m* iterations in the external loop. We observe that with GCIS-1 we perform just one external iteration, that is we solve a linearized FSI problem in a known domain (see (Fernández et al., 2007), (Badia et al., 2008), (Nobile and Vergara 2008), (Crosetto et al., 2009)).

4 NUMERICAL RESULTS

We consider here an application of previous schemes to a real geometry of a patient, namely the human carotid depicted in Figure 1, right. In particular, we want to compare the accuracy of GCIS-1 and GCIS-2 schemes with respect to Double-loop, when three different schemes are used for the time discretization of fluid and structure. In particular we consider *Newmark schemes* for the solid, defined by parameters a_1, a_2, and *theta-methods* for the fluid, depending on the parameter $\theta = a_1/a_2$, see (Nobile et al., 2011). In Table 4 we report the principal features of such temporal schemes. We observe that NWA is obtained by considering *mid-point* for the solid and *Crank-Nicolson* for the fluid. This scheme is globally second order accurate in time, unconditionally stable and less dissipative than other Newmark schemes.

When using NWA, for GCIS-1 we use a suitable extrapolation of the interface quantities and fluid convective term in order to recover a global order 2, whilst for GCIS-2 such extrapolation is not necessary to recover order 2 (Nobile et al., 2011).

We use $P1bubble - P1$ finite elements for the fluid subproblem and $P1$ finite elements for the structure subproblem, and the following data: viscosity $\mu = 0.03 dyne/cm^2$, fluid density $\rho_f = 1 g/cm^3$, structure density $\rho_s = 1.2 g/cm^3$, Young modulus $E = 3\cdot 10^6 dyne/cm^2$, Poisson ratio $\nu = 0.45$, time discretization parameter $\Delta t = 0.001 s$, and elastic coefficient of the surrounding tissue $\alpha_e = 3\cdot 10^6 dyne/cm^2$. This value has been extracted by the experimental results reported in (Liu et al., 2007) and allows to recover a pressure in the physiological range.

For the prescription of the interface continuity conditions, in all the simulations we have considered the RR scheme, with the optimal coefficients proposed in (Gerardo Giorda et al., 2010) and adapted to different temporal schemes in (Nobile et al., 2011).

The results have been obtained with the parallel Finite Element library |LIFEV| developed at MOX—Politecnico di Milano, INRIA—Paris, CMCS—EPF of Lausanne and Emory University—Atlanta.

For the harmonic extension and for the structure, we prescribe at the artificial sections normal homogeneous Dirichlet conditions and tangential homogeneous Neumann conditions, that is we let the domain to move freely in the tangential direction. At the inlet we prescribe the physiological flow-rate depicted in Figure 2, left, through the Lagrange multipliers method (Formaggia et al., 2002), (Veneziani and Vergara 2005). At the outlet, we propose to use the following absorbing boundary condition, obtained by following (Nobile and Vergara 2008), (Nobile et al., 2011):

$$\frac{1}{|\Gamma|}\int_{\Gamma} (\boldsymbol{T}_f \boldsymbol{n})\cdot \boldsymbol{n}\, d\sigma - R_e \int_{\Gamma^n} \boldsymbol{u}\cdot \boldsymbol{n}\, d\sigma = P_{ext} \quad \text{on } \Gamma, \qquad (14)$$

where $R_e = \sqrt{\frac{\rho_f \tau}{2\sqrt{\pi}} \frac{1}{A_0^{3/4}}}, \tau := \frac{EH_s\sqrt{\pi}}{(1-\nu^2)R^2}$, with H_s the structure thickness and R a reference radius. We set $P_{ext} = 0 mmHg$.

We run the simulations on 4 processors for the solution of the fluid problem and on 1 processor for the structure.

We consider a section of the domain, Σ_1 located at $0.7cm$ from the inlet, showed in Figure 2, right. For this section we report in Figure 3, the mean pressure (left) and a zoom of this in the range $t \in (0.45, 0.6)$ (right), obtained with Double-loop by using the three temporal schemes considered. We observe that in all cases the pressure varies in the range $70 - 120 mmHg$, which corresponds to a typical pressure drop in physiological conditions, and the solutions are very similar. However, the solution provided by NWA features some numerical oscillations (Figure 3*a*), which are strongly reduced by using NWB (Figure 3*b*) and disappear completely with NWC (Figure 3*c*).

In Figure 4 we report the fluid velocity (left), the fluid pressure (middle) and the wall shear stress (WSS, right) at the peak instant, obtained with Double loop/NWA scheme. The results obtained with GCIS-1 and GCIS-2 and with NWB and NWC are very similar, so that we do not report them here. In order to quantify the differences, we define the following percentage error with respect to the Double-loop solution

$$E_x = \frac{\| x_{DL} - x_* \|_{L^\infty(\Omega^t)}}{\| x_{DL} \|_{L^\infty(\Omega^t)}} \times 100, \qquad (10)$$

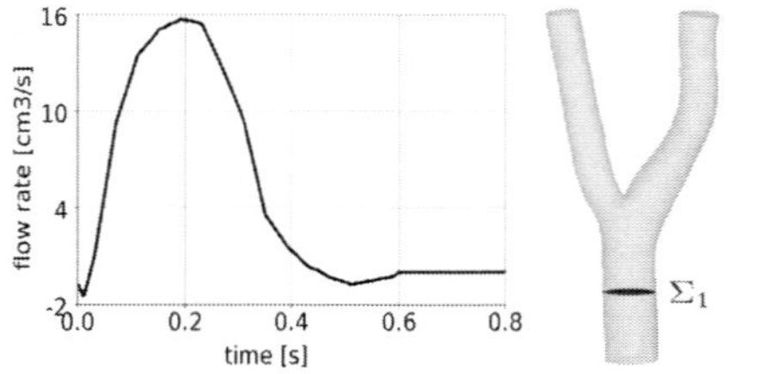

Figure 2. Flow rate waveform prescribed at the inlet of the carotid (left) and fluid domain (right).

Table 1. Parameters of temporal schemes and order of accuracy for the fluid ($\mathcal{F}$-order) and the solid ($\mathcal{S}$-order).

	a_1	a_2	θ	$\mathcal{F}$-order	$\mathcal{S}$-order
NWA	0.25	0.5	0.5	2	2
NWB	0.35	0.5	0.7	1	2
NWC	0.35	0.6	0.75	1	1

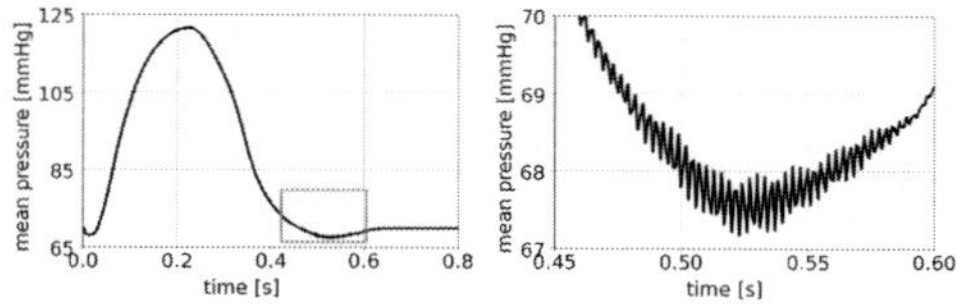

(a) NWA scheme.

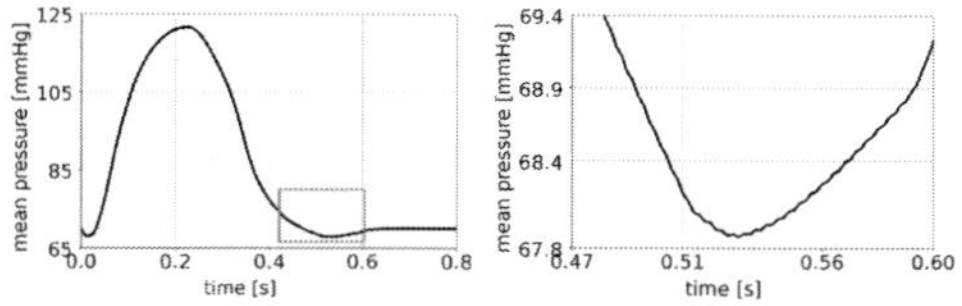

(b) NWB scheme.

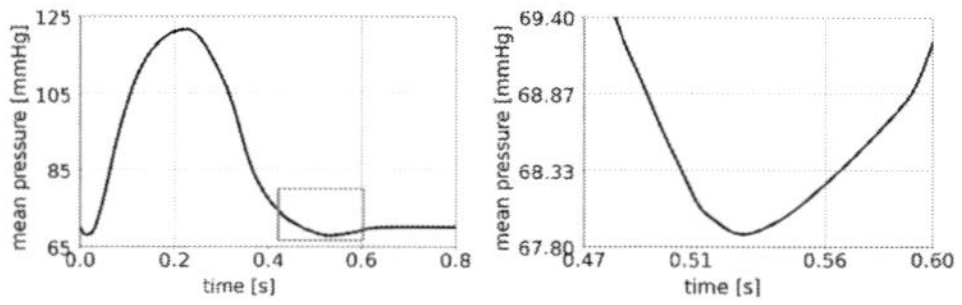

(c) NWC scheme.

Figure 3. Mean pressure [mmHg] (left) and zoom of the box (right).

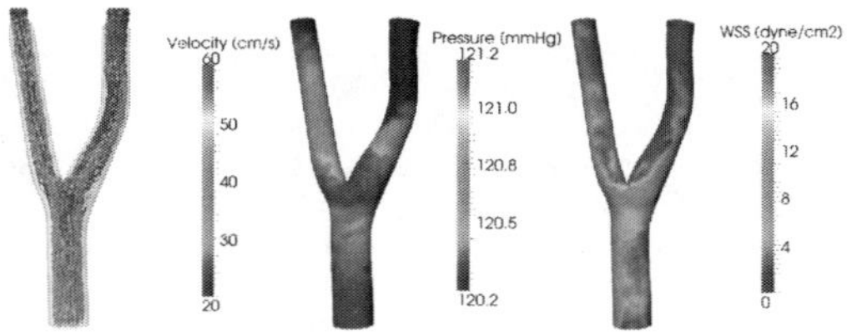

Figure 4. Fluid velocity field (left), fluid pressure (center) and wall shear stress (right) at the peak instant.

where x represents WSS, fluid velocity, fluid pressure or solid displacement, DL stands for Double-loop and * for one of the other scheme. Of course, the errors of GCIS-1 and GCIS-2 obtained by using NWJ scheme, $J = A, B, C$ are computed with respect to the Double Loop solution obtained with NWJ scheme. In Table 2 we report these errors at the peak instant. We observe a good accuracy of such schemes, expecially for GCIS–2.

In Table 3 we report the CPU time normalized with respect to that of Double-loop scheme for Newmark schemes. We observe that Double-loop is about 3 times more expensive than GCIS-1 and more than 2 times more expensive than GCIS-2, and that the normalized CPU time seems to be independent of the temporal scheme.

Table 2. Percentage error of GCIS-1 and GCIS-2 with respect to the Double-loop solution for Newmark schemes. Computation done with 10 at the peak instant.

	E_{WSS}	$E_{\eta s}$	E_{uf}	E_{pf}
GCIS-1/NWA	0.476	0.175	0.286	0.032
GCIS-2/NWA	0.092	0.014	0.035	0.004
GCIS-1/NWB	0.435	0.126	0.400	0.047
GCIS-2/NWB	0.085	0.038	0.022	0.006
GCIS-1/NWC	0.405	0.216	0.241	0.030
GCIS-2/NWC	0.081	0.008	0.021	0.006

Table 3. CPU time normalized with respect to that of Double-loop scheme for NW schemes.

	NWA	NWB	NWC
GCIS-1	0.304	0.316	0.315
GCIS-2	0.458	0.465	0.471

5 CONCLUSIONS

In conclusion, we can state that GCIS-2 scheme is an effective algorithm for the solution of real haemodynamic problems for second order accurate temporal schemes. Indeed, it features a good accuracy with respect to the solution obtained with Double-loop, used here as gold-standard, and a satisfactory improvement in the CPU time (halving the time with respect to Double-loop). These results confirms the nice features of GCIS-2 scheme for real applications highlighted in (Nobile et al., 2011) for a first order temporal scheme.

ACKNOWLEDGMENTS

The authors have been partially supported by the ERC Advanced Grant N.227058 MATHCARD.

REFERENCES

Astorino, M., Chouly, F. and Fernández, M. (2009). Robin based semi-implicit coupling in fluid-structure interaction: stability analysis and numerics. *SIAM J Sc Comp* (31(6)), 4041–4065.

Badia, S., Nobile, F. and Vergara, C. (2008). Fluid-structure partitioned procedures based on Robin transmission conditions. *Journal of Computational Physics 227*, 7027–7051.

Badia, S., Nobile, F. and Vergara, C. (2009). Robin-Robin preconditioned Krylov methods for fluid-structure interaction problems. *Computer Methods in Applied Mechanics and Engineering 198* (33–36), 2768–2784.

Badia, S., Quaini, A. and Quarteroni, A. (2008). Splitting methods based on algebraic factorization for fluid-structure interaction. *SIAM J Sc Comp 30*(4), 1778–1805.

Bazilevs, Y., Calo, V., Zhang, Y. and Hughes, T. (2006). Isogeometric fluid-structure interaction analysis with applications to arterial blood flow. *Computational Mechanics 38*, 310–322.

Causin, P., Gerbeau, J. and Nobile, F. (2005). Added-mass effect in the design of partitioned algorithms for fluid-structure problems. *Computer Methods in Applied Mechanics and Engineering 192 (42–44)*, 4506–4527.

Crosetto, P., Deparis, S., Fourestey, G. and Quarteroni A. (2009). Parallel algorithms for fluid-structure interaction problems in haemodynamics. *EPFL-REPORT-148536*.

Deparis, S., Discacciati, M., Fourestey, G. and Quarteroni, A. (2006). Fluid-structure algorithms based on Steklov-Poincaré operators. *Computer Methods in Applied Mechanics and Engineering 195* (41–43), 5797–5812.

Donea, J. (1982). An arbitrary Lagrangian-Eulerian finite element method for transient dynamic fluid-structure interaction. *Computer Methods in Applied Mechanics and Engineering 33*, 689–723.

Fernández, M., Gerbeau, J. and Grandmont, C. (2007). A projection semi-implicit scheme for the coupling of an elastic structure with an incompressible fluid. *International Journal for Numerical Methods in Engineering* 69(4), 794–821.

Figueroa, C., Vignon-Clementel, I., Jansen, K., Hughes, T. and Taylor, C. (2006). A coupled momentum method for modeling blood flow in three-dimensional deformable arteries. *Computer Methods in Applied Mechanics and Engineering 195*, 5685–5706.

Formaggia, L., Gerbeau, J.-F., Nobile, F. and Quarteroni, A. (2002). Numerical treatment of defective boundary conditions for the Navier-Stokes equation. *SIAM Journal on Numerical Analysis 40(1)*, 376–401.

Formaggia, L., A. Quarteroni, and A. V. (Eds.) (2009). Cardiovascular Mathematics—Modeling and simulation of the circulatory system. Springer.

Forster, C., Wall, W. and Ramm, E. (2007). Artificial added mass instabilities in sequential staggered coupling of nonlinear structures and incompressible viscous flow. *Computer Methods in Applied Mechanics and Engineering* 196(7), 1278–1293.

Gerardo Giorda, L., Nobile, F. and Vergara, C. (2010). Analysis and optimization of robin-robin partitioned procedures in fluid-structure interaction problems. *SIAM Journal on Numerical Analysis 48(6)*, 2091–2116.

Hughes, T. J.R., Liu, W.K. and Zimmermann, T.K. (1981). Lagrangian-Eulerian finite element formulation for incompressible viscous flows. *Computer Methods in Applied Mechanics and Engineering 29*(3), 329–349.

Liu, Y., Charles, C., Gracia, M., Gregersen, H. and Kassab, G.S. (2007). Surrounding tissues affect the passive mechanics of the vessel wall: theory and experiment. *Am J Physiol Heart Circ Physiol 293*, H3290–H3300.

Moireau, P., Xiao, N., Astorino, M., Figueroa, C.A., Chapelle, D., Taylor, C.A. and Gerbeau, J.-F. (2011). External tissue support and fluid-structure simulation in blood flows. *Biomechanics and Modeling in Mechanobiology*.

Nobile, F. (2001). *Numerical approximation of fluid-structure interaction problems with application to haemodynamics*. Ph. D. thesis, École Polytechnique Fédérale de Lausanne. Thesis n° 2458.

Nobile, F., Pozzoli, M. and Vergara, C. (2011). Time accurate partitioned algorithms for the solution of fluid-structure interaction problems in haemodynamics. *MOX Report n. 30/2011*.

Nobile, F. and Vergara, C. (2008). An effective fluid-structure interaction formulation for vascular dynamics by generalized Robin conditions. *SIAM J Sc Comp* 30(2), 731–763.

Perktold, K., Thurner, E. and Kenner, T. (1994). Flow and stress characteristics in rigid walled and compliant carotid artery bifurcation models. *Medical and Biological Engineering and Computing* 32(1), 19–26.

Piperno, S. and Farhat, C. (2000). Design of efficient partitioned procedures for transient solution of aerolastic problems. *Rev. Eur. Elements Finis 9(6–7)*, 655–680.

Tezduyar, T., Sathe, S., Cragin, T., Nanna, B., Conklin, B., Pausewang, J. and Schwaab, M. (2007). Modelling of fluid-structure interactions with the space-time finite elements: arterial fluid mechanics. *International Journal for Numerical Methods in Fluids*, 901–922.

Veneziani, A. and Vergara, C. (2005). Flow rate defective boundary conditions in haemodinamics simulations. *International Journal for Numerical Methods in Fluids*, 803–816.

IX Numerical methods for reactive flows

Numerical Methods for Hyperbolic Equations – Vázquez-Cendón et al. (eds)
© 2013 Taylor & Francis Group, London, ISBN 978-0-415-62150-2

On the use of moving least squares for pressure discretization in low mach number flows

X. Nogueira & I. Colominas
Group of Numerical Methods in Engineering, Department of Mathematical Methods, Universidade da Coruña, Spain

S. Khelladi & F. Bakir
Arts et Métiers ParisTech, Laboratoire de Dynamique des Fluides (DynFluid Lab), Paris, France

J.-C. Chassaing
D'Alembert Institute, UPMC-CNRS, UMR 7190, Univ Paris 6, Paris, France

ABSTRACT: In this work we present a modification of the pressure discretization for low-Mach numerical schemes. We propose using Moving-Least Squares (MLS) approximations to the discretization of the pressure flux for the numerical schemes developed for low-Mach number flows. This simple modification avoids all the problems related with checkerboard and it obtains a very accurate representation of the pressure field. The centered character of MLS approximations ensures the correct scaling of the pressure with the square of the Mach number. We present here the results of the application of the proposed pressure flux discretization to the AUSM-family schemes. In particular we test here the case of the SLAU scheme (E. Shima, K. Kitamura, Comm. Comp. Phys., 2011).

1 INTRODUCTION

Traditionally, two families of finite volume schemes have been developed to compute both compressible and incompressible flows. Thus, density-based solvers (Roe 1981; van Leer 1982; Colella and Woodward 1984) are used for the computation of flows when compressibility effects are important (mainly transonic, supersonic and hypersonic flows), whereas pressure-based solvers (Chorin 1968; Patankar 1980; Peyret and Taylor 1983) are designed to compute incompressible flows. However, it is preferable the development of solvers useful for all the regimes of a flow. This is not only for user's convenience, but also because the importance of flows where low and high Mach regions are present (for example flow past an aerodynamic profile at high angle of attack), or when compressibility effects are important, even in low Mach number flows. Thus, the modification of density or pressure-based solvers to compute all-speed flows is a current active area of research.

Godunov-like schemes are among the most widely used methods for CFD. When Godunov schemes are used in a density-based solver for low-Mach number computations, it is needed a central discretization for the pressure in order to obtain the adequate scaling of pressure with the square of the Mach number (Dellacherie 2010). However, many of the pressure discretization techniques used by the schemes developed for low Mach number, allow the existence of four-field solutions for the pressure (checkerboard). In this work we present a modification in the pressure discretization of low-Mach numerical schemes. We propose using Moving-Least Squares (Lancaster and Salkauskas 1981; Liu et al., 1997) approximations to the discretization of the pressure flux in the AUSM-family numerical schemes developed for low-Mach number flows. This simple modification alleviates the problem related with checkerboard and it obtains a very accurate representation of the pressure field. The centered character of MLS approximations ensures the correct scalingof the pressure with the square of the Mach number (Guillard and Viozat 1999; Dellacherie 2010; Schochet 1994). We present here the results of the application of the proposed pressure flux discretization to the AUSM-family schemes.

2 NUMERICAL METHODS

We present in this section the 2D formulation of the flux proposed in this work. This modification can be apply to any AUSM-family scheme. In this paper we show the results for the Simple Low-dissipation AUSM (SLAU) scheme (Shima and Kitamura 2009; Shima and Kitamura 2011).

2.1 *SLAU scheme*

As it is usual in AUSM-family schemes, we start with a splitting of the numerical flux in convective and pressure parts:

$$\Theta(\boldsymbol{U}^+,\boldsymbol{U}^-)=\Theta_{convection}(\boldsymbol{U}^+,\boldsymbol{U}^-)+\boldsymbol{P}_{flux}. \tag{1}$$

Following (Shima and Kitamura 2011), the expression for $\Theta_{convection}$ is:

$$\Theta_{convection}=\frac{\dot{m}+|\dot{m}|}{2}\Psi^+ + \frac{\dot{m}-|\dot{m}|}{2}\Psi^- \tag{2}$$

where $\Psi=(1,u,\upsilon,H)^T$ and $H=E+p/\rho$ is the enthalpy.

The mass flux function is defined as:

$$\dot{m}=\frac{1}{2}\left[\rho_L\left(V_{nL}+|\bar{V}_n|^+\right)+\rho_R\left(V_{nR}+|\bar{V}_n|^-\right)-\frac{\chi}{\bar{c}}\Delta p\right] \tag{3}$$

with

$$|\bar{V}_n|^+=(1-g)|\bar{V}_n|+g|V_{nL}|$$

$$|\bar{V}_n|^-=(1-g)|\bar{V}_n|+g|V_{nR}|$$

$$|\bar{V}_n|=\frac{\rho_L|V_{nL}|+\rho_R|V_{nR}|}{\rho_L+\rho_R}$$

$$V_{nR}=u_R n_x+\upsilon_R n_y$$

$$V_{nL}=u_L n_x+\upsilon_L n_y$$

$$g=-\max[\min(M_L,0),-1]\cdot\min[\max(M_R,0),1].$$

Where n_x and n_y are the components of the normal vector to the considered face of the volume control and $\Delta()=()_L-()_R$.

As stated by Liou (Liou 2000), the Δ_p term in equation (3) may cause carbuncle phenomena in some hypersonic flows. Thus, the parameter χ is introduced to activate the last term of equation (3) for low Mach number flows (Shima and Kitamura 2009). When the Mach number increases, this term is progressively deactivated. It is defined as:

$$\chi=\left(1-\hat{M}\right)^2$$

$$\hat{M}=\min\left(1.0,\frac{1}{\bar{c}}\sqrt{\frac{{u_L}^2+{v_L}^2+{u_R}^2+{v_R}^2}{2}}\right).$$

On the other hand, the presure flux is:

$$\boldsymbol{P}_{flux}=\left(0,\tilde{p}n_x,\tilde{p}n_y,0\right) \tag{4}$$

In the SLAU scheme, as it is usual in AUSM-family schemes, the pressure flux use the pressure term from the Flux Vector Splitting technique of van Leer to evaluate the flux at the interface of the control volume:

$$\tilde{p}=\beta^+p_L+\beta^-p_R \tag{5}$$

where

$$\beta^+=\begin{cases}\frac{1}{2}\left(1+sign\,M^+\right) & |M^+|\geq 1\\ \frac{1}{4}\left(M^++1\right)^2\left(2-M^+\right) & otherwise\end{cases} \tag{6}$$

$$\beta^-=\begin{cases}\frac{1}{2}\left(1-sign\,M^-\right) & |M^-|\geq 1\\ -\frac{1}{4}\left(M^--1\right)^2\left(2+M^-\right) & otherwise\end{cases} \tag{7}$$

with

$$M^+=\frac{V_{nL}}{\bar{c}} \tag{8}$$

$$M^-=\frac{V_{nR}}{\bar{c}} \tag{9}$$

$$\bar{c}=\frac{c^++c^-}{2} \tag{10}$$

In SLAU, the pressure flux is written as

$$\tilde{p}=\frac{p_L+p_R}{2}+\frac{\beta^+-\beta^-}{2}(p_L-p_R)+(1-\chi)\left(\frac{\beta^++\beta^-}{2}\right)(p_L+p_R) \tag{11}$$

2.2 *The checkerboard problem*

The expression of the pressure flux given by equation (11) leads to the following when the Mach number tends to zero:

$$\tilde{p}=\frac{p_L+p_R}{2}. \tag{12}$$

An asymptotic analysis of the discrete Euler equations (Klainerman and Majda 1982; Guillard and

Viozat 1999; Dellacherie 2010; Schochet 1994; Li and Gu 2008) leads to the following conditions, for both the 0– and 1–order pressure:

$$p_{i-1,j} - p_{i+1,j} = 0, \tag{13}$$

$$p_{i,j-1} - p_{i,j+1} = 0. \tag{14}$$

The discretization given by (11) allows the so called *four field solution (checkerboard)*. As it is shown in Figure 1, equations (13) and (14) consider such a field as a constant field, which evidently is wrong.

In order to cure, or at least, alleviate this problem, different methods have been used. For example, the addition of a pressure stabilization term to the interface fluid velocity in a momentum interpolation (Li and Gu 2008). In this work, we propose a differentapproach that will be explained in the next section.

2.3 *Moving Least Squares (MLS)*

In this section we will present briefly the fundamentals of the Moving Least Squares approximations (Lancaster and Salkauskas 1981; Liu et al. 1997).

Let us consider a function $u(\boldsymbol{x})$ defined in a domain Ω. The idea of the MLS approach is to approximate $u(\boldsymbol{x})$, at a given point $\boldsymbol{x}$, through a weighted least-squares fitting of $u(\boldsymbol{x})$ in a neighborhood of $\boldsymbol{x}$ as

$$u(\boldsymbol{x}) \approx \hat{u}(\boldsymbol{x}) = \sum_{i=1}^{m} p_i(\boldsymbol{x})\alpha_i(\boldsymbol{z})\big|_{z=x} = p^T(\boldsymbol{x})\alpha(\boldsymbol{z})\big|_{z=x}, \tag{15}$$

$\boldsymbol{p}^T(\boldsymbol{x})$ is an m-dimensional polynomial basis and $\alpha(\boldsymbol{z})|_{z=x}$ is a set of parameters to be determined,

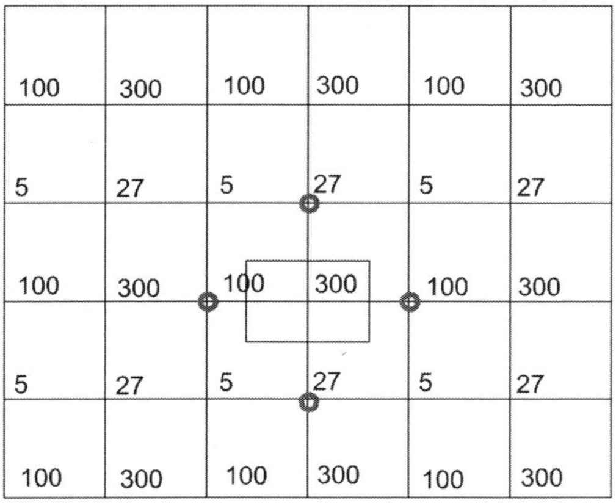

Figure 1. Scheme of the four field solution and the problem of checkerboard.

such that they minimize the following error functional

$$J\left(\alpha(\boldsymbol{z})|_{\boldsymbol{z}=\boldsymbol{x}}\right) = \int_{y\in\Omega_x} W(z-\boldsymbol{y},h)|_{\boldsymbol{z}=\boldsymbol{x}}$$
$$\left[u(\boldsymbol{y}) - \boldsymbol{p}^T(\boldsymbol{y})\alpha(z)|_{\boldsymbol{z}=\boldsymbol{x}}\right]^2 d\Omega_{\boldsymbol{x}},$$

$W(\boldsymbol{z}-\boldsymbol{y},h)|_{\boldsymbol{z}=\boldsymbol{x}}$ is a *kernel* with compact support (denoted by Ω_x) centered at $\boldsymbol{z}=\boldsymbol{x}$, frequently chosen among the kernels used in standard SPH. The parameter h is the smoothing length, which is a measure of the size of the support Ω_x.

Following (Cueto-Felgueroso et al. 2007), the interpolation structure can be identified as

$$\hat{u}(\boldsymbol{x}) = \boldsymbol{p}^T(\boldsymbol{x})\boldsymbol{M}^{-1}(\boldsymbol{x})\boldsymbol{P}_{\Omega_x}\boldsymbol{W}(\boldsymbol{x})\boldsymbol{u}_{\Omega_x}$$
$$= \boldsymbol{N}^T(\boldsymbol{x})\boldsymbol{u}_{\Omega_x} = \sum_{j=1}^{n_x} N_j(\boldsymbol{x})u_j, \tag{16}$$

where the vector $\boldsymbol{u}_{\Omega_x}$ contains the pointwise values of the function u to be reproduced at the n_x particles (nodes) inside $\Omega_{\boldsymbol{x}}$. We define the $(m \times n_{\boldsymbol{x}})$ matrix $\boldsymbol{P}_{\Omega_x} = \left(\boldsymbol{p}(x_1)\ \boldsymbol{p}(x_2)\cdots \boldsymbol{p}(x_{n_x})\right)$ and the $(n_{\boldsymbol{x}} \times n_{\boldsymbol{x}})$ diagonal matrix $\boldsymbol{W} = diag\left[W_i(\boldsymbol{x}-\boldsymbol{x}_i)\right]$ with $i = 1,\&,n_x$. Moreover, $\boldsymbol{M} = P_{\Omega_x}\boldsymbol{W}\boldsymbol{P}_{\Omega_x}^T$ is the moment matrix (see (Cueto-Felgueroso et al. 2007)).

The approximation is written in terms of the MLS "shape functions" $\boldsymbol{N}^T(\boldsymbol{x}) = \boldsymbol{p}^T(\boldsymbol{x})\boldsymbol{C}(\boldsymbol{x})$, where $\boldsymbol{C}(\boldsymbol{x})$ is defined as $\boldsymbol{C}(\boldsymbol{x}) = \boldsymbol{M}^{-1}(\boldsymbol{x})\boldsymbol{P}_{\Omega_x}\boldsymbol{W}(\boldsymbol{x})$.

This technique has been successfully used in the field of finite volume schemes to develop higher-order schemes (Cueto-Felgueroso et al. 2007; Nogueira et al. 2010; Khelladi et al. 2011).

We refer the interested reader to (Lancaster and Salkauskas 1981; Liu et al. 1997; Cueto-Felgueroso et al. 2007; Nogueira et al. 2010) for a deeper explanation of the Moving Least Squares approximations.

2.4 *A MLS-based pressure flux*

The origin of the checkerboard problem is in the centered discretization of the pressure gradient of the Euler equations. However, using a centered discretization of the pressure is needed in order to compute low Mach flows (see (Dellacherie 2010)). Thus, we look for centered discretizations of the pressure in such a way that the checkerboard problem could be avoided.

In this paper we propose to use the following pressure flux

$$\tilde{p} = \chi \cdot p_{MLS} + (1-\chi)\left(\beta^+ p_L + \beta^- p_R\right) \tag{17}$$

where p_{MLS} is the approximation of the pressure at Gauss point using MLS (Lancaster and Salkauskas

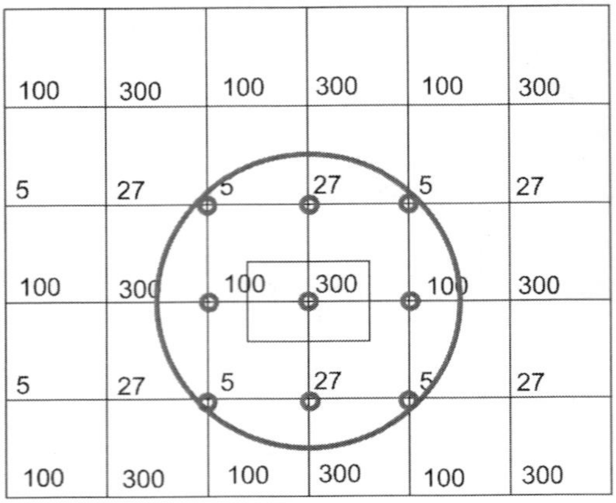

Figure 2. Scheme of the four field solution and the cure for the problem of checkerboard using MLS.

1981; Liu et al. 1997). Note that $\chi \rightarrow 1$ as $M \rightarrow 0$. Thus, we obtain a centered discretization of the pressure flux in the zero-Mach limit, whereas we recover the full-upwind discretization when $M \geq 1$.

Using MLS we obtain a truly multidimensional reconstruction of the pressure field. The stencil for the reconstruction of data using MLS includes points in several directions (Nogueira et al. 2010), including the point (i,j). The discretization using MLS naturally avoids the chekerboard problem, at least for the 0- and 1-order pressure. This is schematically shown in figure 2. However, since we do not modify the 2-order pressure discretization, it still remains the possibility of checkerboard in the order-2 pressure.

The centered character of MLS approximations assures the correct scaling of the pressure with the square of the Mach number (Guillard and Viozat 1999; Dellacherie 2010; Schochet 1994). Note that we can apply this discretization for the pressure to any AUSM-family scheme, not only to SLAU.

3 NUMERICAL EXAMPLES

3.1 *Low mach flow in an unbounded 2D channel with a bump*

Here we compute the flow in an unbounded 2D channel with a bump using a triangular grid. The computational domain is $[0,4]\times[0,1]$. The geometry of the bump is the same than the one in (Dellacherie 2010), that is:

$$y_{bump}(x) = \begin{cases} \frac{1}{10}\left[1-\cos((x-1)\pi)\right] & \text{if } x \in [1,3] \\ 0 & \text{otherwise} \end{cases} \quad (18)$$

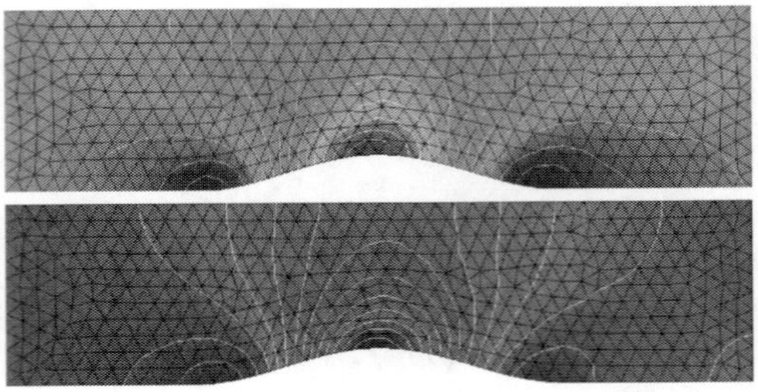

Figure 3. Mach (top) and pressure (bottom) isolines for the $M = 10^{-2}$ flow in an unbounded 2D channel with a bump using the first order SLAU scheme with MLS pressure flux.

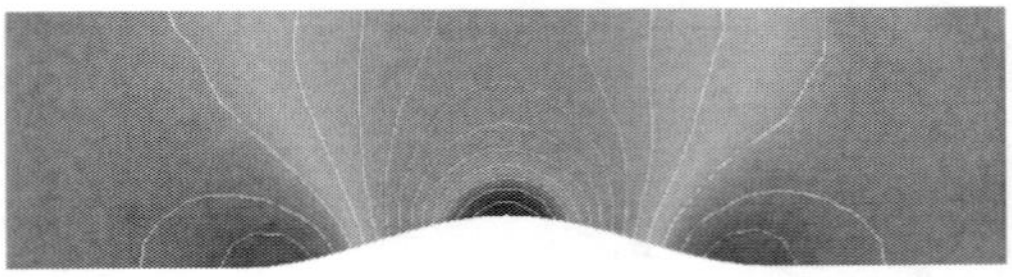

Figure 4. Pressure isolines for the $M = 10^{-2}$ flow in an unbounded 2D channel with a bump using the second order SLAU scheme with MLS pressure flux.

We use the stiffened equation of state for water (Harlow and Amsden 1971; Chang and Liou 2007). This is an equation of state used to compute multiphase flows. The boundary conditions are the following:

$$p = p_0 = 10^5 \text{ Pa}, H = 362.95 \text{ kJ if } x = 0$$

$$p = p_0, \text{ if } x = 4$$

$$u = v = 0, \text{ on } y = 1 \text{ and } y_{bump} \quad (19)$$

The inlet velocity is constant and such as that the Mach number is $M = 10^{-2}$.

We obtain the physical solution of the problem, and no checkerboard modes are observed. However, it would be still possible to find 2-order pressure checkerboard modes.

High-order is easily achieved using MLS for the computation of the derivatives in the reconstruction step of a finite volume scheme (Nogueira et al. 2010). We show the pressure field obtained with the second order scheme. The improvement in the solution is clearly observed.

4 CONCLUSIONS

We have presented a new methodology for the discretization of the pressure in computations of low Mach number flows. When the Mach number tends to zero, we switch the pressure discretization scheme to use a direct, centered MLS approximation at Gauss points of the control volume. This discretization naturally avoids the checkerboard modes for the 0 and 1-order pressure. he pressure.

REFERENCES

Chang, C.-H. & Liou, M.-S. 2007. A robust and accurate approach to computing compressible multiphase flow: Stratified flow model and AUSM+-up scheme. *Journal of Computational Physics*, **225**:840–873.

Chorin, A.J. 1968. Numerical solution of the Navier-Stokes equations. *Mathematics of Computation*, **22**:745–762.

Colella, P., Woodward, P. 1984. The piecewise parabolic method (PPM) for gas-dynamical simulations. *Journal of Computational Physics*, **54**:174–201.

Cueto-Felgueroso, L., Colominas, I., Nogueira X., Navarrina F. & Casteleiro M. 2007. Finite volume solvers and Moving Least-Squares approximations for the compressible Navier—Stokes equations on unstructured grids. *Computer Methods in Applied Mechanics and Engineering*, **196**:4712–4736.

Dellacherie, S. 2010. Analysis of Godunov type schemes applied to the compressible Euler system at low Mach number. *Journal of Computational Physics*, **229**(4):978–1016, 2010.

Guillard, H. & Viozat, C. 1999. On the behaviour of upwind schemes in the low mach number limit. *Computers and Fluids*, **28**:63–86, 1999.

Harlow, F. & Amsden, A. 1971. *Fluid dynamics. Technical Report LA-4700*, Los Alamos National Laboratory.

Khelladi, S., Nogueira, X., Bakir, F. & Colominas, I. 2011. Toward a Higher Order Unsteady Finite Volume Solver Based on Reproducing Kernel Methods. *Computer Methods in Applied Mechanics and Engineering*, **200**:2348–2362.

Klainerman S. & Majda A. 1982. Compressible and incompressible fluids. *Communications on Pure and Applied Mathematics*, **35**:629–651.

Lancaster, P. & Salkauskas, K. 1981. Surfaces generated by moving least squares methods. *Mathematics of Computation*, **37**(155), 141–158.

Li, X.-S & Gu, C.-W. 2008. An All-Speed Roe-type scheme and its asymptotic analysis of low Mach number behaviour. *Journal of Computational Physics*, **227**:5144–5159.

Liou, M.-S 2000. Mass Flux Schemes and Connection to Shock Instability. *Journal of Computational Physics*, **160**:623–648.

Liou, M.-S. 2006. A sequel to AUSM, Part II: AUSM+-up for all speeds. *Journal of Computational Physics*, **214**:137–170, 2006.

Liu, W.K., Hao, W., Chen, Y., Jun, S., Gosz, J. 1997. Multiresolution Reproducing Kernel Particle Methods. *Computational Mechanics*, **20**, 295–309.

Nogueira, X., Cueto-Felgueroso, L., Colominas, I., Khelladi, S. 2010. On the simulation of wave propagation with a higher-order finite volume scheme based on Reproducing Kernel Methods. *Computer Methods in Applied Mechanics and Engineering*, tbf199(23–24):1471–1490.

Patankar, S.V. 1980. *Numerical heat transfer and fluid flow*, McGraw-Hill, New York.

Peyret, R. and Taylor, T.D. 1983. *Computational Methods for fluid flow*, Springer-Verlag, New York.

Roe, P.L. 1981. Approximate Riemann solvers, parameter vectors and difference schemes. *Journal of Computational Physics*,**43**, 357–372, 1981.

Schochet, S. 1994. Fast singular limits of hyperbolic PDEs. *Journal of Differential Equations* **114**:476–512.

Shima, E. & Kitamura, K. 2009. On New Simple Low-Dissipation Scheme of AUSM-Family for All Speeds, *AIAA paper* 2009–136.

Shima, E. & Kitamura, K. 2011. Performance of Low-Dissipation Euler Fluxes and Preconditioned Implicit Schemes in Low Speeds. *Communications in Computational Physics* **10**:90–119.

van Leer, B. 1982. Flux vector splitting for the Euler equations, *Lecture Notes in Physics*, 170. Springer Verlag.

Numerical Methods for Hyperbolic Equations – Vázquez-Cendón et al. (eds)
 ISBN 978-0-415-62150-2

Numerical simulation of a pulverized coal jet

A. Bermúdez & J.L. Ferrín
Departamento Matemática Aplicada, Universidad de Santiago de Compostela, Spain

A. Liñán & L. Saavedra
E.T.S.I. Aeronáuticos, Universidad Politécnica de Madrid, Spain

ABSTRACT: A mathematical model for the group combustion of pulverized coal particles was developed by Bermúdez *et al* in (Bermúdez et al., 2007). It includes the Lagrangian description of the dehumidification, devolatilization and char gasification reactions of the coal particles in the homogenized gaseous environment, followed by the simultaneous group combustion of the three fuels (CO, H_2 and volatiles) produced, by means of the gas phase oxidation reactions, which are considered to be very fast. This model is complemented with an analysis of the particle dynamics, determined principally by the effects of aerodynamic drag and gravity, and its dispersion based on a stochastic model. Two other simpler models for the gasification of the particles were developed: the first one for particles small enough to extinguish the surrounding diffusion flames, and a second one for particles with small ash content where the porous shell of ashes remaining after gasification of the char, non structurally stable, is disrupted.

The goal of this work is to show how these three models can be applied in the numerical simulation of group combustion. Here they are used for the analysis of a simple example of a non-swirling pulverized coal jet with a nearly stagnant air at ambient temperature, with an initial region of interaction with a small annular methane flame. The experiment that we have simulated was done by Hwang *et al* in (Hwang et al., 2006(a); Hwang et al., 2006(b)), where a study of the structure of a turbulent pulverized coal jet flame has been carried out. Firstly, the results obtained with the three versions of the model are compared among them and subsequently we show how the first of the simpler models fits better the experimental results. Although it is shown good agreement between the experimental measurements and our simulations, it can be seen that the particle dispersion is overpredicted. This fact could be explained because that model does not consider changes in particle sizes, due to partial disruption or swelling effects, and also because turbulent dispersion of the particles is overestimated do to the use of the k–ε model and hence larger particles are more dispersed.

Concerning to the numerical methods, to solve the partial differential equations modeling the gas phase, second order modified Lagrange-Galerkin methods are used. These modified methods, analyzed in (Bermejo and Saavedra 2011), have the same accuracy as the standard ones but they are more efficient from a computational point of view.

1 MATHEMATICAL MODELLING

In this section we describe the mathematical models that we use in our simulation. We want to simulate a turbulent mixture of reacting gases where pulverized coal combustion occurs. The main processes involved in the combustion of a coal particle are the evaporation of the moisture, the devolatilization and the char gasification. The char is gasified by three heterogeneous reactions with three species (water, carbon dioxide and oxygen) to generate hydrogen and carbon monoxide that will burn with oxygen in the gas phase.

Summarizing, we consider the following eight reactions in our model; five heterogeneous reactions and three gas phase oxidation reactions:

1. $CO_2 + C_{(s)} \rightarrow 2CO + (q_1)$
2. $\frac{1}{2}O_2 + C_{(s)} \rightarrow CO + (q_2)$
3. $H_2O + C_{(s)} \rightarrow CO + H_2 + (q_3)$
4. $V_{(s)} \rightarrow V_{(g)} + (q_4)$
5. $H_2O_{(s)} \rightarrow H_2O_{(g)} + (q_5)$
6. $CO + \frac{1}{2}O_2 \rightarrow CO_2 + (q_6)$
7. $V_{(g)} + \nu_1 O_2 \rightarrow \nu_2 CO_2 + \nu_3 H_2O + \nu_4 SO_2 + (q_7)$
8. $H_2 + \frac{1}{2}O_2 \rightarrow H_2O + (q_8)$

where q_i is the heat released in the i-th reaction per unit mass gasified. All of them can be obtained from the standard formation enthalpies of the species (the volatiles ones calculated by using the heating value of the coal).

All the volatiles are considered like a single molecule

$$V_{(g)} = C_{\kappa_1} H_{\kappa_2} O_{\kappa_3} S_{\kappa_4}, \tag{1}$$

of molecular mass $M_{\text{vol},}$ where coefficients κ_1, κ_2, κ_3 and κ_4 are deduced from the ultimate analysis of the coal. The stoichiometric coefficients v_i in reaction 7 are calculated in terms of the volatiles composition using these expressions:

$$v_1 = (2\kappa_1 + \kappa_2/2 + 2\kappa_4 - \kappa_3)/2,$$
$$v_2 = \kappa_1, v_3 = \kappa_2/2, v_4 = \kappa_4.$$

To describe the model we must take into account that the two phases are coupled because the solid phase provides sources of mass and energy to the gas phase and the gas phase determines the atmosphere where the coal particles are gasifying. Furthermore, the solid phase is dispersed and what means that volumetric fraction is small compared to the gas phase.

1.1 *Gas phase model*

In order to obtain a simplified mathematical description of the gas phase, we can assume the following hypotheses:

1. All the diffusion coefficients $\mathcal{D}_i$ of chemical species are equal to $\mathcal{D}$.
2. Lewis number is equal to 1, so

   $$k_T = \rho_g \mathcal{D} c_\pi, \tag{2}$$

   where ρ_g is the density of the gas mixture and c_π its specific heat at constant pressure.
3. Fick's law for the diffusion velocities.
4. Low Mach number.

With these hypothesis we can obtain the final equations of the model. First, the state equation

$$\bar{\pi} = \rho_g \mathcal{R} \theta, \tag{3}$$

where θ is the temperature of the gas mixture. Then, the continuity and momentum conservation equations

$$\frac{\partial \rho_g}{\partial t} + \text{div}(\rho_g \mathbf{v}_g) = f^m, \tag{4}$$

$$\frac{\partial(\rho_g \mathbf{v}_g)}{\partial t} + \text{div}(\rho_g \mathbf{v}_g \otimes \mathbf{v}_g) + \text{grad}\pi - \text{div}(T_\tau) = \rho_g \mathbf{g}, \tag{5}$$

where T_τ is

$$T_\tau = (\mu + \mu_t)(\text{grad}\mathbf{v}_g + \text{grad}\mathbf{v}_g^t) - \frac{2}{3}(\mu + \mu_t)\text{div}\mathbf{v}_g I - \frac{2}{3}\rho_g k I, \tag{6}$$

v_g is the gas velocity, $\mathbf{g}$ the acceleration of gravity, μ the molecular viscosity of the gas, μ_t the turbulent viscosity and k the turbulent kinetic energy.

For the turbulence we use the $k - \varepsilon$ standard model:

$$\mu_t = 0.09 \rho_g \frac{k^2}{\in}, \tag{7}$$

$$\frac{\partial(\rho_g k)}{\partial t} + \text{div}(\rho_g \mathbf{v}_g k) - \text{div}\left[(\mu + \mu_t)\text{grad}k\right] = P_k - \rho_g \in, \tag{8}$$

$$\frac{\partial(\rho_g \in)}{\partial t} + \text{div}(\rho_g \mathbf{v}_g \in) - \text{div}\left[\left(\mu + \frac{\mu_t}{1.3}\right)\text{grade} \in\right] = 1.44 \frac{\in}{k} P_k - 1.92 \frac{\in^2}{k}, \tag{9}$$

where P_k is the production of the turbulent kinetic energy and ε is the turbulent dissipation.

The equations for the mass of each species and the energy conservation are

$$\mathcal{L}_g(Y_{O_2}^g) = f_{O_2}^m - \frac{4}{7} w_6 - \frac{32 v_1}{M_{\text{vol}}} w_7 - 8 w_8, \tag{10}$$

$$\mathcal{L}_g(Y_{CO_2}^g) = f_{CO_2}^m + \frac{11}{7} w_6 + \frac{44 v_2}{M_{\text{vol}}} w_7, \tag{11}$$

$$\mathcal{L}_g(Y_{H_2O}^g) = f_{H_2O}^m + \frac{18 v_3}{M_{\text{vol}}} w_7 + 9 w_8, \tag{12}$$

$$\mathcal{L}_g(Y_{SO_2}^g) = f_{SO_2}^m + \frac{64 v_4}{M_{\text{vol}}} w_7, \tag{13}$$

$$\mathcal{L}_g(Y_{CO}^g) = f_{CO}^m - w_6, \tag{14}$$

$$\mathcal{L}_g(Y_V^g) = f_V^m - w_7, \tag{15}$$

$$\mathcal{L}_g(Y_{H_2}^g) = f_{H_2}^m - w_8, \tag{16}$$

$$\mathcal{L}_g(h_T^g) = f^e + q_6 w_6 + q_7 w_7 + q_8 w_8 - \text{div}\mathbf{q}_{rg}, \tag{17}$$

where w_i the rate of the i-th reaction and $\mathcal{L}_g$ is the differential operator

$$L_g(u) = \frac{\partial(\rho_g u)}{\partial t} + \text{div}(\rho_g u \mathbf{v}_g) - \text{div}\left(\left(\rho_g D + \frac{\mu_t}{0.7}\right)\text{grad}u\right). \tag{18}$$

To obtain the composition and the thermal enthalpy of the mixture we have to solve the equations 10–17. In the right hand side of the equations are sources due to the combustion of the particles and terms due to the gas phase oxidation reactions. To eliminate the reaction terms, we are going to assume the Burke-Schumann hypothesis of infinitely fast oxidation reactions, therefore, these reactions occur in a diffusion flame that divides the domain into two regions: Φ_O a region with oxygen and no fuel and Φ_F a region with fuel and no oxygen. In order to obtain equations without the gas phase reaction terms we consider these linear combinations of Shvab-Zeldovich type:

$$X_1^g = Y_{O_2}^g - \frac{4}{7}Y_{CO}^g - \frac{32\nu_1}{M_{vol}}Y_V^g - 8Y_{H_2}^g \tag{19}$$

$$X_2^g = Y_{CO_2}^g + \frac{11}{7}Y_{CO}^g + \frac{44\nu_2}{M_{vol}}Y_V^g \tag{20}$$

$$X_3^g = Y_{H_2O}^g + \frac{18\nu_3}{M_{vol}}Y_V^g + 9Y_{H_2}^g \tag{21}$$

$$X_4^g = Y_{SO_2}^g + \frac{64\nu_4}{M_{vol}}Y_V^g \tag{22}$$

$$H^g = h_T^g + q_6 Y_{CO}^g + q_7 Y_V^g + q_8 Y_{H_2}^g \tag{23}$$

verifying these conservation equations

$$\mathcal{L}_g(X_1^g) = f_{O_2}^m - \frac{4}{7}f_{CO}^m - \frac{32\nu_1}{M_{vol}}f_V^m - 8f_{H_2}^m \tag{24}$$

$$\mathcal{L}_g(X_2^g) = f_{CO_2}^m + \frac{11}{7}f_{CO}^m + \frac{44\nu_2}{M_{vol}}f_V^m \tag{25}$$

$$\mathcal{L}_g(X_3^g) = f_{H_2O}^m + \frac{18\nu_3}{M_{vol}}f_V^m + 9f_{H_2}^m \tag{26}$$

$$\mathcal{L}_g(X_4^g) = f_{SO_2}^m + \frac{64\nu_4}{M_{vol}}f_V^m \tag{27}$$

$$\mathcal{L}_g(H^g) = f^e + q_6 f_{CO}^m + q_7 f_V^m + q_8 f_{H_2}^m - \mathrm{div}\mathbf{q}_{rg}. \tag{28}$$

Once we solve these equations we can recover the composition and the thermal enthalpy of the mixture by using the Burke-Schumann hypothesis.

1.2 *Solid phase model*

To compute the sources of mass and energy from the solid phase we have to solve the solid phase model. We use the BFL combustion model (BFL stands for Bermúdez, Ferrín and Liñán) firstly described in (Bermúdez et al., 2007). This model is based on a Lagrangian computation of the temperature and mass of each coal particle along its trajectory. The velocity of the particle is obtained from:

$$\frac{d\mathbf{v}_p}{dt} = F_A\left(\mathbf{v}_g - \mathbf{v}_p\right) + \mathbf{g}, \mathbf{v}_p(0) = \mathbf{v}_{p0} \tag{29}$$

where F_A is the drag force per unit mass,

$$F_A = \frac{3}{16}\frac{\mu}{\rho_p a^2}C_D \mathcal{R}e_p, \tag{30}$$

C_D is the drag coefficient for a spherical particle and Re_p is the Reynolds number relative to the particle. The position of the particle is given by

$$\frac{d\mathbf{x}_p}{dt} = \mathbf{v}_p, \mathbf{x}_p(0) = \mathbf{x}_{p0}. \tag{31}$$

To take into account the turbulence effect we have coupled the particle motion model with a *discrete random walk model*. The instantaneous fluid velocity is calculated as

$$\mathbf{v}_g = \overline{\mathbf{v}}_g + \left(\xi_1\sqrt{\frac{k}{\varepsilon}}, \xi_2\sqrt{\frac{k}{\varepsilon}}, \xi_3\sqrt{\frac{k}{\varepsilon}}\right) \tag{32}$$

where $\xi_i, i = 1,2,3$ are normally distributed random numbers. The random value of the velocity is kept constant over an interval of time given by the characteristic lifetime of the eddies. We compute the trajectory, the mass, the temperature and the sources to the gas phase for a sufficient number of representative particles ("number of tries") with different sizes dropped from each cell of each inlet.

Once the position of the particle is known, its mass is given by this equation:

$$m_p = m_{H_2O} + m_V + m_C + m_{ash}. \tag{33}$$

The evolution of m_{H_2O}, m_V and m_C, with the radial coordinate r and time t, are given by

$$\frac{dm_V}{dt} = -\upsilon_p w_4, \tag{34}$$

$$\frac{dm_{H_2O}}{dt} = -\upsilon_p w_5, \tag{35}$$

$$\frac{dm_C}{dt} = -\upsilon_p(w_1 + w_2 + w_3), \tag{36}$$

with υ_p the volume of the particle. Therefore, the evolution of the mass is given by the evolution of each of its components and depends on the rates of the heterogeneous reactions. The temperature of the particle is obtained from the equation:

$$m_p c_s \frac{dT_p}{dt} = 4\pi r_p^2 (q_p'' + q_r'') + \int_0^{r_p} \left(\sum_{i=1}^{8} q_i w_i \right) 4\pi r^2 dr, \tag{37}$$

where r_p is the radius of the particle, q_p'' is the heat flux by conduction and q_r'' the heat flux by radiation.

In order to obtain the heterogeneous reaction rates we distinguish the following cases:

1. Large particles with high ash content.
2. Small particles with high ash content.
3. Low ash content particles.

We have developed different models for each of these cases and their equations and how they can been obtained have appeared in (Bermúdez et al., 2011).

2 NUMERICAL SOLUTION

The convection-diffusion equations of the gas phase model can be written as

$$\rho_g \frac{Du}{Dt}(x,t) - \operatorname{div}(\mathcal{D}_t(x,t)\operatorname{grad} u(x,t)) + f_m(x,t)u(x,t) = f(x,t), \tag{38}$$

with $\mathcal{D}_t$ representing the diffusion coefficient or the viscosity depending on the equation.

In order to solve them we use Lagrange-Galerkin methods of second order in space and time. These methods are based on the discretization of the material derivative along the characteristic curves

$$\frac{Du}{Dt}(x,t_{n+1}) \approx \frac{3}{2\Delta t} u^{n+1}(x) - \frac{2}{\Delta t}(u^n \circ \chi^n)(x) + \frac{1}{2\Delta t}(u^{n-1} \circ \chi^{n-1})(x), \tag{39}$$

combined with a Galerkin projection in the framework of finite element methods. The characteristic curve χ is the solution of the initial value problem

$$\frac{d\chi(x,t^{n+1};s)}{ds} = \mathbf{v}_g(\chi(x,t^{n+1};s),s) \tag{40}$$

$$\chi(x,t^{n+1};t^{n+1}) = x \tag{41}$$

The point $\chi^n(x) := \chi(x,t^{n+1};t^n)$, was the position at the instant $t = t^n$ of the fluid particle that is in x at the instant t^{n+1} and that moves with velocity v_g.

These methods have three main advantages:

1. The discretization of the material derivative is a natural way to introduce upwinding from the physical point of view.
2. The resulting linear system of equations is symmetric.
3. The method is unconditionally stable if the Galerkin projection is performed exactly, which allows us to use large Δt in the calculations.

Unfortunately, Lagrange-Galerkin methods have some drawbacks:

1. The calculation of a linear system in every time step, which can have a huge dimension, in particular in 3D problems.
2. The use of high order quadrature rules in order to maintain the good stability properties of the method. These rules have a significant number of points.

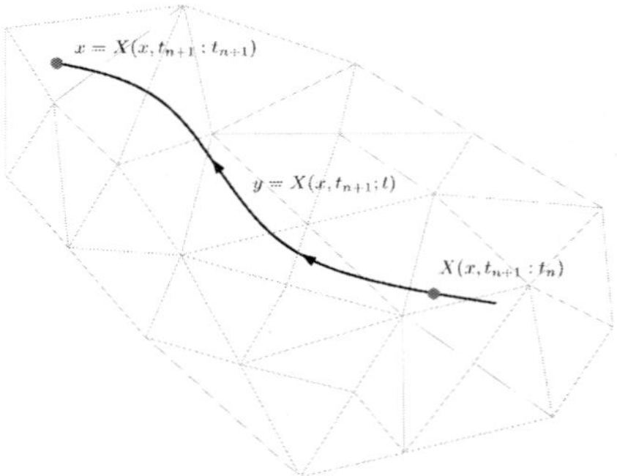

Figure 1. Scheme of a characteristic curve.

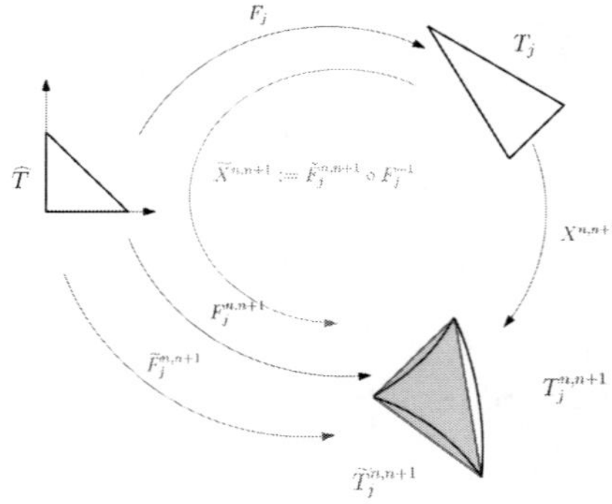

Figure 2. Scheme of Modified Lagrange-Galerkin methods.

3. The calculation in every iteration of the position of each quadrature point at the previous time step.

To overcome these shortcomings we use:

1. Parallelized linear solvers specially designed for Intel processors (MKL library).
2. Modified Lagrange-Galerkin methods. These methods allow us to compute all the quadrature points at the previous step moving only the mesh vertices, so they are more efficient (between 15% and 50%) than the convectional ones maintaining the rate of convergence. A scheme of these methods can be seen in Figure 2 and they are described in (Bermejo and Saavedra 2011).

3 SIMULATION OF A PULVERISED COAL JET FLAME

In (Hwang et al., 2006(a)) and (Hwang et al., 2006(b)) a detailed study of the structure of a coal flame has been done. A laboratory scale burner (Figure 3) was specially manufactured to obtain a pulverized coal jet burning with a flame in ambient air.

The air and the coal are supplied to the main burner and methane to the annular one. Methane is supplied to generate an annular pilot flame needed for the initial heating of the particles to provide the volatiles required for the flame stabilization. The methane flow rate is the minimum needed to form a stable flame. In Table 1 some experimental conditions are shown.

We can distinguish two stages in the experiment: first, the air is supplied to the main burner port and the methane to the annular slit burner and the gas flame due to the air and the methane is formed. Then, when the methane diffusion flame becomes stable the pulverized coal particles are injected.

To make our simulations, first we perform an axisymmetric-steady-state simulation of the first stage with the commercial code Fluent.

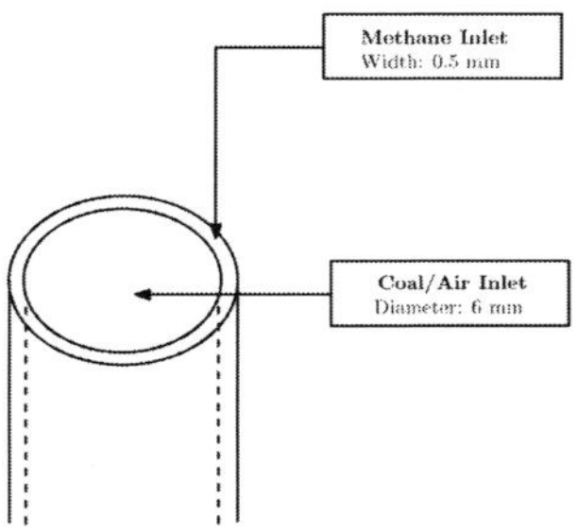

Figure 3. Scheme of the coaxial burner.

Table 1. Experimental conditions.

Air flow rate	$1.80 \times 10^{-4}\,m^3/s$
CH_4 flow rate	$2.33 \times 10^{-5}\,m^3/s$
Pulverized coal flow rate	$1.49 \times 10^{-4}\,kg^3/s$

Table 2. Properties of coal.

Proximate analysis (Dry basis) [wt%]	
Volatile matter	26.9
Fixed carbon	57.9
Ash	15.2
Ultimate analysis (Dry basis) [wt %]	
C	71.9
H	4.4
N	1.58
O	6.53
S	0.39
High heating value	$2.81 \times 10^7\,J/kg$
Density	$1000\,kg/m^3$
Specific heat	$1000\,J/(kgK)$

This simulation has been carried out to calculate profiles for the velocity, the temperature, the composition and the turbulence parameters of the gas mixture near the burner. With these profiles we will carry out a 3D simulation using the BFL combustion model and we compare our results with the experimental data.

The coal used in the experiment is Newlands bituminous coal. Proximate and ultimate analysis of this coal as well as the heating value, density and specific heat are given in Table 2. The mean diameter of the coal particles is 33.3 μm. Thus, taking into account the type of coal used in the experiment, the most appropriate model for the simulation is obtained for small particles with high ash content.

Now we can show some results obtained with our simulations. Figure 4 plots the contours of temperature in an axial plane and Figure 5 shows the 3D diffusion flame. We can notice that the higher temperatures are obtained in the diffusion flame. The contours of mass fraction of O_2 are plotted in Figure 6. As can be seen there is only O_2 in the Φ_O region. Finally, in Table 3 are the percentage of the released volatiles and gasified char. The amounts of volatiles released are very different and to obtain a better result we need to extend our solid phase model to take into account two or more different types of volatiles.

Finally, we are going to compare some of the results obtained with our simulations with the experimental data. In Figure 7 we plot the average

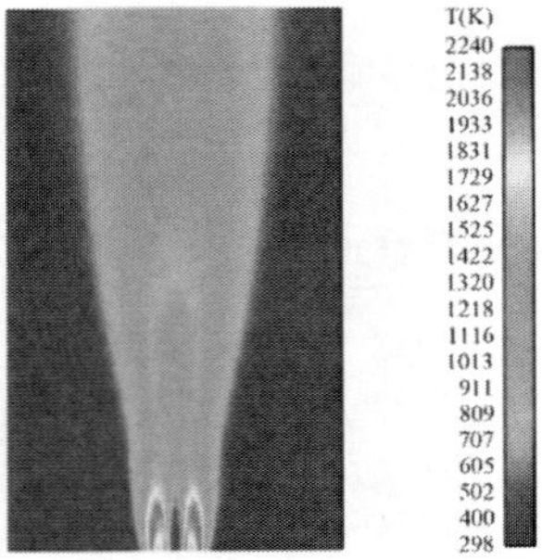

Figure 4. Contours of temperature (K).

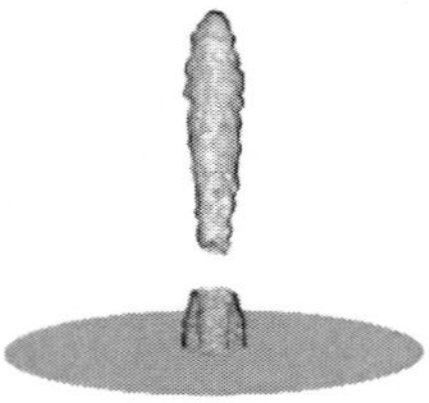

Figure 5. Position of the diffusion flame.

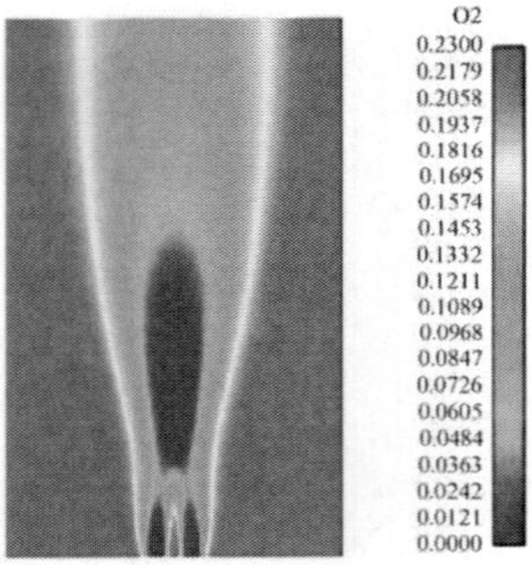

Figure 6. Mass fraction of O_2

Table 3. Volatiles and char gasified.

	Released volatiles (%)	Gasified char (%)
BFL	99.99	11.51
Experiment	44.6	11.3

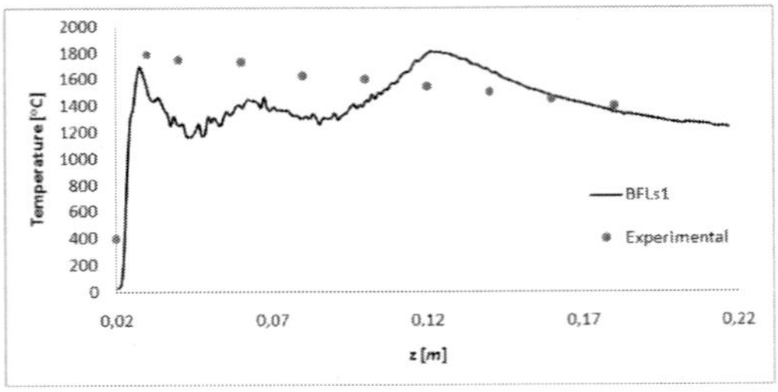

Figure 7. Distribution of particle temperature on the central axis.

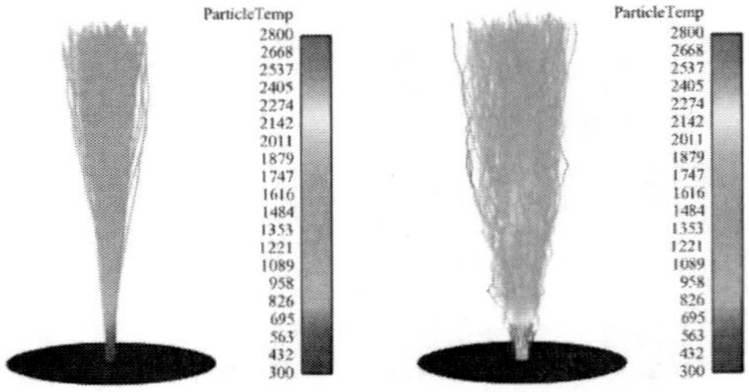

5 μm diameter particles 61 μm diameter particles

Figure 8. Trajectory of coal particles colored by T_p.

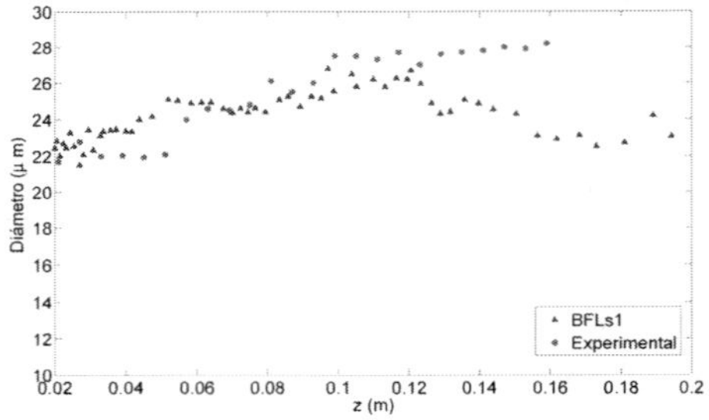

Figure 9. Distribution of mean diameter on the central axis.

of the temperature of the particles that pass through the central axis of the domain. The red dots are the experimental data and the solid line is the numerical result. It can be seen that both results are quite accurate. Figure 8 plots the trajectories of the smallest particles and the biggest ones colored by their temperature. We can notice how the smallest particles are more dispersed than the biggest ones, and that the latter are too dispersed at the top of the domain. In Figure 9 we can see the distribution of mean diameter on the central axis and it can be observed that until $z = 120\text{mm}$ the mean diameter is greater the farther you are from the burner. This is because smaller particles are more dispersed and therefore away from the axis. From $z = 120\text{mm}$ the mean diameter decreases, unlike the experimental data shows. This fact could be explained because the BFL model for coals with high ash content does not consider changes in particle sizes and also because turbulent dispersion of the particles is overestimated.

4 CONCLUSIONS

In this paper we have presented a complete code to simulate a turbulent mixture of reacting gases where pulverized coal combustion occurs. We have included in this code three different models for the gasification of pulverized coal particles depending on its ash content and its size. We have developed new numerical methods more efficient than the standard Lagrange-Galerkin ones maintaining the good properties of the latter ones.

To validate our code we made a simulation of an experimental turbulent jet flame and we have shown that our results are very accurate. In order to obtain better results we must improve the treatment of the turbulence because, as seen in the results presented, the dispersion of the particles is slightly overestimated due to the use of the standard $k-\varepsilon$ model.

REFERENCES

Bermúdez, A., Ferrín, J.L. & Liñán, A. 2007. The modelling of the generation of volatiles, H_2 and CO, and their simultaneous diffusion controlled oxidation in pulverised coal furnaces. *Combustion Theory and Modelling*, 11(6):949–976.

Bermúdez, A., Ferrín, J.L., Liñán, A. & Saavedra, S. 2011. Numerical simulation of group combustion of pulverized coal. *Combustion and Flame*, 158(9):1852–1865.

Hwang, S., Kurose, R., Akamatsu, F., Tsuji H., Makino, H. & Katsuki, M. 2006. Observation of detailed structure of turbulent pulverised-coal flame by optical measurement, (part 1, time-averaged measurement of behavior of pulverised-coal particles and flame structure). *JSME International Journal*, 49(4):1316–1327.

Hwang, S., Kurose, R., Akamatsu, F., Tsuji H., Makino, H. & Katsuki, M. 2006. Observation of detailed structure of turbulent pulverised-coal flame by optical measurement, (part 2, instantaneous two-dimensional measurement of combustion reaction zone and pulverised coal particles). *JSME International Journal*, 49(4):1328–1335.

Bermejo, R. & Saavedra, L. 2011. Modified Lagrange-Galerkin methods of first and second order in time for convection-diffusion problems. *Numerische Mathematik*, doi: 10.1007/s00211-011-0418-8.

Numerical Methods for Hyperbolic Equations – Vázquez-Cendón et al. (eds)
 ISBN 978-0-415-62150-2

A framework for flux-vector splitting schemes

M.E. Vázquez-Cendón
Department of Applied Mathematics, University of Santiago de Compostela, Spain

E.F. Toro
Laboratory of Applied Mathematics, University of Trento, Italy

ABSTRACT: We propose a framework for constructing flux vector splitting schemes and apply the approach to the Zha-Bilgen and the Toro-Vázquez splittings. A key feature of the present approach is the recognition of contact discontinuities. First, the framework identifies a suitable splitting of the flux vector and then considers two resulting systems, the advection and the pressure systems, respectively. For the numerical treatment of the full system we advocate the use of the Riemann problem for the pressure system. This, in addition to the numerical flux for the pressure system itself, provides an advection speed to deal with the advection system. Numerical results for the Euler equations are shown and comparison with classical flux vector splitting schemes is made.

1 INTRODUCTION

A well known limitation of the classical flux vector splitting schemes of Steger-Warming (Steger and Warming 1981) and van Leer (van Leer 1982) is their inability to recognize intermediate characteristic fields, such as contact discontinuities and shear waves, resulting in excessive numerical dissipation. Such failings are not only limited to inviscid problems but is also present in viscous computations, such as for example, when computing shear layers. The more recent flux vector splitting scheme of Liou and Steffen (Liou and Steffen 1993) is however able to recognize contact waves, visibly enhancing accuracy. The splittings of Zha-Bilgen (Zha and Bilgen 1993) and that very recently proposed by Toro and Vázquez (Toro and Vázquez-Cendón 2011) are also, potentially, capable of resolving intermediate characteristic fields, provided an appropriate numerical approach is then used.

In this paper we put forward a framework that contains two aspects, one concerned with flux splitting at the level of the partial differential equations and the other with numerical methods to discretize the resulting systems. We then apply the approach to find numerical fluxes for the Zha-Bilgen and the Toro-Vázquez splittings, leading to useful first-order (monotone) schemes. High-order versions of these are also possible but are not considered here. Numerical experiments for the one-dimensional ideal Euler equations are carried out for which we compare the proposed approach with the classical schemes of Steger-Warming, van Leer and Liou-Steffen.

The rest of this paper is structured as follows. In section 2 we present a framework for constructing flux vector splitting schemes. In section 3 we apply the approach to the Zha-Bilgen and the Toro-Vázquez splittings. In section 3 we show numerical results and conclusions are drawn in section 4.

2 A FLUX-SPLITTING FRAMEWORK

In this section we propose a framework for constructing flux splitting methods that allow the direct use of Godunov-type methods.

2.1 *The framework*

Consider a system of hyperbolic conservation laws, such as the Euler equations, written in differential conservation-law form as

$$\partial_t \mathbf{Q} + \partial_x \mathbf{F}(\mathbf{Q}) = \mathbf{0}, \tag{1}$$

where $\mathbf{Q}$ is the vector of conserved variables and $\mathbf{F}(\mathbf{Q})$ is the flux vector. For solving numerically equations of the type (1) we adopt a conservative method of the type

$$\mathbf{Q}_i^{n+1} = \mathbf{Q}_i^n - \frac{\Delta t}{\Delta x}[\mathbf{F}_{i+\frac{1}{2}} - \mathbf{F}_{i-\frac{1}{2}}], \tag{2}$$

where $\mathbf{F}_{i+1/2}$ is the numerical flux. For background on the Euler equations and conservative schemes of the form (2) see (Toro 2009), for example. We propose to split system (1) via the flux splitting

$$\mathbf{F}(\mathbf{Q}) = \mathbf{A}(\mathbf{Q}) + \mathbf{P}(\mathbf{Q}), \tag{3}$$

so as to end up with the following two hyperbolic systems

$$\partial_t \mathbf{Q} + \partial_x \mathbf{A}(\mathbf{Q}) = 0, \partial_t \mathbf{Q} + \partial_x \mathbf{P}(\mathbf{Q}) = \mathbf{0}, \tag{4}$$

called respectively the *advection system* and the *pressure system*. The aim is then to compute a numerical flux as

$$\mathbf{F}_{i+\frac{1}{2}} = \mathbf{A}_{i+\frac{1}{2}} + \mathbf{P}_{i+\frac{1}{2}}, \tag{5}$$

where $\mathbf{A}_{i+1/2}$ and $\mathbf{P}_{i+1/2}$ are obtained respectively from appropriate Cauchy problems for the advection and pressure systems (4).

The proposed framework can be motivated through the Cauchy problem for the linear advection equation, namely

$$\left.\begin{array}{l} \partial_t q(x,t) + \lambda \partial_x q(x,t) = 0, \, x \in \mathbf{R}, t > 0, \\ q(x,0) = h(x), \end{array}\right\} \tag{6}$$

where λ is a constant wave propagation speed. The exact solution of IVP (6) after a time Δt is

$$q(x, \Delta t) = h(x - \lambda \Delta t). \tag{7}$$

We now decompose the characteristic speed λ as

$$\lambda = \beta\lambda + (1-\beta)\lambda = \lambda_a + \lambda_p, \quad 0 \le \beta \le 1, \tag{8}$$

with definitions

$$\lambda_a = \beta\lambda, \quad \lambda_p = (1-\beta)\lambda, \tag{9}$$

so as to obtain two linear partial differential equations, namely

$$\partial_t q + \lambda_a \partial_x q = 0, \quad \partial_t q + \lambda_p \partial_x q = 0. \tag{10}$$

These are respectively the analogues of the advection and pressure systems in (4). Now consider first the Cauchy problem for the *advection* equation

$$\left.\begin{array}{l} \partial_t \tilde{q} + \lambda_a \partial_x \tilde{q} = 0, \\ \tilde{q}(x,0) = h(x), \end{array}\right\} \tag{11}$$

the solution of which after a time Δt_1 is

$$\tilde{q}(x, \Delta t_1) = h(x - \lambda_a \Delta t_1). \tag{12}$$

Consider now the Cauchy problem for the analogue of the *pressure* system in (4), whose initial condition is the solution (12) of IVP (11), namely

$$\left.\begin{array}{l} \partial_t \bar{q} + \lambda_p \partial_x \bar{q} = 0, \\ \bar{q}(x,0) = h(x - \lambda_a \Delta t_1). \end{array}\right\} \tag{13}$$

The exact solution of IVP (13), after a time Δt_2, is

$$\bar{q}(x, \Delta t_2) = h(x - \lambda_a \Delta t_1 - \lambda_p \Delta t_2). \tag{14}$$

Thus the combined solution of IVPs (11) and (13), solved in succession as shown, for a time $\Delta t_1 = \Delta t_2 = \Delta t$ is

$$\begin{aligned} \bar{q}(x, \Delta t) &= h(x - (\lambda_a + \lambda_p)\Delta t), \\ &= h(x - \lambda \Delta t) = q(x, \Delta t). \end{aligned} \tag{15}$$

The above result can be stated as the following proposition.

Proposition 2.1. The exact solution of the initial value problem (6) can be obtained by solving in sequence the initial-value problems (11) and (13).

We note that in the wave decomposition (8), (9) of the model problem (6) one can choose the characteristic speeds to be arbitrarily different. For example, for $\lambda > 0$, by taking a very small β in (8) we would have $\lambda_a << \lambda_p$, situation that resembles the slow advection waves and the fast pressure waves. From the numerical point of view, Proposition 2.1 suggests a way to compute a numerical flux for IVP (6) by computing numerical fluxes for IVPs (11) and (13).

3 SPLITTINGS FOR THE EULER EQUATIONS

Here we study two splitting schemes that fall within the present framework. We first recall the Euler equations in one space dimension. These, when written as in (1) have

$$\mathbf{Q} = \begin{bmatrix} \rho \\ \rho u \\ E \end{bmatrix}, \quad \mathbf{F}(\mathbf{Q}) = \begin{bmatrix} \rho u \\ \rho u^2 + p \\ u(E+p) \end{bmatrix}. \tag{16}$$

Here ρ is density, u is particle velocity, ρ is pressure and E is total energy given as

$$E = \rho(\frac{1}{2}u^2 + e), \tag{17}$$

where the specific internal energy e is given by an equation of state

$$e = e(\rho, p). \tag{18}$$

For ideal gases $e(\rho, p) = p/\rho(\gamma - 1)$. Here $\gamma = 1.4$.

Now we first recall the flux splitting proposed by Zha and Bilgen (Zha and Bilgen 1993), expressed as

$$\mathbf{A}(\mathbf{Q}) = u\mathbf{Q}, \quad \mathbf{P}(\mathbf{Q}) = p\begin{bmatrix} 0 \\ 1 \\ u \end{bmatrix}. \tag{19}$$

The pressure system is hyperbolic with real eigenvalues

$$\lambda_1 = -C, \quad \lambda_2 = 0, \quad \lambda_3 = C, \tag{20}$$

where

$$C = \sqrt{(\gamma-1)p/\rho}. \tag{21}$$

Note that the system is always subsonic.

To compute numerical fluxes we apply the present numerical approach and solve the linearized Riemann problem for the pressure system, in terms of physical variables, to obtain intercell values for velocity and pressure as follows

$$\left.\begin{aligned} u^*_{i+\frac{1}{2}} &= \frac{\rho_L C_L u_L + \rho_R C_R u_R}{\rho_L C_L + \rho_R C_R} - \frac{(p_R - p_L)}{\rho_L A_L + \rho_R A_R}, \\ p^*_{i+\frac{1}{2}} &= \frac{\rho_R C_R p_L + \rho_L C_L p_R}{\rho_L C_L + \rho_R C_R} - \frac{\rho_L C_L \rho_R C_R (u_R - u_L)}{\rho_L C_L + \rho_R C_R}, \end{aligned}\right\} \tag{22}$$

with

$$C_L = \sqrt{(\gamma-1)p_L/\rho_L} \quad C_R = \sqrt{(\gamma-1)p_R/\rho_R}. \tag{23}$$

Then the advection and pressure fluxes are computed as

$$\mathbf{A}_{i+\frac{1}{2}} = u^*_{i+\frac{1}{2}}\begin{cases} \mathbf{Q}^n_i & \text{if } u^*_{i+\frac{1}{2}} > 0, \\ \mathbf{Q}^n_{i+1} & \text{if } u^*_{i+\frac{1}{2}} \le 0 \end{cases} \tag{24}$$

and

$$\mathbf{P}_{i+\frac{1}{2}} = p^*_{i+\frac{1}{2}}\begin{bmatrix} 0 \\ 1 \\ u^*_{i+\frac{1}{2}} \end{bmatrix}. \tag{25}$$

We now apply the same approach to the recently proposed Toro-Vázquez splitting, (Toro and Vázquez-Cendón 2011), expressed as

$$\mathbf{A}(\mathbf{Q}) = u\mathbf{K}, \quad \mathbf{P}(\mathbf{Q}) = p\begin{bmatrix} 0 \\ 1 \\ u(\rho e/p + 1) \end{bmatrix}. \tag{26}$$

Note that the advection flux $\mathbf{A}(\mathbf{Q})$ contains no pressure terms and may be interpreted as advection of $\mathbf{K} = [\rho, \rho u, \frac{1}{2}\rho u^2]^T$, that is mass, momentum and kinetic energy. Note that for ideal gases $\rho e/p + 1 = \gamma/(\gamma-1)$. The pressure system is hyperbolic with real eigenvalues

$$\lambda_1 = \frac{1}{2}(u - A), \quad \lambda_2 = 0, \quad \lambda_3 = \frac{1}{2}(u + A), \tag{27}$$

where

$$A = \sqrt{u^2 + 4a^2}, \quad a^2 = \frac{\gamma p}{\rho}. \tag{28}$$

Note that again this pressure system is always subsonic. Numerical fluxes for the advection and pressure operators are

$$\mathbf{A}_{i+\frac{1}{2}} = u^*_{i+\frac{1}{2}}\begin{cases} \mathbf{K}^n_i & \text{if } u^*_{i+\frac{1}{2}} > 0, \\ \mathbf{K}^n_{i+1} & \text{if } u^*_{i+\frac{1}{2}} \le 0 \end{cases} \tag{29}$$

and

$$\mathbf{P}_{i+\frac{1}{2}} = p^*_{i+\frac{1}{2}}\begin{bmatrix} 0 \\ 1 \\ \frac{\gamma}{\gamma-1}u^*_{i+\frac{1}{2}} \end{bmatrix}. \tag{30}$$

The intercell values $u^*_{i+1/2}$, $p^*_{i+1/2}$ are found from an approximate solution of the linearized Riemann problem for the pressure system, in terms of physical variables. The result is

$$\left.\begin{aligned} u^*_{i+\frac{1}{2}} &= \frac{C_R u_R - C_L u_L}{C_R - C_L} - \frac{2}{C_R - C_L}(p_R - p_L), \\ p^*_{i+\frac{1}{2}} &= \frac{C_R p_L - C_L p_R}{C_R - C_L} + \frac{1}{2}\frac{C_R C_L}{C_R - C_L}(u_R - u_L), \end{aligned}\right\} \tag{31}$$

with

$$C_L = \rho_L(u_L - A_L); \quad C_R = \rho_R(u_R + A_R), \tag{32}$$

where A_L and A_R are computed from (28) using left and right data.

4 NUMERICAL RESULTS

Here we illustrate the performance of the methods on three representative test problems with reference solutions, for the ideal Euler equations with $\gamma = 1.4$.

Test 1 consists of a stationary isolated contact discontinuity in the spatial interval $0 \le x \le 1$ with initial conditions $\rho(x,0) = 1.4$ if $x < 1/2$ and $\rho(x,0) = 1$ if $x \ge 1/2$; $u(x,0) = 0$ and $p(x,0) = 1$ $\forall x \in [0,1]$. For this test the domain is discretized with 100 cells and results are shown at time $t_{out} = 2$.

Test 2 is a Riemann problem in the spatial interval [0,1], with initial conditions $\rho(x,0) = 1$, $u(x,0) = 3/4$, $p(x,0) = 1$ if $x < 0.3$ and $\rho(x,0) = 0.125$, $u(x,0) = 0$, $p(x,0) = 0.1$ if $x \ge 0.3$. For this test the domain is discretized with 100 cells and results are shown at time $t_{out} = 0.3$.

Test 3 is the blast wave problem of Woodward and Colella (Woodward and Colella 1984) in the spatial interval $0 \le x \le 1$ with initial conditions $\rho(x,0) = 1$ and $u(x,0) = 0$ $\forall x \in [0,1]$; $p(x,0) = 1000$ if $x < 1/10$; $p(x,0) = 0.01$ if $1/10 \le x \le 9/10$, $p(x,0) = 100$ if $x > 9/10$. This problems does not have an exact solution and thus we use a numerically computed reference solution using the most accurate first-order (monotone) method, namely the Godunov method with the exact Riemann solver. For this test the domain is discretized with 3000 cells and results are shown at time $t_{out} = 0.038$.

In all three test problems we show all the cells used for the approximation and the CFL number is set to 0.9, unless otherwise explicitly stated.

Results for Test 1 at time $t_{out} = 2$ are shown in Figures 1 to 6. The schemes of Stewart-Warming and van Leer schemes badly smear the contact wave and such smearing continues to increase at time evolves. As expected the Liou-Steffen scheme

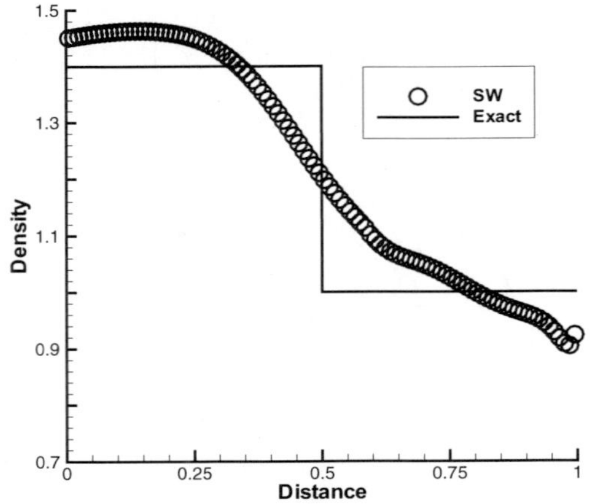

Figure 1. Test 1. Stationary isolated contact. Steger-Warming (SW) and exact solutions.

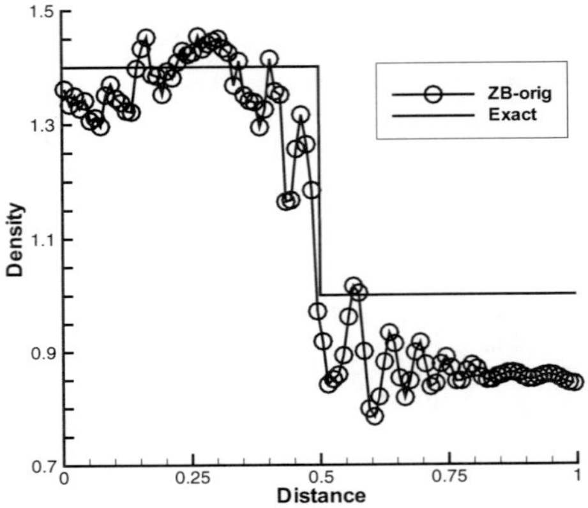

Figure 2. Test 1: Stationary isolated contact. Zha-Bilgen (ZB-orig) and exact solutions.

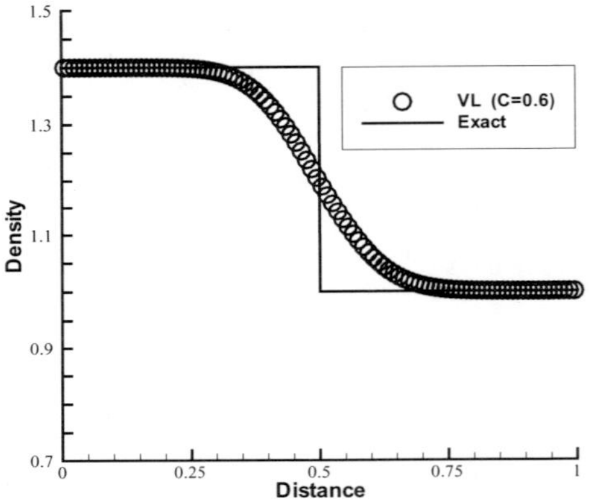

Figure 3. Test 1: Stationary isolated contact. van Leer (VL) and exact solutions.

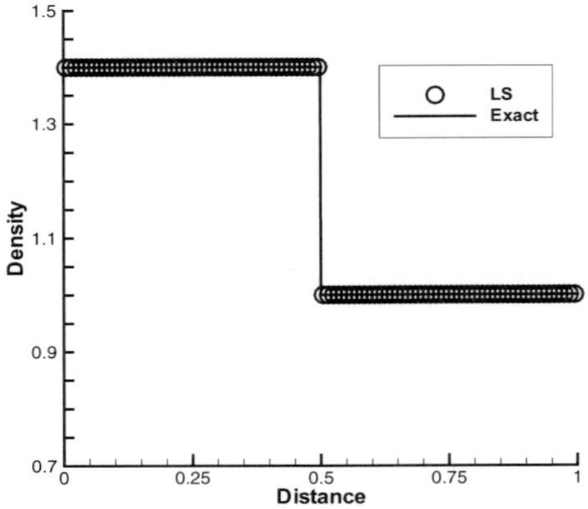

Figure 4. Test 1: Stationary isolated contact. Liou-Steffen (LS) and exact solutions.

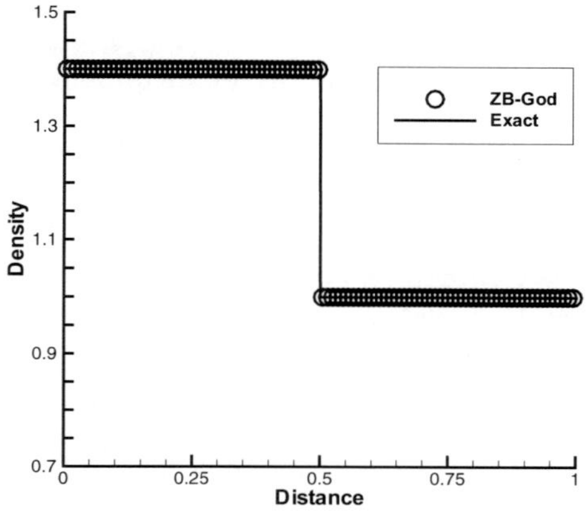

Figure 5. Test 1: Stationary isolated contact. Zha-Bilgen with present approach (ZB-God) and exact solutions.

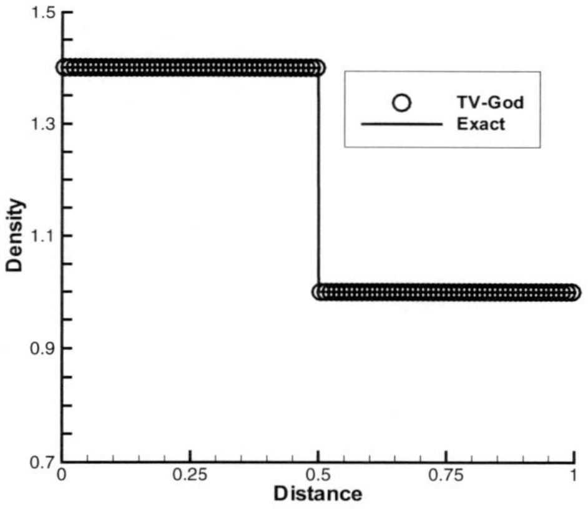

Figure 6. Test 1: Stationary isolated contact. Toro-Vázquez (TV-God) and exact solutions.

recognizes exactly the stationary isolated contact discontinuity. The Zha-Bilgen original splitting and the numerical method they propose can not preserve the contact. This is not due to the splitting but to their numerical method and one can show algebraically that this method cannot preserve the correct solution (Toro and Vázquez-Cendón 2011). However the Zha-Bilgen splitting can be rescued by treating it in the manner proposed in this paper; the modified method reproduces the correct solution. The Toro-Vázquez flux vector splitting along with their proposed numerical approach (Toro and Vázquez-Cendón 2011) also reproduces the correct solution.

Results for Test 2 at time $t_{out} = 0.3$ are shown in Figures 6 to 12. In addition to the right-facing shock and the moving contact wave there is a left-

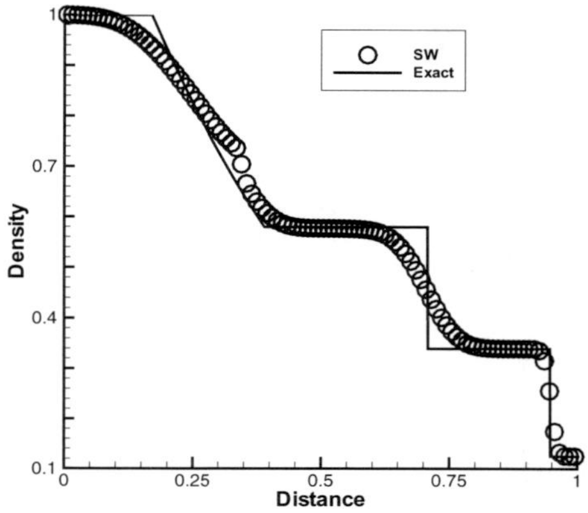

Figure 7. Test 2: Sonic flow. Steger-Warming (SW) and exact solutions.

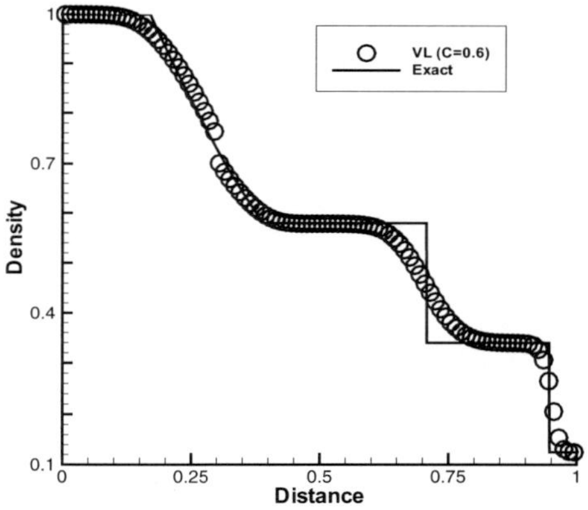

Figure 8. Test 2: Sonic flow. van Leer (VL) and exact solutions.

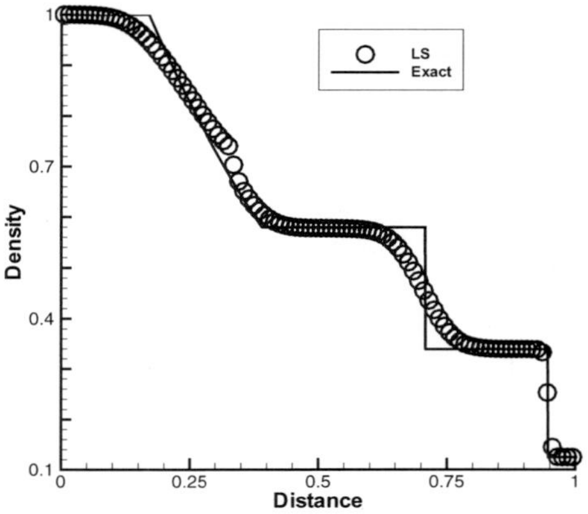

Figure 9. Test 2: Sonic flow. Liou-Steffen (LS) and exact solutions.

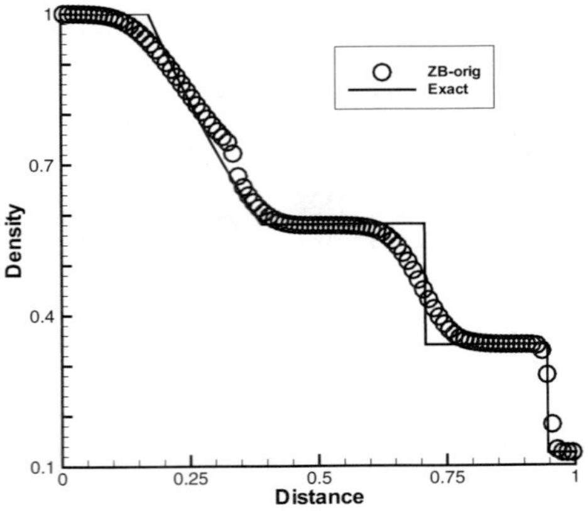

Figure 10. Test 2: Sonic flow. Zha-Bilgen (ZB-orig) and exact solutions.

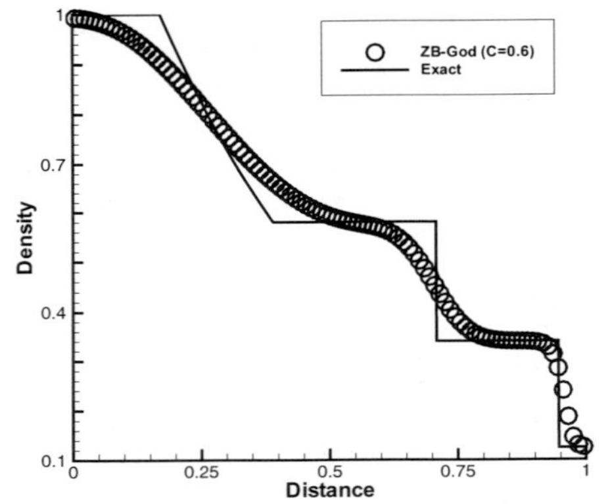

Figure 11. Test 2: Sonic flow. Zha-Bilgen with present approach (ZB-God) and exact solutions.

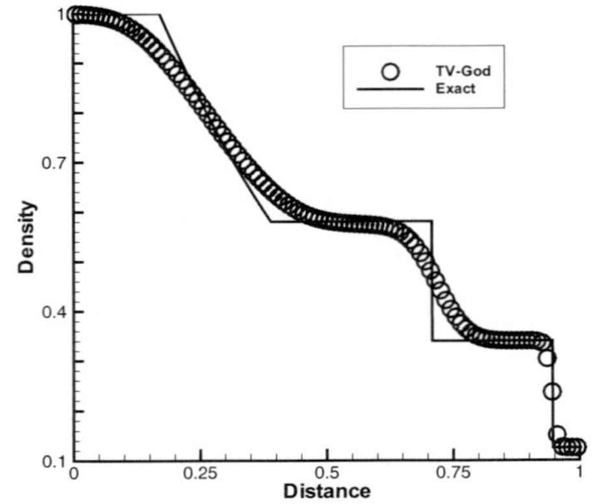

Figure 12. Test 2: Sonic flow. Toro-Vázquez (TV-God) and exact solutions.

facing transonic rarefaction wave. The challenge for numerical methods is to cope with the sonic point. All existing flux vector splitting methods considered here (Stewart-Warming, van Leer, Liou-Steffen, Zha-Bilgen) have considerable difficulty in coping with the sonic point, see Figures 7 to 10. The Zha-Bilgen and Toro-Vázquez splittings with the present numerical approach give a satisfactory resolution of the sonic point. We note, however, that the latter schemes have more numerical diffusion.

Results for Test 3 at time $t_{out} = 0.038$ are shown in Figures 13 to 16. We first note that the choice of the CFL number is crucial for this test. The Liou-Steffen scheme runs only for $CFL = 0.05$ and smaller, but still produces overshoots and has excessive numerical diffusion seen more clearly at peak values. Results for the two versions of the Zha-Bilgen scheme are shown for $CFL = 0.6$. The original Zha-Bilgen scheme does run for larger CFL numbers but results are highly oscillatory in the left part of the domain (not shown) and in the vicinity of the left shock. The modified Zha-Bilgen scheme (ZB-God) following the present approach crushes for $CFL \geq 0.7$ but does run for $CFL \leq 0.6$ and the results are visibly better than those of the original Zha-Bilgen scheme for the same CFL number; it is less diffusive and there are no spurious oscillations. The final result shown is that of the Toro-Vázquez flux vector splitting (TV-God) along with their proposed numerical approach (Toro and Vázquez-Cendón 2011); the scheme is run with $CFL = 0.9$ and shows excellent agreement with the Godunov scheme using the exact Riemann solver. Overall it seems as if the Toro-Vázquez flux vector splitting scheme along with their proposed numerical approach produces the

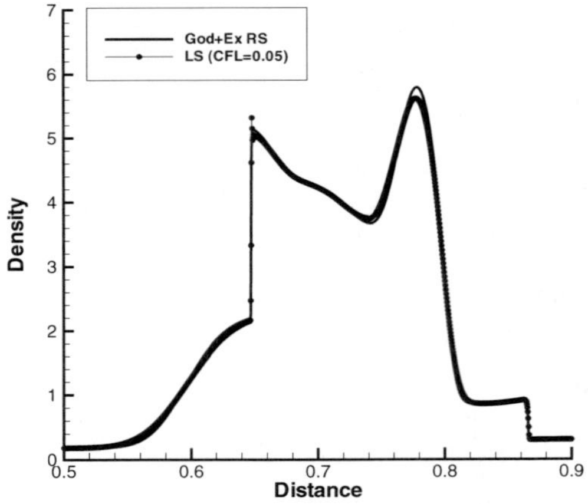

Figure 13. Test 3: Blast wave. Liou-Steffen (LS) and reference solutions.

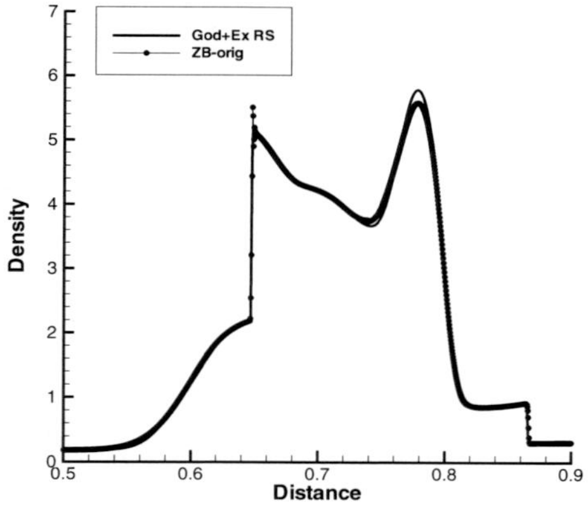

Figure 14. Test 3: Blast wave. Zha-Bilgen (ZB) and reference solutions.

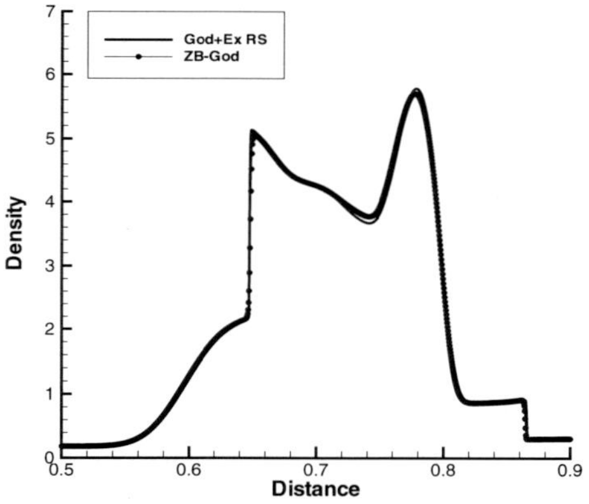

Figure 15. Test 3: Blast wave. Zha-Bilgen with present approach (ZB-God) and reference solutions.

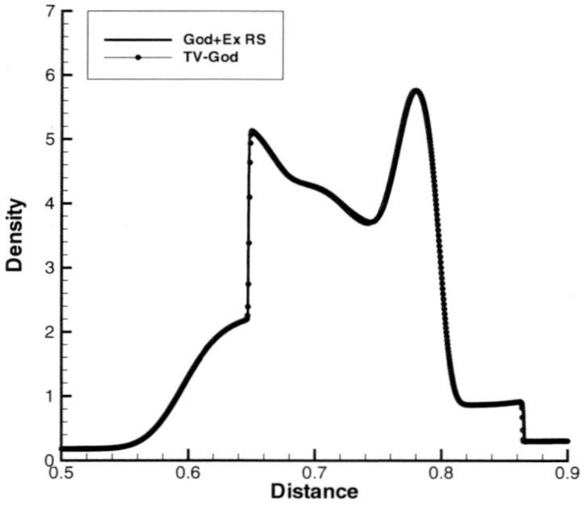

Figure 16. Test 3: Blast wave. Toro-Vázquez (TV-God) and reference solutions.

best results. We speculate that the accuracy and robustness of the Toro-Vázquez scheme comes from the fact that all pressure terms are included in the pressure system.

5 CONCLUSIONS

We have proposed a framework for constructing flux vector splitting schemes and have applied the approach to the Zha-Bilgen and the Toro-Vázquez flux vector splittings for the one-dimensional ideal Euler equations. Numerical results have been shown, which illustrate the benefits of both the splitting approach as well as the numerical methods employed to discretize the advection and pressure systems. Overall, the Toro-Vázquez scheme gives the best results. The present study is limited to the one-dimensional case and the schemes presented are first order accurate. The approach presented can be extended to multidimensional problems, for which we expect considerable gains in accuracy for resolving shear waves and vortical flows. The extension to high order of accuracy can be accomplished through the deployment of any of the existing high-order numerical approaches, such as semi-discrete WENO schemes with Runge-Kutta time stepping or the fully discrete ADER approach. The proposed schemes could also be used in discontinuous Galerkin finite element methods, in semi-discrete of fully discrete form.

REFERENCES

Liou, M.S. and C.J. Steffen (1993). A New Flux Splitting Scheme. *J. Comput. Phys. 107*, 23–39.

Steger, J.L. and R.F. Warming (1981). Flux Vector Splitting of the Inviscid Gasdynamic Equations with Applications to Finite Difference Methods. *J. Comput. Phys. 40*, 263–293.

Toro, E.F. (2009). *Riemann Solvers and Numerical Methods for Fluid Dynamics, Third Edition.* Springer–Verlag.

Toro, E.F. and M.E. Vázquez-Cendón (2011). Flux Splitting Schemes for the Euler Equations. *Computers and Fluids (Submitted)*.

van Leer, B. (1982). Flux–Vector Splitting for the Euler Equations. Technical Report ICASE 82-30, NASA Langley Research Center, USA.

Woodward, P. and P. Colella (1984). The Numerical Simulation of Two–Dimensional Fluid Flow with Strong Shocks. *J. Comput. Phys. 54*, 115–173.

Zha, G.-C. and E. Bilgen (1993). Numerical Solution of Euler Equations by a New Flux Vector Splitting Scheme. *Int. J. Numer. Meth. Fluids 54*, 115–144.

X Poster

Numerical Methods for Hyperbolic Equations – Vázquez-Cendón et al. (eds)

ISBN 978-0-415-62150-2

Extension of fourth-order non-oscillatory central schemes to sediment transport equations

A. Balaguer-Beser, M.T. Capilla & Ll. Gascón
Departamento de Matemática Aplicada, Universidad Politécnica de Valencia, Valencia, Spain

ABSTRACT: We are concerned with the construction of high-order well-balanced non-oscillatory central finite volume schemes for solving the sediment transport equations. Time integration is computed following a Runge-Kutta scheme with a natural continuous extension. Spatial accuracy is obtained in each cell with the three-degree reconstruction polynomials of Balaguer (2011), keeping the local monotonicity of the interpolation data. We have used a treatment for bed slope source term which maintains the established order of accuracy and satisfies the exact conservation property (C-property). In order to do this we present an extension of the procedure described in Gascón and Corberán (2001) to be applied in central schemes. We propose to write the source term in divergence form so that it can be incorporated into the flux vector of the Runge-Kutta scheme. Several standard one-dimensional test cases are used to verify the behaviour of our scheme and its non-oscillatory properties.

1 INTRODUCTION

Shallow water equations are widely used to model flows in rivers and coastal areas. For practical applications the inclusion of a non-flat bottom topography is required. Central schemes allow a simpler reconstruction of the numerical fluxes because the discontinuities in the pointwise solution, produced by the reconstruction algorithm, are located at the center of the staggered control volumes. In this paper we present an extension of the central non-oscillatory Runge-Kutta scheme given in Balaguer (2011) to solve the shallow water equations over a movable non-flat bed described in Caleffi et al. (2007).

We construct a high-order well balanced non-oscillatory numerical scheme for solving the one-dimensional sediment transport problem given by:

$$\frac{\partial}{\partial t}\begin{bmatrix}\eta\\ q\\ Z_b\end{bmatrix}+\frac{\partial}{\partial x}\begin{bmatrix}q+\xi q_b\\ \dfrac{q^2}{\eta-Z_b}+\dfrac{1}{2}g(\eta-Z_b)^2\\ \xi q_b\end{bmatrix} = \begin{bmatrix}0\\ -g(\eta-Z_b)(Z_b)_x\\ 0\end{bmatrix} \tag{1}$$

The above system has been manipulated in order to use the water surface elevation $\eta(x,t)=h(x,t)+Z_b(x,t)$, instead of the water depth $h(x,t)$ (Caleffi et al. (2007)). $q(x,t)=h(x,t)\cdot v(x,t)$ denotes the water discharge; $v(x,t)$ is the water velocity; $Z_b(x,t)$ is the bottom elevation and $g=9.8\ \mathrm{m}/s^2$. Parameter ξ and the bedload sediment transport rate q_b are given by:

$$\xi=\frac{1}{1-\epsilon},\quad q_b(x,t)=Av^3(x,t) \tag{2}$$

where ϵ is the porosity of the riverbed and A is a constant. This vectorial equation may be written as:

$$u_t+f_x(u)=s(u) \tag{3}$$

2 NUMERICAL SCHEME

2.1 *Central finite volume integration*

We integrate the above system over $[x_j, x_{j+1}]\times[t^n, t^{n+1}]$:

$$\bar{u}^{n+1}_{j+\frac{1}{2}}=\bar{u}^{n}_{j+\frac{1}{2}}-\frac{1}{\Delta x}\left[\int_{t^n}^{t^{n+1}} f(u_{j+1}(\tau))d\tau-\int_{t^n}^{t^{n+1}} f(u_j(\tau))d\tau\right]+\int_{t^n}^{t^{n+1}}\bar{s}_{j+\frac{1}{2}}(\tau)d\tau, \tag{4}$$

where $\bar{u}^n_{j+\frac{1}{2}}$ and $\bar{s}_{j+\frac{1}{2}}(t)$ denote the cell averages:

$$\bar{u}^n_{j+\frac{1}{2}}=\frac{1}{\Delta x}\int_{x_j}^{x_{j+1}} u(x,t^n)dx,\qquad \bar{s}_{j+\frac{1}{2}}(t)=\frac{1}{\Delta x}\int_{x_j}^{x_{j+1}} s(u(x,t))dx \tag{5}$$

2.2 *Time integration: Runge-Kutta scheme*

Time flux integrals are evaluated by means of the two-point Gauss quadrature rule:

$$\frac{1}{\Delta x}\int_{t^n}^{t^{n+1}} f(u_j(\tau))d\tau \equiv \frac{\Delta t}{2\Delta x}\left(ff(\hat{u}_j^{n+\beta_0}) + (\hat{u}_j^{n+\beta_1})\right),$$

$$\int_{t^n}^{t^{n+1}} \bar{s}_{j+\frac{1}{2}}(\tau)d\tau \equiv \frac{\Delta t}{2}\left(\bar{s}_{j+\frac{1}{2}}(t^{n+\beta_0}) + \bar{s}_{j+\frac{1}{2}}(t^{n+\beta_1})\right)$$

$\hat{u}_j^{n+\beta_k}, k = 0,1$, are approximated by applying a fourth-order Runge-Kutta scheme coupled with the following Natural Continuous Extension (NCE):

$$\hat{u}_j^{n+\beta_k} \equiv u(x_j, t^n + \beta_k \Delta t) = \hat{u}_j^n + \Delta t \sum_{i=1}^{4} b_i(\beta_k) k_j^{(i)}$$

$k_j^{(i)}, 1 \le i \le 4$ are the Runge-Kutta fluxes, which coincide with a numerical evaluation of $(-f_x + s)$ in the sediment transport problem, and are computed starting from the point-values $\hat{u}_k^{(i)}$, being $\hat{u}_j^{(1)} = \hat{u}_j^n$ and:

$$\hat{u}_j^{(2)} = \hat{u}_j^n + \frac{\Delta t k_j^{(1)}}{2}, \quad \hat{u}_j^{(3)} = \hat{u}_j^n + \frac{\Delta t k_j^{(2)}}{2},$$

$$\hat{u}_j^{(4)} = \hat{u}_j^n + \Delta t k_j^{(3)}$$

$\hat{u}_j^n$ will be defined in section 3 and $b_i(\beta_k)$ are the constants used in Capilla & Balaguer (2010). To calculate $\bar{s}_{j+1/2}(t^{n+\beta_k})$ we use an analytical manipulation of the 2nd component of the source term as in Caleffi (2007) so if we define, $\psi(x) = g \cdot Z_b \cdot \frac{d\eta(x)}{dx}$, then:

$$\begin{aligned}\bar{s}_{j+\frac{1}{2}}^{[2]} &= \frac{-1}{\Delta x}\int_{x_j}^{x_{j+1}} g(\eta - Z_b)\frac{\partial Z_b}{\partial x}dx \\ &= \frac{1}{\Delta x}\int_{x_j}^{x_{j+1}} \psi(x)dx \\ &+ \frac{g}{2\Delta x}\left[\hat{Z}_{b,j+1}^2 - \hat{Z}_{b,j}^2 - 2\hat{\eta}_{j+1}\hat{Z}_{b,j+1} + 2\hat{\eta}_j\hat{Z}_{b,j}\right]\end{aligned} \tag{6}$$

2.3 *Evaluation of the Runge-Kutta fluxes $k_j^{(i)}$*

Considering the point values, $\hat{u}_k^{(i)}$, we compute the flux values:

$$\begin{aligned}K_j(x_k, \hat{u}^{(i)}) &= ff(\hat{u}_k^{(i)}) - (\hat{u}_j^{(i)}) \\ &- \int_{x_j}^{x_k} s(\hat{u}^{(i)}(\psi))d\psi\end{aligned} \tag{7}$$

using a procedure similar to (6) and we apply the three-degree non-oscillatory pointwise interpolating operator $P_j(x; K_j^{(i)})$ described in section 3, to define:

$$k_j^{(i)} = -\frac{dP_j(x; K_j^{(i)})}{dx}, \quad \forall 1 \le i \le 4 \tag{8}$$

3 FOURTH-ORDER NON-OSCILLATORY RECONSTRUCTION

We follow these steps to compute point-values from averages and point-values from point-values:

1. We consider the polynomial, $q_j(x;\bar{u})$, which reconstructs point values based on average values, defined as the mean of the polynomial supported by: $\{\bar{u}_{j-1}, \bar{u}_j, \bar{u}_{j+1}, \bar{u}_{j+2}\}$ and the polynomial defined on $\{\bar{u}_{j-2}, \bar{u}_{j-1}, \bar{u}_j, \bar{u}_{j+1}\}$.
2. We define the point values:

$$w_{j\pm1} = \hat{u}_{j\pm1}, \quad w_{j-\frac{1}{2}}^r = q_j(x_{j-\frac{1}{2}};\bar{u}),$$

$$w_j = \hat{u}_j, \quad w_{j+\frac{1}{2}}^l = q_j(x_{j+\frac{1}{2}};\bar{u})$$

where

$$\hat{u}_j \equiv \bar{u}_j - \left(\frac{\bar{u}_{j-1} - 2\bar{u}_j + \bar{u}_{j+1}}{24}\right) = u(x_j, t^n) + O(\Delta x)^4$$

3. We consider the polynomial, $W_j(x;w)$, defined as the mean of the polynomial supported by the point values $\{w_{j-1/2}^r, w_j, w_{j+1/2}^l, w_{j+1}\}$ and the polynomial defined by $\{w_{j-1}, w_{j-1/2}^r, w_j, w_{j+1/2}^l\}$. Then:

$$\frac{1}{\Delta x}\int_{x_{j-\frac{1}{2}}}^{x_{j+\frac{1}{2}}} W_j(x;w)dx = \bar{u}_j, \quad \text{and}$$

$$W_j(x;w) = u(x,t) + O(\Delta x)^4, \quad \forall x \in \left[x_{j-\frac{1}{2}}, x_{j+\frac{1}{2}}\right]$$

4. Point values $\{w_{j-1/2}^r, w_j, w_{j+1/2}^l\}$ should satisfy the following non-oscillatory conditions:

(Ave1) If $\bar{u}_{j-1} \le \bar{u}_j \le \bar{u}_{j+1}$ then $w_{j-\frac{1}{2}}^r \le w_j \le w_{j+\frac{1}{2}}^l$

(Ave2) If $\bar{u}_{j-1} \ge \bar{u}_j \ge \bar{u}_{j+1}$ then $w_{j-\frac{1}{2}}^r \ge w_j \ge w_{j+\frac{1}{2}}^l$

These conditions are not fulfilled when,

$$6\min\left\{\left|\overline{UR}_j\right|, \left|\overline{UL}_j\right|\right\} + 5\left|\overline{UC}_j\right| < Sign(\overline{UC}_j)\overline{UC2}_j$$

where: $\overline{UC}_j = \bar{u}_{j+1} - \bar{u}_{j-1}$, $\overline{UR}_j = \bar{u}_{j+1} - \bar{u}_j$,

$\overline{UL}_j = \bar{u}_j - \bar{u}_{j-1}$, $\overline{UC2}_j = \bar{u}_{j+2} - \bar{u}_{j-2}$

In these cases we apply a correction so:

$$w_{j-\frac{1}{2}}^r = q_j(x_{j-\frac{1}{2}};\bar{u}) + K_{j-\frac{1}{2}}^r, \quad w_j = \hat{u}_j + K_j,$$

$$w_{j+\frac{1}{2}}^l = q_j(x_{j+\frac{1}{2}};\bar{u}) + K_{j+\frac{1}{2}}^l$$

and $K^r_{j-1/2}+4K_j+K^l_{j+1/2}=0$. Such constants are defined in Balaguer (2011).

5. We use the parameter $\theta^{\bar{u}}_j \in [0,1]$ defined by Liu and Tadmor (1998) so $\forall x \in [x_{j-\frac{1}{2}}, x_{j+\frac{1}{2}}]$:

$$R_j(x;\bar{u}) = \theta^{\bar{u}}_j \cdot W_j(x;w) + \left(1-\theta^{\bar{u}}_j\right)\cdot \bar{u}_j \quad (9)$$

satisfying the following non-oscillatory properties:

(ExI) If $\bar{u}_j > \bar{u}_{j+1}$ then $R_j(x_{j+\frac{1}{2}};\bar{u}) \geq R_{j+1}(x_{j+\frac{1}{2}};\bar{u})$

(ExII) If $\bar{u}_j < \bar{u}_{j+1}$ then $R_j(x_{j+\frac{1}{2}};\bar{u}) \leq R_{j+1}(x_{j+\frac{1}{2}};\bar{u})$

(ExIII) If $\bar{u}_j = \bar{u}_{j+1} \rightarrow R_j(x_{j+\frac{1}{2}};\bar{u}) = R_{j+1}(x_{j+\frac{1}{2}};\bar{u})$

However in the cells with a maximum or minimum we must impose additional conditions:

(ExIV) If $\bar{u}_{j-1} < \bar{u}_j > \bar{u}_{j+1}$ then

$$W_j(x_{j\pm\frac{1}{2}};w) \geq \min\left\{\frac{\bar{u}_j + \bar{u}_{j\pm1}}{2}, W_{j\pm1}(x_{j\pm\frac{1}{2}};w)\right\}$$

(ExV) If $\bar{u}_{j-1} > \bar{u}_j < \bar{u}_{j+1}$ then

$$W_j(x_{j\pm\frac{1}{2}};w) \leq \max\left\{\frac{\bar{u}_j + \bar{u}_{j\pm1}}{2}, W_{j\pm1}(x_{j\pm\frac{1}{2}};w)\right\}$$

which are fulfilled with the choice of $\left\{K^r_{j-1/2}, K_j, K^l_{j+1/2}\right\}$ described in Balaguer (2011).

6. Using $R_j(x;\bar{u}^n)$, the cell-averaged values of u on the staggered grid, $\bar{u}^n_{j+\frac{1}{2}}$, can be approximated as:

$$\frac{1}{\Delta x}\left[\int_{x_j}^{x_{j+\frac{1}{2}}} R_j(x;\bar{u}^n)dx + \int_{x_{j+\frac{1}{2}}}^{x_{j+1}} R_{j+1}(x;\bar{u}^n)dx\right]$$

We also calculate the point-values on the non-staggered grid: $\hat{u}^n_j \equiv R_j(x_j;\bar{u}^n)$

7. A similar procedure leads to the definition of $P_j(x;K^{(i)}_j)$ using the point-values $K_j(x_k,\hat{u}^{(i)})$ given in (7) (see Balaguer (2011)).

8. The function $\psi(x)$ in (6) is also computed by the polynomial used in step (7) and then it is integrated exactly by means of this procedure:

   a. We start with the point-values $\hat{\eta}_j$ of step (6), and then we compute the reconstruction polynomials, $P_j(x;\hat{\eta})$, which are used to approximate the point-values of $\psi(x)$:

$$\hat{\psi}_j = g\hat{Z}_{b,j}\frac{d\hat{\eta}_j}{dx} = g\hat{Z}_{b,j}\frac{dP_j(x_j;\hat{\eta})}{dx} \quad (10)$$

   b. We compute the point-value reconstruction polynomials, $P_j(x;\hat{\psi})$, using the point-values, $\hat{\psi}_j$, and finally we evaluate the following integral:

$$\int_{x_j}^{x_{j+1}} \psi(x)dx = \int_{x_j}^{x_{j+\frac{1}{2}}} P_j(x;\hat{\psi})dx + \int_{x_{j+\frac{1}{2}}}^{x_{j+1}} P_{j+1}(x;\hat{\psi})dx \quad (11)$$

9. Integrals $\int_{x_k}^{x_j} g(\eta - Z_b)\frac{\partial Z_b}{\partial x}dx$ given in (7) are evaluated with the same procedure, (6), (10) and (11).

10. In the following time step we compute the averages $\bar{u}^{n+2}_j$, using the following equation:

$$\bar{u}^{n+2}_j = \bar{u}^{n+1}_j - \frac{1}{\Delta x}\left[\int_{t^{n+1}}^{t^{n+2}} f(u_{j+\frac{1}{2}}(\tau))d\tau - \int_{t^{n+1}}^{t^{n+2}} f(u_{j-\frac{1}{2}}(\tau))d\tau\right] + \int_{t^{n+1}}^{t^{n+2}} \bar{s}_j(\tau)d\tau \quad (12)$$

4 NUMERICAL RESULTS

4.1 *Test 1: Unsteady flow over a sinusoidal bump*

We will consider that $\xi = 0$ with the following bottom function and initial conditions:

$$Z_b(x,0) = \sin^2(\pi x); \quad q(x,0) = \sin(\cos(2\pi x));$$
$$\eta(x,0) = 5 + e^{\cos(2\pi x)} + Z_b(x,0),$$

with $x \in [0,1]$ and periodic boundary conditions.

Figure 1 shows the free surface level computed at time $t = 0.1$. There is very good agreement between the reference solution (12800 cells) and the numerical solution computed with 100 cells and 4000 time steps.

4.2 *Test 2: The dam-break problem over a rectangular bump*

We solve equations (4) considering that $\xi = 0$, using the following fixed discontinuous bottom profile:

$$Z_b(x,0) = \begin{cases} 8, & \text{if } |x - 750| \leq 1500/8, \\ 0, & \text{otherwise,} \end{cases} \quad (13)$$

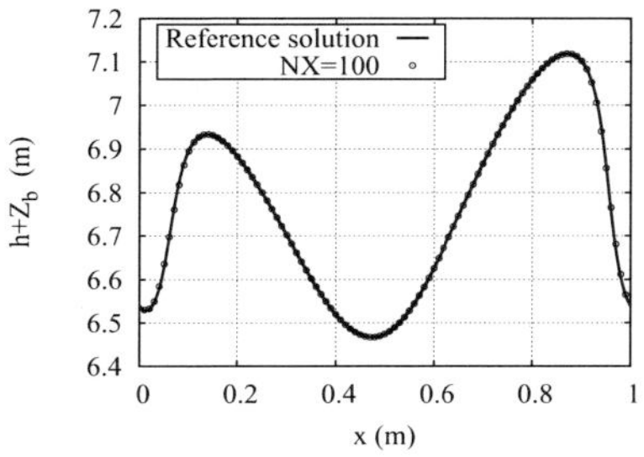

Figure 1. Unsteady flow over a sinusoidal bump.

for $0 \le x \le 1500$ m. The initial conditions are:

$$q(x,0) = 0 \, and \, \eta(x,0) = \begin{cases} 20, & \text{if } x \le 750, \\ 15, & \text{otherwise.} \end{cases} \quad (14)$$

The numerical solution is computed with $NX = 500$ cells and it is compared with a reference solution computed using $NX = 5000$ cells. Both solutions are shown in Figure 2 for $h(x) + Z_b(x)$ at time $t = 15$.

Numerical solution showed a good agreement with solutions presented in Capilla & Balaguer (2010) and Caselles et al. (2009).

4.3 *Test 3: Intense sediment transport problem*

The initial conditions for $0 \le x \le 1000$ are given by $\eta(x,0) = 10$, $q(x,0) = 10$ and:

$$Z_b(x,0) = \begin{cases} \sin^2\left(\dfrac{\pi(x-300)}{200}\right), & \text{if } 300 \le x \le 500 \\ 0, & \text{otherwise,} \end{cases} \quad (15)$$

We consider that $A = 1$ and $\epsilon = 0.2$ ($\xi = 1.25$). In Figure 3 water depth is plotted at the times $t = 200$ and $t = 700$ with 250 cells. Numerical solutions agree with results presented by Caleffi et al. (2007).

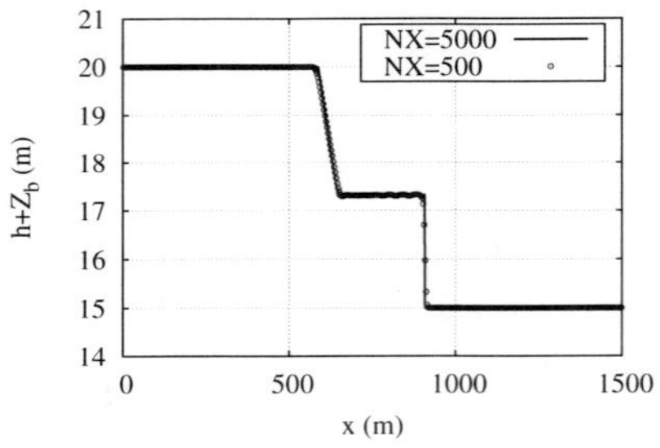

Figure 2. Dam-break problem. Numerical solution at $t = 15$ using 500 and 5000 cells.

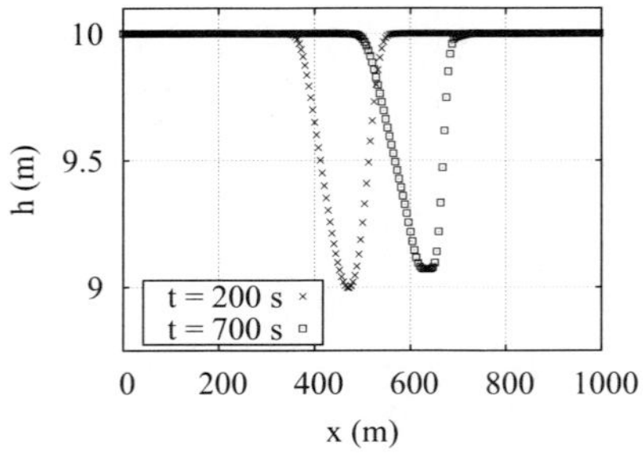

Figure 3. Intense sediment transport problem.

5 CONCLUSIONS

We have presented a new fourth-order central scheme to solve the sediment transport equations. The treatment that we apply for the source term spatial integration preserves the time and space accuracy and results in a well-balanced scheme. We have used a reconstruction algorithm that defines a polynomial in each cell $[x_{j-1/2}, x_{j+1/2}]$. based on the point values $\{w^r_{j-1/2}, w_j, w^l_{j+1/2}\}$. Taking a fourth-order approximation a correction of these point values is required in order to the reconstruction to be of the non-oscillatory type. Numerical results have shown its non-oscillatory behavior. The exact conservation property and the fourth-order of accuracy of our scheme may also checked in a similar way to Capilla & Balaguer (2010) or Capilla et al. (2011).

ACKNOWLEDGEMENTS

The authors agree the financial support provided by the Universidad Politécnica de Valencia (PAID-06-10) and the Spanish Ministry of Education and Science (Projects CGL2009-14220, CGL2010-19591 and DPI2010-20333).

REFERENCES

Balaguer-Beser, A. 2011. A new reconstruction procedure in central schemes for hyperbolic conservation laws. *Int. J. Numer. Meth. Engng*, 86, 1481–1506. DOI: 10.1002/nme.3105.

Caleffi, V. & Valiani, A. & Bernini, A. 2007. High-order balanced CWENO scheme for movable bed shallow water equations. *Advances in water resources*, 30, 730–741.

Capilla, M.T. & Balaguer-Beser, A. 2010. High-Order Non-Oscillatory Central Schemes for Shallow Water Equations. In B.H.V. Topping, J.M. Adam, F.J. Pallarés, R. Bru, M.L. Romero, (editors), *Proceedings of the Seventh International Conference on Engineering Computational Technology*, Civil-Comp Press, Stirlingshire, UK, Paper 165.

Capilla, M.T. & Balaguer-Beser, A. & Gascón, Ll. 2011. Reconstrucción de los flujos Runge-Kutta en esquemas centrados de alto orden que resuelven sistemas hiperbólicos no homogéneos. *Actas XXII CEDYA / XII CMA*. Mallorca. Spain.

Caselles, V. & Donat, R. & Haro, G. 2009. Flux-gradient and source-term balancing for certain high resolution shock-capturing schemes. *Computers and fluids*, 38, 16–36.

Gascón, Ll. & Corberán, J.M. 2001. Construction of second-order TVD schemes for nonhomogeneous hyperbolic conservation laws. *J. Comput. Phys*, 172, 261–297.

Liu, XD. & Tadmor, E. 1998. Third order nonoscillatory central scheme for hyperbolic conservation laws. *Numer. Math*, 79, 397–425.

Numerical Methods for Hyperbolic Equations – Vázquez-Cendón et al. (eds)
© 2013 Taylor & Francis Group, London, ISBN 978-0-415-62150-2

First approach to the application of operational 4DVAR data assimilation to the regional ocean model system

Marcos Cobas-García, Andrés Gómez-Tato & Carmen Cotelo-Queijo
Centro de Supercomputación de Galicia, Santiago de Compostela, Spain

M. Elena Vázquez-Cendón
Applied Mathematics Department, University of Santiago de Compostela, Spain

Pablo Carracedo-García, Pedro Costa & Breogán Gómez-Hombre
Meteogalicia, Santiago de Compostela, Spain

Joaquín Triñanes-Fernández
Systems Laboratory, University of Santiago de Compostela, Spain

Manuel Ruiz-Villarreal
Instituto Español de Oceanografía, A Coruña, Spain

ABSTRACT: We present in this contribution an initial study of the application of the 4DVAR data assimilation method to the Regional Ocean Modelling System (ROMS), within the framework of the OCEANO project. One of the objectives of this project is to run an operational configuration of the ROMS model in the North-West Iberian coast, and to perform buoy and satellite data assimilation in the area. Two sets of experiments were run, in order to test the influence of the algorithm in the correction of temperature an salinity, as a function of the spatial correlation scale parameter, and the amount and type of data provided. The results reproduce the sensitivity with respect to the scs parameter, and show that salinity data constitute the most important dataset for the correction of these two variables.

1 INTRODUCTION

As a consequence of the uncertainties present in the application of models to real-life problems (parametrization of small-scale phenomena, simplifications made in the development of the models, accuracy of the numerical methods involved, etc.), even models initialized with accurate initial conditions will eventually present a drift from the actual evolution of the phenomena being modellised. This is actually the case for models that simulate the motion of Earth fluids, like the Regional Ocean Model System (ROMS). The data assimilation procedure has been described by (Talagrand 1997) as "the process through which all the available information is used to determine as accurately as possible the state of the atmospheric or oceanic flow", and this way minimise the forementioned drift. The 4DVAR data assimilation achieves this by optimization of a quadratic function along multiple dimensions, to find the model initial state whose evolution minimises the distance to the observations, within a given time window, and taking into account the spatial correlation scales (scs) of the variables being corrected. The Regional Ocean Model System (ROMS) is a state-of-the-art, three dimensional, free-surface ocean model, with a wide comunity of users (www.myroms.org), that includes this data assimilation technique.

An initial study of the application of the 4DVAR algorithm to the ROMS model was conducted, as a starting point for a future implementation in an operational run of the model. The application of this algorithm can be very sensitive to the data provided and the settings, and therefore the overall purpose of these experiments is to test the correct application of the method, with experiments that could confirm our intuition with respect to these two elements. Two sets of experiments were run: on the one hand, the sensitivity of the model to the settings of the spatial correlation scale (scs) parameter was tested, and on the other hand the impact in the outcome of the algorithm of the variability in the amount and type of data. The setup of the two series of experiments is explained in section 2, and the results and conclusion will be examined in section 3.

2 MODEL SETUP

The 4DVAR data assimilation algorithm is based on the minimization of a funcional that accounts for the difference between the model and the observations, which takes the following form:

$$J = \frac{1}{2}[Hx - y]R^{-1}[Hx - y] + \frac{1}{2}[x - x_b]B^{-1}[x - x_b] \quad (1)$$

In the previous equation, x, y and x_b stand for the vector of model values interpolated to the position of the observations, the vector of observations, and the vector of background values respectively. H denotes the "observation operator", that performs the interpolation of values of the model variables located in the model grid to the observation locations. R and B denote the observations and background error covariance matrices respectively. In other words, the functional computes the differences of the model with respect to the observations and to a background state, or "first guess", taking into account the confidence in both of them, expressed in the R and B matrices. The R matrix is easily obtained from the specifications of the measuring instruments, but the B matrix can not be computed, owing partly to its huge size, and therefore has to be modelised, based on the scs parameter and statistical computations performed on a standard forward run of the model (Weaver and Courtier 2001).

Therefore, before any assimilation cycle was performed, a conventional forward run of the model was carried out. The grid of this forward run expands between 41 and 44.5 degrees North and 8.5 and 13 degrees West, with a horizontal resolution of 1/15 degrees, and 30 vertical levels. The time step was set to 900 seconds, with a resulting barotropic Courant number of 0.53. For the purpose of this test, no river discharges were configured. It was forced with atmospheric data from the HIRLAM model, with 0.2 degrees resolution, and data from a climatological run were provided as boundary conditions.

Other applications present in the literature (Broquet et al. 2009), (Powell et al. 2008), use a much longer period to gather statistical data. We are aware that a shorter period can reduce the variability of the data, and this could produce better results in our case than would in a realistic case. For a future application to an operational run, a longer period is being considered but, in any case, this should be taken into account when interpreting the results of the present study.

The typical scenario for the application of a 4DVAR data assimilation consists of a set of initial conditions (ic) for the model values and a number of observations posterior to these ic, in such a way that, in general, there would be a misfit between the evolution of these ic and the observations. The role of the 4DVAR algorithm is to modify these ic so that after the application of the 4DVAR the misfit between model and observations is reduced.

The assimilation experiments were carried out to mimic this typical scenario, in the following way: First, temperature, salinity, sea surface elevation and velocities were extracted from the previous conventional forward run of the ROMS model at a certain time, to be used as ic. Aditional data were extracted at subsequent neighbouring times, that will play the role of observations to be assimilated. This way, the evolution of these ic would match exactly the "observations" extracted. Then, several types of perturbation were introduced into the surface salinity and SST of the ic (hereinafter "initial perturbations"), so that there would be a misfit between the observations and the evolution of the perturbed ic. Given the high spatial correlation between these two variables (around 0.9), the salinity field was perturbed respecting this correlation. Finally, the observations were assimilated into the perturbed ic, and corrected ic were generated by the algorithm. Ideally, the assimilation cycle should reduce the initial perturbations to zero, but in practice this will not be achieved, and after the assimilation cycle there will be a difference with the exact solution (i.e., the unperturbed ic). Hereinafter, we will refer to this difference as the "final perturbation". Since by construction we know the exact solution, we can assess the performance of the algorithm by comparing the initial and final perturbations. In this sense, we will present plots of the initial and final perturbations, the corrections made by the alrotithm, and computations of the average absolute values of the perturbations.

Following this strategy, two sets of data assimilation cycles were built, to test the importance of the scs parameter and the quality of data. A central assimilation cycle was run with conventional settings and idealistic data quality, that will serve as a "canonical" run, around which these two sets of assimilation cycles were run. The following subsections explain the detailed setup of these two sets of experiments.

2.1 *Spatial correlation scale*

In this section we focus on the sensitivity of the model to the scs parameter, which takes part in the modelization of the B matrix. By introducing a misfit between the model and data scs, the importance of the fine tunning of this parameter becomes apparent. Initially, the scs was set to 0.588 degrees (lat/lon) in the 4DVAR algorithm, as in (Broquet et al. 2009), whereas the initial conditions were perturbed with noises with scs of 1.0, 0.588, 0.3 and 0.15 degrees. Each perturbation consists of an accumulation of gaussian functions centered in each of the grid points, with the same scs, but different amplitudes (top of each group of three plots in Figure 1). Finally the model is reset for the case of the biggest misfit, 0.15 degrees, using this scs for the model and the perturbation.

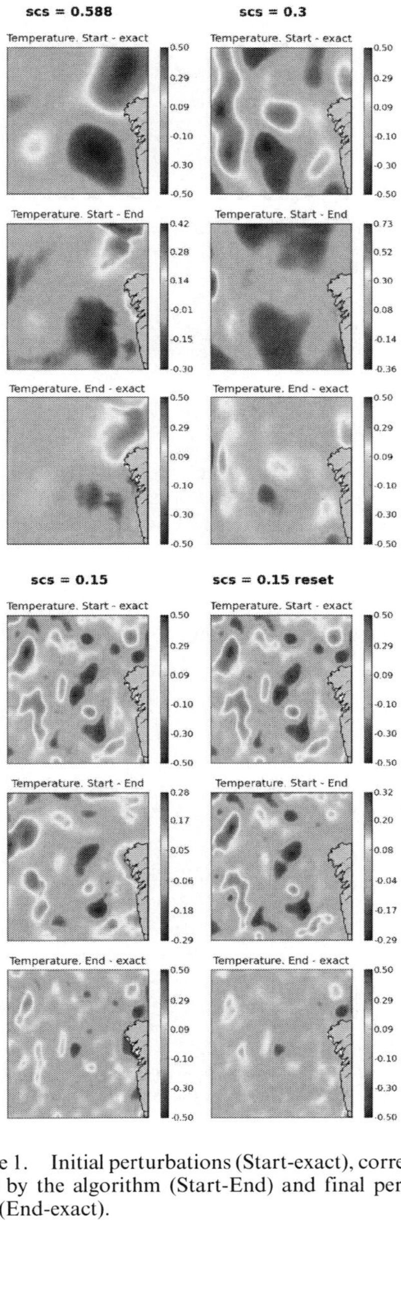

Figure 1. Initial perturbations (Start-exact), corrections made by the algorithm (Start-End) and final perturbations (End-exact).

2.2 *Data quality*

The first set of experiments was run with an idealistically abundant amount of data, to discard uncertainties with regard to the data quality, and focus on the settings of the algorithm. In a second series of experiments, we will focus on the datasets. The model was provided with different categories of data: full datasets, data with missing datasets (no SST, no salinity,...), scarce or incomplete data of one kind (e.g. incomplete SST fields, reduced salinity stations, etc.), and finally with a more realistic scenario of data, consistent of along-track SSH data, "cloudy SST" (i.e., resembling the data obtained under cloudy conditions) and a few salinity stations near the coast, to test the impact of the quality of data. Each of the datasets were tested separately, and in pairs, and compared with the central assimilation experiment, in which full SST, full SSH and 200 randomly distributed salinity stations were employed. The scs for these runs was set to the original setting of the previous section, 0.588 degrees both in the perturbation and the model configuration, which presented a good ratio of corrections.

3 RESULTS AND CONCLUSIONS

The aim of the first part of the study was to confirm the correct application of the algorithm, and assess its theoretical performance. The algorithm behaved as expected: the more resemblance between the data and model scs's, the better the correction, as can be seen in Table 1. A certain threshold of misfit is tolerated by the model, above which the performance degrades, as the "0.15 scs" row in Table 1 shows. The fact that the resetting of the model scs to the data scs ("0.15_{rs}" row in Table 1) improves the average correction rates, points to this misfit as the cause for the degradation, although other factors may be involved, since

Table 1. Spatial correlation scale sensitivity. Average perturbations before and after the assimilations.

	scs	Before	After	*After/before*
Temp	1	0.243	0.114	0.47
	0.588	0.168	0.085	0.50
	0.3	0.169	0.085	0.50
	0.15	0.126	0.083	0.66
	0.15_{rs}	0.126	0.063	0.50
Salt	1	4.925	2.341	0.47
	0.588	3.403	1.777	0.52
	0.3	3.431	2.069	0.60
	0.15	2.557	2.071	0.81
	0.15_{rs}	2.557	1.756	0.68

the improvement in the assimilation of the salinity field is not as conspicuous. Other tests made during this study suggest that the ratios of correction increase with the magnitude of the initial perturbation, which is another possible cause of this difference. Figure 1 shows the initial perturbation, corrections made by the algorithm and final perturbations in some of the runs. This figure shows the matching patterns of the corrections and the perturbations, especially in the top two columns. This match is somewhat diffused in the bottom-left column, in which the misfit is the widest, and recovered once the model scs is reset to the data scs (bottom-right column).

The assimilation of temperature and salinity is greatly affected by the quality of the data, as was expected. Figure 2 summarizes the results of this section. The principal conclusion of the second set of experiments is that the most important dataset for the assimilation of temperature and salinity is formed by the salinity dataset. SSH data has little impact on the correction of these two variables, at least if SST and salinity data are provided. The availability of SST data improves the correction of temperature, but not salinity, despite the high correlation between both variables. The use of "cloudy" SST, or even no SST at all, has little impact on the assimilation of temperature, as long as salinity data are provided. On the contrary, the absence or scarcity of salinity data degrades the correction of temperature and, to a greater extent, salinity, even to the point of increasing the initial perturbation. The final experiments, which try to resemble a realistic scenario, in which only along-track SSH data, cloudy SST and just a few salinity stations near the coast were employed, suggest that the initial correction rates are not to be expected in a realistic application of the algorithm. The two last executions of this configuration, adding 20 supplementary salinity stations in each of them (Realistic+20, Realistic+40), emphasize the importance of the salinity dataset. The addition of

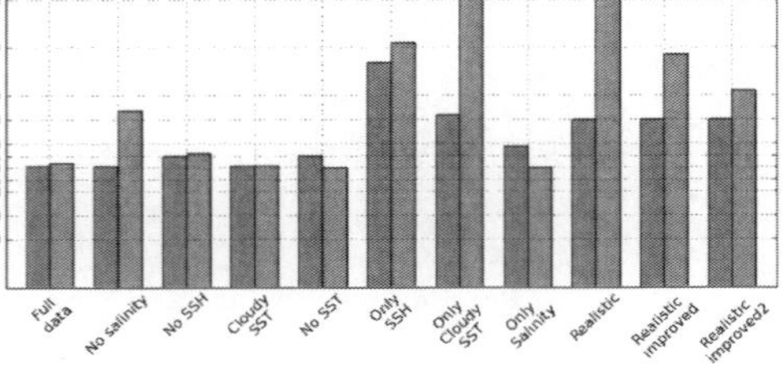

Figure 2. Summary of ratios of reduction *After*/*Before* for temperature and salinity.

Table 2. Data quality experiments. Average perturbations before and after the assimilations.

	Name	Before	After	*After*/*before*
Temp	Full data	0.168	0.085	0.51
	No salinity	0.168	0.085	0.51
	No SSH	0.168	0.092	0.55
	Cloudy SST	0.168	0.086	0.51
	No SST	0.168	0.092	0.55
	Only SSH	0.168	0.158	0.94
	Only SST	0.168	0.121	0.72
	Only salinity	0.168	0.099	0.59
	Realistic	0.168	0.118	0.70
	Realistic+20	0.168	0.118	0.70
	Realistic+40	0.168	0.118	0.70
Salt	Full data	3.403	1.777	0.52
	No salinity	3.403	2.505	0.74
	No SSH	3.403	1.915	0.56
	Cloudy SST	3.403	1.748	0.51
	No SST	3.403	1.704	0.50
	Only SSH	3.403	3.474	1.02
	Only SST	3.403	4.829	1.42
	Only salinity	3.403	1.688	0.50
	Realistic	3.403	4.819	1.42
	Realistic+20	3.403	3.306	0.97
	Realistic+40	3.403	2.807	0.82

20 more stations improves dramatically the correction of salinity, but not temperature, with respect to the original perturbation, with further improvements adding 20 more stations.

ACKNOWLEDGEMENTS

This study was funded by the regional government of Galicia, as part of the OCEANO project, code number 09MDS009CT.

REFERENCES

Broquet, G., C. Edwards, A. Moore, B. Powell, M. Veneziani, and J. Doyle (2009). Application of 4d-variational data assimilation to the california current system. *Dynamics of Atmospheres and Oceans 48*, 69–92. Modeling and Data Assimilation in Support of Coastal Ocean Observing Systems.

Powell, B., H. Arango, A. Moore, E.D. Lorenzo, R. Milliff, and D. Foley (2008). 4dvar data assimilation in the intra-americas sea with the regional ocean modeling system (roms). *Ocean Modelling 23*, 130–145.

Talagrand, O. (1997). Assimilation of observations, an introduction. *Jour. Met. Soc. Japan 75* (1B), 191–209.

Weaver, A. and P. Courtier (2001). Correlation modelling on the sphere using a generalized diffusion equation. *Quarterly Journal of the Royal Meteorological Society 127*(575), 1815–1846.

Numerical Methods for Hyperbolic Equations – Vázquez-Cendón et al. (eds)
 ISBN 978-0-415-62150-2

Changes in buoyancy-driven instabilities using a reaction-diffusion system

D.M. Escala, J. Guiu-Souto, J. Carballido-Landeira & A. Pérez-Muñuzuri
Non Linear Physics Group, Faculty of Physics, University of Santiago de Compostela, Spain

M.E. Vázquez-Cendón
Applied Mathematics Department, University of Santiago de Compostela, Spain

ABSTRACT: At the interface generated by mixing two miscible fluids, instabilities may be induced due to difference between the fluids densities and/or diffusion coefficients. These instabilities generate characteristic patterns that affect the mass transport between the two species. The Belousov-Zhabotinsky reaction is a chemical reaction in which, due to autocatalysis of its intermediaries and differences between diffusion coefficients, chemical oscillations and waves are generated that result in pattern formation when the reaction is carried out in a two-dimensional media. The aim of this study was to analyze the influence of a chemical reaction system on the instabilities when two fluids with different densities and diffusion coefficients react in a Hele-Shaw cell. All models involved in these phenomena were solved using commercial software Flow-3D® v9.3.2.

1 INTRODUCTION

1.1 *Hydrodynamic instabilities*

Hydrodynamic instabilities are physical phenomena involving fluids. The most common situation consists on two different fluids—with different thermodynamic properties—interact under an external force. That difference induces different patterns and behaviors at the interface between the two fluids. Some examples of these instabilities are the very well know Rayleigh-Taylor instability, double diffusion, salt fingering, viscous fingering, etc.

1.2 *Oscillating chemical reactions*

Oscillating chemical reactions are systems where an intermediate specie of the reaction oscillates in time. The most studied oscillating system is the Belouzov-Zabotinsky reaction. The oscillating behavior of this reaction is produced by the reduction and oxidation of the color indicator by an organic and inorganic compound respectively. These oscillating reactions are examples of non equilibrium system and shows a non linear complex behavior. In a stirred and homogeneous reactor, the oscillating behavior is demonstrated as a change of the catalyzer color with time. However, when this reaction is put in a Petri dish, the diffusion phenomenon and autocatalysis produces the generation of chemical waves and different complex pattern formation like spirals and target points. It is known that this kind of mechanism of pattern formation is involved in different processes in Nature.

Our aim is to simulate two of the classic hydrodynamical instabilities (Rayleigh-Taylor and double diffusion) using commercial software Flow-3D® (Flow Science inc) to compare the result with experimental data. Then we will simulate the effect of the inclusion of a chemical reaction on a hydrodynamic instability. Two cases will be analyzed. First, we will analyze the inclusion of a simple bimolecular chemical reaction like $A + B \rightarrow C$. Then we will analyze the effect of an oscillating chemical reaction. In last case, we will not compare the simulated results with experimental data, but we keep them for future reference.

2 PHYSICAL SYSTEM AND MATHEMATICAL MODELS

2.1 *Bouyancy driven instabilities*

The physical system involved in this work is compound by two fluids disposed in a Hele-Shaw cell as shown in Figure 1. For the Rayleigh-Taylor case (initially unstable), the denser fluid is located on the top and the less dense one is put on the bottom of the system. For the double diffusion case (initially stable), the situation is the inverse. Independently of these two situations, the mathematical model involved for both is the same.

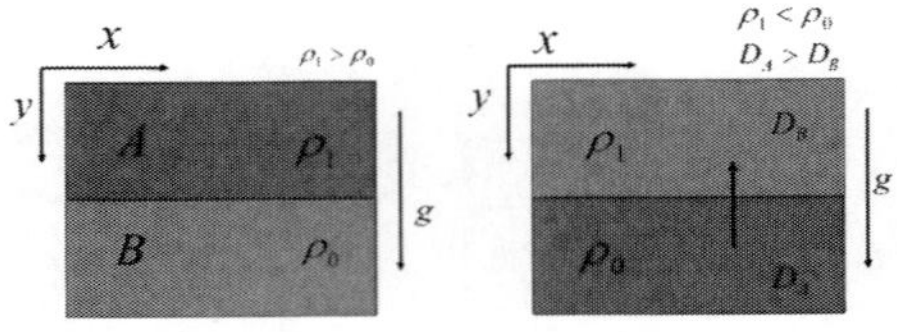

(a) Initially unstable case (Rayleigh-Taylor)
(b) Initially stable case (double diffusion)

Figure 1. Non reactive reference system.

At first instead, according with the Figure 1 and based on previous research (Trevelyan et al. 2011, Zalts et al. 2008), the non-reactive system is governed by the following diffusion-advection system which employ the Darcy's law and Bousinessq's approximation for pressure and density calculation:

$$\nabla p = -\frac{\mu}{k}\mathbf{u} + \rho(A, B)\mathbf{g}, \tag{1a}$$

$$\nabla \cdot \mathbf{u} = 0, \tag{1b}$$

$$\phi\frac{\partial A}{\partial t} + \mathbf{u}\cdot\nabla A = \phi D_A \nabla^2 A, \tag{1c}$$

$$\phi\frac{\partial B}{\partial t} + \mathbf{u}\cdot\nabla B = \phi D_B \nabla^2 B, \tag{1d}$$

where the height coordinate, y, is positive in the direction of gravity field, $\mathbf{g}$. $A = A(x, y)$, $B = B(x, y)$ are the fluid species, p and $\mathbf{u}$ are the pressure and velocity fields, μ and κ are the kinematic viscosity and the permeability respectively and φ is the porosity that it is equal to 1 in Hele-Shaw cells. D_A and D_B are the diffusion coefficients of fluids A and B respectively. ρ is the density and depends on the involved species in the following manner:

$$\rho(A, B) = \rho_0(1 + \alpha_A A + \alpha_B B), \tag{2}$$

where ρ_0 is the density of pure solvent, (water in this work), and α_A and α_B are the solutal expansion coefficients defined by:

$$\alpha_A = \frac{1}{\rho_0}\frac{\partial \rho}{\partial A}, \quad \alpha_B = \frac{1}{\rho_0}\frac{\partial \rho}{\partial B} \tag{3}$$

2.2 *Chemically driven instabilities—non oscillating system*

Like the non reactive system that was explained before, we will define the mathematical model involved in the physical system.

As is shown on Figure 2(a), the product of the reaction, the species C, is created by the contact at interface of the other two species A and B. In this case, the initial fluid configuration at time $t = 0$ is stable.

The mathematical description of this system is given by the following reaction-diffusion-advection system

$$\nabla p = -\frac{\mu}{\kappa}\mathbf{u} + \rho([A],[B],[C])\mathbf{g}, \tag{4a}$$

$$\rho = p_0\left(1 + \alpha_A[A] + \alpha_B[B] + \alpha_C[C]\right) \tag{4b}$$

$$\nabla \cdot \mathbf{u} = 0, \tag{4c}$$

$$[A]_t + \mathbf{u}\cdot\nabla[A] = D_A\nabla^2[A] - k[A][B], \tag{4d}$$

$$[B]_t + \mathbf{u}\cdot\nabla[B] = D_B\nabla^2[B] - k[A][B], \tag{4e}$$

$$[C]_t + \mathbf{u}\cdot\nabla[C] = D_C\nabla^2[C] - k[A][B], \tag{4f}$$

where $[A]$, $[B]$ and $[C]$ are the molar concentrations of species A, B and C respectively and k is a rate constant.

2.3 *Chemically driven instabilities—oscillating system*

In this case the situation is similar to the previous one in which both species will interact at the interface in an initially stable configuration. We will use the Lokta-Volterra equations because it is the most simple model for an oscillating chemical reaction. The chemical model is given by:

$$A + X \xrightarrow{k_1} 2X, \tag{5a}$$

$$X + Y \xrightarrow{k_2} 2Y, \tag{5b}$$

$$Y \xrightarrow{k_3} B. \tag{5c}$$

where the equations 5a and 5b models the autocatalytic behavior of the system. Furthermore, in this case A represent the organic substrate, X and Y the

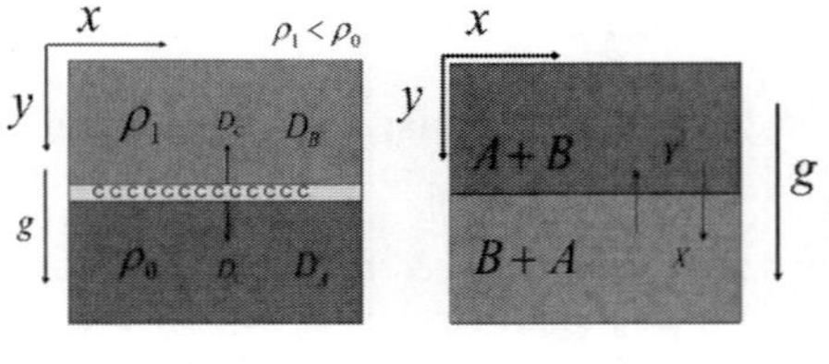

(a) Non oscillating case
(b) Oscillating case

Figure 2. Reactive reference system.

oscillating intermediaries, B the product, and k_1, k_2 and k_3 are rate constants. Using the mass action law to calculate the rate equations considering a fixed amount of substrate and product in the system and supposing that the density variation only depends on the intermediates species, we obtain the mathematical model for the system 2(b):

$$\nabla p = -\frac{\mu}{\kappa}\mathbf{u} + \rho([X],[Y])\mathbf{g}, \tag{6a}$$

$$\rho = p_0\left(1 + \alpha_X[X] + \alpha_Y[Y]\right) \tag{6b}$$

$$\nabla \cdot \mathbf{u} = 0, \tag{6c}$$

$$[X]_t + \mathbf{u}\cdot\nabla[X] = D_X\nabla^2[X] - k_2[X][Y] + k_1[A][X], \tag{6d}$$

$$[Y]_t + \mathbf{u}\cdot\nabla[Y] = D_Y\nabla^2[Y] - k_3[Y] + k_2[X][Y]. \tag{6e}$$

2.4 *Simulation parameters*

All simulations were performed using one simple rectangular mesh with 10000 nodes. The boundary conditions were set as non-slip wall on x coordinate and an outflow boundary on y coordinate. The simulation time depends on every specific case.

3 RESULTS

3.1 *Bouyancy driven instabilities*

For this case, we have obtained very accurate results for both system, unstable and stable, in comparison with experimental data. These results are shown on Figures 3 and 4.

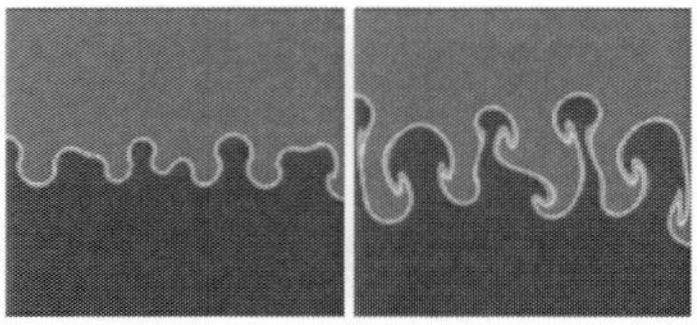

(a) Fluid fraction map obtained with Flow-3D®

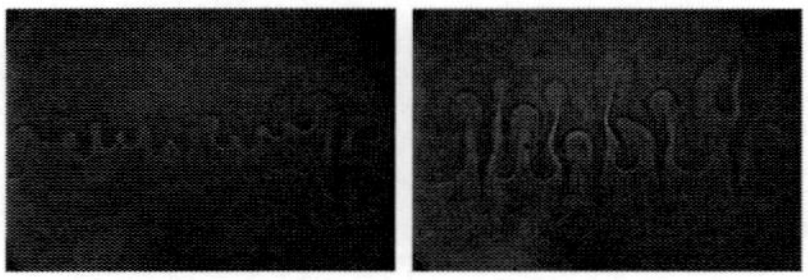

(b) Experimental results

Figure 3. Initial conditions unstable.

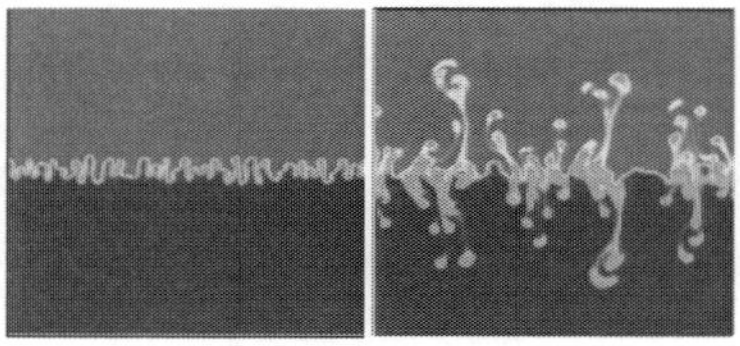

(a) Fluid fraction map obtained with Flow-3D®

(b) Experimental results

Figure 4. Initial conditions stable.

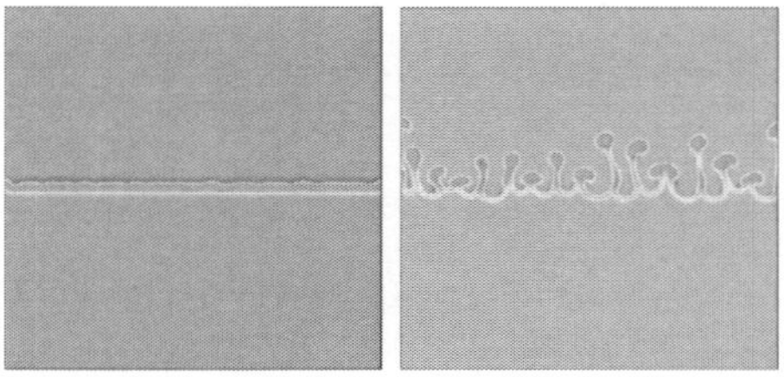

(a) Product C is less dense than reactants

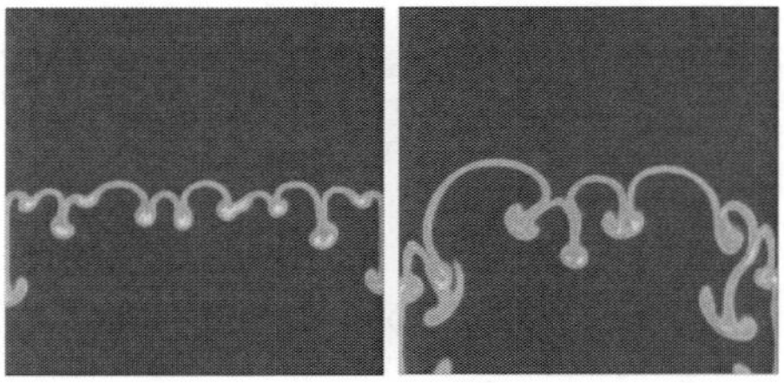

(b) Product C is more dense than reactants

Figure 5. Two different density maps obtained from Flow-3D®.

All this simulations are based on results found on bibliography. (Trevelyan et al. 2011, De Wit 2004).

3.2 *Chemically driven instabilities*

For the non oscillating system, the results obtained are shown on Figure 5. In these cases, we obtained

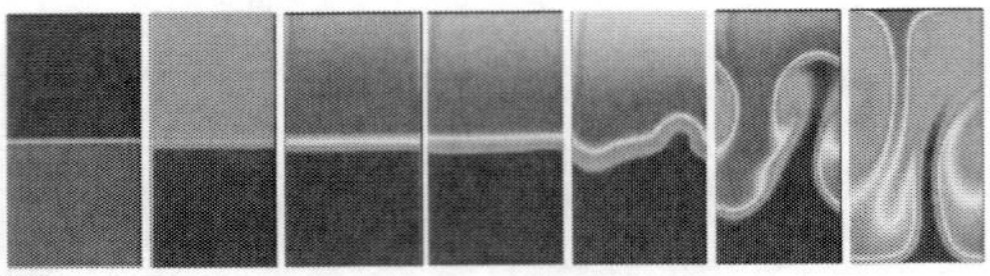

Figure 6. Simulation results obtained for oscillating system using Flow-3D®. The figure shows different states of density map along time.

two results depending on the density of the product *C*. The behavior obtained is consistent with the chemical characteristics of the reaction. These results can be compared directly with previous works. (Almarcha et al. 2010, Zalts et al. 2008, APS web page).

For oscillating system, we only performed simulation experiments. The results are shown in Figure 6.

In this situation, it can be appreciated when the system begins in a stable configuration, the variation on the intermediaries concentration leads to a change in the system density. This generates a general instability to the entire system.

4 CONCLUSIONS

The theoretical model has been simulated with acceptable results and compared with previous experimental work. Both, classic, (Rayleigh-Taylor and double diffusion) and complex (chemically driven) instabilities, shows good correlation with experimental work and simulations.

In future research on models with oscillating reactions we must get a compromise between the reaction parameters and the hydrodynamic scenario.

We hope to see a new behavior on system stability. The coupling between diffusive and hydrodynamic phenomena can lead us to obtain important conclusions about the system dynamic and stability. Flow-3D® has shown very good results considering speed calculation, accuracy and easy of use. It will be an alternative to take account in our next work.

ACKNOWLEDGEMENTS

The authors thank MICINN by funding through research projects, FIS2010-21023, and MAT2010-17336, Xunta de Galicia through projects INCITE09206020PR and for European Regional Development Funds (research project 2010/50) and Fundación Ramón Areces for additional financial support.

REFERENCES

Almarcha, C., P. Trevelyan, P. Grosfils, and A. De Wit (2010). Chemically driven hydrodynamic instabilities. *Physical Review Letters 104*(4), 44501.

APS web page. American physical society. http://physics.aps.org/story/v25/st3#videos.

De Wit, A. (2004). Miscible density fingering of chemical fronts in porous media: Nonlinear simulations. *Physics of Fluids 16*, 163.

FlowScience inc. Flow-3d® user manual v9.3. http://www.flow3d.com.

Trevelyan, P., C. Almarcha, and A. De Wit (2011). Buoyancy-driven instabilities of miscible two-layer stratifications in porous media and hele-shaw cells. *Journal of Fluid Mechanics 670*(1), 38–65.

Zalts, A., C. El Hasi, D. Rubio, A. Urena, and A. D'Onofrio (2008). Pattern formation driven by an acid-base neutralization reaction in aqueous media in a gravitational field. *Physical Review E 77*(1), 015304.

Numerical Methods for Hyperbolic Equations – Vázquez-Cendón et al. (eds)
© 2013 Taylor & Francis Group, London, ISBN 978-0-415-62150-2

A mean gradient method to solve shallow flows

Jaime Fe, Fermín Navarrina & Luis Cueto-Felgueroso
Grupo de Métodos Numéricos en Ingeniería, Universidade da Coruña, Spain

ABSTRACT: The present work consists in the implementation of a second order Finite Volume Method for the numerical resolution of the Two-Dimensional Shallow Water Equations (2D-SWE) with turbulent term. The model achieves second order accuracy in space by making use of the mean gradient of the variables in a cell and second order in time by using the second order Henn Method. The eddy viscosity values are obtained from the Rodi's depth-averaged $k - \varepsilon$ method. The model has been applied to the Cavity Flow Problem (for different Reynolds numbers) and to a real problem to compare the numerical results with some experimental measurements.

1 INTRODUCTION

The 2D-SWE are a hyperbolic system of non linear conservation laws. They derive from the Navier-Stokes equations through some simplifying hypothesis, the most important of which is the hydrostatic pressure distribution (Vreugdenhil 1994).

This set of equations describes the behaviour of incompressible fluids when the ratio of the depth to the horizontal dimensions is small, a feature commonly found in the flow in channels, rivers and estuaries and even in sea currents near the shore.

The effects of turbulence are considered in the 2D-SWE both by the friction and the turbulent source terms. In many cases the influence of using the turbulent term is not significant, for instance when we only need an estimate of energy losses. For this reason 2D-SWE are frequently solved in a simplified form, which does not include the turbulent term. Its inclusion, however, may become critical in the simulation of flows in which recirculation zones play an important role.

2 THE SHALLOW WATER EQUATIONS

The 2D-SWE system in conservative form is expressed as

$$\frac{\partial \mathbf{U}}{\partial t} + \frac{\partial \mathbf{F}_1}{\partial x} + \frac{\partial \mathbf{F}_2}{\partial y} = \mathbf{G} \tag{1}$$

being $\mathbf{U} = (h, hu, hv)^T$ the vector of unknowns, being the flux terms

$$\mathbf{F}_1 = \begin{pmatrix} hu \\ hu^2 + \frac{1}{2}gh^2 \\ huv \end{pmatrix}, \quad \mathbf{F}_2 = \begin{pmatrix} hv \\ huv \\ hv^2 + \frac{1}{2}gh^2 \end{pmatrix} \tag{2}$$

and being the source term

$$\mathbf{G} = \begin{pmatrix} 0 \\ gh(S_{0x} - S_{fx}) + S_{t1} \\ gh(S_{0y} - S_{fy}) + S_{t2} \end{pmatrix}. \tag{3}$$

In the above expressions h is the fluid depth, u and v are the horizontal velocity components and g is the gravity acceleration. S_{0x} and S_{0y} are the geometric slopes. S_{fx} and S_{fy} are the friction slopes, that are calculated from the Manning formula

$$S_{fx} = \frac{n^2 u\sqrt{u^2 + v^2}}{h^{4/3}}, \qquad S_{fy} = \frac{n^2 v\sqrt{u^2 + v^2}}{h^{4/3}}, \tag{4}$$

being n the Manning coefficient. Finally, the turbulent terms S_{t1}, S_{t2} are

$$S_{t1} = \frac{\partial}{\partial x}\left(2\nu_t h \frac{\partial u}{\partial x}\right) + \frac{\partial}{\partial y}\left(\nu_t h\left[\frac{\partial v}{\partial x} + \frac{\partial u}{\partial y}\right]\right), \tag{5}$$

$$S_{t2} = \frac{\partial}{\partial x}\left(\nu_t h\left[\frac{\partial v}{\partial x} + \frac{\partial u}{\partial y}\right]\right) + \frac{\partial}{\partial y}\left(2\nu_t h \frac{\partial v}{\partial y}\right), \tag{6}$$

where ν_t is a variable called turbulent kinetic viscosity or eddy viscosity.

3 THE FINITE VOLUME DISCRETIZATION

The finite volumes used here are of cell-vertex type. They are defined starting from a previous triangular mesh, the nodes of which are the centers of the finite volumes mesh (see Fig. 1). To form the boundary of cell C_i we take the barycenters of all the triangles that have the common vertex I as well as the midpoints of the edges that meet at I. Γ_{ij} represents the common boundary of cells C_i and C_j and η_{ij} is the outward normal vector to Γ_{ij}. By $\|\eta_{ij}\|$ we represent the length of the common boundary η_{ij} and $\tilde{\eta}_{ij} = (\tilde{\alpha}_{ij}, \tilde{\beta}_{ij})$ is the corresponding unit vector. The subcell T_{ij} is the union of triangles AMI and MBI.

4 THE MEAN GRADIENT RECONSTRUCTION

To discretize the hydrodynamic equations, our first choice was to implement the first order finite volume method proposed by Bermúdez et al. in (Bermúdez et al. 1998), which assumes that the variables values are constant at both sides of every cell edge and atevery time step.

A key point to get good results with this method is to properly calculate the numerical flux at the cell edges and, as it has been shown by these authors, the upwinding of the flux term proves to be a useful stabilizing technique.

This approach, however, has the drawback of producing a considerable amount of numerical diffusion, which led us to develop a method second order accurate in space. It makes use of the mean gradient of the variables in a cell, which is calculated from their values at the cell edges, that in turn depend on the values at the adjacent cells. In this way, to calculate the values at the edges of a cell, the considered cell and the ones that surround it are involved.

If Γ_{ij} is the edge between cells C_i and C_j, and u_i, u_j are the uniform values of variable u, the values of the reconstructed variable u^* at each side of Γ_{ij} are

$$\text{Side } C_i: \quad u_i^* = u_i + \left(\overline{\nabla u}\right)_{C_i} \cdot (\mathbf{r}_M - \mathbf{r}_I) \tag{7}$$

$$\text{Side } C_j: \quad u_j^* = u_j + \left(\overline{\nabla u}\right)_{C_j} \cdot (\mathbf{r}_M - \mathbf{r}_J) \tag{8}$$

being $M \in \Gamma_{ij}$ the midpoint between I and J. We are then employing a linear approximation for the variables within each cell, instead of the constant values assumption, thus reducing the value of the upwinding term (see Fig. 2). The time derivatives are obtained with the Henn method, which is second order accurate.

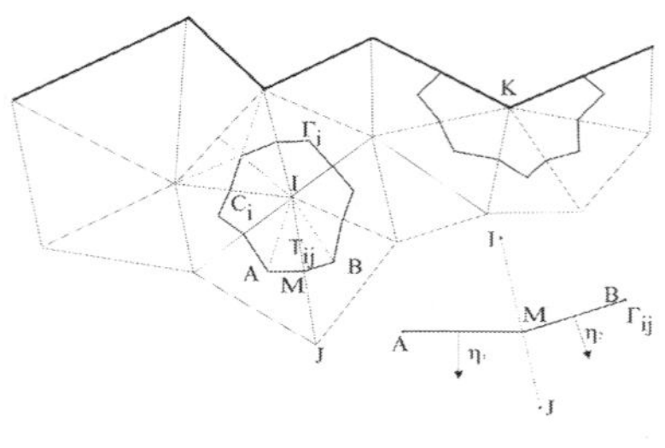

Figure 1. Finite volume mesh.

5 THE CAVITY FLOW PROBLEM

To test this technique, the Cavity Flow problem has been solved. It is a classical benchmark for the two dimensional Navier-Stokes equations and its results show the ability of the proposed model to accurately represent viscous flows.

The problem consists in obtaining the velocity field in a square domain of $1 \times 1\, m^2$. A 81×81 non-uniform mesh (finer near the corners) has been employed. The boundary conditions, of Dirichlet type, are: $u = 1, v = 0$ at the upper side; $u = v = 0$ (no-slip condition) at the other three.

Taking $V = L = 1$ for the velocity and length scales, the resulting Reynolds number is

$$Re = \frac{VL}{\nu} = \frac{1}{\nu} \tag{9}$$

and we can simulate different Reynolds numbers by varying the viscosity value.

Calculations have been made for viscosity values corresponding to Reynolds numbers up to 20,000. In Figures 3–6 we represent the resulting streamlines obtained with Reynolds numbers of 1000, 5000, 10,000 and 20,000. It can be noticed that the Cavity Flow problem is not commonly solved for Reynolds numbers over 10,000, over

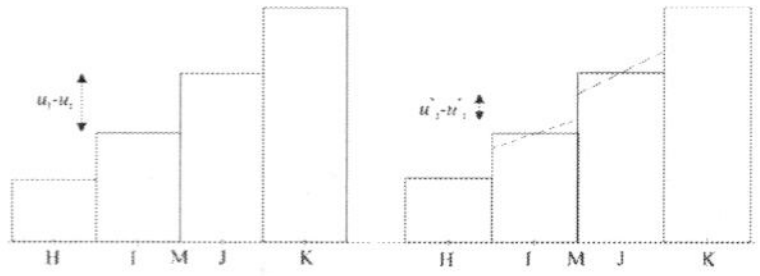

Figure 2. a) First order scheme, b) Second order scheme.

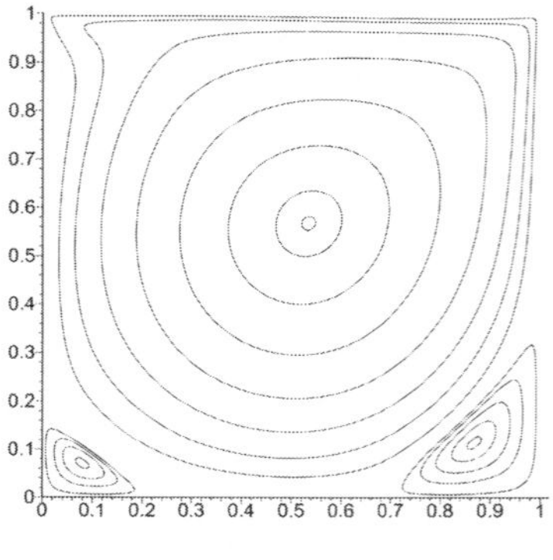

Figure 3. Streamlines. $Re = 1000$.

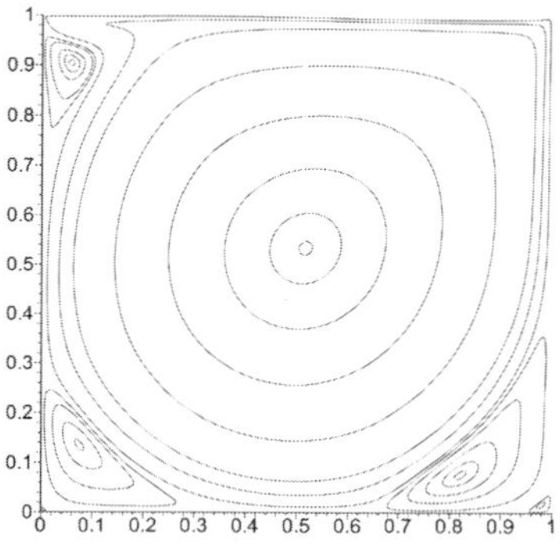

Figure 4. Streamlines. $Re = 5000$.

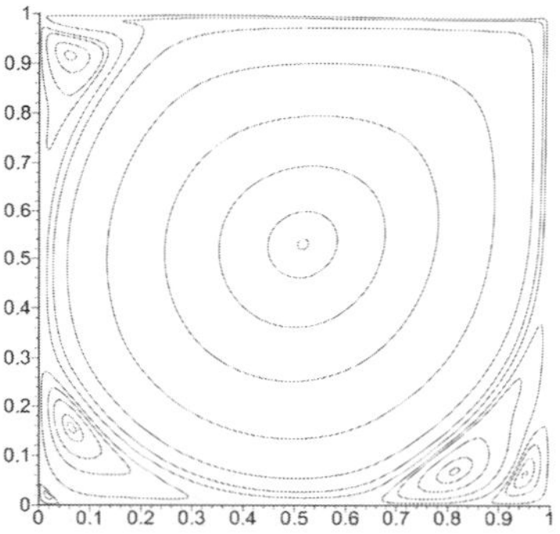

Figure 5. Streamlines. $Re = 10000$.

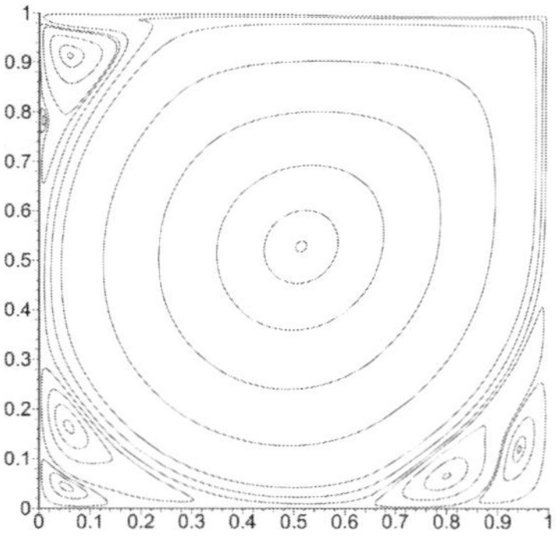

Figure 6. Streamlines. $Re = 20000$.

which limit it is said to become unstable (Vellando et al. 2002) unless a very fine mesh is used, as Erturk et al. have shown (Erturk et al. 2005).

6 THE TURBULENCE MODEL

In a subsequent stage, to deal with practical problems, we needed to calculate the turbulent viscosity at every point, therefore a turbulence model was required. To this aim, the depth-averaged $k - \varepsilon$ model, described in (Rodi 1993) was chosen for our work.

The turbulent source terms, S_{t1}, S_{t2} (see section 2) contain the eddy viscosity ν_t. This coefficient is calculated from the turbulent kinetic energy k and the dissipation rate per unit mass ε as

$$\nu_t = c_\mu \frac{k^2}{\varepsilon}, \tag{10}$$

being k and ε the turbulence kinetic energy and the dissipation rate per unit mass respectively.

The values of magnitudes k and ε are obtained with Rodi's depth-averaged $k - \varepsilon$ turbulence model. They are given by the transport equations

$$\frac{\partial k}{\partial t} + u\frac{\partial k}{\partial x} + v\frac{\partial k}{\partial y} = \frac{\partial}{\partial x}\left(\frac{\nu_t}{\sigma_k}\frac{\partial k}{\partial x}\right) + \frac{\partial}{\partial y}\left(\frac{\nu_t}{\sigma_k}\frac{\partial k}{\partial y}\right) + P_h + P_{kV} - \varepsilon, \tag{11}$$

$$\frac{\partial \varepsilon}{\partial t} + u\frac{\partial \varepsilon}{\partial x} + v\frac{\partial \varepsilon}{\partial y} = \frac{\partial}{\partial x}\left(\frac{\nu_t}{\sigma_\varepsilon}\frac{\partial \varepsilon}{\partial x}\right) + \frac{\partial}{\partial y}\left(\frac{\nu_t}{\sigma_\varepsilon}\frac{\partial \varepsilon}{\partial y}\right) + c_{1\varepsilon}\frac{\varepsilon}{k}P_h + P_{\varepsilon V} - c_{2\varepsilon}\frac{\varepsilon^2}{k}, \tag{12}$$

where

$$P_h = \nu_t\left[2\left(\frac{\partial u}{\partial x}\right)^2 + 2\left(\frac{\partial v}{\partial y}\right)^2 + \left(\frac{\partial v}{\partial x} + \frac{\partial u}{\partial y}\right)^2\right], \tag{13}$$

$$P_{kV} = \frac{1}{\sqrt{c_f}}\frac{U^{*3}}{h} \tag{14}$$

and

$$P_{\varepsilon V} = c_{\varepsilon\Gamma}\frac{c_{2\varepsilon}}{c_f^{3/4}}\sqrt{c_\mu}\frac{U^{*4}}{h^2}. \tag{15}$$

P_h is the production of k due to the interaction of the turbulent stresses with the horizontal mean-velocity gradients. P_{kv} and $P_{\varepsilon v}$ are the productions of k and ε due to the vertical velocity gradients and they are related to the friction velocity U^*.

These equations have been numerically solved in conservative form, coupled with the 2D-SWE. A full description of the process, as well as details on the values used for the different coefficients involved, many of them empirical, can be found at Fe et al.(2009) (Fe et al. 2009).

7 FLOW IN SETTLING TANKS

To test the numerical model with a real problem, experimental measurements were made at the Hydraulics Laboratory of the Civil Engineering School of A Coruña (Bonillo 2000).

The data were obtained using SONTEK Micro Acoustic Doppler Velocimeters that produce a small distortion of the velocity field. They are highly accurate (10^{-3} m/s) and they take between 80 and 250 measurements per second. As the output values have a maximum frecuency of 50 Hz, every one represents an average of several measurements.

The numerical results have been obtained with a simplified model that applies the mean gradient method only to the velocity components. This simplification improves stability significantly, with little loss of accuracy.

The experimental domain is composed of two flat bottom settling tanks. The bottom level of both tanks is situated 12 cm below the level of the channels that arrive to and start from them. In Figure 7, experimental values of the velocity modulus and streamlines are shown. Figure 8 represents the numerical values of the velocity modulus and Figure 9 the calculated streamlines as well as the levels of the eddy viscosity.

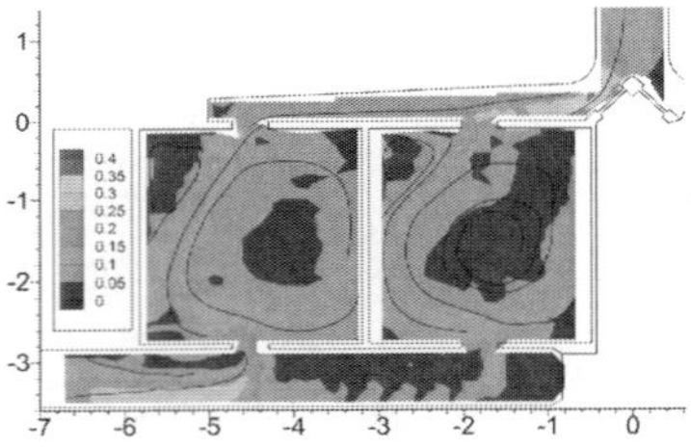

Figure 7. Experimental streamlines and velocity modulus.

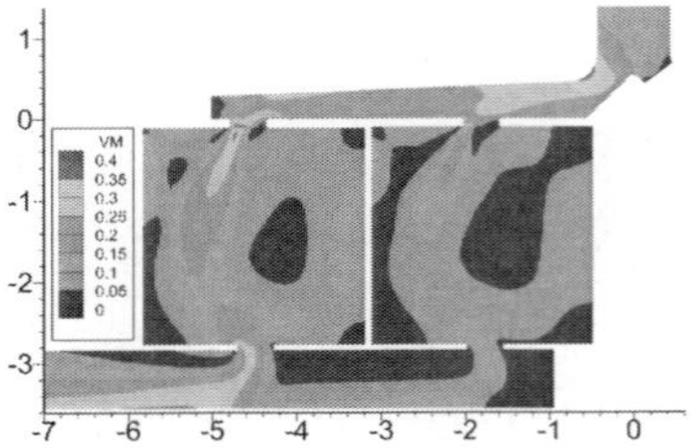

Figure 8. Numerical velocity modulus.

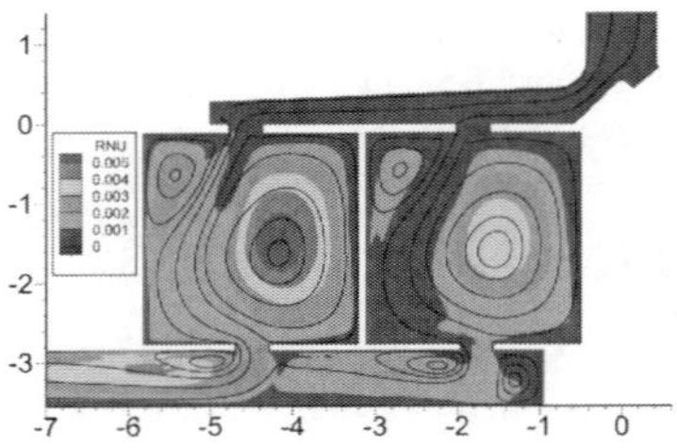

Figure 9. Numerical streamlines and eddy viscosity.

The results show a very good agreement between numerical and experimental velocity modulus and streamlines. Numerical values for the eddy viscosity are also displayed in Figure 9 but, as this magnitude cannot be directly measured, no experimental validation is shown.

8 CONCLUSIONS

In this work 1) we have described a second order finite volume method for the resolution of the 2D-SWE with turbulent term; 2) we have shown the results of applying the model to the Cavity Flow problem, for uniform values of the viscosity, with Reynolds numbers up to 20,000; and 3) we have applied the model to a real case, calculating the turbulent viscosity values from the depth-averaged $k - \varepsilon$ model, and comparing the computational results with experimental measurements. Good agreement has beenobtained between the presented results and the validation tests.

ACKNOWLEDGEMENTS

This work has been partially supported by the *Ministerio de Ciencia e Innovación* (grant #DPI2009-14546-C02-01 and #DPI2010-16496), by R&D projects of the *Xunta de Galicia* (grants #PGDIT09-MDS00718PR and #PGDIT09REM005118PR) cofinanced with FEDER funds, and the *Universidade da Coruña*.

REFERENCES

Bermúdez A., Dervieux A., Desideri J. & Vázquez M.E. 1998. Upwind schemes for the two dimensional shallow water equations with variable depth using unstructured meshes, *Computer Methods in Applied Mechanics and Engineering*; 155, 49–72.

Bonillo, J.J. 2000. Un modelo de transporte de sustancias solubles para flujos turbulentos en lámina libre, Doctoral Thesis, University of A Coruña.

Erturk, E., Corke, T.C. & Gokcol, C. 2005. Numerical solutions of 2-D steady incompressible driven cavity flow at high Reynolds numbers, *International Journal for Numerical Methods in Fluids*; 48, 747–774.

Fe, J., Navarrina, F., Puertas, J., Vellando P. & Ruiz, D. 2009. Experimental validation of two depth-averaged turbulence models, *International Journal for Numerical Methods in Fluids*; 60(2), 177–202.

Rodi W. 1993 Turbulence models and their application in hydraulics. A state-of-the-art review, Iahr monograph (3rd. ed.). Rotterdam: Balkema.

Vellando, P., Puertas, J. & Colominas, I. 2002. Supg stabilized finite element resolution of the Navier-Stokes equations. Applications to water treatment engineering. *Computer Methods in Applied Mechanics and Engineering*; 191, 5899–5922.

Vreugdenhil C.B. 1994. *Numerical methods for shallow-water flow*. Dordrecht: Kluwer Academic Publishers.

Numerical Methods for Hyperbolic Equations – Vázquez-Cendón et al. (eds)
© 2013 Taylor & Francis Group, London, ISBN 978-0-415-62150-2

Upwind schemes for non-homogeneous hyperbolic conservation laws with stationary solutions

Ll. Gascón, A. Balaguer-Beser & M.T. Capilla
Departamento de Matemática Aplicada, Universidad Politécnica de Valencia, Spain

ABSTRACT: In a previous work, Balaguer & Conde (2005) describe a fourth-order nonoscillatory scheme for solving homogeneous hyperbolic conservation laws, which is based on a reconstruction procedure where the point values of the solution are calculated from the cell averages by avoiding the increase in the number of solution extrema at the interior of each cell. They applied the ideas developed by Liu & Tadmor (1998) to define third order nonoscillatory schemes. In Gascón & Corberán (2001), the authors describe a strategy to extend well-known schemes for the homogeneous case to non-homogeneous conservation laws, guaranteeing the balance of the flux and source terms at steady states of a natural way. In this work, we present a revision of the procedure described in Gascón & Corberán (2001) in order to obtain an extension for non-homogeneous hyperbolic conservation laws of the fourth-order scheme developed in Balaguer & Conde (2005). Numerical results are presented.

1 INTRODUCTION

In this research we use the method developed in Gascón & Corberán (2001) to extend some well-known high-order schemes to hyperbolic conservation laws with source terms in the scalar case. The main idea consists in the transformation of the non-homogeneous problem into homogeneous through the definition of a new flux formed by the physical flux and the primitive of the source term. This transformation allows to apply schemes already known for the homogeneous case to non-homogeneous problems and allows us to include correctly the source term as a divergence term providing the balance between flux and source terms of a natural way. For this, the formulation must be such that all the differences that appear in the schemes are expressed as flux-differences including source terms and not as flow variable differences. In the following we present a revision of this technique.

We consider a problem of initial values associated with a non-homogeneous hyperbolic conservation law

$$u_t + f(u)_x = s(x) \tag{1}$$

that models the evolution of the variable $u(x,t)$ and where $f(u)$ represents the flux and $s(x)$ a source term that we assume first dependent only on x. To assure a correct discretization of the stationary equation associated to the non-homogeneous conservation law, it seems convenient the description of schemes suggesting the same type of discretization for the fluxes as for the primitive of the source terms (Gascón & Corberán 2001). With this aim, we denote

$$u_t + g(x,u)_x = 0 \quad \text{where} \quad g(x,u) = f(u) - \int_a^x s(y)dy$$

This strategy has been applied to extend the second-order Lax&Wendroff scheme and the TVD schemes of Harten in Gascón & Corberán (2001), the flux limiters of Sweby (Gascón & García 2001, Gascón et al. 2003) and ENO schemes (Caselles et al. 2009). Following the technique described in Gascón & Corberán (2001) to approximate (1) and denoting

$$b_{i,k} = -\int_{x_i}^{x_k} s(y)dy, \quad \lambda = \frac{\Delta t}{\Delta x} \quad \text{and} \quad \alpha = \lambda \frac{\partial f}{\partial u}$$

we propose write

$$u_j^{n+1} = u_j^n - \lambda\left[\hat{g}_{j+1/2} - \hat{g}_{j-1/2} + b^n_{j-1/2,j} + b^n_{j,j+1/2}\right]$$

where

$$\hat{g}_{j+1/2} = \frac{1}{2}\Big[f_j^n + f_{j+1}^n - b^n_{j,j+1/2} + b^n_{j+1/2,j+1} - \Phi(\alpha_{j+1/2})(f_{j+1}^n - f_j^n + b^n_{j,j+1})\Big]$$

being $\Phi(x) = x$ for the Lax&Wendroff extension (Gascón & Corberán 2001), $\Phi(x) = \text{sign}(x)$ for the first-order upwind extension (Gascón & Corberán 2001) and $\Phi(x) = \text{sign}(x) + \chi(r)(x - \text{sign}(x))$ for the flux limiters extension (Gascón & García 2001, Gascón et al. 2003). Finally, the extension of the TVD of Harten, described in Gascón & Corberán (2001), can be given by

$$\hat{g}_{j+1/2} = \frac{1}{2}\big[f_j^n + f_{j+1}^n - b_{j,j+1/2}^n + b_{j+1/2,j+1}^n - \Phi(\alpha_{j+1/2})(f_{j+1}^n - f_j^n + b_{j,j+1}^n)\big] + \Psi_{j+1/2}$$

where

$$\Psi_{j+1/2} = s_{j+1/2}\max\left\{0, \min\left\{s_{j+1/2}\left|\phi_{j-1/2}\right|, s_{j+1/2}\left|\phi_{j+3/2}\right|, \phi_{j+1/2}\right\}\right\}$$

being

$$\phi_{j+1/2} = \frac{1}{2}(\Phi(\alpha_{j+1/2}) - \alpha_{j+1/2})(f_{j+1}^n - f_j^n + b_{j,j+1}^n)$$
$$s_{j+1/2} = \text{sign}(\phi_{j+1/2})$$

and the function $\Phi(x) = \text{sign}(x)$. The extension to systems has been accomplished by local linearization.

2 FOURTH-ORDER NON-OSCILLATORY INTERPOLATION

By integrating Equation 1 over the control volume, we obtain

$$\bar{u}_j^{n+1} = \bar{u}_j^n - \frac{1}{\Delta x}\int_{t^n}^{t^{n+1}}\big[g(x_{j+1/2}, u(x_{j+1/2}, \tau)) - g(x_{j-1/2}, u(x_{j-1/2}, \tau))\big]d\tau \tag{2}$$

where the cell average for u is defined as

$$\bar{u}_j^n = \frac{1}{\Delta x}\int_{x_{j-1/2}}^{x_{j+1/2}} u(x, t_n)\, dx$$

and

$$g(x, u) = f(u) - \int_a^x s(y)dy \tag{3}$$

in order to secure a similar discretization for the fluxes and the source terms, preserving steady states.

The integrals with respect to the time variable in Equation 2 are approximated using a two-point Gauss quadrature. Thus

$$\int_{t^n}^{t^{n+1}} g(x_{j\pm1/2}, u(x_{j\pm1/2}, \tau))d\tau \approx \frac{\Delta t}{2}\big[\hat{g}_{j\pm1/2}^{n+\beta_0} + \hat{g}_{j\pm1/2}^{n+\beta_1}\big]$$

where

$$\beta_0 = \frac{1 - 1/\sqrt{3}}{2} \quad \text{and} \quad \beta_1 = \frac{1 + 1/\sqrt{3}}{2}$$

and to approximate the pointvalues of g at the time steps we may use a Taylor expansion, replacing the time derivatives of the solution by spatial derivatives

$$\hat{g}_{j+1/2}^{n+\beta_i} = g_{j+1/2}^n - (\Delta t \cdot \beta_i)\left(\frac{\partial f}{\partial u}\right)\Bigg|_{j+1/2}^n \frac{\partial g}{\partial x}\Bigg|_{j+1/2}^n + \cdots$$

where the source terms have been included (by Equation 3) as a divergence term providing the balance between flux and source terms of a natural way.

To estimate the pointvalues of g and its spatial derivatives, we propose to use a three-degree non-oscillatory reconstruction polynomial, following a procedure similar to Balaguer & Conde (2005).

Initially we consider the polynomial of degree 3 that verifies these conditions

$$q_j(x_j; g^n) = g_j^n; \quad q_j(x_{j-1}; g^n) = g_{j-1}^n;$$
$$q_j(x_{j+1}; g^n) = g_{j+1}^n; \quad \Delta x\frac{dq_j}{dx}(x_j; g^n) = d_j^n$$

This polynomial can be expressed as

$$q_j(x; g^n) = g_j^n + A_j(x; g^n)$$
$$A_j(x; g^n) = d_j^n\left(\frac{x - x_j}{\Delta x}\right) + \left(\frac{g_{j-1}^n - 2g_j^n + g_{j+1}^n}{2}\right)\left(\frac{x - x_j}{\Delta x}\right)^2 + \left(\frac{-g_{j-1}^n - 2d_j^n + g_{j+1}^n}{2}\right)\left(\frac{x - x_j}{\Delta x}\right)^3$$

where the slope d_j has been defined as in Balaguer & Conde (2005). Numerical results in this paper have been obtained using d_j such that q_j coincides with the centered polynomial defined as the average value between two conservative piecewise polynomials,

$$\{g_{j-1}, g_j, g_{j+1}, g_{j+2}\} \text{ and } \{g_{j-2}, g_{j-1}, g_j, g_{j+1}\}$$

We have considered a convex combination of the polynomial q_j and the linear piecewise interpolant L_j, as in Kurganov & Levy (2002). Then,

$$p_j(x; g^n) = (1 - \theta_j^n)L_j(x; g^n) + \theta_j^n \cdot q_j(x; g^n)$$

being

$$L_j(x;g^n) = g_j^n + s_j^n\left[\frac{x - x_j}{h}\right]$$
$$s_j^n = \min\mathrm{mod}(g_{j+1}^n - g_j^n\ ,\ g_j^n - g_{j-1}^n)$$

and the parameter θ_j, which is in [0,1], has been modified according to Liu & Tadmor (1998) and Kurganov & Levy (2002) in order to guarantee the nonoscillatory behaviour of the polynomial reconstruction p_j. Then,

$$M_j = \max\left\{\ q_j(x_{j+1/2}),\ q_j(x_{j-1/2})\right\}$$

$$m_j = \min\left\{\ q_j(x_{j+1/2}),\ q_j(x_{j-1/2})\right\}$$

$$M_{j\pm1/2} = \max\left\{\frac{1}{2}\left(L_j(x_{j\pm1/2}) + L_{j\pm1}(x_{j\pm1/2})\right),\ q_{j\pm1}(x_{j\pm1/2})\right\}$$

$$m_{j\pm1/2} = \min\left\{\frac{1}{2}\left(L_j(x_{j\pm1/2}) + L_{j\pm1}(x_{j\pm1/2})\right),\ q_{j\pm1}(x_{j\pm1/2})\right\}$$

and θ_j is defined as

$$\min\left\{\frac{M_{j+1/2} - L_j(x_{j+1/2})}{M_j - L_j(x_{j+1/2})},\ \frac{m_{j-1/2} - L_j(x_{j-1/2})}{m_j - L_j(x_{j-1/2})},\ 1\right\}$$

when $g_{j-1} < g_j < g_{j+1}$. The parameter θ_j is equal to

$$\min\left\{\frac{M_{j-1/2} - L_j(x_{j-1/2})}{M_j - L_j(x_{j-1/2})},\ \frac{m_{j+1/2} - L_j(x_{j+1/2})}{m_j - L_j(x_{j+1/2})},\ 1\right\}$$

when $g_{j-1} > g_j > g_{j+1}$, and otherwise, θ_j is equal to 1.

Finally, the scheme proposed is given by

$$\bar{u}_j^{n+1} = \bar{u}_j^n - \frac{\lambda}{2}\left[\hat{g}_{j+1/2}^{n+\beta_0} + \hat{g}_{j+1/2}^{n+\beta_1} - (\hat{g}_{j-1/2}^{n+\beta_0} + \hat{g}_{j-1/2}^{n+\beta_1})\right]$$

where

$$\hat{g}_{j+1/2}^{n+\beta_i} \approx p_k^n(x_{j+1/2};g^n) - (\Delta t \cdot \beta_i)\left(\frac{\partial f}{\partial u}\right)\Bigg|_{j+1/2}^{n} \frac{\partial p_k}{\partial x}\Bigg|_{j+1/2}^{n} + \cdots$$

being

$$k = \begin{cases} j & \text{if } \alpha_{j+1/2} \geq 0 \\ j+1 & \text{if } \alpha_{j+1/2} < 0 \end{cases}$$

The cell average for u in Equation 2 has been obtained by a fourth–order interpolation similar to that used previously for g, but considering

$$p_j(x;\bar{u}^n) = (1 - \theta_{u,j}^n)L_j(x;\bar{u}^n) + \theta_{u,j}^n\left(q_j(x;\bar{u}^n) - \frac{1}{24}(\Delta x)^2\frac{d^2q_j(x;\bar{u}^n)}{dx^2}\right)$$

to secure the fourth-order spatial accuracy.

3 NUMERICAL RESULTS

In order to examine the capacity of the above scheme to capture steady states of non-homogeneous conservation laws, we show numerical experiments for the test proposed in Greenberg et al. (1996). The second problem shows the accuracy of the scheme.

3.1 *The Leroux test*

Let us consider the problem

$$u_t + \left[\frac{u^2}{2}\right]_x = c'(x),\ c(x) = \begin{cases}\cos^2(\pi x/2), & -1 \leq x \leq 1 \\ 0, & \text{otherwise}\end{cases}$$
$$u(x,0) = 0$$

Figures 1 and 2 show numerical solutions for this problem calculated with the fourth-order scheme, described previously, at t = 0.2 and t = 3 seconds, respectively, using 240 nodes.

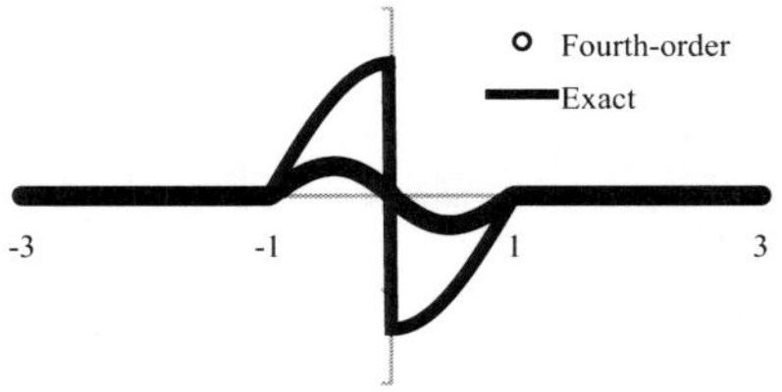

Figure 1. Numerical solution for the Leroux test (at t = 0.2 sec).

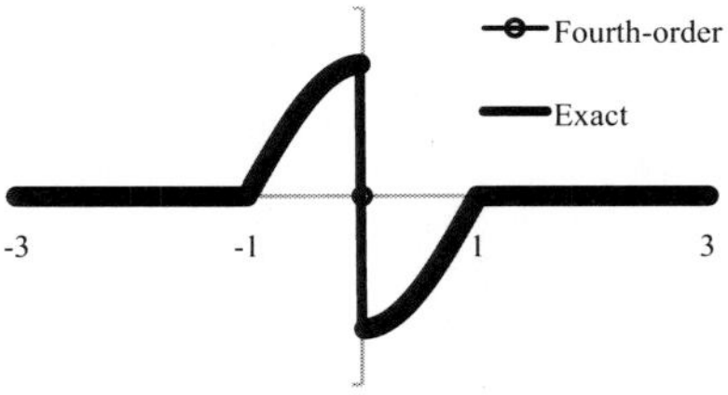

Figure 2. Numerical solution for the Leroux test (at t = 3 sec).

3.2 *The Toro test*

In order to compare the accuracy of the scheme described in Section 2 with previous schemes revised in the introduction, we consider the following test

$$u_t + u_x = \left(\frac{1+x}{x}\right)u, \quad u_0(x) = \begin{cases} x, & 0.3 < x < 0.7 \\ 0, & \text{otherwise} \end{cases}$$

Figures 3–5 show a comparison of the exact solution (solid line) with the numerical results calculated at t = 1 second, using 300 nodes (L = 3 m), by the extension of the upwind first-order (Fig. 3), the TVD extension of Harten and minmod flux limiter (Fig. 4), described in the introduction, and the fourth-order scheme presented in Section 2 (Fig. 5).

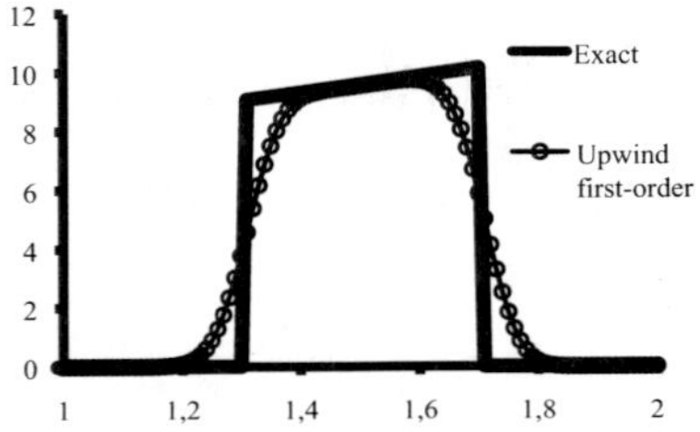

Figure 3. Comparison of the exact solution (solid line) with the numerical results calculated at t = 1 s for the Toro problem, by the extension of the upwind first-order.

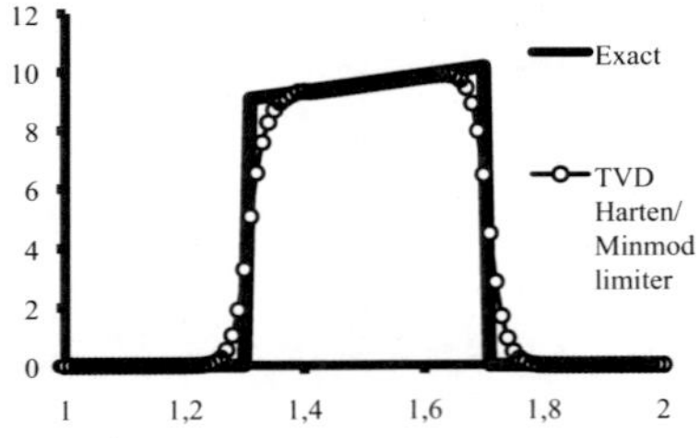

Figure 4. Comparison of the exact solution (solid line) with the numerical results calculated at t = 1 s for the Toro problem, by the TVD extension of Harten and minmod flux limiter described in the introduction.

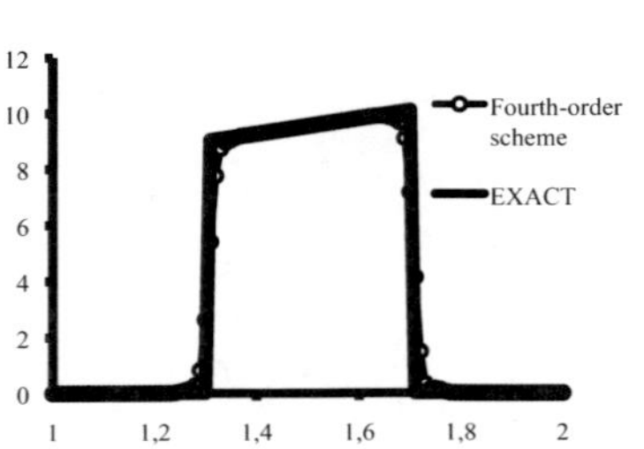

Figure 5. Comparison of the exact solution (solid line) with the numerical results calculated at t = 1 s for the Toro problem, by the fourth-order scheme described in Section 2.

4 CONCLUSIONS

In this paper, the technique described in Gascón & Corberán (2001) has been used to construct a new well balanced scheme. Fourth-order spatial accuracy has been obtained using a non-oscillatory three-degree reconstruction polynomial similar to that described in Balaguer & Conde (2005), applied to the addition of the source's primitive and the physical flux, in order to secure the balance between the discretizations of the flux and the source terms. Preliminary numerical results, calculated with second-order temporal and fourth-order spatial accuracy, as compared with exact solutions, show that the proposed numerical scheme is superior to existing well-balanced schemes.

ACKNOWLEDGEMENTS

The authors agree the financial support provided by the Universidad Politécnica de Valencia (PAID-06-10) and Spanish Ministery of Science and Innovation (Projects CGL2009-14220-C02-01, CGL2010-19591/BTE and DPI2010-20333).

REFERENCES

Balaguer A. & Conde C. 2005. Fourth-order non-oscillatory upwind and central chemes for hyperbolic conservation laws. *SIAM Journal on Numerical Analysis* 43: 455–473.

Caselles V., Donat R. & Haro G. 2009. Flux-gradient and source term balancing of certain high resolution shock-capturing schemes. *Computers and Fluids* 38: 16–36.

Gascón Ll. & Corberán J.M. 2001. Construction of second-order TVD schemes for nonhomogeneous hyperbolic conservation laws. *Journal of Computational Physics* 172: 261–297.

Gascón Ll. & García, J.A. 2001. Esquemas de alta resolución aplicados a sistemas hiperbólicos no-homogéneos con soluciones estacionarias, *Actas del XVII CEDYA, Salamanca, 24–28 Septiembre 2001.*

Gascón Ll., Corberán J.M. & García, J.A. 2003. Construction of High-order Schemes for Hyperbolic Equations involving Source Terms. *New and Advanced Numerical Strategies in Forming Processes Simulation; Proc. 6th International ESAFORM Conference. Salerno, 2003,* Vol I: 683–686.

Grennberg, J.M., Leroux, A.Y., Baraille, R. & Noussair, A. 1997. Analysis and approximation of conservation laws with source terms. *SIAM Journal on Numerical Analysis* 34(5): 1980–2007.

Kurganov A. & Levy D. 2002. Central-upwind schemes for the Saint-Venant system. *Numerical Modeling and Numerical Analysis* 36: 397–425.

Liu X.D. & Tadmor E. 1998. Third Order Nonoscillatory Central Scheme for Hyperbolic Conservation Laws. *Numerical Mathematics* 79: 397–425.

Numerical Methods for Hyperbolic Equations – Vázquez-Cendón et al. (eds)
 ISBN 978-0-415-62150-2

Viscous fingering instabilities in reactive miscible media

Jacobo Guiu-Souto, Darío M. Escala, Jorge Carballido-Landeira & Alberto Pérez-Muñuzuri
Group of Non Linear Physics, University of Santiago de Compostela, Spain

Elena Martín-Ortega
Department of Mechanical Engineering, University of Vigo, Spain

ABSTRACT: In living organisms, fingering structures are often observed in open systems such as growing bacterial colonies, or in porous media flows as oil and gas reservoir. Viscous fingering phenomenon classically occurs when a less viscous fluid displaces a more viscous one in a porous media. We investigate numerically how the instability can be triggered by a simple reaction-diffusion flow. In order to perform our simulations we employed two solutions with different viscosities reacting in the interface and generating a more viscous product. The properties of the fingering pattern observed in the region where the less viscous reactant pushes the more viscous product are studied as a function of the relevant parameters of the problem.

1 INTRODUCTION

Viscous fingering phenomena is a hydrodynamic instability occurring when a less viscous fluid displaces another more viscous in a porous medium (Homsy 1987). The interface between the two fluids becomes unstable and adopts a finger-like pattern. The triggering mechanism is always the same, the difference of viscosities in the system. This topic has long been studied in non reactive systems (Tan & Homsy 1988). However, recently the coupling between fingering phenomena and chemical reactions has been investigated (Nagatsu & Udea 2003). In this case, the fingering dynamics may be influenced by the chemical reactions.

Fingering patterns are often observed in open systems as bacterial colonies (Tokita *et al.* 2009) and especially in porous media flows as oil and gas reservoir (Brailovsky *et al.* 2006) or in contaminated groundwater sites (Zhan & Smith. 2002).

We focus on the miscible viscous fingering in a reactive system. The reactant A is displacing another reactant B with identical viscosity in a miscible solution. The instability is triggered by the following reaction $A + B \rightarrow C$ where a more viscous specie C is produced at the interface. Thus, the pure 1D flow destabilized by the viscous fingering instability is studied by the finite element package Comsol Mult-physics from a mobile reference system solider with the fluid displacement.

This paper is organized as follows. In Sec. II we introduce the equations which are used to model the reactive system. Sec. III describes in detail the numerical procedures. The results are presented in Sec. IV. Finally, Sec. V collects the conclusions and discussion.

2 MODEL

We consider an isotropous and homogeneous porous media with a rectangular geometry (see Fig. 1) subjected to Hele-Shaw cell conditions. In the system, a solution of reactant A is injected (with a viscosity, μ_0, close to the water viscosity since the solutions are very diluted) from the left boundary with speed U along the x direction into a solution of reactant B with the same viscosity. A chemical reaction $A + B \rightarrow C$ takes places

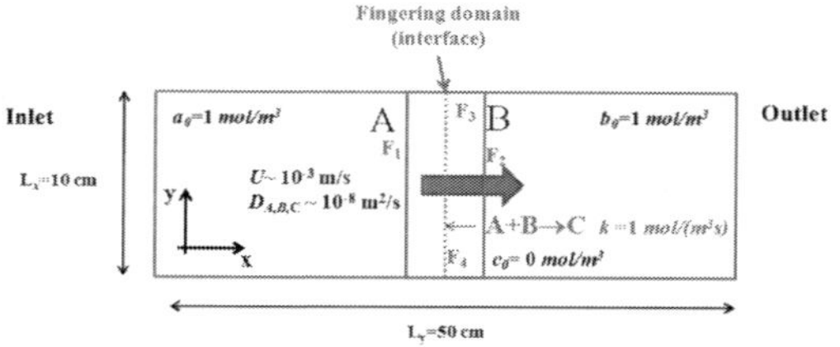

Figure 1. (Color online) Scheme of the modeling system. The reactant A (B) is allocated in the left (right) side. The motion of the fluids is from left to the right faces with a uniform velocity of $U = 10^{-3}$ m/s in the x direction. The initial concentrations of the species A, B and C are respectively a_0, b_0 (both with a value of 1 mol/m^3) and c_0 (0 mol/m^3). *Idem* for the diffusion coefficients $D_{A,B,C}$ (all with a value of 10^{-8} m^2/s). The red box at the interface represents the domain where fingering phenomena takes place.

(Podgorski *et al.* 2007, Cornell 1995) when A and B get in contact yielding the more viscous product C whose viscosity is governed by equation (8). The dynamics of the system can be described by the following set of equations:

$$\nabla \cdot \mathbf{u} = 0 \tag{1}$$

$$\nabla p = -\frac{\mu(c)}{k}\mathbf{u} \tag{2}$$

$$\mu(c) = \mu_0 \exp\left(R\frac{c}{a_0}\right) \tag{3}$$

where μ_0 (water viscosity) and a_0 are a reference viscosity and concentration respectively. The R parameter represents the log-mobility ratio. It is important to note that we consider an incompressible flow (Eq. 1). Equation (2) is the Darcy's law that relates the pressure gradient ∇p, the bidimentional velocity field $\mathbf{u}$ and the permeability κ. The Reaction-Diffusion-Convection (RDC) model results:

$$\frac{\partial a}{\partial t} + \mathbf{u} \cdot \nabla a = D_A \nabla^2 a - kab \tag{4}$$

$$\frac{\partial b}{\partial t} + \mathbf{u} \cdot \nabla b = D_B \nabla^2 b - kab \tag{5}$$

$$\frac{\partial c}{\partial t} + \mathbf{u} \cdot \nabla c = D_C \nabla^2 c + kab \tag{6}$$

where a, b and c denote, respectively, the concentrations of the two reactants, A and B, and the product C with k being the kinetic constant. Also $D_{A,B,C}$ are the diffusion coefficients of the species A, B and C, respectively.

At this point, we switch to a reference frame moving with velocity U, solidarity with the motion of the fluid. In addition, we nondimensionalize the equations by the following characteristics parameters, the velocity U, the hydrodynamical time $\tau_h = D_C/U^2$, the length $L_h = D_C/U$ and the concentration a_0. The dimensionless concentrations result $a' = a/a_0$, $b' = b/a_0$ and $c' = c/a_0$. Viscosity is normalized as $\mu' = \mu(c)/\mu_0$ and pressure as $p' = p\kappa/(\mu_0 D_C)$. Moreover, we introduce the following ratios between the diffusion coefficients $\delta_{A,B} = D_{A,B}/D_C$, as well as, the dimensionless Damköler number $D_{am} = D_C k a_0/U^2$ which expresses the ratio between the hydrodynamic and chemical times. After all these considerations the dimensionless equations in the moving frame result:

$$\nabla' \cdot \mathbf{u}' = 0 \tag{7}$$

$$\nabla' p' = -\mu'(c)[\mathbf{u}' + \mathrm{e}_x] \tag{8}$$

$$\frac{\partial a'}{\partial t'} + \mathbf{u}' \cdot \nabla' a' = \delta_A \nabla'^2 a' - D_{am} a' b' \tag{9}$$

$$\frac{\partial b'}{\partial t'} + \mathbf{u}' \cdot \nabla' b' = \delta_B \nabla'^2 b' - D_{am} a' b' \tag{10}$$

$$\frac{\partial c'}{\partial t'} + \mathbf{u}' \cdot \nabla' c' = \delta_A \nabla'^2 c' - D_{am} a' b' \tag{11}$$

In our dimensionless scales, the initial dimensions of our domain (L_x and L_y) correspond with the Peclet numbers (Pe_x and Pe_y, respectively). Pe_x controls the maximum time of the simulations and Pe_y the number of fingers present across the domain.

3 NUMERICAL PROCEDURE

We perform a numeric simulation of the Equations (7)–(11) by using COMSOL Multiphysics which is based in the finite elements method. In the implementation of the model it is important to observe the dimensionless character of our variables, as well as, acclimatizing the boundaries F_1, F_2, F_3 and F_4 (see Fig. 1). We have considered a wall condition for the boundaries F_3 and F_4. However for F_1 (F_2), we have applied a zero velocity inlet (outlet) condition because our domain is in the moving frame of reference. In addition, we have used constant concentrations a_0 and b_0 at the boundaries F_1 and F_2, respectively. The reason is these boundaries are considered to be sufficiently far from the interface and the concentrations can be considering unalterable. For the initial pressure we have considered the following initial condition by integrating the adimensional Darcy law (8):

$$p_0' = -\mu_0' Pe_x + x' \tag{12}$$

where μ_0' is the viscosity at initial time, x' the dimensionless longitudinal component and Pe_x the Peclet number in the x-direction.

At the interface, we introduce the following random function in order to trigger the instability:

$$c_{i,\mathrm{inter}}(x,y) = \frac{\varepsilon(x,y)}{2}\left[\begin{array}{l} c_{i0,left} + c_{0,right} \\ + \left(c_{i0,left} - c_{0,right}\right)\cos\dfrac{\pi(x-x_0)}{l_{gap}} \end{array}\right] \tag{13}$$

where $\varepsilon(x,y)$ is a random function, and $c_{i,\mathrm{inter}}$ with $i = a,b$ are the concentrations at the interface of the species A and B, respectively. Analogously, $c_{i0,\mathrm{left\,(right)}}$ represents the concentration of each specie at the

Table 1. Initial parameters.

Initial concentration specie A (a_0)	1
Initial concentration specie B (b_0)	1
Initial concentration specie C (c_0)	0
Diffusion ratio specie A (δ_A)	1
Diffusion ratio specie B (δ_B)	1
Damkhöler number (D_{am})	1
Mobility ratio (R)	3
Longitudinal Peclet number (Pe_x)	600
Transversal Peclet number (Pe_y)	300

left (right) side of the interface. The parameters l_{gap} and x_0 represent the thickness of the interface and its starting point along x-coordinate.

A table of all previous initial values can see below.

The computational domain consists in a mapped mesh of 30.000 elements. To solver the Equations (6–10) we considered first order Lagrange elements and we segregate in two groups the preceding equations. Both of them were solved by using the UMFPACK method. The first group (stationary) contains the Darcy law and the second group (transient) the RDC equations. The relative and absolute applied tolerances were 0.01 and 0.001, respectively.

4 RESULTS

The different stages of viscous fingering can be seen in Figure 2, the red color represents the concentration of the specie A, and the blue color is associated with the specie B. Note that due to the mobile reference frame the instability is generated around the center of the domain. Initially, we observe the random initial condition in the centre of the domain, and also the initial concentration of each species, the A at the left side and B at the right. When the reaction takes place and product C is generated at the interface, the viscosity is increased in this region and, therefore, the front begins to deform. At the final stages of the simulation, we observe clearly the corresponding fingers.

The velocity field and the corresponding stream lines are shown in Figure 3. We observed four zones of recirculation corresponding with fingers development. It is important to note as in these zones the velocity takes the highest values (see the arrows in the figure). The concentration of product C, in gray scale, represents the changes in the viscosity according with (3) and therefore, being the cause of the front's deformation, i.e. fingering phenomena.

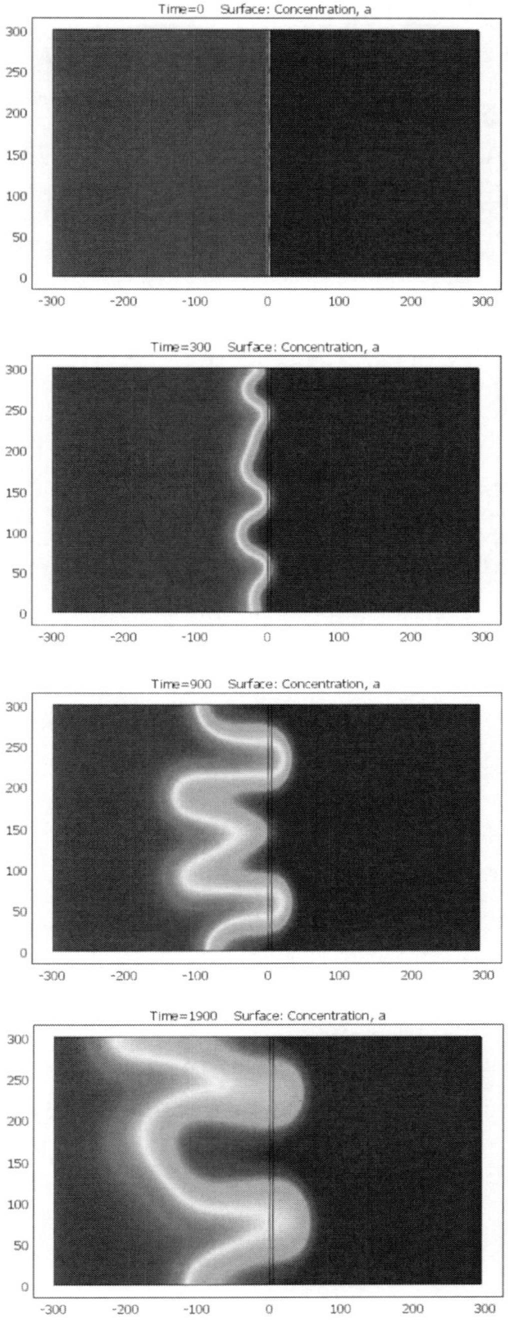

Figure 2. (Color online) Finger development. The concentration of specie A is plotted by jet color map, i.e. the red (blue) represents the maximum (minimum) values. The length of adimensional domain corresponds with the values of the Peclet numbers from Table 1.

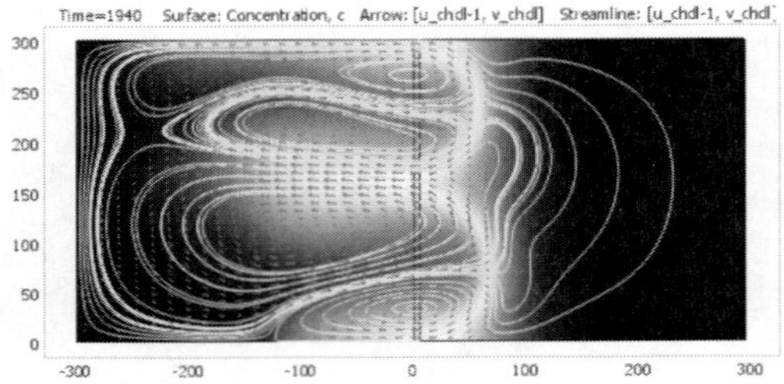

Figure 3. (Color on line) Velocity field and concentration of product *C*. The streamlines are plotted in yellow, and their corresponding velocity arrows in red. While the concentration of product *C* is represented in gray scale.

5 CONCLUSIONS

The performed simulations by Comsol Multiphysics for viscous fingering reproduce satisfactorily the phenomena. The use of a mobile frame solitary with the motion of the fluid gives us two important advantages. In the one hand, it reduces the computational domain at only those interesting zones where fingering can occur. And on the other hand, it allows us to observe the behavior of velocity field with its recirculation.

Finally, authors want to indicate that this work contributes to improve the knowledge of the hydrodynamics instabilities in presence of chemical reactions.

This work has been supported by the DGI (Spain) under project No. FIS2010-21023.

REFERENCES

Brailovsky, I., Babchin, A., Frankel, M., Sivashinsky, G. 2006. *Transport in Porous Media* 63, 363.

Cornell, S., Koza, Z., & Droz, M.1995 *Phys. Rev E* 52, 3500.

Homsy, G.M. 1987. Annu. Rev. Fluid. Mech. 19, 271.

Nagatsu, Y. & Udea, T. 2003. *AIChE J*. 49, 789.

Podgorski, T., Sostarecz, M.C., Zorman, S. & Belmonte, A. 2007. *Phys. Rev. E* 76, 016202.

Tan, C.T. & Homsy, G.M. 1988. *Phys. Fluid.* 31, 1330.

Tokita, R., Katoh, T., Maeda, Y., Wakita, J., Sano M., Matsuyama T. & Matsushita M. 2009. *Journal of Physics Society of. Japan*. 78, 074005.

Zhan, Z.F., Smith, J.S. 2002. *Transport in Porous Media* 48, 41.

Numerical Methods for Hyperbolic Equations – Vázquez-Cendón et al. (eds)
© 2013 Taylor & Francis Group, London, ISBN 978-0-415-62150-2

Author index